Hyperspectral Satellites and System Design

Hyperspectral Satellites and System Design

By

Shen-En Qian

CRC Press
Taylor & Francis Group
Boca Raton London New York

CRC Press is an imprint of the
Taylor & Francis Group, an **informa** business

CRC Press
Taylor & Francis Group
6000 Broken Sound Parkway NW, Suite 300
Boca Raton, FL 33487-2742

Printed on acid-free paper

International Standard Book Number-13: 978-0-367-21790-7 (Hardback)

Visit the Taylor & Francis Web site at
http://www.taylorandfrancis.com

and the CRC Press Web site at
http://www.crcpress.com

Contents

Preface

Since the beginning of this millennium, spaceborne hyperspectral imaging has emerged as a new generation of remote sensing satellites due to its expanded capability of acquiring hundreds of contiguous and narrow spectral bands for each pixel in the scene. Such distinct spectral signatures or "fingerprints" of the pixels can provide direct identification of the surface materials. Hyperspectral imaging has a wide range of remote sensing applications in various disciplines including agriculture, forestry, environment, geology, ocean, atmosphere, climate change, defence and security, and law enforcement. Fascinating detailed spectral information provided by hyperspectral imagery often provides results not possible with multispectral or other types of imaging technologies.

For over 30 years, I have focused my work on the development of optical satellites. In the last 25 years, as a senior scientist with the Canadian Space Agency (CSA) in optical payloads, I have been dedicating my efforts to the development of hyperspectral satellite technology and missions. In particular, as the technical lead in multiple hyperspectral missions in Canada and in Canadian joint ventures with international partners, including NASA, the US Naval Research Laboratory, and the Italian Space Agency, I have been fortunate enough to collaborate with and draw upon the knowledge of many highly qualified teams in the space industry and academia, during different development phases of hyperspectral satellite missions. This experience has provided crucial firsthand know-how and insight into the key aspects of hyperspectral satellite system design and implementation.

While a number of books have been published on hyperspectral remote sensing and on the general concepts of satellite design, I'm not aware of a book that specifically addresses hyperspectral satellites and their system design. With hyperspectral satellites still in their relative infancy and likely to become more common in the next decade or so, a book on this topic certainly seems beneficial and timely for current and future professionals in this field. Therefore, having written two books on the subject of optical satellite signal processing (published by SPIE Press), authored chapters in five others, and edited a book on optical payloads for space missions (published by John Wiley & Sons), I have now systematically selected and organized my firsthand knowledge and know-how on hyperspectral satellite system design into this book.

This book consists of 14 chapters and is written in a way to be an 'A to Z' of all aspects of hyperspectral satellites and their designs. Chapters 1–3 provide background information on hyperspectral imaging, including introduction to hyperspectral satellites, an overview of hyperspectral sensors on orbit, and an overview of hyperspectral imaging applications in Earth observation. Chapters 4–12 form the main body of the book devoted to system design. These chapters start with a discussion on the mission concept and trade-off study, then move on to subsystem design including optical system and design, focal plane arrays, system performance modeling, thermal and mechanical design, in-flight calibration, instrument control and onboard data handling, and finally provide information on the ground segment, and on-ground calibration and characterization. The last two chapters (Chapters 13 and 14) are about data processing on ground, including radiometric conversion and data correction, and atmospheric correction.

The design of a hyperspectral satellite is undeniably complex and involves a process that is still not well-established and documented. Therefore, this book, which is written to cover all the aspects of this process and draws upon many years of experience and knowledge, should be of significant value as a reference or guideline for engineers and decision makers at space agencies, industry professionals, and academics and students in the aerospace sector and related fields. Academics in the broader field of remote sensing would likely find the chapters on background and data processing to be the most beneficial part of this book.

The contents of the book are intended for readers with a university education in an engineering domain. No prerequisite knowledge is needed. The book could also be used as a reference for students who are interested in satellites and plan to build cube-satellites, or as a teaching aid for senior undergraduate or graduate level courses. Such courses might include aerospace engineering or specialist courses that have an emphasis on space science and technology.

I am proud to be a Canadian government scientist working at the CSA contributing towards the advancement of its hyperspectral satellite program. Canada is one of the leading countries in the development of hyperspectral technology. Canada built its first hyperspectral sensor in 1984 when the Fluorescence Line Imager (FLI), an airborne imaging spectrometer, was delivered to the Canadian government as the first step towards a hyperspectral satellite. Canadian researchers and engineers pioneered this field, together with their American colleagues at NASA's Jet Propulsion Laboratory in building the FLI. The FLI flew over a wide variety of land and ocean sites to acquire imagery with spectral resolution and sensitivity higher than that of other comparable sensors at the time. The Compact Airborne Spectrographic Imager (CASI) has been fabricated and marketed commercially by ITRES Research Ltd. based in Calgary, Canada since 1989. The Canada Centre for Remote Sensing developed the first Canadian shortwave infrared (SWIR) imaging spectrometer—SWIR Full Spectrographic Imager (SFSI)—in 1992. Since joining CSA in 1994, I have participated in and led multiple Canadian hyperspectral missions. In the early 2000s, CSA initiated the Hyperspectral Environment and Resource Observer (HERO) mission. At the end of Phase A development of the HERO mission, CSA teamed with the Italian Space Agency (ASI) to develop the Italy-Canada joint hyperspectral satellite mission in 2006. Due to some uncontrollable reasons both the HERO mission and the Italy-Canada joint mission were terminated in 2008. However, in 2012, my colleagues at the agency and I initiated the Canadian Hyperspectral Mission (CHM), a new hyperspectral mission, to meet the needs of government stakeholders. I led the Canadian industry teams and academia in the CHM mission concept studies, following the completion of the two feasibility studies for different payload concepts. Since 2014, I have been the technical lead for the Canadian coastal and inland water color hyperspectral satellite (referred to as WaterSat). In 2015, CSA and the US Naval Research Laboratory jointly developed a concept of a hyperspectral payload called Coastal Ocean Color Imager (COCI) for deployment as a complement payload to NASA's PACE Mission to benefit the coastal and inland water user community. As the Technical Authority of government contracts, I am leading the Canadian industry to build prototypes for both the WaterSat mission and the COCI payload. The first prototype, referred to as WaterSat Imaging Spectrometer Experiment (WISE), has been built and is now operating onboard aircraft to serve the Canadian water color research community. The development of the second prototype, called Dual Imaging spectrometer COCI Experiment (DICE) with enhanced performance, is in progress and is to be completed in the first half 2020.

During the course of writing this book, I received strong support and assistance from my colleagues, experts, and scholars in Canada. From this perspective, the book can be considered a collective effort of a Canadian team, enriched by three decades of experience and accomplishments of all the team members. I would like to thank my colleague Oleg Djazovski for providing me the material used in Chapter 6 (Focal Plane Arrays). I thank my partners at Canadian government departments, including Canada Centre for Remote Sensing (CCRS), Agriculture and Agri-Food Canada (AAFC), Defence Research and Development Canada (DRDC), and Pacific Forestry Centre (PFC), for providing their research results and figures on hyperspectral applications and data processing. I also thank Canadian industry teams—ITRES Research Ltd. based in Calgary, ASL Environmental Sciences (former Borstad Associates Ltd.) based in Victoria, and GHGSat Inc. based in Montreal—for providing me the material and figures included in the book. I would particularly like to thank Louis Moreau at ABB Inc., Canada based in Quebec City and Stephen Achal at ITRES Research Ltd. (now with 4 Pi Lab) for reviewing my manuscript and providing constructive comments for the improvement of the book. I thank my colleagues at the agency Martin Bergeron, Andrew Gibson (now with ESR Technology in the United Kingdom), and Michael Maszkiewicz for reviewing the chapters of the book. I thank Karl Staenz at Lethbridge University (retired), who was the chair of the User and Science Team of the HERO hyperspectral mission when he was with

CCRS, for his careful review of the manuscript and detailed comments. I thank my former colleagues Allan Hollinger at the agency and David Goodenough at PFC for their leadership in the development of the HERO mission, and Peter Oswald and Ron Buckingham for their consultation service. I would also like to thank my colleagues Jean-Pierre Ardouin, Josée Lévesque, and Vincent Roy at the DRDC Valcartier for reviewing the chapters and providing me the material and figures used in the book. I thank Canadian industry teams, including MDA, COM DEV, ABB Canada (Quebec), MPB Communication, former EMS Technologies, former Routes AstroEngineering Ltd., Spectral Applied Research, and EHR Optical Systems, for their contributions toward the development of Canadian hyperspectral missions.

I would also like to thank the four anonymous reviewers for their strong endorsement of this book and the constructive comments leading to the improvement of the book. Many thanks as well to the CRC Press/Taylor & Francis staff Jonathan Plant, Kyra Lindholm, and Bhavna Saxena for their professional service and turning my manuscript into this book.

Finally, I would like to thank my wife Nancy and daughter Cynthia for their help and support. They provided great encouragement and assistance during the many hours of my spare time in writing this book.

Shen-En Qian (錢神恩), PhD
Senior Scientist
Canadian Space Agency
Montreal, Canada, September 2019

Author Biography

Dr. Shen-En Qian is a senior scientist, a scientific authority of government contracts, and a technical lead of space missions at the Canadian Space Agency. He is an internationally recognized expert in optical spacecraft payloads, space technologies for satellite missions and deep space exploration, remote sensing, satellite signal processing and enhancement, onboard satellite data compression, data handling, and establishment of international standards for spacecraft data systems. He has over 30 years of experience in these areas.

Dr. Qian received his Ph.D. degree in Telecommunication and Electronic Systems in 1990. He has written three books on optical satellites and their signal processing, has authored chapters in five others, and has edited a book on optical payloads for space missions. Dr. Qian has published over 120 articles. He holds 35 granted patents in United States, Europe, Canada, and Australia developed in Canadian government laboratory. These technologies have helped overcome the technical challenges in the development of the space missions and reduce their cost. The Governor General of Canada presented him with the Canadian national award, "Public Service Awards of Excellence," in 2016. He was the only individual recipient in the Scientific Contribution category. He also received the IEEE Canada Outstanding Engineer Award and a silver medal in 2019.

Dr. Qian is a fellow of the Canadian Academy of Engineering, a fellow of the International Society of Optics and Photonics Engineering (SPIE), and a senior member of IEEE. He is an Associate Editor of the Journal of Applied Remote Sensing. He is an adjunct professor at York University.

1 Introduction to Hyperspectral Satellites

1.1 SPACEBORNE SPECTROSCOPY AND IMAGING

While multispectral satellites, such as Landsat and SPOT satellites, have been in regular use since the 1970s, hyperspectral satellites have emerged as a new generation of remote sensing satellites since the beginning of this millennium. Imaging spectrometry, also known as hyperspectral imaging, is a combination of the traditional spectroscopy technology and the modern imaging system (Qian 2013).

Spectroscopy is a key analytical method used to investigate material composition and related processes through the study of the interaction of light with matter. Energy is absorbed by the matter, creating an excited state. The interaction creates some form of electromagnetic waves. Isaac Newton described the spectral nature of light. Spectroscopy or spectrometry deals with the measurement of a specific spectrum for identification of matters. By using a spectrometer, one can determine the level of excitement in the matter's atoms to identify what kind of material it is. It is incredibly difficult to make the distinction just with the naked eye, as human eyes are not able to see the fine details. Joseph von Fraunhofer invented the spectroscope in 1814 and used it to characterize the optical properties of glass for the development of more powerful telescopes. He also identified the dark lines in the solar spectrum. Kirchhoff and Bunsen used spectroscopy to investigate the composition of the solar atmosphere by establishing the connection between the solar Fraunhofer lines and the spectroscopic signatures of elements observed in the laboratory. Determining composition remotely, without physical contact, is one of the most valuable capabilities of spectroscopy. Since its beginning, spectroscopy has evolved and been used to enable a broad range of scientific discoveries by Edwin Hubble to deduce the expanding nature of the universe.

Spectroscopy is widely used in laboratories in the disciplines of physics, chemistry, and biology to investigate material properties. Spectroscopic data are often represented by an emission spectrum, a plot of the response of interest as a function of wavelength or frequency. For example, colorimetry used for the investigation of cholesterol or blood sugar in chemical laboratories is a form of spectroscopy. Spectrometry has also been used for determining blood alcohol levels, checking automobile emissions, and monitoring smokestack pollution.

A modern imaging system converts the visual characteristics of an object, such as a physical scene or the interior structure of an object, into digital signals and creates digitally encoded representations that are processed by a processor or a computer and give output in the form of a digital image. Imaging systems typically consist of a camera and an imaging lens, along with an illumination source. Depending on the system setup, imaging systems magnify or enhance observed objects to ease the viewing or inspection of small or unclear details. Computers are becoming more and more powerful with increasing capacities for running programs of any kind, especially digital imaging software. Software is becoming both smarter and simpler at a fast pace.

In the late 1970s, detector, optical, and computer technologies advanced sufficiently to enable a combination of spectroscopy with an imaging system. The combination of the spectroscopy technology and the modern imaging system is referred to as *imaging spectrometry*, now also called *hyperspectral imaging*. It could measure a spectrum for every pixel in an image. This provides revolutionary ways of observing the Earth by collecting information of each pixel in a scene across the electromagnetic spectrum. Imaging spectroscopy operating in the solar-reflected spectrum senses objects on the ground in detail spectrally and spatially. Molecules and particles of the land, water and atmosphere interact with solar energy in the 400–2500 nm spectral region through absorption, reflection, and scattering processes. Imaging spectrometers in the solar-reflected spectrum measure spectra of the ground objects as images in some or the entire portion of the spectra. These spectral measurements are used to determine constituent composition through the physics and chemistry of spectroscopy for scientific research and applications over the regional scale of the image.

The primary advantage of hyperspectral imaging is that, because an entire spectrum is acquired for each pixel of the acquired imagery, an operator needs no prior knowledge of the sample, and post-processing allows all available information from the dataset to be exploited (Chang 2013). Hyperspectral imaging can also take advantage of the spatial relationships among the different spectra in a neighborhood, allowing more elaborate spectral-spatial models for a more accurate segmentation and classification of the image (Grahn and Geladi 2007). The story of hyperspectral imaging is closely tied to advances in digital electronics and computing capabilities due to its complexity for acquiring and processing large data volume.

The term *hyperspectral imaging* was originally defined by Goetz et al. (1985) as "the acquisition of images in hundreds of contiguous, registered spectral bands such that for each pixel a radiance spectrum can be derived." It sets this type of spectral remote sensing apart from *multispectral imaging* by requiring the spectral bands to be contiguous, so that no gaps occur through which precious spectral information might slip undetected. This original definition, however, did not explicitly mention the bandwidth of the spectral bands with respect to that of multispectral imaging. It did implicitly define the bandwidth of the hyperspectral bands as being much narrower, because of hundreds of bands in hyperspectral imaging within the same wavelength range rather than a few bands in multispectral imaging. The bandwidth of hyperspectral sensors is typically 10 nm or less, which is much narrower than that of multispectral sensors whose bandwidth is about 100 nm or so. The narrowness of the spectral bands in hyperspectral imaging is more important than the number of spectral bands being in hundreds. An imaging sensor that produces spectral images with a bandwidth of 10 nm or less should be classified as a hyperspectral sensor even if its total number of bands is less than a hundred. The original definition of hyperspectral imaging could be updated as "the acquisition of *many* images of contiguous, *narrow*, registered spectral bands such that for each pixel a radiance spectrum can be derived" to explicitly mention the bandwidth of the spectral bands.

Figure 1.1 shows the concept principle of a hyperspectral satellite. It acquires images of a given scene on the ground in hundreds (or tens sometimes) of continuous and narrow spectral bands over wavelengths that range from the near-ultraviolet to the shortwave infrared. Each image, often referred to as spectral image or band image, corresponds to a particular wavelength or spectral band number. The collected "datacube" contains both spatial and spectral information of the materials within the scene. Each element or pixel in the scene is sampled across hundreds (or tens) of narrow band images at a particular spatial location in the datacube, resulting in a one-dimensional (1D) spectrum. It is a plot of wavelength versus radiance or reflectance. The spectrum for a single pixel acquired by a hyperspectral satellite appears similar to a spectrum collected by a spectrometer in a spectroscopy laboratory. The spectrum can be used to identify and characterize the particular feature of the pixels within the scene based on the unique spectral signatures or "fingerprints." Hyperspectral imagery can provide direct identification of the surface materials and has been used in a wide range of remote sensing applications, including geology, agriculture, forestry, environment, ocean, atmosphere, climate change, defence and security, and law enforcement. The fascinating detailed spectral information provided by hyperspectral imagery often provides results not possible with multispectral or other types of imagery (Qian 2016).

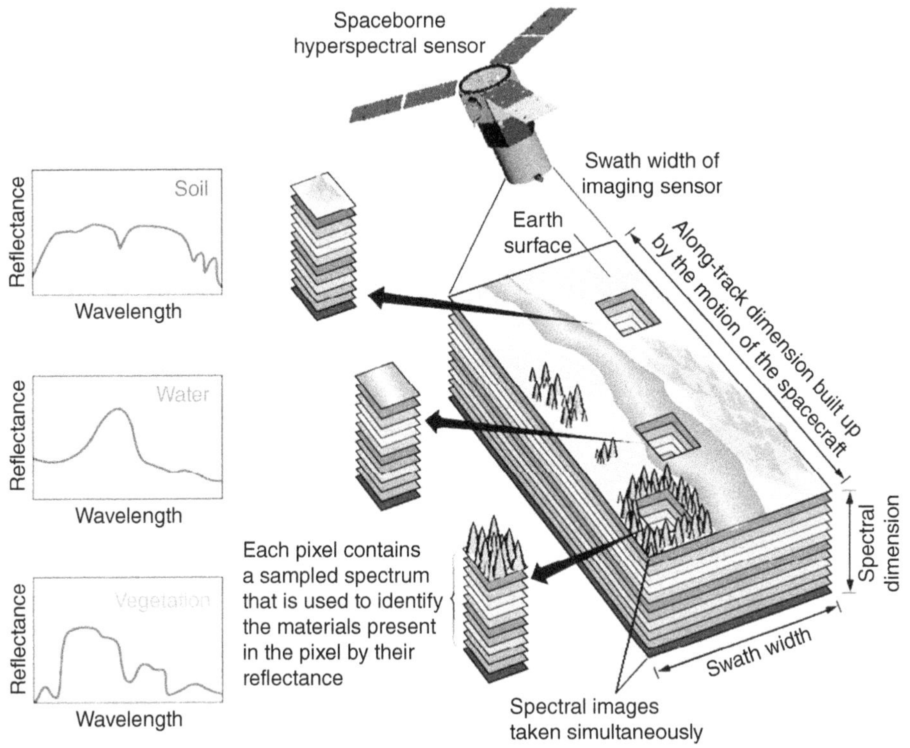

FIGURE 1.1 Concept principle of a hyperspectral satellite. A datacube and spectrum of each pixel is generated by a hyperspectral satellite.

1.2 HYPERSPECTRAL IMAGING APPROACHES

Hyperspectral sensors acquire both spatial and spectral information of a scene and generate a datacube for the scene. There are at least three different approaches to acquiring the hyperspectral data in terms of the type of imaging spectrometers:

1. Dispersive elements based approach
2. Spectral filters based approach
3. Snapshot hyperspectral imaging

1.2.1 Dispersive Elements Based Approach

Figure 1.2 shows an example of the concept of a dispersive element based hyperspectral sensor with a dispersing element and a two-dimensional (2D) detector array. In this approach, a spectrometer separates the radiation light from an object into spectral components, or dispersion, by the use of an optical element possessing a known functional dependence on wavelength, specifically prisms or diffraction gratings. A dispersive element is inserted into a collimated beam of the instrument and then creates a dispersed spectrum centered on the object's location in the instrument's field of view (FOV). A grating or a transparent prism can be used to break up the radiation light into its constituent monochromatic components. The grating-dispersed monochromatic components are in linear distribution. The monochromatic components dispersed using a prism are in nonlinear distribution. A grism can also be used to break radiation light up into its constituent monochromatic components. A grism is a combination of a grating and a prism (also called a grating prism) so that radiance light at a chosen central wavelength passes straight through.

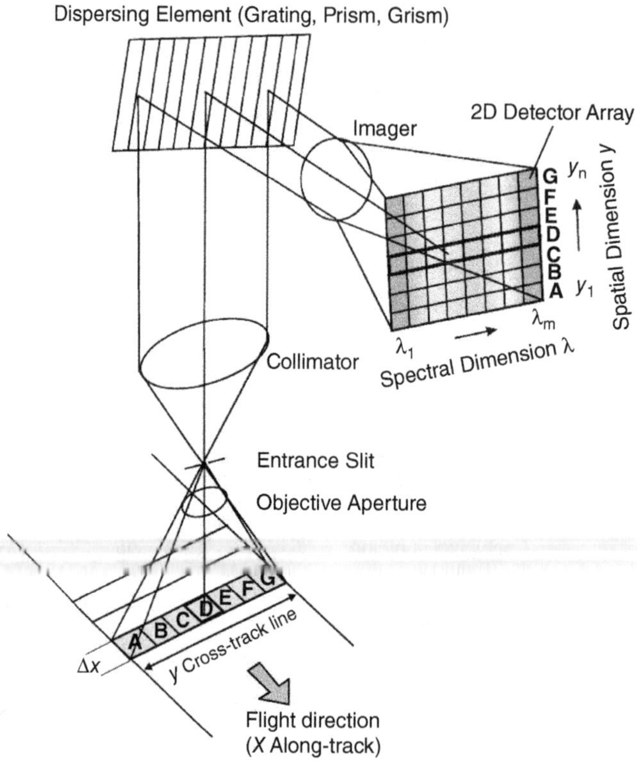

FIGURE 1.2 An example of the concept of a dispersive elements based hyperspectral sensor.

Dispersive element based hyperspectral sensors are the most popular ones for both airborne and spaceborne remote sensing. This type of hyperspectral sensors needs to scan the scene on the ground either by a dedicated scanner or by the entire instrument with the aircraft or satellite flight motion to obtain the spatial coverage. There are two operating modes in terms of scanning of the scene: whiskbroom and pushbroom.

1.2.1.1 Whiskbroom Mode

For the early airborne hyperspectral sensors, the whiskbroom operating mode was often used, such as the Airborne Visible/Infrared Imaging Spectrometer (AVIRIS) developed by NASA's Jet Propulsion Laboratory (JPL) (Green et al. 1998) (the recent AVIRIS still uses the whiskbroom mode). This was because linear detector arrays were used that could record the monochromatic components of the spectrum of only one ground pixel (or the sampling cell). The instrument needs to scan the ground sampling cells one after another in a cross-track line using the whiskbroom mode, as shown in Figure 1.3. After completion of scanning the entire current cross-track line, the instrument starts to scan the ground sampling cells of the next cross-track line when the aircraft or satellite flies forward in the flight direction (also referred to as along-track direction), and so on. The advantages of the whiskbroom operating mode are:

1. Simple design of the instrument.
2. Wide FOV, as there is no constraint imposed by the available number of pixels of a 2D detector array in the spatial direction thanks to using only 1D linear detector array.
3. Easy calibration, since all the spectra of the ground sampling cells within the scene are generated by the same linear detector array and the same optics, which have the identical spectral characteristics. This type of hyperspectral sensor has no spatial distortion (also referred to as keystone) as does a pushbroom hyperspectral sensor.

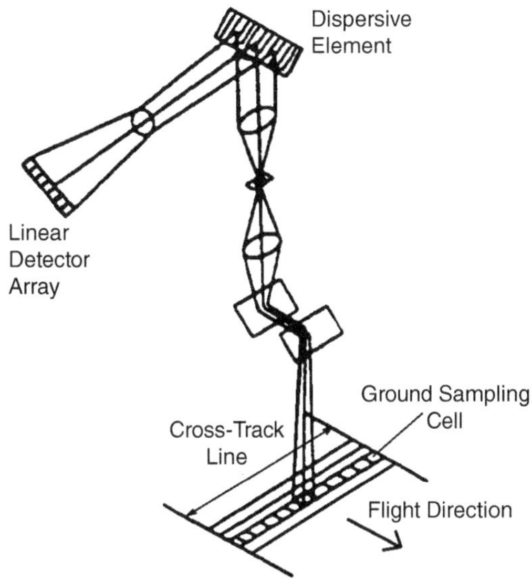

FIGURE 1.3 Concept of a hyperspectral sensor operating in the whiskbroom mode with a linear detector array.

The disadvantages of the whiskbroom operating mode are:

1. Requirement of a mechanical scanner, which contains moving parts in a vacuum chamber.
2. Requirement of post-processing for spatial incongruence.
3. Constraints of high spectral and spatial resolution requirements due to low integration time.

1.2.1.2 Pushbroom Mode

Almost all the spaceborne hyperspectral sensors use the pushbroom operating mode. As shown in Figure 1.2, in the pushbroom mode all the ground sampling cells in an entire cross-track line are imaged into a 2D detector array simultaneously. One dimension of the detector array corresponds to the spatial direction of the cross-track line and another dimension corresponds to the spectral extension of the ground sampling cells. The advantages of the pushbroom operating mode are:

1. No moving parts.
2. Congruence spatial images.
3. Longer integration time for each ground sampling cell, because each of them is sensed simultaneously by a row of elements of the 2D detector array (e.g., rows A, B, C, … G in Figure 1.2) instead of one cell after another, which omits the time sharing scanning of all the ground sampling cells in a cross-track line. Longer integration time means more photos are collected and results in higher signal-to-noise ratio (SNR).

The disadvantages of the pushbroom operating mode are:

1. Complex optical design and complex focal plane.
2. Narrow FOV, which is constrained by the available number of pixels of the 2D detector array in the spatial direction.
3. Complex calibration.
4. Spectral distortion (also referred to as smile) and spatial distortion (also referred to as keystone). The hyperspectral data collected by a pushbroom hyperspectral sensor need to be sufficiently corrected for smile and keystone distortion before being distributed to users for downstream applications (Qian 2013).

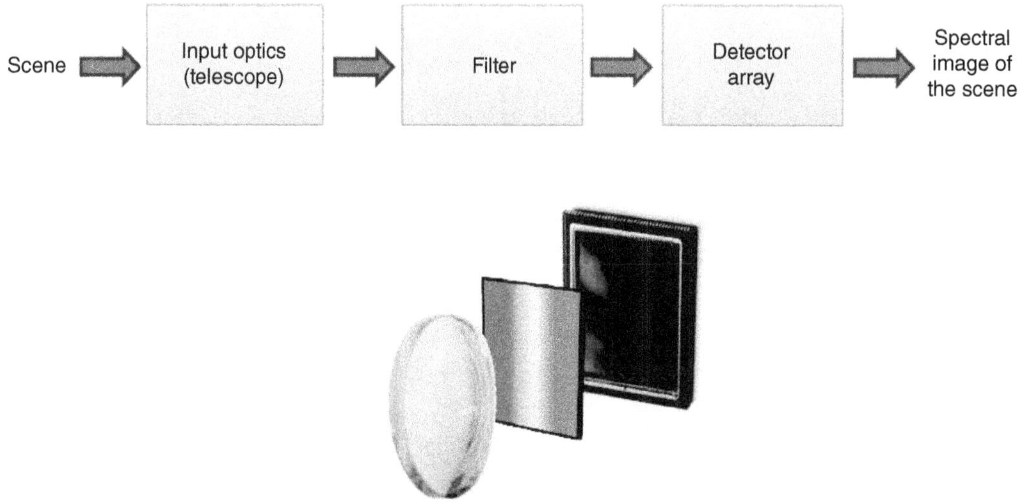

FIGURE 1.4 A block diagram of composition of a spectral filter based hyperspectral sensor.

1.2.2 Spectral Filters Based Approach

Spectral filters based hyperspectral sensors are less popular and appear more recently. For this type of hyperspectral sensors, one or more spectral filters, such as an absorption or interference filter, are used to transmit the selected spectral bands of interest. As the beam passes through a filter, some of its spectral components are blocked through an absorption or interference process, while the desired spectral components are transmitted. Various interference filters, from the ultraviolet through the far-infrared region, in various dimensions, are available as commercial-off-the-shelf (COTS) products. Electronically tunable filter (ETF) is another kind of filters that transmit desired spectral bands by controlling voltage, or acoustic signal, etc. (Gat 2000).

The basic principle of a spectral filter based hyperspectral sensor is similar to that of a dispersive element based sensor. However, the main difference is that the radiation light from a telescope, instead of being collimated and then dispersed by a grating or a prism in a traditional spectrometer, is directly focused on a 2D filter. The filter has the same size as the 2D detector array and is assembled in front and close to the detector's sensitive surface. It distributes spectral content of the ground samples in the spectral dimension (Qian 2013). Figure 1.4 shows a block diagram of the composition of a filter-based hyperspectral sensor. It is essentially composed of three parts: a telescope, a filter, and a detector array. A traditional bulky spectrometer is omitted and replaced by a tiny optical filter.

Figure 1.5 shows an example of how the volume and mass of a spectral filter based hyperspectral sensor are reduced in comparison to a conventional grating-based hyperspectral sensor (Mouroulis et al. 2008). In this case, a linear variable filter (LVF) based hyperspectral sensor is chosen. Both designs use the same front-end optical elements. The elimination of the spectrometer results in a much more compact instrument. In the figure, an optical filter based imaging spectrometer replaces the traditional spectrometer assembly with a simple optical filter mounted in front of the detector array at the focal plane (Qian et al. 2013). A filter-based hyperspectral sensor can yield a much more compact design when compared to the conventional approaches using Dyson (1959) or Offner (1973) type spectrometers. However, this simplified optical design comes with several limitations.

1.2.2.1 Linear Variable Filters

A LVF is a glass coated with an interference filter of increasing thickness along one direction, as shown in the upper left part in Figure 1.6. This direction is referred to as the spectral direction. The spectral position of the peak of the transmission curve varies with the thickness of the deposition. Hence, the

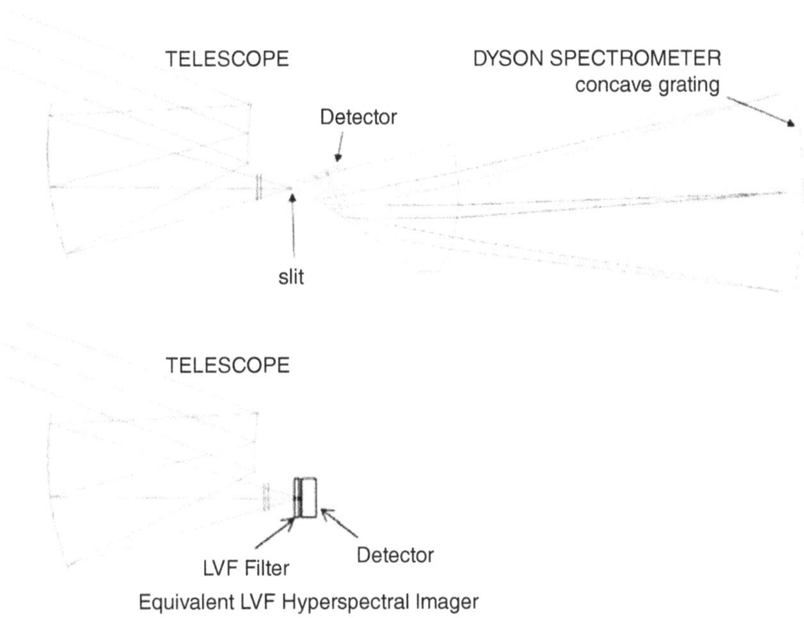

FIGURE 1.5 An example of how the volume and mass of a filter-based hyperspectral sensor is reduced compared to a traditional dispersive element based approach.

spectral position of the peak of the transmission curve varies linearly along one physical dimension of the filter. The slope of that curve is called the LVF's spectral gradient (upper right in Figure 1.6).

At a given position on the LVF, the transmittance has a certain shape. That shape shifts with the position. Figure 1.7 shows the transmittance curves of a typical LVF for a few positions along the spectral direction of the filter. Because the change is gradual, a LVF in the visible region exhibits a gradual rainbow pattern when seen in transmittance, as shown in the photo in the lower part in Figure 1.6. The shape of the transmittance curve also varies along the spectral dimension of the LVF. For example, Figure 1.7 shows the full width at half maximum (FWHM) of a typical LVF at different positions. Typical LVFs have spectral gradient between 10 nm/mm and 200 nm/mm.

The LVF exists in the spectral range from visible to the thermal infrared region, typically within the visible and near-infrared (VNIR) regions (e.g., 450–900 nm). The spectral bandwidth of an optical filter varies as a function of wavelength. The longer the wavelength, the wider is the spectral bandwidth.

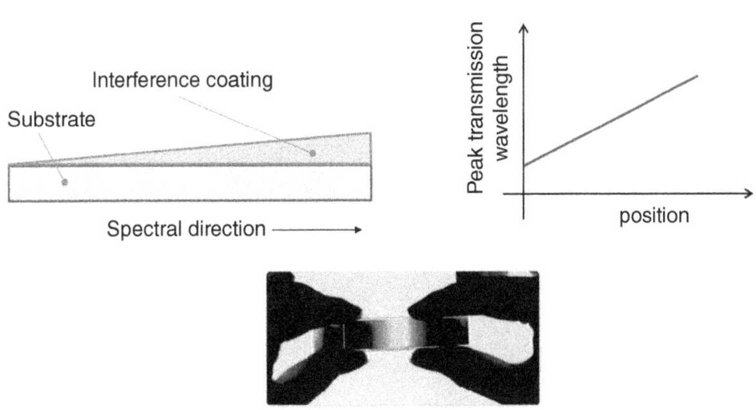

FIGURE 1.6 A linear variable filter and its spectral gradient.

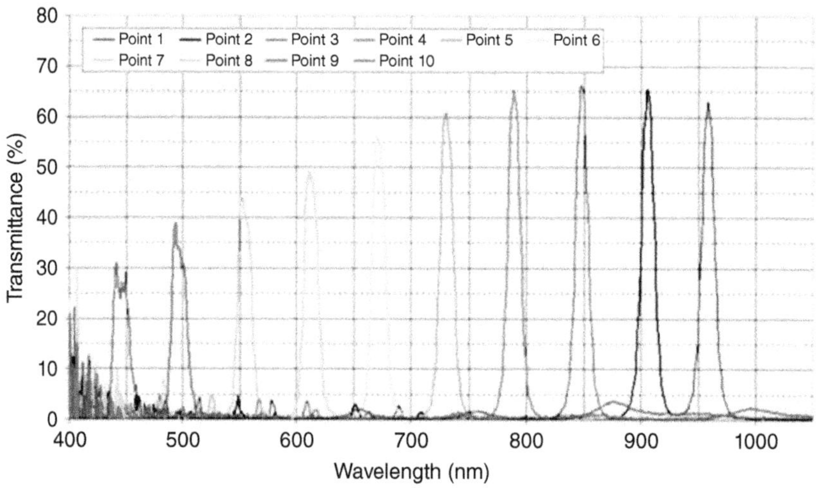

FIGURE 1.7 Transmittance curves of a typical linear variable filter.

The spectral bandwidth is limited by its coating thickness, glass index, quality of the coatings, the f/# of the optics, and the wavelength of the light. A typical band-pass (i.e., FWHM) ranges from ~7 nm at 450 nm to ~12 nm at 900 nm. Typical LVFs have a FWHM variation 1–2% of the center wavelength or more. The LVF transmission degrades as the FWHM is narrowed. The throughput ranges from ~40% to ~66% over the spectral range.

Figure 1.8 illustrates the concept of a LVF-based hyperspectral sensor. A simple spectrometer can be built by placing a LVF in front of a 2D detector array after the telescope collects the input light of a ground scene. A pushbroom hyperspectral sensor can then be built based on this type of a spectrometer. The 2D detector array of the instrument "sees" the complete scene at once. As described in Section 1.2.1, a dispersive element based hyperspectral sensor working in the

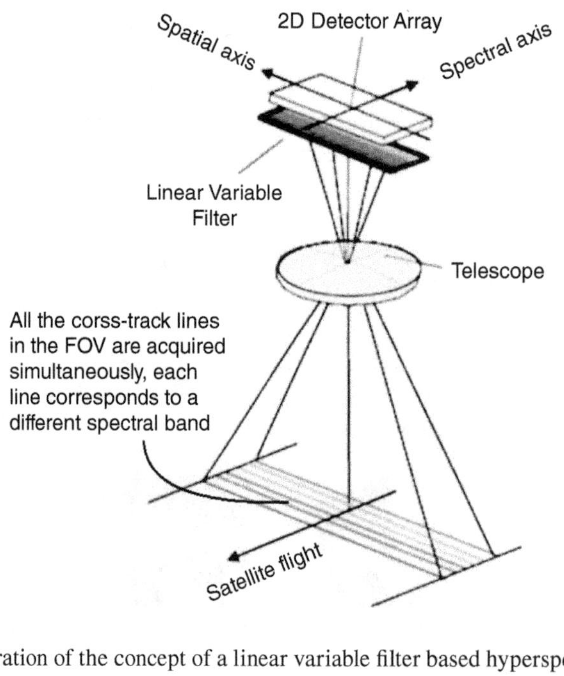

FIGURE 1.8 An illustration of the concept of a linear variable filter based hyperspectral sensor.

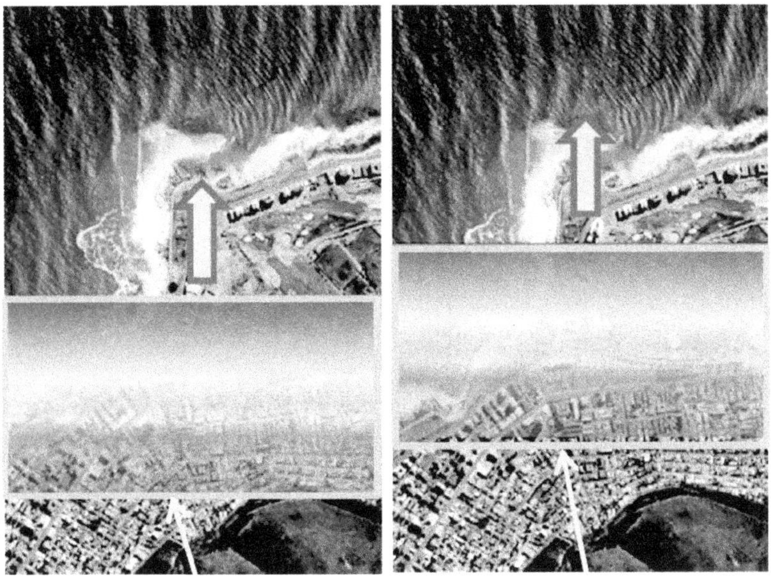

FIGURE 1.9 Two field of views of a linear variable filter based hyperspectral sensor passing over a ground track when a satellite flies at the moment $T = 0$ and $T = t$.

pushbroom mode acquires an entire cross-track line on the ground at a moment (see Figure 1.2), whereas a dispersive element based hyperspectral sensor working in the whiskbroom mode acquires a single sampling cell in a cross-track line at a moment (see Figure 1.3). Unlike a dispersive element based hyperspectral sensor, a LVF-based hyperspectral sensor acquires simultaneously all the cross-track lines in the FOV, this is because the FOV of a LVF-based hyperspectral sensor is not limited by a slit to only one cross-track line in the along-track direction. Each row of the detector array images a corresponding cross-track line of the scene but in a different waveband than the neighboring lines. Hence, every moment of time a detector frame is imaged, a complete 2D scene is acquired. Each line of the scene is acquired at a different wavelength. Figure 1.9 illustrates the two FOVs of a LVF-based hyperspectral sensor passing over a ground track when a satellite flight at the moment $T = 0$ and $T = t$. As the satellite flies over the scene, each filter row passes over the scene. Once the complete FOV has passed over the scene, each different filter row has sensed all cross-track lines of ground sampling cells in the scene and acquired all spectral components. A spectrum of each ground sampling cell can be reconstructed by reorganizing these images acquired at different moments.

Like a grating- or prism-based spectrometer, the throughput of an optical filter based spectrometer is a complex function that includes a principle wavelength as well as a variety of cross-talk wavelengths. For a single detector row, there is a principle wavelength, and up to 15% of the total throughput corresponds to cross-talk from other wavelengths. Removal of the spectral cross-talk needs to be carefully done as part of the image post-processing. Trying to correct too many cross-talk wavelengths, for example, will introduce excess noise and degrade the resulting product SNR. The optimal approach is to correct only the key cross-talk wavelengths.

At the time of this book being written, there were two spaceborne hyperspectral imagers based on a LVF. The Hyperspectral Imager (HySI) onboard Indian Mini Satellite-1 (IMS-1) launched in April 2018 used a LVF, which covers a wavelength range from 400 nm to 900 nm with a bandwidth of 10 nm dispersed over 512 spectral elements. This hyperspectral satellite is described in Section 2.12. Another LVF-based spaceborne hyperspectral imager is hyperspectral nanosatellite HyperScout.

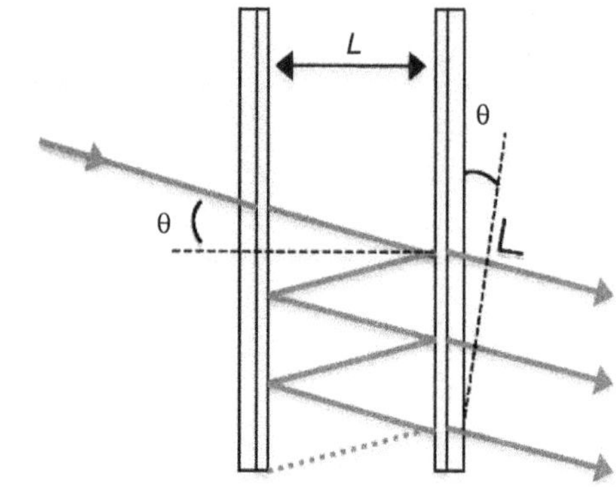

FIGURE 1.10 The principle of a Fabry-Pérot filter.

It is a 3U CubeSat with the hyperspectral imager being 1U (10 cm × 10 cm × 10 cm). The LVF covers a wavelength range from 450 nm to 900 nm with a spectral resolution of 10 nm (Conticello et al. 2016). This spaceborne hyperspectral imager is described in Section 2.20.

1.2.2.2 On-Chip Stepped Fabry-Pérot Filters

Fabry-Pérot interferometer-based filters have been adopted to transmit the selected spectral bands of interest. It is named after Charles Fabry and Alfred Pérot, who developed the filter in 1899 (Fabry and Pérot 1899). A Fabry-Pérot filter consists of two tightly spaced parallel mirror surfaces, i.e., an etalon and an interferometer, between which incoming light reflects and interferes (Figure 1.10). This interference allows some wavelengths to pass, while others are filtered out. Its transmission spectrum, as a function of wavelength, exhibits peaks of large transmission corresponding to resonances of the etalon. By carefully controlling the distance between the reflecting mirror surfaces, a Fabry-Pérot filter can be designed to transmit only a narrow-wavelength band wanted.

A practical solution for fast, compact, user-friendly and low-cost hyperspectral imaging sensors can be achieved by monolithically integrating optical interference filters on top of complementary metal-oxide-semiconductor (CMOS) based image sensors. These integrated filters can be made by traditional semiconductor fabrication tools, enabling mass production and, therefore, low-cost manufacturing. In order to integrate Fabry-Pérot filters on top of a CMOS detector array, the design of the filters requires the selection of CMOS-compatible materials and suitable fabrication steps. However, the process steps and materials used for post-processing the filter structures must be carefully selected. They are crucial for the quality of the Fabry-Pérot filters but may not damage the underlying CMOS detector arrays.

Geelen et al. (2013) reported that Fabry-Pérot filters in a wavelength range of VNIR (400–1000 nm) have been processed directly on top of the wafer that contains silicon-based CMOS image sensor chips. Figure 1.11 shows the Fabry-Pérot filters for a line scan imager. Each row of a CMOS detector array is integrated with a row of same band-pass Fabry-Pérot filters. Each Fabry-Pérot filter has a fixed distance between the two reflecting mirror surfaces for a particular band-pass. The use of monolithically integrated Fabry-Pérot filters on top of a detector array is an innovative approach to combining spectroscopy with imaging technology for hyperspectral sensors. This can result in low cost, compactness, and high speed. As a natural evolution of this innovative technology, the wavelength range of the CMOS-based Fabry-Pérot filters has been extended to shortwave infrared (SWIR): 1000–1700 nm, together with InGaAs-based detector arrays in 2018 (Imec 2018).

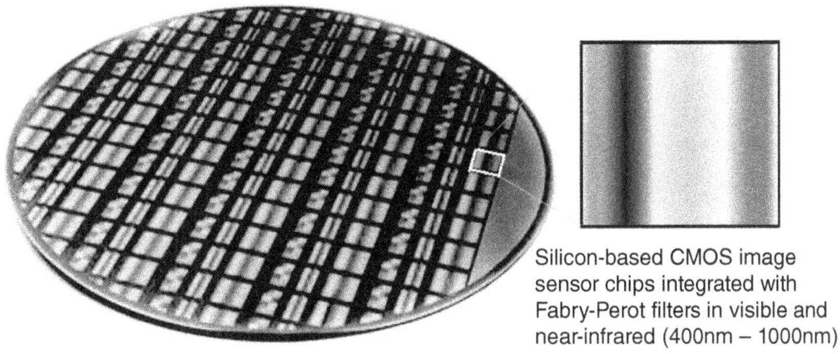

FIGURE 1.11 Fabry-Pérot filters in a wavelength range of visible and near-infrared (400–1000 nm) integrated on top of silicon-based CMOS image sensor chips at the wafer level.

This new integration approach is unique and has attractive advantages: First, combining the production of filters and detector arrays into one CMOS-compatible process leads to an overall simplification and cost reduction, and enables massive production. Second, the monolithic integration induces less cross-talk between neighboring bands and reduces stray light in the system. This also has a positive effect on the system's sensitivity and speed.

The upper part of Figure 1.12 shows a silicon-based CMOS detector array of format 1088 rows by 2014 pixels working in a wavelength range of VNIR. The same band-pass Fabry-Pérot filters are integrated on top of the detector pixels in a particular row of the detector array when the detector array was fabricated at the wafer level. The spectral band-pass of the filters is controlled by the distance between the two parallel mirror surfaces (i.e., the thickness). The same band-pass Fabry-Pérot filters can be deployed on multiple rows of the detector array. In the figure, the same band-pass filters (having the same thickness) are integrated on top of every six adjacent rows of pixels. These multiple rows of pixels have the same thickness (or height) of deposited filters and look like a step of a stair. This is why this kind of Fabry-Pérot filters on top of a detector array is also referred to as on-chip stepped filters (OCSF), as illustrated in the lower part of Figure 1.12.

The configuration of the OCSF is a series of filter rows disposed one behind the other in the along-track direction that allows to create a pushbroom hyperspectral imager same as in a LVF-based hyperspectral imager. Most importantly, this configuration provides the choice to sense objects on the ground using multiple detector rows (such as 6 rows) of the same spectral band-pass to boost the SNR using the on-chip time-delayed integration (TDI) or off-chip data binning in the spectral dimension.

The VNIR detector array with the OCSF shown in Figure 1.12 has 181 spectral bands spanning a wavelength range from 450 nm to 960 nm with a spectral sampling interval of 5 nm. The FWHM of each spectral band is about 15 nm. The width of a step is 6 detector rows (these rows are deposited the same Fabry-Pérot filters).

The OCSF can be adjusted to the center pixel to an accuracy of about ±1 nm. The instrumental line shape is a Lorentzian function. The FWHM of the line shape is approximately between 7 nm and 20 nm with an accuracy of about ±1 nm. The OCSF shown in Figure 1.12 covering the spectral range from 450 nm to 960 nm, this spectral range has been extended to 420–1000 nm. Typical transmittance efficiencies are between 80% and 95% when not considering the detector quantum efficiency.

Figure 1.13 illustrates the concept of data acquisition of spectral components using an OCSF-based hyperspectral sensor versus using a conventional dispersive element based hyperspectral sensor. In the case of the conventional dispersive element based hyperspectral sensor (assuming working in the pushbroom mode), all spectral components (λ_1, λ_2, ..., λ_n) of the ground sampling cells within a cross-track line (x_1) are acquired simultaneously at the moment $T = t_1$. All spectral

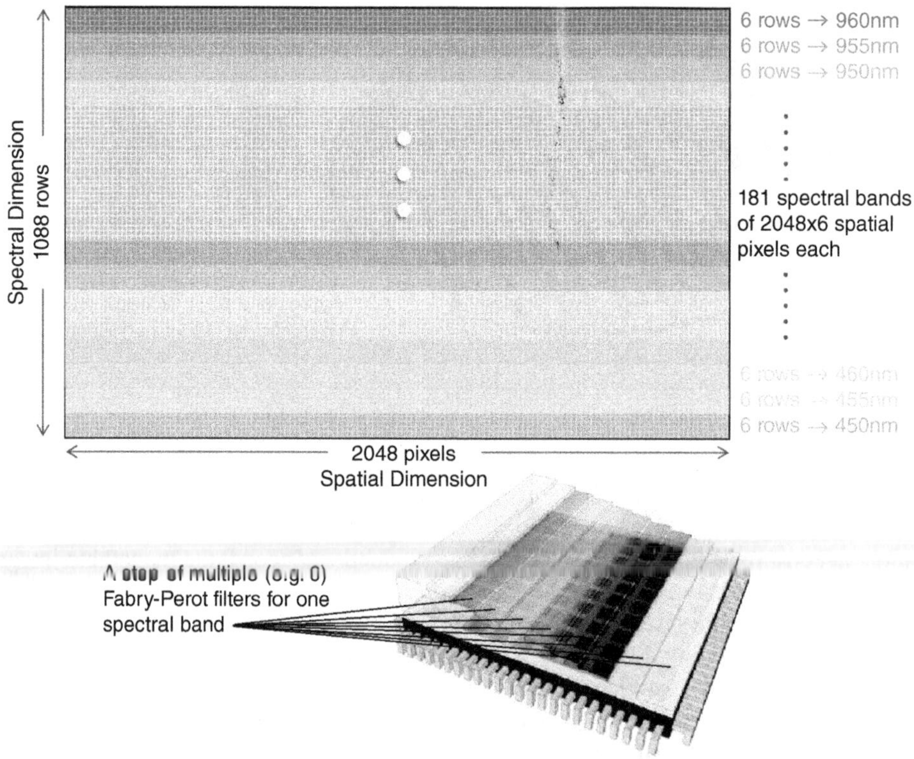

FIGURE 1.12 On-chip stepped filters covering a wavelength range of visible and near-infrared (400–1000 nm), each step having 6 rows with the same spectral band-pass filters.

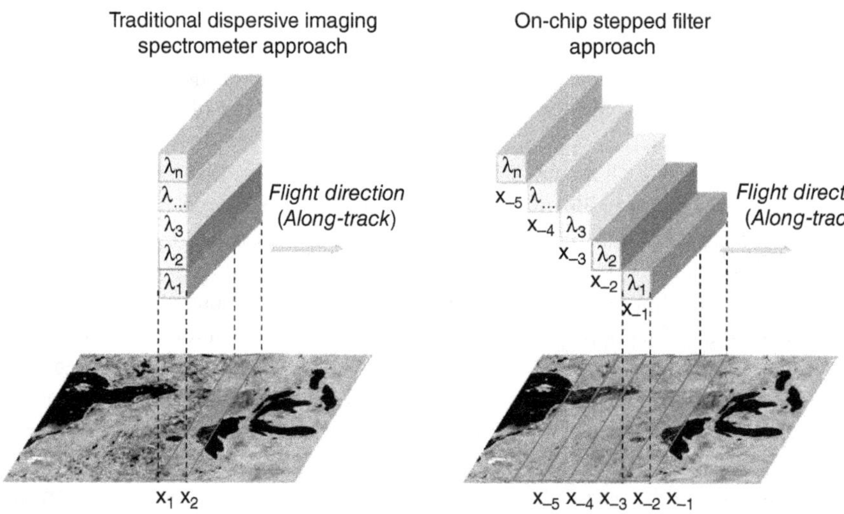

FIGURE 1.13 Illustration of acquisition of spectral components using an on-chip stepped filter based hyperspectral sensor versus using a conventional dispersive element based hyperspectral sensor hyperspectral sensor.

components (λ_1, λ_2, ..., λ_n) of the ground sampling cells within the next cross-track line (x_2) are acquired simultaneously at the moment $T = t_2$ when the satellite flies in the along-track direction, and so on, until a sufficient amount of cross-track lines are acquired to generate a datacube.

In the case of the OCSF-based hyperspectral sensor, at the moment $T = t_l$, the ground sampling cells in a total of n cross-track lines are acquired, but only for one corresponding spectral component, i.e., the ground sampling cells in cross-track line x_1 are acquired for spectral component λ_1, in cross-track line x_2 for spectral component λ_2, ..., and in cross-track line x_n for spectral component λ_n. In order to generate a datacube, the ground sampling cells in each cross-track line need to be scanned by all the filter steps and acquired. Let's use cross-track line x_1 as an example to see how it works. At the moment $T = t_l$, the ground sampling cells in cross-track line x_1 have been acquired with spectral component λ_1. At the moment $T = t_2$, with the satellite flight in the along-track direction, the detector array with filter step λ_2 chipped in over cross-track line x_1, the ground sampling cells in cross-track line x_1 are acquired with spectral component λ_2. At the moment $T = t_3$, the ground sampling cells in cross-track line x_1 are acquired with spectral component λ_3, and so on, until at the moment $T = t_n$, the ground sampling cells in cross-track line x_1 are acquired with spectral component λ_n. After n moments, the ground sampling cells in each cross-track line are acquired with all spectral components, and thus a datacube can be constructed.

The main difference between an OCSF-based hyperspectral sensor and a dispersive element based hyperspectral sensor is that at one moment the former acquires a 2D image of a scene containing n cross-track lines, each of which obtained only one spectral component, whereas the later acquires the ground sampling cells in only one cross-track line, but obtained all the spectral components simultaneously.

Figure 1.14 shows two COTS hyperspectral cameras that were fabricated by Interuniversitair Micro-Electronica Centrum (IMEC) using an OCSF. These hyperspectral sensors consist of a fore-optics and a silicon-based CMOS detector array with integrated OCSF, plus imaging controlling electronics. There are no spectrometers in the hyperspectral cameras. That is why their volume and mass are all small. The volume is about 6 cm × 6 cm × 8 cm and the mass is around 0.6 kg. The hyperspectral camera shown in the lower part of Figure 1.14 uses IMEC's first-generation OCSF covering a spectral range 600–1000 nm with about 100 spectral bands and 5-nm bandwidth. The hyperspectral camera shown in the upper part of Figure 1.14 uses IMEC's second-generation OCSF covering a spectral range 470–960 nm with about 150 spectral bands.

FIGURE 1.14 Two on-chip stepped filter based hyperspectral cameras.

The author of this book worked with his federal government partners at the Department of National Defence (DND) and the members of Canadian industry teams and conducted studies on the OCSF-based hyperspectral sensors. An OCSF-based hyperspectral imaging breadboard was built using an IMEC hyperspectral camera. The breadboard hyperspectral sensor was characterized in an optical laboratory. Hyperspectral images were collected using a translation stage setup to acquire images and analyze spatial alignment and spectral fidelity of the reconstructed cubes.

The characterization of the breadboard was carried out on various targets with interested or known spectra to verify the capability of the OCSF in identifying the spectrum of materials. Five rare earth doped spectralon targets were used because their spectra are well known and often used as standards to validate spectrometers. Figure 1.15 shows the rare earth holmium (WA17A-1622) doped spectralon target used in the tests. Figure 1.16 presents spectra of the rare earth holmium doped spectralon acquired by using a field portable spectrometer manufactured by the Analytical Spectral Devices (ASD) and the Fabry-Pérot filter based hyperspectral camera with the second-generation OCSF (covering a wavelength range from 470 nm to 960 nm). In the figure, the spectrum acquired by the OCSF-based hyperspectral camera is the mean of multiple single-pixel spectra selected manually in a region of interest. Two spectra of the standard deviation upper and lower bound (+std, −std) are also shown. It can be seen from the figure that the spectrum of the rare earth holmium doped spectralon acquired by the OCSF-based hyperspectral camera is quite similar to that acquired by the ASD spectrometer, especially for the absorption features at around 540 nm, 640 nm, 760 nm, and 880 nm. The narrow absorption features at 650 nm and 665 nm are not well captured by an OCSF-based hyperspectral camera. This is understandable because the FWHM of the Fabry-Pérot filters integrated on top of the silicon-based CMOS detector array is around 15 nm, even though their spectral sampling interval is 5 nm in the wavelength range from 450 nm to 960 nm.

Following the success of the laboratory tests, hyperspectral images were acquired outdoors using the breadboard in parallel with a conventional dispersive element based hyperspectral instrument. In-scene targets included white and dark panels, rare earth targets, an orange cone, and people. The reconstructed datacubes from the OCSF-based hyperspectral sensor were compared to those from both the reference ASD spectrometer and the dispersive element based hyperspectral sensor to evaluate the spectral output of the OCSF-based hyperspectral sensor.

FIGURE 1.15 Rare earth holmium (WA17A-1622) doped spectralon.

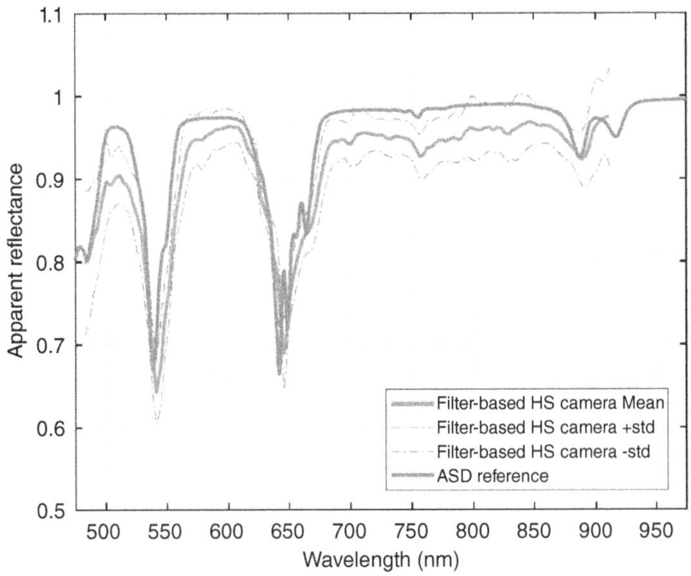

FIGURE 1.16 Spectra of the rare earth holmium doped spectralon acquired by an ASD spectrometer and an on-chip stepped filter based hyperspectral camera.

In order to demonstrate the capability of an OCSF-based hyperspectral sensor, the breadboard hyperspectral camera underwent two airplane flight campaigns together with the conventional dispersive element based hyperspectral instrument. Figure 1.17 shows an example of the airplane flight campaigns. The airborne setup consisted of a platform for airborne measurement with the dispersive element based hyperspectral instrument (HySpex) in which the OCSF-based hyperspectral breadboard was integrated. A scene near a baseball field in Saint-Henri, Quebec, Canada was selected. Ten regions of interest (ROIs) were identified in the scene, as listed at the bottom of Figure 1.17. The airplane flew at an altitude of 10,000 feet at a speed of about 100 knots. In the figure, the spectra of the material in ROI 10 acquired by both OCSF-based hyperspectral breadboard and the dispersive element based hyperspectral instrument are shown. These are the spectra of at-sensor radiance after processing of spatial alignment and spectral calibration. It can be seen that they are quite similar and this technology shows great promise. The team continues to work on the processing methodology to improve the spectral performance.

1.2.2.3 Electronically Tunable Filters

An ETF-based hyperspectral sensor uses a filter that is mounted in front of a monochrome camera by electronically tuning its spectral transmission (i.e., band-pass) to produce a stack of image slices at a sequence of wavelengths. An ETF is a device whose spectral transmission can be electronically controlled by applying voltage, acoustic signal, etc. (Gat 2000). The advantage of an ETF-based imaging spectrometer is that an entire 2D spatial image of a spectral band is formed instantly when the filter tunes to a particular band-pass wavelength. Unlike dispersing element based and LVF/OCSF-based hyperspectral sensors, there is no need to observe multiple cross-track lines by the satellite flight motion to obtain the second spatial dimension or to accumulate the spectral dimension. This advantage is at the cost of the additional time required to tune the filter to cover the whole wavelength range before the satellite moves its FOV to observe the next FOV on the ground.

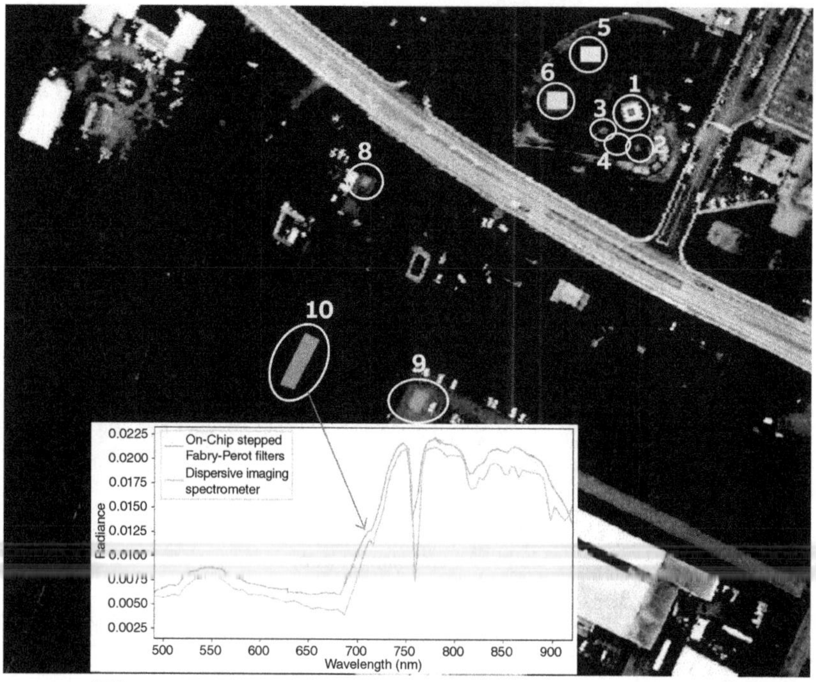

ROIs:

1. Grey calibration panel
2. Black calibration panel
3. Blue tarp
4. Infield (dirt)
5. Outfield green
6. Outfield yellow
7. N/A
8. Blue roof
9. Parking Asphalt
10. Field green

FIGURE 1.17 An example of the airplane flight campaigns to demonstrate the capability of an OCSF-based hyperspectral sensor.

There are typically three classes of ETFs:

1. Liquid crystal tunable filter (LCTF), which uses electronically controlled liquid crystal elements to transmit a desired wavelength of light and block others. A LCTF has high image quality and is relatively easy to be integrated into an optical system. A disadvantage of a LCTF is its lower peak transmission values in comparison with conventional fixed-wavelength optical filters due to the use of multiple polarizing elements. A LCTF is also effected by temperature. Higher temperature can decrease the transition time for the molecules of the liquid crystal material to align themselves and for the filter to tune to a particular wavelength. Lower temperature can increase the viscosity of the liquid crystal material, and thus increase the tuning time of the filter from one wavelength to another.
2. Acousto-optic tunable filter (AOTF), which is based on the principle of diffraction. An acousto-optic modulator, also called a Bragg cell, uses the acousto-optic effect to diffract and shift the frequency of light. Compared with a LCTF, an AOTF has faster tuning speed (microseconds versus milliseconds) and a wider wavelength range. A disadvantage of an AOTF is its relative poor imaging quality due to the acousto-optic effect of sound waves to diffract and shift the frequency of light.
3. Interferometer-based filters. For example, a Fabry-Pérot filter is an interferometer-based filter that has been described in Section 1.2.2.2.

Abdlaty et al. (2018) reported a comparison of performance between an AOTF-based and LCTF-based hyperspectral imagers in medical applications for the purpose of highlighting the leverage points of the two type of filters to facilitate their selection in hyperspectral imager design. In their experiments, three parameters were examined: spectral resolution, out-of-band suppression (spectral cross-talk), and image quality in the sense of spatial resolution. The experimental results demonstrated that AOTF-based hyperspectral imager showed superiority in spectral resolution, out-of-band suppression and random switching speed between wavelengths, whereas LCTF-based hyperspectral imager had better performance in terms of the spatial image resolution, both horizontal and vertical, and high definition quality. They concluded that an efficient design of a hyperspectral imager is application-dependent. For medical applications, for instance, if the tissue of interest required more spectral information for undefined optical properties, or contains close spectral features, AOTF might be the better option. Otherwise, LCTF is more convenient and simpler to use, especially if the tissue chromophore's spatial mapping is needed.

The Visible and Near-infrared Imaging Spectrometer (VNIS) aboard Chinese Chang'E-3 lunar spacecraft is an AOTF-based hyperspectral imager. Chang'E-3 achieved lunar orbit on December 6, 2013 and landed on December 14, 2013, becoming the first spacecraft to soft-land on the moon since the Soviet Union's Luna 24 in 1976. The VNIS is described in detail in Section 2.15.

At the time of this book being written, to the best knowledge of the author there was no spaceborne hyperspectral imager based on a LCTF.

Figure 1.18 shows a concept drawing of a Fabry-Pérot filter based hyperspectral sensor. The Fabry-Pérot filter is mounted in front of the imaging optics and lets pass only light that is at the resonance condition $\lambda = 2d$ (first-order). Each plate separation generates a 2D image at a wavelength with spectral extent given by the FWHM of the transmission response. The way a Fabry-Pérot filter is used here is different from that in the on-chip Fabry-Pérot filter based hyperspectral sensor. Fabry-Pérot filters described in Section 1.2.2.2 are deposited on the top of a CMOS detector array at the wafer fabrication level with fixed cavity for a fixed band-pass, whereas Fabry-Pérot filters used here are tunable for band-pass by electronically controlling the width of the cavity.

At the time of this book being written there were at least two spaceborne hyperspectral imagers that are based on the technology of electronically tunable Fabry-Pérot filters. The Greenhouse Gas Satellite-Demonstrator (GHGSat-D) microsatellite, which was launched in June 2016 (Germain et al. 2016), uses an electronically tunable Fabry-Pérot filter operating in a wavelength region between 1600 nm and 1700 nm with a spectral resolution on the order of 0.1 nm. This wavelength range and spectral resolution were selected for the presence of spectral features for greenhouse gases methane and carbon dioxide, as well as relatively little interference from other

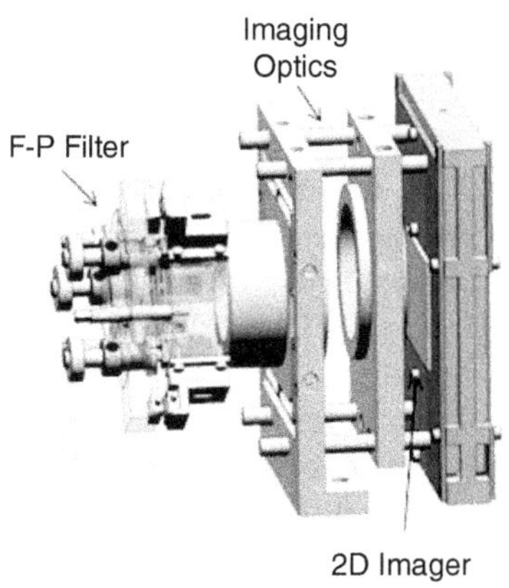

FIGURE 1.18 A concept drawing of an electronically tunable Fabry-Pérot filter based hyperspectral sensor.

atmospheric species, H_2O in particular. This miniaturized ETF-based hyperspectral imager has a mass of 5.4 kg and a volume of 36 cm × 26 cm × 180 cm. This spaceborne hyperspectral imager is described in Section 2.17.

Another spaceborne hyperspectral imager is called Aalto-1 Spectral Imager (AaSI) onboard the Aalto-1 3U CubeSat. A tunable Fabry-Pérot filter is electronically controlled in a closed capacitive feedback loop by three different piezo actuators to cover a spectral range from 500nm to 900 nm with a spectral resolution of 6–20 nm. For detailed description of this spaceborne hyperspectral imager, refer to Section 2.18.

1.2.3 Snapshot Hyperspectral Imagers

Both dispersive element based hyperspectral imagers (Section 1.2.1) and spectral filters based hyperspectral imagers (Section 1.2.2) require a scanning to sequentially generate a hyperspectral datacube by either whiskbroom (point-scanning), pushbroom (line-scanning), or wavelength scanning (using spectral filters). A snapshot hyperspectral imager generates a hyperspectral datacube during a single integration time of a detector array. No scanning is required. It incorporates specialized components to distribute a scene's 3D spatial-spectral information onto a 2D detector array. Snapshot hyperspectral imaging typically requires detector arrays with a high number of pixels. The elimination of moving parts means that motion artifacts can be avoided.

The system design of snapshot hyperspectral imagers is generally more complex than scanning hyperspectral imagers. They makes use of recent technology, such as large format detector arrays, high-speed data transmission, advanced manufacturing methods and precision optics. A snapshot hyperspectral imager can often offer much higher light collection efficiency than equivalent scanning instruments. However, this advantage can be fully benefited only by tailoring the design to particular applications, such as in the professional astronomical community. One of the main reasons for the popularity of the snapshot technology in the astronomical community is that it offers large increases in the light-collection capacity of a telescope when performing hyperspectral imaging (Hagen et al. 2012).

Hagen and Kudenov (2013) reviewed the snapshot spectral imaging technologies comprehensively. In the review, they defined the spectral imaging as a measurement attempting to obtain an $I(x, y, \lambda)$ datacube of a scene, in which the spectral dimension is sampled by more than three elements. This section briefly describes five popular snapshot spectral imaging technologies that show more promise for hyperspectral imaging.

1.2.3.1 Multi-Aperture Filtered Camera

A multi-aperture filtered camera (MAFC) is composed of an array of imaging elements, such as a monolithic lens array, with a different filter placed at each element of the detector array to collect a desired spectral band, as shown in Figure 1.19. There are typically three different versions of implementation.

Figure 1.19a and b show two versions of implementation that consist of an array of lenses co-registered to an array of spectral filters, with the entire set coupled to a monolithic detector array. The difference between these two implementations is the location of the filters. The filters are placed in front of the lens array in Figure 1.19a (Shogenji et al. 2004), whereas the filters are placed in front of the detector array in Figure 1.19b (Imec 2013).

Figure 1.19c shows another implementation using a light field camera with a filter array. A lenslet array is placed at the focal plane of the objective lens, and the detector array lies at a pupil plane as imaged by the lenslets. The image behind each lenslet is an image filtered by the filter array and modulated by the scene's average spectral distribution across the lenslet. This implementation has an advantage of being able to use a variety of objective lenses, so that zooming, refocusing, and changing focal lengths are easier to achieve at the cost of more complex and less compact than the other two implementations (Levoy et al. 2006, Horstmeyer et al. 2009).

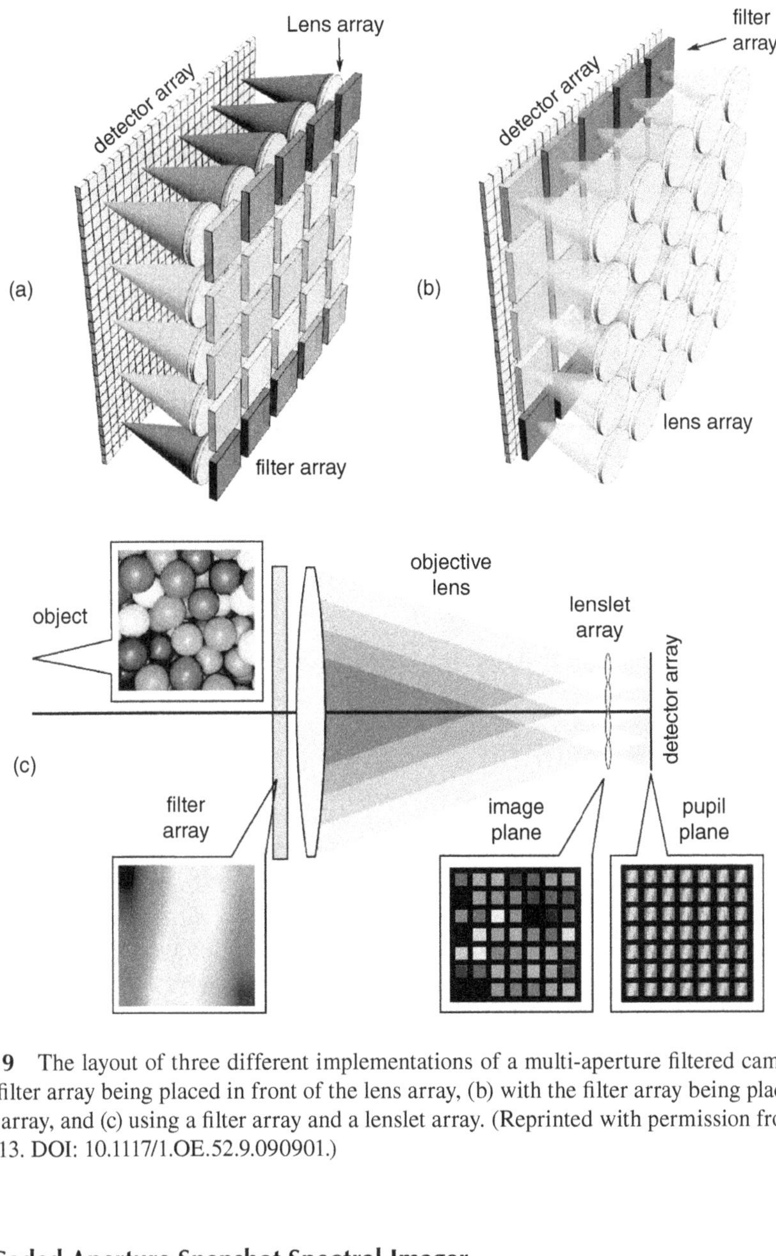

FIGURE 1.19 The layout of three different implementations of a multi-aperture filtered camera approach, (a) with the filter array being placed in front of the lens array, (b) with the filter array being placed in front of the detector array, and (c) using a filter array and a lenslet array. (Reprinted with permission from Hagen and Kudenov 2013. DOI: 10.1117/1.OE.52.9.090901.)

1.2.3.2 Coded Aperture Snapshot Spectral Imager

A coded aperture snapshot spectral imager (CASSI) replaces the entrance slit of a dispersive spectrometer with a much wider field stop. A binary-coded mask (typically an S-matrix pattern or a row-doubled Hadamard matrix, see the lower part in Figure 1.20) is inserted inside the field stop. The mask attempts to create a transmission pattern at each column within the slit such that each column's transmission code is orthogonal to that of every other column (Mende et al. 1993, Wagadarikar et al. 2007, Wagadarikar et al. 2009). The encoded light, transmitted by the coded mask within the field stop, is then passed through the collimating lens, disperser, lens, and detector array of a standard spectrometer. Because the columns of the coded mask are orthogonal, when they are smeared together by the disperser and multiplexed on the detector array, they can be de-multiplexed during post-processing. The setup allows the imager to collect light over a wide aperture without sacrificing the spectral resolution that one would lose by opening wide the slit of

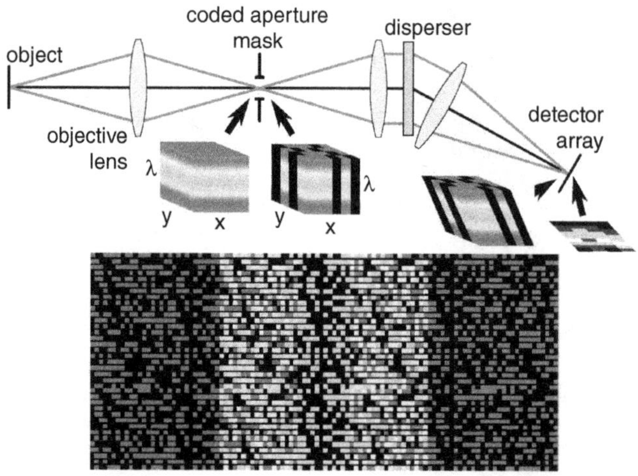

FIGURE 1.20 Top: The layout of a coded aperture snapshot spectral imager, showing only the single-disperser configuration. Bottom: the pattern on the detector array. (Reprinted with permission from Hagen and Kudenov 2013. DOI: 10.1117/1.OE.52.9.090901.)

a standard slit spectrometer. The trade-off is a factor of two in light loss at the coded mask, and in some noise enhancement due to the signal processing.

1.2.3.3 Image Mapping Spectrometry

An image mapping spectrometry (IMS) based snapshot spectral imager uses an array of micro-faceted mirrors placed at an image plane. These mirror facets share the same tilt angle, so that multiple slices of the image are mapped to each individual pupil plane. The resulting pattern, imaged at the detector array, resembles that of a scene through a picket fence. If there are nine individual pupils in the imager system, then the spaces between the fence's slats are 1/9th the slat widths. Figure 1.21 shows the concept layout of an IMS-based snapshot imager. In the figure, nine individual pupils are used as an example. There are nine facet images on the detector array, each representing the image shifted by 1/9th of a slat width relative to the others. Figure 1.21a displays the nine sub-images with the lorikeet being imaged. Figure 1.21b displays the nine sub-images with the lorikeet being imaged when the prism array has been removed from the system. Figure 1.21c displays the nine sub-images when a spectrally and spatially uniform object is being imaged through the full system. Assembling all nine sub-images thus allows one to replicate the original scene with the spectrum dispersed.

The IMS approach allows each pupil to be shared among many mirror facets; the system design becomes much more compact and allows for higher spatial resolution. The primary drawback of the IMS approach is the need for very high precision for cutting the mapping mirror, which is the central element of the system. Current ultraprecision lathes have advanced to the point where it is possible to make these elements on monolithic substrates, though considerable care is involved.

Gao et al. (2009) reported the first IMS-based snapshot spectral imager (they referred it to as an image slicing spectrometer or ISS at that time). The demonstrated prototype can simultaneously acquire 25 spectral images spanning a spectral range of 140 nm with a spectral resolution of 5.6 nm for fluorescence spectroscopy. The generated hyperspectral datacube has a size of 100 pixels × 100 pixels × 25 spectral-bands using a large-format CCD array. Two years later, they improved the datacube size to 350 × 350 × 46 (Gao et al. 2011).

The IMS-based snapshot spectral imaging technology has been pushed to remote sensing applications, such as monitoring vegetation, urban development, and lightning observation. Dwight et al. (2018) reported a compact snapshot image mapping spectrometer (SNAP-IMS) for unmanned aerial vehicle (UAV) hyperspectral imaging in collaboration with a NASA team. The SNAP-IMS

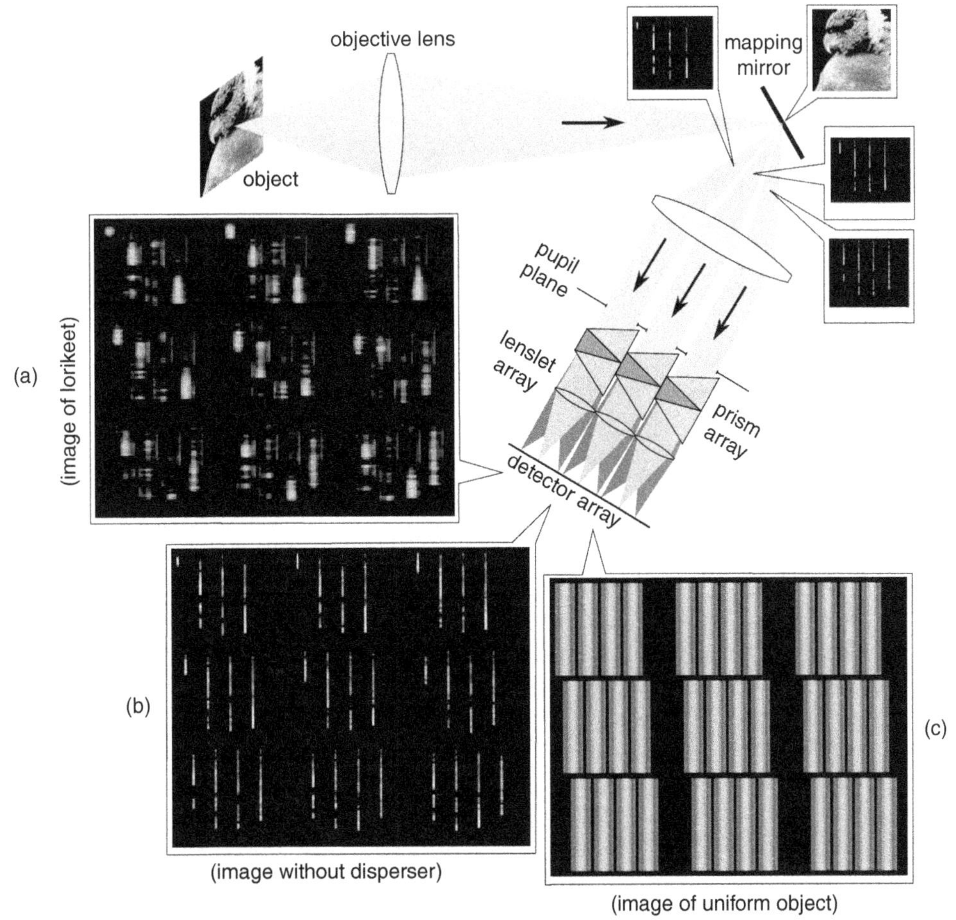

FIGURE 1.21 The concept layout of an image mapping spectrometry based snapshot spectral imager with nine individual pupils. (Reprinted with permission from Hagen and Kudenov 2013. DOI: 10.1117/1.OE.52.9.090901.)

instrument has a FOV of 10.6° and an instantaneous field of view (IFOV) of 0.03°. It can acquire a hyperspectral datacube of size 350 × 400 × 55 (x, y, λ) within a single camera frame. These 55 spectral bands are acquired simultaneously covering a wavelength range from 470 nm to 670 nm in the visible spectral region. This wavelength range is limited only by the choice of the detector array, band-pass filter, and dispersion optics. A hyperspectral datacube can be acquired at 1/500 sec to 1/100 sec, eliminating motion artifacts associated with the platform motion. The instrument has a very small volume of 28.8 cm × 15.3 cm × 16.3 cm and a mass of 3.6 kg. It has been integrated with an Octocopter UAV. The payload's power consumption is marginal as there are no mechanical scanning components. The existing power requirements are dedicated exclusively to CCD frame acquisition. Experimental testing included several flights onboard the Octocopter UAV. The small size and mass, and low power consumption can offer longer and higher flights at a smaller drone, which is attractive for nano- or micro-satellite missions.

1.2.3.4 Snapshot Hyperspectral Imaging Fourier Transform Spectrometer

A snapshot hyperspectral imaging Fourier transform (SHIFT) spectrometer is based on a birefringent polarization interferometer behind a lenslet array (Kudenov and Dereniak 2010, 2012). As depicted in Figure 1.22, the interferometer contains a pair of Nomarski prisms (NP), NP_1 and NP_2, each of which consists of two birefringent crystal prisms with wedge angle α. There is also a half-wave plate (HWP) between NP_1 and NP_2. An N × M lenslet array images a scene through the generating polarizer (G),

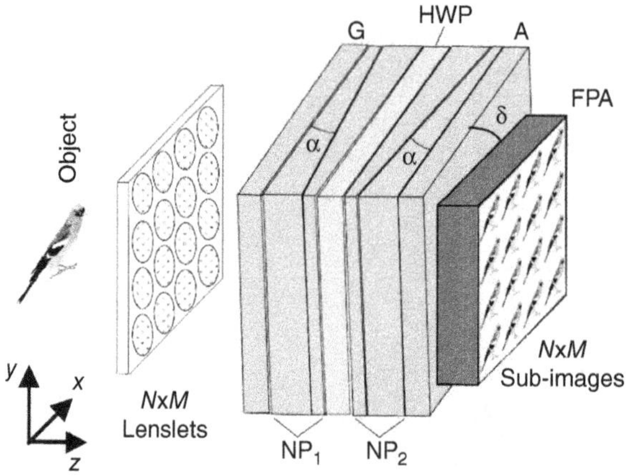

FIGURE 1.22 The layout for a snapshot hyperspectral imaging Fourier transform spectrometer. (Reprinted with permission from Hagen and Kudenov 2013. DOI: 10.1117/1.OE.52.9.090901.)

a pair of Nomarski prisms and the analyzer polarizer (A). Thus, N × M sub-images are formed on a detector array. Rotating the interferometer by a small angle δ relative to the detector array enables each one of the sub-images to be exposed to a different optic path difference (OPD). Therefore, a 3D interferogram cube can be assembled by sequentially extracting each one of the sub-images. Fourier transformation, along the OPD axis of the interferogram cube, enables reconstruction of the 3D datacube. This prism-based design allows for a reduced system volume and an improved robustness to vibration. A SHIFT spectrometer performs its spectral measurement in the time domain. It is similar to MAFC but uses a Fourier-domain approach without any filters.

Kudenov and Dereniak (2012) reported a laboratory prototype of the SHIFT spectrometer. A 10 × 10 lenslet array is placed behind a 10 × 10 aperture array. Each lenslet has a 2.4-mm radius of curvature and a 4.6-mm focal length. The 50-mm focal length objective and collimating lenses are identical and have focal ratios of 1.8. The focal plane array is a Kodak Megaplus camera with a 2048 × 2048 pixel and pitch size of 7.4 × 7.4 μm. Three emission sources were imaged to verify the SHIFT spectrometer's spectral calibration. These included mercury (Hg), neon (Ne), and high pressure sodium (Na) gas discharge lamps. Acquired spectra from the SHIFT spectrometer were compared to those obtained using an Ocean Optics USB 2000 spectrometer with a full-width spectral resolution of approximately 25 cm^{-1} at 633 nm in a wavelength range of 425–750 nm. The spectra produced by the prototype of the SHIFT spectrometer is reasonably similar to those measured using the Ocean Optics spectrometer.

Additional test data from the SHIFT proof-of-concept sensor have been collected for outdoor targets, including several moving automobiles at a traffic light intersection and grass. These data were acquired with an integration time of 1/40th of a second in the late afternoon on a clear day. Notable in the spectrum of the grass is the increased signal for vegetation past 714 nm due to chlorophyll.

1.2.3.5 On-Chip Fabry-Pérot Filters

The IMEC in Belgium reported an innovative snapshot hyperspectral imaging technology (Imec 2013). The key concept of their snapshot hyperspectral imager is the use of Fabry-Pérot filters that are directly post-processed at the wafer level on top of a CMOS detector array. A snapshot hyperspectral imager based on this concept allows spectral information of a scene to be captured at one discrete point in time. It is simply composed of three parts: (1) a fore-optics (e.g., an objective lens or a telescope) that forms an image of the scene; (2) an optical subsystem that duplicates the scene onto each filter tile; and (3) a filter array organized in a tiled configuration monolithically integrated on top of a CMOS detector array, where each filter is designed to sense only one narrow band of wavelengths.

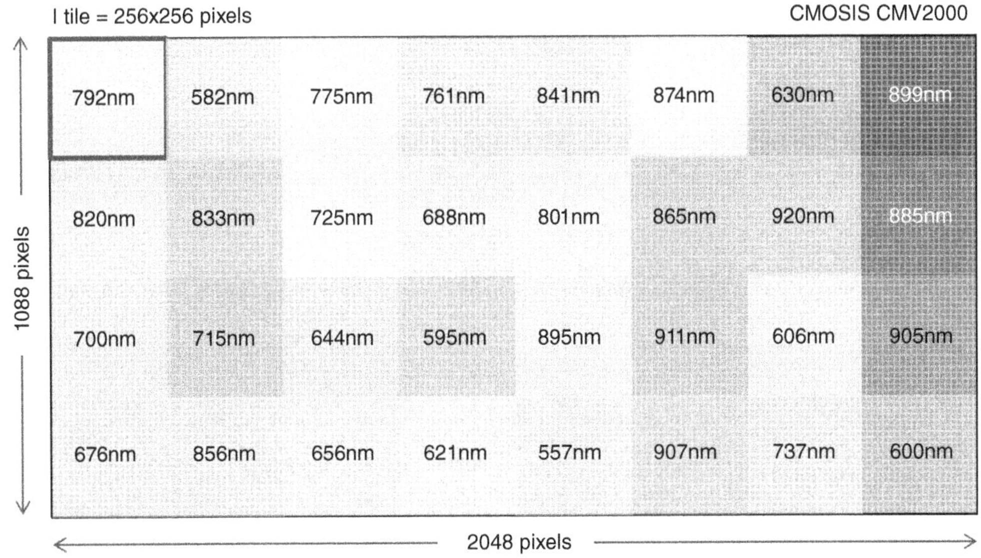

FIGURE 1.23 Fabry-Pérot filters monolithically integrated on top of a CMOS detector array in a tiled configuration for snapshot hyperspectral imaging. Each tile contains 256 × 256 pixels for a narrow (10–15 nm) band of wavelengths as shown in the tile. (Courtesy of IMEC.)

Figure 1.23 shows a layout of the Fabry-Pérot filters monolithically integrated on top of a CMOS detector array in a tiled configuration. There are 4 × 8 = 32 tiles spanning a wavelength range of 557–920 nm. Each tile contains 256 × 256 pixels for a narrow (FWHM≈10–15 nm) band of wavelengths as shown in the tile. These simultaneously acquired spectral bands selected by the filters in the tiles of a scene can be easily reconstructed into a datacube of size 256 × 256 × 32 (x, y, λ). The acquisition speed can be up to 340 datacubes per second.

Figure 1.24 exhibits a mosaic configuration of the same format array of Fabry-Pérot filters. In this configuration, a block of 5 × 5 detector pixels is integrated with 5 × 5 different Fabry-Pérot filters on

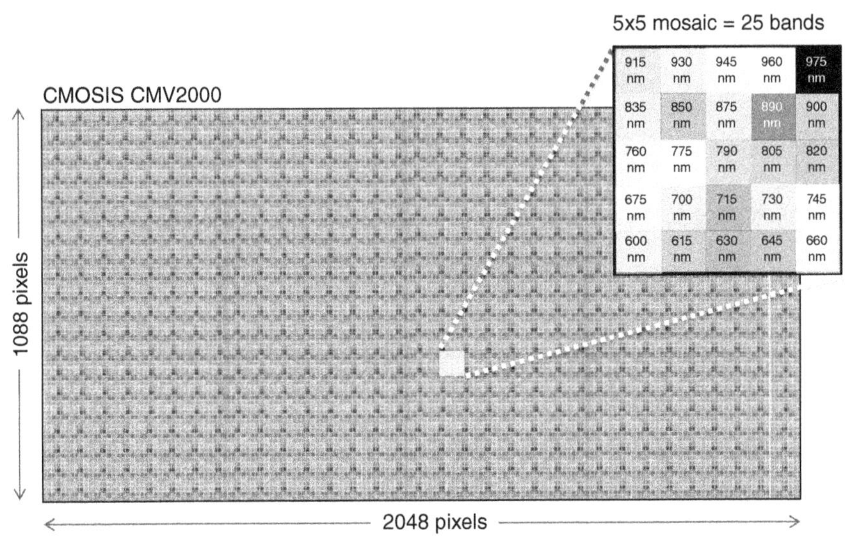

FIGURE 1.24 Fabry-Pérot filters monolithically integrated on top of a CMOS detector array in a 5 × 5 mosaic configuration for snapshot hyperspectral imaging. (Courtesy of IMEC.)

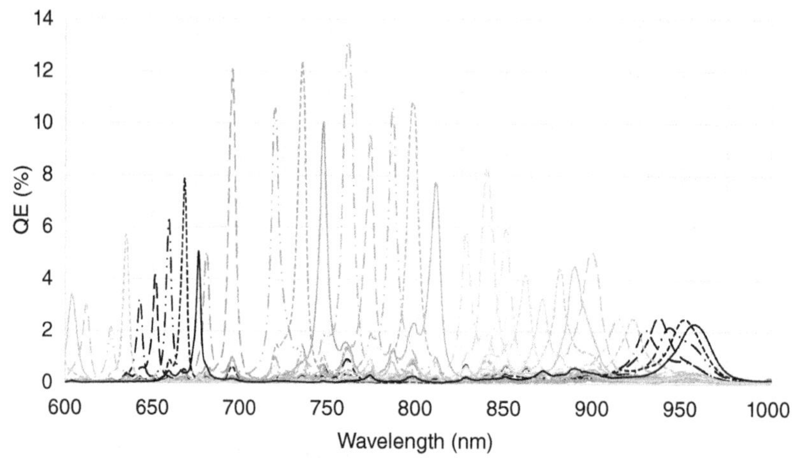

FIGURE 1.25 Shape and bandwidth of the 5 × 5 Fabry-Pérot filters in the 600–1000 nm near-infrared region. (Courtesy of IMEC.)

top of them covering a spectral range of 600–975 nm with a FWHM bandwidth of 10–15 nm. With this mosaic configuration of the Fabry-Pérot filters, there is no need of an optical subsystem that duplicates the scene onto each filter tile, thus the snapshot imager becomes even simpler. The simultaneously acquired spectral bands selected by the filters in the mosaic blocks of a scene can be easily reconstructed into a datacube of size 410 × 218 × 25 (x, y, λ). The acquisition speed can also be up to 340 datacubes per second. Figure 1.25 depicts the shape and bandwidth of the Fabry-Pérot filters in the near infrared (NIR) region (600–1000 nm) without applying the micro-lenses. It can be seen from the figure that the quantum efficiency (QE) is relatively low at the two ends of the wavelength range. There are also high-order band-passes for some filters, thus special treatments are required to remove them.

There is a difference between a tile configuration based snapshot imager and a mosaic configuration based snapshot imager in acquiring spectral information of ground cells in the scene. In the case of a mosaic configuration based snapshot imager, the ground area sensed by a 5 × 5 pixels mosaic block should contain an object of the same material. In other words, the area of an object of the same material should not be smaller than the area sensed by a mosaic block of detectors. In this way, all 25 spectral bands of the same material are collected. Otherwise a mosaic block of detectors may collect spectral information for more than one material. There is no such problem in the tile configuration based snapshot imager due to the optical subsystem that duplicates the scene onto each filter tile.

1.3 HYPERSPECTRAL IMAGING FROM AIRCRAFT TO SPACECRAFT

1.3.1 Scientific Rationale for Hyperspectral Remote Sensing

As the level of sophistication in optical remote sensing instruments and techniques is growing, more applications are being found that require imaging with high spectral resolution. Imaging spectrometers, which record a full spectrum for each pixel of an image, are ideal sensors for this purpose. They allow optimal mapping of targets with well-defined optical signatures, flexible band placement for minimizing and estimating the effects of atmospheric scattering and absorption, and accurate simulation of the responses of other sensors.

The term imaging spectroscopy has been used more frequently from the perspective of instrumentation, especially in the beginning when it emerged as a new technology. Now, it is often referred to as hyperspectral imaging and has a long history of development. The development of hyperspectral remote sensing began in the 1970s from field spectral measurements in support of

NASA's Landsat-1 data analysis. The need for hyperspectral imaging came from the recognition based on laboratory and field spectral measurements, primarily of minerals and soils, that multispectral imaging in four broad spectral bands with the Landsat's Multispectral Scanner (MSS) was not adequate to discriminate among, much less identify, minerals on the earth's surface that were important in resource exploration and environmental assessment.

The scientific rationale for hyperspectral imaging is evident given its strong needs from diversity of disciplines. In the 1970s, when geologists used Landsat MSS images to identify minerals, they found that many of the morphological clues of the minerals were absent in the four spectral band data and that the spectral reflectance was the major key to mapping geological units. Field measurements with portable spectrometers covering the 0.4–2.5 μm region showed that spectral reflectance could be used to identify the mineralogy of exposed units, particularly if an OH-bond was present. Laboratory spectra of OH-bearing minerals clearly showed that spectrometers with a resolution of 10 nm FWHM or less were necessary to identify individual minerals. The Landsat MSS had 100–200 nm wide bands, which are too wide to identify the individual minerals (Goetz 1995).

The need for imaging spectroscopy in vegetation analysis has spurred renewed interest in the potential for hyperspectral remote sensing. For example, the rates of ecosystem function reflected in the lignin-to-nitrogen ratio are important to understanding the land contribution to the carbon cycle. Moreover, imaging spectroscopy provides the only means of assessing the biochemistry of forest canopies remotely.

Hyperspectral imaging is, for example, helpful in understanding the productivity and nutrient cycling in oceans for modeling the global carbon cycle. Phytoplankton distribution and variability is an important determinant of food-web structure and trophodynamics, which ultimately controls the flux of carbon in the oceans. Light in the visible region of the spectrum penetrates the ocean and lake water from a few centimeters to tens of meters, the depth depending on the amount of absorbing and suspended material in the water. The upwelling spectra can be decomposed into abundances of different phytoplankton groups: colored dissolved organic matter, detritus, and suspended sediments. Imaging spectrometry is a powerful tool in determining the composition of suspended matter when all the above components are present.

Remote sensing of snow and ice also needs hyperspectral imaging. For example, the extent of snow and ice cover changes dramatically with the season, and because of its high reflectance, it has a large effect on the Earth's albedo. Snow reflectance depends on the grain size, illumination angle, contaminants, and depth when the thickness is small. In the NIR region, ice is moderately absorbing and the grain size has a large effect on the reflectance. In turn, the grain size is indicative of the thermodynamics of the snow and can give an indication of the beginning of melt.

1.3.2 History of Development of Airborne Hyperspectral Imagers

1.3.2.1 First Airborne Hyperspectral Imager—AIS

The history of terrestrial hyperspectral remote sensing began with the development of the first hyperspectral imager for remote sensing—the Airborne Imaging Spectrometer (AIS). At that time neither the detectors, optics, electronics nor computers for rapid data analysis were readily available. The AIS was built in the early 1980s as an engineering test bed using the first hybrid infrared 2D detector arrays, as well as a tool for research in hyperspectral remote sensing (Goetz et al. 1985, Vane and Goetz 1988).

In 1979, the first hybrid detector array became commercially available. This 2D detector array consisted of a matrix of mercury-cadmium-telluride (HgCdTe) detectors bonded to a matched silicon CCD readout array. Although this first detector array contained only 32 × 32 elements, it enabled the construction of an imaging spectrometer that covered the spectral region of 1.2–2.4 μm beyond the 1.1-μm cutoff range of silicon detector arrays.

At the heart of the AIS was a 32 × 32 element HgCdTe array. The optical layout is shown in Figure 1.26. Mirrors M1 and M2 comprise the telescope, which is in a concentric Schwarzchild

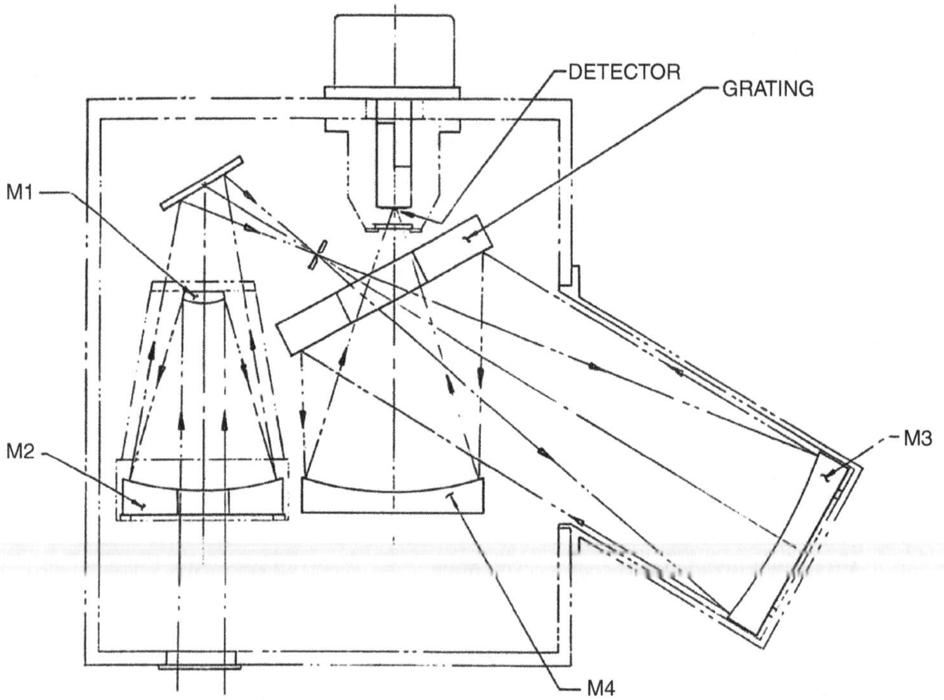

FIGURE 1.26 The optical layout of the Airborne Imaging Spectrometer (AIS) instrument. (Courtesy of NASA/JPL.)

configuration. The focal length is 70.7 mm with an effective focal ratio of f/3. The telescope images a cross-track line on the ground onto the entrance slit of the spectrometer that in turn disperses the light onto the 2D array detector. The AIS is composed of M3 and M4 mirrors and the grating. Mirror M3 is a 300 mm effective-focal-length f/3 paraboloid and M4 is a 150 mm effective-focal-length f/1.5 paraboloid. The resulting magnification of 0.5 combined with the telescope, which has a magnification of 2, provided a system focal length of 35 mm and IFOV of 1.9 mrad per pixel.

The AIS spectral sampling interval was 9.6 nm per pixel. As the detector array had 32 × 32 pixels, only a wavelength range of 32 × 9.6 nm = 307 nm could be covered. In order to expand the wavelength range, the grating of the AIS was rotated through 4 positions within the 40 ms integration time it took the aircraft to move forward one cross-track line width. Each rotation was 0.42 degrees apart for moving the dispersed image across the detector array such that wavelength bands of 1.2–1.5 μm, 1.5–1.8 μm, 1.8–2.1 μm, and 2.1–2.4 μm are presented sequentially for a readout. This technique was only partially successful, because of the time required for the grating to settle in a new position.

The AIS was a pushbroom imaging spectrometer covering a spectral range of 1.2–2.4 μm. It flew aboard the NASA C-130 aircraft over a number of geological targets in the western United States, and over geobotanical targets and natural and agricultural vegetation across the United States. The AIS provided the first direct identification of minerals on the earth's surface using a remote imaging system. Hyperspectral data of ground scenes were acquired remotely first time onboard an aircraft. This allowed subtle spectral features to be observed that could not be seen in the multispectral data.

The experiments of AIS flights and analyses of the acquired hyperspectral data provided valuable knowledge and useful experience for the development of hyperspectral imagers. The best data were acquired when the AIS was operating in the fixed grating position covering the 2.03–2.32 μm wavelength region, approximately the range covered by the single Landsat Thematic Mapper Band. The first successful hyperspectral data were obtained from a flight over the Cuprite Mining District,

NV, in August 1983. After the individual band images were processed by normalizing each pixel to the scene average to bring out band-to-band differences, spectral reflectance variations became apparent. It was also the first time to see that the hyperspectral band images reconstructed from the raw data showed a major problem associated with the use of a 2D detector array. There were vertical strips in the band images caused by dead pixels or poorly calibrated pixels in the detector array. Each band image had a different set of stripes depending on the location of the bad pixels. To deal with this problem, a strategy of removing the stripes was developed. The bad pixels were removed by replacing them with the average of each of its neighbors in the spectral direction. It was historic that the first hyperspectral image acquired by the AIS showed unambiguous evidence of mineral identification, in particular the minerals alunite and kaolinite. The spectral reflectance curves overlay those taken from laboratory spectra of samples taken from the same field locations.

A second AIS flight over the Cuprite Mining District was undertaken to double-check the capability of the hyperspectral remote sensing. An unknown spectrum was fortuitously observed, which turned out to be the mineral buddingtonite, an ammonium feldspar, after comparing with the known spectral libraries in the laboratory (Goetz and Srivastava 1985). It was thought that buddingtonite might become a pathfinder mineral for gold. The success story of mineral identification, in particular the buddingtonite discovery, which became part of a Public Broadcasting System (PBS) documentary, led to a greater interest within NASA to pursue further hyperspectral sensor development.

The AIS had several shortcomings, including lack of full spectral coverage in the VNIR and a very narrow FOV, analogous to observing the Earth from an aircraft through a soda straw. A later version of the instrument, AIS-2, incorporated a second-generation 64 × 64 element detector array, and covered the 0.8–2.4 μm region. The promising hyperspectral imaging results, and the demonstrated need, led to the development of an imaging spectrometer program within NASA. In 1984, the initial program contained the next-generation aircraft system, the AVIRIS.

1.3.2.2 Airborne Imaging Spectrometer Development Between the 80's and 90's

Following the success of the AIS, in 1983 Alexander F. H. Goetz and his colleagues at Jet Propulsion Laboratory (JPL) proposed to design and develop the AVIRIS. The goal was to introduce the newly available solid-state detector technology into remote sensing applications and to enable the acquisition of images in hundreds of contiguous spectral bands (from visible to shortwave infrared) such that a complete radiance spectrum is available for each pixel (Goetz 1995).

AVIRIS development began in 1984 and the instrument first flew aboard a NASA ER-2 aircraft at a 20-km altitude in 1986, first science data was delivered in 1987, and it has been fully operational since 1989. Since then, it has gone through several major upgrades as technology has advanced in detector arrays, electronics, and computing. The AVIRIS is arguably the finest, best-calibrated, airborne hyperspectral imager ever flown. It is regarded as the first operational hyperspectral instrument and has been the primary provider of airborne imaging spectrometer data to the research community.

The AVIRIS is a whiskbroom imaging spectrometer using 1D linear detector arrays. It covers a wavelength range of 0.4–2.45 μm with 224 spectral bands. The instantaneous FOV is 1.0 mrad, equivalent to a 20-m footprint on the ground from a 20-km altitude. The swath (the width of a cross-track line) is covered by 614 pixels, or approximately 12 km (Vane et al. 1993, Green et al. 1998). AVIRIS is the most accurate spectral imaging system that is radiometrically calibrated, making it possible to apply atmospheric models to obtain apparent surface reflectance. AVIRIS is described in detail in Section 1.3.4.

In the late 1980s, several commercial hyperspectral imagers entered the market. In 1989, ITRES Research Limited based in Calgary, Canada, marketed the Compact Airborne Spectrographic Imager (CASI), an area CCD array based pushbroom imaging spectrometer, which covers a wavelength range from 0.45 μm to 0.90 μm with a bandwidth of 1.8 nm and 512 active pixels in the spatial imaging direction (Babey and Anger 1989, Weale et al. 1990). ITRES has been in commercial airborne hyperspectral imagers since then and extended their product line to cover SWIR and thermal regions of the spectrum. The CASI is described in detail in Section 1.3.4.

TABLE 1.1
Parameters of the Multispectral Infrared and Visible Imaging Spectrometer (MIVIS)

Spectrometer	Wavelength Range (µm)	Bandwidth (nm)	Number of Bands
Visible	0.43–0.83	20	20
Near-IR	1.15–1.55	50	8
Mid-IR	1.983–2.478	8	64
Thermal-IR	8.18–12.70	340–540	10

Geophysical Environmental Research of Millbrook, NY, introduced the Digital Airborne Imaging Spectrometer (DAIS), which is an airborne imaging spectrometer with 79 spectral bands covering the visible and SWIR regions as well as the thermal infrared region (Richter 1996).

Daedalus Inc. manufactured the Multispectral Infrared and Visible Imaging Spectrometer (MIVIS) with a total of 102 spectral bands. It was a modular instrument with 4 spectrometers designed to collect radiation from the Earth's surface in the visible, near-IR, mid-IR and thermal-IR imaging spectrometer. Table 1.1 lists the parameters of the MIVIS. It was a second-generation imaging spectrometer developed for use in environmental remote sensing studies across a broad spectrum of scientific disciplines (Bianchi et al. 1995).

An airborne imaging spectrometer called Hyperspectral Data and Information Collection Experiment (HYDICE) was developed and flown under the sponsorship of the Naval Research Laboratory (NRL) (Rickard et al. 1993). This imaging spectrometer was designed around a prism dispersion concept and a single, hybrid HgCdTe 2D array to cover the 0.4–2.5 µm wavelength range. It had an IFOV of 0.5 mrad, equivalent to a 3-m footprint on the ground from a 6-km altitude. The instrument was mounted on a stabilized platform. Although results were preliminary, SNRs considerably exceeded those of the early version AVIRIS. Based on results from data analysis, HYDICE data should make possible the identification of materials that cover only 1–2% of a pixel, provided that their reflectance spectra are unique. The HYDICE began as a dual-use program, it soon reverted to an all Department of Defense (DoD) program.

Other commercial sensors in this wavelength range were available. The one most similar to AVIRIS is the Australian sensor Hyperspectral Mapper (HyMap) built by the HyVista Corporation, which markets a full-range (0.4–2.5 µm) hyperspectral imaging service on a global basis. The design of HyMap was modular and this led to an ability to configure the spectral and spatial characteristics to suit the customer's specific requirements. The visible (VIS) module includes a silicon CCD detector array operated at ambient temperature, and the NIR and shortwave infrared-1 (SWIR1) and shortwave infrared-2 (SWIR2) modules incorporate InSb arrays cooled by liquid nitrogen to 77 K. Each spectrographic module provides 32 spectral bands. Table 1.2 lists parameters of the HyMap sensor. Its FOV is 60° (512 pixels) corresponding to a swath of 2.3 km at 5-m IFOV or 4.6 km at 10-m IFOV (Cocks et al. 1998).

TABLE 1.2
Parameters of the HyMap Imaging Spectrometer

Spectral Region	Wavelength Range (µm)	Bandwidth (nm)
Visible (VIS)	0.45–0.89	15
Near-infrared (NIR)	0.89–1.35	15
Shortwave Infrared-1 (SWIR1)	1.40–1.80	13
Shortwave Infrared-2 (SWIR2)	1.95–2.48	17

1.3.2.3 Early Imaging Spectrometer Development in Canada

In 1992, Staenz (1992) conducted a review on a decade's worth of research and development of imaging spectrometry in Canada. The development of imaging spectrometers in Canada began in the early 1980s when the Fluorescence Line Imager (FLI), an airborne imaging spectrometer, was designed and constructed for Canada's Department of Fisheries and Oceans (DFO) in 1981 as the first step in developing an advanced satellite sensor based on CCD detector arrays for remote sensing of the Earth's surface. Canadian researchers and engineers pioneered this field, together with their American colleagues at NASA's Jet Propulsion Laboratory, and built the FLI, one of the first imaging spectrometry sensors. Starting in 1983, the FLI flew over a wide variety of land and sea targets to acquire imagery with higher spectral resolution and sensitivity than other comparable sensors (Gower et al. 1987, 1992).

A compact, single-camera version of the instrument, the Compact Airborne Spectrographic Imager (CASI), had been marketed commercially since 1989, as mentioned earlier (Babey and Anger 1989). The Canada Centre for Remote Sensing developed the first Canadian SWIR imaging spectrometer–the SWIR Full Spectrographic Imager (SFSI) in 1992 (Neville and Powell 1992).

These instruments use the flexibility of the imaging spectrometer design to provide high-sensitivity pushbroom imaging in the visible, NIR, and SWIR regions with programmable spectral band configurations.

The FLI was designed as a prototype instrument, with its main goal to map ocean and coastal phytoplankton concentrations by imaging the radiation from solar-stimulated fluorescence of chlorophyll *a*. The detection of phytoplankton from an aircraft by this method was first reported by Neville and Gower (1977). Following the success of several phytoplankton-mapping exercises using the in vivo fluorescence technique in the Arctic (Borstad and Gower 1984, Borstad 1985) and off British Columbia, Canada (Gower 1980, Borstad et al. 1980, Gower and Borstad 1981, Gower et al. 1984), the Canadian Department of Fisheries and Oceans began investigating the possibility of mapping chlorophyll fluorescence from space.

A feasibility study, completed in March 1981, concluded that the required sensitivity and spectral resolution could be achieved using an imaging spectrometer based on 2D CCD detector arrays. The study also noted that an airborne prototype for this instrument would be a significant improvement over other available systems, providing greatly increased sensitivity and spectral resolution.

Development of the FLI airborne prototype by Moniteq Ltd. of Toronto and ITRES Research Ltd. of Calgary began in 1981, and the instrument was delivered in May 1984. Construction was funded jointly by the Department of Fisheries and Oceans and the Inter-departmental Committee on Space (the predecessor of the Canadian Space Agency), which was at the time responsible for Canada's space program. The eventual goal was to produce a sensor that could be launched as part of a satellite payload (Gower et al. 1987).

The FLI was operated by Moniteq Ltd. on a total of about 100 flight campaigns in the period from 1984 to 1990. Projects included the mapping of phytoplankton and benthic vegetation, and the measurement of water depths in lakes and coastal areas. The FLI has successfully mapped chlorophyll fluorescence, thereby fulfilling its design goal (Gower and Borstad 1990). Observations also covered a wide variety of land targets and included projects in North America and Europe. In many of these cases, it was the same target molecule, the photosynthetic pigment chlorophyll *a*, whose fluorescence or absorption effects were being studied on land and in water.

The CASI was designed to make use of the rapid technical developments in computer hardware and CCD detector arrays during the period from 1984 to 1989. Design of more recent units has continued to evolve. In the early 1990s, about 10 CASI instruments were in operation. These have increased the number of applications of imaging spectrometry by providing a sensitive, flexible, and economical source of airborne hyperspectral imagery. For example, the CASI's small volume had made it possible to fly on operational fish-counting surveys in light aircraft for fisheries management and on airborne water-depth mapping surveys with the hydrographic lidar.

Besides FLI and CASI that work in VNIR spectral region, the Canada Centre for Remote Sensing (CCRS) had undertaken to develop an imaging spectrometer in the SWIR region of the spectrum. The objective was to achieve both high spectral resolution (10 nm) and high spatial resolution (0.5 m) from an airborne platform and to record the full spectrum for all image pixels. The SFSI covers a spectral range from 1.2 μm to 2.4 μm in 122 contiguous, 10-nm wide spectral bands. It uses a 2D PtSi detector array and provides a full-image cube with an area coverage of 512 × 512 pixels.

Canadian design and applications experience from the FLI and CASI were used to define the specifications for European Space Agency's (ESA's) Medium Resolution Imaging Spectrometer (MERIS) onboard ESA's ENVISAT, which was launched in 2002.

The FLI, CASI, and SFSI were developed in the mid-1980s to early 1990s, a time of rapid development in the field of electro-optics. Several other instruments were also constructed in the same timeframe or were planned for the near future. Table 1.3 compares the specifications of the FLI, CASI, and SFSI with AVIRIS and MERIS in early years of hyperspectral technology development between 1980s and early 1990s.

1.3.3 Planned NASA Orbiting Imaging Spectrometers in the 1990s

In the late 1970s, JPL developed the Shuttle Multispectral Infrared Radiometer (SMIRR). It was scheduled to fly on the second flight of Space Shuttle (STS-2) in 1979 but was delayed until November 1981. SMIRR was installed in the payload bay of STS-2 alongside the Shuttle Imaging Radar-A (SIR-A). The development of SMIRR could be considered as the beginning of the development

TABLE 1.3
Comparison of Specifications of Selected Imaging Spectrometers in Early Years Between the 1980s and Early 1990s

Parameters	FLI	CASI	SFSI	AVIRIS	MERIS
Operation year	1984	1989	1992	1987	2002
Operation mode	Pushbroom	Pushbroom	Pushbroom	Whiskbroom	Pushbroom
Detector array format	385 × 288	612 × 288	512 × 488	1 × 32 (S1) 1 × 64 (S2–S4)	770 × 576
Number of detector arrays	5[1]	1	1	4[2]	5[3]
Spectral range (nm)	430–805	418–926	1200–2400	410–700 (S1) 680–1270 (S2) 1250–1860 (S3) 1840–2450 (S4)	400–1050
Number of spectral elements	288	288	122	224	520
Number of spectral bands formed	8	72	122	224	15
Spectral sampling interval (nm)	1.3	1.8	10	10	1.25
Number of spatial elements	1925	512	512	614	4440
Field of view (degrees)	70	35	11.7	30	82
IFOV (mrad)	0.65	1.2	0.4	1.0	0.33
Swath (km)	4.8	1–5	0.26	11	1460
Ground sampling distance (m)	2.5	2–5	0.5	20	260
Nominal platform altitude (km)	2	2	2	20	820
Platform	Aircraft flexible	Aircraft flexible	Aircraft flexible	Aircraft ER-2	Spacecraft
Data rate (Mbps)	7	0.875		17	22.3
Digitization	12	12	12	10	12

[1] The five detector arrays were used to cover the full swath of 1925-pixel wide.
[2] Four detector arrays associated with four spectrometers were used to cover the full spectral range (S1–S4).
[3] Five detector arrays associated with five hyperspectral cameras were used to cover the full swath of 4440-pixel wide.

of spaceborne imaging spectrometers. The purpose of the experiment was to test the feasibility of direct mineral identification from the Earth's orbit. The instrument was designed as a profiler to acquire radiance values in 10 spectral bands covering the VNIR and SWIR regions along a 100-m wide track beneath the spacecraft. The instrument consisted of a spare Mariner Venus-Mercury mission telescope with a rotating filter wheel in front of an HgCdTe detector array in the focal plane. Two 16-mm fighter-aircraft gun cameras recorded the ground track. The experiment was successful and resulted in the first direct identification of soil minerals from space through the use of the five closely spaced spectral channels in the 2.2–2.5 μm region (Goetz et al. 1982). The minerals kaolinite and calcite, in the form of limestone, were identified in an Egyptian dry lake and the surrounding hills and were confirmed based on returned samples.

The promising results coming from AIS and SMIRR and the demonstrated needs led to the development of an imaging spectrometer program within NASA. In 1984, the initial program contained the next-generation aircraft hyperspectral imager AVIRIS, the Shuttle Imaging Spectrometer Experiment (SISEX), and the free-flyer High Resolution Imaging Spectrometer (HIRIS), as shown in Figure 1.27. The AVIRIS successfully reached operational status, while SISEX was canceled after the space shuttle Challenger disaster in 1986 when the Shuttle was destroyed by a booster rocket malfunction shortly after liftoff.

In response to the Earth Observing System (EOS) flight proposal opportunity, the HIRIS concept was developed further in 1988 (Dozier 1988). The Science Team of the HIRIS mission consisted of fourteen individuals from all areas of Earth observation science. The mission goal was to deploy on the EOS platform an imaging spectrometer that was pointable covering the wavelength range 0.4–2.5 μm with a 10-nm bandwidth for 192 contiguous spectral bands and spanning a 50-km swath from the 705-km orbital altitude (Goetz and Herring 1989). The designed ground sampling distance was 30 m × 30 m (Barrett and Curtis 1997). The HIRIS proposal was accepted

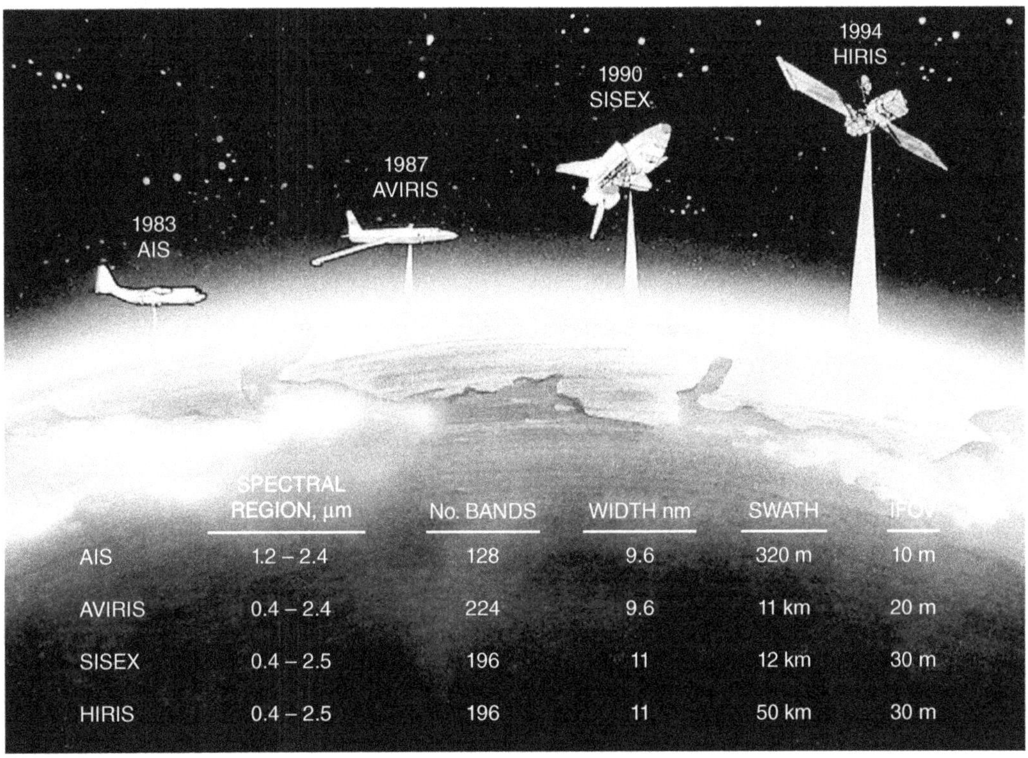

FIGURE 1.27 NASA's plan of early time imaging spectrometer program. (Courtesy of NASA/JPL.)

for the EOS; however, HIRIS did not fly as the overall program was scaled back to approximately one-fourth of its original size.

It became clear at that time that the Earth observation community was not sufficiently exposed to the advantages of imaging spectrometry and they had not worked with high-quality hyperspectral datasets in order for them to become more familiar with the hyperspectral satellites. In addition, they were not aware of what scientific questions could be addressed uniquely with imaging spectrometry and, therefore, the community was not in a position to become a strong advocate for the satellite.

In retrospect, at the time of the demise of HIRIS only a handful of investigators had actually worked with hyperspectral data mainly because there existed only a few hyperspectral datasets, and they were of poor quality compared to what is available today. Additionally, there were not any readily available software tools to process the datasets that were orders of magnitude larger than the multispectral datasets from Landsat, and desktop and powerful computing systems were not yet up to the task. Therefore, it is not surprising that HIRIS was not included in the EOS.

Ultimately, in November 2000, NASA launched Hyperion, a spaceborne imaging spectrometer built by TRW, aboard the Earth Observing-1 (EO-1) satellite (Ungar et al. 2003). If one compares the successful Hyperion with HIRIS, it can be easily seen that Hyperion bears some of the HIRIS instrument characteristics. EO-1 was created to develop and validate a number of instrument and spacecraft bus breakthrough technologies. It was intended to enable the development of future Earth imaging observatories that would have a significant increase in performance while also having reduced cost and mass. EO-1 was decommissioned in March 2017 after over 16 years of successful operation. Hyperion onboard EO-1 is described in detailed in Section 2.5.

In the beginning of 1990s, NASA also worked with TRW to develop a spaceborne spectrometric system called HyperSpectral Imager (HSI) for the LEWIS mission as a technology demonstration as part of its Small Spacecraft Technology Initiative (SSTI) program (De Long et al. 1995). HSI was designed to shoot for 128 bands in the spectral range 0.4–1.0 μm and another 256 bands in the 0.9–2.5 μm range, for a total of 384 bands. The spectral sampling intervals in both spectral ranges were 5.0 nm and 6.5 nm, respectively, which are still very competitive to the present-day spaceborne hyperspectral sensors. The swath width was 7.7 km and ground sampling distance was 30 m. Unfortunately, HSI was unsuccessful. Three days after the launch on 23 August 1997, the control of the satellite was lost and subsequently entered the Earth atmosphere in September 1997 (Lewis (SSTI 1) 2014).

1.3.4 Two Major Airborne Hyperspectral Sensor Developments Since the Beginning

The development of airborne hyperspectral sensors is a precursor to the development of spaceborne hyperspectral sensors. Firsthand experience acquired and lessons learned during the course of development of airborne projects will significantly enhance the chances of success of the following space mission. Airborne instrument development helps elevate technical readiness level of spaceborne hyperspectral missions and significantly reduce the technical risk of the payloads.

This section first describes the difference between airborne and spaceborne hyperspectral imaging, and then provides a detailed description of two major airborne hyperspectral imagers: AVIRIS and CASI. These are the two airborne hyperspectral imagers developed since the very beginning in the history of terrestrial hyperspectral imaging. They are good examples and considered being representative. AVIRIS is the first operational hyperspectral instrument and is widely used by the scientific community, whereas CASI is a commercial airborne hyperspectral imager with the longest history in the market and has been extended to cover other spectral regions.

Before getting into the detailed description of the two typical airborne hyperspectral imagers, it is worth the effort to briefly discuss the difference between airborne and spaceborne hyperspectral imagers from the perspective of system design and operation.

1.3.4.1 Difference Between Airborne and Spaceborne Hyperspectral Imaging

The design consideration and development of a spaceborne hyperspectral imager are different from those for an airborne hyperspectral imager. The design and development of spaceborne instruments are very demanding tasks and need to focus on assurance of long-term reliability and stability in the harsh space environment, where the instruments are expected to be exposed to hard ionization and proton radiation, extreme temperature excursions, ultrahigh vacuum, and very strong vibrations and accelerations during the launch with a space carrier. Moreover, hard limitations are incurred with regard to mass, volume, and available power supply. A relevant limitation is related to the narrow band for data download to ground reception stations, which requires data preprocessing and compression onboard the spacecraft.

Given the high cost of space missions and an inherent impossibility of satellite maintenance on orbit, an extreme reliability of the space instruments is of great importance. The basic means to assure the reliability includes usage of space-qualified components and materials, qualified and certified technological procedures, and qualification testing of finished spaceborne devices. Also important is a redundant multi-level design, from individual functional blocks up to full dual redundancy (i.e., two identical systems).

Since the AIS is the first airborne hyperspectral imager, its design and development were to prove the concept of imaging spectrometer technology. It was based on the best technology available at that time, such as the 32 × 32 HgCdTe detector array. Some design strategies were used to match the available technology or overcome their shortages. For example, due to the extremely small number (32) of pixels of the detector array in the spectral dimension, the wavelength range that could be covered was only 307 nm. In order to expand the wavelength range, the AIS was designed to rotate the grating through 4 positions to cover 4 segments to reach the final wavelength range of 1.2–2.4 μm. This strategy was only partially successful because of the reduction of integration time and the additional time required for the grating to settle in a new position. With the technology advancement of detector arrays, large format detector arrays became available, and this design strategy was retired automatically.

The design of AVIRIS has four key features: (1) whiskbroom-scan architecture, (2) four separate spectrometers, each covering a portion of the entire spectral range, (3) very fast optics (f/1), and (4) large detector pitch size (200 μm). An advantage of the whiskbroom scan permits the use of 1D linear detector arrays instead of 2D area detector arrays. This yields easier and more accurate calibration while minimizing the number of artifacts that bother the design using area array detectors. The penalty of the whiskbroom-scan design is its limited integration time for ground sampling cells, because the total integration time for a cross-track line on the ground is shared by the 614 ground sampling cells when the spectrometers scan each of them before the aircraft moves in the along-track direction to image the next cross-track line (see Figure 1.3). Maximizing the throughput of the spectrometers is necessary in order to counteract the limited integration time. The last three features of AVIRIS were aimed at maximizing the signal by providing maximum optical throughput and transmittance including grating efficiency. The point-scan (i.e., one ground sampling cell per scan) design requires a large aperture, which in the case of AVIRIS is 200 mm in diameter. The whiskbroom-scan architecture dominated the design of AVIRIS and yielded a large volume of the instrument.

Unlike AIS whose overall spectral range was achieved by rotating the grating through four positions, the AVIRIS design used four separate spectrometers, each covering a portion of the entire spectral range. This design eliminated the moving mechanism of the grating. It also removed the need of additional time used for settling the grating at each position and led to allocation of more time for integration. Obviously, this design of using four separate spectrometers is a negative factor for reducing the volume and mass of the system. Present available 2D detector arrays have sufficient number of pixels in both dimensions for covering an entire range of wavelength and spanning a wide swath in the cross-track direction. The design strategy of using four separate spectrometers for covering an entire spectral range is no longer being adopted.

A pushbroom imaging spectrometer images an entire line of ground sampling cells in the cross-track direction, while the platform flies forward in the along-track direction (see Figure 1.2). The line image (1D) of the cross-track line, formed on the slit of the spectrometer by the fore-optics, is then dispersed by a dispersive element onto a 2D detector array, which provides spectral information along one axis and spatial information along the other. This architecture effectively integrates as many acquired spectra as there are from the ground sampling cells in a cross-track line into a single instrument. Therefore, it permits substantial mass and volume savings, a definitive advantage especially in resource-constrained space missions.

Other potentially significant advantages of pushbroom hyperspectral imagers include an increase in the integration time, and hence SNR and avoidance of moving parts, which leads to a reduction of polarization sensitivity. In addition, the large increase in integration time permits the use of smaller pitch size of the detector arrays. The pitch size of the detector arrays that AVIRIS uses is 200 μm, which is about 10 times larger than the typical pitch size of the detector arrays widely used in airborne and spaceborne instruments. Longer integration time offered by pushbroom hyperspectral imagers also allows the use of spectrometers with a slower optical speed than the extremely demanding f/1 of AVIRIS.

A potential disadvantage of pushbroom hyperspectral imagers is the requirement for a wide-field telescope. However, many designs of wide-field telescope have been proposed that provide a wide linear field with excellent correction. Another disadvantage is the use of 2D array detectors, which causes problems of uniformity, calibration, readout artifacts, and striping. A systematic approach for reducing or removing artifacts has been proposed and developed (Neville et al. 2003, 2004, 2008, Sun et al. 2006, Qian et al. 2006, Qian 2013). Detector and calibration artifacts can typically be handled with post-processing of data.

In the design of spaceborne instruments, the volume and mass of the instruments are the key parameters that need to be taken into account. The experience learned from the AVIRIS indicates that a whiskbroom-scan hyperspectral imager that uses 1D detector arrays normally has a large volume and heavy mass. The weight of the AVIRIS is around 327 kg (720 lbs), and the volume is around 1.57 m^3 (33 in. wide × 63 in. long × 46 in. high). A pushbroom hyperspectral imager that uses the 2D detector array technology normally yields relatively small volume and mass. For example, Hyperion is a pushbroom hyperspectral imager onboard NASA's EO-1 satellite. The mass of Hyperion is 49 kg and the volume is 0.19 m^3 (0.75 m × 0.39 m × 0.65 m). That is the reason why almost all the spaceborne hyperspectral imagers, which had been in orbit, are currently flying, or have to be flown, are of pushbroom design. (Table 2.1 provides a summary of spaceborne hyperspectral sensors to date.)

From the perspective of operations, a basic difference between airborne and spaceborne hyperspectral imaging is the use of their platform. Airborne hyperspectral imaging uses aircrafts as a platform and has some distinct advantages for remote sensing systems. Airborne remote sensing missions are often carried out as one-time operations, whereas Earth observation satellites offer the possibility of continuous monitoring of the Earth. Aircrafts can fly at very low altitudes and gain relatively strong input signals. Airborne hyperspectral imagers admit a sub-meter spatial resolution. The airborne sensors can be used more flexibly, because of their easy adaption to variation of altitude and the forward scan velocity. Therefore, spatial resolution and swath width can be easily adapted to the task requirements in the course of a flight. Aircraft can easily change the flight schedule to avoid weather problems, such as clouds, which may block a passive FOV of the imaging spectrometer, and to allow for sun illumination. Instrument maintenance, repair, and configuration changes are easily applicable to aircraft platforms. There are no bounds for an aircraft but the country borders. The disadvantages are for a low-flying aircraft with a narrow FOV; a number of flybys are needed to cover a large ground area due to the small coverage area. Aircrafts are less stable than a spacecraft, thus the image quality of the instrument onboard suffers. The cost of per unit area of ground coverage is high. The turnaround time it takes to get the data to the user can be long due to necessity of landing the aircraft before transferring the raw imagery to the data processing facility.

Spaceborne hyperspectral imaging can cover a wide area of the Earth's surface and provide periodic monitoring of the same area of interest during the satellite revisits. For satellite hyperspectral imaging, there are no territorial borders, so they can sweep any part of the globe, regardless of the altitude of the countries the onboard imaging flies over. The basic disadvantage of the spaceborne hyperspectral imaging is its multi-hundred million dollars cost to develop the system as well as ground support facilities, while keeping in mind the spacecraft's relatively short operating life of about 5 years or less. Decrease of the satellite mass can result in substantial cost saving, primarily due to less powerful and low-cost launchers to be involved in putting the satellite into orbit. This is why the concept of small satellite (100–500 kg) based hyperspectral imaging was developed and implemented by means of micro-satellites.

1.3.4.2 AVIRIS and Its Next Generation

Airborne Visible/Infrared Imaging Spectrometer (AVIRIS) is the best known airborne hyperspectral imager in the community of terrestrial hyperspectral remote sensing. It is the first hyperspectral imager that measures the solar reflected spectrum from 400 nm to 2500 nm at 10-nm spectral sampling intervals. It offers best calibrated hyperspectral data resulting from its very high SNR and carefully designed calibration system and well-implemented calibration procedure. As an operational airborne imaging spectrometer, AVIRIS is the primary provider of hyperspectral data to the research community since 1987. JPL has organized over 10 workshops on AVIRIS since 1988. Numerous scientific papers related to AVIRIS have been published.

The AVIRIS system has been upgraded and improved since the beginning in a continuous effort to meet the requirements of investigators using AVIRIS spectral images for science research and applications. These improvements have been directed towards the AVIRIS sensor, calibration, data system, and flight operations. In parallel with the sensor, the science research and applications pursued with AVIRIS have diversified and evolved (Green et al. 1998).

1.3.4.2.1 Key Science Requirements

The key science requirements from the AVIRIS investigators drove the determination of performance parameters for the AVIRIS instrument. Some compromise was made after a trade-off study to match the performance available from the key components chosen (e.g., detector arrays) and to accommodate the environmental constraints imposed by the U2 aircraft.

The spectral sampling requirement for the AVIRIS instrument was determined by two key science requirements: the desire to detect shifts in the chlorophyll spectrum on the order of 10–40 nm at 0.7 μm, and the desire to resolve spectral features as narrow as the kaolinite doublet at 2.2 μm. Previous work with the AIS instrument had determined that the 10-nm sampling chosen for the AVIRIS would be adequate. The instrument's longwave cutoff point of 2.5 μm was chosen to avoid viewing thermal emissions (Wallace and Enmark 1987). The AVIRIS is considered a second-generation imaging spectrometer by JPL in comparison to the AIS. It covers the entire spectral region in the VNIR and SWIR from 0.4 μm to 2.5 μm with its 224 contiguous spectral bands over a swath 614 pixels wide.

The SNR requirements were determined by analysis to be at least 100:1 at 0.7 μm for the detection of the chlorophyll shift and at least 50:1 for the detection of the kaolinite doublet. These requirements were for assuming a surface albedo of 0.5, viewed through a standard LOWTRAN atmosphere with 23-km visibility.

The FOV was chosen to be 30°. The ground instantaneous field of view (GIFOV) was chosen to be 20 m. A GIFOV of 10 m would have been achieved by increasing the scan rate and shortening the detector integration time. This provides 614 spatial pixels in the scan direction and a swath width of over 10 km when the altitude of the aircraft was 20 km.

1.3.4.2.2 Description of the Initial AVIRIS Instrument

AVIRIS is modular in construction, consisting of six optical subsystems and five electrical subsystems. In order to achieve the required spectral range, the AVIRIS design used four separate spectrometers (A–D), as shown in Figure 1.28. Spectrometer A covers 0.4–0.7 μm; Spectrometer B covers 0.65–1.25 μm; Spectrometer C covers 1.20–1.82 μm; and Spectrometer D covers 1.78–2.40 μm.

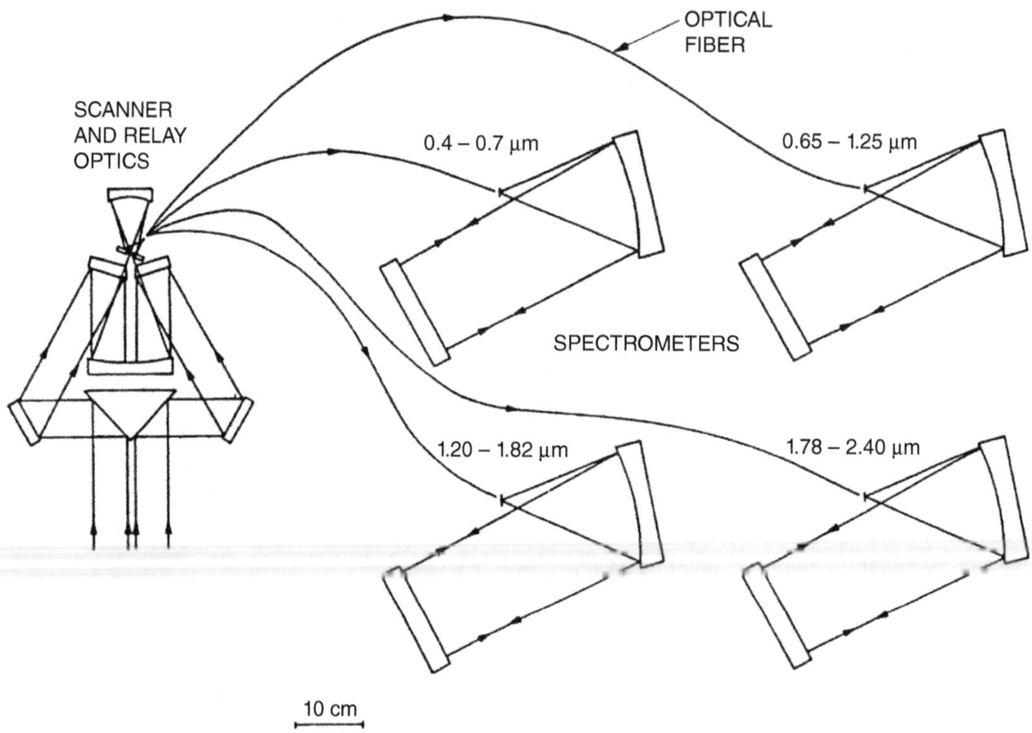

FIGURE 1.28 A schematic of AVIRIS optics. (Courtesy of NASA/JPL.)

An all-reflective decentered aperture Schmidt design was selected for the spectrometers. All four spectrometers have the same basic optical layout and only the grating tilt changes significantly from Spectrometers A to D, as required by the spectral range covered by each of them. A silicon 32-element line detector array was selected for Spectrometer A, while three indium antimonide 64-element line detector arrays were used for Spectrometers B, C, and D. All the four detector arrays have a pitch size of 200 μm × 200 μm.

In addition to four spectrometers, the optical subsystems also include a scanner and a calibration source. They are coupled together through optical fibers. The use of optical fibers to interconnect optics is unique in the AVIRIS system design. This was necessitated by the need to incorporate four separate spectrometers while keeping each spectral band less than one octave wide, to avoid spectral contamination from grating orders overlap. This design provided additional benefits in that it greatly simplified the mechanical layout of the instrument and allowed the various subsystems to be aligned and tested independently of each other.

The fore-optics exhibits a FOV of 33° and active FOV of 30°. The telescope has an effective focal length of 19.76 cm and effective pupil diameter of 14.5 cm. The fore-optics assembly mounted on top of the scanner collects the light from the ground sampling cells being viewed and sends it to the four spectrometers via four optical fibers. The four optical fibers, each 200 μm in diameter (one pixel), lie in a row looking at adjacent pixels on the ground. The fibers carrying the VNIR part of the spectrum to their respective spectrometers (A and B) are made of silica glass. The other two fibers, which carry the SWIR to their respective spectrometers (C and D), are made of fluoride glass.

In each spectrometer, the radiant energy emerging from the fiber optics is collimated by a spherical mirror, which is used in double pass in this AVIRIS configuration. A reflective diffraction grating is located on the double-pass Schmidt corrector at the center of the curvature of the collimating mirror. The grating mask forms the aperture stop of the spectrometer. The dispersed spectrum is imaged on the cooled linear detector array of the silicon detectors for the 0.4–0.7 μm spectral

TABLE 1.4
Parameters of the Four AVIRIS Spectrometers

Parameter	Spectrometer A	Spectrometer B	Spectrometer C	Spectrometer D
Wavelength range (μm)	0.41–0.70	0.68–1.27	1.25–1.86	1.84–2.45
Spectral sampling interval (nm)	9.7	9.5	9.8	9.8
Number of bands	31	63	63	63
Fiber optics material	Silica	Silica	Fluoride	Fluoride
Detector material	Silicon	Indium antimonide	Indium antimonide	Indium antimonide
Detector elements	32	64	64	64
Signal chain gain	235	375	385	750

range (Spectrometer A) and on indium antimonide detector array for the 0.7–2.45 μm spectral range (Spectrometers B–D). Table 1.4 lists the parameters of the four spectrometers.

An onboard reference source is also connected by optical fibers to the spectrometers, providing information on spectral and radiometric calibration.

The four spectrometers are heated to maintain proper focus in the aircraft environment. Each line array detector of the spectrometers is mounted in a Dewar.

The five electrical subsystems include the signal chain, the digital control, data buffers, the roll correction gyro, and the power supplies. The instrument is operated under the control of the digital control subsystem that is based on an Intel 8085A-2 microprocessor. The control subsystem also interfaces with the aircraft's navigation computer to receive flight parameter data, which are recorded along with the science and calibration data. Data pertaining to the operation of the instrument are also recorded.

The aircraft flight parameters and the spatial imaging requirements defined the detector integration time to be 87 μsec. This short integration time produces a small output signal on the order of only a few percent of what the detector arrays are capable of. Because of this, the signal chain is very sensitive to pixel-to-pixel variation and dark current of the detector arrays. Onboard processing was adopted to correct the dark current offset of each pixel of the detector arrays. The Dewars operating at liquid nitrogen temperature were used to maintain the temperature change of the detector arrays within one degree of centigrade.

The development of the initial AVIRIS instrument was completed in 1986 and the first flight was in 1987. It worked within all of its design requirements except for its SNR performance. Both the spectral performance and geometric performance met the design requirements. It achieved a SNR of 150:1 at the spectral band located at 0.7 μm, which exceeded the designed requirement of 100:1, whereas the SNR achieved at 2.2 μm was 30:1, which was lower than the requirement of 50:1.

1.3.4.2.3 Upgrades of the AVIRIS System

Since the development of its initial version, AVIRIS has undergone many modifications and upgrades over the years as technology has advanced in detector arrays, electronics, and computing, which has resulted in performance far exceeding original expectations. These upgrades have been reported in the literature (Porter et al. 1990, Chrien et al. 1991, 1992, 1993, Green et al. 1993, Sarture et al. 1995, Eastwood et al. 2000).

Improvements to the focal planes cover five areas: multiplexer upgrades, signal chain noise reductions, photodiode array material optimization, blocking filter improvements, and an operating temperature change for the silicon detector array. These improvements significantly contribute to the increase of SNR of the AVIRIS. Detailed description on improvements to the AVIRIS focal planes can be found in the literature (Eastwood et al. 2000). Below are some examples of the modifications and upgrades.

In the AVIRIS 1992 engineering and maintenance cycle, the SNR was improved by reducing the noise contribution of the preamplifier circuitry in each of the four spectrometers. This activity resulted in a reduction in noise from approximately 1.8–1.1 digital numbers (DNs). Fixed pattern noise was reduced through increased shielding of the timing and signal chain cables. These noise reduction efforts led to significant improvements in the SNR performance (Green et al. 1993).

During this maintenance cycle, a number of transmissive elements in the AVIRIS instrument were modified. Spectrometer spherical mirrors were recoated to repair accumulated corrosion damage. Antireflection coatings were added to the hatch window, the fore-optics window and the fore-optics termination of the fiber harness. These coatings provided incremental improvements in the AVIRIS throughput. The fore-optics window was tilted to minimize the effects of unwanted internal reflections. All these efforts to improve signal throughput contributed to an increase in the SNR.

To further improve the signal throughput, new spectral order blocking filters were installed in the AVIRIS spectrometers in the 1993 engineering and maintenance cycle. These filters provided both higher throughput in the band-pass and improved blocking of unwanted signals. As an example, the improvement in Spectrometer B ranges between 20% and 50%. In Spectrometer C, a 5–30% increase in the throughput was achieved. Better long-wavelength energy blocking was achieved in Spectrometer D. Reduction in this background energy decreased the noise in the Spectrometer D by as much as 30%.

A set of new fiber harnesses was installed in the AVIRIS. The new fiber material was manufactured to have improved strength and resistance to humidity degradation. Problems with both breakage and humidity damage had been encountered with the vintage fiber material used in the initial version. Improvements were made to the fiber optic spectrometer connector to improve repeatability and stability of spectral alignment. The old connector may have been responsible for some change in the spectral calibration of the Spectrometer B during the 1992 flight season.

An additional improvement to the onboard calibrator was implemented in the 1993 engineering and maintenance cycle as well. In the past, the onboard calibrator signal was reflected into the optical system from one of two sides of a shutter. The side of the shutter measured was an unpredictable function of the flight line acquisition timing. A slight difference in the reflectance of the two sides of the shutter was measured in 1992. A change was implemented in 1993 to measure both sides of the shutter at the beginning and end of each flight line. This eliminated any ambiguity in the measured onboard calibrator signal.

Prior to the 1994 flight season, a new set of detector arrays was integrated into AVIRIS to improve reliability and performance. These detector arrays were based on the updating detector and multiplexor technology. These new detector arrays significantly decreased detector noise with respect to the technology of the detector arrays in 1986. As a consequence of the decreased noise anticipated in the new detector arrays, the AVIRIS digitization was increased from 10 bits to 12 bits to fully encode this improved precision.

Figure 1.29 shows the curves of SNR of the AVIRIS instrument as a function of wavelength for the initial instrument (1987) and after upgradation in years 1994, 1997, and 2000, respectively. It can be seen that the improvements are significant and encouraged. The AVIRIS team worked very hard to achieve these performances and bring the instrument to provide high-quality hyperspectral remote sensing data. There are many useful experiences, know-how, and lessons learned from the development of this high-end airborne instrument. These heritages of the airborne hyperspectral imager development will greatly benefit the development of spaceborne hyperspectral imagers.

1.3.4.2.4 Next-Generation AVIRIS

On the basis of the input received from the research community and potential partners, the JPL team began to conceptualize the next generation of an airborne AVIRIS-like sensor (referred to as AVIRISng) in 2007 to support NASA's Terrestrial Ecology Program and the planned Hyperspectral Infrared Imager (HyspIRI) mission. The AVIRISng instrument was built and integrated in 2010. The volume of the instrument is 83 cm high × 57 cm in diameter plus electronic boxes and racks. Its mass is 465 kg. Major components include a telescope and spectrometer benches, a detector, and a source for the onboard calibrator, as shown in Figure 1.30. The AVIRISng flew for the first time in April 2012 and became operational in early 2013.

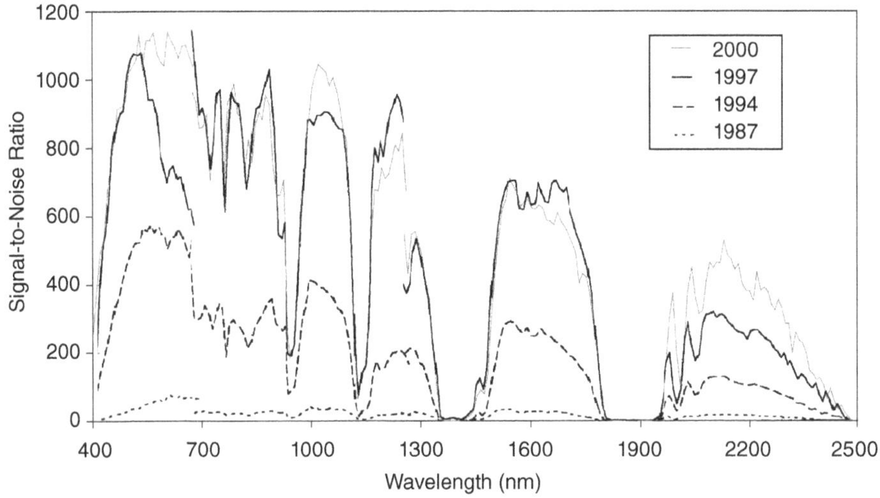

FIGURE 1.29 Evolution of the improvement of signal-to-noise ratio (SNR) of the AVIRIS instrument. (Courtesy of NASA/JPL.)

FIGURE 1.30 A photo of the AVIRIS next-generation assembly. (Image credit: NASA/JPL.)

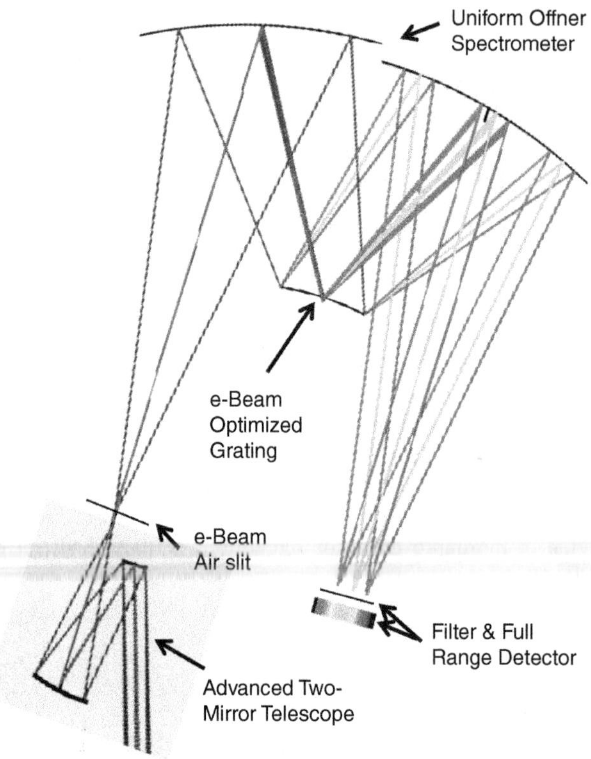

FIGURE 1.31 The optical layout of the Next-Generation AVIRIS. (Courtesy of NASA/JPL.)

The design of AVIRISng incorporated a number of design advantages, including a two-mirror anastigmatic telescope for simplified alignment and a high throughput, advanced technology of detector arrays, as well as a concentric, multi-blazed grating for tailored broadband efficiency. It covers the entire solar-reflected spectrum in the 380–2500 nm range with a single 2D detector array and a single grating operating in first order using a single spectrometer, rather than four spectrometers in the case of AVIRIS (Bender et al. 2010) Figure 1.31 shows the optical layout of the AVIRISng. Table 1.5 lists the parameter characteristics of AVIRISng in comparison to the AVIRIS classic.

AVIRISng has a completely different design compared to that of the AVIRIS classic and has achieved superior performance. It is a pushbroom imaging spectrometer and covers the VNIR and SWIR regions using one spectrometer with a single 2D detector array. This significantly simplifies the hyperspectral imagers and eliminates the gap between the VNIR and SWIR regions caused by using two separate spectrometers, which is the case for most imaging spectrometer covering a spectral range of 0.4–2.5 µm. In addition, the instrument design provided sufficient optical tolerancing that allows to assess the spectrometer and the telescope individually. The successful AVIRISng flights demonstrated a thermal, actively cooled design that limits opto-thermal-mechanical distortions to <0.01% of a pixel. AVIRISng also demonstrated a new alignment methodology for optical instruments including the alignment of a HgCdTe detector array at an ambient temperature.

1.3.4.3 CASI and Its Spectrally and Spatially Extended Siblings

1.3.4.3.1 History of Development

In the late 1980s, ITRES Research Limited based in Calgary, Canada, undertook to develop a general-purpose hyperspectral imaging instrument for airborne remote sensing. The objective was to provide an imaging spectrometer that was sufficiently flexible, portable, and low cost based on an area CCD detector array and readout system to provide the remote sensing community with an

TABLE 1.5
Performance Characteristics of the Next-Generation AVIRIS-Next Generation in Comparison with AVIRIS-Classic

Parameter	AVIRIS-Classic	AVIRIS-Next Generation
	General	
Number of spectrometers	4 (each covering a portion of the full spectral range)	1
Spectrometer type	all-reflective decentered aperture Schmidt	Offner form
f/#	1	3
Telescope	all-reflective modified Kennedy Scanner	2-mirror anastigmatic
Detector array format	32 × 1, 64 × 1	480 (spectral direction) × 640 (cross-track)
Detector pitch size	200 μm × 200 μm	27 μm × 27 μm
	Spectral Performance	
Wavelength range	0.38–2.5 μm	0.38–2.51 μm
Spectral sampling interval (SSI)	10 nm	10 nm
Spectral resolution (FMHM)	1~1.5 SSI	1~1.5 SSI
	Spatial Performance	
Field of view	34°	34°
Instantaneous field of view (IFOV)	1 mrad	1 mrad
Spatial resolution (FMHM)	1~1.5 IFOV	1~1.5 IFOV
Ground sampling distance (GSD)	4–20 m	0.3–20 m
	Radiometric Performance	
Signal-to-noise ratio	>1000:1 @ 0.6 μm >400:1 @ 2.2 μm	>2000:1 @ 0.6 μm >1000:1 @ 2.2 μm
Radiometric accuracy	>90%	>95%
Linearity	≥99% characterization	≥99% characterization
	Uniformity	
Spectral cross-track	>98% across FOV	>95% across FOV
Spectral IFOV variation	>98% Spectral direction	>95% Spectral direction
Mass	327 kg	465 kg
Volume	1.57 m^3 0.84 m (W) × 1.6 m (L) × 1.17 m (H)	0.83 m (H) × 0.57 m (Dia.) plus electronics boxes and racks

affordable tool useful for a variety of applications. Its capabilities were intended to fall in the range between satellite imaging instruments and aerial photography in terms of spatial resolution, but to give users full flexibility in specifying their spectral information, together with excellent SNRs, wide dynamic range, a very friendly user interface, and a direct digital output.

The design of Compact Airborne Spectrographic Imager (CASI) made use of the experience gained in constructing and operating the FLI instrument (see Section 1.3.2.3). In 1988, ITRES built a proof-of-concept flyable prototype based on an earlier Space Shuttle instrument and deployed it on an airplane for summer flight campaigns. The results from the flights were very encouraging in the areas of surface fish school surveys and spectral signature analysis (Nakashima et al. 1989, Royer et al. 1989).

A commercial prototype was then built and operated in the summer of 1989 in a number of programs, including further capelin fish surveys, algae bloom and oil spill surveys, and vegetation test site overflights. The first four production units of the CASI instrument were delivered to customers in Canada and Europe in the spring of 1990.

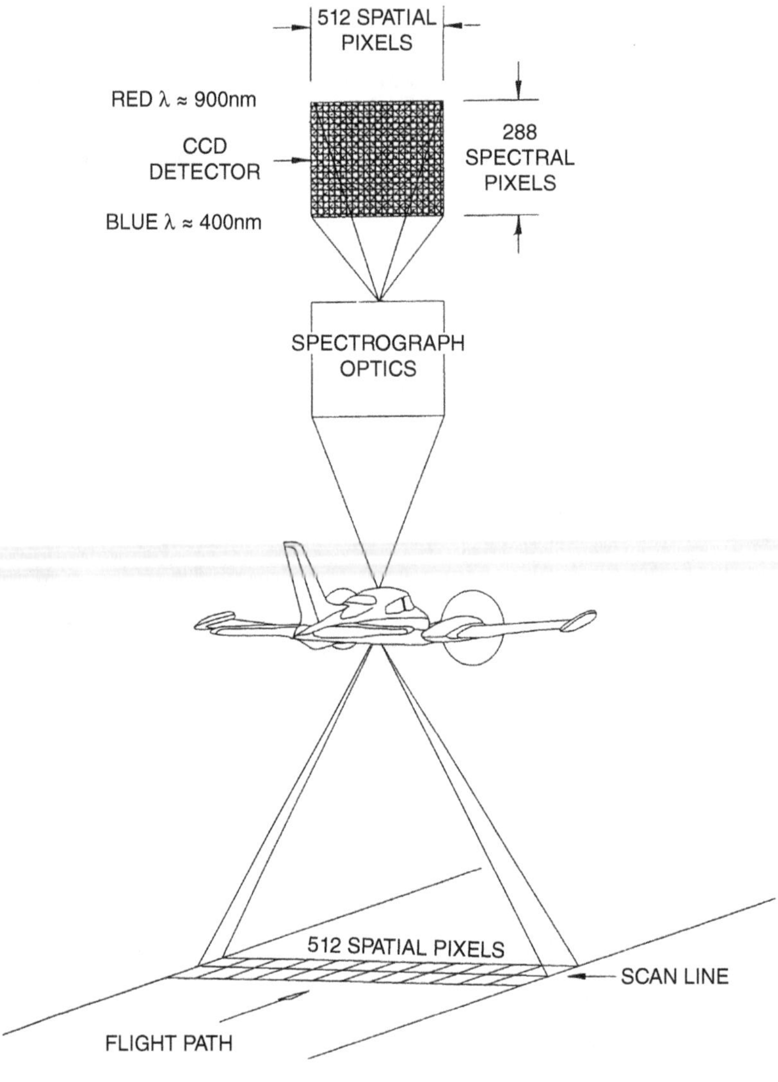

FIGURE 1.32 The concept of CASI instrument operation. (Courtesy of ITRES Research.)

The CASI is a pushbroom hyperspectral imager using a grating to disperse the spectrum. It operates by looking down in a fixed direction and images successive scan lines of the scene under the aircraft, thereby building up a 2D image as the aircraft moves forward. The instrument employs an imaging spectrograph with an area CCD detector array to capture both spatial and spectral information of ground sampling pixels in each scan line, as shown in Figure 1.32. This is done without any scanning or other mechanical motions except for the forward motion of the aircraft platform, resulting in a very compact and lightweight instrument. Because of the low noise, high sensitivity, and wide dynamic range of the instrument and its readout system, and because each individual sensing element integrates its signal over the full time that the instrument looks at each scan line in the scene, the CASI has a very high sensitivity.

1.3.4.3.2 Early Version CASI

The early version CASI employed an e2v area CCD array of size 612 × 576 pixels. It collects 288 spectral bands covering a spectral range from 400 nm to 926 nm with a spectral sampling interval (SSI) of 1.8 nm. It has 512 active pixels in the spatial imaging direction. Integration times are

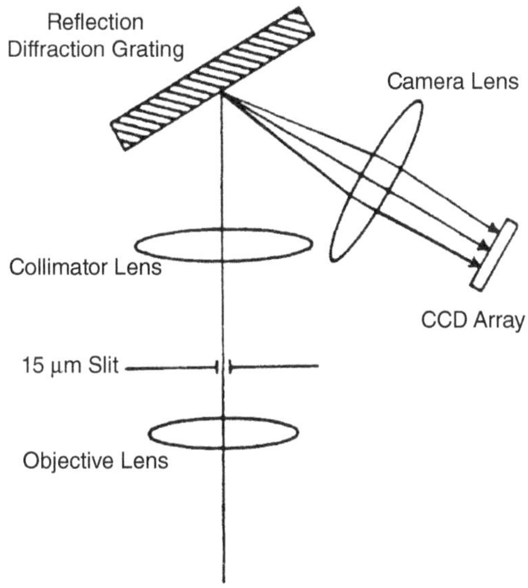

FIGURE 1.33 The optical layout of the CASI spectrometer. (Courtesy of ITRES Research.)

variable from 10 ms up to several seconds. A fiber-optic tap into the spectrograph slit enables a real-time spectral calibration to be recorded with the data (Anger et al. 1990).

The spectrometer is a reflection grating based pushbroom imaging spectrograph with a fast optical throughput (f/2). Figure 1.33 shows the optical layout of the CASI spectrometer.

The spectrum is sampled by the CCD detector array in contiguous 1.8-nm sampling intervals. This results in an over-sampling of the spectrum and gives a nominal overall average resolution of approximately 2.9 nm at FWHM. Because the grating dispersion varies slightly as a function of wavelength, the spectral bandwidth also varies somewhat with the wavelength. Thus, the actual spectral resolution increases from 2.8 nm at 450 nm to 2.9 nm at 850 nm. An operator may choose bandwidths as multiples of the basic 1.8-nm SSI.

CASI has a FOV of 35° corresponding to 578 pixels in a scan line in cross-track direction, typically reduced to 512 pixels for ease of data processing. This gives 1–5 km swath width on the ground depending on the altitude of the aircraft. The IFOV is 1.2 mrad, which corresponds to the ground sampling distance (GSD) of typically 2–5 m depending on the altitude of the aircraft, ground speed, and integration time.

Spectral calibration is determined by performing a third-order polynomial fit to the laboratory spectral line sources, interference filters, and atmospheric absorption lines. The calibration stability is ±1 nm. Radiometric calibration is obtained using an integrating sphere traceable to the National Institute of Standards and Technology (NIST). The overall calibration accuracy is ±5%. Recalibration is performed at the ITRES laboratory and is recommended once per season. The calibration software corrects for offset variations, dark current, and nonuniformities across the FOV, introduced by slit imperfections and pixel-to-pixel sensitivity variations, which are of the order of a few percent.

In the early '90s, the data storage was the cartridge tape recorder that limited the rate at which data could be recorded. In order to achieve acceptable ground resolution during aircraft operation, data were normally collected in one of two modes: spatial mode or spectral mode. The spatial mode digitizes and records up to 15 spectral bands of information while maintaining full spatial resolution of 612 pixels, of which the imaging FOV comprises 512 pixels. Each band is summed on-chip and may include any number of adjacent, nonoverlapping spectral rows. The spectral mode maintains full spectral elements 288 pixels encompassing the wavelength range specification of 400–926 nm. In this mode, up to 39 look directions, each of which corresponds to a single column on the CCD, can be defined for data acquisition. An individual single-row band at full spatial resolution, the Scene Recovery Channel (SRC),

TABLE 1.6
Specifications of the Early Version CASI Instrument

Optics	Reflection Grating Spectrometer
f/#	2
Detector	612 × 576 pixels e2v UT104, 512 of the 612 elements are recorded 288 of the 576 elements are used
Spectral coverage	400–926 nm sampled by 288 detector elements Sampling interval 1.8 nm Spectral resolution (FWHM) 2.9 nm
Spatial coverage	35° swath 612 pixels in cross-track direction Ground sampling (IFOV) 1.2 mrad Spatial resolution (FWHM) 1.6 mrad
Spectral mode	39 spectra of the full 400–926 nm range are recorded with 2.9 nm spectral resolution, from 39 different footprints across the swath
Spatial mode	Spectral pixels are grouped to form up to 15 bands (15 pushbroom images each 512 pixels wide). The number of bands governs the integration time
Integration times	50 msec, typical in the spatial mode 100 msec, typical in the spectral mode
Data recording	Exabyte digital recording to 8-mm tapes (1.1 Gbytes/tape). 280 kbit/sec
Mass	55 kg
Volume	Head: 15 cm × 22 cm × 22 cm Control unit: 42 cm × 48 cm × 24 cm

is also recorded in order to create a high-resolution monochromatic reference image. Lately, a third mode (full-frame mode) has been added to the CASI instrument software. This mode digitizes and records the entire data frame of 612 × 288 pixels. The recording of this amount of data takes approximately 2 seconds per frame, making this mode useful mainly in laboratory situations for the acquisition of radiometric calibration data. It has also been used in the field with a computer-controlled platform to synchronize the scanning motion of the platform with the acquisition of hyperspectral data frames. Table 1.6 summarizes the specifications of the early version CASI instrument.

1.3.4.3.3 Follow-Up Development

In the early 2000s, the Joint Airborne Lidar Bathymetry Technical Center of Expertise (JALBTCX), under a charter from the US Army Corps of Engineers (USACE), urged ITRES to develop a wide-swath hyperspectral imager with superb sensitivity from 365 nm to 1050 nm. Their main application was to perform operations, research, and development in airborne Lidar bathymetry and hyperspectral technologies to support the mapping of US coastlines. ITRES, subsequently, embarked on a program to develop an updated version of CASI with 1500 spatial pixels and covering an extended spectral range from 365 nm to 1050 nm, referred to as the "CASI-1500"—to satisfy JALBTCX's needs.

The CASI-1500 is a 1500 spatial-pixel, 288 spectral-band hyperspectral imager with a high throughput and sensitivity covering a spectral range of 365–1050 nm. It employs an ITRES' custom-designed CCD detector array specifically for use in hyperspectral applications. The CCD has a relatively large pitch size (20 μm × 20 μm) and provides a relatively large throughput and full-well capacity. The CCD's image region has been sectioned into three image areas—blue sensitive (thin polysilicon with some bare silicon exposed), red sensitive (regular thickness polysilicon), and far near-infrared sensitive (deep depletion p-doped Silicon). Note that back-illuminated CCDs have good blue sensitivity, but possess poor far near-infrared sensitivity.

CASI-1500's spectrometer is an ITRES-designed all-reflective Offner-type. It is optically fed by a very low distortion all-refractive telecentric apochromatic fore-optic to facilitate efficient ortho-rectification. The combination yields a low spectral distortion (a.k.a. smile) and spatial distortion (also referred to as keystone). Both distortions are <0.35 pixels (peak to valley) across the entire spectral range and FOV. Thanks to its outstanding performance, the CASI-1500 replaced the initial version CASI and became the main product of airborne imaging spectrometers of the company. Around 40 units have been sold worldwide.

The CASI-1500's spectrometer is superbly adept for low-albedo target mapping applications, such as water color remote sensing. More than one-third of all CASI-1500s sold are used for water applications. For example, the US Naval Research Laboratory (NRL) owns a CASI-1500 spectrometer. NRL then used a slightly modified CASI-1500 spectrometer in their Hyperspectral Imager for Coastal Ocean (HICO) payload. The HICO payload successfully operated on the International Space Station from 2009 until its demise in late 2014. More than 10,000 images of ocean and coastal areas were collected by HICO and distributed to the scientific community.

Following the success of the CASI-1500, ITRES further developed a hybrid variant of the CASI-1500, called CASI-1500H, by integrating the separate instrument controller unit with the sensor body. CASI-1500H maintains the performance edge of the original CASI-1500, but it is smaller and lighter. It has been reduced in volume by ~70% from the original CASI-1500 sensor plus controller, in addition to being ~20% lighter. The performance specifications of the instrument have not been degraded. All system control, optical, readout, and recording functions have been consolidated into a single portable unit to further simplify installation in smaller aircrafts.

During the same period of time, ITRES has developed a line of airborne hyperspectral imager products by extending the spectral range from VNIR to SWIR, mid-wave infrared (MWIR) and thermal infrared (TIR), as well as widening the swath to span more spatial pixels. In addition to the scientific-grade hyperspectral imagers, ITRES has also developed three micro-version commercial-grade hyperspectral imagers for use in unmanned aerial vehicles (UAVs), desktop, and industrial imaging. These small and portable systems complement their full-size, scientific-grade, high-performance siblings. Figure 1.34 outlines all the products of the airborne hyperspectral and thermal imagers fabricated by ITRES, which indicate their spectral range and number of spatial pixels in a cross-track line. Table 1.7 summarizes the key parameters and performance of the airborne hyperspectral and thermal imaging sensors.

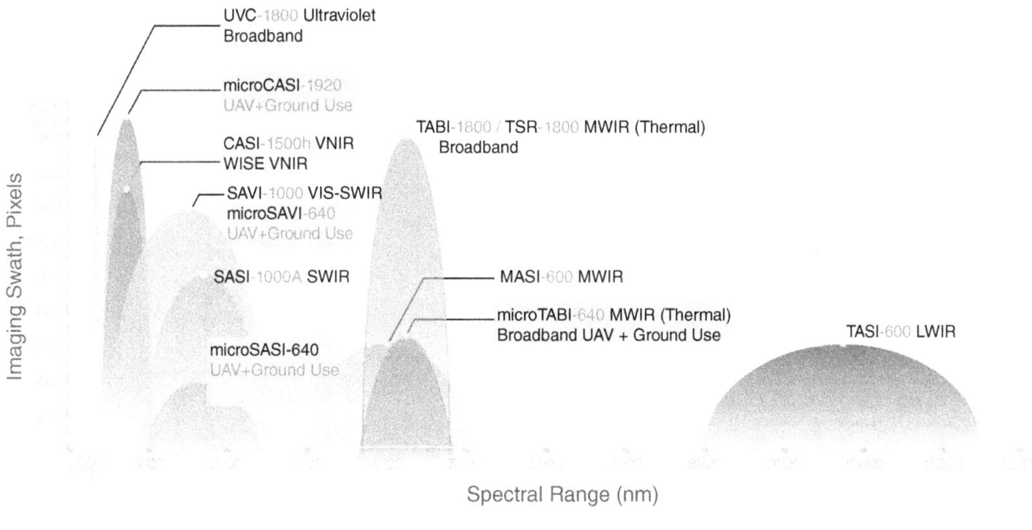

FIGURE 1.34 Spectral range and imaging swath of ITRES airborne hyperspectral and thermal imaging sensor products in 2019. (Courtesy of ITRES Research.)

TABLE 1.7
List of ITRES Airborne Imaging Spectrometers and Their Key Performance Parameters

Parameters	CASI-1500H	μCASI-1920	SASI-1000A	μSASI-640	SAVI-1000	MASI-600	TABI-1800	μTABI-640	TASI-600
# Cross-track Pixels	1500	1920	1000	640	1000	600	1800	640	600
Spectral range (μm)	0.38–1.05	0.4–1.0	0.95–2.45	1.0–2.5	0.4–2.5	3.0–5.0	3.7–4.8	3.7–4.8	8.0–11.5
# Spectral bands	288	288	200	256	256	64	1	1	32
Field of view (°)	40	36.6	40	40	40	40	40	40	40
IFOV (mrad)	0.49	0.36	0.698	1.8	0.698	1.2	0.405		1.21.2
f/#	f/3.5	f/2.5	f/2.4	f/2.5	f/2.4	f/2.0	f/2.0	f/2.0	f/2.0
Spectral sampling interval (nm)	2.4	2.1	7.5	5.9	8.2	32	-	-	110
Spectral resolution (FWHM) (nm)	<3.5	<5.0	<12	<8.5	<12	32	N/A	N/A	110
FPA pitch size (μm)	20 × 20	5.86 × 5.86	30 × 30	15 × 15	24 × 32		15 × 15	15 × 15	30 × 30
Detector full well capacity (Me)	>1.0	0.325	>1.0	≥1.0	>1.0	≥16.0	≥4.0	≥4.0	≥300
Spectral distortion (pixel)	±0.35	<0.5	<0.2	<0.2	<0.2	±0.33	N/A	N/A	±0.35
Spatial distortion (pixel)	±0.35	<0.5	<0.2	<0.2	<0.2	±0.33	-	-	±0.35
Frame rate (Hz)	up to 333	≥90	up to 100	≥150	>100	60	≥90	Up to 120	Up to 200
Volume (W × H × D cm³)	30 × 53 × 38	19 × 19 × 10	85 × 20 × 50 (SHU) 85 × 18 × 53 (ICU)	10 × 23 × 25	35 × 86 × 19	48 × 18 × 52 (ICU)	36 × 61 × 40	10 × 23 × 25	30 × 85 × 20 (SHU) 48 × 18 × 53 (ICU)
Mass (kg)	21	2.5	25 (SHU) 16 (ICU)	3.8	35	16 (ICU)	31	3.8	40 (SHU) 16 (ICU)

CASI: Compact Airborne Spectrographic Imager
CASI-1500H: Hybrid CASI VNIR Pushbroom Sensor
SASI: Shortwave infrared Airborne Spectrographic Imager
SAVI: Shortwave and Visible Spectrographic Imager
MASI: Mid-wave IR Airborne Spectrographic Imager
TABI: Thermal Airborne Broadband Imager (Cooled HgCdTe)
TASI: Thermal Airborne Spectrographic Imager
μ: micro-imagers

1.3.4.3.4 Development Toward Spaceborne Hyperspectral Imagers

In 2013, the Canadian Space Agency (CSA) initiated its microsatellite program for providing a more systematic implementation with an objective of sustainability. Five concepts of candidate microsatellites were selected through a whole-of-government approach, where stakeholders were invited to support the missions that best fit their interests. In this context, microsatellites are considered to have a spacecraft mass, including payloads within 75–150 kg. Development time should be within 3–4 years with a minimum mission lifetime of 2 years. They would not likely have propulsion and would be intended to orbit in low Earth orbit (LEO) (Bergeron et al. 2014).

The Canadian Coastal and Inland Waters Satellite (WaterSat) was one of the five selected candidates. This was to respond to the needs to improve the monitoring and better understand the quality and productivity of coastal and inland waters, as they have an important impact on the health of the population and on economic activities in Canada. The WaterSat mission aims to (1) provide ecological information of coastal waters and medium to large freshwater bodies, (2) monitor hazards, discharges, effluents, and pollution events, (3) assess productivity of marine coastal ecosystems, (4) characterize harmful algal blooms (HAB), and (5) monitor the quality of medium and large freshwater bodies. WaterSat is expected to be implemented as a pushbroom hyperspectral sensor covering the wavelength range from 350 nm to 900nm with a 5-nm spectral resolution and 100-m spatial resolution over a 250-km swath.

ITRES was selected by the CSA as the winner of a Canadian government contract to build a prototype for WaterSat payload under CSA's Space Technology Development Program (STDP) after a tender evaluation process. Funded by the CSA contract, ITRES completed the development of an airborne hyperspectral imager for a water color microsatellite called "WaterSat Imaging Spectrometer Experiment (WISE)" in early 2018.

The WISE prototype was developed based on the specification of the WaterSat payload listed in Table 1.8. In order to validate the design concept and demonstrate the performance achieved, a fore-optics with a FOV of 39° and IFOV of 0.25 mrad for aircraft flight operation was designed and built and was integrated into an airborne instrument enclosure. The delivered WISE instrument is ruggedized and can sustain the aircraft environment for flight operations.

The WISE instrument consists of three key subsystems: (1) fore-optics, (2) imaging spectrometer, and (3) CCD with its readout, control electronics, and software. WISE's fore-optics is a novel wide-angle, compact unobstructed diffraction-limited two-mirror telecentric system. WISE's imaging spectrometer is an innovative compact modified-Dyson design based on an ITRES' patent technology. WISE also incorporates a custom-designed high-speed low-noise CCD with superb sensitivity and dynamic range (Achal et al. 2018).

This WISE instrument yields a high-performance hyperspectral imaging system. The SNR is a factor of 2 higher than CASI 1500 based on the laboratory test. Both spatial distortion (keystone) and spectral distortion (smile) are <0.1 pixel. The WISE instrument acquires 1500 spatial pixels spanning a FOV of 39.46°, up to 288 spectral bands covering a spectral range of 358–992 nm with a spectral sampling interval of 2.20 nm. It occupies a volume of <0.06 m^3, weighs <10 kg and consumes <70 W. Comparing to the specification of the WaterSat payload listed in Table 1.8, the performance of WISE met the requirements and exceeded them in many items. WISE now becomes a member of the ITRES airborne imaging spectrometers family. Figure 1.35 shows outlook picture of the WISE instrument.

ITRES has conducted two airplane test flights of the WISE instrument and demonstrated all the performance parameters. High-quality hyperspectral images were acquired and confirmed the SNR improvement by at least a factor of two. The first WISE instrument, as an asset of the government of Canada, has been transferred to the Flight Research Laboratory at the National Research Council (NRC) of Canada as a new member of their airborne instrument fleet for serving the Canadian water color research community. In the summer 2019, a user team including researches from four Canadian universities and four federal government departments flew the WISE instrument over a coastal zone in eastern Canada (Manicouagan Peninsula, Quebec). They assessed the potential of airborne hyperspectral imagery to quantify seagrass beds, retrieved bathymetry and water column optical properties in optically complex waters using the WISE hyperspectral data.

TABLE 1.8
Specifications of Water Color Hyperspectral Imagers for Canadian Microsatellite WaterSat and the COCI Payload (Items for COCI Are Highlighted (shaded) Enhancement/Change Compared to Those of WaterSat)

Parameter	WaterSat (WISE)	COCI (DICE)	Note
Orbit	675 km	675 km	
FOV	11°	19.6°	In cross-track direction
Swath width	150 km (half swath)	240 km	Can be achieved by two sub-swaths
GSD at nadir	100 m	100 m	
Spatial resolution @ FWHM	<1.2*GSD	<1.2*GSD	
Spectral Range	380–970 nm	360–910 nm	350–1000 nm as a goal
Spectral sampling interval	≤5 nm	1.25 nm	Native instrumental SSI
Nominal spectral sampling	≤5 nm	5 nm	After binning
Number of spectral bands	118	440	Before binning
Spectral resolution	≤10 nm	≤6 nm	
Smile	<0.2 [illegible]	<0.1 [illegible]	
Keystone	<0.2 pixel	<0.1 pixel	
Stray light	3.0%	1.0%	
Polarization sensitivity	<5%	<3.0% <0.2%	without depolarizer with depolarizer
Peak SNR	350:1	950:1	@550 nm, 100 m GSD, 10 nm SSI, and 5% albedo
Minimum SNR	100:1	300:1	From 360 nm to 700 nm
Saturation level	no saturation	85%	For maximum scene
Dyson spectrometer	Yes	Yes	
Controller, software and GUI	Yes	Yes	
Structure and mechanics	Yes	Yes	
Fore-optic for airborne flight validation and demonstration	Yes	Yes	39° FOV, 0.25 mrad IFOV
Ruggedization for airborne test	Yes	Yes	
Airplane flight assessment	Yes	Yes	
Mass	<10 kg	<35 kg	For ruggedized airborne prototype including fore-optics and electronics
Volume	<0.10 m^3	0.20 m^3	
Power	<70 W	65 W	

In early 2016, CSA in collaboration with the US NRL and NASA initiated a joint venture study and proposed a Coastal Ocean Color Imager (COCI) as a complement payload to the NASA's Plankton, Aerosol, Cloud, ocean Ecosystem (PACE) mission (NASA 2012). The proposed COCI payload will have a GSD of 100 m with a swath width of 240 km covering a spectral range from 360 nm to 910 nm with a 5-nm spectral sampling. The peak SNR will be >950:1 (for a reference albedo of 5% spectrally resampled at 10 nm) to meet users' needs for allowing the retrieval of accurate geophysical products. It will avoid saturation from relatively bright surfaces such as sea ice or clouds (Qian et al. 2017).

To comply with the demanding requirements of accurately retrieving water leaving radiance for coastal and inland waters, stray light and polarization sensitivity of the instrument should be minimized and atmospheric correction bands in the SWIR region need to be included. A ±20 degrees tilting and ±45 degrees cross-track pointing capacity are also needed to maximize coverage and avoid sun glint and to allow for targeted imaging. The planned specification of COCI payload is listed in Table 1.8. The items highlighted (shaded) are the enhancement/change compared to those of WaterSat.

FIGURE 1.35 Appearance of the WaterSat Imaging Spectrometer Experiment (WISE) instrument, a prototype of the Canadian microsatellite WaterSat.

In order to elevate the Technology Readiness Level (TRL) of the COCI payload, CSA issued new Request for Proposals (RFP) for Canadian industry to develop a prototype COCI instrument. ITRES Research once again has been selected as the winner to build the prototype. The contract was kicked off in 2018 and is currently under development and expected to be completed by the first half 2020.

REFERENCES

Abdlaty, R., et al. 2018. Hyperspectral imaging: Comparison of acousto-optic and liquid crystal tunable filters. *Proceedings of SPIE* 10573:1057332P_01–09.

Achal, S., S.-E. Qian, M. Bergeron, P. Liu, A. U. Diaz, and Rita Leung. 2018. WaterSat imaging spectrometer experiment (WISE) for Canadian microsatellite mission. *Proceedings of Ocean Optical.*

Anger, C. D., S. K. Babey, and R. J. Adamson. 1990. New approach to imaging spectroscopy. *Proceedings of SPIE Imaging Spectroscopy of the Terrestrial Environment* 1298:72–86.

Babey, S. K. and C. D. Anger. 1989. A compact airborne spectrographic imager (CASI). *Proceedings of IEEE IGARSS'89/Twelfth Canadian Symposium on Remote Sensing* 2:1028–1031.

Barrett, E. C. and L. F. Curtis. 1997. *Introduction to Environmental Remote Sensing*, 3rd ed. Singapore: Chapman & Hall.

Bender, H. A., P. Mouroulis, R. O. Green, and D. W. Wilson. 2010. Optical design, performance and tolerancing of next-generation airborne imaging spectrometers. *Proceedings of SPIE* 7812:0P-1–12.

Bergeron, M., et al. 2014. Feasibility study for the next generation of Canadian microsatellites. *Proceedings of IEEE International Geoscience and Remote Sensing Symposium (IGARSS'2014)* 5207–5210.

Bianchi, R., et al. 1995. MIVIS airborne hyperspectral remote sensing in Europe. *Proceedings of International Symposium on Spectral Sensing Research*—ISSSR. Melbourne, Australia.

Borstad, G. A. 1985. Water colour and temperature in the southern Beaufort sea: Remote sensing in support of ecological studies of the bow head whale. *Canadian Technical Report of Fisheries and Aquatic Sciences* 1350:1–69.

Borstad, G. A., R. M. Brown, and J. F. R. Gower. 1980. Airborne remote sensing of sea surface chlorophyll and temperature along the outer British Columbia Coast. *Proceedings of the Sixth Canadian Symposium on Remote Sensing* 541–547.

Borstad, G. A. and J. F. R. Gower. 1984. Phytoplankton distribution in the eastern Canadian Arctic. *Arctic* 37:224–233.

Chang, C.-I. 2013. *Hyperspectral Data Processing*. Hoboken: John Wiley & Sons.

Chrien, T. G. 1992. AVIRIS: Recent instrument maintenance, modifications and 1992 performance. *Summaries of the Third Annual JPL Airborne Geoscience Workshop*. Pasadena, CA: JPL.

Chrien, T. G., M. L. Eastwood, C. M. Sarture, R. O. Green, and W. M. Porter. 1991. Current instrument status of the airborne visible/infrared imaging spectrometer (AVIRIS). *Proceedings of the Third AVIRIS Workshop*, 91–128. pp. 302–313 San Diego, CA: JPL Pub.

Chrien, T. G., R. O. Green, C. M. Sarture, C. Chovit, M. L. Eastwood, and B. T. Eng. 1993. Airborne visible/infrared imaging spectrometer (AVIRIS): Recent improvements to the sensor. *Summaries of the 4th Annual JPL Airborne Geoscience Workshop. Volume 1: AVIRIS Workshop* 27–30.

Cocks, T., R. Jenssen, A. Stewart, I. Wilson, and T. Shields. 1998. The HyMap airborne hyperspectral sensor: The system, calibration and performance. *Proceedings of the 1st EARSeL Workshop on Imaging Spectroscopy*, eds. M. Schaepman, D. Schläpfer, and K. I. Itten, 37–43, 6–8 October 1998, Zurich, Paris: EARSeL.

Conticello, S. S., et al. 2016. Hyperspectral imaging for real time land and vegetation inspection. *The 4S Symposium 2016*.

De Long, R. K., J. Marmo, D. P. Bowler, and T. E. Romesser. 1995. Airborne and satellite imaging spectrometer development at TRW. *Proceedings of SPIE International Symposium on Aerospace/Defense Sensing and Control and Dual-Use Photonics* 2480: 287–294 .

Dozier, J. 1988. HIRIS The high resolution imaging spectrometer. *Proceedings of SPIE 0924 Recent Advances in Sensors, Radiometry, and Data Processing for Remote Sensing*: 23-30.

Dwight, J. D., et al. 2018. Compact snapshot image mapping spectrometer for unmanned aerial vehicle hyperspectral imaging. *Journal of Applied Remote Sensing* 12(4):044004.

Dyson, J. 1959. Unit magnification optical system without Seidel aberrations. *Journal of the Optical Society of America* 49:713–716.

Eastwood, M. L., et al. 2000. Recent improvements to the AVIRIS sensor: Flight season 2000. *Proceedings of Airborne Visible/Infrared Imaging Spectrometer (AVIRIS) 2000 Workshop*, Pasadena, US.

Fabry, C. and A. Pérot. 1899. Theorie et applications d'une nouvelle methode de spectroscopie interferentielle. *Annales de Chimie et de Physique* 16(7):115–144.

Gao, L., R. T. Kester, and T. S. Tkaczyk, 2009. Compact image slicing spectrometer (ISS) for hyperspectral fluorescence microscopy. *Optics Express* 17(15):12293–12308.

Gao, L., et al. 2011. Depth-resolved image mapping spectrometer (IMS) with structured illumination. *Optics Express* 19(18):17439–17452

Gat, N. 2000. Imaging spectroscopy using tunable filters: A review. *Proceedings of SPIE* 4056:50–64.

Geelen, B., N. Tack, and A. Lambrechts. 2013. New multi- and hyperspectral cameras cover diverse applications. *Photonics Spectra*, (March) https://www.photonics.com/Articles/New_Multi-_and_Hyperspectral_Cameras_Cover/a53223.

Germain, S., et al. 2016. Global monitoring of greenhouse gas emission. *Proceedings of 30th Annual AIAA/USU SmallSat Conference* (August), Logan, UT, USA.

Goetz, A. F. H. 1995. Imaging spectrometry for remote sensing: Vision to reality in 15 years. *Proceedings of SPIE* 2480:1–13.

Goetz, A. F. H. and M. Herring. 1989. The high resolution imaging spectrometer (HIRIS) for EOS. *IEEE Transactions on Geoscience and Remote Sensing* 27:136–144.

Goetz, A. F. H., L. C. Rowan, and M. J. Kingston. 1982. Mineral identification from orbit: Initial results from the shuttle multispectral infrared radiometer. *Science* 218:1020–1024.

Goetz, A. F. H. and V. Srivastava 1985. Mineralogical mapping in the Cuprite mining district, Nevada. *Proceedings of the Airborne Imaging Spectrometer Data Analysis Workshop JPL Publication* 85(41):22–31.

Goetz, A. F. H., U. Vane, J. Solomon, and B. N. Rock. 1985. Imaging spectrometry for Earth remote sensing. *Science* 228:1147–1153.

Gower, J. F. R. 1980. Observations of in situ fluorescence of chlorophyll *a* in Saanich inlet. *Boundary Layer Meteorology* 18:235–245.

Gower, J. F. R. and G. A. Borstad. 1981. Use of the *in vivo* fluorescence line at 685 nm for remote sensing surveys of surface chlorophyll *a*, In *Oceanography from Space*, ed. J. F. R. Gower, 329–338. New York: Plenum Press.

Gower, J. F. R. and G. A. Borstad. 1990. Mapping of phytoplankton by solar-stimulated fluorescence using an imaging spectrometer. *International Journal of Remote Sensing* 11:313–320.

Gower, J. F. R., G. A. Borstad, C. D. Anger, and H. R. Edel. 1992. CCD-based imaging spectroscopy for remote sensing: The FLI and CASI programs. *Canadian Journal of Remote Sensing* 18(4):199–208.

Gower, J. F. R., G. A. Borstad, L. H. Gray, and H. R. Edel. 1987. The fluorescence line imager: Imaging spectroscopy over water and land. *Proceedings of the Eleventh Canadian Symposium on Remote Sensing*, Waterloo, Ontario, June 22–25, 689–697.

Gower, J. F. R., S. Lin, and G. A. Borstad. 1984. The information content of different optical ranges for remote chlorophyll estimation in coastal waters. *International Journal of Remote Sensing* 5:349–364.

Gower, J. F. R., and G. A. Borstad. 1990. Mapping of Phytoplankton by Solar-stimulated Fluorescence Using an Imaging Spectrometer. *International journal of Remote Sensing* 11:313–320.

Grahn, H. and P. Geladi. 2007. *Techniques and Applications of Hyperspectral Image Analysis*. Hoboken: John Wiley & Sons.

Green, R. O., et al. 1993. Airborne visible/infrared imaging spectrometer (AVIRIS): Recent improvements to the sensor and data facility. *Proceedings of SPIE* 1993:180–190.

Green, R. O., et al. 1998. Imaging spectroscopy and the airborne visible/infrared imaging spectrometer (AVIRIS). *Remote Sensing of Environment* 65(3):227–248.

Hagen, N., R. T. Kester, L. Gao, and T. S. Tkaczyk. 2012. Snapshot advantage: A review of the light collection improvement for parallel high-dimensional measurement systems. *Optical Engineering* 51:111702.

Hagen, N. and M. W. Kudenov. 2013. Review of snapshot spectral imaging technologies. *Optical Engineering* 52(9):090901.

Horstmeyer, R., et al. 2009. Flexible multimodal camera using a light field architecture. *IEEE International Conference on Computational Photography* 1–8.

Imec. 2013. New multi- and hyperspectral cameras cover diverse applications. *Photonics Media*. https://www.photonics.com/Articles/New_Multi-_and_Hyperspectral_Cameras_Cover/a53223

Imec. 2018. Meadia Release: Imec demonstrates shortwave infrared (SWIR) range hyperspectral imaging camera. SPIE Photonics West in San Francisco.

Kudenov, M. W. and E. L. Dereniak. 2010. Compact snapshot birefringent imaging Fourier transform spectrometer. *Proceedings of SPIE* 7812:781206.

Kudenov, M. W. and E. L. Dereniak. 2012. Compact real-time birefringent imaging spectrometer. *Optics Express* 20(16):17973–17986.

Levoy, M., et al. 2006. Light field microscopy. *ACM Transactions on Graphics* 25(3):924–934.

Lewis (SSTI 1). 2014. (available at http://space.skyrocket.de/doc_sdat/lewis.htm, accessed on December 5, 2018).

Mende, S. B., et al. 1993. Hadamard spectroscopy with a two-dimensional detecting array. *Applied Optics* 32(34):7095–7105.

Mouroulis, P., R. O. Green, and D. W. Wilson. 2008. Optical design of a coastal ocean imaging spectrometer. *Optics Express* 16(12):9087–9096.

Nakashima, B. S., G. A. Borstad, D. A. Hill, and R. C. Kerr, 1989. Remote sensing of fish schools: Early results from a digital imaging spectrometer. *Proceedings of IEEE IGARSS'89/12th Canadian Symposium on Remote Sensing, Vancouver, B.C.* 4:2044–2046.

NASA. 2012. Pre-Aerosol, Clouds, and ocean Ecosystem (PACE) Mission Science Definition Team Report. https://pace.oceansciences.org/docs/pace_sdt_report_final.pdf.

Neville, R. A. and J. F. R. Gower. 1977. Passive remote sensing of phytoplankton via chlorophyll a Fluorescence. *Journal of Geophysical Research* 82:3487–3493.

Neville, R. A. and I. Powell. 1992. Design of SFSI: An imaging spectrometer in the SWIR. *Canadian Journal of Remote Sensing* 18(4):210–223.

Neville, R. A., L. Sun, and K. Staenz. 2003. Detection of spectral line curvature in imaging spectrometer data. *Proceedings of SPIE* 5093:144–154.

Neville, R. A., L. Sun, and K. Staenz. 2004. Detection of keystone in image spectrometer data. *Proceedings of SPIE* 5425:208–217.

Neville, R. A., L. Sun, and K. Staenz. 2008. Spectral calibration of imaging spectrometers by atmospheric absorption feature matching. *Canadian Journal of Remote Sensing* 34(S1):S29–S42.

Porter, W. M., T. G. Chrien, E. G. Hansen, and C. M. Sarture. 1990. Evolution of the airborne visible/infrared imaging spectrometer (AVIRIS) flight and ground data processing system. *Proceedings of SPIE* 1298:11–17.

Offner, A. 1973. Unit power imaging catoptric anastigmat. *U.S. Patent No. 3, 748, 015.*

Qian, S.-E. 2013. *Optical Satellite Signal Processing and Enhancement*. Bellingham: SPIE Press.

Qian, S.-E. 2016. Review of spaceborne optical payloads. In *Optical Payloads for Space Missions*, ed. S.-E. Qian, 1–25. West Sussex, UK: John Wiley & Sons.

Qian, S.-E., R. Girard, and G. Kroupnik. 2013. Development of Canadian hyperspectral imager onboard micro-satellites. *Proceedings of IEEE International Geoscience and Remote Sensing Symposium (IGARSS'2013)* 3506–3509.

Qian, S.-E., J. Lévesque, and R. A. Neville. 2006. Effect of removing random noise of radiance data using smoothing on data compression onboard a hyperspectral satellite. *WSEAS Transactions on Systems* 1(5):219–224.

Qian, S., et al. (other 18 coauthors). 2017. A spaceborne coastal and inland water color hyperspectral imager. *Proceedings of IEEE International Geoscience and Remote Sensing Symposium (IGARSS'2017)* 447–450.

Richter, R. 1996. Atmospheric correction of DAIS hyperspectral image data. *Computers and Geosciences* 22:785–793.

Rickard, L. J., R. Basedow, E. Zalewske, P. R. Silvergiate, and M. Landers. 1993. HYDICE: An airborne system for hyperspectral imaging. *Proceedings of SPIE* 1937:173–179.

Royer, A., N. T. O'Neill, D. Williams, P. Cliche, and R. Verreault. 1989. System De mesures De reflectances pour les spectrometres imageurs. *Proceedings of IEEE IGARSS'89/12th Canadian Symposium on Remote Sensing, Vancouver* 4:2097–2099.

Sarture C. M., T. G. Chrien, R. O. Green, M. L. Eastwood, H. K. Johnson, C. J. Chovit, P. Hajek, J. L. Holbrook, J. J. Raney, and M. A. Hernandez. 1995. Airborne visible/infrared imaging spectrometer (AVIRIS): Sensor improvements for 1994 and 1995. *Proceedings of the Fifth Annual JPL Airborne Earth Science Workshop* (January), 1:23–26.

Shogenji, R., et al. 2004. Multispectral imaging using compact compound optics. *Optics Express* 12(8):1643–1655.

Staenz, K. 1992. A decade of imaging spectrometry in Canada. *Canadian Journal of Remote Sensing* 18(4):187–197.

Sun, L., R. Neville, and K. Staenz. 2006. Automatic destriping of Hyperion imagery based on spectral moment matching. *Canadian Journal for Remote Sensing* 34(S1):S68–S81.

Ungar, S. G., J. S. Pearlman, J. A. Mendenhall, and D. Reuter. 2003. Overview of the Earth observing one (EO-1) mission. *IEEE Transactions on Geoscience and Remote Sensing* 41:1149–1153.

Vane, G. and A. F. H. Goetz. 1988. Terrestrial imaging spectroscopy. *Remote Sensing of Environment* 24:1–29.

Vane, G., et al. 1993. The airborne visible/infrared imaging spectrometer (AVIRIS). *Remote Sensing of Environment* 44:127–143.

Wagadarikar, A. A., M. E. Gehm, and D. J. Brady. 2007. Performance comparison of aperture codes for multimodal, multiplex spectroscopy. *Applied Optics* 46(22):4932–4942.

Wagadarikar, A. A., N. P. Pitsianis, X. Sun, and D. J. Brady. 2009. Video rate spectral imaging using a coded aperture snapshot spectral imager. *Optics Express* 17(8):6368–6388.

Wallace, M. P. and H. T. Enmark. 1987. A system overview of the airborne visible/infrared imaging spectrometer (AVIRIS). *Proceedings of SPIE Imaging Spectroscopy II* 834:22–31.

Weale, G. P., S. B. Achal, and C. D. Anger. 1990. Application of CASI, a commercial remote-sensing instrument, to astronomy. *Proceedings of SPIE* 1235.

2 Overview of Hyperspectral Sensors on Orbits

2.1 SPACEBORNE HYPERSPECTRAL SENSORS AT A GLANCE

Thanks to the wealthy information obtained in both spatial and spectral dimensions of the observing targets, hyperspectral imaging is regarded as a diagnostic monitoring technology. It can play a decisive role in obtaining accurate information for better understanding the observing targets, their identification information, risks, and consequences of the changes. Airborne hyperspectral imaging has been widely utilized because of its easy adaption to variation of altitude, flight schedule to avoid weather problems, such as clouds, and flexibility of adaption to the task requirements in the course of flight. However, a low-flying aircraft imaging has a narrow field of view (FOV), a number of flybys are needed to cover a large ground area due to the small coverage swath. Aircrafts are less stable, thus the image quality obtained from the instrument onboard suffers.

The ever-growing needs for timely coverage of wide area and the demand for specific information for the remote and inaccessible areas on Earth have driven the development of the satellite hyperspectral sensors. Since the beginning of this millennium, the research and development of spaceborne hyperspectral sensors has been grown exponentially. With the advancement of technologies for focal plane arrays (FPAs), optical materials manufacture and diamond tuning, data acquisition, data storage, computation, and telemetry, it is made possible to build high-performed hyperspectral sensors and reduce the cost of development of such space systems. Satellite hyperspectral sensors are now more readily available not only for scientific applications but also for mandate operations. Spaceborne hyperspectral sensors proved their capability to provide critical information in numerous application areas as of civilian origin as of military.

The author of the book conducted a survey on a list of spaceborne hyperspectral sensors to date and found that there exist at least 25 spaceborne hyperspectral sensors that have been deployed to orbits of Earth, moon, Mars, Venus, and comet (3 slightly variant VIRTIS sensors have flown in three planetary missions). The list of spaceborne hyperspectral sensors since the beginning is tabulated in Table 2.1. In the table, the spaceborne hyperspectral sensors are listed in the order of years chronologically. This list may miss some spaceborne hyperspectral sensors. It is worth to note that during a short period from 2016 to 2020, close to 10 spaceborne hyperspectral sensors have been launched or scheduled to be launched to space. There is a leap for the number of spaceborne hyperspectral sensors launched in 2018. More spaceborne hyperspectral sensors have been planned and will come up soon.

It can be seen from the table that the earliest spaceborne hyperspectral sensor was the Ultraviolet and Visible Imagers and Spectrographic Imagers (UVISI) onboard the Midcourse Space Experiment (MSX) mission of the US Department of Defense (DoD), which was launched in 1996. It consisted of five spectrographic imagers (SPIMs) covering a wavelength range from ultraviolet (UV) to visible and near-infrared (VNIR) regions. It is not popular because of its large ground sampling distance (770m) and the nature as a military satellite. This spaceborne hyperspectral sensor is described in Section 2.2.

The second earliest spaceborne hyperspectral sensor was HyperSpectral Imager for the LEWIS mission launched in 1997 (De long et al. 1995) as a technology demonstration under the NASA's Small Spacecraft Technology Initiative (SSTI) program. Unfortunately, HyperSpectral Image did not reach the orbit. Three days after the launch on August 23, 1997, the control of the satellite was lost and it subsequently entered the Earth atmosphere in September 1997 (Lewis 2014).

TABLE 2.1
List of Spaceborne Hyperspectral Sensors

No.	Hyperspectral Sensor	Satellites/ Platform	Launch Year	Demission Year	Number of Bands	Spectral Range(μm)	Spectral Sampling Interval(nm)	Ground Sampling Distance(m)	Swath Width(km)	Orbit	Spectral Imaging Technique
1	SPIMs 1-5	MSX	1996	2008	272	0.11–0.9	0.5–4.3	770	15	LEO	Grating, pushbroom
2	HSI	LEWIS	1997	1997	128 256	0.4–1.0 0.9–2.5	5.0 6.5	30	7.7	LEO	Grating, pushbroom launched, but the satellite lost control
3	MODIS	Terra Aqua	1999 2002	Active Active	36	0.41–14.4	10–50	250: b1-b2 500: b3-b7 1000: b8-b36	2330	LEO	Band-pass filters, whiskbroom
4	Hyperion	EO-1	2000	2017	220	0.4–2.5	10	30	7.7	LEO	Grating, pushbroom
5	CHRIS	PROBA	2001	Active	19–62	0.4–1.0	1.25–11	25–50	13	LEO	Prism, pushbroom, multi-viewing
6	MERIS	ENVISAT	2002	2012	520 (transmit 15)	0.39–1.04	1.25	300	1150	LEO	Grating, pushbroom, onboard bandwidth selection
7	VIRTIS	Rosetta Venus Express NASA-Dawn	2004 2005 2007	2016 2015 2018	432 432	0.28–1.10 1.05–5.13	1.89 9.47	0.014°× 0.014° (IFOV)	3.67 × 3.67° (FOV)	Comet 67P Venus Vesta, Ceres	Grating, slit scan
8	CRISM	MRO	2005	Active	455	0.37–3.92	6.55	18	10.8	Mars orbit (300 km)	Grating, pushbroom, onboard compression
9	M3	Chandrayaan-1	2008	2009	260	0.43–3.0	10	70	40	Lunar orbit	Grating, pushbroom
10	FTHSI	HJ-1A	2008	Active	115	0.45–0.95	4	100	50	LEO	Fourier interferometer
11	HySI	IMS-1	2008	2012	64	0.40–0.95	8	500	130	LEO	Linear variable filter (LVF)
12	ARTEMIS	TacSat-3	2009	2012	400	0.4–2.5	5	4	4	LEO	Grating, pushbroom
13	HICO	ISS	2009	2014	128	0.35–1.08	5.7	90	51	ISS orbit	Grating, pushbroom
14	VNIS	Change'E	2013	2015	100	0.45–0.95 0.9–2.4	5	-	8.5° × 8.5° (FOV)	Lunar rover	Acousto-optic tunable filter (AOTF)
15	OLCI	Sentinel-3A	2016	Active	520 (transmit 21)	0.39–1.04	1.25	300	1270	LEO	Grating, pushbroom, onboard bandwidth selection
16	MHRIS	GHGSat-D	2016	Active	512	1.6–1.7	0.2	50	15	LEO	Tunable Fabry–Pérot filter, 25U nanosat
17	AaSI	Aalto-1	2017	Active	6–20	0.50–0.90	7–10	192	97	LEO	Tunable Fabry–Pérot filter, 3U nanosat
18	DESIS	ISS	2018	Active	235	0.40–1.0	2.55	30	30	ISS orbit	Grating, pushbroom
19	HyperScout	GomX-4B	2018	Active	45	0.4–1.0	15	50	200	LEO	Linear variable filter (LVF), 3U nanosat

20	AHSI	Gaofen-5	2018	Active	330	0.40–2.50	5 VNIR 10 SWIR	30	60	LEO	Grating, pushbroom
21	HysIS	IMS-2	2018	Active	256	0.40–2.4	10	30	30	LEO	Dispersive(?)
22	PRISMA	PRISMA	2019	Active	237	0.40–2.51	12	30	30	LEO	Prism, pushbroom
23	HISUI	ISS	2019	Active	185	0.40–2.50	10 VNIR 12.5 SWIR	30	30	LEO	Grating, pushbroom
24	EnMAP	German HS	2020	N/A	244	0.42–2.50	5, 10	30	30	LEO	Prism, pushbroom
25	MAJIS	JUICE	2022	N/A	508 508	0.50–2.35 2.55–5.54	3.6 6.4	75	30	Ganymede orbit (500 km)	Grating, slit scan

Acronyms list

Active: when the book was written on December 28, 2019
AaSI: Aalto-1 Spectral Imager
AHSI: Advanced Hyperspectral Imager on GaoFen-5 satellite
ARTEMIS: Advanced Responsive Tactically Effective Military Imaging Spectrometer
CHRIS: Compact High-Resolution Imaging Spectrometer
CRISM: Compact Reconnaissance Imaging Spectrometer for Mars; MRO: Mars Reconnaissance Orbiter
DESIS: DLR Earth Sensing Imaging Spectrometer
EnMAP: Environmental Mapping and Analysis Program
FTHSI: HyperSpectral Imager on HJ-1A satellite
HICO: Hyperspectral Imager for Coastal Ocean
HISUI: Hyperspectral Image Suite
HSI: HyperSpectral Imager; LEWIS mission
Hyperion: Hyperspectral imager; EO-1: Earth Observing-1 Mission
HyperScout: A 1U (1 litre) hyperspectral camera
HySI: Hyperspectral Imager onboard Indian Mini Satellite-1 (IMS-1)
HysIS : Hyperspectral Imaging System onboard Indian Mini Satellite-2 (IMS-2)
M3: Moon Mineralogy Mapper
MAJIS: Moons and Jupiter Imaging Spectrometer onboard spacecraft of JUpiter ICy moons Explorer
MERIS (MEdium Resolution Imaging Spectrometer); ENVISAT: ESA's Environmental Satellite
MHRIS: Miniature high-resolution imaging spectrometer
MODIS: Moderate Resolution Imaging Spectroradiometer
OLCI: Ocean and Land Color Imager
PRISMA: PRecursore IperSpettrale della Missione Applicativa
SPIMs 1–5: Spectrographic Imagers 1–5; MSX: Midcourse Space Experiment satellite
VIRTIS: Visible and Infrared Thermal Imaging Spectrometer
VNIS: Visible and Near-infrared Imaging Spectrometer aboard Chang'E 3 Spacecraft

The Moderate Resolution Imaging Spectroradiometers (MODIS) were deployed onboard NASA's Terra (EOS AM) satellite and Aqua (EOS PM) satellite, which were launched in 1999 and 2002, respectively. MODIS uses band-pass filters to disperse spectrum of the object scene. This is the technology used for multispectral sensors (e.g., Landsat, SPOT). From this perspective, MODIS should be classifies as a multispectral sensor. However, many of its bands are narrow enough (10 nm; see Table 2.5), which fall in the definition of a hyperspectral sensor "the acquisition of many images of contiguous, narrow, registered spectral bands such that for each pixel a radiance spectrum can be derived" described in Section 1.1. That is why it is treated as a hyperspectral sensor and included in this chapter.

Hyperion is well known in the remote sensing community and is often regarded as the first spaceborne hyperspectral sensor for the reasons described in the paragraphs above. It was onboard NASA's Earth Observing-1 (EO-1) satellite launched in 2000 as part of the New Millennium Program to develop and validate new technologies for future Earth imaging observatories. Hyperion had continuously acquired hyperspectral data for scientific research and user community until its retirement in 2017, although it was designed for a 1-year life. This was a great success story of spaceborne hyperspectral sensors.

Regarding the operation mode, among the 25 spaceborne hyperspectral sensors listed in Table 2.1, all of them use two-dimensional (2D) detector arrays and operated or are operating in pushbroom mode, except MODIS that use 1D linear detector arrays and operates in whisk-broom mode.

With respect to spectral dispersion means, 17 spaceborne hyperspectral sensors use dispersive elements, either gratings or prisms. One (MODIS) uses band-pass filters. Two of them use linear variable filters (LVFs) to disperse spectrum. These two sensors are Hyperspectral Imager (HySI) onboard the Indian Mini Satellite-1 (IMS-1) and HyperScout on European Space Agency's GomX-4B nanosatellite. Three hyperspectral sensors utilize electronically tunable filters, of which two sensors use tunable Fabry-Perot filters and one sensor uses acousto-optic tunable filter (AOTF). There is also one spaceborne hyperspectral sensor that uses Fourier transform interferometer to disperse spectrum. One sensor's (Hyperspectral Imaging System, HysIS) dispersion means is unknown at the time of writing this book.

In terms of platforms and orbits of these spaceborne hyperspectral sensors, majority of them are aboard satellites on low Earth orbits (LEOs). Three of them are deployed on the International Space Station (ISS). There are also six hyperspectral sensors out of Earth orbits, one (Compact Reconnaissance Imaging Spectrometer for Mars [CRISM]) on a Mars orbit, one (M3) on a lunar orbit, and one (Visible and Near-Infrared Imaging Spectrometer [VNIS]) on a lunar rover for in situ observation. The Visible and Infrared Thermal Imaging Spectrometer (VIRTIS) and its two slight variants were deployed onboard the space probes of three planetary missions on orbits of a comet, Venus and two protoplanets.

In Table 2.1, two upcoming spaceborne hyperspectral sensors are also listed. These sensors are Environmental Mapping and Analysis Program (EnMAP), and Moons and Jupiter Imaging Spectrometer (MAJIS).

This chapter describes these spaceborne hyperspectral sensors from the perspective of instrument. The applications of hyperspectral satellites are discussed in Chapter 3.

2.2 ULTRAVIOLET AND VISIBLE IMAGERS AND SPECTROGRAPHIC IMAGERS

Ultraviolet and Visible Imagers and Spectrographic Imagers (UVISI) was deployed on the MSX satellite, which was launched on April 24, 1996 by US DoD, for military applications (Paxton et al. 1996). It constituted a great leap forward by developing a first spaceborne hyperspectral imager and

TABLE 2.2
Characteristics of Spectrographic Imagers SPIM 1–5

Characteristics	SPIM 1	SPIM 2	SPIM 3	SPIM 4	SPIM 5
Spectral range (nm)	110–170	165–258	151–387	381–589	581–900
Number of spectral bands	272	272	272	272	272
Spectral sampling interval (nm)	0.8, 0.5	1.2, 0.9	1.8, 1.5	2.8, 2.1	4.3, 3.3
Sensitivity (photo/cm^2/sec)	5	2	3	1	1

demonstrating its application from a satellite. The MSX satellite contains a combined instrument suite, in addition to UVISI, including also the Spatial Infrared Imaging Telescope III (SPIRIT III) and the Space-Based Visible (SBV) experiment. SPIRIT III was an interferometer and not designed for imaging. SBV was a VNIR imager covering a spectral range of 0.55–1.0 μm.

The UVISI is composed of five UV and visible hyperspectral imagers (called SPIMs) and four UV and visible multispectral imagers (MSIs). These nine imaging sensors provided hyperspectral and multispectral capabilities from 110 nm to 900 nm (Carbary 1994).

The five SPIMs share an off-axis optical design in which selectable slits alternate FOVs (1.00° × 0.10° or 1.00° × 0.05°) and spectral resolutions between a wide slit and a narrow slit. The SPIMs have a programmable number of spectral bands of 68, 136, or 272 pixels across each individual spectral band, and a programmable spatial dimension with 5, 10, 20, or 40 pixels across the 1° slit length in cross-track direction. A scan mirror sweeps the slit through a second spatial dimension to generate a 1° × 1° spectrographic image once every 5, 10, or 20 sec, depending on the scan rate.

The FPA of each SPIM utilized is intensified CCD (charge-coupled device) detectors that have an intrascene dynamic range of ~10^3 and an interscene dynamic range of ~10^5; neutral-density filters provide an additional dynamic range of ~10^{2-3}. The detector array uses an automatic gain control that permit the SPIMs to adjust to scenes of varying intensity. The five SPIMs have common boresights and can operate separately, simultaneously, or synchronously.

The five SPIM 1–5 covers a wavelength range 110–170, 165–258, 251–387, 381–589, and 581–900 nm, respectively, with a varying spectral resolution from 0.5 nm to 4.3 nm, depending on wavelength range and data mode. Each SPIM generated up to 272 spectral bands. The UVISI records data in 1360 spectral bands simultaneously, with a spatial resolution of 770 m at nadir and a swath about 15-km wide. The SPIM 1–3 provides UV imaging capability and SPIM 4 and 5 are visible/near-IR SPIMs. Table 2.2 lists the characteristics of SPIM 1–5.

The four imagers provide narrow-field (1.59° × 1.28°) and wide-field (13.1° × 10.5°) viewing. Each imager has a six-position filter wheel that selects various spectral regions and neutral densities. The characteristics of the four imagers are listed in Table 2.3.

TABLE 2.3
Characteristics of Four Multispectral Imagers 1–4

Characteristics	IUN Imager (Narrow FOV)	IUW Imager (Wide FOV)	IVN Imager (Narrow FOV)	IVW Imager (Wide FOV)
Bandwidth (nm)	180–300	115–180	300–900	380–900
Field of view	1.59° × 1.28°	1.31° × 10.5°	1.59° × 1.28°	13.1° × 10.5°
Clear aperture (cm^2)	130	25	130	25
Ground sampling distance (m)	80	800	80	800

MSX has the ability to view a scene from a variety of angles. It is a powerful tool for understanding the surface features or structure and composition of the atmosphere and its constituents or the properties of ocean. The MSX also carried panchromatic (PAN) imagers. The combination of UVISI's UV/VNIR hyperspectral imagers, MSIs, and PAN imager was a powerful means in satellite-based remote sensing since the UVISI sensors provided detailed spectral information, while the imagers set the stage for the observation by providing a broader context for the data.

2.3 HYPERSPECTRAL IMAGER (HSI) FOR THE LEWIS MISSION

In the beginning of 1990s, NASA's Office of Space Access and Technology initiated the Small Spacecraft Technology Initiative (SSTI) to advance the state of technology and reduce program costs associated with the development and operation of small satellites. Under SSTI program in July 1994, Thompson Ramo Wooldridge Inc. (TRW) began the development of LEWIS satellite and a spaceborne spectrometric system called HyperSpectral Imager (HSI) as a technology demonstration (Delong et al. 1995).

HSI was based on the technology of TRWIS III, an airborne grating-based hyperspectral imager developed by TRW. The optical design of HSI utilized shared fore-optics to generate three line images separately slightly in field, two of which are for the VNIR and shortwave infrared (SWIR) regions, the third is for a PAN channel. The VNIR and SWIR regions together cover a spectral range from 0.4 μm to 2.5 μm. The VNIR region has 128 bands in the spectral range of 0.4–1.0 μm with a spectral sampling interval (SSI) of 5.0 nm, whereas the SWIR region has 256 bands in the spectral range of 0.9–2.5 μm with a SSI of 6.5 nm. This resulted in 384 spectral bands. There is an overlap from 0.9 μm to 1.0 μm between the two regions, which were co-aligned to have virtually identical FOV. The refractive elements in the spectrometer were designed and coated specifically to the respective wavelength regions. The spectrometer was designed to very low distortion and for IFOV matching to maintain tight spatial co-registration of the spectral bands.

The FPA used in the VNIR region is a four-ported, split frame transfer CCD detector array that is thinned and backside illuminated for high quantum efficiency and to avoid the fringing associated with the gate structure in front side illuminated CCD arrays. The CCD detector array has 768 pixels for spatial and 384 pixels for spectral with a pitch size of 20 μm. The full well capacity of the CCD array is 1.2 million electrons. In order to increase the signal-to-noise ratio (SNR), the detector pixels in the VNIR spectrometer were aggregated 3 × 3 to produce 256 pixels in spatial direction and 128 pixels in spectral direction with 60-μm macro pixels.

The SWIR FPA used is a 2.45-μm wavelength cutoff mercury-cadmium-telluride (MCT or HgCdTe) detector array based on the technology of Near Infrared Camera and Multi-Object Spectrometer (NICMOS). The format of the MCT detector array is 256 × 256 pixels, each of which has a pitch size of 60 μm. The readout and multiplexer is four-ported and uses a capacitive feedback transimpedance amplifier in each unit cell to provide the sensitivity and linearity at the low photo fluxes characteristic of the operation. The integration capacitors are selectable for different applications.

The HSI used both solar and in-flight calibration sources for absolute radiometric calibration. The MCT detector array was cooled to 115 K using a TRW miniature pulse tube cryocooler. The CCD array was cooled to 273 K to reduce dark current noise.

HSI also had a PAN channel. The PAN imager covers a spectral range in visible region 0.48–0.75 μm. It has a ground sampling distance (GSD) of 5 m in the cross-track direction and 2592 pixels in a cross-track line, which is equivalent to a swath of width close to 13 km. The PAN image allows for spatial resolution sharpening of the 30-m hyperspectral image using ground post processing.

The HSI has an envelope of roughly 43-cm wide × 94-cm high × 69-cm long and a mass of 40 kg including control and power electronics. HSI satellite could conduct pointing in cross-track

TABLE 2.4
Characteristics of HyperSpectral Imager (HSI)

Characteristics	Value
Orbit altitude (km)	525
Swath width (km)	7.68
Ground Sampling Distance (GSD) (m)	30
Spectral range (μm)	0.4–2.5 (Overall) 0.4–1.0 (VNIR) 0.9–2.5 (VNIR)
Spectral Sampling Interval (SSI) (nm)	5 (VNIR) 6.38 (SWIR)
Number of spectral bands	384 (Overall) 128 (VNIR) 256 (SWIR)
Field of view (FOV) (mrad)	15.4
Instantaneous Field of View (IFOV) (mrad)	0.06
Aperture (cm)	12.5
Focal length (cm)	104.8
f/#	8.3
Spectral dispersion	Grating
Focal plane arrays	CCD (VNIR) MCT (SWIR)
CCD detector array size	768 × 384
CCD pitch size (μm)	20
MCT detector array size	256 × 256
MCT pitch size (μm)	60
Mass (kg)	40
Volume (W × H × L)	43 cm × 94 cm × 69 cm
Power (W)	66 (average)

direction up to 20° off nadir. The swath width is 7.68 km and GSD is 30 m at the orbit altitude of 525 km. The characteristics of HSI are summarized in Table 2.4.

2.4 MODERATE RESOLUTION IMAGING SPECTRORADIOMETER ON TERRA AND AQUA SATELLITES

Moderate Resolution Imaging Spectroradiometer (MODIS) is a whiskbroom (cross-track scanning) imaging spectrometer. There are two MODIS sensors. The first MODIS sensor is on the NASA Earth Observing System (EOS) "Terra" satellite launched on December 18, 1999. The second MODIS sensor is on the "Aqua" EOS satellite launched on May 4, 2002. Both of them have continued to work quite successfully (as of December 2019) for 20 years in the case of the Terra MODIS instrument, and over 17 years for the Aqua MODIS instrument.

MODIS was designed and developed to collect continuous global data for studies of both short- and long-term changes in the Earth's land, ocean, and atmosphere systems and to help the science community assess the impact of global environmental and climate changes. It observes the Earth with a very wide swath of 2330 km, and therefore produces a complete global coverage in <2 days.

MODIS has 36 spectral bands covering a wavelength range from 0.41 μm to 14.2 μm. There are three different GSD: 250 m for bands 1 and 2, 500 m for bands 3–7, and 1 km for bands 8–36.

Table 2.5 is a summary of MODIS spectral bands, their center wavelengths, bandwidths, and primary science applications. MODIS bands 1–19 and 26, covering a wavelength range within 0.4 to 2.2 μm, are the reflective solar bands (RSB) and bands 20–25 and 27–36, covering wavelengths from 3.75 μm to 14.24 μm, are the thermal emissive bands (TEB) (Xiong et al. 2009).

MODIS acquires data by scanning in cross-track direction over a scan angle range of ±55° relative to instrument nadir via a double-sided scan mirror. The scan of each mirror side produces a swath of 10 km (nadir) in along-track direction by 2330 km in cross-track direction for

TABLE 2.5
MODIS Bands, Bandwidth, and GSD

Band	Center Wavelength (μm)	Bandwidth (nm)	Ground Sampling Distance (m)	Primary Science Applications
1	0.645	50	250	Land/cloud/aerosol
2	0.858	35	250	
3	0.469	20	500	Land/cloud/aerosol properties
4	0.555	20	500	
5	1.240	20	500	
6	1.640	24	500	
7	2.130	50	500	
8	0.412	15	1000	Ocean color/phytoplankton/ biogeochemistry
9	0.443	10	1000	
10	0.488	10	1000	
11	0.531	10	1000	
12	0.551	10	1000	
13	0.667	10	1000	
14	0.678	10	1000	
15	0.748	10	1000	
16	0.869	15	1000	
17	0.905	30	1000	Atmospheric water vapor
18	0.936	10	1000	
19	0.940	50	1000	
26	1.375	30	1000	Cirrus clouds
20	3.75	180	1000	Surface/cloud temperature
21	3.96	60	1000	
22	3.96	60	1000	
23	4.05	60	1000	
24	4.47	70	1000	Atmospheric temperature
25	4.52	70	1000	
27	6.72	360	1000	Water vapor
28	7.33	300	1000	
29	8.55	300	1000	Cloud properties
30	9.73	300	1000	Ozone
31	11.03	500	1000	Surface/cloud temperature
32	12.02	500	1000	
33	13.34	300	1000	Cloud top altitude
34	13.64	300	1000	
35	13.94	300	1000	
36	14.24	300	1000	

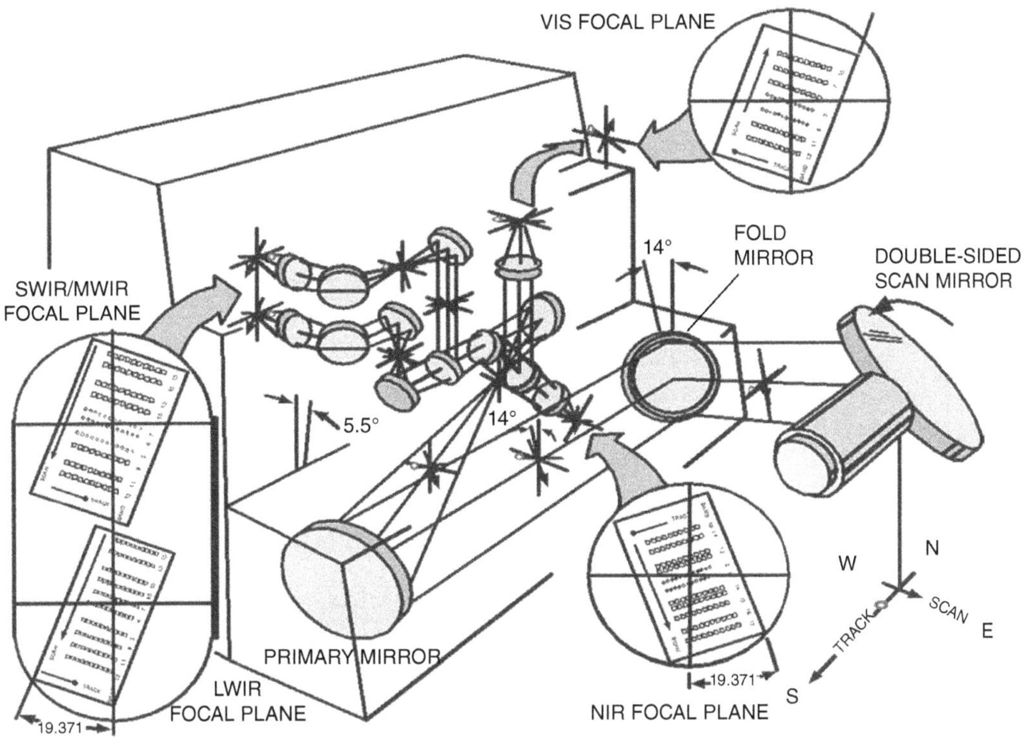

FIGURE 2.1 Layout of MODIS optical system. (Courtesy of NASA.)

each spectral band. The radiant flux reflected from the scan mirror is directed by a fold mirror to the off-axis telescope, consisting of a primary and a secondary mirror as illustrated in Figure 2.1. The aft optics includes 3 beam-splitters, 4 objective assemblies, and various blocking and spectral band-pass filters.

MODIS uses band-pass filters to separate spectrum of the observed scene. Blocking filters and spectral band-pass filters are placed in front of the linear detector arrays to select the desired spectral bands as illustrated in Figure 2.2. The filters for the 36 spectral bands are located according to their wavelengths on 4 focal plane assemblies (FPAs):

- Visible (VIS), for bands 3, 4, 8, 9, 10, 11, 12.
- Near-infrared (NIR), for bands 1, 2, 13, 13′, 14, 14′, 15, 16, 17, 18, 19.
- Shortwave infrared and mid-wave infrared (SWIR/MWIR), for bands 5, 6, 7, 20, 21, 22, 23, 24, 25, 26.
- Long-wave infrared (LWIR), for bands 27, 28, 29, 30, 31, 32, 33, 34, 35, 36.

As shown in Figure 2.2, the spectral bands are aligned in the along-scan direction and detector arrays are aligned in each spectral band in the along-track direction.

The material of the linear detector arrays for VIS and NIR FPAs is silicon. The VIS and NIR detector arrays are custom silicon P-I-N photovoltaic (PV) hybrid CMOS (Complementary Metal Oxide Semiconductor) FPAs covering spectral range 0.4 to 1.0 μm. The material for the SWIR/MWIR and LWIR FPAs is HgCdTe. The SWIR/MWIR detector arrays are PV hybrid CMOS FPAs covering a spectral range of 1.2–4.5 μm. The LWIR FPA includes a six-band

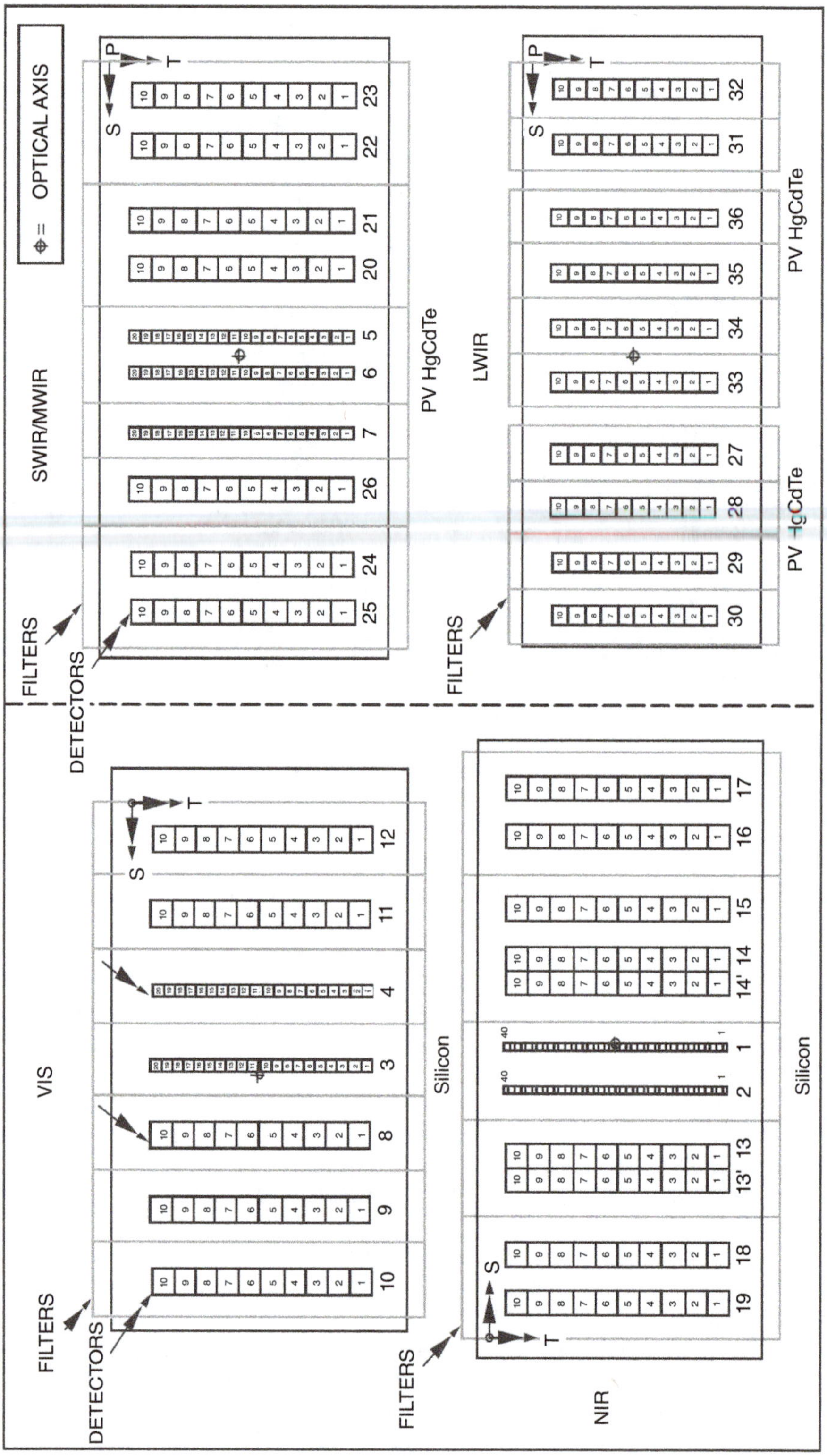

FIGURE 2.2 Layout of band-pass filters and linear detector arrays of the MODIS bands on the four FPAs. (Courtesy of NASA.)

photo-conductive (PC) detector arrays for the wavelengths beyond 10 μm. The pitch size of the detector arrays ranges from 135 μm to 540 μm square. Each detector array contains an array of indium bumps to increase the interconnection reliability and to supply mechanical support. The detector arrays are mated to readout integrated circuits (ROICs), which provide signal pre-amplification. The signals are then multiplexed and sent off-chip via 1–3 output lines per ROIC. Readout circuit design features include redundant bias, auto clock, shift registers, and an output amplifier to improve reliability and minimize single-point failures. Capacitive transimpedance amplifier (CTIA) readout unit cell preamplifiers provide customized gains for each of the multiple bands within a single readout.

The detector arrays have different number of elements for achieving different GSDs within a fixed width of 10 km in along-track direction. The detector arrays for bands 1 and 2 have 40 elements, each of which corresponds to a GSD of 250 m. The detector arrays for bands 3 to 7 have 20 elements, each of which corresponds to a GSD of 500 m. The detector arrays for remaining bands have 10 elements, each of which corresponds to a GSD of 1 km. As an exception, both bands 13 and 14 have a pair of 10-element detector arrays for high- and low-gain observations. The outputs of these detector arrays are summed in the scan direction, which is called time-delay integration (TDI). Table 2.6 summarizes the main performance parameters of MODID instrument.

The temperatures of the VIS and NIR FPAs are not controlled and float with the instrument, whereas the temperatures of SWIR/MWIR and LWIR FPAs are controlled nominally at 83 K via a passive radiative cooler that are referred to as the cold FPA (CFPA).

One of the major improvements of the MODIS instrument over its heritage sensors was its stringent calibration requirements. In order to achieve and maintain high-quality calibration requirements, MODIS was designed and built with state of the art on-board calibrators (OBCs), which include a solar diffuser (SD), a solar diffuser stability monitor (SDSM), a blackbody, and a spectro-radiometric calibration assembly (SRCA). In addition, a space view port is part of the OBCs, which enables measurements of instrument background and detector offsets (Xiong et al. 2016). Details on onboard calibration are described in Chapter 9.

2.5 HYPERION ONBOARD EO-1 MISSION

Hyperion hyperspectral sensor was onboard the EO-1 Mission, part of NASA's New Millennium Program to develop and validate a number of instrument and spacecraft bus breakthrough technologies designed to enable the development of future Earth imaging observatories. Hyperion is a pushbroom hyperspectral sensor with a 7.65-km wide swath. The ground footprint size is 30 m × 30 m. The 30-m size in the along-track direction was obtained by basing the frame rate on the velocity of the spacecraft for a 705-km orbit. An entire 7.65-km wide swath is obtained in a single frame. Each image contains data for a 7.65-km wide in cross-track direction by 185-km long in along-track direction (Pearlman et al. 2003).

Hyperion was designed as a technology demonstration and provided high-quality calibrated data for evaluation of hyperspectral applications (Pearlman et al. 2001). It had a fast-track schedule and was delivered to NASA Goddard Space Flight Center (GSFC) for spacecraft integration in <12 months. To achieve this goal, the developer TRW used focal planes and associated electronics remaining from the HSI for LEWIS mission under NASA Small Satellite Technology Initiative (SSTI) program. HSI for LEWIS mission is described in Section 2.3.

Hyperion has a single telescope and two spectrometers: one VNIR spectrometer and one SWIR spectrometer. The telescope is a three-mirror anastigmat (TMA) design with a 12-cm primary aperture and an effective f/# of 11. Both VNIR and SWIR spectrometers are of three-reflector Offner form design using convex gratings. The telescope images the scene on ground onto a slit that defines an instantaneous field of view (IFOV) of 0.6240° × 0.0024°, which corresponds to a ground

TABLE 2.6
MODIS Main Performance Parameters

Parameters	Value
Orbit altitude	705 km
Equator-cross time	10:30 AM descending (Terra) 1:30 PM ascending (Aqua)
Swath width	2330 km (cross-track) × 10 km (along-track)
Imaging scan mode	Whiskbroom
Telescope	Two mirror off-axis
Field of view (FOV)	±55°
Ground sampling distance (GSD)	B1–B2: 250m B3–B7: 500m B8–B36: 1000m
Wavelength range	0.41–14.2 μm
Spectral bandwidth	10–500 nm
Number of spectral bands	36
Spectral dispersion element	38 spectral band-pass filters
FPA	–18 Si linear detector arrays, –14 photovoltaic HgCdTe linear detector arrays, –6 photoconductive HgCdTe linear detector arrays
Format of detector arrays	40 × 1, 20 × 1, and 10 × 1 in each of detector array for 250, 500, and 1000 m GSD bands
Detector pitch size	135–540 μm square
Radiometric accuracy	±5% (reflective solar bands) ±0.5–1.0% (thermal emission bands)
Signal-to-noise ratio (SNR)	VNIR bands: 74:1–1087:1 (SNR)
Noise equivalent temperature difference (NEΔT)	SWIR/LWIR bands: 0.05–0.35 (NEΔT)
Digitization	12 bits
Data rate	10.6 Mbit/sec (peak daytime) 6.1 Mbit/sec (orbital average)
Power	162.5 W
Mass	228.7 kg
Volume	1.62 m^3 (1.0 × 1.6 × 1.0)

cross-track line of 7.65-km long (swath width) with 30-m wide in the satellite flight direction from an orbit of 705 km altitude. This slit image of the ground scene is relayed at a magnification of 1.38:1 to two focal planes of the VNIR and SWIR spectrometers. A dichroic filter (beam-splitter) in the system reflects the spectrum from 400 nm to 1000 nm to the VNIR spectrometer and transmits the spectral from 900 nm to 2500 nm to the SWIR spectrometer. The SWIR overlaps with the VNIR from 900 nm to 1000 nm and allows cross-calibration of the two spectrometers. There is an order-sorting filter in the VNIR spectrometer.

The VNIR spectrometer uses a 128 × 256 CCD detector array, only a section of 70 (spectral) × 256 (spatial) pixel is used. The SWIR spectrometer uses a MCT detector array, and has 256 × 256 pixels of 60-m pitch and a custom pixel readout. Only a 172 pixel (spectral) × 256 pixel (spatial) section is used. The two spectrometers produce 242 bands.

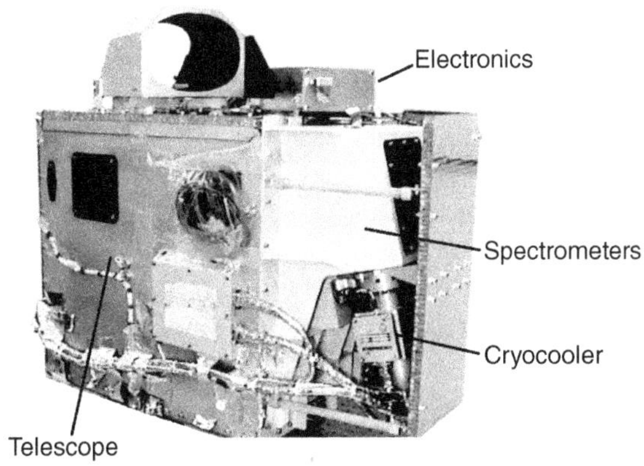

FIGURE 2.3 Hyperion sensor assembly. (Source of NASA.)

The Hyperion instrument consisted of the following three physical units:

1. the Hyperion sensor assembly (HSA),
2. the Hyperion electronics assembly (HEA), and
3. the Cryocooler electronics assembly (CEA).

These units were placed on the nadir deck of the spacecraft with the viewing direction along the major axes of the spacecraft. The HSA included the optical systems, cryocooler, in-flight calibration system, and the focal plane electronics as shown in Figure 2.3. The HEA and the CEA contained the data and control electronics for the sensor and the cryocooler. The HSA enclosure is 38.6-cm wide × 75-cm long × 64.6-cm high. The HSA enclosure controls the optics thermal environment, and the housing is maintained at 293 K ± 2 K for precision imaging and alignment. The VNIR spectrometer FPA is passively cooled by a radiator and operate at 283 K. The SWIR spectrometer FPA is actively cooled by the cryocooler with a thermal head set to 110 K. Table 2.7 summarizes the general, spectrometer, and instrument characteristics of Hyperion.

The SNR of Hyperion was both modeled and measured under an assumption of 30% uniform albedo, a 60° solar zenith angle, an instrument f/# of 11, a 10 nm spectral bandwidth, and a 224-Hz frame rate. As listed in Table 2.7, the measured SNR is between 140:1 and 190:1 in VNIR 550–700 nm, 96:1 at 1225 nm, and 38:1 at 2125 nm.

The use of a 2D detector array for pushbroom configuration versus a traditional linear detector array based on whiskbroom configuration requires new approaches to calibration. Hyperion adopted both prelaunch calibration and on-orbit radiometric calibration. Prelaunch calibration included extensive laboratory characterization and tests using both lamp-based and solid-state detector measurements. The lamp-based and solid-state detector-based calibrations showed a 5–15% difference in absolute values but similar spectral response profiles. Solar, lunar, and Earth surfaces observing "vicarious" measurements were used for the on-orbit calibration. On-orbit radiometric calibration was also performed with internal reference sources (lamps) mounted inside Hyperion.

2.6 COMPACT HIGH-RESOLUTION IMAGING SPECTROMETER (CHRIS) ON PROBA SATELLITE

The Compact High Resolution Imaging Spectrometer (CHRIS) was developed to provide remote sensing data for land applications and aerosol measurements as well as coastal zone monitoring. It is the main instrument payload on the European Space Agency's (ESA) Project for Onboard

TABLE 2.7
Characteristics of Hyperion

	Characteristics	Value
General	Orbit altitude	705 km
	Swath width	7.65 km
	Ground sampling distance	30 m
	Field of view	0.624°
	Instantaneous Field-of-view	0.0024° (44.25 µrad)
	Wavelength range	0.4–2.5 µm
	Spectral sampling interval	10 nm
	Number of spectral bands	242
	Number of spectral bands processed	198
Spectrometer	Fore-optics	TMA telescope
	Aperture	12 cm
	f/#	11
	Spectral dispersive element	Grating
	VNIR spectrometer	Offner form
	VNIR spectral range	0.4–1.0 µm
	VNIR Focal Plane Array	128 × 256 CCD
	SWIR spectrometer	Offner form
	SWIR spectrometer	0.9–2.5 µm
	SWIR Focal Plane Array	256 × 256 MCT
	SNR (VNIR 550-700 nm)	140:1–190:1
	SNR (SWIR ~1225 nm)	96:1
	SNR (SWIR ~2125 nm)	38:1
Instrument	Frame rate	223 Hz
	Digitization	12 bits
	Mass	49 kg
	Volume (L × W × H)	75 × 39 × 65 cm³
	Power	51 W (average) 126 W (peak)

Autonomy (PROBA) satellite launched on October 22, 2001. The primary objective of PROBA was to test a number of innovations in platform design, such as attitude control and recovery from errors, autonomous operation with minimal intervention from the ground (Barnsley et al. 2004). The longevity of CHRIS is quite impressive. It is still running after 18 years orbiting (as of December 2019).

The CHRIS sensor acquires data in the VNIR region of the electro-magnetic spectrum. It uses a 2D CCD detector array to combine high spatial resolution (17–20 m or 34–40 m) with a multi-angle viewing capability and programmable hyperspectral bands. It acquires up to 62 spectral bands at 5–15 nm spectral bandwidth in a wavelength range of 415–1050 nm. CHRIS has five formal operating modes, each of which has varied nominal number of bands, wavelength range, spectral bandwidth, and the nominal GSD, with GSD decreasing as spectral bandwidth increases as listed in Table 2.8. At perigee, CHRIS provides a GSD of 17 m, over typical image areas 13 km square. The instrument provides sets of images of selected target areas, at different pointing angles, forming up to 5 images of each target in a single overpass. Table 2.9 summarizes the characteristics of CHRIS instrument.

The CHRIS instrument consists of a telescope and a spectrometer attached to a 2D CCD array detector. The telescope is a two-mirror catadioptric configuration design that provides the required

TABLE 2.8
CHRIS Five Operating Modes

Key Parameters	Mode 1	Mode 2	Mode 3	Mode 4	Mode 5
Number of bands	62	18	18	18	37
Spectral range (nm)	406–992	406–1003	438–1035	486–788	438–1003
Bandwidth (nm)	6–20	6–33	6–33	6–11	6–33
GSD at nadir (m)	34	17	17	17	17

spectral range without using aspheric or off-axis elements. The focal length of the telescope is set at approximately 74.6 cm and the aperture diameter is set at 12 cm corresponding to an f/# of 6. A large meniscus lens at the entrance pupil of the instrument is employed to correct for spherical aberration. This also provides a convenient method for mounting the secondary mirror, which is cemented to the inner face of the meniscus. The telescope optics are completed by two small lenses in the converging beam in the entrance slit assembly, which correct for some minor telescope aberrations and allow the telescope to be approximately telecentric (Cutter 2000).

The spectrometer is an Offner configuration design with two curved prisms and three mirrors. The two prisms with curved surfaces are integrated into a modified Offner relay. The design uses only spherical surfaces and only one material (fused quartz) for the prisms. The spectrometer mirrors were made up of common optical glass. The dispersion of the spectrometer varies from approximately 1.3 to 12 nm across the spectrum with the highest dispersion at 415 nm and the lowest dispersion in the near-infrared at 1050 nm.

The two-mirror telescope design presents some challenges in terms of stray light control. The main source of stray light error is low-angle scatter at optical surfaces, arising from imperfections in the polish and coatings. In particular, some light from the scene can reach the entrance slit by transmission through the three lens elements, without reflection at either mirror. An oversized secondary mirror and a sequence of baffles were deployed to mitigate these stray lights.

The FPA used is a frame transfer CCD area array with 576 × 770 pixels. It is thinned, back-illuminated, and has a single-layer antireflection coating, which is uniform over the image area, and

TABLE 2.9
Characteristics of CHRIS Instrument

Characteristics	Value
Orbit altitude	615 km (550-670 km)
Swath width	13 km
Ground sampling distance	17 m, 34 m
Wavelength range	0.415 – 1.05 μm
Spectral sampling interval	5-15 nm
Number of spectral bands	up to 62
Fore-optics	Two-mirror catadioptric telescope
Aperture	12 cm
f/#	6
Spectrometer	Offner configuration
Spectral dispersive element	Grating
Focal Plane Array	576 × 770 pixels CCD
Mass	14 kg
Volume	79 × 26 × 20 cm^3

quarter-wave effective thickness at 1000 nm. This gives high quantum efficiency, including good performance in the deep blue spectral region (20% at 400 nm) and better than 7% at 1000 nm. The detector pitch size is 22.5 μm. The CCD array is capable to make radiometric measurements in the spectral range 0.4–1.05 μm with a SSI that varies from 1.25 nm to 11.25 nm across the spectral range.

Since CHRIS is able to acquire hyperspectral data from five different viewing angles, its data can potentially improve image classification, the quantification of vegetation structure and function. It can also provide information about sun-target-sensor geometries from which a measure of the Bidirectional Reflectance Distribution Function (BRDF) can be derived.

2.7 MEDIUM-RESOLUTION IMAGING SPECTROMETER ONBOARD ESA'S ENVISAT

The Medium Resolution Imaging Spectrometer (MERIS) was onboard ESA's ENVISAT satellite launched on March 1, 2002, from Europe's spaceport in Kourou, French Guiana. The ENVISAT mission initially planned for 5 years and ended after 10 years of operation on April 8, 2012, following an unexpected loss of contact with the satellite (Bézy et al. 2016).

MERIS was developed for observing the color of oceans, both in the open ocean and in coastal zones to study the oceanic biology and marine water quality of the global carbon cycle and the productivity of these regions and the atmosphere and land surface related processes. It allows accurate determination of oceanic constituents, such as chlorophyll, suspended matter, and dissolved organic material, thereby providing vital information about the water's quality and its productivity. The ENVISAT flew on a sun-synchronous orbit with a mean altitude of 799.8 km and an inclination of 98.55°. The orbit period is 100.6 min with a repeat cycle of 35 days. The MERIS covers the global of the Earth in 3 days (Gortl and Huot 2003).

MERIS is the first large swath spaceborne imaging spectrometer with high spectral and radiometric accuracy. It is a nadir-looking sensor and operates in a pushbroom mode. MERIS has a wide FOV with a swath width of 1150 km measuring the solar radiation reflected by the Earth in 15 spectral channels covering a wavelength range from 412.5 nm to 900 nm. The spatial sampling distance (GSD) varies in the cross-track direction, between 260 m at nadir and 390 m at swath extremities. Figure 2.4 shows MERIS instrument configuration. It consists of five identical imaging spectrometers, calibration mechanism, power and control electronics, sun baffle, radiator, and interface panel with satellite. The five imaging spectrometers (also referred to as hyperspectral cameras) are mounted in a fan-out configuration on the optical bench each covering one-fifth of the wide swath.

Fifteen spectral channels provided by MERIS can be changed in width (variable between 1.25 nm and 30 nm) and position by ground command. By design, the sensor could record 520 wavebands within the instrument spectral range 390 to 1040 nm in terms of the instrument native SSI of 1.25 nm. However, the MERIS was restricted by its downlink capability and transmitted only 15 channels, where each channel is an average taken over 8–10 elements of the detector array (Rast and Bezy 1990). Table 2.10 tabulates the band center and bandwidth of the 15 channels and their main applications.

The ocean signal, i.e., water leaving radiance from ocean, is very small, typically about 5% albedo. It is very challenge for the design of an ocean color satellite sensor. The radiometric performance is one of the most crucial requirements for MERIS because the signals coming from the ocean are weak and thus is most difficult to detect and quantify. MERIS also has to encompass a large dynamic range to cover these low-level signals as well as signals emanating from bright targets such as clouds and land surfaces, throughout its spectral range. This imposes a rather demanding requirement on the MERIS design for radiometric performance. Ocean color applications require extremely accurate absolute and inter-band radiometric calibration to support the atmospheric correction. This also entails low instrument sensitivity to polarization to cope with the large degree of polarization of the backscattered atmospheric radiation.

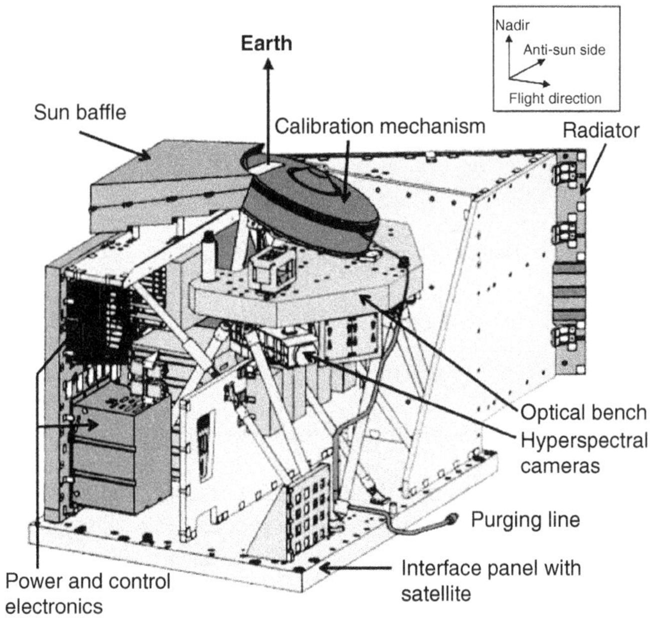

FIGURE 2.4 MERIS instrument configuration, showing the locations of the subsystems. (Courtesy of ESA.)

Each of the five identical imaging cameras has a FOV of 14°. These five cameras cover an overall FOV of 68.6°, with 0.4° overlap between adjacent cameras. These cameras are mounted on an optical bench in a fan-shaped configuration. The temperature of the optical bench is controlled to 20°C ± 1°C. These cameras view the Earth through five depolarizing windows. This modular design of MERIS ensures high optical image quality over the large FOV. The output of each camera is processed separately in an analogue and digital processing unit.

TABLE 2.10
Specification of the 15 Channels of the MERIS Sensor

Channel Number	Band Center (nm)	Bandwidth (nm)	Applications
1	412.5	10	Yellow substance and pigments detritus
2	442.5	10	Chlorophyll absorption maximum
3	490	10	Chlorophyll and other pigments
4	510	10	Suspended sediment, red tides
5	560	10	Chlorophyll absorption minimum
6	620	10	Suspended sediment
7	665	10	Chlorophyll absorption and fluorescence reference
8	681.25	7.5	Chlorophyll fluorescence peak
9	708.75	10	Fluorescence reference, atmospheric corrections
10	753.75	7.5	Vegetation, cloud
11	761.75	3.75	Oxygen absorption R-branch
12	778.75	15	Atmosphere corrections
13	865	20	Vegetation, water vapor reference
14	885	10	Atmosphere corrections
15	900	10	Water vapor, land

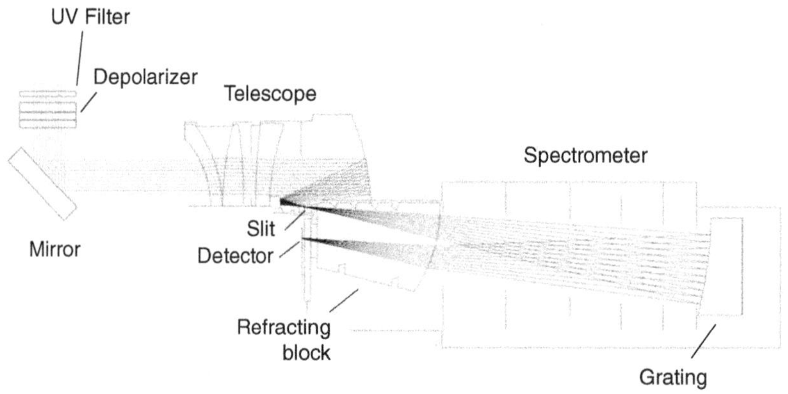

FIGURE 2.5 Optical layout of MERIS spectrometer. (Courtesy of ESA.)

The optics of each camera consists of a depolarizer, a telescope, and a spectrometer as shown in Figure 2.5. The first optical surface of the camera is an uncoated bulk absorption UV filter to protect the rest of the optical parts from solar UV radiation. Moreover, the external face of the first lens of the fore-optics has an inverse filter to equalize the instrument spectral responsivity (including optical transmission, diffraction grating efficiency, and CCD quantum efficiency) for radiometric performance.

A depolarizer is used to significantly reduce the polarization sensitivity of the instrument to meet the demanding requirements of ocean color sensing. It was made of three cemented wedges, two in quartz and one in fused silica for chromatic correction. The depolarizer is positioned at the entrance pupil. Quartz has the property to alter the polarization state of the transmitted light, the change in polarization state depending on the thickness of quartz. The depolarizing effect is a function of wedge angles. Larger angles produce more cycles in change of polarization state across an optical aperture of given dimensions. The depolarizer made the instrument insensitive to the Earth spectral radiance polarization status with a radiometric error lower than 0.3%.

The telescope is an off-axis catadioptric form design with a 67.3-mm focal length made of three large lenses in fused silica, a concave primary mirror, a convex secondary mirror, and a field lens in fused silica and located in the image plane. The lenses have antireflection coating optimized for the spectral range 390–1040 nm. The diameter of entrance pupil area is 50-mm of which an approximately rectangular area of 20 mm × 32 mm is used as the entrance pupil.

The design of MERIS spectrometers is of Dyson type. A spectrometer consists of a refracting Dyson block and a concave grating as shown in Figure 2.5. A Dyson type of spectrometer is selected for its compact size, lower polarization, and high throughput. The grating combines the collimating, spectral dispersion and imaging functions of a classical dispersive spectrometer. It is a low-groove density holographic grating with a spectral dispersion of 1321 p/mm.

With the aperture stop of the camera at the diffraction grating, the spectrometer is telecentric at both the object and image planes. Due to off-axis optical design, the beam is tilted at 11°. The Dyson block is made of fused silica to improve the correction of the spectrometer aberrations of the dispersed image. The spectrometer entrance slit is located near the diffraction grating center of curvature, on a flat face of the Dyson block. The opposite face of the block is spherical and is concentric with the diffraction grating. A blocking filter is included in the construction of the Dyson block in order to suppress the second-order spectrum of the diffraction grating.

The spectrometer works at "unit magnification," which means that a square of side 22.5 μm at the entrance slit (corresponding to a 260 m × 260 m area on Earth) is imaged as a square of side 22.5 μm on the detector array. The grating disperses a 1.25-nm spectral interval across one 22.5-μm detector element. The entrance slit is straight. It is imaged as a straight line in each wavelength on a detector element in spectral direction. Five hundred twenty (520) detector elements are used to cover the nominal spectral range 390–1040 nm.

MERIS instrument reduced the stray light to the minimum by careful baffling of the zero and higher diffraction orders, including grooves in the corrector block. Ghosts are reduced by the use of wide band high-quality antireflection coatings on all surfaces.

The FPA of the hyperspectral cameras is an area CCD array with 780 × 576 imaging elements custom-designed for MERIS. The pitch size of the detector element is 22.5 μm × 22.5 μm. The CCD arrays are thinned and backside illuminated to avoid the absorption and reflection in the blue by the electrode structure located at the front face of the device and they offer the required high responsivity in the blue region of the spectral range. The quantum efficiency between 450 nm and 750 nm is better than 80%. A permanent protective window is applied to protect the detector antireflection coating from degradation. The window, outside the Imaging Zone, is gold coated to optimize the thermal interfaces with the surrounding optics. A graded antireflection coating in the spectral direction has been deposited on the CCD array by adjusting the coating thickness across the device to match the wavelength diffracted by the spectrometer and to locally meet the condition of minimum reflection. This enabled to mitigate the optical ghosts generated by reflection between the CCD and its window.

For the 780 columns × 576 rows in imaging area of the CCD array, only 740 × 520 are used as imaging elements. The central 740 columns correspond to the spatial footprints in a ground cross-track line at a given wavelength. The 520 rows correspond to the spectral bands of a footprint at all wavelengths in the 390–1040 nm range. Five masked columns are used on both sides of the imaging area to monitor the variation of the detector dark current and offset. Twenty additional columns are implemented between the imaging area and the masked columns to account for possible misalignment with the mask. Additional rows are used on both sides of the imaging area to account for possible shift of the spectrum with respect to the CCD to account for possible misalignment with the mask and to protect the smear band against charge contamination. The CCD arrays are cooled using Peltier cooler, each of them is associated with one cooler to –22.5°C ± 0.01°C to reduce dark current. The Peltier coolers are referenced through heat pipes to a deep sky radiator.

MERIS is equipped with an onboard calibration unit based on flat plate sun-illuminated diffusers to meet demanding radiometric and spectral accuracy requirements.

The calibration unit is mounted on the optical bench and includes the calibration wheel and three baffles: (1) a sun baffle to limit illumination of the diffusers during calibration to only that of the sun; (2) a protection baffle covering the calibration wheel and including a dry nitrogen purges in positive pressure throughout assembly, integration, and launch; and (3) a camera baffle reducing the along track scatter both in observation and calibration. The calibration wheel has five disc positions: two radiometric diffusers, one wavelength diffuser, one Earth diaphragm, and one Earth shutter. When not required to perform the calibration, the diffusers are stayed in a cavity to protect them from contamination and UV radiation.

The reason of having two radiometric diffusers is for monitoring the ageing of the diffuser. A calibration diffuser has been exposed to the Sun for a total cumulative period of about 6.8 h during MERIS's lifetime at an operation frequency of once every 2 weeks. Degradation is expected due to vacuum UV radiation and particle exposure. A second identical diffuser is equipped to evaluate the degradation. The second diffuser is used infrequently (about once every 3 months) and thus does not degrade at the same rate as the first diffuser. Diffuser ageing is monitored by comparing the data acquired with both diffusers.

Wavelength calibration is achieved by using the wavelength diffuser featuring well known and stable absorption peaks. MERIS spectral bands are reprogrammed to sample the absorption features with the highest possible SSI (1.25 nm). From this calibration, the spectral position of any spectral band can be derived. Use of the solar Fraunhofer absorption lines and the O_2-A absorption spectra had also been exercised as a complement to the spectral diffuser.

MERIS can be operated in either full resolution (FR) imaging mode with a spatial resolution of 300 m or reduced resolution (RR) mode after onboard spatial binning of 4 × 4 FR pixels for a spatial resolution of 1200 m. The wide swath of 1150 km ensured the global Earth coverage within 3 days, which is required by oceanographic and atmospheric users.

TABLE 2.11
MERIS Main Performance Parameters

Parameters	Value
Swath width	1150 km
Field of view (FOV)	68.6°
Ground sampling distance (GSD)	260 m × 300 m (Full resolution) 1040 m × 1200 m (Reduced resolution)
Telescope type	Off-axis catadioptric
Spectrometer type	Dyson form
Spectral dispersive element	Grating
Wavelength range	390–1040 nm
Spectral sampling interval (SSI)	1.25 nm
Spectral resolution (FWHM)	1.5 nm
Spectral channels downlinked	15 (programmable in center band and width)
FPA	Back-illuminated frame-transfer 2D CCD array
Detector array format	780 × 576
Detector pitch size	22.5 μm × 22.5 μm
Relative spectral accuracy	0.05%
Absolute radiometric accuracy	2%
Signal-to-noise ratio (SNR)	150:1–1500:1
Modulation transfer function (MTF)	>0.3
Polarization sensitivity	<0.3%
Data rate	24 Mbit/sec (Full resolution) 1.6 Mbit/sec (Reduced resolution)
Mass	200 kg
Volume	1.62 m^3 (1.6 × 0.9 × 1.0)

The MERIS instrument has a mass of 200 kg and volume of 1.6 m^3, draws on average approximately 200 W and delivers 24.0 Mbits/sec data rate in FR mode and 1.6 Mbits/sec in RR mode. The main performance parameters are summarized in Table 2.11.

2.8 VISIBLE AND INFRARED THERMAL IMAGING SPECTROMETER FOR ROSETTA, VENUS-EXPRESS, AND NASA-DAWN PLANETARY MISSIONS

The Visible and Infrared Thermal Imaging Spectrometer (VIRTIS) was originally built for ESA's Rosetta cometary mission (Coradini et al. 2007). It was the third cornerstone mission of the ESA's Horizon 2000 programme. Rosetta was a space probe, including a lander Philae, launched on March 2, 2004, for studying comet 67P/Churyumov–Gerasimenko (67P). The spacecraft reached the comet on August 6, 2014. During its journey to the comet, the spacecraft flew by Mars and the asteroids 21 Lutetia and 2867 Šteins.

The scientific payloads of Rosetta mission were designed to obtain the information of the comet by combining in situ analysis of comet material obtained by the small lander Philae and by a long-lasting and detailed remote sensing of the comet obtained by instrument onboard the orbiting spacecraft. The combination of remote sensing with in situ measurements increases the scientific return of the mission. VIRTIS is one of the scientific payloads of the Rosetta Orbiter to detect and characterize the evolution of specific signatures—such as the typical spectral bands of minerals and molecules—arising from surface components and from materials dispersed in the coma. The identification of spectral features is a primary goal of the Rosetta mission as it allows identification of the nature of the main constituent of the comets. Moreover, the surface thermal evolution during comet approach to sun is also studied.

Thanks to its suitability for planetary exploration, VIRTIS was also selected as a key instrument for ESA's Venus Express and NASA-Dawn missions. This makes the VIRTIS a great success story for three planetary missions.

Venus Express mission was aimed to study the Venusian atmosphere and clouds in detail, and to study the plasma environment and the surface characteristics of Venus from orbit. It was launched on November 9, 2005, and entered its target Venus orbit at apoapsis on May 7, 2006 (Piccioni et al. 2007).

Dawn is a space probe launched on September 27, 2007, for studying two of the three known protoplanets of the asteroid belt, Vesta and Ceres. The space probe entered orbit around Vesta on July 16, 2011, and completed a 14-month survey mission before leaving for Ceres in late 2012. It then entered orbit around Ceres on March 6, 2015 (Russell et al. 2007).

The VIRTIS was developed in cooperation among three countries Italy, France, and Germany. Although with some modifications, the VIRTIS instruments for Venus Express and NASA-Dawn missions are essentially identical to the instrument carried by Rosetta mission (Piccioni et al. 2007). It is an imaging spectrometer with three focal planes in two channels. The mapping channel, referred to as VIRTIS-M, has two 2D focal planes covering the visible region (0.28–1.1 μm) and infrared (IR) region (1.05–5.13 μm). The spectroscopic channel, referred to as VIRTIS-H, has a single aperture covering a wavelength range 1.84–4.99 μm with high SSI. VIRTIS-M generates 432 spectral band images of size 256 × 256 pixels in the wavelength range 0.28–4.99 μm with SSI of 1.89 and 9.49 nm in visible and IR regions, respectively. There are eight different operational modes by binning the acquired data spectrally and spatially to reduce the data rate. The FOV of VIRTIS-H is centered in the middle of the VIRTIS-M image. VIRTIS-H generates spectra with high SSI in the small portion of the image. Figure 2.6 shows the Optical Module of VIRTIS payload before being integrated on the spacecraft. Table 2.12 summarizes the performance characteristics of the VIRTIS payload.

The hyperspectral imager VIRTIS-M consists of a Shafer-type telescope and an Offner spectrometer that serve both the visible and IR regions. Instead of two separate spectrometers, one spectrometer forms two focal planes to cover both the visible and IR regions. This is a rather unique

FIGURE 2.6 Optical Module of the VIRTIS payload before installation of the multilayer insulation and integration on the spacecraft. The VIRTIS-M is at right top; the VIRTIS-H is at left top. (Courtesy of ESA.)

TABLE 2.12
VISTIS Performance Parameters

Parameters	VISTIS-M Visible	VISTIS-M Infrared	VISTIS-H
Platform	Rosetta mission spacecraft Venus-Express mission spacecraft NASA-Dawn mission spacecraft		
Spectral range (μm)	0.28–1.10	1.05–5.13	1.84–4.99
Max Spectral Sampling Interval (nm)	1.89	9.47	0.6
Field of view (°)	3.67 × 3.67		0.025 × 0.077
IFOV (Max spatial sampling distance) (°)	0.014 × 0.014		-
Image size (pixel)	256 × 256		-
Telescope	Shafer type		Off-axis parabolic mirror
Aperture (mm)	47.5		32
f/#	5.6	3.2	2.04
Slit dimension (mm)	0.038 × 9.53		0.029 × 0.089
Spectrometer form	Offner		Echelle
Spectral dispersion	Grating		Grating
Detector array	CCD	HgCdTe	HgCdTe
Detector format	508 × 1024	270 × 436	270 × 436
Pixel pitch size (μm)	19	38	38
Detector cut-on and cut-off wavelength (μm)	0.25–1.05	0.95–5.0	0.95–5.0
Readout noise (e-)		<300	<300
Mean dark current	<1 e$^-$/sec	<2 fA @ 90K	<2 fA @ 90K
Operating temperature (K)	150–190	65–90	65–90
Mass (kg)		33	
Power (W)	50 (during science operation), 70 (peak)		

design. In addition, this optical configuration eliminates the need for a beam-splitter, collimators, and camera objectives, thereby simplifying fabrication and minimizing volume and mass. However, a grating spectrometer that does not rely on a collimator and camera objective requires perfect matching with its collecting telescope. Not only must they have matching F-numbers, but the telescope must be telecentric or have its exit pupil positioned on the grating. The Shafer-type telescope is matched to the Offner spectrometer because both are telecentric. The entrance pupil is imaged on the slit of the spectrometer.

Since the pupil optics conjugate is on the grating, the splitting of visible and IR spectral regions is achieved by the grating, which was made having three-circle zones with different groove densities and different depths (Piccioni et al. 2000). The two inner zones, which make up the central 30% of the conjugate pupil area, have higher groove density for generating the higher spectral resolution for the visible region. The external circle zone has lower groove density, makes up 70% of the pupil area for compensating the low IR solar irradiance, for the IR region.

The visible FPA is a Thomson-CSF TH7896 CCD detector. It uses a full-frame image sensor with 1024 × 1024 sensitive elements, two registers, and four outputs. It is used as a frame-transfer device, shielding half the sensitive area that works as a memory section. Only one horizontal register and one output are actually used. In order to meet the requirement of the IFOV of 0.014°, 2 × 2 binning is implemented to achieve a pixel size of 38 μm, same size as the pixel of the IR detector array.

The IR FPAs used in VIRTIS-M and VIRTIS-H are photovoltaic HgCdTe detector arrays. The arrays were formed through hybridization of HgCdTe material with dedicated Si CMOS. The dimension of the array is 270 × 436 pixels with a pitch size of 38 μm, a wavelength range from 0.95 μm up to 5.0 μm, and an operating temperature of 80 K.

2.9 COMPACT RECONNAISSANCE IMAGING SPECTROMETER FOR MARS

The Compact Reconnaissance Imaging Spectrometer for Mars (CRISM) is a visible near-infrared and infrared (VNIR+IR) hyperspectral imager onboard the NASA spacecraft of Mars Reconnaissance Orbiter (MRO), which was launched on August 12, 2005, and attained Martian orbit on March 10, 2006. The primary science objectives of the MRO mission are (1) to search for evidence of aqueous and/or hydrothermal activity; (2) to map and characterize the composition, geology, and stratigraphy of surface deposits; and (3) to characterize seasonal variations in dust and ice aerosols and water content of surface materials, recovering science lost with the failure of the Mars Climate Orbiter (MCO). MRO also has two secondary objectives: (1) to provide information on the atmosphere complementary to the reflown MCO investigations, and (2) to identify new sites with high science potential for future investigation (Murchie et al. 2009).

CRISM is one of six major science instruments on MRO. It is used to produce detailed maps of the surface mineralogy of Mars. CRISM is being used to identify minerals and chemicals indicative of the past or present existence of water on the surface of Mars. These materials include iron, oxides, phyllosilicates, and carbonates, which have characteristic patterns in their visible-IR energy.

The CRISM measures visible and IR electromagnetic radiation from 362 nm to 3920 nm with a SSI of 6.55 nm. It operates in three modes: multispectral mode, targeted mode, and atmospheric mode. In the multispectral mode, the CRISM points at planet nadir and uses multispectral means to reconnoiter Mars with 72 of its 544 measurable spectral bands at a spatial resolution of 100–200 m per pixel. Nearly the entire planet can be mapped in this fashion. The objective of this mode is to identify new scientifically interesting locations that could be further investigated.

In targeted mode, the CRISM uses hyperspectral means to detect Mars. The imaging spectrometer measures energy reflected from Mars surface in all 544 spectral bands. When the MRO spacecraft is at an altitude of 300 km, the CRISM detects a region of interest at full spatial and spectral resolution (15–19 m/pixel, 362–3920 nm at 6.55 nm/channel). Ten additional abbreviated, spatially binned images are taken before and after the main image, providing an emission phase function of the site for atmospheric study and correction of surface spectra for atmospheric effects.

In atmospheric mode, only the emission phase function is acquired. Global grids of the resulting lower data volume observations are taken repeatedly throughout the Martian year to measure seasonal variations in atmospheric properties. Raw, calibrated, and map-projected data are delivered to the community with a spectral library to aid in interpretation. Table 2.13 tabulates the performance parameters of the CRISM instrument.

CRISM consists of three units: (1) Optical Sensor Unit (OSU), which includes the optics, gimbal, focal planes cryocoolers, radiators, and focal plane electronics; (2) Gimbal Motor Electronics (GME), which commands and powers the gimbal motor and encoder, and analyzes data from the encoder in a feedback loop, and (3) Data Processing Unit (DPU), which accepts and processes commands from the spacecraft and accepts and processes data from the OSU and communicates it to the spacecraft (Murchie et al. 2007).

Figure 2.7 shows the layout of the optical design of the OSU. The telescope of CRISM is a Ritchey-Chretien on-axis design with a 441-mm focal length and 100-mm aperture. It focuses incoming light onto a slit. Both the primary and secondary mirrors are coated aluminum, and are baffled to block out-of-field paths to the slit. The secondary mirror is mounted by a spider and obscures 29% of the aperture. The telescope is protected by a composite baffle with flexure mounts to the optical bench.

TABLE 2.13
CRISM Performance Parameters

Parameters	Value	Comment
Platform	Mars Reconnaissance Orbiter (MRO)	
Orbit altitude	255–320 km	
Fore-optics	On-axis Ritchey-Chretien	
Spectrometer type	Offner form	
Spectral dispersive element	Grating	
Swath width	9.4–11.9 km	From 255 km to 320 km Martial orbit
Spatial sampling	15.7 to 19.7 m/pixel	From 255 km to 320 km Martial orbit
Number of spectrometers	2	One for VNIR, one for IR
Overall spectral range	362–3920 nm	>3600 nm allows greater sensitivity to carbonate
VNIR Spectrometer range	362–1053 nm	Overlap with IR spectrometer
IR Spectrometer range	1002–3920 nm	
Spectral sampling interval	6.55 nm	
Spectral resolution	7.9–10.1 nm VNIR, 9.0–19.0 nm IR	Measured at FWHM
Number of spectral bands	544	
Field of view	2.12°	
Instantaneous FOV	61.5 mrad	Pixel angular size
Focal length	441 mm	
Aperture	100 mm	
Detector array format	640 × 480 pixels	
Detector pixel pitch	27 μm × 27 μm	
Signal-to-noise ratio	425:1 at 2300 nm, >100:1 at 400 nm and 3600 nm	
System MTF	0.73 (VNIR) 0.4 (IR)	At 400 nm and 3600 nm, for average material at 30° phase angle
Spectral distortion	<1.2 pixel	
Spatial distortion	<±0.4 pixel	
Stray light	<2%	
Pointing	±60°	Along track
Scan jitter	25 μrad	
Design lifetime	>4 years	
Mass	32.92 kg	
Power	44–47 W	During normal operations
	16 W	During standby with subsystems off

The slit is made of nickel and its telescope facing side is gold plated, both to resist effects of heating and to dissipate incident solar energy in the event of a direct view of the sun. The slit is 27-μm wide and 16.3-mm long and is mounted in an assembly for fastening to the optical bench.

Following the slit is the spectrometer optics. A wedged dichroic beam splitter is used to split light by reflecting the VNIR while transmitting the IR to the VNIR and IR spectrometers and FPAs. The spectrometers are of modified Offner form. The wedge directs internal reflections in the transmitted IR light out of its nominal path, so that the reflections can be blocked by the order-sorting filter on the detector array.

In the VNIR spectrometer, the first and third mirrors (VNIR M1 and M3) are prolate ellipsoids. The grating (M2) is mounted on a spherical surface. In the IR spectrometer, IR M1 and IR M2 are spherical and IR M3 is a generalized asphere approximating an oblate ellipsoid.

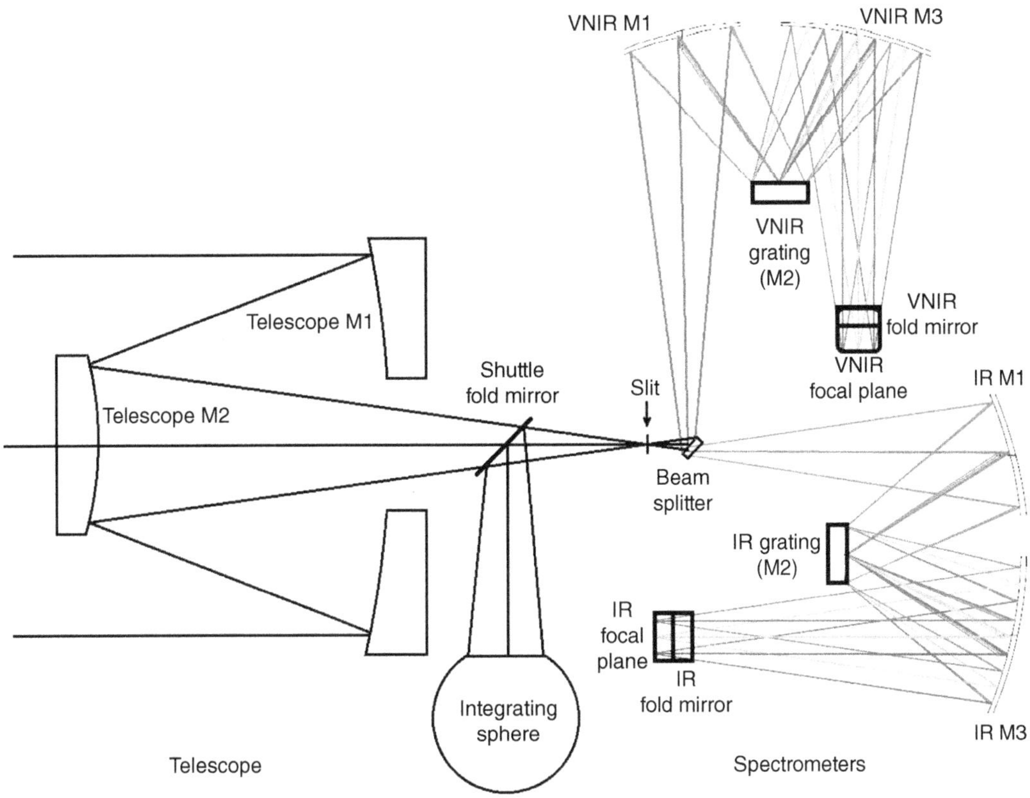

FIGURE 2.7 Optical layout of CRISM imaging spectrometer.

Both spectrometers account for the curved slit, thereby creating a flat, well-corrected slit image at their detector arrays. Each spectrometer also has a fold mirror after M3 that directs the light out of plane to focus on the FPA mounted on the side of the spectrometer for thermal control. Nominally each spectrometer has unity magnification, but due to manufacturing tolerances the angular sizes of the two FOVs differ by 1.2%.

The diffraction grating of the spectrometers disperses the light and focuses it onto their respective focal planes. The gratings are an aluminized polymer manufactured using an electron beam process (Wilson et al. 2003). Each is dual zone, with each zone blazed to maximize efficiency in either the longer or shorter wavelength parts of the VNIR or IR spectral range. The areas of each zone are sized to balance SNR in their two wavelength ranges. The first-order diffracted light is used. Higher orders from the gratings are blocked by order-sorting filters mounted on the detectors.

The VNIR focal plane is a 640 columns × 480 rows detector array. It is silicon photodiode detector array indium bump-bonded to a TCM 6604A multiplexer. The pitch size of the detector is 27 mm. Full well capacity is approximately 780,000 electrons (e^-), and response is quasi-linear up to about 93% of full well. The readout noise is approximately 180 e^-, and gain is 80 e^- per 14-bit digital number (DN). There are four quadrants each 160 columns in width. Readout occurs one row (spectral direction) at a time, and each quadrant's output is sent to a separate analog-to-digital converter that digitizes it to 14 bits.

The IR focal plane is a HgCdTe detector array indium bump-bonded to a TCM 6604A multiplexer, the same as used in the VNIR array. Read noise is slightly lower at 140 e^-, but gain and full well capacity are similar to the VNIR array. The fixed-mounted filter is a three-zone interference filter designed to block not only higher orders from the grating, but also thermal background to which the detector responds at wavelengths <4250 nm. Zones 1 and 2 are band-pass filters that transmit wavelengths of 1000–1810 nm and 1580–2840 nm, respectively. Zone 3, covering longer wavelengths, is a linearly variable filter (LVF) with an 80-nm band-pass to match the dispersion of the IR spectrometer. All three zones overlap in order to eliminate leaks of thermal background to the IR detector array. The temperature of the IR detector array is maintained at <120 K using an active cryogenic cooler system to minimize dark current.

CRISM has several internal calibration capabilities that allow monitoring bias, dark current and thermal background, detector nonuniformity, and responsivity of the parts of the system behind the slit. An integrating sphere and a shutter are built into the optical bench. The integrating sphere as shown in Figure 2.7 provides a smooth, near flat field of dispersed light as viewed through each of the spectrometers. It samples all of the optics except the telescope, and is intended as the primary in-flight reference for radiometric calibration. Illumination is provided by either of two small incandescent lamps, one controlled by the VNIR focal plane electronics and another is controlled by the IR focal plane electronics. The VNIR-controlled bulb is the primary bulb for in-flight calibration. The lamps' current level can be commanded either under open loop or closed loop control. The shutter is an aperture-filling vane with a polished aluminum rear surface, attached to a 33-position of a stepper motor. In its closed position, the shutter enables measurement of bias, dark current, and thermal background for the IR detector array.

2.10 MOON MINERALOGY MAPPER

The Moon Mineralogy Mapper (M3) is a NASA's contributed imaging spectrometer to the India's first mission to the moon, Chandrayaan-1, launched on October 22, 2008. The M3 is the first high-resolution imaging spectrometer to map the entire lunar surface spatially and spectrally. The measured information will help provide clues to the early development of the solar system and guide future astronauts to store precious resources. The Chandrayaan-1 mission was cut short at 10 months in August 2009, when contact was lost with the spacecraft. Despite the abbreviated mission, M3 was able to meet its mission requirements: collecting >95% of the moon in Global Mode along with a small number of Target Mode images (Green et al. 2011).

The M3 instrument is a hyperspectral sensor operating in pushbroom mode. It generates images of moon surfaces in long narrow strips in a wavelength range from 400 nm to 3000 nm (blue to IR light) with a SSI of 10 nm. This forms 260 spectral images for a scene of the lunar surface. The swath width (i.e., width of a line scene) is 40 km on the moon's surface at a moment. This line scene is imaged onto 600 detector pixels, with each pixel representing a footprint of size 70 m × 70 m on the surface. The second spatial dimension of the scene is obtained by the flight of the spacecraft along the flight direction. The circumference of the moon is 10,930 km. With overlap, it takes >274 image swaths to completely map the moon.

M3 instrument was designed to include a number of key enabling elements to achieve the science requirements and overcome the additional constrains imposed by the Chandrayaan-1 mission, such as low mass (<10 kg), compact volume, limited power, and limited downlink capacity. A high uniformity and high throughput imaging spectrometer optical design was chosen, which is both compact and comparatively simple. In order to measure the full spectral range from 430 nm to 3000 nm with a single spectrometer, an all-reflective Offner spectrometer design was selected (Mouroulis et al. 2000).

Figure 2.8 shows the optical layout of the M3 instrument. Light from the moon passes through a pair of baffles and is reflected from a fold mirror to a compact TMA telescope. The telescope provides a FOV of 24° in the cross-track direction and an IFOV of 0.7 mrad in the along-track direction, thus supporting the required 40-km swath and 70-m spatial sampling distance from a nominal 100-km lunar orbit. Light from the telescope is imaged on a uniform open slit. Light selected by the slit is passed to the surface of the spectrometer mirror where it is reflected to the efficiency-tuned diffraction grating. Light is spectrally dispersed with optimized efficiency in the –1 order over the wavelength range from 430 nm to 3000 nm. The spectrally dispersed light from the diffraction

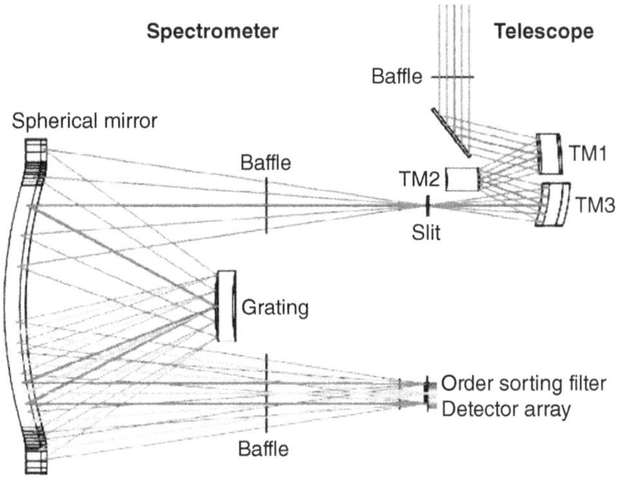

FIGURE 2.8 Optical design of M3 imaging spectrometer. (Courtesy of NASA/JPL.)

grating is reflected for the second time by the spectrometer mirror and selectively transmitted by the order sorting filter and focused on the HgCdTe area detector array. The order sorting filter is a three-zone filter with a nominal cut on at 425 nm and zone boundaries at 815 and 1565 nm. At the detector array, the dispersed light is converted to an electronic signal and passed to the electronic signal chain for amplification, digitization, compression, formatting, and storage prior to Chandrayaan-1 transmission to Earth (Green et al. 2011). Table 2.14 lists the key performance parameters of M3 imaging spectrometer.

2.11 FOURIER TRANSFORM HYPERSPECTRAL IMAGER ONBOARD CHINESE ENVIRONMENT PROTECTION SATELLITE HJ-1A

The minisatellite constellation Huan Jing (HJ) is a national program under the National Committee for Disaster Reduction and the State Environmental Protection Administration (NCDR/SEPA) of China to construct a network of Earth observing satellites. The overall objective is to establish an operational Earth observing system for disaster monitoring and mitigation using remote sensing technology and to improve the efficiency of disaster mitigation and relief. *Huan Jing* in Chinese means *environment*.

The first phase of the program implementation is referred to as HJ-1 (or Environment-1). The HJ-1 constellation includes three minisatellites (2 + 1 constellation). The satellites of the constellation are referred to as HJ-1A, HJ-1B, and HJ-1C. HJ-1A/B satellites were launched on September 6, 2008, in Taiyuan Satellite Launch Center, China, with the technology of "one rocket for two satellites."

The primary goal of HJ-1A/B is to validate new technologies of spaceborne instruments and to provide remotely sensed data to the user community for environment and disaster monitoring. The three primary payloads onboard HJ-1A/B are a hyperspectral imager, a CCD-based VNIR multispectral imager (MSI) and an IR MSI. Table 2.15 lists the payloads on the two satellites and their main performance parameters.

The hyperspectral imager onboard HJ-1A satellite is a Fourier transform based hyperspectral imager (FTHSI). It is a kind of spatially modulated imaging interferometer, which has been developed in the 1990s (Rafert et al. 1995, Smith and Hammer 1996). Unlike dispersive element based imaging spectrometers, a Fourier transform based imaging spectrometer produces interferometric data in Fourier transform domain, which need to be processed before obtaining radiometric data.

TABLE 2.14
Key Performance Parameters of M3 Imaging Spectrometer

Parameter	Values
Platform	Lunar orbiter
Orbit altitude	100 km
Swath width on moon	40 km
Spatial sampling distance	70 m
Imaging scan mode	Pushbroom
Wavelength range	0.43–3.0 μm
Spectral sampling interval	10 nm
Number of spectral bands	260
Spectrometer type	All-reflective Offner
Spectral dispersive element	Grating
Telescope type	Three-mirror anastigmat
FOV	24°
IFOV	0.7 mrad
Detector array material	HgCdTe (substrate-removed with spectral range extended to visible wavelengths)
Detector array format	640 × 480 pixels
Full well capacity	650,000 e
Readout noise	100 e
Detector pitch size	27 μm × 27 μm
Signal-to-noise ratio	>400:1 at equatorial reference radiance >100:1 at polar reference radiance
Power	<20 W
Mass	<10 kg
Volume	<50 × 50 × 50 cm^3

TABLE 2.15
Three Primary Payloads on Board HJ-1A/B Satellites and Their Main Performance Parameters

Satellite	Payload	Number of Bands	Spectral Range (μm)	Ground Sampling Distance (m)	Swath Width (km)
HJ-1A	VNIR multispectral imager	4	B1: 0.43–0.52 B2: 0.52–0.60 B3: 0.63–0.69 B4: 0.76–0.90	30	360
	Fourier transform hyperspectral imager	115	0.45–0.95	100	50
HJ-1B	VNIR multispectral imager	4	B1: 0.43–0.52 B2: 0.52–0.60 B3: 0.63–0.69 B4: 0.76–0.90	30	360
	IR multispectral imager	4	B1: 0.75–1.10 B2: 1.55–1.75 B3: 3.50–3.90 B4: 10.5–12.5	15	720

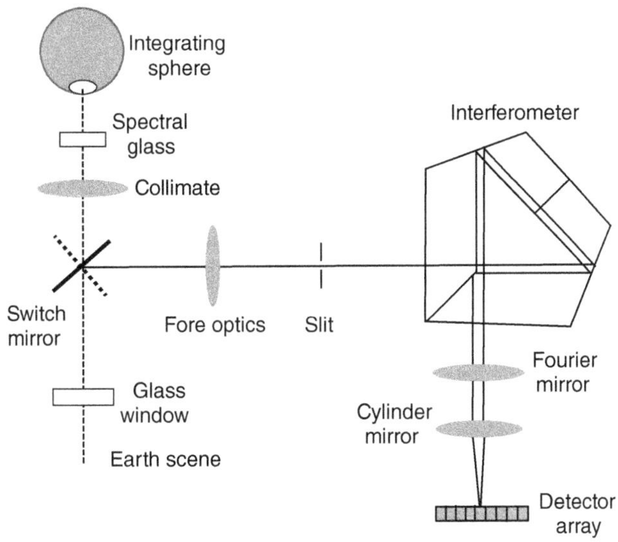

FIGURE 2.9 Construction diagram of the FTHSI instrument.

The FTHSI instrument has 115 bands covering a spectral range from 0.45 μm to 0.95 μm after processing of the raw Fourier transform data and returning to spectral domain. Zhao et al. (2010) reported their work on processing and calibration of the FTHSI instrument data.

Figure 2.9 shows the construction diagram of the FTHSI instrument. It consists of a fore optics, Fourier interferometer, and calibration subsystem. In the fore optics, a switch mirror is equipped to select the incoming light of the interferometer for observation or calibration. When the mirror is turned to the observation position, the incoming light from a scene on ground is directed toward the interferometer. When the mirror is turned to the calibration position, the incoming light from the calibration subsystem is directed toward the interferometer. Table 2.16 tabulates the key performance parameters of the FTHSI instrument.

2.12 HYPERSPECTRAL IMAGER ONBOARD INDIAN MINI SATELLITE-1

The Indian Mini Satellite-1 (IMS-1), which was originally called third world satellite (TWSAT), was launched on April 28, 2008. It carried a HyperSpectral Imager (HySI) and a miniature multispectral imager. The main goals of this mission were to design, build, and operate a 3-axis stabilized

TABLE 2.16
Key Parameters of the FTHSI Instrument

Instrument Type	Imaging Fourier Interferometer
Scan mode	Pushbroom
Orbit altitude	650 km
Swath width	50 km
Ground sampling distance	100 m
Spectrometer type	Fourier transform interferometer
Wavelength range	0.45–0.95 μm
Spectral resolution	98.5 cm^{-1} (8.1 nm)
Number of bands	115
Digitization	12 bits

remote sensing satellite providing easy access of data to students and scientists in developing countries (Kumar and Samudraiah 2016).

The IMS1-HySI instrument was a 64-band VNIR imaging spectrometer with spatial sampling distance of about 500 m and swath width of about 130 km. It was aimed at validating the design of hyperspectral imager and also to generate data cubes for experimental studies of applications on ocean, atmosphere, etc. It was also aimed at providing hands on experience for users and scientists of the hyperspectral applications.

IMS1-HySI was an optical filter based hyperspectral imager. It used a wedge filter, also known as a linear variable filter (LVF), to disperse spectrum. Based on the best knowledge of the author, HySI is the first spaceborne hyperspectral imager that uses a LVF as the dispersive element. The operation principal of an optical filter based hyperspectral imager is described in Section 1.2.2. The HySI imager acquired the images of Earth in pushbroom mode as shown in Figure 1.8. It consisted of a collecting optics, an area array detector with wedge filter mounted very close to it and the associated front-end electronics. The HySI imager was mounted on the IMS1 platform, which was a 3-axis stabilized satellite, such that its optical axis oriented toward nadir.

As shown in Figure 1.8, a frame of image is generated at a moment by the instantaneous projection of a ground scene through the wedge filter onto the detector array. The telescope and the detector pixels respond to all wavelengths, while the spectral transmission of the wedge filter varies linearly with narrow spectral band-pass from one end to other, and hence a row of elements in the detector array receives only a small particular portion of wavelengths. Accordingly, different rows correspond to different wavelengths (i.e., spectral bands). The satellite movement (in along-track direction) is used to cover the scene on the ground and also the spectrum. The exposure time of the HySI imager was kept less than the time required to cross a footprint on the ground (dwell time) to minimize the smear in the along-track direction. The frame rate (integration time) was selected such that there was one frame per each dwell time of a footprint. The figure shows violet (shortest wavelength) at one end and red (longest wavelength) at the other end. As the satellite moved forward, the same ground line was swept by different rows of the wedge filter and detector array and thus creating images with complete spectrum (i.e., all spectral bands). The resolution and swath were dictated by focal length, pixel size, and the number of elements in the detector array.

Figure 2.10 shows the optical configuration of the IMS1-HySI instrument. The fore-optics collected the light from the scene and focused it onto the LVF and detector assembly. The fore-optics had FOV of 26° with a f/# of 4. It was a telescope with a seven-element telecentric lens assembly with a thermal filter at the front as shown in the figure. The telecentric design, with chief ray at each

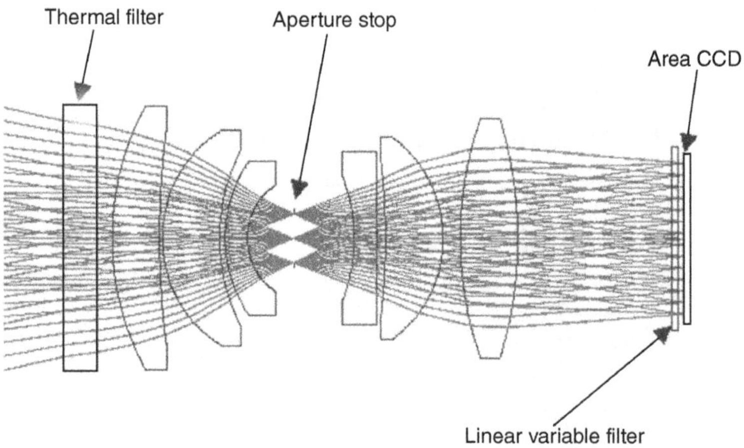

FIGURE 2.10 Optical configuration of HySI. (Courtesy of ISRO.)

image filed parallel to the optical axis, was adopted to minimize the spectral shift with look angle. All the lens elements had spherical surface profiles. The design placed the lens elements very close to each other to make it compact. The thickness of each of the lens elements was minimized to the extent possible to reduce the mass of the instrument.

A LVF was placed in front of the detector array to disperse the spectrum. This approach was selected as it helped to achieve a simplest form of hyperspectral imager because of the omission of collimator optics and focusing optics, which are generally associated with grating- or prism-based spectral dispersion systems, and accordingly it made the system very compact. A wedge filter is an interference filter, where only a narrow band of wavelengths is allowed to pass at any point of wavelength and the other wavelengths in the spectrum of the input signal are blocked. The central wavelength of this pass-band varies linearly from one end to the other end of the filter in one dimension of the filter frame. As the filter characteristics are sensitive to incidence angle, the imaging optics was designed as a telecentric lens and the LVF was assembled very close to the detector array. With wedge filter bandwidth of about 10 nm, system-level bandwidth was about 20–25 nm. The filter dimensions were selected in such a way that 400–950 nm was dispersed over 512 rows of the detector array. This resulted in oversampling at about 1 nm spectral interval. Considering the application requirements and the limitations of data rate, 8 rows binning was implemented. After binning, the SSI works out to be about 8 nm. This arrangement of spectral oversampling provides ample possibility of constructing better shaped bands using the oversampled signal.

The focal plane was CMOS 2D detector array 512 × 256 elements with a pitch size of 50 μm × 50 μm. The 512 elements side was used for along-track direction that dispersed spectral components, whereas the 256 elements side used for cross-track deciding the swath. It incorporated pixel-level charge to voltage converter and amplifier, column-level amplifiers, programmable amplifier, 12-bit pipelined analog-to-digital converter, timing and control logic, data serializer, memory for configuration selection, etc. This made the system compact and consumed very low power.

The test results showed that HySI achieved a linear spectral dispersion of about 8.4 nm/band in the central wavelength. The spectral instability performance was found be <1 pixel all through the environmental tests. The out of band suppression was 3 orders of magnitude. The typical smile distortion was <1 pixel, indicating that all the errors/aberrations at various stages of design, fabrication, and assembly were very less. The SNR was measured to be >820:1 as against the specification of ≥400:1. The dark noise was about 25 DNs in 15 bit system (max count of about 32,000 DNs). It was observed that the system noise was predominantly dictated by detector readout noise. The key performance parameters of IMS1-HySI are summarized in Table 2.17.

2.13 ADVANCED RESPONSIVE TACTICALLY EFFECTIVE MILITARY IMAGING SPECTROMETER ONBOARD TACSAT-3

The Advanced Responsive Tactically Effective Military Imaging Spectrometer (ARTEMIS) was onboard TacSat-3 satellite, which was the third in a series of US DoD military reconnaissance satellites and launched on May 19, 2009. The TacSat satellites are all designed to demonstrate the ability to provide real-time data collected from space to combatant commanders in the field. In addition to ARTEMIS hyperspectral imager, TacSat-3 also carried other two distinct payloads: the Ocean Data Telemetry Microsatellite Link and the Space Avionics Experiment (Lockwood et al. 2006).

The ARTEMIS was to demonstrate a hyperspectral imaging capability from space direct to the tactical warfighter within 10 min of a collection opportunity. TacSat-3 has provided key insights into hyperspectral imaging capabilities hosted on a small satellite platform. This mission has given insights into new concepts of operations in the tactical employment of satellites and the balance between onboard processing, automation, and performing these functions on the ground.

The ARTEMIS hyperspectral sensor uses a single 2D detector array covering both VNIR and SWIR spectral region from 0.4 μm to 2.5 μm at a uniform spectral resolution of 5 nm with 4 m spatial

TABLE 2.17
Key Performance Parameters of IMS1-HySI

Parameter	Value
Orbit altitude	632 km
Swath width	130 km
FOV	26°
Number of pixels in a cross-track line	256
Ground sampling distance	505 m
Fore-optics	7-telecentric lens telescope
f/#	4
Spectrometer	Omission
Spectral dispersive element	LVF
Spectral range	410–965 nm
Spectral sampling interval	8.4 nm
Number of spectral bands	64
Spectral distortion (Smile)	<1 pixel
Detector array format	512 × 256
Detector pitch size	50 μm × 50 μm
SNR at saturation	≥820:1
Quantization	15 bits
Mass	4 kg
Power	2.6 W

resolution and a swath width of 4 km. It consists of a telescope, a spectrometer, a real-time processor, and onboard health monitor (OBHM). Table 2.18 tabulates the key parameters of ARTEMIS payload.

The telescope is a standard Ritchey-Chrétien form having an aperture of 35 cm. It is telecentric as is required to meet the spectral and spatial uniformity goals of the imaging spectrometer. Additionally, the secondary mirror has a built-in focus mechanism for on-orbit optimization.

The spectrometer is of the basic Offner form consisting of two powered reflecting surfaces comprising the primary and tertiary elements (Offner 1987). The secondary mirror is replaced by a curved grating for dispersion of radiation light and is the limiting stop of the system. This form has the merit of being simple, compact, and both spatially and spectrally uniform (Mouroulis et al. 2000). The spatial and spectral uniformity is critical to the operational performance of imaging spectrometers as it enables robust exploitation of data products. Additionally, the design has <5% spatial and spectral nonuniformity. The slit is reticulated with small apertures at the top and bottom to aid in alignment and testing.

The grating is dual-angle blaze that was selected largely due to its superior performance in reducing the effect of obscuration at the grating stop. To make the SNR performance approximately equal at all wavelengths, the grating was also designed to optimize its optical efficiency. This was done by suppressing the optical efficiency at the blue wavelengths corresponding to the peak of the solar Planck function while increasing the efficiency from 1.4 μm to 2.5 μm, where the solar illumination is over a factor of 20 times lower.

The FPA is a substrate-removed MCT 2D detector array that extends its sensitivity into the blue wavelengths to cover the full spectral range. The quantum efficiency of the MCT detector array is better than 70% at all wavelengths and the array is equipped with a three-zone blocking filter for order sorting. This single focal plane eliminates the co-registration issues associated with multiple FPA systems.

The real-time onboard processor provides reprogrammable digital signal processing and derives surveillance information onboard and downlinks it in-theater for tactically effective military applications.

The OBHM was equipped to establish and monitor on-orbit functionality and to evaluate spectral calibration performance. The OBHM consists of a small blackbody source with a color temperature

TABLE 2.18
Key Parameters of ARTEMIS Payload

Parameter	Value
Satellite platform	TacSat-3
Orbit altitude	467 km
Wavelength range	0.4–2.5 μm
Spectral sampling interval	5 nm
Number of spectral bands	400
Number of spectrometers covering the spectral range	1
Spectral nonuniformity	<5%
Telescope	Standard Ritchey-Chrétien, telecentric
Aperture	35 cm
Spectrometer	Offner form
Spectral dispersive element	Dual-angle blaze grating
Spatial nonuniformity	<5%
FPA	HgCdTe 2D detector array
Detector QE	>70%
Onboard health monitor	Monitoring spectral, spatial, and radiometric performance
Mass	170 kg

of about 2200 K, an elliptical reflector, and a spectral filter. The OBHM source is placed at the center of the secondary mirror of the telescope and is within the shadow of the obscuration. The OBHM is not intended to be a radiometric source as its irradiance will vary on orbit and is not verifiable. The spectral filter is used to confirm spectral calibration. It is a composite of a filter of NIST (National Institute of Standards and Technology) standard reference material (SRM 2035) and Mylar, and exhibits spectral features across the spectral range of interest. The OBHM does not completely fill the imaging spectrometer limiting the illumination of the grating to the center. Performance is improved by the grating design trade result, as a multi-zone grating would not be as efficient at all wavelengths as is a dual-angle blaze grating.

The ARTEMIS payload successfully implemented a number of design and test decisions to meet the program's challenging cost and schedule. These include (1) significant use of COTS and tactical-grade electronic components with minimal redundancy; (2) use of a single spectrometer FPA to expedite laboratory alignment and achieve stringent spectral/spatial uniformity; (3) use of an onboard health monitor for trending spectral, spatial, and radiometric performance; (4) implementation of a focus mechanism to achieve on-orbit focus of the sensor; and (5) vicarious techniques for on-orbit spectral and radiometric calibration. These design decisions enabled the successful development and delivery of the ARTEMIS sensor by significantly reducing the cost of hardware components and duration of pre-launch ground testing. ARTEMIS payload has also shown lessons in key areas of improving responsive space goals (Straight et al. 2010).

2.14 HYPERSPECTRAL IMAGER FOR THE COASTAL OCEAN ONBOARD THE INTERNATIONAL SPACE STATION

The Hyperspectral Imager for the Coastal Ocean (HICO) was installed on the ISS on September 23, 2009. It was the first spaceborne imaging spectrometer designed for coastal ocean research (Lucke et al. 2011). Sponsored by the Office of Naval Research (ONR) as an innovative naval prototype, HICO was developed to demonstrate improved coastal remote sensing products, including bathymetry, bottom types, water optical properties, and on-shore vegetation maps. It was built within a short period (about 18 months) from non-space-hardened, commercial off-the-shelf (COTS) parts. Having met all its navy goals in the first year, HICO was granted a 2-year operations extension from the ONR and then NASA stepped in to sponsor this ISS-based sensor, extending HICO's operations for another 2 years. It stopped operation in September 2014. During its 5 years life, it collected approximately 10,000 hyperspectral scenes of the Earth. The data have enabled ocean color scientists and managers to assess data quality and apply the imagery to a variety of scientific and societal problems (Kappus et al. 2016).

HICO was a pushbroom imaging spectrometer working in VNIR region. It covered a wavelength range of 380–960 nm with a SSI of 5.7 nm. It had a swath width of 51 km when the ISS altitude was 420 km. Its GSD was 100 m, which was much smaller than that of other spaceborne ocean color sensors (300 m of MERIS and 1000 m of MODIS ocean color bands). Even with such small GSD comparing to other ocean color sensors, it still achieved reasonable high SNR: maximum SNR 470: 1 at 480 nm, SNR > 200: 1 in spectral range 400–600 nm.

The HICO instrument consisted of a fore-optics, a spectrometer, and a FPA camera. The fore-optics was a telescope with a five-element telecentric lens assembly. The spectrometer was an Offner-form grating spectrometer. It dispersed spectrum of the incoming signal from a cross-track line of pixels. It had a high grating efficiency, about 80% at the blaze wavelength of 400 nm, and low spectral (smile) and spatial (keystone) distortion. The size of the slit was 8.2-mm long × 16-μm wide, which was reimaged at 1:1 magnification onto the 16-μm square pitch size of the CCD detector array. This slit image corresponded to 512 pixels of footprint size 100 m × 100 m in a ground cross-track line.

The camera was a COTS unit designed for laboratory use, not built to operate in vacuum, and was therefore housed in a hermetic enclosure containing nitrogen gas, with a fused silica window to admit light from the spectrometer. The main modifications for spaceflight were to conformally coat the printed circuit (PC) boards and add stiffening to their mountings. The thermoelectric cooler for the FPA was deactivated because the heat from its 80 W power consumption could not be removed from the camera due to the limited thermal conductivity from the rotating camera enclosure. Without the cooler the temperature of the detector was not held constant and the dark counts changed as the temperature changed.

The size of the CCD detector array was 512 × 512 pixels, with 16-μm pitch size. It is a thinned, backside-illuminated silicon CCD, with high quantum efficiency in the blue wavelengths, which are important for the retrieval of aquatic biophysical products. It is a frame-transfer CCD, in which charges are moved along one direction (spectral) for being read out along one edge of the array. Charges continue to accumulate during the transfer process, and this effect is referred to as frame transfer smear and must be characterized in advance and removed during post processing. All 512 pixels in one dimension were used to cover the swath width of 51 km. Only the first 384 pixels in another dimension were used to cover the wavelength range of 350–1080 nm, other 128 pixels were not used. The discarded wavelengths were longer than the silicon sensitivity wavelength (approximately 1000 nm), and discarding them enabled operation of the camera at a faster frame rate than if all pixels were recorded.

Spreading 1080 – 350 = 730 nm of spectrum over 384 pixels meant that each pixel covered SSI of 1.9 nm. In HICO's normal mode of operation, pixels 1–384 were binned spectrally by three at readout to yield 128 spectral bins, each covering a SSI of 5.7 nm. Thus, one complete frame of data contains 512 (spatial) × 128 (spectral) = 65,536 data samples. Each sample is digitized to 14 bits and read out as a 16-bit word. Since the spectral range covered more than a factor of two in wavelength, second-order light from wavelengths 350–540 nm falls onto the same pixels as first-order light from 700 nm to 1080 nm.

HICO instrument was mounted on a rigid optical bench that was attached by silicon-rubber isolators to a cradle that was rotated about a horizontal axis to direct the line of sight (LOS) to the desired off-nadir direction as shown in Figure 2.11.

The swath and GSD of HICO depended on the ISS altitude (approximately 420 km) and the imaging frame rate. For an ISS altitude of 420 km, GSD at nadir is 100 m in the cross-track direction

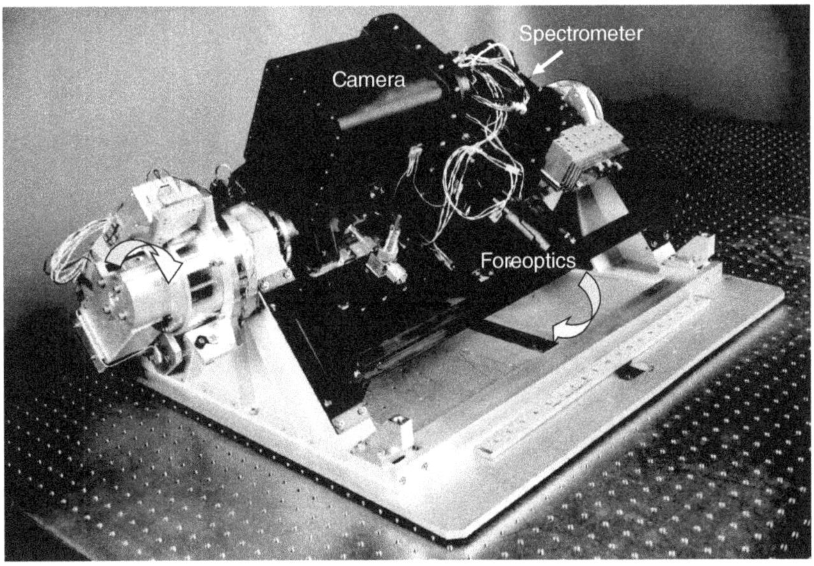

FIGURE 2.11 HICO flight instrument. (Courtesy of NRL.)

and larger when HICO imaged off-nadir. HICO's GSD is 95 m in along-track direction, which is a function of imaging frame rate in addition to altitude. Each HICO image contains 512 spatial pixels cross-track and 2000 frames along-track, resulting in image dimensions of 51 km × 190 km on the ground for an ISS altitude of 420 km.

The HICO's spectral distortion (smile) was smaller than 0.3 pixel, whereas spatial distortion (keystone) was smaller than 0.4 pixels. This was characteristic of a well-aligned Offner spectrometer system, and a primary reason for choosing the Offner spectrometer for HICO. Further details of HICO performance, including spatial and spectral resolution, methods for measuring spectral smile and spatial keystone, polarization sensitivity, optical distortion, and calibration were given by Lucke et al. (2011). Primary operating parameters and other information for the HICO sensor are provided in Table 2.19.

HICO mission has demonstrated that a low-cost hyperspectral sensor, built mainly from COTS parts can produce high-quality data to scientific community to improve understanding of the complex coastal ocean. Meanwhile, it provided many useful lessons learned for spaceborne hyperspectral sensors that use non-space-qualified parts and fast-paced development, as well as using the ISS as the platform.

The most serious impact of non-space-qualified parts on HICO was that the non-hardened computer suffered from frequent lockups, presumably due to single event radiation upsets.

Hundreds of lockups were resolved over the years by simply rebooting the computer. However, in September 2014 HICO did not recover from the computer upset, which caused HICO's termination of its operation life.

The initial spectrometer design incorporated an order-sorting filter and a zero-order beam dump. However, mounting the filter posed a significant risk and so it was not installed. This second-order effect was measured in the laboratory with the intent of colleting the information to make corrections during data post processing. The lack of the second-order filter was a setback in the construction of HICO. Difficulties in calibration confirmed the value of having a physical filter, rather than relying on software corrections for grating spectrometer instruments.

The lack of an FPA cooler allowed the FPA temperature to increase during the imaging period, so that an empirical procedure to account for the rising temperature effect on dark noise had to be established. This procedure appears to be effective for the HICO demonstration.

Not having an on-orbit calibration system had created additional challenges for HICO calibration, in addition to the complicated calibration and characterization process.

TABLE 2.19
HICO Instrument and Operating Parameters

	Parameters	Value
Spatial	Platform	International Space Station
	Nominal orbital altitude	420 km
	Swath width	51 km
	Ground sampling distance (GSD)	100 m (cross-track) 95 m (along track)
	Spatial resolution (FWHM)	1.6 GSD typical
Telescope	Telescope type	Five-element telecentric lens assembly
	Focal length	6.72 cm
	Aperture	0.19 cm
	f/#	3.5
	FOV	6.92°
	IFOV	0.24 mrad
Spectrometer	Spectrometer form	Offner
	Spectral dispersive element	Grating (blaze wavelength 400 nm)
	Wavelength range	0.35–1.08 μm (nominal) 0.38–0.96 μm (best data)
	Number of spectral bands	128
	Spectral sampling interval	1.9 nm (instrument native) 5.7 nm (3-pixel binning)
	Spectral resolution (FWHM)	1.6 SSI typical
	Slit size (long × wide)	8.2 mm × 16 μm
FPA	Detector array	Thinned, backside illuminated silicon CCD
	Detector array format	512 × 512 pixels
	Pitch size	16 μm × 16 μm
	Dark noise	3.8 DN
Radiometric	Spatial distortion (Keystone)	0.4 pixel
	Spectral distortion (smile)	0.3 pixel
	Polarization sensitivity	<4%
	SNR	Peak 470:1 at 480 nm >200:1 over 400–600 nm
	Digitization	14 bits
Operation	Off-nadir pointing	45° to port 30° to starboard
	Frame rate	72.7 Hz
	Scene Area (W × L)	51 km × 190 km
	Scene time	27.5 sec
	Mass	41 kg

The ISS is an unusual platform for a hyperspectral sensor. A big advantage is the cost savings compared to a separate satellite. Other advantages include the availability of power; Guidance, Navigation, and Control (GNC) services; and a spacecraft control team. On the ISS, there are opportunities to collect imagery in a variety of viewing conditions, opening the way to observing phenomena not usually visible in typical sun-synchronous viewing geometries. Disadvantages include limited viewing opportunities compounded by irregular changes to the orbit and interruptions caused by other ISS operations. In addition, the ad hoc orbital maneuvers, along with only triweekly updates to the ephemeris, complicates mission planning and sometimes causes a payload to miss the desired target.

2.15 VISIBLE AND NEAR-INFRARED IMAGING SPECTROMETER ABOARD CHANG'E 3 SPACECRAFT

The Visible and Near-infrared Imaging Spectrometer (VNIS) is one of the main scientific payloads of China's lunar rover Yutu ("Jade Rabbit" in Chinese) of the Chang'E 3 mission, which reached lunar orbit on December 6, 2013, and landed on the moon on December 14, 2013. The Chang'E 3 mission included a lander and a lunar rover Yutu, each of them carried scientific payloads. After soft landing on the moon, Chang'E 3 carried out a lunar survey and scientific exploration activities, including (1) survey of lunar surface for topography and geological structure, (2) lunar surface material composition and available resource exploration, and (3) detection of the Earth's plasma layer and moon-based optical astronomical observation (Ye and Peng 2006, Dai et al. 2014).

The objective of VNIS was to make in situ measurements of the composition and resources of the lunar surface via imaging and spectrometry in the VNIR and SWIR regions. As a passive optical instrument, the VNIS measures the radiance diffusely reflected from the moon's surface of the solar illumination. Mounted on the platform of lunar rover Yutu in the front, the VNIS detected the lunar surface objects with a 45° view angle and acquire spectral and geometric data for determining the lunar surface mineral composition and performing comprehensive analysis of the chemical composition (He et al. 2011).

The VNIS instrument consisted of two separate parts: a spectrometer probe located outside of the rover, and a logical control and AOTF radio frequency (RF) driver module, called Remote Electronic Control Box, located inside the rover. These two parts are connected by cables (Wang et al. 2016).

The VNIS instrument includes a VNIR imaging spectrometer covering a wavelength range of 0.45–0.95 μm and a SWIR spectrometer covering a wavelength range of 0.9–2.4 μm. Acousto-optic tunable filters were used as the dispersive components of the two spectrometers.

Figure 2.12 shows a block diagram of the optical systems of VNIS instrument. Both the VNIR and SWIR spectrometers were composed of the fore-optics, AOTF, and aft-optics. The fore-optics includes objective lens, field diaphragm, and collimating lens. The aft-optics includes imaging lens and detector array.

The VNIS spectrometer used a CMOS area array detector, and the SWIR spectrometer used an InGaAs single element detector. Both spectrometers used noncollinear AOTFs as light dispersive devices. The AOTF is based on the acousto-optic interaction. When a RF signal generated by the AOTF RF driver is applied to the transducer of the AOTF crystal, the electrical signal is converted into an ultrasonic vibration. Then a coupling, quasi-monochromatic wavelength is diffracted at a given separation angle by momentum matching between the collimated light and ultrasound vibration at a given frequency. The spectrally tuned light reached at the surface of the detector for spectral imaging. The key performance parameters of the VNIS are listed in Table 2.20.

The VNIS has two operating modes: detection and calibration. In detection mode, VNIS acquires scientific data from lunar surface objects. The default SSI is 5 nm. In addition, both the VNIR and

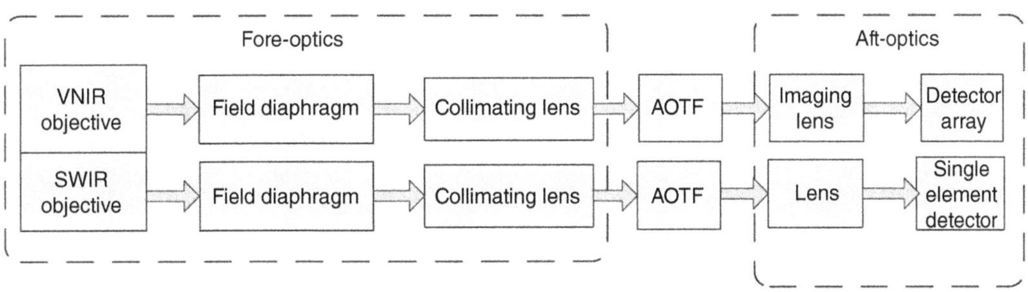

FIGURE 2.12 Block diagram of VNIS optical systems.

TABLE 2.20
Key Performance Parameters of the VNIS Instrument

Parameter	Spectrometer VNIR	Spectrometer SWIR
Platform	Lunar rover Yutu	
Wavelength range (nm)	450–950	900–2400
Spectral sampling interval (nm)	5	5
Spectral resolution (nm)	2–7	3–12
Number of spectral band	100	300
FOV (degree)	8.5 × 8.5	3.6
Spectral dispersive element	Acousto-optic tunable filter	
Detector format	256 × 256	1 × 300
Detector material	CMOS	InGaAs
SNR (dB)	≥31	≥32
Digitization (bit)	10	16
Power (W)	19.8	
Mass (kg)	1.7 (Spectrometer probe) 0.7 (Electronic Control Box)	

the SWIR spectrometers acquire 20 extra frames of dark current for dark current subtraction of data processing for data recovery. By sending instruction codes, VNIS can shift the central wavelength of the detected band-pass, so that it can acquire a spectral image or data in a specified band-pass.

In calibration mode, using solar radiation as the calibration source, the diffusing calibration panel of the calibration unit is set to a horizontal position to allow calibration of the instrument. The workflow for the calibration mode is identical to that for the detection mode.

The VNIS had carried out several in-orbit calibrations and lunar surface measurements since it was first successfully operated on the moon on December 23, 2013, which was the first in situ spectral imaging detection on the lunar surface. The high resolution and effective spectral imaging data obtained by VNIS has provided valuable hyperspectral data for lunar scientific applications.

2.16 OCEAN AND LAND COLOR IMAGER (OLCI) ON SENTINEL-3A

The Ocean and Land Color Imager (OLCI) is a VNIR pushbroom imaging spectrometer. It is the successor of MERIS onboard ESA's ENVISAT, which was out of service since Aril 2012. OLCI is one of the seven instruments onboard ESA's Sentinel-3A launched on February 16, 2016. Sentinel-3 is an Earth observation satellite constellation, including two satellites A and B. It is developed by ESA as part of the Copernicus Program of the European Union. The Sentinel-3 mission's main objective is to measure sea-surface topography, sea- and land-surface temperature and color with accuracy in support of ocean forecasting systems, and for environmental and climate monitoring (Donlon et al. 2012).

OLCI was designed to provide global and regional measurements of ocean and land surface at a high level of accuracy. It is based on the heritage design from MERIS. OLCI has a spatial resolution of 300 m over both land and water surfaces and slightly wider swath of 1270 km than that of MERIS. Its SNR has been improved. It has also improved instrument characterization including stray light, camera coverage overlap, and calibration diffusers. Its revisit times with global coverage have been reduced to 3 days, instead of around 15 days of MERIS (Nieke et al. 2016).

OLCI transmits to ground more spectral channels, 21 channels, compared to the 15 on MERIS. Table 2.21 summarizes the 21 channels. The six new spectral channels provide the means for improved

TABLE 2.21
Specification of the 21 Channels of the OLCI Sensor

Channel Number	Band Center (nm)	Bandwidth (nm)	Comparison to MERIS Band	Applications
1	400	15	New	Aerosol correction, improved water constituent retrieval
2	412.5	10	Same	Yellow substance and pigments detritus
3	442.5	10	Same	Chlorophyll absorption maximum
4	490	10	Same	Chlorophyll and other pigments
5	510	10	Same	Suspended sediment, red tides
6	560	10	Same	Chlorophyll absorption minimum
7	620	10	Same	Suspended sediment
8	665	10	Same	Chlorophyll absorption and fluorescence reference
9	673.75	7.5	New	For improved fluorescence retrieval and to better account for smile together with bands 665 nm and 681 nm
10	681.25	7.5	Same	Chlorophyll fluorescence peak
11	708.75	10	Same	Fluorescence reference, atmospheric corrections
12	753.75	7.5	Same	Vegetation, cloud
13	761.75	2. 5	Same	Oxygen absorption R-branch
14	764.375	3.75	New	Atmospheric correction
15	767.5	2.5	New	O_2A used for cloud top pressure, fluorescence over land
16	778.75	15	Same	Atmosphere corrections
17	865	20	Same	Vegetation, water vapor reference
18	885	10	Same	Atmosphere corrections
19	900	10	Same	Water vapor, land
20	940	20	New	Water vapor absorption, Atmospheric/Aerosol correction
21	1020	40	New	Atmospheric/Aerosol correction

water constituent retrieval (400 nm and 673 nm), improved parameter retrieval in the O_2A-band (767–770 nm) and atmospheric correction (940 nm and 1020 nm).

A lesson learned from MERIS is the negative impact of the direct solar reflection at sea surface to the sensor, which is referred to as sun glint. To minimize the impact of sun glint, OLCI adopted an asymmetric swath with respect to the satellite ground track to avoid sun glint. The amount of tilt was defined by the need to minimize the maximum observation zenith angle (OZA) at the outer border of the swath and at the same time guaranteeing global coverage. A cross-track tilt of 12.6° of the overall FOV is used that results in a maximum OZA slightly above 55°.

Same as MERIS, OLCI instrument has five identical fan-arranged Dyson spectrometers (also called cameras) with five FPAs mounted on a temperature-controlled optical bench to cover a wide swath. OLCI instrument includes the following components:

1. An optical bench.
2. A calibration mechanism.
3. A depolarizer assembly.
4. Five fan-arranged camera optical subassemblies.
5. Five FPAs.
6. Five video acquisition modules containing the whole analogue imaging chain down to the digital conversion.
7. A OLCI electronic unit (OEU) managing all the instrument functions.
8. A calibration assembly allowing a radiometric and spectral calibration.
9. A heat pipe networks insuring the thermal control of the video acquisition modules.

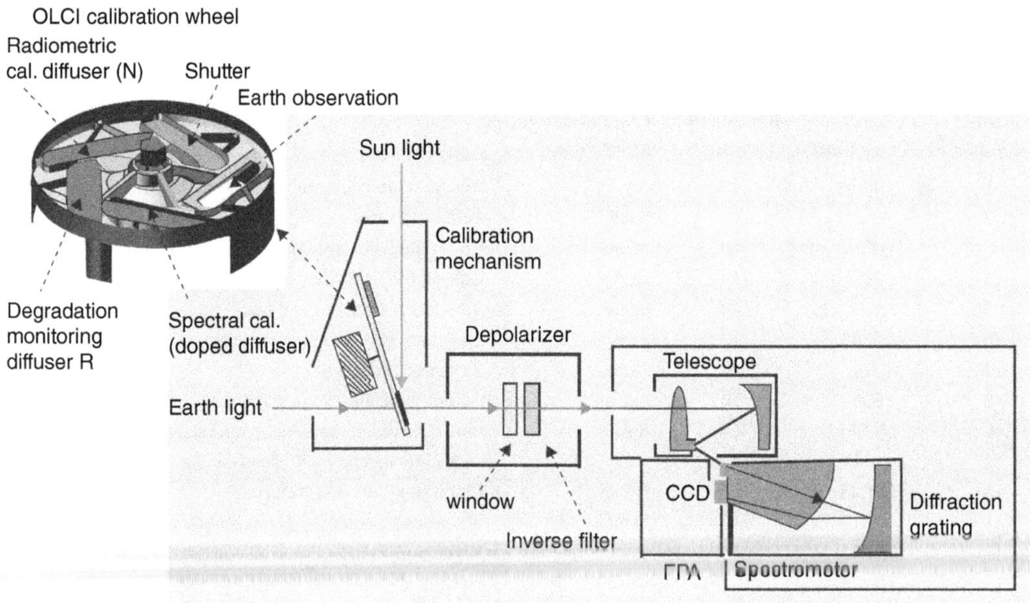

FIGURE 2.13 OLCI optical configuration. (Courtesy of ESA.)

Figure 2.13 shows the optical layout of the OLCI sensor. An off-axis catadioptric telescope collects the light through the calibration mechanism, either from a scene on Earth or the sun-illuminated diffuser, and a depolarizer (including the depolarizer window and inverse filter). The collected light is focused onto the entrance slit of the Dyson spectrometer, which includes an off-axis concave holographic diffraction grating and a co-centric refractive Dyson block. Then the grating spectrometer generates a dispersed image of the slit on a 2D CCD array: one dimension of the array is the spatial extension of the slit, and the other dimension the spectral dispersion of the slit image in the range between 390 nm and 1040 nm. The calibration mechanism allows a view of the earth surface or one of several onboard calibration targets through a slit window by rotating each target mounted on a calibration wheel into the FOV of the instrument.

OLCI calibration and validation processes are critical to the quality of the data. The calibration and validation processes include the following three phases: the pre-launch phase (C/D), the commissioning phase (E1), and the exploitation phase (E2) (Nieke et al. 2010). Table 2.22 summarizes the characteristics of OLCI.

2.17 MINIATURE HIGH-RESOLUTION IMAGING SPECTROMETER ON GHGSAT-D

The miniature high-resolution imaging spectrometer (MHRIS) is an electronically tunable filter based SWIR hyperspectral imager for monitoring targeted greenhouse gas emitters, such as area fugitive sources (tailing ponds and landfills) and stacks emissions such as flaring and venting. It is onboard the Greenhouse Gas Demonstration Satellite (GHGSat-D) owned by GHGSat Inc., a commercial venture based in Montreal, Canada. GHGSat's mission is to become the global reference for remote sensing of greenhouse gas and air quality gas emissions from industrial sites using satellite technology. GHGSat-D was launched on June 22, 2016, as a secondary payload to ISRO's CartoSat-2C spacecraft (Germain 2016).

Two new high-resolution greenhouse gas monitoring satellites GHGSat-C1 and GHGSat-C2 are under development and are planned to be launched in 2020. The MHRIS on each satellite integrates lessons learned from GHGSat-D and are expected to lower the methane detection threshold

TABLE 2.22
OLCI Performance Characteristics

Parameters	Value
Swath	1270 km
FOV	68.6°
Cross-track tilt of the FOV	12.6° (to avoid sun-glint)
Ground sampling distance (GSD)	300m (for both land and water scenes)
Telescope type	Off-axis catadioptric
Spectrometer type	Dyson form
Spectral dispersive element	Grating
Wavelength range	390–1040 nm
Spectral sampling interval (SSI)	1.25 nm
Spectral resolution (FWHM)	1.5 nm
Spectral channels downlinked	21 (6 bands more than MERIS)
FPA	Back-illuminated frame-transfer 2D CCD array
Detector array format	780 × 576
Detector pitch size	22.5 μm × 22.5 μm
Relative spectral accuracy	0.05%
Absolute radiometric accuracy	<2.0%
Relative radiometric accuracy	0.5%
Radiometric stability	0.1%
Signal-to-noise ratio (SNR)	190:1–2450:1
Modulation transfer function (MTF)	>0.3
Polarization sensitivity	<0.3%
Mass	150 kg
Volume	1.3 m^3

by 7–10 times compared to GHGSat-D. Data acquired by the GHGSat will enable industries to monitor and quantify their own carbon dioxide and methane emissions in order to adjust and better manage and reduce their impact on air quality.

The GHGSat is a nanosatellite based on a low-cost and high-performance nanosat bus NEMO-AM made by University of Toronto, Institute for Aerospace Studies/Space Flight Laboratory (UTIAS/SFL) (Zee 2016.). It has a launch mass of 15 kg and a volume of approximately 25 U (20 cm × 30 cm × 42 cm) plus a mezzanine of size 7 cm × 18 cm × 42 cm on one side (-X) as shown in Figure 2.14.

The MHRIS on GHGsat-D and GHGsat-C1 was designed and developed by MPB Communications Inc., based in Pointe Claire, Québec, Canada. The MHRIS on GHGsat-C2 is built by ABB Inc. of Quebec City. It consists of beam-folding mirrors, lens assemblies, and a tunable filter-based spectrometer. The beam-folding mirrors are used to fit the telescope into the NEMO bus. The first lens assembly is a large telescope doublet. The second lens assembly is a collimator to provide the magnification required by the system. The third lens assembly is to form an image of the target scene on the FPA.

The tunable filter-based spectrometer is a Fabry-Pérot interferometer. It is an optical resonator consisting of a single plate with two parallel reflecting surfaces. Light passing into the spectrometer can only pass through when its wavelength corresponds to the resonances of the etalon that creates a narrow-band spectrum on the focal plane that is precisely tuned to the desired wavelengths in order to create a high-resolution spectrum of the backscattered signal. The Fabry-Pérot interferometer restricts the incident spectral passbands within a narrow wavelength region between 1600 nm and 1700 nm selected for the presence of spectral features for methane and carbon dioxide, as well as

FIGURE 2.14 Miniature high-resolution imaging spectrometer aboard GHGSat-D nanosatellite. (Courtesy of GHGSat Inc.)

relatively little interference from other atmospheric species, H_2O in particular. The spectral resolution is on the order of 0.1 nm.

The FPA selected for GHGSat-D is a high-sensitivity InGaAs SWIR detector array with enhanced dynamic range. The detector array format is 640 × 512 pixels, of which GHGSat-D masks the area outside the central 512 × 512 array. The selected InGaAs array has heritage on a NASA mission.

The MHRIS instrument has a spatial resolution of under 50 m, covering a 15 km ground swath. It has a compact size of about 36 cm × 260 cm × 180 mm, including baffle, a mass of 5.4 kg. It was designed to work from −40°C to +80°C.

Power generation is provided with body-mounted solar cells. Three 26 W/h Li-Ion batteries deliver a nominal operating voltage of 12.3 V regulated to 5 and 3.3 V for distribution to the satellite subsystems. Reaction wheels and magnetic torque rods are employed for precise attitude control to ensure accurate pointing to observation targets.

In addition to MHRIS, GHGSat-D also contains a secondary instrument for cloud and aerosol detection, called Cloud and Aerosol Instrument (C&A Instrument), to enhance retrievals from the primary instrument. Figure 2.15 shows the photo of the GHGSat-D nanosat on a test bed before launch. Table 2.23 summarizes the key parameters of the GHGSat-D satellite.

2.18 AALTO-1 SPECTRAL IMAGER ON A 3U NANOSATELLITE

Aalto-1 Spectral Imager (AaSI) is an electronically tunable filter (ETF) based imaging spectrometer. It is the main payload of the Aalto-1 nanosatellite that was launched on June 23, 2017, by PSLV-C38 rocket from India. The Aalto-1 nanosatellite is built mainly by students at Aalto University in Finland and coordinated by the Department of Radio Science and Engineering and supported by Space Technology teaching. The goals of the Aalto-1 project are to: (1) design, build and operate first Finnish Earth observation nanosatellite, (2) demonstrate a technology of a very small spaceborne imaging spectrometer for Earth observation, (3) demonstrate a technology of a very small radiation detector for future satellites, (4) develop and demonstrate a deorbiting device for nanosatellites based on e-sail concept and measurement of its performance, and (5) promote engineering education in Finland with the aid of a satellite project. The nanosatellite is based on a 3U CubeSat with a volume of 34 cm × 10 cm × 10 cm and a mass of ~4 kg. The design life is 2 years. It has an average power production of 4.8 W (Praks et al. 2011).

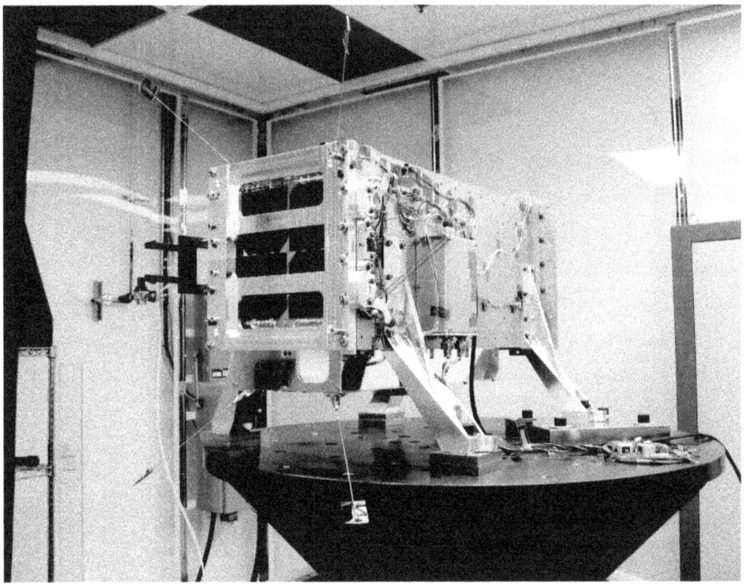

FIGURE 2.15 GHGSat-D satellite on a test bed before launch. (Courtesy of GHGSat Inc.)

AaSI is based on a spectrally tunable Fabry-Pérot interferometer and developed by VTT Technical Research Centre of Finland. It consists of two camera modules, the primary part is the spectral camera that includes a tunable Fabry-Pérot interferometer and a CMV4000 CMOS detector array. The second part is a visible camera, which is a normal red-green-blue (RGB) camera with a wider FOV than the spectral camera. The visible camera has the same CMOS detector array as the spectral camera. It is combined with COTS optics. Both cameras share the common control electronics in order to minimize the system complexity. The objective of the visible camera is to confirm

TABLE 2.23
Miniature High-Resolution Imaging Spectrometer on GHGSat-D Nanosat

Parameters	Value
Satellite platform	GHGSat-D
Orbit altitude	512 km
Swath width	15 km
Ground sampling distance	50 m
Fore-optics	Folding mirrors and lens assemblies
Spectrometer type	Fabry-Pérot interferometer
Spectral dispersive element	ETF (Fabry-Pérot filter)
Spectral range	1600–1700 nm
Spectral sampling interval	0.1 nm
Number of spectral bands	>300
FPA	InGaAs SWIR detector array
Detector array format	640 × 512 pixels (center 512 × 512 pixels used)
Instrument Mass	5.4 kg
Instrument Volume	16.7U (36 cm × 26 cm × 18 cm)
Nanosat bus mass	15 kg
Nanosat bus volume	25.2U (20 cm × 30 cm × 42 cm) plus mezzanine 5.3U (7 cm × 18 cm × 42 cm)

TABLE 2.24
Parameters of Aalto-1 Spectral Imager

Parameter	Value
Satellite platform	Aatlo-1 nanosatellite
Orbit altitude	550 km
FOV	10° × 10°
Swath width	97 km (@ 550km altitude)
IFOV	0.02° (0.34 mrad)
Spectrometer type	Fabry-Pérot interferometer
Spectral dispersive element	ETF (Fabry-Pérot filter)
Ground sampling distance	192 m (@ 550 km altitude)
Spectral range	500–900 nm
Spectral sampling interval	1 nm (spectral step)
Spectral resolution @ FWHM	10–30 nm
Number of spectral bands	6–20 (over 60 possible)
f/#	3.6
Focal length	3.2 cm
FPA	CMOS detector array
Detector array format	2048 × 2048 pixels, binned to 512 × 512
SNR	>50 (@3-ms integration, 20-nm bandwidth, 30% albedo, 60° latitude)
Instrument Mass	0.6 kg
Instrument Volume	0.45 U (9.7 cm × 9.7 cm × 4.8 cm)
Power	<4 W (peak)
Nanosat bus mass	4 kg
Nanosat bus volume	3.4U

the location of the AaSI imagery, and to determine whether it is sensible to downlink the high rate data, due to, for example, cloud cover in the target area (Praks et al. 2015).

The spectral camera is able to record 2D spatial images at selected spectral bands by electronically tuning the Fabry-Pérot interferometer. The interferometer consists of two highly reflecting surfaces separated by a tunable air gap. The spectral camera is controlled in a closed capacitive feedback loop by three different piezo actuators. With these actuators the air gap can be adjusted from 0.5 μm to 3.0 μm, with a spectral range from 500 nm to 900 nm. Filter apertures of 7 mm or even 19 mm can be reached with the piezo-actuated Fabry-Pérot interferometer with a spectral resolution of 7–10 nm. Table 2.24 lists the performance parameters of the AaSI and Aalto-1 nanosatellite.

2.19 DLR EARTH SENSING IMAGING SPECTROMETER ON THE INTERNATIONAL SPACE STATION

The DLR Earth Sensing Imaging Spectrometer (DESIS) is a pushbroom hyperspectral imager spectrally sensitive over the VNIR range from 400 nm to 1000 nm with a minimum SSI of 2.55 nm. It has a ground swath width of 30 km with a ground footprint size of 30 m. It is hosted on the Multi-User System for Earth Sensing (MUSES) mounted on the ISS. The DESIS instrument was launched on June 29, 2018, as part of the SpaceX CRS-15 logistics flight to the ISS and was installed to the exterior of the ISS on August 27, 2018.

MUSES is a commercial Earth-imaging platform on the ISS. It is designed, built, owned, and operated by Teledyne Brown Engineering based in Huntsville as part of the company's new commercial space-based digital imaging business to increase the Space Station's research capabilities.

MUSES provides position sensing, data downlink, and other core services for each payload attitude control (Perkins et al. 2016).

The DESIS instrument is the first commercially available, production-class, spaceborne hyperspectral sensor capable of delivering near-global coverage with high quality, high spectral resolution data in VNIR region. This will enable significant new research, expand the dimensions of humanitarian crisis response, and provide improved large-scale commercial spectral analytic applications. DESIS has multiple experimental modes to support additional research, including as an off-nadir along-track pointing mirror for BRDF investigations, forward ground motion compensation studies, and stereo imaging in the hyperspectral domain. These modes allow development of new scientific methods and research applications (Muller et al. 2016).

DESIS satellite data with high spectral information and high global revisit time will increase the value of the derived information for humanitarian aid. Applications will include monitoring of ecosystems, habitat restoration and remediation, vegetation development trends, water quality of coastal zones and oceans, as well as raw materials and minerals inventories and snow and ice cover assessment. The high spectral, medium spatial, and medium temporal resolution of the DESIS instrument will support commercial applications where fine VNIR spectral measurements performed at intervals of weeks to months over moderate to large geographic areas will provide enhanced value. Potential commercial markets include assessments of medium- to large-scale crops, forests, and terrestrial environments, as well as marine, ocean, and inland fresh-water monitoring.

DESIS instrument is a VNIR imaging spectrometer based on a modified Offner design. Its telescope is based on a TMA design. DESIS uses a 2D back-illuminated CMOS detector array of format 1056 (for spatial dimension) × 256 (for spectral dimension). It has a swath width of 30 km with a ground sampling distance of 30 m at the nominal ISS altitude. It uses convex grating to disperse spectrum within a range of 400–1000 nm, which provides an instrument native SSI of 2.55 nm with 235 spectral bands. The design of DESIS instrument is similar to conventional hyperspectral imagers. The main difference between DESIS design and most of the hyperspectral design is that DESIS is equipped with a pointing mirror in front of the entrance slit. It can point in forward direction and back direction up to ±15°. It can operate in either the static mode with 3° angle steps or the dynamic mode with up to 1.5° change in viewing direction per seconds. When operating in the static mode, it allows acquiring experimental data to produce BRDF products or stereo images. When operating in the dynamic mode, it allows continuous observations of the same targets with ground motion compensation to further improve SNR of the acquired imagery. Figure 2.16 shows the configuration of the DESIS instrument.

The peak SNR of DESIS at 550 nm is 205:1 when SSI is 2.55 nm and 406:1 when SSI is aggregated to 10.21 nm modeled based on Modtran with standard mid latitude summer atmosphere with 30% albedo and 0.2 nm sampling. This SNR performance is much improved compared to early spaceborne hyperspectral sensor, such as Hyperion (see Section 2.5).

DESIS has a mass of 93 kg and the volume of the spectrometer is 430 mm × 190 mm × 135 mm. It is integrated in one of the large containers of the MUSES platform. Two gimbals allow a rotation of the whole MUSES platform around two axes resulting in ±25° forward/backward view, 45° backboard (port) view, and 5° starboard view. The pointing accuracy is smaller than 30 arc sec, which corresponds to about 60 m on ground from 400-km altitude. Together with the pointing unit of the DESIS instrument, a ±40° along-track viewing is achievable. Table 2.25 lists the performance parameters of DESIS instrument.

2.20 HYPERSCOUT HYPERSPECTRAL CAMERA ON A 6U NANOSATELLITE (GOMX-4B)

HyperScout is a miniaturized hyperspectral camera of size 1U (10 cm × 10 cm × 10 cm) developed by Cosine Research in the Netherlands. It is onboard ESA's nanosatellite GomX-4B, which is one of a pair of two nanosatellites (GomX-4A and GomX-4B). They were launched at the same time

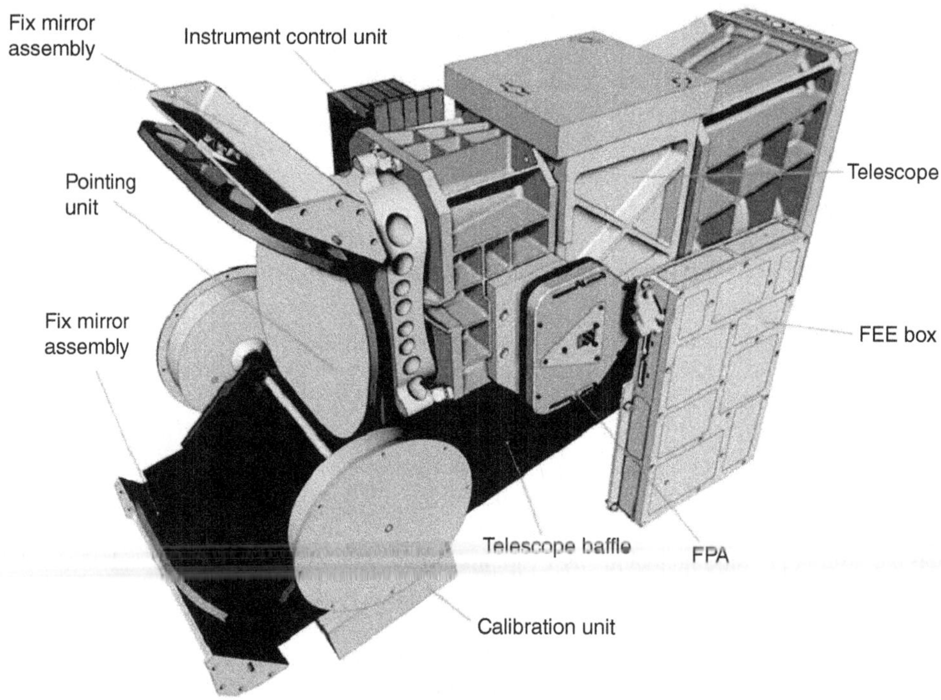

FIGURE 2.16 DESIS instrument configuration. (Courtesy of DLR.)

as secondary payloads on February 2, 2018, on a Long March 2D vehicle from Jiuquan Satellite Launch Center (JSLC), China.

GomX-4 contains two 6U nanosatellites with the objective to demonstrate key technologies to handle large satellite formations. Like its predecessor GomX-3, the GomX-4 mission is a collaboration between ESA and GomSpace ApS of Aalborg, Denmark, to demonstrate miniaturized technologies, preparing the way for future operational nanosatellite constellations. GomX-4B used its butane cold gas propulsion system to manoeuvre away from its twin, flying up to 4500 km away in a fixed geometry—a limit set by Earth's curvature, and representative of planned CubeSat constellation spacing—to test intersatellite radio links allowing the rapid transfer of data from Earth between satellites and back to Earth again.

GomX-4B carries 5 demonstration payloads onboard: the 6U propulsion module from NanoSpace, the innovative Inter-Satellite Link (ISL) from GomSpace, the Chimera board developed by ESA, the HyperScout hyperspectral camera from Cosine, and the Star Tracker from Innovative Solutions In Space (ISIS).

HyperScout is a LVF based hypespectral imager and consists of four components: a telescope, a FPA, an instrument control unit, and onboard data handling unit. The telescope is a TMA design. It comprises three powered mirrors, which focus the incoming radiance of the scene in the FOV on the FPA, and the opto-mechanical system, which provides a stable support to the optical and electronic units. A LVF is used to separate the different wavelengths before the radiance reaching on a 2D CMOS detector array, which is then read by the read out electronics (ROE). The instrument control unit contains the control software and provides electrical interface to the spacecraft. It distributes power, clocks, telemetry, and commands between the units, controls the detector through the ROE, and merges the data acquired with the platform ancillary information creating L0 data, which is then stored in the data storage subsystem (Contocello et al. 2016).

Because HyperScout is a spectral filter based hyperspectral imager using LVF to disperse spectrum, the wavelength separation is performed in the along-track direction, with a constant wavelength in the cross-track direction. This means, ground footprints in each cross-track line it observes

TABLE 2.25
Performance Parameters of DESIS Instrument

Parameter	Value
Platform	ISS
Altitude	400 km (nominal)
FOV	4.4°
Swath width	30 km
Number of pixels in cross-track	1024
IFOV	0.004°
Ground sampling distance	30 m
Wavelength range	0.4–1.0 μm
Spectral sampling interval (SSI)	2.55 nm
Spectral bands	235 117 (after 2-band binning) 78 (after 3-band binning) 58 (after 4-band binning)
Off-nadir pointing	± 15° in along-track direction
Peak SNR	205:1 at 2.55 nm SSI 406:1 at 10.21 nm SSI (after 4-band binning)
Telescope	Three-mirror anastigmat (TMA)
Focal length	320 mm
f/#	2.8
Spectrometer	Offner configuration
Spectral dispersive element	Convex grating
Spectrometer size	430 × 190 × 135 mm^3
Radiometric linearity	>95%
FPA	CMOS 2D detector array
Detector array format	1056 × 256 pixels
Detector pitch size	24 μm × 24 μm
Maximum frame rate	232 Hz
Digitization	13 bits
Mass	93 kg

is seen at a different wavelength from 400 nm to 1000 nm, with the onward movement of the satellite allowing a complete hyperspectral image to be built up rapidly. As shown in Figure 1.8, the 2D detector array is used in pushbroom mode: the full hyperspectral datacube generation requires the acquisition of a series of subsequent frames, so that each cross-track line on ground is imaged in all wavelengths and can then be used to reconstruct the hyperspectral datacube. Table 2.26 lists the parameters of hyperspectral camera HyperScout.

Funded by the European Space Agency, Cosine Research has integrated thermal infrared (TIR) technologies into a miniaturized VNIR hyperspectral imager to fit the combined spectral channels in a volume of less than 2U. The imager is named HyperScout-2 as it uses the HyperScout platform, as building block to further integrate spectral channels. HyperScout-2, as shown in Figure 2.17, is based on an athermal telescope with free-form reflective elements, shared by both VNIR and TIR channels. It is equipped with a hybrid processing platform composed of a CPU, GPU, and vision processing unit (VPU). HyperScout-2 will be used as an in-orbit test bed to benchmark the performance of a miniaturized class of systems as well as to perform hands-on investigations to forecast the benefits of combining frequent co-registrated measurements in the VNIR and TIR from nanosatellites, with less frequent but very accurate measurements performed by institutional satellites such as the Copernicus fleet (Esposito and Marchi 2018, Esposito et al. 2019).

TABLE 2.26
Parameters of Hyperspectral Camera HyperScout

Parameters	Value
Satellite platform	6U nanosatellite
Orbit altitude	500 km, sun synchronous
Swath width	200 km
FOV	23° × 16° (cross-track × along-track)
Ground scene size	200 km × 150 km (cross-track × along-track)
Ground sampling distance	70 m
Wavelength range	0.4–1.0 μm
Spectral resolution	15 nm
Number of spectral bands	45
Telescope form	Three-mirror anastigmat (TMA)
Spectrometer	Omission
Spectral dispersion element	Linear variable filter (LVF)
FPA	2D CMOS detector array
SNR	50:1–100:1
Digitization	12-bit
Power	11 W
Instrument mass	1.1 kg
Instrument volume	1 U (10 × 10 × 10 cm^3)

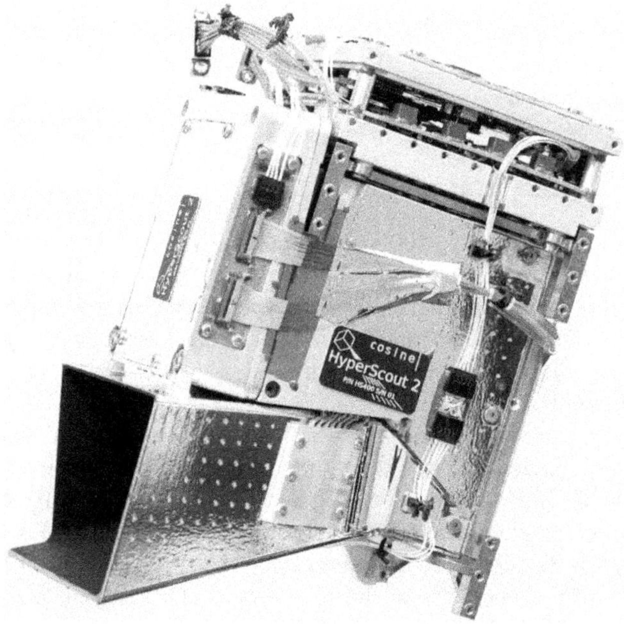

FIGURE 2.17 Hyperspectral camera HyperScout-2. (Courtesy of Cosine Research.)

2.21 ADVANCED HYPERSPECTRAL IMAGER (AHSI) ON CHINESE GAOFEN-5 SATELLITE

The Advanced Hyperspectral Imager (AHSI) is the main payload of the Chinese Gaofen-5 (GF-5) satellite, which is a remote sensing satellite for scientific research on the Earth's atmosphere and terrestrial observation launched on May 8, 2018. The satellite carries six payloads, a hyperspectral payload, and a multispectral payload for terrestrial Earth observation, along with four atmospheric observation payloads. These payloads will enable researches to study greenhouse gases, pollution, air quality, climate change, and map geological resources and crop production, among other tasks. The objectives of AHSI are to address many key science questions and operational needs using remote sensing technology, such as ecological and environmental monitoring, investigation of geology and mineral resources, land and resources, disaster monitoring, precision agriculture, forestry management, precision animal husbandry, and urban planning.

The AHSI was designed and built by the Shanghai Institute of Technical Physics, Chinese Academy of Science. It is China's first spaceborne hyperspectral imager that uses a convex grating to disperse spectrum. AHSI has 330 spectral bands covering a wavelength range from 0.4 μm to 2.5 μm. The SSI is 5 nm in VNIR (0.4~1.0 μm) region and 10 nm in SWIR (1.0~2.5 μm) region. The ground sampling distance of AHSI is 30 m, which is the same as that of Hyperion, whereas the swath width of AHSI is 60 km, which is about 8 times wider than that of Hyperion. Figure 2.18 shows a photo of the AHSI payload before being launched (Liu 2018).

Table 2.27 lists the key parameters of the specification of the AHSI payload. The AHSI consists of a wide-field telescope, a field splitter, a slit, two Offner spectrometers with convex gratings, an ensemble of FPAs, a baffle, and an onboard calibration subsystem, as well as subsystems such as components, drivers, and signal acquisition and communication control and information processing electronics. The telescope is an off-axis TMA. The field splitter separates the input light from the telescope into VNIR and SWIR portions to fill the two corresponding spectrometers. The slit limits the radiation light to the spectrometers. The convex gratings of the spectrometers disperse the input light and image the spectrum onto the focal planes of the spectrometers. The 2D CCD detector array and the 2D HgCdTe detector array mounted on the focal planes of the VNIR and SWIR spectrometers sense the spectra and convert them to electronic signals.

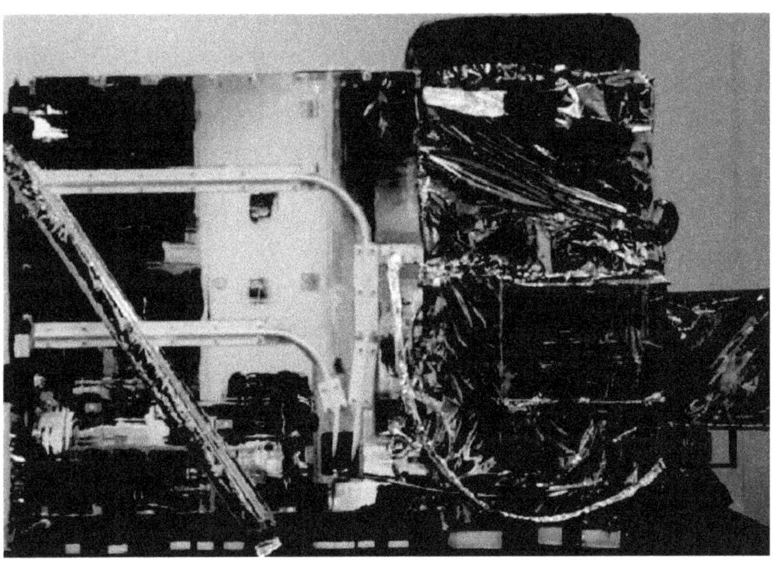

FIGURE 2.18 Photo of the AHSI payload before launch. (Courtesy of Shanghai Institute of Technical Physics, Chinese Academy of Science.)

TABLE 2.27
Key Parameters of the Specification of the AHSI Payload

Parameter	Value
Orbit altitude	705 km
Swath width	60 km
Ground sampling distance	≤30 m
Wavelength range	0.40–2.5 μm
Number of spectral bands	330
Spectral Sampling Interval	≤5 nm (VNIR) ≤10 nm (SWIR)
Spectral error	≤1.0 nm
Signal-to-noise ratio	≥200:1 (0.40–0.90 μm) ≥150:1 (0.90–1.75 μm) ≥100:1 (0.75–2.50 μm)
Telescope	TMA
Spectrometers	Offner configuration
Detector arrays	CCD (VNIR) HgCdTe (SWIR)

To achieve the requirements of the large FOV, a design of full reflection off-axis TMA was adopted for the telescope. On the basis of the traditional Offner configuration, a convex grating is added as a correction lens. The radiation lights pass through the grating twice prior entering the Offner structure and after leaving the Offner structure, respectively. The spectral curvature (smile) and spatial distortion (keystone) caused by the long slit are corrected by the different incident angles of the slit center and edge lights to the grating. The AHSI is also equipped with an onboard calibration subsystem to ensure the stability and quantification of acquired image data. This includes by imaging the onboard LED calibration components, combining the occultation to observe the atmospheric profile for spectrometer on-orbit spectral calibration, and by introducing sunlight to illuminate the diffuse panel to calibrate the spectrometer while using a separate diffuser to monitor the attenuation of the main diffuse panel.

Table 2.28 reports the pre-launch characterization and test results of the ASHI payload witnessed by the representatives of the client and the mission management team. The test results demonstrate that the flight model of the AHSI payload met and exceeded all the required specifications. Compared to Hyperion hyperspectral sensor, the AHSI has a higher SNR (3–4 times), a wider swath width (around 8 times), and more spectral bands (over 100 more; Liu 2018). This kind of progress of spaceborne hyperspectral sensors is encouraged and expected by the hyperspectral user community for almost two decades since the launch of Hyperion in 2000.

2.22 ITALIAN HYPERSPECTRAL SATELLITE PRISMA

PRISMA (PRecursore IperSpettrale della Missione Applicativa) is a preoperative Italian hyperspectral satellite, aiming to qualify the technology, contribute to develop applications, and provide products to institutional and scientific users for environmental observation and risk management. It was launched on March 22, 2019, on a Vega launch vehicle from the European base of Kourou in French Guyana into a sun synchronous orbit. It focuses primarily on the European area of interest, enabling the download of the data on two ground stations located in Italy (Candela et al. 2016).

PRISMA instrument is composed of a hyperspectral imager and a PAN camera. The instrument is the core of the PRISMA mission, fully funded by the Agenzia Spaziale Italiana (ASI), and the prime

TABLE 2.28
Pre-Launch Test Results of the ASHI Payload

Parameter	Requirement	Measured Result
Swath width (km)	60 (1.00 ± 1%)	61.127 (VNIR) 60.159 (SWIR)
Ground sampling distance (m)	≤30	29.67 (VNIR) 29.70 (SWIR)
Wavelength range (μm)	0.40–2.5	0.388–2.518
Spectral sampling interval (nm)	5 (VNIR) 10 (SWIR)	<4.47 (VNIR) <8.60 (SWIR)
Spectral error (nm)	≤1.0	<0.829 (VNIR) <0.747 (SWIR)
Absolute radiation accuracy	<5%	2.63–2.93% (VNIR) 3.45–4.31% (SWIR)
Relative radiation accuracy	<3%	2.10% (VNIR) 2.24% (SWIR)
Spectral registration accuracy (nm)	0.5 (VNIR) 1.0 (SWIR)	0.39 (VNIR) 0.65 (SWIR)
Static MTF	≥0.25	>0.45
Signal-to-noise ratio	≥200:1 (0.40–0.90 μm) ≥150:1 (0.90–1.75 μm) ≥100:1 (0.75–2.50 μm)	654:1 (500 nm) 341:1 (900 nm) 380:1 (1100 nm) 397:1 (1700 nm) 191:1 (2400 nm)

contractor is a consortium of Italian companies. SELEX ES is responsible to the development of the hyperspectral imager, including level 0–Level 1 (L0–L1) product algorithms (Meini et al. 2012).

The hyperspectral imager operates in pushbroom mode. It is made up of a VNIR spectrometer and a SWIR spectrometer to cover spectral bands ranging from 400 nm to 1010 nm and from 920 nm to 2505 nm. It provides hyperspectral images of the Earth with 30-m ground sample distance (GSD), 30-km swath width, and spectral bands at an SSI of 12 nm. The PAN camera provides black-and-white images at spatial resolution of 5 m within a spectral range of 400–700 nm, co-registered to the hyperspectral images, so as to allow images fusion to sharpen the spatial resolution of the hyperspectral images (Meini et al. 2016).

The Optical Head Unit houses a common telescope, a double-channel imaging spectrometers operating in VNIR and SWIR regions, and a PAN camera. It collects the input radiation from a scene on the ground by a telescope common to the hyperspectral imager and PAN camera, disperses the radiation by the prisms of the two spectrometers, converts photons to electrons by means of appropriate detector arrays, and amplifies the electronic signal and converts it into digital data stream. Figure 2.19 shows the optical layout of the PRISMA hyperspectral imager. The Main Electronics Unit controls the instrument and handles the bit stream representing the spectral images up to the interface with the spacecraft transmitter.

The telescope is a TMA design that assures excellent optical quality with a minimum number of optical elements. This solution is very compact and without obstruction. The TMA telescope optics layout is shown on the left in Figure 2.19. The shape of the three mirrors is aspherical with only conic constants. The secondary mirror is almost on-axis. The off-axis values of the primary mirror and of the tertiary mirror are designed in order to facilitate the mirror manufacturing. The position of the aperture stop lies on surface M2. The telescope optical system is telecentric with respect to the entrance pupil. The stray light effects have been extensively analyzed and addressed to define

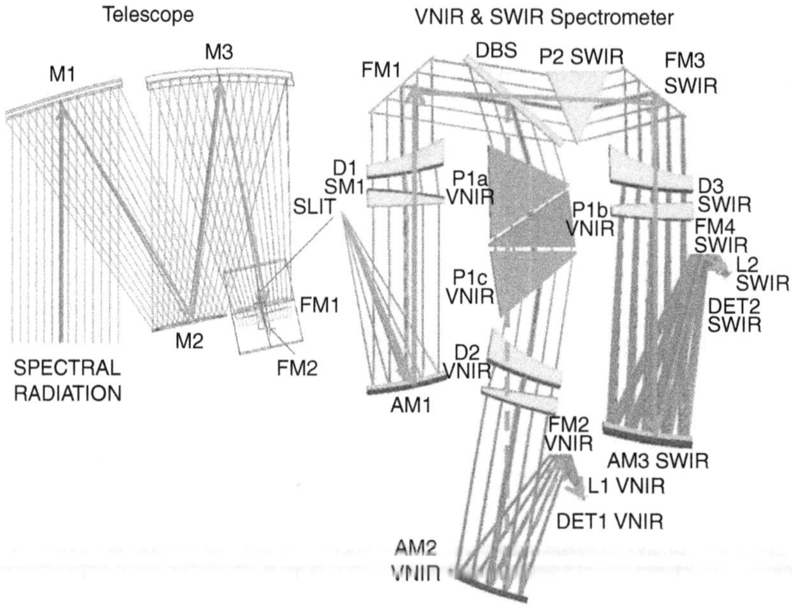

FIGURE 2.19 Optical layout of the PRISMA hyperspectral imager. (Courtesy of SELEX ES.)

the particular requirements on the optics design in terms of element dimensions, mechanical profiling, finishing, scratch and digs, coatings, contamination, and baffling.

The spectrometers are of collimator-prism-imager configuration. A spectrometer consists of a collimator common to both VNIR and SWIR channels, a dispersing prism, and an objective (different for the two channels). The telescope images the spectral radiation of a cross-track line on Earth on the entrance slit of the VNIR and SWIR spectrometers. The dimension of the slit is 30 mm × 30 μm, with the 30 mm orientation toward cross-track direction and 30 μm to along-track direction. The overall input spectral radiation (400–2505 nm) is split into two channels (VNIR and SWIR) by a dichroic beam splitter (DBS). The collimator images the slit image at infinity, then the prism disperses radiation spectrum reaching on its surface, the objective focuses the chromatic images on the dedicated detector array placed on the corresponding VNIR and SWIR focal planes as detailed in Figure 2.19.

The main advantages of this kind of design are the same FOV for both VNIR and SWIR channels (i.e., same entrance slit) and the use of several common optical elements for both channels. The VNIR channel covers a wavelength range of 400–1010 nm with 66 spectral bands, while the SWIR channel covers a wavelength range of 920–2505 nm with 171 spectral bands. The overlap between the VNIR and SWIR channels ranges from 920 nm to 1010 nm. This overlap allows a cross-calibration between the two channels, increasing the confidence in the calibration process.

The two spectrometers use prisms as the spectral dispersive elements. This prism-based solution has advantage of obtaining higher efficiency and lower polarization sensitivity than those achievable by grating-based spectrometers. The high efficiency allows reducing the instrument dimension and mass with less demanding resources to the spacecraft and less criticalities for the optics design. The disadvantage is that the spectral dispersion is not constant with respect to the wavelength.

Both VNIR and SWIR channels have a magnification to match the detector array of 1000 × 256 pixels, with pitch size of 30 μm × 30 μm. The instrument design guarantees the spectral distortion (Smile) and the spatial distortion (Keystone) effects to be maintained within 10% of the pixel for both VNIR and SWIR detector plane arrays.

The PAN channel is obtained by separating the main beam coming from the TMA telescope by an in-field separator (FM2 in the telescope) that allows the use of a common fore-optics for

TABLE 2.29
Performance Parameters of PRISMA Instrument

Parameter	VNIR	SWIR	PAN
Platform	PRISMA satellite		
Altitude	615 km		
FOV	2.77°		
Swath width	30 km		
IFOV	48.34 μrad		
Telescope	Three-mirror anastigmat (TMA)		
Telescope focal length	62 cm		
Telescope aperture	21 cm		
Telescope f/#	2.95		
Spectrometer	Collimator-prism-imager configuration		
Spectral dispersive element	Prism		
Ground sampling distance	30 m	30 m	5 m
Number of pixels in cross-track	1000	1000	6000
Wavelength range	0.4–1.01 μm	0.92–2.505 μm	400–700 nm
Spectral sampling interval (SSI)	≤12 nm	≤12 nm	-
Spectral bands	66	171	1
Detector format	1000 × 256 pixels	1000 × 256 pixels	6000 × 1 pixels
Detector pitch size	30 μm × 30 μm	30 μm × 30 μm	6.5 μm × 6.5 μm
SNR	Peak 500:1 @0.65 μm 200:1 @0.4–1.0 μm	Peak 400:1 @1.55 μm 200:1 @1.0–1.75 μm 100:1 @1.95–2.35 μm 200:1 @2.1 μm	240:1
MTF @Nyquist frequency	>0.8	>0.7	>0.2
Spectral distortion	0.1 pixel	0.1 pixel	
Spatial distortion	0.1 pixel	0.1 pixel	
Absolute radiometric accuracy	>5%		
Frame rate	230 Hz		
Digitization	12 bits		
Instrument power	110 W (average)		
Instrument mass	200 kg		
Instrument volume	1.0 × 1.01 × 1.65 m^3		

both hyperspectral imager and PAN camera, greatly simplifying the overall instrument design and products co-registration. The in-field separation effect is a constant offset in terms of geo-location between hyperspectral and PAN images, which will be taken into account by image processing algorithms, when co-registering the hyperspectral and PAN images.

PRISMA instrument is also equipped with in-flight calibration unit to allow operations of absolute and relative radiometric calibration as well as geometric and spectral calibrations. Table 2.29 tabulates the performance parameters of PRISMA instrument.

2.23 HYPERSPECTRAL IMAGE SUITE ABOARD THE INTERNATIONAL SPACE STATION

Hyperspectral Imager Suite (HISUI) includes a hyperspectral imager (HSI) and a multispectral imager (MSI) onboard the Japan Experiment Module (JEM) in the ISS. HISUI is developed by Japanese Ministry of Economy, Trade, and Industry (METI) for space demonstration to see if the sensor works in orbit on JEM of the ISS. The HISUI-Exposed Payload (HISUI-ExP) and HISUI-Mission Data Recorder—Pressurized Module (MDR-PM) are launched from US Cape Canaveral Air Force Station to the ISS by SpaceX's Falcon-9 Dragon cargo rocket on December 6, 2019. HISUI-ExP will be deployed on JEM Exposed Facility (EF) as a nadir-viewing instrument. MDR-PM will be installed in JEM-PM. HISUI also has support sensors such as a gyro, two star trackers, GPS receivers, and a mission data processor (Matsunaga et al. 2017).

Due to the constraint of data downlink capacity of the ISS, the hyperspectral and multispectral data generated by HISUI will be partially transmitted to ground stations (≈10 GB/day ≈ 30,000 km^2). The rest (≈ max. 300 GB/day ≈ 900,000 km^2) will be recorded in mass memory storages that will be shipped back to Earth by cargo ships three or four times a year.

The HSI and MSI are fabricated in two separate boxes and operate independently or simultaneously. The basic specifications of two imagers are summarized in Table 2.30. The swath of HSI is 20 km, which is one-third of that of the MSI (60 km) due to the constraint of optical design and the availability of large form 2D detector arrays for HSI. To fill the gap between the swaths of two

TABLE 2.30
Specifications of HISUI Hyperspectral Imager and Multispectral Imager

Parameter	Hyperspectral Imager	Multispectral Imager
Platform	ISS	
Altitude	400 km (nominal)	
Swath	20 km	60 km
Number of pixels in a cross-track line	1000	18,000
Ground sampling distance	20 m (cross-track) × 30 m (along-track)	3.3 m (cross-track) × 5m (along-track)
Wavelength range	0.40–2.5 μm	0.45–0.9 μm
Spectral sampling interval	2.5 nm (VNIR) 6.25 nm (SWIR)	N/A
Spectral resolution	10.0 nm (VNIR) 12.5 nm (SWIR)	60–110 nm
Number of spectral bands	185 (57 VNIR + 128 SWIR)	4
Telescope type	TMA	TMA
Telescope aperture	30 cm	
Telescope f/#	2.2	
Pointing capacity	±5° (±35 km) in cross-track direction	N/A
Spectrometer configuration	Offner	
Spectral dispersive element	Grating	Band-pass filters
FPA	2D Si-CMOS (VNIR) 2D HgCdTe (SWIR)	1D Si-CMOS
SNR (30% albedo)	≥450:1@620 nm ≥300:1@2100 nm	≥200:1
MTF	≥0.2	≥0.3
Spectral distortion (smile)	<1 nm (VNIR) <2.5 nm (SWIR)	N/A
Spectral accuracy	0.2 nm (VNIR) 0.625 nm (SWIR)	N/A
Absolute radiometric accuracy	±5%	N/A
Digitization	12 bits	12 bits
Onboard data compression	Lossless (70% reduction)	Lossless (70% reduction)
Data rate	300 GB/day (transmitted 10 GB/day)	
HISUI Exp Mass	550 kg (incl. 240 kg for HSI)	
HISUI Exp Volume	2.3 × 1.5 × 1.6 m^3	

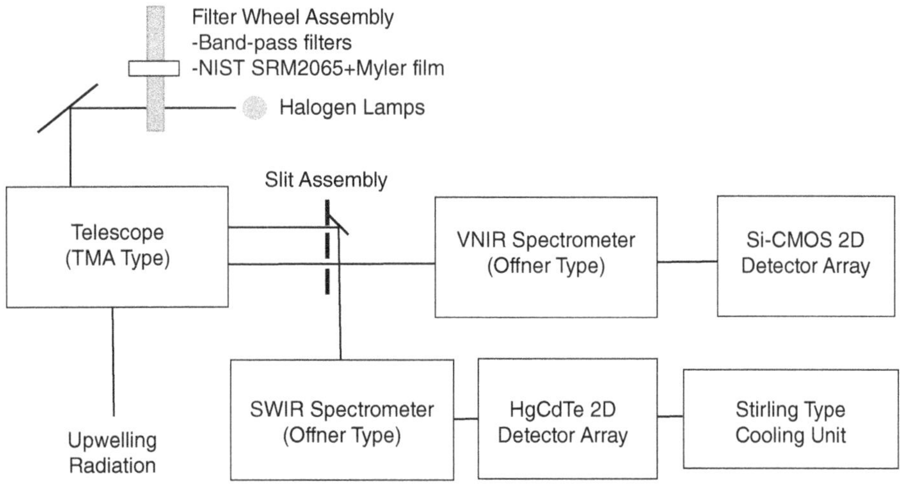

FIGURE 2.20 Block diagram of composition of HISUI Hyperspectral Imager.

imagers, HSI is equipped with a pointing mechanism, which can tilt HSI for ±5° in cross-track direction to match the swaths of the imagers.

As shown in Figure 2.20, the HSI is composed of a TMA-type reflective telescope and two Offner form grating based spectrometers that cover the VNIR and SWIR regions. Each spectrometer consists of a grating to disperse the spectrum and a 2D detector array. The SWIR spectrometer has a Stirling cooler for the SWIR detector array. The MSI consists of a TMA type telescope, 4 linear detector arrays, and 4 band-pass filters (Matsunaga et al. 2016).

HISUI instrument is also equipped with an onboard data correction and calibration mechanism for HSI. It is based on the instrument data measured on the ground. These data are recorded in the onboard memory prior to launch. Raw data from detector array are radiometrically corrected, including corrections of nonlinearity and offset as well as photo response nonuniformity (PNRU). After radiometric correction, the raw data are binned in the spectral direction. Every 4 pixels in VNIR and 2 pixels in SWIR are binned to produce the required spectral band width. Smile correction is applied using weight functions. Weighted radiance data from adjacent 6 (VNIR) or 4 (SWIR) detector pixels are added. Then, the center of the wavelength of the spectral band after binning is corrected.

The onboard calibration uses internal light sources, vicarious calibration at selected sites across the world, cross calibration with other remote sensing instruments, and lunar calibration (Yamamoto et al. 2012, Kouyama et al. 2014). The HSI has a partial aperture calibration unit that includes halogen lamps and a filter wheel, which accommodates multiple filters with known and stable spectral characteristics. Calibration data will be periodically acquired in the night time with an interval of several tens of days.

2.24 GERMAN HYPERSPECTRAL IMAGER FOR ENVIRONMENT MAPPING AND ANALYSIS PROGRAM (ENMAP)

The Environmental Mapping and Analysis Program (EnMAP) is a German hyperspectral satellite mission scheduled to be launched in 2020. It aims at monitoring and characterizing the Earth's environment on a global scale. The German Aerospace Center (DLR) is responsible for the mission management and operation. Helmholtz Centre Potsdam German Research Centre for Geosciences (GFZ) is the scientific principal investigator and OHB System AG is the industrial prime contractor for the payload, spacecraft and launch.

The scientific objectives of EnMAP are to 1) provide high-quality hyperspectral data that are not achievable by the currently available spaceborne hyperspectral sensors for advanced remote sensing analyses, 2) Obtain diagnostic geochemical, biochemical and biophysical parameters that describe the status and dynamics of various ecosystems to improve our understanding of complex environmental processes, 3) Provide information products that can serve as input for advanced ecosystem models, 4) Foster and develop novel methodologies that improve the accuracy of currently available remote sensing information and provide additional science-driven information products, 5) Significantly contribute to environmental research studies in the fields of ecosystem functions, natural resource management, natural hazards and Earth system modeling, and 6) Develop new concepts and techniques for data extraction and assimilation to achieve synergies with other sensors (Kaufmann et al. 2016).

EnMAP will be a high-performed spaceborne hyperspectral imager. It is a prism based imaging spectrometer operating in pushbroom mode. It has 242 spectral bands covering a wavelength range from 420 nm to 2450 nm with a SSI of 6.5 nm for VNIR bands and 10 nm for SWIR bands. Its ground swath width is 30 km with a ground sampling distance of 30m × 30m. EnMAP is designed to achieve better SNR than the available spaceborne hyperspectral imagers. The SNR will be greater than 500:1 for a 10 nm equivalent bandwidth of the spectral band at 495 nm. In the SWIR region, an SNR of more than 150:1 will be reached (Sang et al. 2008, Stuffler et al. 2009). Table 2.31 tabulates the performance parameters of EnMAP.

The EnMAP hyperspectral imager is composed of a telescope, two spectrometers, FPAs, onboard calibration and control electronics. Figure 2.21 shows the layout of the optical system of the EnMAP hyperspectral imager. The telescope is a standard off-axis unobscured TMA without intermediate focus locations. Its optical speed is f/3. Similar TMA telescope designs have been used in multiple Earth observation missions. The aperture stop is located at the secondary telescope

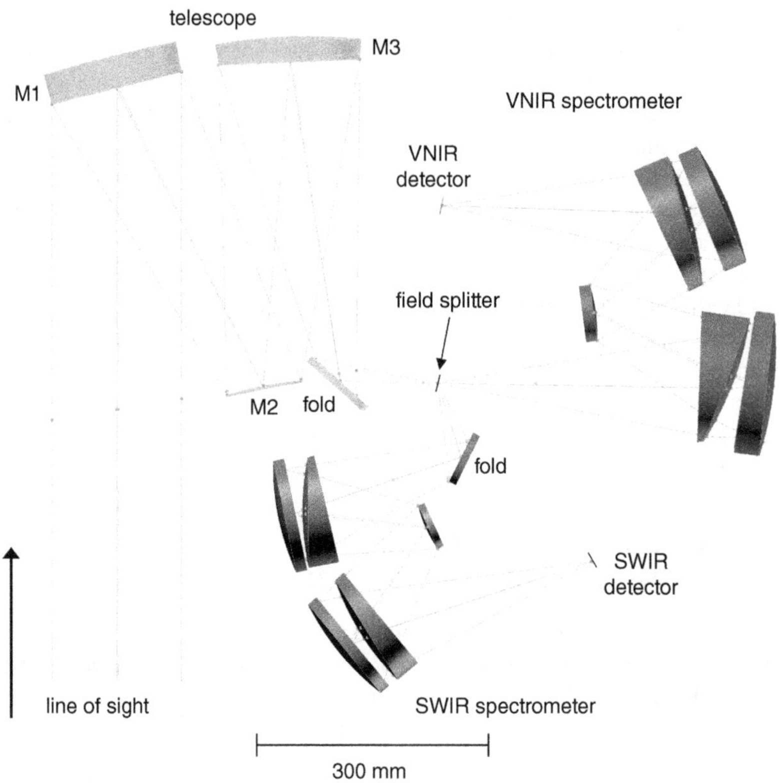

FIGURE 2.21 Layout of the optical system of the EnMAP hyperspectral imager. (Courtesy of German Aerospace Center Space Agency.)

TABLE 2.31
Performance Parameters of EnMAP

Parameter	VNIR	SWIR
Platform	EnMAP satellite	
Altitude	653 km	
Scanning type	Pushbroom	
Swath width	30 km	
Ground sampling distance	30 m	
Number of pixels in cross-track	1024	
Off-nadir pointing	±30° in cross-track direction	
Telescope	Three-mirror anastigmat (TMA)	
FOV	2.63°	
Telescope focal length	52.2 cm	
Telescope aperture	17.4 cm	
Telescope f/#	3	
Spectrometer	Offner form	
Spectral dispersive element	Curved prism	
Wavelength range	0.42 – 1.0 μm	0.90–2.45 μm
Spectral sampling interval (SSI)	6.5 nm	10 nm
Spectral bands	88	154
Spectral calibration accuracy/stability	0.5 nm	1.0 nm
Absolute radiometric accuracy	>5%	>5%
Spectral distortion (smile)	0.2 pixel	0.2 pixel
Spatial distortion (keystone)	0.2 pixel	0.2 pixel
Detector array material	Si-CMOS	HgCdTe
Detector array format	1024 × 108 pixels	1024 × 256 pixels
Detector pitch size	24 μm × 24 μm	24 μm × 32 μm
Full well capacity	1 Me-	1.2 Me- (low gain) 300 Ke- (high gain)
Detector readout noise	200 e- (low gain) 70 e- (high)	290 e- (low gain) 160 e- (high)
SNR	>500:1 @0.495 μm	>150:1 @ 2.2 μm
MTF @ Nyquist frequency	>0.25 @ 16.6 cyc/km for all wavelength	
Polarization sensitivity	<5%	
Frame rate	230 Hz (4.3 ms integration time)	
Digitization	14 bits	
Data rate	866 Mbit/sec science data, 650 Gbit uncompressed, 400 Gbit compressed	
Instrument Power	<300 W (peak) 237 W (standby)	
Instrument Mass	369 kg	
Instrument Volume	1.51 m^3 (1.8m × 1.2m × 0.7 m)	

mirror, generating a telecentric imaging situation at the field stop and matching the entrance pupil location of the spectrometers.

The required SSI for the VNIR region is 6.5 nm. This requirement is a compromise between resolving power and keeping the SNR as well as the data volume at acceptable levels. For the SWIR region, the required SSI is 10 nm on average, which is sufficient to resolve the typical mineralogical features around 2000 nm, guaranteeing a good SNR in the region where solar irradiation is low. Due to different spectral sampling requirements for the VNIR and SWIR regions, a dual spectrometer approach was selected to cover the required spectral range from 420 nm to 2450 nm. Based on the demanding polarization sensitivity requirement as well as the high optical throughput necessary to achieve the requested SNR performance, a prism was selected as a spectrum disperser.

In order to align the images generated by the VNIR and SWIR spectrometers, an overlap between the images is required. The common approach to generate this overlap is to use a single entrance slit combined with a DBS to separate the spectrometer bands. This method allows good co-registration between the spectrometer images, but is problematic with respect to the polarization sensitivity and will result in reduced SNR in the overlap region. EnMAP uses a dual aperture split FOV concept to overcome these difficulties. The two spectrometers are coupled to a common telescope using a field-splitting unit that features two closely spaced entrance slit apertures and a beam separating optic. Therefore, both spectrometers deliver full SNR performance in the region of spectral overlap, thereby permitting the data sets to be merged with high precision and without the drawbacks of increased dichroic-induced polarization sensitivity. Section 5.5.1 describes the design of the dual aperture slit in details.

The VNIR and SWIR spectrometers are a novel design, which combines an Offner design with a curved prism disperser as shown in Figure 2.21. Following the symmetry of the Offner design, a pair of two prisms is introduced into an Offner relay, both of which are used in a double pass configuration for increase of dispersive power. The optimal design of the spectrometers inherits the low distortion properties of the Offner configuration and exhibits good imaging performance in a compact design with all spherical surfaces and minimum volume.

A prism-based spectrometer inherently suffers from nonlinear spectral dispersion (see Section 5.4.3 for details). To alleviate this problem, different materials were chosen for the pair of the two prisms. For the VNIR spectrometer, the strong dispersion of glasses in the UV region of the spectrum dictates the use of two compensating glass types. A combination of fused silica and a flint glass materials was selected. The SSI varies from 4.8 nm to 8.2 nm with an average of 6.5 nm over the full VNIR spectral range. For the SWIR spectrometer, a fused silica disperser was chosen based on the dispersion characteristics and the good properties of this material with respect to the space radiation environment. The dispersive behavior results in an average SSI of 10 nm with variations of +20% and –25% over the SWIR spectral range. For both spectrometers, the spectral resolution of a band as defined by the full width at half maximum (FWHM) value of the corresponding spectral response function (slit function) is similar to the local SSI, deviating from this value by <10% with a low smile distortion.

As shown in Figure 2.21, the telescope images a cross-track line on ground and focuses it via a fold mirror onto the field splitter. Radiation light transmitted through the dual spectrometer entrance slits is directed into the VNIR and SWIR spectrometers, which form spectrally resolved images of the slits on the detector arrays. A silicon-based 2D detector array was selected for the VNIR region from 420 nm to 1000 nm, and a HgCdTe detector array was selected for SWIR region from 900 nm to 2450 nm.

The FPA of the VNIR spectrometer is a high-speed silicon CMOS 1024 × 108 pixel detector array featuring on-chip correlated double sampling to provide low noise in snapshot mode, including stare-and-scan capabilities. It is optimized for a large full well capacity (1 Me^-), high quantum efficiency, high resolution, high speed operation, low power consumption, and low noise. The dual column amplifiers divide the detector array into two areas (bottom, top) with dual low and high gain. A thermoelectric cooler provides thermal stabilization during operation to a temperature of 21°C ± 0.05°C.

The FPA of the SWIR spectrometer is a HgCdTe 1024 × 256 pixel photovoltaic array. With indium bumps, the photovoltaic array is attached to the silicon ROIC chip forming the so-called IR-hybrid. The hybrid is optimized with respect to quantum efficiency and sensitivity. The amplification provides two integration capacitors, which can be selected line-by-line individually for gain adjustment. The SWIR array must operate at a nominal temperature of 150 K or lower to reduce thermal noise and dark current. The operating temperature is supplied by a split Stirling cryocooler with a pulse tube cold finger and the cooler electronic.

Several onboard facilities for instrument calibration allow consistent monitoring of instrument response and enable a high data quality and reliability to be achieved after on-ground processing.

2.25 MOONS AND JUPITER IMAGING SPECTROMETER OF ESA'S JUPITER ICY MOONS EXPLORER

The Moons And Jupiter Imaging Spectrometer (MAJIS) is a spaceborne hyperspectral imager covering wavelength range from VNIR to IR. It is selected as one of the scientific payloads by ESA in February 2013 for its Jupiter Icy Moons Explorer (JUICE) mission intended to explore Jupiter and three of its icy moons: Europa, Callisto, and Ganymede. It is scheduled to be launched in 2022. The spacecraft of the JUICE mission is targeted to fly by Callisto, Ganymede, and Europa, then a 1-year orbital phase around Ganymede (Langevin et al. 2014).

MAJIS will perform imaging spectroscopy required to achieve many of the mission scientific objectives. These include investigation of the nature and location of chemical compounds, especially organic and non-ice constituents on the surfaces of the Galilean moons. In addition, it will characterize the Galilean moons' exospheres and monitor their peculiar aspects, for example, Io and Europa tori or Io's volcanic activity. It will also study Jupiter's atmosphere and spectral characterization of the whole Jupiter system.

MAJIS is a dual-grating spectrometer design with the VNIR spectrometer covering a spectral range from 0.5 µm to 2.35 µm and the IR spectrometer covering a spectral range from 2.25 µm to 5.54 µm. The two spectrometers share a common TMA telescope. A dichroic beam-splitter separates the light between the two spectrometers.

The optical design of MAJIS payload benefits from heritage of other imaging spectrometers in the visible and IR spectral range designed and developed by the same builder in the past years for other planetary missions, namely VIRTIS for Rosetta (Coradini et al. 2007), VIRTIS for Venus Express (Piccioni et al. 2007), VIR spectrometer for Dawn (De Sanctis et al. 2011), and JIRAM for Juno (Adriani et al. 2017), although the combination of solutions adopted for MAJIS optical design is unique for spectral range, cooling strategy, and structural design.

The spectral and spatial resolution of MAJIS takes advantage of up-to-date developments of detector technology with 2 times 508 spectral bands from 0.5 µm to 5.5 µm at SSI of 3.6 nm and 6.4 nm over 400 spatial pixels. The SNR will exceed 100 over most of the spectral range, except for deep ice absorption bands such as those observed for Europa above 2.8 µm. The IFOV of 150 µrad of the instrument corresponds to a footprint size of 75 m on Ganymede from a 500-km circular orbit over Ganymede and to a footprint size of 150 km for observations of the atmosphere of Jupiter when flies by it. Spatial and spectral binning will be implemented, in combination with an effective onboard compression scheme, so as to provide extensive spatial coverage of the icy Galilean moons at medium resolution (1–5 km/pixel) as well as time evolution sequences for the atmosphere of Jupiter and the exospheres of the moons. Table 2.32 lists the parameters of the MAJIS instrument.

The TMA telescope is constituted of two off-axis (M1 and M3) and one on-axis (M2) mirrors with focal length 24 cm and an equivalent aperture of 7.5 cm, which results in a f/3.2. It is telecentric in image space with the aperture stop placed in correspondence of the M2 mirror, defining the pupil shape. Along the optical path, there are four folding mirrors in order to match with the mechanical design. The first folding mirror FM1, which is part of a scan unit, reflects the radiation from the scan mirror to M1, while folding mirrors FM2 and FM3 guide the light toward the two spectrometers. This kind of configuration guarantees a good optical quality inside the whole FOV (Guerri et al. 2018).

In order to avoid any defocus due to the large range of operative temperature and the excursion between the room temperature (at which the telescope will be aligned) and the operative temperature, the material of all the mirrors of the telescope is aluminum RSA 6061, which has the same coefficient of thermal expansion (CTE) as the material of the optical bench (Al6061). The only

TABLE 2.32
Performance Parameters of MAJIS Hyperspectral Imager

Parameter	VNIR	IR
Platform	JUICE spacecraft	
Orbit Altitude	500 km orbiting Ganymede	
Scanning type	Pushbroom, slit scan	
Swath width	30 km on Ganymede	
Ground sampling distance	75 m on Ganymede	
Number of pixels in cross-track	400	
Telescope	Three-mirror anastigmat (TMA)	
FOV	3.4°	
IFOV	150 µrad	
Spatial resolution @FWHM	≤225 µrad	
Telescope focal length	24 cm	
Telescope aperture	7.5 cm	
Telescope f/#	3.2	
Spectrometer	Schmidt off-axis collimator	
Spectral dispersive element	Grating	
Wavelength range	0.50–2.35 µm	2.25–5.54 µm
Spectral sampling interval (SSI)	3.6 nm	6.5 nm
Spectral resolution @FWHM	≤5.5 nm	≤10.0 nm
Number of spectral bands	508	508
Detector array material	HgCdTe	HgCdTe
Detector array format	400 × 508 pixels	400 × 508 pixels
Detector pitch size	36 µm × 36 µm	36 µm × 36 µm
SNR	>100:1	>100:1
Spectral distortion (smile)	1.0 pixel	1.0 pixel
Spatial distortion (keystone)	1.0 pixel	1.0 pixel
Stray light	≤1.0% (out-of-field), ≤0.5% (in-field)	
Scan angle of the scan mirror	±4° (in the direction perpendicular to the slit)	

exception is the scan unit that mounts a flat mirror in beryllium. The combination between the beryllium CTE and the CTE of the scan axis allows the mirror to remain blocked at room temperature and be free to rotate at cryogenic temperatures, ensuring the maximum stability of the mechanism during the launch.

The scan mirror is equipped to scan the line of sight in a direction perpendicular to the slit at operative scan angle ±4° from boresight to image a fixed target or to increase the dwell time on a moving target (i.e., ground motion compensation). The maximum excursion by the scan unit is ±19° to allow also the rotation necessary to direct the optical axis in the direction of the internal calibration unit (ICU).

The ICU is a subsystem integrated inside the baffle designed to perform in-flight calibration during the entire operative life of the instrument. It foresees two different calibrated sources: an incandescent lamp for the VNIR spectrometer and a blackbody that illuminates a common diffuser for the IR spectrometer.

Figure 2.22 shows the optical layout of the MAJIS spectrometers. The entrance slit of size 14.4 mm × 36 µm is placed on the telescope focal plane, which defines the MAJIS instrument FOV. Consequently, the light is collimated by a Schmidt off-axis collimator with a specular corrector plate placed in correspondence of its pupil.

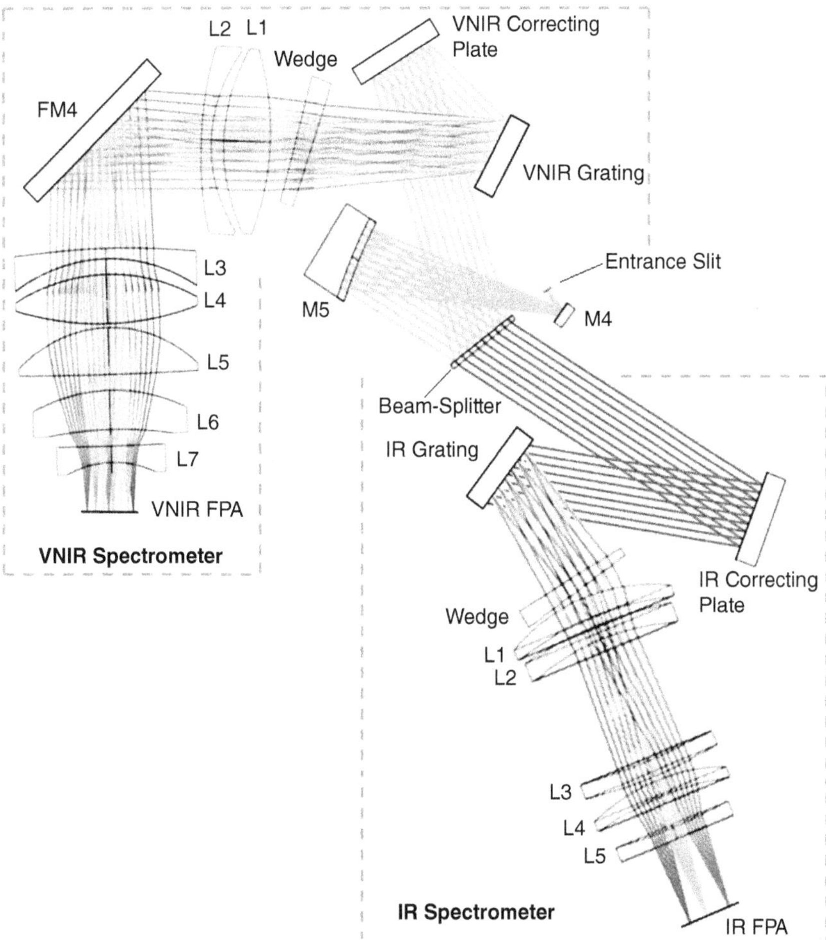

FIGURE 2.22 Optical layout of MAJIS spectrometers.

As shown in Figure 2.22, the light separation for the VNIR and IR spectrometers is implemented by a dichroic beam-splitter inserted between the collimator primary mirror (M5) and the IR correcting plate to permit the adjustment of a different corrector plate for each spectrometer in terms of aspheric coefficients. This solution guarantees a very good correction for the spherical aberration and the coma.

Same as the telescope mirrors, all collimator mirrors are made in aluminum RSA 6061 for the same CTE. During the collimator alignment only, the primary mirror M5 is expected to be adjusted, while all the other elements are mounted at mechanical tolerances. In each spectrometer channel, the collimated light is then reflected by a flat ruled grating that disperses the light, and finally it crosses a completely dioptric objective. A wedge on the top of each objective compensates the pupil distortion introduced by the grating. Having both objectives the same focal length of the collimator and f/#, the two spectrometers share the same 1× magnification.

The two gratings have different groove densities (85.9 grooves/mm for the VNIR, 49 grooves/mm for the IR), because the spectral bandwidths of their spectrometers to be dispersed are 3.6 nm and 6.5 nm. Their profiles are optimized to maximize the efficiency of the first order: higher orders are rejected by an order-sorting filter placed in front of the detector array. In particular, the order-sorting filter for the IR spectrometer is a band-pass filter (able to suppress also part of the thermal background) and the order-sorting filter for the VNIR spectrometer is a high-pass filter. The zero order is suppressed by the dedicated diaphragms along the optical path inside the barrels.

As depicted in Figure 2.22, the VNIR objective consists of one wedge and seven lenses, two of them (L2 and L3) with one aspherical surface. The wedge compensates the pupil distortion introduced by the grating. A folding mirror (FM4) is used after the first two lenses (L1 and L2) to allow detector array to be mounted at a location which minimizes cable length to the external connector. The whole objective (also including the wedge and folding mirror) is mounted inside a barrel in titanium. The folding mirror FM4 is in BK7, since its CTE is closer to the titanium CTE than aluminum RSA 6061.

The IR objective consists of five spherical lenses, four in Silicon and one in Germanium, as shown in Figure 2.22. Same as for the VNIR channel, pupil distortion is kept under control using a wedge, while grating's keystone is corrected by introducing the right amount of lateral color in the optical design.

As reported by Guerri et al. (2018), the experiments results show that design performances of the instrument meet requirements with enough margin for manufacturing, mounting, alignment, and environmental effects.

REFERENCES

Adriani, A., et al. 2017. JIRAM, the Jovian infrared auroral mapper. *Space Science Reviews* 213(1–4):393–446.

Barnsley, M. J., et al. 2004. The PROBA/CHRIS mission. A low-cost smallsat for hyperspectral, multi-angle, observations of the earth surface and atmosphere. *IEEE Transactions on Geosciences and Remote Sensing* 42:1512–1520.

Bézy, J.-L., J.-P. M. Huot, S. M. Delwart, L. Bourg, R. Bessudo, and Y. Delclaud. 2016. Medium resolution imaging spectrometer for ocean colour onboard ENVISAT. In *Optical Payloads for Space Missions*, ed. S.-E. Qian, 91–120. West Sussex, UK: John Wiley & Sons.

Candela, L., R. Formaro, R. Guarini, R. Loizzo, F. Longo, and G. Varacalli. 2016. The PRISMA mission. *Proceedings of IEEE Geoscience & Remote Sensing Symposium (IGARSS2016)* 2016:253–256.

Carbary, J. F. 1994. Ultraviolet and visible imaging and spectrographic imaging instrument. *Applied Optics* 33(19):4201–4213.

Contocello, S. S., et al. 2016. Hyperspectral imaging for real time land and vegetation inspection. *Proceedings of Small Satellites, System & Services Symposium* (4S Symposium 2016).

Coradini, A., et al. 2007. VIRTIS: An imaging spectrometer for the Rosetta mission. *Space Science Reviews* 128(1–4):529–559.

Cutter, A. M. 2000. Compact high resolution imaging spectrometer (CHRIS). *Proceedings of RTO MP-61 Symposium on Space-Based Observation Technology*, 16–18 October 2000.Island of Samos, Greece.

Dai, S. W., J. Wu, H. X. Sun, B. M. Zhang, J. F. Yang, G. Y. Fang, J. Y. Wang, H. Y. Wang, and J. S. An. 2014. Chang'E-3 lunar rover's scientific payloads. *Chinese Journal of Space Science* 34(3):332–340.

De Long, R. K., J. Marmo, D. P. Bowler, and T. E. Romesser. 1995. Airborne and satellite imaging spectrometer development at TRW. *Proceedings of SPIE international Symposium on Aerospace/Defense Sensing & Control and Dual-Use Photonics* 2480:287–294.

De Sanctis, M. C., et al. 2011. The VIR spectrometer. *Space Science Reviews* 163(1–4):329–369.

Donlon, C., et al. 2012. The global monitoring for environment and security (GMES) sentinel-3 mission. *Remote Sensing of Environment* 120:37–57.

Esposito, M., S. S. Conticello, M. Pastena, and B. Carnicero Domínguez. 2019. In-orbit demonstration of artificial intelligence applied to hyperspectral and thermal sensing from space. *Proceedings of SPIE* 11131:C1–C9.

Esposito, M. and A. Z. Marchi. 2018. In-orbit demonstration of the first hyperspectral imager for nanosatellites. *Proc. of Int. Conference on Space Optics*, Chania, Greece.

Germain, S. 2016. GHGSat—Global Emissions Monitoring GHGSat, Inc., January 2016 (available at https://3c.gmx.net/mail/client/attachment/download/tatt0_1—tmai14429e73031cba9f/;jsessionid=F156C9863C2317CE6D8886D39CA98B96-n4.bs51b).

Gortl, P. and J. P. Huot. 2003. Overview of the Envisat MERIS and AATSR data quality, calibration and validation program. *Proceedings of IEEE Geoscience and Remote Sensing Symposium (IGARSS '03)*3:1588–1590.

Green, R. O., et al. 2011. The moon mineralogy mapper (M3) imaging spectrometer for lunar science: Instrument description, calibration, on-orbit measurements, science data calibration and on-orbit validation. *Journal of Geophysical Research* 116(E00G19):1–31.

Guerri, I., et al. 2018. The optical design of the MAJIS instrument on board of the JUICE Mission. *Proceedings of SPIE* 10690(1L):1–21.

He, Z. P., R. Shu, and J. Y. Wang. 2011. Imaging spectrometer based on AOTF and its prospects in deep-space exploration application. *Proceedings of SPIE International Symposium on Photoelectronic Detection and Imaging* 8196: 819625-1–819625-7.

Kappus, M. E., et al. 2016. Hyperspectral imager for the coastal ocean on the international space station. In *Optical Payloads for Space Missions*, ed. S.-E. Qian, 29–49. West Sussex UK: John Wiley & Sons.

Kaufmann, H., et al. 2016. Environmental mapping and analysis program—A German hyperspectral mission. In *Optical Payloads for Space Missions*, ed. S.-E. Qian, 161–181. West Sussex UK: John Wiley & Sons.

Kouyama, T., Y. Ishihara, R. Nakamura, S. Tsuchida, T. Matsunaga, F. Sakuma, Y. Yokota, and S. Yamamoto. 2014. Comparison of A New Lunar Radiometric Model based on SELENE/SP with Satellite Observing Lunar Images. *45th Lunar and Planetary Science Conference* 1302.pdf.

Kumar, A. S. K. and D. R. M. Samudraiah. 2016. Hyperspectral imager onboard Indian mini satellite-1. In *Optical Payloads for Space Missions*, ed. S.-E. Qian, 141–160. West Sussex UK: John Wiley & Sons.

Langevin, Y., et al. 2014. The MAJIS VIS-NIR imaging spectrometer for the JUICE mission. *Proceedings of 45th Lunar and Planetary Science Conference*, 17–21 March 2014, 1777:2493. The Woodlands, TX.

Lewis. 2014. Lewis (SSTI 1) (available at http://space.skyrocket.de/doc_sdat/lewis.htm, accessed on December 5, 2018).

Liu, Y. 2018. Visible-shortwave infrared hyperspectral imager of GF-5 satellite. *Spacecraft Recovery & Remote Sensing* 39(3):25–28 (in Chinese).

Lockwood, R. B. et al. 2006. Advanced responsive tactically-effective military imaging spectrometer (ARTEMIS) design. *Proceedings of IEEE Geoscience and Remote Sensing Symposium IGARSS* 2006: 1628–1630.

Lucke, R. L., et al. 2011. Hyperspectral imager for the coastal ocean: Instrument description and first images. *Applied Optics* 50:1501–1516.

Matsunaga, T., et al. 2016. Hyperspectral imager suite (HISUI): Japanese spaceborne hyperspectral imager for resource and environmental mapping. In *Optical Payloads for Space Missions*, ed. S.-E. Qian, 215–222. West Sussex UK: John Wiley & Sons.

Matsunaga, T., et al. 2017. Current status of hyperspectral imager suite (HISUI) onboard international space station (ISS). *Proceedings of IEEE International Geoscience and Remote Sensing Symposium (IGARSS 2017)* (available at https://ieeexplore.ieee.org/abstract/document/8126989).

Meini, M., F. Battazza, R. Formaro, M. Dami, E. Fossati, and L. Giunti. 2012. Hyperspectral Payload for PRISMA Mission. *Small Satellites Systems and Services—The 4S Symposium, Portoroz, Slovenia*, 4–8 June 2012.

Meini, M., A. Bini, L. Giunti, E. Fossati, and R. Formaro. 2016. Hyperspectral Payload for Italian PRISMA Programme. In *Optical Payloads for Space Missions*, ed. S.-E. Qian, 183–213. West Sussex UK: John Wiley & Sons.

Mouroulis, P., R. O. Green, and T. G. Chrien. 2000. Design of pushbroom imaging spectrometers for optimum recovery of spectroscopic and spatial information, *Applied Optics* 39(13):2210–2220.

Muller, R., et al. 2016. The new hyperspectral sensor DESIS on the multi-payload platform MUSES installed on the ISS. *International Archives of the Photogrammetry, Remote Sensing and Spatial Information Sciences* XLI-B1:461–467.

Murchie, S., et al. 2007. Compact reconnaissance imaging spectrometer for Mars (CRISM) on Mars reconnaissance orbiter (MRO). *Journal of Geophysics Research* 112(E05S03):1–57.

Murchie, S., et al. 2009. Compact reconnaissance imaging spectrometer for Mars investigation and data set from the Mars reconnaissance orbiter's primary science phase. *Journal of Geophysics Research* 114(E00D07):1–15.

Nieke, J., P. Gory, C. Donlon, and B. Berruti. 2010. GMES sentinel-3: Mission overview and aspects of the Calval Plan for its optical payload. *Proceedings of SPIE* 7862: 786203-1–786203-10.

Nieke, J., C. Mavrocordatos, C. Donlon, B. Berruti, T. Garnier, J.-B. Riti, and Y. Delclaud. 2016. Ocean and land color imager on sentinel-3. In *Optical Payloads for Space Missions*, ed. S.-E. Qian, 223–245. West Sussex UK: John Wiley & Sons.

Offner, A. 1987. Annular field systems and the future of optical microlithography. *Optical Engineering* 26:294–299.

Paxton, L. J., C.-I. Meng, D. E. Anderson, and G. J. Romick. 1996. MSX—A multiuse space experiment. *Johns Hopkins APL Technical Digest* 17(1) (available at https://www.jhuapl.edu/Content/techdigest/pdf/V17-N01/17-01-Paxton.pdf).

Pearlman, J., P. Barry, C. Segal, J. Shepanski, D. Beiso and S. Carman. Hyperion, a space-based imaging spectrometer. 2003. *IEEE Transactions on Geoscience and Remote Sensing* 41(6):1160–1173.

Pearlman, J., S. Carman, C. Segal, P. Jarecke, P. Barry, and W. Browne. 2001. Overview of the Hyperion imaging spectrometer for the NASA EO-1 mission. *Proceedings of IEEE IGARSS* 6:3504–3506.

Perkins, R., D. Krutz, R. Mueller, E. Carmona, R. Griffin, L. Graham, R. Miller. 2016. Teledyne's Multi-User System for Earth Sensing (MUSES). *Joint Agency Commercial Imagery Evaluation (JACIE) Workshop*, Fort Worth, Texas, USA, 11–15 April (available at https://calval.cr.usgs.gov/wordpress/wp- content/uploads/2016_JACIE-TDY_MUSES-3.pdf).

Piccioni, G., S. Amici, S. Fonti, A. Coradini, and M. Dami. 2000. Efficiency measurements of the VIRTIS-M grating. *Planetary and Space Science* 48(5):411–417.

Piccioni, G., et al. 2007. VIRTIS: The visible and infrared thermal imaging spectrometer. *ESA-SP-1295, ESA Publications Division* SP-1295, Noordwijk, The Netherlands.

Praks, J., A. Kestila, M. Hallikainen, H. Saari, J. Antila, P. Janhunen, and R. Vainio. 2011. Aalto-1—An experimental nanosatellite for hyperspectral remote sensing. *Proceedings of IEEE International Geoscience and Remote Sensing Symposium (IGARSS)*:4367–4370.

Praks, J., A. Kestilä, T. Tikka, H. Leppinen, O. Khurshid, and M. Hallikainen. 2015. Aalto-1 earth observation cubesat mission—Educational outcomes. *Proceedings of IEEE International Geoscience and Remote Sensing Symposium IGARSS'15* (available at https://ieeexplore.ieee.org/document/7326023).

Rafert, J. B., R.G. Sellar, and J. H. Blatt. 1995. Monolithic Fourier transform imaging spectrometer. *Applied Optics* 34(31):7228–7230.

Rast, M. and J. L. Bezy. 1990. ESA's medium resolution imaging spectrometer (MERIS): Mission, system and applications. *Proceedings of SPIE* 1298:114–126.

Russell, C. T., et al. 2007. Dawn mission to Vesta and Ceres. *Earth Moon Planet* 101:65–91.

Sang, B., et al. 2008. The EnMAP hyperspectral imaging spectrometer: Instrument concept, calibration and technologies. *Proceedings of SPIE* 7086: 708605-1–708605-15.

Smith, W. H. and P. D. Hammer. 1996. Digital array scanned interferometer: Sensors and results. *Applied Optics* 35(16):2902–2909.

Straight, S. D., et al. 2010. Tactical satellite-3 mission overview and initial lessons learned. Proceedings of the 24th Annual AIAA/USU Conference on Small Satellites (available at https://www.semanticscholar.org/paper/Tactical-Satellite-3-Mission-Overview-and-Initial-Straight-Doolittle/e796b69aa76ec1103808d910b5d20b0b5fdf2f70).

Stuffler, T., K. Förster, S. Hofer, M. Leipold, B. Sang, H. Kaufmann, B. Penné, A. Mueller, and C. Chlebek. 2009. Hyperspectral imaging—An advanced instrument concept for the EnMAP mission (Environmental Mapping and Analysis Program). *Acta Astronautica* 65:1107–1112.

Wang, J.-Y., et al. 2016. Visible and near-infrared imaging spectrometer aboard Chinese Chang'E 3 spacecraft. In *Optical Payloads for Space Missions*, ed. S.-E. Qian, 121–139. West Sussex UK: John Wiley & Sons.

Wilson, D., P. Maker, R. Muller, P. Mouroulis, and J. Backlund. 2003. Recent advances in blazed grating fabrication by electron-beam lithography. *Proceedings of SPIE* 5173:115–126.

Xiong, X. X., B. N. Wenny, and W. L. Barnesc. 2009. Overview of NASA earth observing systems Terra and Aqua moderate resolution imaging spectroradiometer instrument calibration algorithms and on-orbit performance. *Journal of Applied Remote Sensing* 3:032501.

Xiong, X.-X., et al. 2016. Moderate resolution imaging spectroradiometer on Terra and Aqua missions. In *Optical Payloads for Space Missions*, ed. S.-E. Qian, 53–90. West Sussex UK: John Wiley & Sons.

Yamamoto, H., R. Nakamura, and S. Tsuchida. 2012. Radiometric calibration plan for the hyperspectral imager suite (HISUI) instruments. *Proceedings of SPIE* 8527:85270V-1–85270V-11.

Ye, P. J. and J. Peng. 2006. Deep space exploration and its prospects in China. *Engineering Sciences* 8(10):13–18.

Zee, R. E. 2016. Space flight laboratory: Disruptively capable smaller satellites. *Proceedings of Canadian SmallSat Symposium*, Toronto, Canada, 2–3 February, 2016.

Zhao, X., Z. Q. Xiao, Q. Kang, Q. Li, and L. Fang. 2010. Overview of the Fourier transform hyperspectral imager (HSI) boarded on HJ-1A satellite. *Proceedings of IEEE Geoscience & Remote Sensing Symposium IGARSS* 2010:4272–4274.

3 Overview of Applications of Hyperspectral Satellites in Earth Observations

3.1 REMOTE SENSING APPLICATIONS FROM MULTISPECTRAL TO HYPERSPECTRAL IMAGING

Earth observation satellites have formed the basis for acquiring data remotely for many years and are viewed as a time- and cost-effective way to undertake large-scale monitoring (Okin et al. 2001) of the Earth, which can be used to determine the quality, quantity, and geographic distribution of its resources. Multispectral imagery has been used as the data source for remote sensing from airborne and satellite systems since the 1960s (Landgrebe 1999).

Multispectral sensors collect data generally in 4–10 broad spectral bands in a single observation from the visible and near-infrared up to the infrared region of the electromagnetic spectrum, as shown in Figure 3.1. Detailed spectral signatures of objects are averaged out due to the wide bandwidth. These broad spectral bands of the reflected and emitted energy from the Earth are the primary limiting factor of multispectral sensors. Over the past three decades, advances in sensor technology have overcome this limitation of Earth observation systems with the development of hyperspectral sensor technologies. Hyperspectral systems have made it possible for the collection of tens up to hundreds fine spectral bands within an imagery, thus producing many more detailed spectral data.

Remotely sensed images are characterized by their spatial and spectral resolutions. The terms spatial and spectral resolution represent pixels of an image displayed in a geometric relationship to one another and variations within pixels as a function of wavelength.

Spectral resolution refers to the width of the portions of the electromagnetic spectrum measured by an imaging sensor. A sensor may be sensitive to a large range of the electromagnetic spectrum, but have poor spectral resolution if it captures a small number of wide bands. A sensor that is sensitive to the same range of the electromagnetic spectrum, but captures many narrowbands within that range, would have greater spectral resolution. The objective of fine spectral sampling is to enable the human or computer, to distinguish between elements in a scene. Detailed information about how individual elements in a scene reflect or emit electromagnetic energy increases the probability of identifying unique characteristics for a given element, which allows for better distinction from other elements in the scene. Multispectral remote sensing sensors record energy over several separate broad spectral bands at various spectral resolutions. Hyperspectral sensors acquire hundreds of contiguous narrow spectral bands throughout the visible and near-infrared (VNIR), to shortwave-infrared (SWIR) range of the electromagnetic spectrum. A distinct advantage of their very high-spectral resolution facilitates fine discrimination between different targets based on their spectral response in each of the narrowbands.

Spatial resolution defines the level of spatial detail depicted in an image. This may be described as a measure of the smallness of objects on the ground that may be distinguished as separate entities in the image, with the smallest object necessarily being larger than a single pixel of the sensor. In this sense, spatial resolution is directly related to pixel size of an image. The spatial property of an image is a function of the design of the sensor in terms of its ground sampling distance (GSD), field-of-view (FOV), and the altitude at which the sensor operates above the surface. Each of the detector

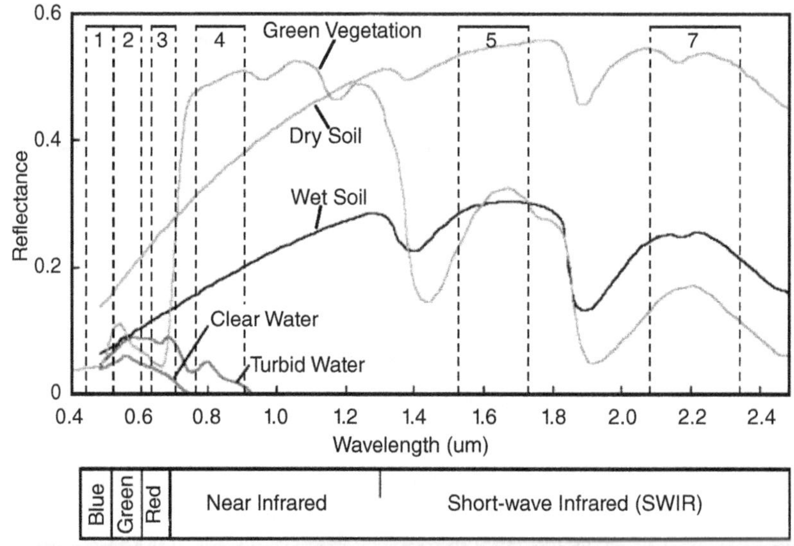

FIGURE 3.1 Reflectance spectra of three popular surface materials (vegetation, soil, and water) measured by a hyperspectral sensor in comparison with the broad multispectral bands of Landsat-7.

elements of an imaging sensor measures the energy received from a finite patch on the ground surface. The smaller the individual patches are, the higher the spatial resolution and the more detail one can spatially interpret from the image. The GSD of the current spaceborne hyperspectral sensors is typically in the order of 20–30 m, whereas airborne hyperspectral sensors generally have a higher spatial resolution usually in the order of 5 m or finer.

Understanding of spectral signatures is essential in the interpretation of a remotely sensed image. Different materials are discriminated by wavelength-dependent absorptions and the reflected solar energy of the materials, called spectral reflectance, is known as the spectral signature. It is a ratio of the reflected energy to incident energy as a function of wavelength. The graph of the spectral reflectance of an object as a function of wavelength is termed as the spectral reflectance curve (spectrum). Spectral signatures obtained from multispectral images are discrete rather than contiguous. Contiguous spectral signatures obtained from hyperspectral images allow for the detection of surface materials through detailed analysis of their spectral signature and abundances, as well as inferences of biological and chemical processes.

Figure 3.1 shows the reflectance spectra of three popular surface materials on Earth measured by a hyperspectral sensor: green vegetation, bare soil (dry and wet), and water body (clear and turbid). For the sake of comparison, the six multispectral bands of Landsat-7 in the same wavelength range are also shown (Qian 2013). For the green vegetation, the absorptions (valleys) in the visible region of the spectrum are determined by the pigmentation of the plant. For example, chlorophyll absorbs strongly in the blue (0.45 μm) and red (0.67 μm) regions, also known as the chlorophyll absorption bands. Chlorophyll is the primary photosynthetic pigment in green plants. There is a strong absorption of the red and blue wavelengths and a strong reflection of the green wavelengths. This is the reason for the human eye perceiving healthy vegetation as green. When the plant is subjected to stress that hinders normal growth and chlorophyll production, there is less adsorption in the red and blue regions and the amount of reflection in the red waveband increases.

The spectral reflectance signature illustrates a dramatic increase in the reflection for healthy vegetation at around 0.70 μm, which is referred to as red-edge. In near-infrared (NIR) between 0.70 μm and 1.30 μm, a plant leaf typically reflects between 40% and 50% of incident radiation, the rest of the radiation is transmitted, with only about 5% being adsorbed. Structural variability in leaves in this range allows human to differentiate between species, even though they might look the same in the visible region (Lillesand and Kiefer 1999). Beyond 1.30 μm the incident energy on the vegetation is largely absorbed or reflected with very little transmittance of energy. Two strong water absorption bands are noted at 1.40 μm and 1.90 μm within this spectral range.

The spectrum for bare soil shows far less variation in reflectance compared to that of green vegetation. The variation of the soil reflectance is mainly due to moisture content, soil texture, surface roughness, presence of minerals, and organic matter content. The dry soil reflects more than the wet soil, but the absorption bands are located at the same wavelengths. The influence soil reflectance is complex, variable, and interrelated.

The most distinctive characteristic of clear water is the absorption in the NIR and beyond. There are various conditions of water bodies that manifest themselves in the visible wavelengths. There are complex energy-matter interactions at these wavelengths that depend on a number of factors. These factors include the interaction with the water surface and material suspended in the water. Clear water reflects the greatest at wavelengths around 0.60 μm. The presence of organic and inorganic materials greatly influences the transmittance of the water and, therefore, the reflectance is dramatically affected. For example, a water body with high amounts of suspended sediments reflects more than clear water. An increase in chlorophyll decreases the reflectance in the blue wavelengths and increases in the green. This characteristic can be useful in the detection of algae in water using remote sensing (Lillesand and Kiefer 1999).

The ability of hyperspectral imaging to provide detailed spectral information for identifying materials makes it ideal for a wide array of Earth observation applications, such as agriculture, forestry, environment and resource management, marine, coastal and inland waters, geology and mineral exploration, defense and security. Its application has now spread into fields, such as widespread as biology, medical and law enforcement, as well as historical manuscript research, such as the imaging of the Archimedes Palimpsest. This technology is continually becoming more available to the public.

In this chapter, the hyperspectral applications covered are not a complete set, rather a selected list to illustrate the power and potential of hyperspectral imaging. In addition, the sections describing the individual applications may not cover the whole spectrum within the individual application area.

3.2 AGRICULTURE AND PRECISION FARMING

The spectral signatures of vegetation in general exhibits several characteristic features, namely, the green peak, the chlorophyll well, the red-edge, the NIR plateau, and water absorption features (Zhang et al. 2002). Hyperspectral measurement can provide detailed spectral information through across numerous contiguous narrow spectral bands to facilitate detailed analysis of crop classification, monitoring plant diseases, pests and invasive plant species, estimation of crop yield, quantifying soil property variability, etc.

Precision agriculture is an excellent application of improving crop yields and assisting management decisions using remotely sensed technology and analysis tools (NRC 1997). Hyperspectral remote sensing can play an important role in crop diagnosis and soil physical and chemical characteristics. The combination of hyperspectral remote sensing imagery and local modeling and validation can improve farmers' efficiency if the soil and plant maps are provided within a short time period for some applications such as for irrigation purposes. For example, prediction of yield, plant stress, nutrient deficiencies, and weed and insect infestations require timely applications of water, fertilizers, and pesticides. Combined with meteorological data, remote sensing on local and regional levels can improve forecasting of the response of crops to management practices (Whiting et al. 2006).

The goal of precision agriculture is to assist farmers in the management of their business. The effective use of hyperspectral imagery in precision agriculture not only helps farmers to effectively use resources, such as herbicides and pesticides, but it also provides insight into the current stage of crop development and health. For example, hyperspectral remote sensing can provide insight into the existence of weeds in a small and localized region of a wheat field. The traditional farmer's approach to this problem would be to spray the entire field with herbicides ensuring the weed is eradicated from the field. However, with the data from the hyperspectral analysis in hand, the farmer could apply the herbicide to a localized region of the field saving a great deal of time and money.

3.2.1 Crop Stress Mitigation and Site Specific Management

All crops require adequate macro- and micronutrients to grow and produce (yield). Plants draw upon the nutrients available in the soil to satisfy this need. Growth and development of crops can be adversely affected by deficiencies or excesses of any one of a number of key nutrients. Knowledge of the nutrient status of the soil and the crop is required to mitigate these effects.

In an ideal environment, crops would be healthy throughout their life cycle and provide the maximum yield. In reality, however, shortages of essential nutrients and water or external pressures create stress on the crops and reduce their potential yield. Farmers attempt to mitigate these effects by applying crop inputs, such as fertilizer and herbicides, at the appropriate time and place. Information that can improve farm management is highly desirable. Hyperspectral data is applicable in the following four situations:

1. Nutrient deficiencies
2. Water availability
3. Weed mapping
4. Mapping of crop disease and insect damage

3.2.1.1 Nutrient Deficiencies

Nitrogen (N) is the most important nutrient controlling canopy development, and an essential constituent of cell walls, cytoplasmic proteins, nucleic acids, and chlorophyll, in addition to many other components. Therefore, it plays an important role in the rate of leaf expansion, final leaf size and longevity, tiller or branch formation and survival, and ultimately productivity.

Leaf chlorophyll content is one of the parameters first affected by nitrogen stress and observable by remote sensing. A quantitative knowledge of both the intensity and the location of pigment loss is extremely valuable information for both the real-time management of inputs as well as for the a posteriori interpretation of production factors that can explain crop productivity limitations.

The challenge, from a remote sensing point-of-view, is to isolate the effects of leaf chlorophyll pigment content on the observed spectral reflectance from those due to other canopy attributes such as plant biomass and soil, at crop stages where the canopy is not closed. For that purpose, the hyperspectral approach has the potential to isolate chlorophyll from other factors, which currently cannot be achieved using broadband multispectral technologies.

As shown in Figure 3.1 for healthy vegetation, such as crop canopy, chlorophyll *a* and *b* absorb energy in a very narrow range that can be readily measured by a hyperspectral sensor. When the chlorophyll is reduced, due to a lack of nitrogen for instance, a crop's spectral reflectance will differ from a healthy crop.

Figure 3.2 shows a map of the amount of chlorophyll for an agricultural field derived from the data acquired by a hyperspectral imager. With this map a user can determine where problems exist and the magnitude of the problem.

Hu et al. (2004) retrieved crop chlorophyll content and leaf area index (LAI) from hyperspectral data for applications in precision agriculture. The same retrieval process was also applied to decompressed hyperspectral data for the purpose of evaluating the effect of near lossless data compression on remote sensing applications. Two hyperspectral data sets acquired by a CASI imaging spectrometer were used. The first hyperspectral data set was acquired over the corn fields at the L'Acadie experimental farm (Agriculture and Agri-Food Canada), Saint Jean-sur-Richelieu, Quebec, during the summer of 2000. The second data set was acquired over corn, soybean, and wheat fields at

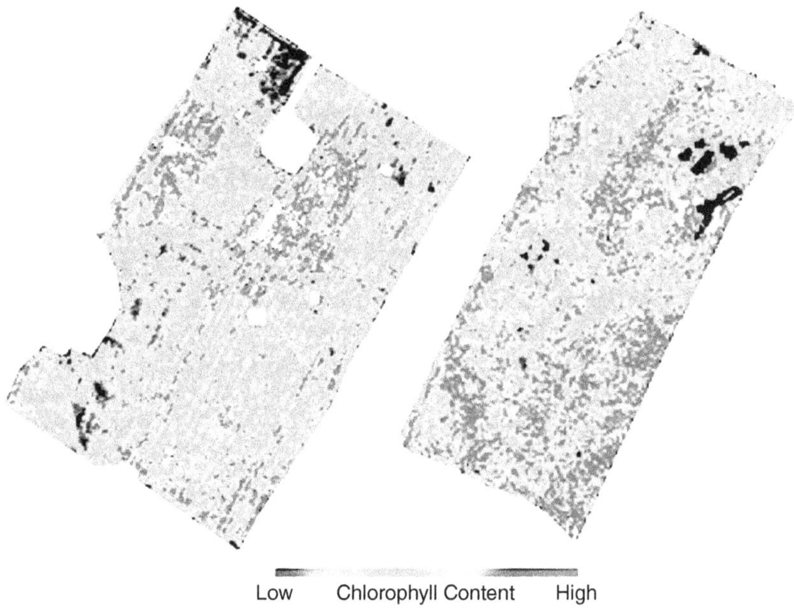

FIGURE 3.2 Maps of chlorophyll content for a bean field (left) and a corn field (right) derived from hyperspectral Probe 1 data. (Courtesy of Canada Centre for Remote Sensing, Natural Resources Canada.)

the former Greenbelt farm (Agriculture and Agri-Food Canada), Ottawa, in three intensive field campaigns (IFC) during the summer of 2001. The two CASI data sets have spectral bands with a spectral resolution of 7.6 nm in the range of 408–947 nm and spatial resolution of 2 m × 2 m.

Figure 3.3a shows the color composite image of the first CASI hyperspectral data set with band 53 (775 nm) printed in red, band 33 (631 nm) in green, and band 20 (538 nm) in blue. The study

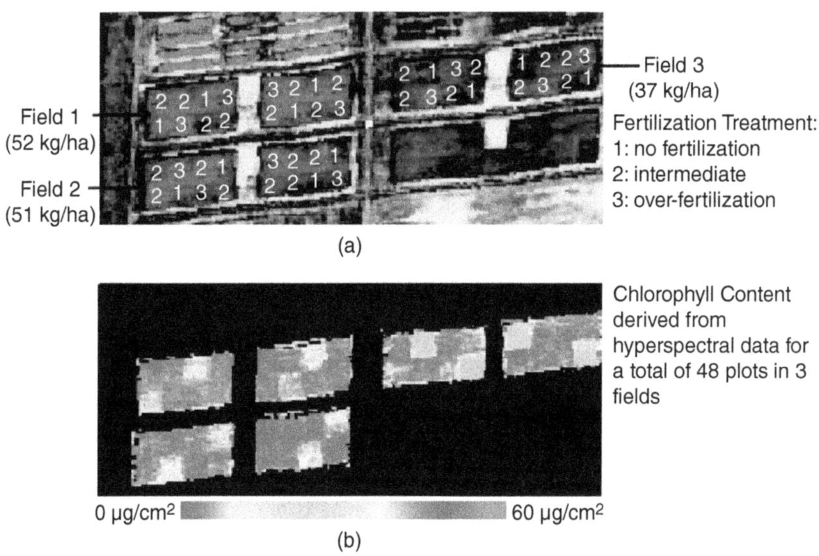

FIGURE 3.3 Chlorophyll content retrieved from hyperspectral data over three corn fields with different background nitrogen content before seeding and three different nitrogen fertilization treatments of 16 plots in each filed. (a) CASI color composite image with band 53 (775 nm) printed in red, band 33 (631 nm) in green, and band 20 (538 nm) in blue; (b) chlorophyll content derived from the hyperspectral dada for 16 plots in each field.

area shown in the figure includes three adjacent corn fields denoted as Field 1, Field 2, and Field 3. In each field, there are two blocks, each containing eight 20 m × 20 m plots. In total, there are 48 plots in the study area. The three fields had various cropping histories and, thus, the nitrogen content in the soil varied before seeding. There was a larger amount of nitrogen in the soils in Fields 1 and 2 (52 kg/ha and 51 kg/ha, respectively) than in Field 3 (37 kg/ha). Three nitrogen fertilization treatments were supplied to the study area: (1) no fertilization, (2) intermediate fertilization, and (3) over-fertilization.

Figure 3.3b shows chlorophyll content derived from the hyperspectral data for 48 plots in the 3 fields (16 plots in each field). The difference in the nitrogen content in the soil before seeding among the fields and the nitrogen fertilization treatments applied to the plots led to the differences in the growth of the corn crop among plots. The corn canopies growing on the plots with more nitrogen content tend to show increased vigor and to be more homogeneous compared with those growing on the plots with less nitrogen content. As a result, it can be seen from the figure that in each field, the mean value of chlorophyll content increases and the spatial variation decreases in order from plots with 1: no nitrogen fertilization to 2: intermediate nitrogen fertilization and to 3: nitrogen over-fertilization. It also can be seen that the plots in Field 3 (low nitrogen content in the soil before seeding time) have the lower mean chlorophyll content and higher spatial variations than the plots in Fields 1 and 2 (high nitrogen content in the soil before seeding time) for 1: no nitrogen fertilization.

The LAI retrieved from the second CASI hyperspectral data set acquired over crop fields at the former Greenbelt farm during three field campaigns in the summer of 2001 (IFC-1, June 13; IFC-2, June 26; and IFC-3, July 19) is shown in Figure 3.4. The top row shows color composite images. The three campaigns were planned to coincide with different phenological development stages of the crops. The phenological changes of the crops are clearly shown in these three images. By the first campaign (IFC-1), the crops had started to grow; the corn crop had three to six fully expanded leaves. The second campaign (IFC-2) occurred during the active growing season of the crops; as a result, the IFC-2 image has much more red tone than the IFC-1 image in response to the increased NIR reflectance. During the last campaign (IFC-3), the crops were in the reproductive period of the growing season. The corn canopy had grown 8–13 fully expanded leaves with tassels emerged in productive zones; the soybean canopy had set pods; and the wheat crop reached its maturity dominated by extensive heading. Due to the heads of the wheat canopy, less red tone is observed in the IFC-3 image than in the IFC-2 image of the wheat fields, while more red tone is observed in the IFC-3 image than in the IFC-2 image in both the corn and soybean fields. Some spatial variability of the crops, especially during IFC-1, can also be observed. For instance, the crop in the northeastern portion of the wheat field showed more vigor than the wheat crop in the northwestern portion. The spatial patterns in the wheat field are a result of variations in soil conditions. The wheat field in the northeastern portion is in a productive soil area with low clay and high organic matter content, while the northwestern portion of the field slopes toward the water creek flowing between the wheat and corn fields, with very low organic matter content due to erosion.

The LAI images of the second data set corresponding to three field campaigns are depicted on the bottom row in Figure 3.4. It can be observed that the northeastern area of the wheat field has higher LAI than the northwestern area of the wheat field for IFC-1 and IFC-2; and the reverse trend is observed for IFC3 in relation to wheat maturity. These trends are caused by soil conditions and topography difference between the two areas. By comparing the LAI images, the temporal variation of the LAI can be identified. It can be observed that the LAI values of corn and soybean crops showed the expected increase through the growing season from IFC-1 to IFC-3. Conversely, LAI values of wheat crops, as expected, increased from IFC-1 to IFC-2, then decreased from IFC-2 to IFC- 3.

3.2.1.2 Water Content in Crop Canopy and Soil

Water is an essential factor in plant growth, and the dominant factor in the development of canopy leaf area. Too much or too little moisture results in a progressive decline in the rates of cell and leaf

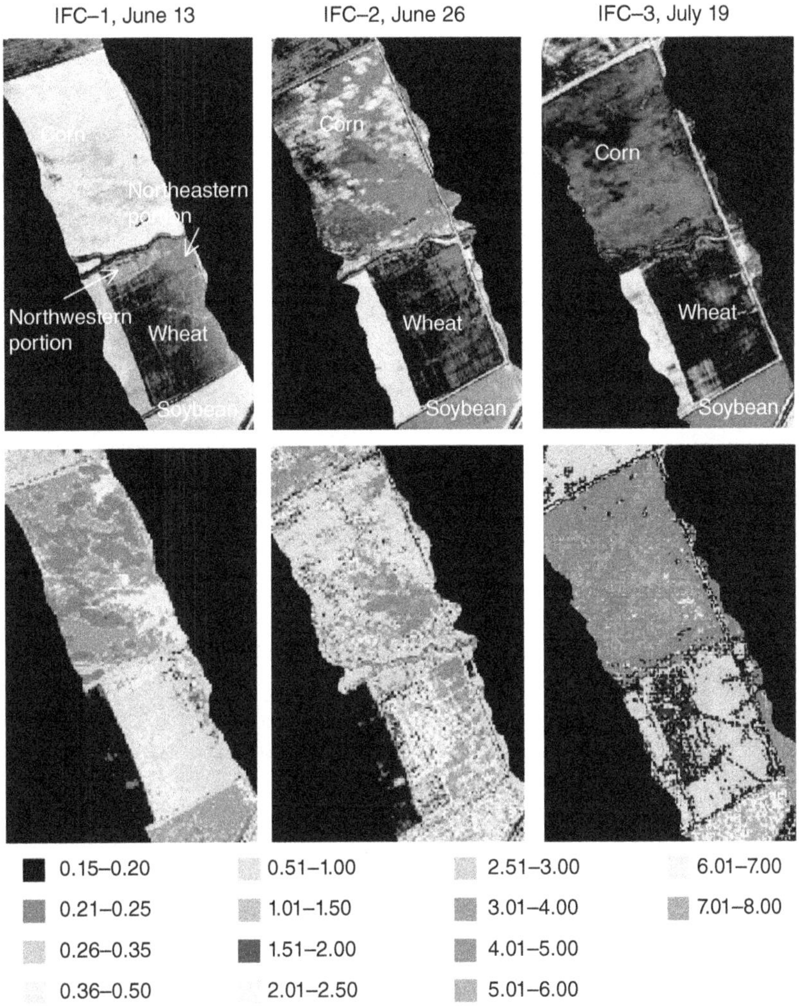

FIGURE 3.4 The hyperspectral images and the LAI images derived from CASI datacubes collected in three intensive field campaigns (IFC) during the crop growing season of the corn, wheat, and soybean. Top row: The color images are composed by band 41 (708 nm) printed in red, band 24 (580 nm) in green, and band 8 (460 nm) in blue; bottom row: LAI images corresponding to three field campaigns.

expansion, resulting in lower leaf area and green leaf duration, limiting total dry matter accumulation and yield potential.

Water is becoming a scarce resource and water use management is important for sustainable agriculture. Irrigation draws a significant amount of water and is used extensively for high-value crops. Knowledge of the water content in the crop canopy and in the soil is an important information for precision agriculture. This can assist in irrigation scheduling and optimize the use of water. In addition, water availability can be used in combination with other parameters measured with a hyperspectral sensor and meteorological stations to predict yield.

The water content in the canopy of crops can be measured using a hyperspectral sensor. Canopy water content (CWC) is the amount of water within the stalks, leaves, and fruit of the crop canopy. Hyperspectral remote sensing has demonstrated great potential for accurate retrieval of CWC. A canopy equivalent water thickness (EWT) can be derived by selecting the 850–1250 nm wavelength range, including the 940 nm and 1130 nm atmospheric water vapor absorption bands and liquid water absorption minima (Champagne et al. 2003). The EWT is functionally equivalent to the

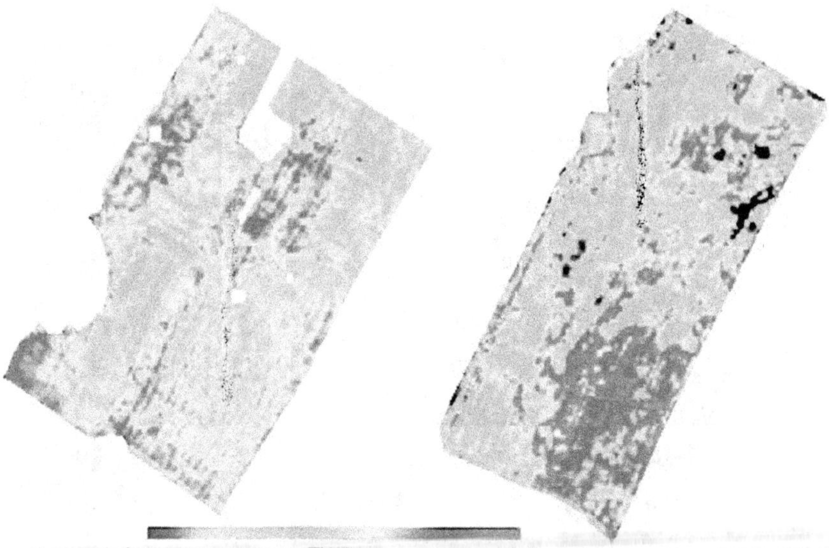

FIGURE 3.5 Canopy equivalent water thickness (EWT) for a bean field (left) and a corn field (right) derived from hyperspectral Probe 1 data. (Courtesy of Canada Centre for Remote Sensing, Natural Resources Canada.)

volume of water in the canopy and is a measure of plant water status. A map of the canopy water over an agriculture field can be readily made by calculating the EWT for each pixel location, as shown in Figure 3.5.

Clevers et al. (2010) reported retrieval of CWC using hyperspectral data, in particular the spectral information provided by the canopy water absorption feature at 970 nm. They estimated and predicted CWC using in situ spectroradiometric measurements and a modeling approach. The relationship of the first derivative at the right slope of the 970 nm water absorption feature with CWC was investigated with a radiative transfer model (referred to as PROSAIL) and tested for field spectroradiometer measurements on two test sites. The first site was a heterogeneous floodplain with natural vegetation like grasses and various shrubs. The second site was an extensively grazed fen meadow. PROSAIL simulations showed a linear relationship between the first derivative over the spectral range 1015–1050 nm and CWC ($R^2 = 0.97$).

3.2.1.3 Weed Detection and Mapping

Weeds have a detrimental effect on the yield of various agricultural crops. They compete with the crop for space, nutrients, water, and light. Thus, weed detection, discrimination, and determination of their abundance are useful information for farmers. Weed infestations are extremely variable in terms of spatial extent of weed patches and in terms of type of infestation. Traditional farming approaches call for a blanket application of herbicides over the entire field. Since weeds tend to grow in patches, a blanket approach has much of the herbicide applied to locations where no weeds are growing. The excessive application of herbicides can contaminate surrounding ecosystems when these chemicals are leached to groundwater or when they are carried into surface waterways in the process of surface runoff and soil erosion. Because weed species vary across a field, the type of herbicide applied should be appropriate for the type of infestation. Site-specific management also means that expenses associated with farm chemicals are reduced and, thus, farm profits can be positively affected.

Many agricultural service providers offer crop-scouting services. These scouts visit or scout client fields periodically throughout the growing season. They watch for infestations of weeds, insects, and fungus and provide recommendations on treating detected infestations. Scouts generally provide hand-written sketches of the general location of the weeds, and detailed information on weed species. The weed species information is used to determine the type of herbicide application required. However, since the weed locations are very generalized, this information is not sufficient for site-specific application of herbicides.

Mapping weeds using hyperspectral imagery prior to crop emergence is relatively straightforward. Weeds can be detected by the presence of biomass when normally there would be bare soil. Once a significant crop cover is present, separating crop plants from weed plants becomes more challenging. Hyperspectral imagery, with a large number of narrow and contiguous bands, is able to exploit small differences in reflectance among weed species, and between weeds and crops. Hyperspectral sensors can uniquely detect and identify specific weed species without the need for ground reference information, because each plant species possesses a unique spectral *fingerprint* based on the unique biochemical makeup of a particular species. Figure 3.6 shows the spectral

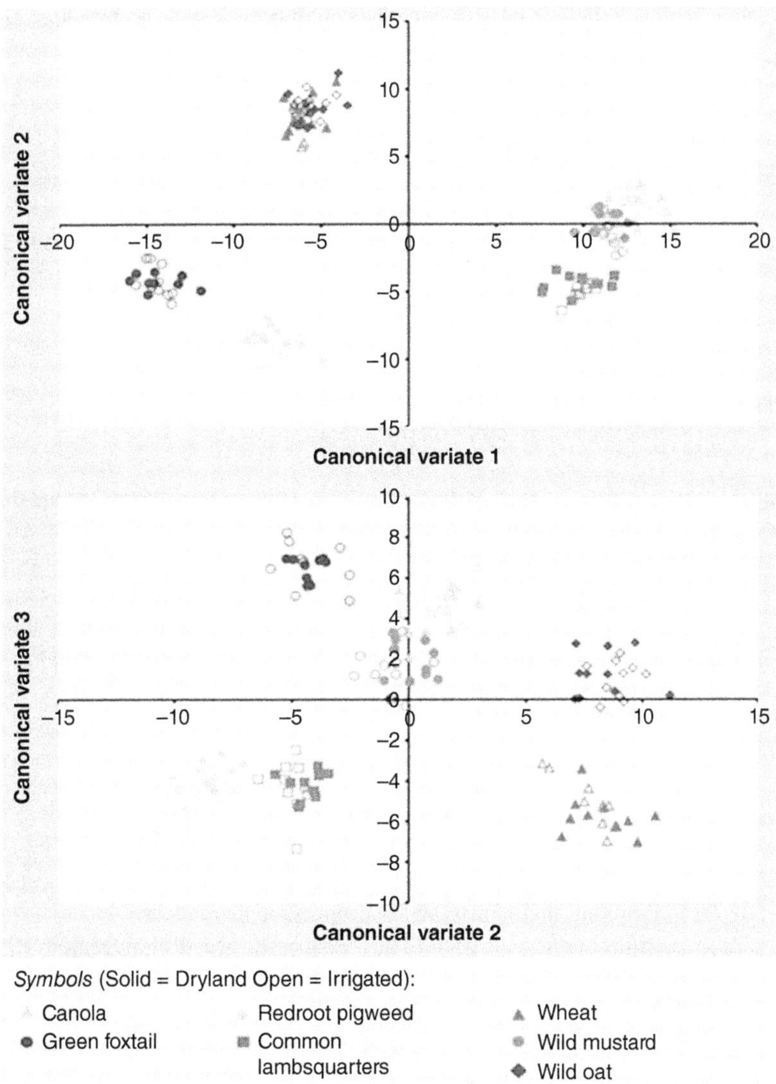

FIGURE 3.6 Spectral separation of different weeds. (Courtesy of Agriculture and Agri-Food Canada.)

separation of different weeds using statistical techniques on hyperspectral data collected with an ASD field spectroradiometer. It is evident that some weeds such as green foxtail, redroot pigweed, and common lamb's-quarters are clearly separable from wheat and canola.

The potential capabilities of hyperspectral remote sensing for plant species discrimination arise directly from its high spectral resolution. The basis for plant species discrimination relies critically on whether different plant species are sufficiently different biochemically and that these differences are manifested in their reflectance spectra. The 0.4–2.5 μm spectra of vegetation contain week absorption bands associated with a very wide range of biochemical properties (lignin, cellulose, etc.) (Curran 1989). These absorption bands are actually associated with specific molecular species, such as CH_2, NH_3, as well as many others, and their wavelength positions, widths, and depths are a function of many factors such as the molecular state, and abundance of the particular functional group. Thus, if the biochemical properties that affect these absorption band properties differ between different plant species, and if the spectral resolution of the hyperspectral sensor is sufficiently high, plant species discrimination should be tractable. The ability to obtain detailed absorption band information suggests that plant species discrimination should be possible using spectral variations such as relative and absolute absorption band depths and widths, and variations in the wavelength positions of specific absorption bands. Once this capability is demonstrated, automated algorithms can be applied to hyperspectral data to derive the necessary products. This level of analysis is not available for multispectral data.

Weed maps identify the location, density, and type of weed. Products (maps) should be referenced for precision farming applications and geo-referenced so that the information can be integrated with other precision agriculture technologies such as GPS and variable rate applicators. Once the areas of weed infestations are identified, a geo-referenced weed map can be used on variable rate sprayers to locate the weed patches and apply the appropriate type and amount of herbicide, resulting in benefits such as:

- More efficient crop scouting
- Reduced herbicide use, reducing costs to the producer
- Lower impact from excess herbicide run off into groundwater

Eddy et al. (2013) reported weed and crop discrimination using hyperspectral image data. A hyperspectral instrument covering a wavelength range of 400–1000 nm at 10 nm spectral sampling interval with 61 spectral bands was used to collect image data over field plots to discriminate weeds (redroot pigweed, wild oats) from crops (canola, wheat, and pea). The hyperspectral instrument was mounted on an extendable boom affixed to a flat-bed truck and centered 1 m above each field plot target with nadir viewing, resulting in a GSD of 1.25 mm × 1.25 mm. Artificial neural network (ANN) was used to classify the crops and weeds from the hyperspectral images. They achieved classification accuracies of 89–95% when the ANN classification applied to the full 61-band hyperspectral datacube.

When the ANN classification is applied to a reduced 7-band dataset of the original datacube derived through principal component analysis (PCA), the classification accuracies of the crop and weeds are between 88% and 94%, which are only 1% less compared to those of the classification on the full 61-band hyperspectral datacube. This large reduction in data volume (54 bands fewer) has significantly less computational overhead, an important factor when envisioning these procedures applied to real-time precision agriculture, site-specific weed management (SSWM) techniques. Figure 3.7 shows weed and crop maps derived from hyperspectral imagery using the ANN classification.

Another study to evaluate the potential of hyperspectral imagery for the detection of dicotyledonous weeds in durum wheat during weeding period was reported by Hadoux et al. (2012). A pushbroom hyperspectral instrument (HySpex VNIR 1600-160) fitted on a tractor-mounted motorized rail. The hyperspectral instrument has a spectral range from 0.4 μm to 1.0 μm with a spectral

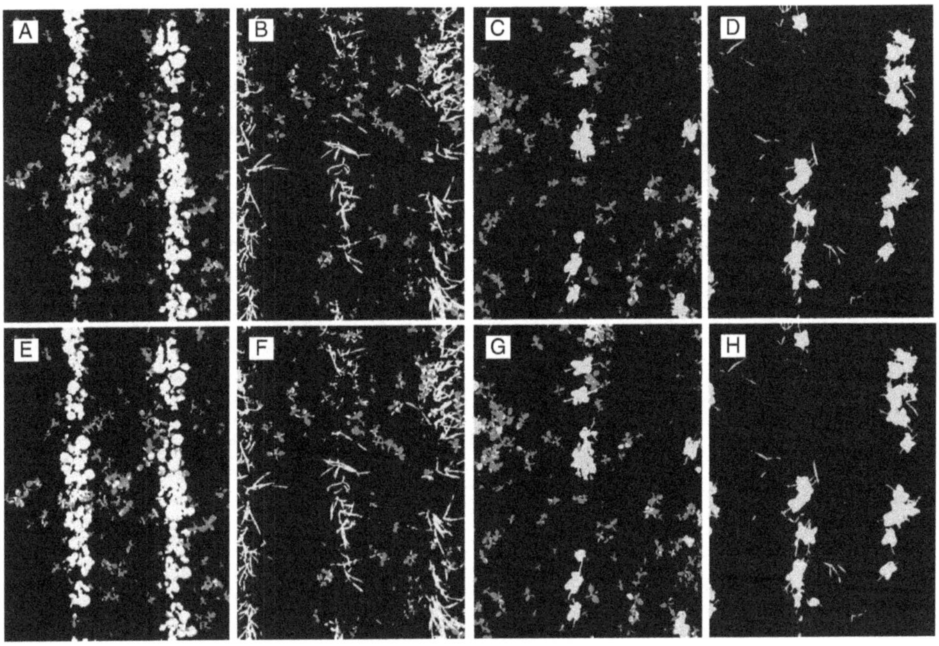

FIGURE 3.7 Weed and crop discrimination using hyperspectral imagery and artificial neural network classification (weeds: redroot pigweed in red, wild oat in orange; crops: canola in yellow, wheat in cyan, pea in green). Images a, b, c, and d are classification maps with classification applied on the full 61-band datacube. Images e, f, g, and h are classification maps with classification applied on a reduced dataset of 7-band derived through principal component analysis. (Reprinted from Eddy et al. 2013, © 2014 Government of Canada.)

resolution of 3.7 nm. The spatial dimension of the instrument has 1600 pixels in the cross-track direction and the spectral dimension has 160 bands. Each image represents about 0.30 m in cross-track direction by 1.50 m in along-track direction acquired at 1 m above the canopy with the lens and the viewing angle being fixed. The spatial resolution in cross-track direction is 0.2 mm, while the spatial resolution along-track direction depends on the motion speed. Spectral discrimination between weeds and crop appears quite efficient (91.6%) with a 8% prediction error on an independent test set.

A study of detection of weed species in crop fields using the combination of spectral, shape, and textural features was reported by Lin et al. (2017). This study used a field imaging spectrometer that supports micro-scale plant feature analysis by acquiring high-resolution hyperspectral images of corn and a number of weed species in a laboratory. The imaging spectrometer has 344 spectral bands covering a spectral range of 446–912 nm. The spectral sampling interval is about 1.4 nm. The object-oriented classification with segmentation and decision-tree algorithms was utilized on the hyperspectral images to extract shape and texture features of eight species (corn plus 7 weeds) of plant leaves, while the spectral identification characteristics of different species were determined through sensitive waveband selection and using vegetation indices calculated from the sensitive band data of the images.

Two categories of weeds, monocotyledonous and dicotyledonous, were studied. Monocotyledonous weeds include foxtail (*Setaria viridis L.*, SETVI) and goose grass (*Eleusine indica L.*, ELEIN). Dicotyledonous weeds include round leaf pharbitis (*Pharbitis purpurea L. Voigt*, PHAPU), lobed leaf pharbitis (*Pharbitis nil L. Choisy*, PHANI), redroot amaranth (*Amaranthus retroflexus L.*, AMARE), purslane (*Portulaca oleracea L.*, POROL) and lambs quarters (*Chenopodium album L.*, CHEAL). These seven weeds are widely distributed in North China, and are major weeds that impact corn yield. Whole plants were collected in the field and sent to the laboratory as quickly as possible to minimize analytical errors resulting from the wilting of the leaf blade. In the laboratory, fully expanded leaves were clipped and wiped clean for measurement by the imaging spectrometer. Results of the model evaluation showed that the global accuracy and the kappa coefficient of the model were both over 95%.

3.2.1.4 Mapping of Crop Disease and Insect Damage

Insect infestation, plant viruses, fungi, and bacteria are among the most challenging environmental hazards that farmers face. Successful mitigation depends on early detection and regional tracking. The use of hyperspectral data can aid in the collection of data useful in early detection of plant damage/defoliation, and potentially allow the agricultural industry to respond to damage resulting from insect infestation and disease.

Hyperspectral data can enhance the detection and mapping of crop disease. Pest and disease infestations change leaf pigment content, which can be detected with a hyperspectral sensor. Products relevant to monitoring crop disease and insect damage include maps showing variation of leaf chlorophyll, LAI, and biomass. Other potential products include the mapping of leaf senescing pigments, and generation of disease maps.

Early detection and mapping of diseases reveals the extent and magnitude of the problem. Temporal data could show the origin and progression of the disease or pest infestation. Mapped products would be of great benefit in the implementation of more cost-effective mitigation strategies that could result in cost benefits and benefits to the environment (precision application of fungicides, pesticides, etc.).

For example, root rot of peas and lentils is a crop disease, which is caused by a complex of diseases that affect the belowground portion of the developing plant, leading to poor performing pulse crops. The organisms that cause the disease are seed or soil-borne and can infect the plant at any stage. Unfortunately, once root rot has set in, there is nothing that can be done. Understanding the disease, identifying the risks for root rot infection, and thorough planning for prevention are the only current options. Figure 3.8 shows root rot fraction map of a bean field.

Apan et al. (2004) reported the usage of spaceborne hyperspectral data acquired by Hyperion to map *orange rust* disease in sugarcane crops. Orange rust is a fungal disease in sugarcane that produces orange leaf lesions (pustules) and tends to be grouped in patches. The ruptured leaves let water escape from the plant leading to moisture stress (Croft et al. 2000). Orange rust occurs in summer/autumn and is favored by humid warm conditions. Apan et al. (2004) studied a set of spectral indices based on their correlations with the discriminant function. They concluded that narrowband indices derived from the Hyperion image data were able to discriminate sugarcane crops severely affected by orange rust disease. The indices used spectral bands that are known to be sensitive to changes in leaf pigments, internal leaf structure and moisture content. They underlined that the best spectral indices allowing one to determine the affected sugarcane were those combining the VNIR bands with the 1660-nm band sensitive to moisture.

Disease detection of mustard crop using hyperspectral data acquired by Hyperion was reported by Dutta et al. (2006). Diseased mustard crop have been discriminated from healthy crop using hyperspectral data. Five spectral indices for disease water stress proposed by Apan et al. (2004) were used. A disease score was introduced. The spectral response of the diseased crop obtained from the Hyperion data were compared to the disease scores obtained through ground reference data. Their study showed that among the five spectral indices the index 1660/680 nm was found to be best correlated with the disease score with $R^2 = 0.68$.

Hyperspectral data acquired by Hyperion was also used for categorical and quantitative mapping of salinity stress in sugarcane fields to estimate salinity stress of sugarcane crops (Hamzeh et al. 2016). The salinity and water stress index (SWSI) derived from the Hyperion data provided a prediction of salinity stress with an R^2 of 0.68 for quantitative mapping of salinity stress in sugarcane fields.

FIGURE 3.8 Root rot fraction map for a bean field derived from hyperspectral Probe-1 data. (Courtesy of Canada Centre for Remote Sensing, Natural Resources Canada.)

Apan et al. (2005) studied the feasibility to detect the effects of insect pest and disease in vegetable crops using hyperspectral data. They used a handheld spectrometer to collect data of the tomato crops affected by a fungal "early blight" disease (*Alternaria solani*) and eggplant leaves exhibited skeletal interveinal damage ("holes") caused by 28-spotted ladybirds (*Epilachna vigintioctopunctata*) in field. A partial least squares (PLS) regression was used to assess the predictive power of the relationship between disease incidence and reflectance data. PLS regression is a bilinear modeling method for relating the variations in one or several response variables (*Y*-variables) to the variations of several predictors (*X*-variables). PLS performs particularly well when the various *X*-variables have high correlation in hyperspectral data. Information in the original X-data is projected onto a small number of underlying variables called PLS components. The cross-validated results showed that the incidence of disease on tomato and pest on eggplant could be predicted using hyperspectral data with an accuracy of 82%. The most significant spectral bands for the tomato disease prediction corresponded to the reflectance red-edge (690–720 nm), the visible region (400–700 nm), and part of near-infrared wavelengths (735–1142 nm). For the eggplant, the bands between 732 nm and 829 nm were equally significant as the red-edge (694–716 nm) in the prediction. Their results confirmed the utilization of hyperspectral data to diagnose pests and diseases of vegetables. This approach can improve detection speed and provide an opportunity for nondestructive sampling.

Detecting and discriminating winter wheat aphids using hyperspectral imagery was reported by Huang et al. (2012). Wheat aphid is one of the most destructive pests in agricultural systems, especially in temperate climates of the northern and southern hemispheres. Wheat aphids appear annually in the wheat planting area, causing great economic damage to crops as a result of their direct feeding activities. With high enough density, wheat aphids can remove plant nutrients, and potentially reduce the number of heads, the number of grains per head, and overall seed weight.

The damage is especially high when wheat aphids occur in the flowering and filling stage of wheat. It is reported that average density over 20 aphids per plant can cause substantial losses of yield and quality of wheat (Basky and Fónagy 2003). There are also indirect damages including excretion of honeydew from aphids and as a vector of viruses. To prevent the occurrence and prevalence of aphids, large amounts of insecticides are used, causing environment pollution. Therefore, large-scale, real-time prediction and monitoring of wheat aphid incidence and damage degree using remote sensing technology are extremely important.

The study carried out by Huang et al. (2012) aimed to identify spectral characteristics of wheat leaf and canopy infected by aphids and find the sensitive spectral bands to detect aphid damage at canopy level in filling stage of wheat, and to establish an aphid damage hyperspectral index (ADHI) based on those sensitive spectral bands for detecting aphid damage in wheat canopy level in filling stage of wheat. The experimental results showed that the reflectance of wheat leaf infested by aphids was higher in the visible and SWIR region and lower in NIR region than that that of the uninfested leaf. Compared to the canopy spectra of the healthy wheat, the canopy reflectance of aphid infested wheat was gradually decreased in the range from 350 nm to 1750 nm, especially in the NIR region. The sensitive bands were selected based on the relevance analysis between reflectance and aphid damage levels. The most sensitive bands to aphids were 551 nm with $R^2 = 0.741$ in the visible, 823 nm with $R^2 = 0.865$ in the NIR, and 1651 nm with $R^2 = 0.668$ in the SWIR, respectively. An ADHI was established based on these most sensitive bands from hyperspectral data. It was concluded that the ADHI exhibited high relationship with aphid damage levels ($R^2 = 0.839$) by correlation analysis between ADHI and aphid-damage level using 25 investigation points.

3.2.2 Crop Productivity and Yield

Crop productivity is the quantitative measure of crop yield in a given measured area of a field. Crop yield is one of the most important pieces of information for crop management in precision agriculture. It integrates the effects of various variables, such as soil properties, topographic attributes, tillage, plant population, fertilization, irrigation, and pest infestations. Remote sensing provides continuous spatial coverage and complements meteorological based data.

In the United States, for example, the agricultural yield survey provides farmers with reported survey data of expected crop yields used to forecast and estimate crop production levels throughout the growing season. The agricultural yield survey is conducted in all states except Alaska and Hawaii. Samples of farm operators are selected from the March crops/stocks survey (for small grains) and the June crops/stocks survey (for late season crops and tobacco). Farmers reporting acreage of at least one commodity of interest are included in the monthly data collection to forecast crop yields.

Crop yield prediction is one of the main outputs of remote sensing for agri-businesses. Potential yield estimation is also of interest in precision farming and for food processors. Crop yield forecasting monitors the meteorological and development parameters (plant growth) from the start of the growing season to the present day, then uses historical data (i.e., normals) for the remainder of the year to predict yield. The three most common approaches to predicting yield involve:

1. Correlating the vegetation health with historical yield data
2. Assimilating meteorological and/or remote sensing derived data into a crop yield model
3. Using remote sensing derived products to update or adjust crop models

Imagery from traditional multispectral satellites, such as the Landsat and SPOT satellites, has long been used to monitor crop-growing conditions and to estimate crop yields over large geographic areas. With the advancement of hyperspectral satellites having large area coverage, hyperspectral imagery can be used for these approaches. The first approach could use hyperspectral vegetation measures, which include:

- Leaf chlorophyll (link to nitrogen)
- Leaf area index
- Biomass
- Percent of green vegetation cover

Hyperspectral products will improve crop yield forecasting through more accurate vegetation assessments (e.g., LAI estimations). A typical graph of the amount of a vegetative measure over time shows an increase during the vegetative growth stage until it reaches a peak then decreases during the ripening phase. The length of time it takes to reach the peak and the magnitude of vegetative measure will vary from year to year depending on the environmental conditions. A comparison with a reference year gives an indication of the relative crop productivity. Historical relationships between the yield and vegetative measures can also be used to estimate the yield potential for the current year. A variety of spectral products have been developed that are highly related to crop yield potential.

Vegetation indices (VIs) derived from remotely sensed imagery are commonly used to estimate crop yields. Figure 3.9 shows an example of the map of the photochemical reflective index (PRI) at 570 nm derived from hyperspectral CASI data and the yield monitor map for crop in a soybean field, which shows their correlation.

The second approach uses crop and soil derived parameters as inputs to a crop growth model. In addition to the vegetative health products, other hyperspectral products that can be used are:

- Leaf water content
- Leaf senescing pigments

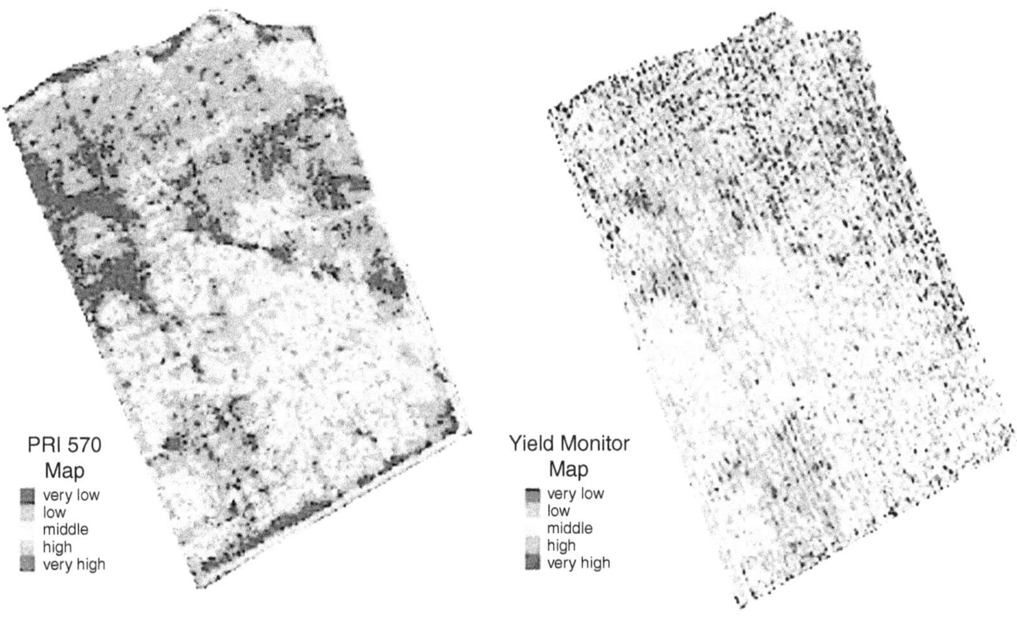

FIGURE 3.9 Map of photochemical reflective index at 570 nm (PRI 570) derived from hyperspectral CASI data and its correlation with the yield monitor map for crop in soybean field. (Courtesy of J. Miller, York University, and E. Pattey, Agriculture and Agri-Food Canada.)

Crop models predict growth and development from inputs, such as daily minimum and maximum temperatures, solar radiation, and precipitation. These models are generally constructed around a growth function that places LAI as a core parameter driving the crop ability to intercept light and transform it into biomass.

The third approach uses periodical checking of the predicted versus actual LAI. When necessary, model discrepancies with the observed reality are corrected within-season.

Yield maps not only help identify within-field spatial variability for variable rate applications, but also enable farmers to evaluate the economic returns of different farming management strategies. Estimates of relative yield variation within the growing season are critical to address the crop problems, such as nutrient deficiencies, water stress, or pest infestations, which should be managed during the growing season. In addition, yield maps are important for field-level improvements, such as drainage, land leveling, irrigation, fencing, and for off-field information use.

Yang et al. (2004) reported a study of estimating yield variability using airborne hyperspectral imagery for grain sorghum and the correlations between the maps derived from the yield monitor data and hyperspectral imagery. An airborne hyperspectral imager (Yang et al. 2003) with a swath of 640 pixels and 128 spectral bands covering a spectral range from 457.2 nm to 921.7 nm at 3.63-nm spectral sampling interval was used for image acquisition. The first 5 bands and the last 21 bands were removed from each image due to poor data quality. The remaining 102 bands covering a spectral range from 475 nm to 845 nm were used for the analysis. The hyperspectral image was acquired around the time of peak plant development of the crop. To eliminate the redundancy in the hyperspectral image, the 102-band image was transformed into a set of unrelated and independent principal components. Stepwise regression results for relating grain yield to the first ten principal components show that five of the ten principal components were found to be significant and explained 80% of the yield variability. To identify the spectral bands or combinations of the bands that were particularly responsive to yield variability, stepwise regression was performed directly on the yield data and the 102-band hyperspectral image data. Seven of the 102 bands were identified to be significant and explained about 82% of the variability in yield. Four multispectral bands were also simulated using the Landsat ETM+ broad bands. Based on the stepwise regression analysis, these four broad bands accounted for 76% of the yield variability, compared to 82% of the variability explained by the seven significant narrow hyperspectral bands. Therefore, the hyperspectral image data provided better yield estimation than the broad band multispectral image.

When linear spectral unmixing is applied to a hyperspectral image, it produces a set of abundance images (or fraction images), each of which corresponds to an endmember (i.e., pure material component). Each fraction image shows the spatial distribution of the spectrally defined component as an NDVI (normalized difference vegetation index) image does. The fraction of crop plants determined from linear spectral unmixing is a more direct measure of plant cover than an NDVI value. Yang et al. (2007) applied this technique to hyperspectral imagery for mapping the yield variation in two grain sorghum fields and their results indicate that plant fraction images can be used as relative yield maps. They also examined how variations of endmember spectra affect the results and found that correlation coefficients between yield and unconstrained plant fractions are not sensitive to the selection of plant and soil endmembers, even though the correlation coefficients between yield and constrained plant fractions are affected by the choice of endmember spectra.

The spectral angle mapper (SAM) provides an alternative approach to quantifying the spectral differences among all pixels in a hyperspectral image and, therefore, has the potential for mapping yield variability. Yang et al. (2008) reported a study to map yield variability using the SAM technique. Airborne hyperspectral imagery was acquired from two grain sorghum fields, and yield data were collected using a grain yield monitor. SAM images were generated from the hyperspectral images based on six reference spectra extracted directly from the hyperspectral images and four reflectance spectra measured on the ground. A statistical analysis showed that the ten SAM images for each field produced similar correlation coefficients with yield. The SAM images based on the soil reference spectra provided higher correlation coefficients with yield of 75% and 92%

for two separate fields (fields 1 and 2), respectively. A SAM image can be easily generated from a hyperspectral image to characterize the spatial yield variability. Moreover, a SAM image based on a single reference spectrum can be a good representation of yield variability if actual yield data are not available. The results from the study indicate that the SAM technique can be used alone or in conjunction with other VIs for yield estimation from hyperspectral imagery.

A study of yield estimation of grass crops using simulated EnMAP hyperspectral data and Sentinel-2A multispectral data was reported by Ancin-Murguzur et al. (2019). A total of 254 reference squares in size of 0.25 m^2 in-grass crop fields in northern Norway during four years (2014–2017) were acquired using a FieldSpec 3 portable spectrometer (made by ASD Inc., Boulder, Colorado) in a spectral range of 350–2500 nm, with a 1-nm interpolated spectral resolution. Thirteen (13) Sentinel-2A multispectral bands and 240 EnMAP hyperspectral bands were simulated from the FieldSpec datasets. For that purpose, binning was performed on the original spectra following the EnMAP specifications to achieve a bandwidth of 6.5 nm in the VNIR region (420–1000 nm) and 10 nm in the SWIR region (1000–2450 nm). In addition, a model using only the VNIR region between 420 nm and 900 nm was developed to avoid the region most affected by water absorptions and compare it with the results of the model with the full spectral range. The grass-crop yield was modeled using the full spectral range of the simulated data (i.e., 420–2450 nm) and the VNIR region to estimate how accurate remote-sensed yield models might be using EnMAP data compared to those involving Sentinel-2A data. The experimental results show satisfactory predictability of grass-crop yields ($R^2 > 0.55$), even when captured under suboptimal conditions. EnMAP hyperspectral bands showed their potential applicability to measure yields in northern latitudes with a comparable reliability to the portable spectrometer. The spectral region of 350–900 nm showed to be more robust than the full spectrum (350–2500 nm) against influencing factors such as moisture, facilitating the development of reliable models to measure yield of agricultural fields in high latitudes. The study showed that models resulting from hyperspectral measurements can help optimize the management of fodder production in high latitudes.

3.2.3 Soil Quality/Erosion

Soil properties and tillage practices (percent residue cover) have a significant impact on soil quality and levels of soil erosion. Soil erosion reduces soil productivity and impacts agricultural runoff and contaminates surrounding ecosystems with sediment, fertilizer, and pesticide loadings. Crop residue cover and soil tillage also impact the flux of greenhouse gases from agricultural land. Knowledge of crop residue cover can provide valuable information for identifying areas susceptible to erosion.

Agencies responsible for monitoring farm practices require:

- Information on residue and tillage to establish erosion rates and, thus, target conservation programs and prescribe conservation approaches
- A benchmark to evaluate the effectiveness of soil conservation adoption programs initiated by both regional and national governments
- A tool to determine if the country is meeting its national and international commitments for safe drinking water, control of environmental contamination, and reduction in greenhouse gases

Classification images showing percent residue cover would be the primary product. Figure 3.10 shows an example of a map of residue fraction of a white bean field derived from hyperspectral image data using spectral unmixing. Residue mapping with hyperspectral data and vegetation indices is relatively straightforward. In most cases, only one spring estimate is required. Some standards require the estimation of percent residue cover after spring planting.

Residue cover estimates from optical sensors are based on the fact that crop senescence results in significant changes in the spectral response of vegetation throughout the 0.45–2.5 μm spectral

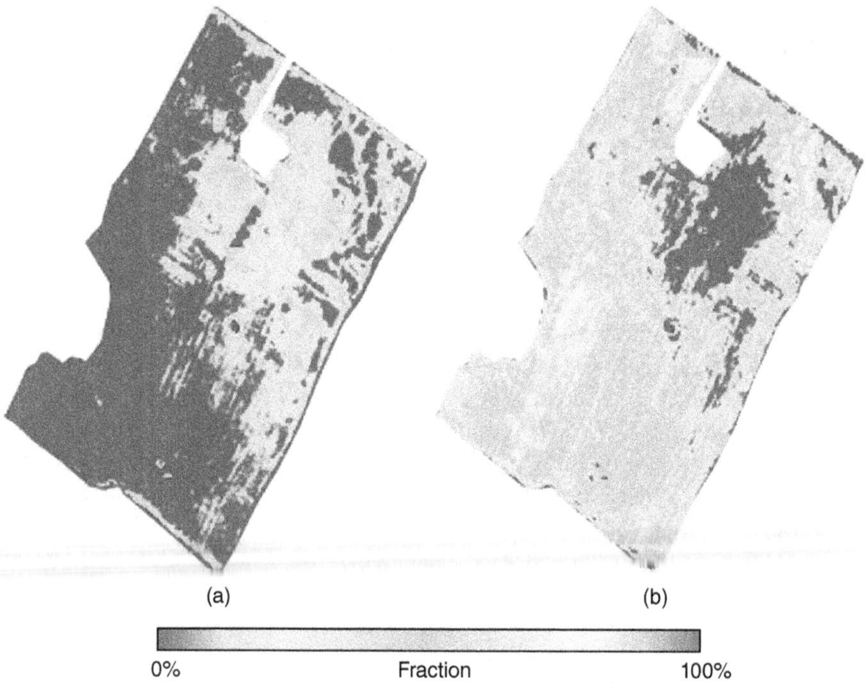

FIGURE 3.10 Residue map derived from hyperspectral imagery; (a) bare soil fraction, and (b) residue fraction of a white bean field. (Courtesy of Canada Centre for Remote Sensing, Natural Resources Canada.)

range. The response is dependent on which spectral range is being observed, visible, or NIR. Figure 3.11 shows spectra of bare soil and crop residues measured using a GER 3700 spectrometer. Crop residues are brighter than soils and have a distinct feature in the SWIR (around 2100 nm) related to cellulose absorption.

Accurate assessment of bare soil and crop residue will allow the identification of areas at high risk of soil erosion. This knowledge will allow mitigation measures, such as altered tillage practices, to be incorporated in farm management practices. Commercial companies could use information on farm practices to focus the development and sale of fertilizer, seed, chemicals, and farm implements. Governments could use this for monitoring soil conservation practices.

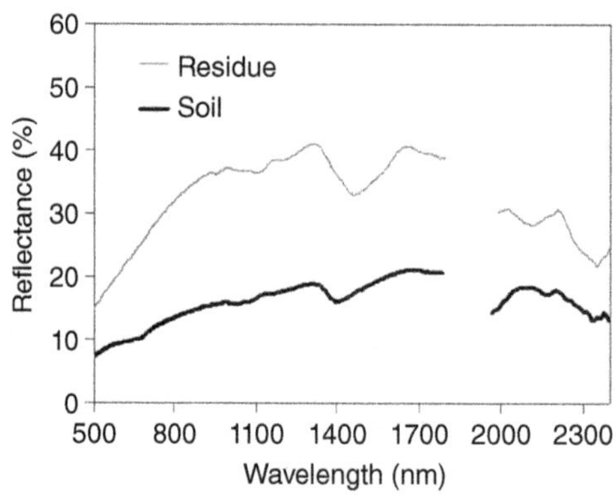

FIGURE 3.11 Spectra of bare soil and crop residues. (Courtesy of Canada Centre for Remote Sensing, Natural Resources Canada.)

Soil quality assessment using airborne hyperspectral data was reported by Paz-Kagan (2015). The primary goal of the study was to prove and demonstrate the ability of hyperspectral data to evaluate soil properties and quality across anthropogenically induced land-use changes. A spectral soil quality index (SSQI) using hyperspectral data was proposed. Thirteen (13) soil physical, biological, and chemical properties and their derived soil quality index (SQI) were measured. Two different SPECIM's AISA hyperspectral airborne sensors were used. The first one is a hyperspectral sensor with a single optics covering the VNIR and SWIR regions (380–2500 nm) with 448 spectral bands and a spectral resolution of 3.5–5.5 nm (at FWHM). The sensor was mounted onboard a small aircraft flown at an altitude of 762 m, resulting in a GSD of 1 m × 1 m. The second hyperspectral sensor combines a VNIR (400–970 nm) sensor and a SWIR (970–2500 nm) sensor with different optics covering the overall 420–2450 nm spectral range with 367 spectral bands and a spectral resolution of 4.5 nm in the VNIR and 6.3 nm in the SWIR regions. Correlations between the laboratory spectral values and the calculated SQI, the coefficient of determination (R^2) and ratio of performance to deviation (RPD) were $R^2 = 0.84$, RPD = 2.43 and $R^2 = 0.78$, RPD = 2.10 for the two study sites, respectively. The partial least squares-discriminate analysis model that was used to develop the SSQI showed high classification accuracy for both sites. The correlations between the SSQI and the SQI were $R^2 = 0.71$ and $R^2 = 0.70$ for the two study sites, respectively. It was concluded that soil quality can be effectively monitored using the spectral-spatial information provided by the hyperspectral technology. Hyperspectral-based classification of soils can provide the basis for a spatially explicit and quantitative approach for monitoring soil quality and as a function at a local scale.

Arable soil quality indicators, such as microbial biomass-C (MBC) and hot water-extractable C (HWEC), complemented by organic carbon (OC) and nitrogen (N) were assessed by Vohland et al. (2017) using laboratory and airborne hyperspectral data. MBC is a measure of the mass of the living component of soil organic matter. It underpins any biologically related property of soil systems and is essentially linked to nutrient cycles, humification, physical structure formation, degradation of contaminants, and soil fertility (Bastida et al. 2008, Ritz et al. 2009). Thus, MBC and its species and functional diversity are recently gaining significance as bioindicators of soil quality (Velasquez et al. 2007, Ritz et al. 2009, Pulleman et al. 2012), leading to a huge output of studies throughout the last decade (Bastida et al. 2008, Havlicek 2012). Compared to total OC, MBC is substantially more dynamic and quickly responds to agricultural management changes, such as tillage practices, organic amendments, and incorporation of cover crops, or to soil moisture and temperature changes. MBC depends on the amount of available soil carbon substrate, so that the MBC-to-OC ratio indicates the substrate availability to soil microorganisms (Böhme et al. 2005, Fließbach et al. 2007). With selected variables, quantification accuracies improved markedly for MBC (RPD increased to 2.35 from 1.80) and OC (RPD increased to 2.20 from 1.94). The study confirmed that OC, which is one key soil characteristic, could be estimated with success using airborne hyperspectral data. The results are well in line with results of other studies, for example, Steinberg et al. (2016) ($R^2 = 0.74$, RPD = 1.90) or Hbirkou et al. (2012) ($R^2 = 0.83$, RPD = 2.45).

3.2.4 Agro-Environmental Health Monitoring and Forecasting

National governments require information related to national objectives on climate change, water quality and biodiversity, including:

1. Land use, crop rotation, and management maps
2. Environmental monitoring

Some of this information is common to the various issues in environmental health. Information required for climate change includes:

- Evaluation of crop damage caused by frost (mainly for forage crops, such as alfalfa in Canada), flooding, high winds (cereal lodging), water/heat stress, and diseases

- Phenology and yield forecasting
- Data in support of quantification of greenhouse gas emissions from agro-ecosystems, such as surface condition (crop rotation, tillage), location and extent of vegetation, duration of snow cover, water equivalent of snow, flooded field area and duration, crop biomass, chlorophyll content for estimating residual soil nitrogen, amount of crop residue
- Calculation of carbon credits as a result of soil and crop management practices, in connection with the Kyoto commitment on greenhouse gas emissions: soil organic matter content, biomass production and chemical composition for estimating carbon sink, crop residues, annual and perennials to verify models and soil databases

Hyperspectral products described in the previous subsections for detecting and identifying the status of crops all have applications in agro-environmental monitoring. Information on crop condition and stress, crop productivity, potential yield, and soil properties are useful inputs to an agri-environmental management strategy.

In addition, the high-spectral resolution of hyperspectral data improves the discrimination between crop types as well as increases classification accuracy. This is important for identifying crop rotation practices and management practices.

Monitoring and forecasting the environmental health of agricultural regions can be used to determine the impact of changes in land use, climate change, and farming practices on the agricultural environment.

3.3 ENVIRONMENT AND SUSTAINABILITY

The environment is a multidisciplinary field covering a large number of applications of hyperspectral remote sensing. This section reviews only a selected number of environmental applications that benefit from the hyperspectral remote sensing data. For each application area, the advantages of using hyperspectral data are presented. In addition, potential benefits to the users and/or the environment are outlined. The following selected topics are described:

1. Wetlands
2. Mine waste management/reclamation
3. Land degradation and soils mapping
4. Climate change

3.3.1 WETLAND APPLICATIONS

Wetlands are among the most productive, life-supporting ecosystems on Earth (Mitsch and Gosselink 1993). They provide habitat for fish, wildlife, and waterfowl; hydrological benefits by regulating water levels; water quality protection and several economic benefits through recreation and tourism (Klemas et al. 1993). Wetlands play a key role in the global environment and are critically linked to many major global environmental issues such as climate change, water quality, the hydrological and carbon cycles, and wildlife habitat (Steeger et al. 2001, Johnson and O'Neil 2001).

The users' needs of wetland specialists can be listed as follows: (1) mapping wetland locations; (2) delimiting wetland boundaries; (3) undertaking inventories, vegetation (type and density) mapping; (4) land-use mapping and planning; (5) ecological land classification; (6) mapping water quality; (7) mapping the substrate; and (8) developing baseline data.

Particularly in coastal wetlands, water levels can change considerably from year to year and storms can destroy large areas of vegetation. Thus, wetland specialists require to have an annual update of changes in wetland extent, but this may be precluded by cost for all except the most critical areas.

There is interest in determining the health of wetlands using remote sensing techniques. This would be a particularly suitable task for hyperspectral sensors, but certainly not an easy one. However, it is anticipated that hyperspectral imagery could greatly increase the information on both emergent and submergent species within a wetland environment compared to the traditional methods, including multispectral remote sensing.

Depending on the nature of the information required by a user, a variety of geocoded maps can be produced.

1. Wetland Vegetation Maps: These would provide information needed for legislative or regulatory requirements; determination of the location of seasonally inundated and permanently flooded areas; decisions on sites for agricultural, residential, or industrial development; wildlife management and habitat assessment; development of recreational opportunities; monitoring change in vegetative cover and wetland acreage (change detection maps). Figure 3.12 shows a wetland map using hyperspectral CASI data.
2. Wetland Substrate Maps: In areas of relatively clear water, it is possible to see and ultimately map differences in the substrate. This is important as the substrate often provides a food source for aquatic life.

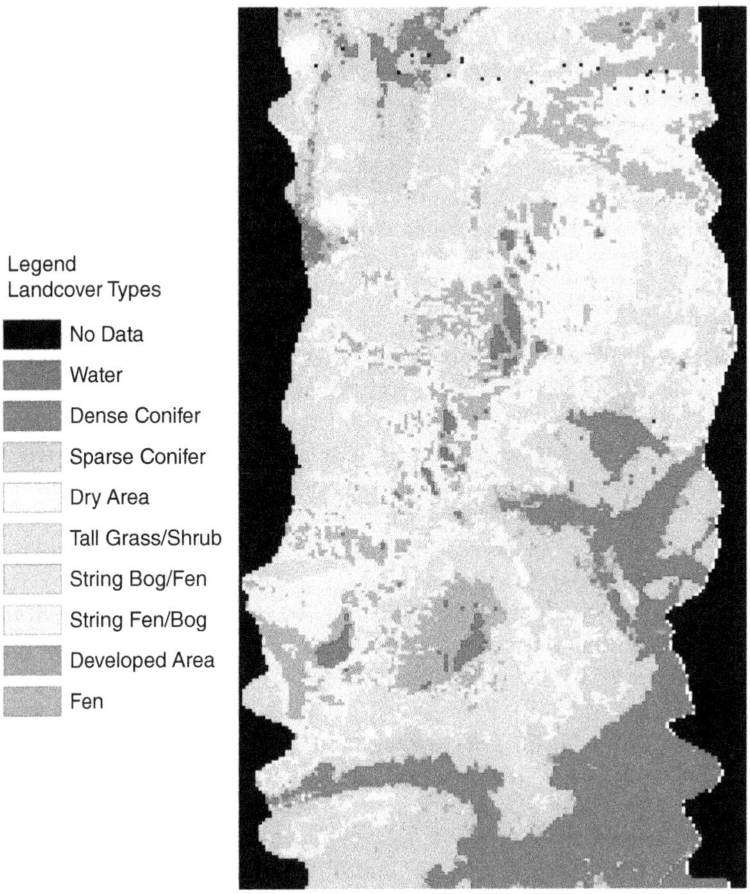

FIGURE 3.12 Wetland mapping using airborne CASI hyperspectral data in Labrador. (Image courtesy of Canada Centre for Remote Sensing, Natural Resources Canada.)

3. Land-Cover and Land-Use Maps: Thematic maps showing land cover and land use in wetland areas are used for wetland inventory, wetland evaluation and assessment, and as a base for monitoring wetlands. In addition to maps that show wetland classes, it is also important that the land-cover and land-use maps show the landscape surrounding the wetlands. The surrounding landscape provides habitat for wildlife and can have a considerable effect on the health and viability of the wetland.

Closely related to the above types of mapping products are maps that show water characteristics (e.g., suspended sediment distributions, algal distribution, and abundance maps). These are described in Section 3.4.

The primary benefit of using a hyperspectral sensor would be a significant increase in classification accuracy for surface and submerged vegetation and general land cover. Due to the increased spectral resolution water quality issues can also be addressed more readily (mapping of phytoplankton, suspended sediments, etc.).

Mapping vegetation of wetlands is challenging. This is because wetland plants and their properties are not as easily detectable as terrestrial plants. First, herbaceous wetland vegetation exhibits high spectral and spatial variability, because of the steep environmental gradients, which produce short ecotones and sharp demarcation between the vegetation units (Adam and Mutanga 2009). Hence, it is often difficult to identify the boundaries between vegetation types. Second, the reflectance spectra of wetland vegetation canopies are often very similar and are combined with reflectance spectra of the underlying soil, hydrologic regime, and atmospheric water vapor. The most important factors affecting the spectral reflectance among wetland vegetation are the biochemical and biophysical parameters of a plant, such as chlorophyll *a* and *b*, carotene, and xanthophylls. Wetland species appear to vary greatly in chlorophyll and biomass reflectance as a function of plant species and hydrologic regime. Spectral behavior of wetland vegetation is also influenced by leaf water content, which determines the absorption of the NIR and SWIR region. LAI is also a key variable in the canopy reflectance of the wetland vegetation. The canopies with a high LAI reflect more than the canopies with medium or low LAI. High-spectral resolution of hyperspectral remote sensing allows in-depth examination and discrimination of vegetation types, which would be lost with broadband multispectral imagers (Schmidt and Skidmore 2003, Govender et al. 2006).

A number of attempts have been successfully made to discriminate and classify wetland species based on their fresh leaf reflectance at laboratory levels with the view to scaling it up to airborne remote sensing (Vaiphasa et al. 2005, 2007) and field reflectance at canopy scale (Best et al. 1981, Penuelas et al. 1993, Schmidt and Skidmore 2003, Rosso et al. 2005). The earliest effort on spectral discrimination of wetland species was reported by Anderson (1970) who attempted to evaluate the discrimination of 10 marshplant species, which dominated a wetland in Chesapeake Bay using a spectroradiometer. He concluded that the spectral difference between the species is minor in the visible spectrum, but significant in the near-infrared. The variation in the spectral reflectance with the changing seasons was also reported in the study.

Rosso et al. (2005) assessed the separability of five marsh species (*Salicornia, S. foliosa, S. foliosa, S. alterniflora,* and *Scirpus*) using AVIRIS data. Spectral mixture analysis (SMA) and multiple endmember spectral mixture analysis (MESMA) were applied on the AVIRIS imagery data. Using both SMA and MESMA, it was possible to distinguish between the species with higher classification accuracies. The MESMA technique appeared to be more appropriate, because it could incorporate more than one endmember per class.

Similar work was also reported by Li et al. (2005). They used AVIRIS imagery to discriminate three salt marsh species (*Salicornia, Grindelia,* and *Spartina*) in China and in the San Pablo Bay of California, United States. They developed a model that mixed the spectral angle (SA) together with physically meaningful fraction and the rms. In a fraction image, the percentage coverage of pixels in a polygon that are classified as the species corresponding to the endmember was used as the measurement of the discrimination of the species. For the two marsh vegetation species Spartina and Salicornia, which occupied 93.8% of the marsh area, the percentage coverage in the corresponding fraction image was up to 71% and 64%, respectively. These results were satisfactory in discriminating these two species. However, it was difficult to discriminate Grindelia from Spartina and Salicornia due to the spectral similarity between the species.

Wang et al. (2007) attempted to map highly mixed vegetation in salt marshes in the lagoon in Venice in Italy. Six significant bands of Compact Airborne Spectral Imager (CASI) hyperspectral imagery data were selected using spectral reconstruction (SR). The results showed that the accuracy of Vegetation Community based Neural Network Classifier (VCNNC) can be used effectively in the situation of mixed pixels; thus, it yielded an accuracy higher (91%) than the neural network classifier (84%).

3.3.2 Environmental Geology

Associated with mining activities are environmental issues that need to be addressed, primarily how to manage waste products. Part of this process involves the identification and mapping of mine tailings and examining potential impacts on the surrounding environment and local population.

A major problem associated with active and abandoned mines is the oxidation of mine waste that can produce acidic water, which can affect adjacent ecological systems and human drinking water supplies. For example, the estimated costs for clean-up liabilities of mine tailings site in Canada alone amount to $2-5 billion in 2003.

The spectral resolution of spaceborne hyperspectral data allows the accurate assessment of rehabilitation efforts within larger mine tailings sites. Hyperspectral data also has great benefits in mapping mineral fractions within mine tailing areas, as they can be used to map acid mine drainage potential, and monitor potential environmental impacts. A major constraint for using these products is the spatial resolution. A large number of the mine tailings were generated from small mines; spaceborne sensors will likely be best suited to monitor larger mine tailing sites.

Hyperspectral remote sensing has been applied successfully for the detection of mine waste due to its unique capability to resolve mineral absorption features (Clark et al. 1998, Shang et al. 2000, Swayze et al. 2000, Neville et al. 2001). The mapping of individual minerals (e.g., pyrite, jarosite, goethite, hematite) provides information on the degree of acid mine drainage that occurs on mine tailings (Lévesque et al. 1997, 1999, Shang et al. 2000).

Hyperspectral remote sensing is also an excellent tool to monitor and temporally assess the rehabilitation of mine tailing sites (Lévesque et al. 1999, Staenz et al. 1999). It uses its ability to map the proportional surface cover, which can be referred to percent ground cover of surface materials (vegetation).

There are two areas where spaceborne hyperspectral remote sensing can be highly beneficial:

- Monitoring mine tailings rehabilitation
- Mapping acid mine drainage, whether from tailings dam failures or from abandoned mines

A major by-product of mining activities is the waste produced from the extraction of sulfide-rich ore. The principal effect of sulfide mine waste is an increase in soil acidity. For this reason, lime is often used to help the revegetation process. Products related to the monitoring of mine tailing rehabilitation would include temporal maps showing fractions of green vegetation compared to fractions of ground cover or mine tailing fractions (oxidized tailings and lime).

Figure 3.13 shows fraction images for monitoring mine tailings revegetation at Copper Cliff Mine, Sudbury, Ontario, Canada derived from hyperspectral images acquired with CASI in 1996 and 1998 and acquired with Probe-1 in 1999 over the tailings impoundment area. A commercial company was tasked to rehabilitate these tailings. Following various lime treatments, they plant various type of bluegrass, ryegrass, and fescue. The revegetation process did not always work so that a monitoring tool such as hyperspectral remote sensing was very useful to locate areas where more attention was required. Hyperspectral data allow identification of new vegetation (grass) and native vegetation that

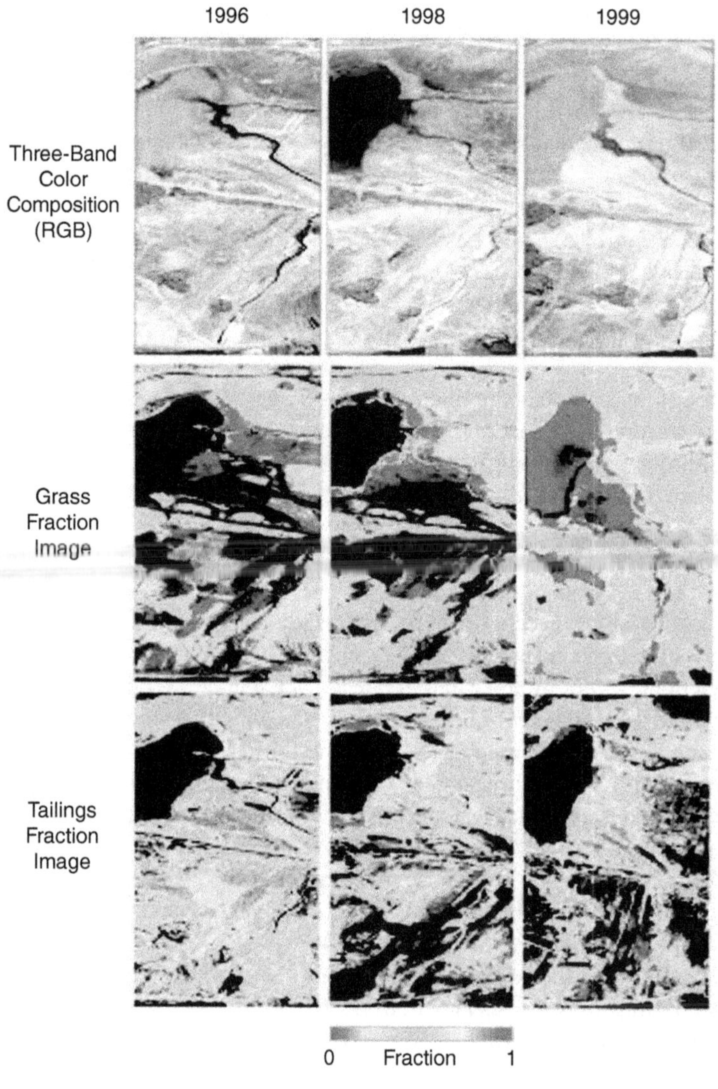

FIGURE 3.13 Fraction images derived from hyperspectral images acquired with CASI in 1996 and 1998 and acquired with Probe-1 in 1999 for monitoring mine tailings revegetation at the Copper Cliff Mine, Sudbury, Ontario, Canada. (Courtesy of J. Lévesque).

is not part of the revegetation (native trees outside the tailings area). Results indicate that advances in reclamation of the site have been made over the 3-year period (Lévesque et al. 1997, 1999).

Acid mine drainage (AMD) potential maps will aid identifying sources of AMD and detect changes over the year in the vicinity of the AMD source. Such changes can include vegetation die-off, and can indicate such problems as leakage and seepage from tailings impoundment dams. Tailings are a product of the extraction of metals from rocks. Many of them contain a considerable amount of reactive sulfides. When exposed to oxygen and water, oxidization of the tailings occurs, acid waters are produced and heavy metals are liberated into the environment. A result is the production of a range of iron-bearing secondary minerals. The type of mineral that occurs is dependent on the degree of oxidation, the pH, the moisture content and the composition of the solution (Swayze et al. 2000). The typical pattern is for the fresh, unoxidized tailings to be dominated by pyrite and pyrrhotite. With progressive oxidation, these weather into jarosite and then goethite as

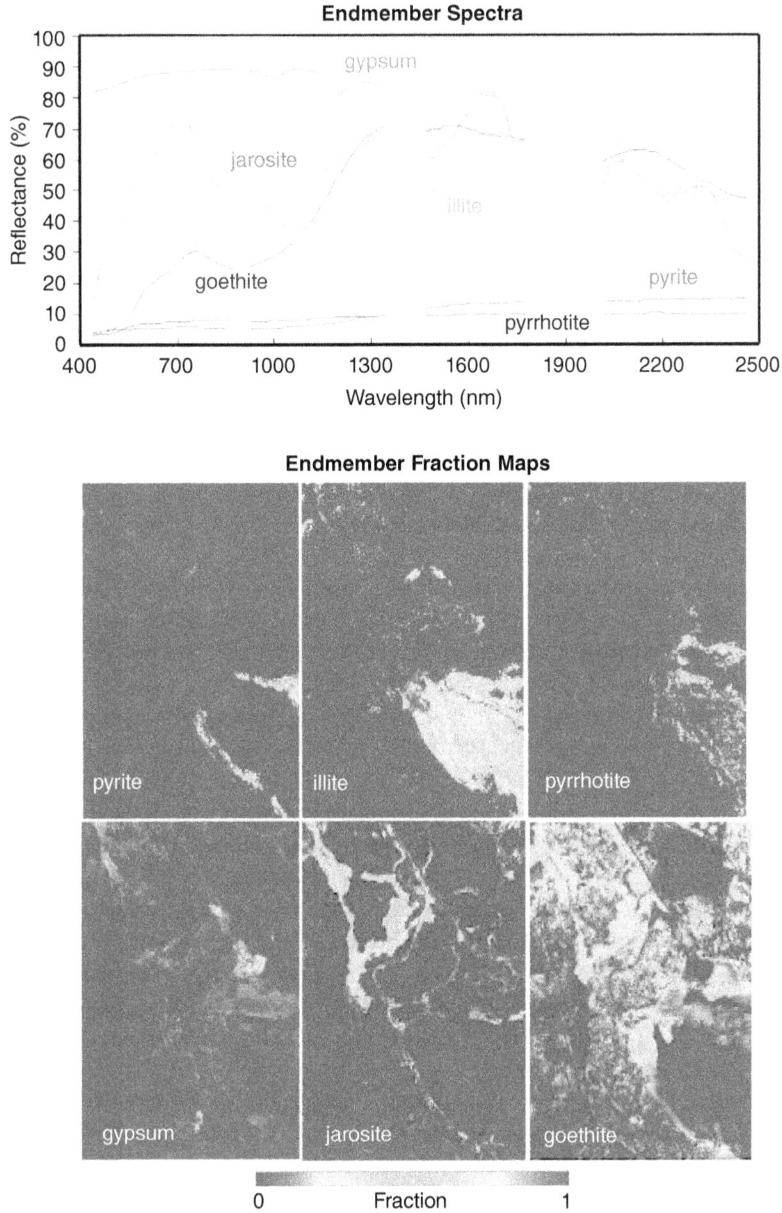

FIGURE 3.14 End members and corresponding fraction images for mapping mine tailings for AMD potential. (Courtesy of Canada Center for Remote Sensing.)

the pH increases. The spatial distribution of these minerals identifies those areas with low pH and the occurrence of heavy metals. Figure 3.14 shows mine tailing mineral fraction images for AMD potential. This will provide valuable information when designing proper reclamation procedures.

3.3.3 Land Degradation and Soils Mapping

It was estimated that arid and semi-arid regions of the world occupy approximately one-third of the continental land mass. These regions are primarily located between 15° and 35° north and south of the equator in areas that are dominated by the climatological subtropical zones of high pressure.

Land degradation is of great concern in arid and semi-arid environments. Evidence of this is the fact that almost 200 nations have ratified the "UN Convention to Combat Desertification."

The major land degradation processes are deterioration of soil structure, soil erosion by wind and water, salinization due to the upward movement of salts within the soils or near-surface materials, and depletion of nutrients by over-cultivating or leaching (Sujatha et al. 2000). Costs in terms of lost production are considerable. For example, in the Murray-Darling Basin of Australia, the annual cost of lost production due to saline soils was estimated to be US$200 million (Dehaan and Taylor 2002). For planning reclamation strategies and establishing sustainable agricultural practices, information is needed on the nature, extent, severity, and geographical distribution of degraded lands.

Hyperspectral data would be of great benefit due to its ability to reliably and efficiently identify unique absorption features associated with varying degrees of salinization/degradation and associated vegetative communities.

Hyperspectral products would include geocoded maps showing saline and degraded soils and their associated vegetation in arid and semi-arid environments. Depending on the nature and characteristics of the surface features, varying degrees of salinity can be mapped based on absorption features (Dehaan and Taylor 2002). Spectra from several salt-tolerant plants have also been measured and display distinctive signature. Maps identifying vegetative species would provide additional information regarding the degree and extent of salinization.

Weng et al. (2010) used hyperspectral data acquired by the Hyperion sensor to quantitatively map soil salinity in the Yellow River Delta region of China. A soil salinity spectral index (SSI) was constructed from continuum-removed reflectance at 2052 nm and 2203 nm to analyze the spectral absorption features of the salt-affected soils. They found that there existed a strong correlation ($R^2 = 0.91$) between the SSI and soil salt content (SSC). A model for estimation of SSC with SSI was established using univariate regression. Validation of the model yielded a root mean square error (RMSE) of 0.986 and an R^2 of 0.873. The model was applied to a Hyperion reflectance image on a pixel-by-pixel basis and the resulting quantitative salinity map was validated successfully with a RMSE = 1.921 and $R^2 = 0.627$. These results demonstrate that the satellite hyperspectral data had the potential for predicting SSC in a larger area than airborne data.

Time-series data could also be used to monitor change in soil characteristics and vegetative communities and efforts associated with soil sustainability and soil reclamation.

Application of hyperspectral remote sensing represents a promising solution in the assessment of soil degradation by erosion at a regional and global scale. Soil erosion and accumulation affect chemical and physical properties of upper layer of soils. Thus, spectral reflectance is varied, and eroded soils have different spectral response from non-eroded healthy soils (Demattê and Focht 1999, Schmid et al. 2012, 2016). Hyperspectral image analysis of soil spectral characteristics has a great potential for soil erosion assessment (Ben-Dor et al. 2009, Ben-Dor and Demattê 2015) thanks to the high-spectral resolution and, therefore, allows for more efficient quantification of spectral features in comparison with multispectral data. However, there are still challenges in the use of hyperspectral data for soil erosion assessment. On one hand, there are restraints connected to data collecting, such as atmospheric attenuation, signal-to-noise ratio, spatial resolution of the images and, above all, the bidirectional reflectance distribution function (BRDF) effect (Ben-Dor and Demattê 2015). BRDF effect depends on the off-nadir angles of the FOV. It is more severe for airborne data acquisition than satellite-based systems unless they have off-nadir capabilities. On the other hand, the challenges are often caused by surface characteristics, such as soil covered with vegetation, litter, dust, or soil crust, which make the direct soil assessment difficult. Moreover, the spectral information is considerably influenced by soil properties (e.g., soil moisture, soil roughness, soil texture or size of soil aggregates) and heterogeneity of environmental settings. This is valid mainly in an erosion relief, where the combination of soil loss and redeposition may lead to similar soil surface properties in eroded and accumulated soils (Hill et al. 1995, Zádorová et al. 2013). A successful application of hyperspectral data then requires a very thorough planning of data acquisition flight campaigns, precise image corrections and a sufficient amount of reference soil data.

In temperate climate, a multi-temporal approach must be considered when assessing larger areas, as in each period some of the soils are covered with vegetation.

Several studies have employed hyperspectral image data for soil erosion assessments (Hill et al. 1994, 1995, Chabrillat et al. 2003, Ben-Dor et al. 2004, Haubrock et al. 2004, Haubrock et al. 2005). Hill et al. (1994, 1995) used linear spectral mixture analysis and k-means clustering for classification of erosion stages based on skeleton content determination. These studies dealt with the monitoring of specific soil properties related to soil erosion, such as development of soil crust (Ben-Dor et al. 2004) or changes in soil moisture (Haubrock et al. 2004, 2005). Chabrillat et al. (2014) distinguished erosion classes using support vector machine classification method and image-derived endmembers for each erosion class. Lin et al. (2013, 2014) analyzed spectral response of different eroded soils in subtropical region in China. They compared a direct identification of erosion stages by analyzing laboratory hyperspectral data with an indirect identification using prediction of soil properties. The latter performed better due to mixing and crossover effect among erosion groups.

A study is reported by Žížala et al. (2017) to analyze the possibilities of using aerial hyperspectral data for the assessment of soil degradation by erosion on arable land in temperate agriculture regions with different dominant soil units. The objectives of the study were to: (1) quantify the prediction performance of selected soil surface properties derived from hyperspectral data in the context of various soil environments and spectral preprocessing and regression methods; and (2) determine erosion classes that represent soils differently impacted by soil redistribution by applying the predicted soil surface properties as site-specific indicators and input data into classification models of erosion classes. The influence of site-specific conditions on prediction of soil properties and classification of erosion classes was assessed. The prediction accuracy measured by R^2 of the best performing models predicting the soil properties varies in the range from 0.80 to 0.91 for soil organic carbon content, 0.21–0.67 for sand content, 0.40–0.92 for silt content, 0.38–0.89 for clay content, 0.73–089 for Fe_{ox}, 0.59–0.78 for F_{ed} and 0.82 for $CaCO_3$.

3.3.4 Climate Change

A wide range of international programs concerned with issues of climate change now make use of remote sensing data to provide the synoptic, repetitive satellite views of the Earth's surface and atmosphere that are of importance when developing and testing ecosystem and climate models (Donoghue 2002). Of particular interest is the issue of carbon exchange between the biosphere and the atmosphere.

To establish operational systems, there is a need to provide improved land-cover information (DeFries and Belward 2000). This will firstly meet the needs of the scientific community for studies related to carbon fluxes and global climate models, and secondly, for some of the policy objectives of international agreements, such as the Kyoto Protocol, provide accurate land-cover change information.

Since the 1980s, data from the Advanced Very High Resolution Radiometer (AVHRR) have provided land-cover information at continental scales. Various algorithms have been developed to produce these results. Reevaluation of some of these data had indicated inaccuracies in the interpretation of land-cover change due to errors in atmospheric corrections and the influences of soil reflectance.

It was suggested that high-spectral resolution sensors provide a better link between ecosystem processes and remote observations (Donoghue 2002). Work done by Ustin and Xiao (2001) and Goodenough et al. (2002) showed that greatly increased accuracies could be achieved with hyperspectral imagery when generating vegetation maps of forested areas compared to conventional multispectral satellites.

Hyperspectral products would consist of geocoded maps showing detailed land-cover information, showing vegetation types, quantitative estimates of biomass, and carbon. Products showing land-cover change would be important to establish significant trends, for example, afforestation and deforestation.

High-resolution hyperspectral sensors with the ability of species identification can be used to develop improved land-cover mapping at coarse resolutions, such as the methods for deriving the fractional cover types within grid cells (DeFries and Belward 2000).

In addition to land-cover maps as input to climate models, hyperspectral data are essential to study the impacts of climate change on ecosystems, for example, to quantify subtle changes in ecosystems before drastic changes occur. In Canada, for example, as the largest federal land manager, Parks Canada has an ongoing need for basic biophysical mapping over large areas. It has a legal requirement under the National Parks Act of Canada to provide up-to-date detailed biophysical inventories of ecosystems in national parks. In the province of British Columbia, the Ministry of Environment, Lands and Parks has a similar need for vegetation inventories and believes that it could obtain more detailed information from hyperspectral imagery than from the Landsat TM data.

Hyperspectral data showed contrasting results for land-cover mapping applications. The monitoring of changes in vegetation is particularly useful for land-degradation assessments (Huete et al. 2003). Xu and Gong (2007) compared the capacity of Hyperion with the multispectral ALI sensor to discriminate different land use and land cover in California. They showed that for various dry grass vegetation classes and impervious land use categories (new residential lots), the Hyperion data produced better results than the ALI data. Petropoulos et al. (2012) emphasized the relevance of Hyperion imagery for mapping and monitoring land degradation and desertification. They showed the performance of Hyperion data for land-use classification in Mediterranean environments. However, despite a good overall accuracy, some classes were more difficult to identify, such as sparsely vegetated areas, heterogeneous agricultural areas, burnt areas or transition between land uses. Ben-Arfa et al. (2015) used Hyperion imagery for mapping oases dynamics caused by competition for land and water between different user groups in the Gabes area in Tunisia. They emphasized the efficiency of the cellulose index, vegetation mask, and water presence index for the discrimination of land use.

3.4 MARINE, COASTAL, AND INLAND WATERS

The monitoring and maintenance of water quality and ecological quality of coastal zones and inland waters is a growing global concern. Over 60% of the human population lives in the coastal zone. Runoff and delivery of nutrients and sediments to lakes and coastal waters are modified through human activities, such as land clearing, agriculture, mining, urban and industrial development. Coastal development can lead to modification of foreshores, loss of key habitats, changes to flushing rates, resuspension of sediments, and direct inputs of nutrients and toxicants.

Coastal waters are of particular significance in terms of the global fishery. Human recreational activities and tourism are concentrated in coastal waters and often depend on the maintenance of high water quality. The managing of impacts of urbanization and industrialization on the coastal zone ecology has become a high priority for many nations and, hence, the need to develop better methods for monitoring and predicting change in coastal systems.

Spaceborne hyperspectral data will contribute significantly to the monitoring and mapping of:

- Coastal and inland water phytoplankton biomass and blooms, and river plumes
- Coastline and shoal mapping, beach and sea-weed surveys, and wetlands
- Ships, fishing activity, and pollution events

Let's use the author's country, Canada, as an example. Canada is often called a "water-rich" nation, with 9% of the world's renewable freshwater supply and the longest ocean coastlines of any country, as well as the second largest continental shelf along its coasts. However, in a global context,

Canada is the second highest user of water in our day-to-day lives. Its high per capita water usage, coupled with other stressors such as population growth, introduces pressures on Canada's freshwater resources. Also, while Canadians enjoy one of the highest standards of clean water in the world, pollution remains an important problem in some of its waters. In some areas, people cannot swim or eat the fish they catch.

Since Confederation in 1867, Canada has lost 65% of Maritime tidal marshes and 70% of wetlands in some areas (e.g., Great Lakes—St. Lawrence basin, Fraser River Delta). Understanding and quantifying the physical, biogeochemical, and ecological processes that occur within its natural aquatic environments are critical components of regional water resource management, ranging from protecting and monitoring water quality, carbon stocks and fluxes, near-shore/wetland habitats, fisheries and aquaculture, public health, to navigation and shipping, security, oil spill/pollution event response, and impacts from episodic flood and storm events.

Water resource management concerns the allocation and demand for water resources. The overall aim is to ensure that current and future water uses are met in the most economically efficient manner possible. Management involves the collection of data and information on water use and pricing in both the municipal and the industrial sectors of the nation's economy, the analysis of demand patterns and the factors that influence them as well as the forecasting of national, regional, and basin level trends. Geospatial information of the type provided by Earth observation satellites is obviously a key component of all of these activities.

3.4.1 Applications in Marine Environments

The optical properties of natural bodies of water are influenced by three main components. These are:

1. *Phytoplankton*—includes phytoplankton and other microscopic free-floating organisms found in the illuminated surface layers of the ocean. They are the single-celled plants that form the base of the aquatic food web, and are an important component of the global carbon cycle. The concentration of the main phytoplankton pigment, chlorophyll *a*, is often taken as an index of phytoplankton biomass.
2. *Suspended material* (inorganic)—includes all inorganic particulate material that is not included in the phytoplankton component. In coastal regions variations of suspended material are a function of out flows from rivers.
3. *Yellow substances*—these are colored, dissolved organic substances, also *detrital* particulate material; for example, from the degradation of phytoplankton cells and other organic particles.

Whereas the optical properties of deep off-shore waters are primarily a function of phytoplankton concentration, coastal waters and inland waters represent a more complex optical environment. The optical properties in these areas are due to a mix of phytoplankton, suspended sediments, and yellow substances, which can vary independently from each other.

Spaceborne hyperspectral data is ideally suited to determine the nature of the optical properties of a water body and, thus, derive the relevant information products for use by resource managers. The following subsections outline the marine applications that will benefit from the availability of spaceborne hyperspectral remote sensing data. For each application area, selected hyperspectral products are identified, and the potential benefits of hyperspectral data are outlined. There is a wide range of possible application areas. This section summarizes some of the aspects of the following areas:

1. Impacts of land use and river runoff
2. Eutrophication/harmful algae

3. Fisheries applications
4. Land-related water applications
5. Coastal zone change detection

3.4.2 Impacts of Land Use and River Runoff

Anthropogenic activities in watersheds, such as including agriculture, forestry, mining, irrigation, and urban and industrial development, can result in major changes in the volume and pattern of runoff, and result in increased nutrient and sediment loading of rivers and subsequent impacts on downstream estuarine and coastal habitats. Changes in runoff may have significant impacts on local and regional biochemical cycles, resulting in loss of benthic communities due to:

- Direct disturbance
- Increases in turbidity and sediment deposition
- Nuisance algal blooms due to eutrophication

Hyperspectral sensors are best suited to derive quantitative measures of the above parameters, especially in optically complex waters. A high-spatial resolution hyperspectral sensor is ideally suited to map the spatial extent of impacts associated with major flood events and point source pollutants. Concentrations of suspended sediments and yellow substances can be used to track plume distribution and dilution, and sedimentation, to calibrate or validate dispersion models, and identify potential interactions with benthic ecosystems, such as fish habitat, coral reefs, etc.

Spaceborne hyperspectral products could include suspended sediments, Gelbstoff (yellow substances), and chlorophyll *a* concentration. Figure 3.15 shows three water constituents chlorophyll, sediment and Gelbstoff over the German Bight derived from MOS-IRS (Modular Optoelectronic Scanner on the IRS-P3 satellite) data using the principal components approach. This set of images illustrates the decomposition of the individual components from the multivariate data set.

Most of the analyzed studies dealing with water resources management focused on quality assessment and component classification. Giardino et al. (2007) assessed the water quality of the Garda Lake in Italy using the Hyperion data. Their results were satisfactory for deriving chlorophyll *a*, while the results were moderate for detecting the tripton that sedimented in the lake. Devred et al. (2013) reviewed water resource applications of the future HyspIRI (now also referred to as Surface Biology and Geology, or SBG) sensor, such as water quality, species succession in estuaries, nutrients loads and changes in fisheries. The spatial and spectral resolutions of the HyspIRI sensor with 60 m and 10 nm, respectively, represent good advantages for water studies.

3.4.3 Eutrophication and Harmful Algae Blooms in Lakes, Rivers and Coastal Areas

Eutrophication is the enrichment of a water body through increased loads of nutrients. Eutrophication can result from land-use changes within a watershed, which increase river inputs, and from direct urban inputs (sewage treatment plants, industry, and storm water drains). Eutrophication generally results in increases in algal biomass. In naturally occurring oligotrophic (nutrient-poor) systems, it often results in the replacement of indigenous aquatic vegetation by benthic macro-algae and then phytoplankton grows extremely rapidly (blooms). This also holds true for coastal waters and inland lakes. Eutrophication represents one of the most severe and widespread environmental problems for coastal zone and freshwater managers.

Harmful algal blooms (HABs) are phytoplankton blooms that produce harmful toxins. In both coastal and inland waters, certain plankton species can grow explosively, forming massive blooms, commonly referred to as "red tides," even though they are sometimes caused by blue-green algae.

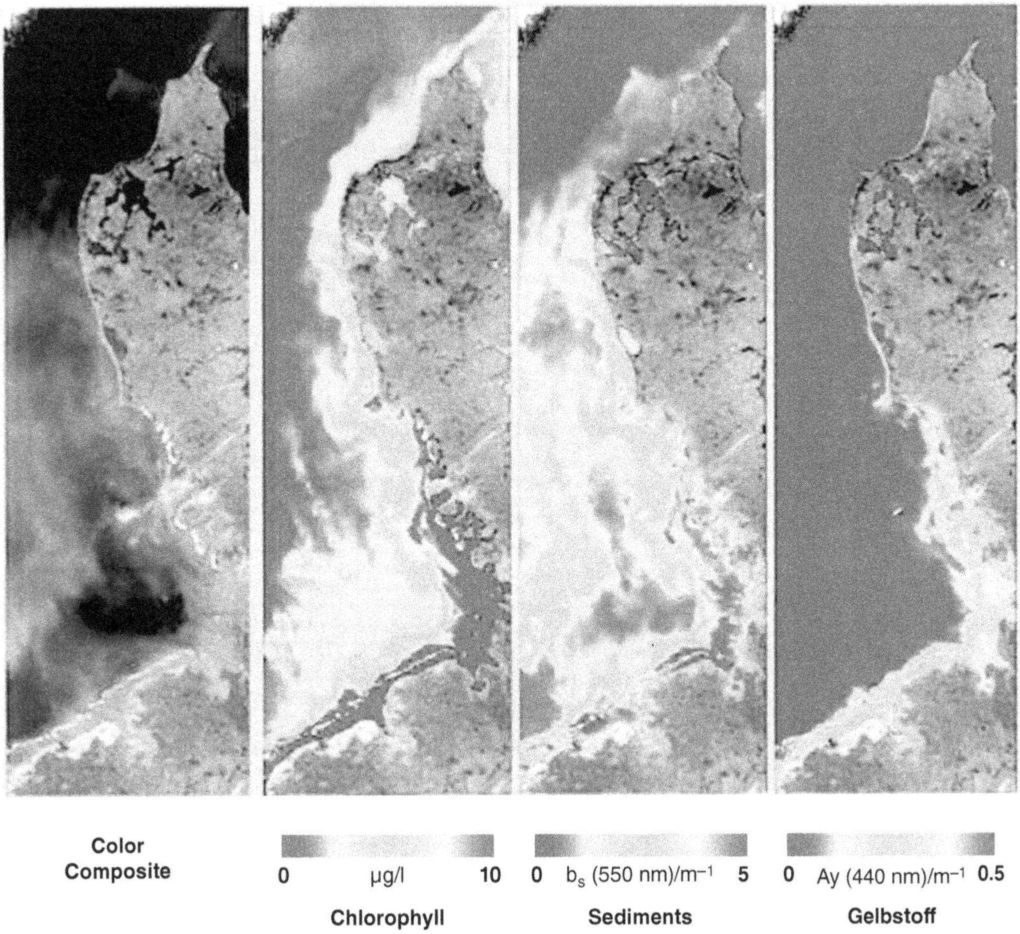

FIGURE 3.15 Water constituents over the German Bight derived from hyperspectral data. (Courtesy of German Aerospace Centre, DLR.)

Figure 3.16 shows an intense phytoplankton bloom near Tofino, British Columbia in Canada from airborne CASI imagery. The remnant red edge peak at 710 nm is displayed as the red channel in this RGB image, with bands on either side of the peak as green and blue. Some coastal blooms can kill fishes in net pens. Hepato and neural toxins from blue-green algae have killed cattle and other animals in lakes in Manitoba, Canada. Sadly, elsewhere in the world, humans have even died from freshwater blooms.

The primary product would be geocoded thematic maps showing concentration of chlorophyll *a* and algal type. Time series data would be used to examine the dynamics of eutrophication and algae blooms. Figure 3.17 shows near-shore chlorophyll mapped using 4 m × 4 m airborne CASI hyperspectral data over the Bay of Quinte, Ontario, Canada to simulate a MERIS fluorescence line height (FLH) image. Phytoplankton chlorophyll can be quantified based on solar-stimulated fluorescence at 685 nm. In this figure, the spatial resolution has been degraded to match the spatial resolution of the MERIS sensor. For optimal results of near-shore and coastal inlets the spatial resolution of spaceborne hyperspectral sensors needs to be higher than that of MERIS or OLCI.

Hyperspectral satellites can provide a cost-effective method to monitor chlorophyll levels on a regular basis and would have substantial benefits for the management of eutrophication, especially in remote areas. Accordingly, hyperspectral data would have an important role in detection and

FIGURE 3.16 Red tide observed on the west coast of Vancouver Island in September 1990 using an airborne CASI imaging spectrometer. (Image Courtesy of Borstad Associates.)

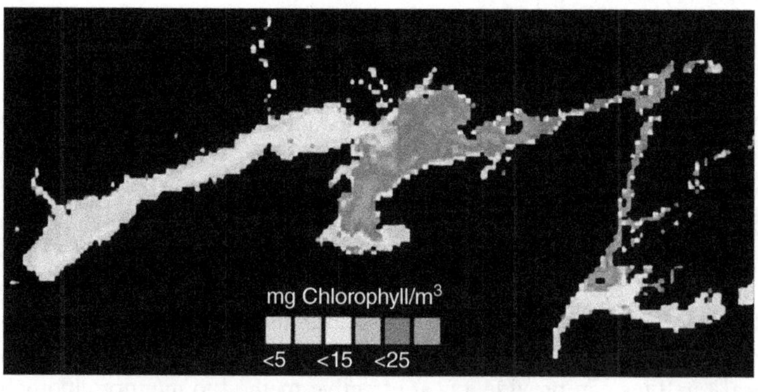

FIGURE 3.17 Near-shore chlorophyll mapped using airborne CASI hyperspectral data over the Bay of Quinte, Ontario, Canada. (Image Courtesy of Borstad Associates.)

monitoring of chlorophyll concentrations well before blooms occur. Time series data would be of particular use in identifying initiating sites and bloom dynamics.

The ability of using hyperspectral data such as Hyperion, CHRIS/Proba and HyspIRI to detect phycocyanin concentrations through chlorophyll *a* content was evaluated by Ogashawara et al. (2013). They simulated spectral bands of these existing and future spaceborne hyperspectral sensors. They evaluated the algorithms that detect the phycocyanin absorption feature in the water reflectance spectra between 600 nm and 625 nm. Results showed that the algorithms for retrieving chlorophyll *a* of phycocyanin were less sensitive for the Hyperion-simulated data than for the CHRIS-simulated data. Studies reported by Devred et al. (2013) and Kudela et al. (2015) show that future HyspIRI data is able to map cyanobacterial blooms if the data is taken on the optimal day.

3.4.4 Fisheries Applications

Fish stocks are an important high-protein food source for mankind. During the past a few decades the fisheries productivity of the world has been declining due to pressures from overfishing, habitat change, pollution, and climate change. Sustainable use of marine resources requires effective monitoring and management of the world's fish stocks. Remote sensing techniques are being used to help manage fisheries at sustainable levels, while also guiding fishing fleets to locate fish schools more efficiently. Fish tend to aggregate in ocean areas that exhibit conditions favored by specific fish species. Some of the relevant oceanographic conditions, such as sea surface temperature, ocean color (productivity) and oceanic fronts, which strongly influence natural fluctuations of fish stocks, can be observed and measured by spaceborne and airborne sensors. The remotely sensed data are provided in near-real time to help fishermen save fuel and ship time during their search for fish; to modelers who produce fisheries forecasts; and to scientists who help develop strategies for sustainable fisheries management.

In coast zones, bays, and estuaries, the water becomes quite turbid and contains suspended sediment, dissolved organics, and other substances, in addition to chlorophyll. Increasing emphasis has to be placed for understanding processes affecting stock recruitment and the conservation of nursery habitats. Coastal fisheries managers concern about degradation of coastal water quality as it relates to increased nutrient and sediment loading, and coastal development. Coastal fisheries include a number of issues. To identify each substance in this complex mixture of Case II waters requires higher spatial resolution, hyperspectral sensors and more sophisticated algorithms than the empirical regression models used in Case I waters for the open ocean.

The sea surface temperature and ocean color have long been used to help locate high fish densities and have been used in the off-shore fishery. With hyperspectral data, fish stock distribution models could be extended to a wider variety of coastal and shelf fisheries.

The aquaculture industry has become an important component in the coastal fishery. Key issues as they relate to this industry are the identification of suitable sites for expansion and the management of environmental impacts of and on the industry. Of prime concern is the monitoring and maintenance of water quality. There is a need to monitor potentially harmful algae blooms and particle-bound contaminates. There are also concerns related to the effects of intensive aquaculture on the water quality of adjacent ecosystems.

Hyperspectral products for the wild fisheries include geocoded maps showing the distribution and concentration of coastal phytoplankton and suspended solids. Hyperspectral products for the aquaculture industry would include maps of phytoplankton concentration (including phytoplankton blooms), phytoplankton type, and light attenuation.

In addition, hyperspectral products provide valuable time series data to monitor fish habitat and distribution, especially in the optically complex coastal waters. Water quality products provide fisheries managers with valuable insight into algae bloom dynamics and early indications of water quality degradation for aquaculture.

3.4.5 Land-Related Water Applications

Land-related water targets include wetlands, bogs, fens, and salt marshes. Generally, these targets have spatial scales and time rates of change defined by land topography and vegetation growth and, therefore, belong to land targets.

Wetlands provide habitat for fish, wildlife, and waterfowl, hydrological benefits by regulating water levels and runoff, water quality protection and several economic benefits through recreation and tourism. Information needs for wetlands include mapping of wetland locations and extent, vegetation mapping (wetland classification), and the establishment of baseline data on wetlands such that quantitative assessments of wetland change (either natural or anthropogenic) and associated impacts can be made.

Land-related coastal applications include coastline, shoal and coral reef mapping, and beach, seaweed, and other surveys. Hyperspectral products for land-related water applications include:

- True-color images
- Coastlines at different tide elevations and water levels in wetlands
- Wetland vegetation maps (surface and subsurface aquatic vegetation)

Wetlands are extremely difficult to map using conventional techniques, since they often have too much water to move about on foot, and not enough water to move about in a boat. Hyperspectral data would be of great benefit in mapping coastal and inland wetland vegetation due to its high radiometric and spectral resolution. Good correlations between water quality and vegetation characteristics have been demonstrated.

A hyperspectral satellite would provide the necessary capability to monitor change due to short-term climatic events (storm-induced coastal changes) or longer-term changes due to different climate regimes and changes in land use.

Figure 3.18 shows a thematic map of wetland vegetation in the vicinity of Ladner Marsh, in the Fraser River Estuary south of Vancouver, British Columbia, Canada derived from 4 m × 4 m airborne CASI hyperspectral data. The CASI bands were defined to optimally match the spectral reflectance features of the marsh vegetation.

Turpie et al. (2015), for example, assessed the prospective of the hyperspectral sensor HyspIRI to observe fine-scale structure in coastal wetlands and concluded that the HyspIRI sensor can help to study coastal environment, properties, biochemical cycling, and water quality. Moreover, supported by its thermal infrared (TIR) instrument, HyspIRI is also able to follow hydrologic and hydrometeorological processes.

3.4.6 Coastal Zone Environment and Change Detection

Coast is a unique environment in which atmosphere, hydrosphere and lithosphere contact each other. Coastline is one of the most important linear features on the Earth's surface, which display a dynamic nature. Coastal zone, and its environmental management requires the information about coastlines and their changes. There is an ongoing need to monitor water bodies to detect episodic point source pollution events. Targets included in this category encompass events seen as water color changes compared to normal water targets. Such water color changes may be aperiodic, and due to known and continuing point sources such as vessel, municipal or industrial discharges, or unknown nonpoint sources from agricultural or other runoff. Some water color changes are also caused by dissolved materials in river outflow or land runoff.

At sea, combining the optical data with SAR data will aid in the separation of spill signatures from naturally occurring water color changes due to dissolved organic materials. The low backscatter areas resulting from oil slick damping would be one example. The spatial resolution and number of spectral bands required for these targets make them more suitable for hyperspectral missions,

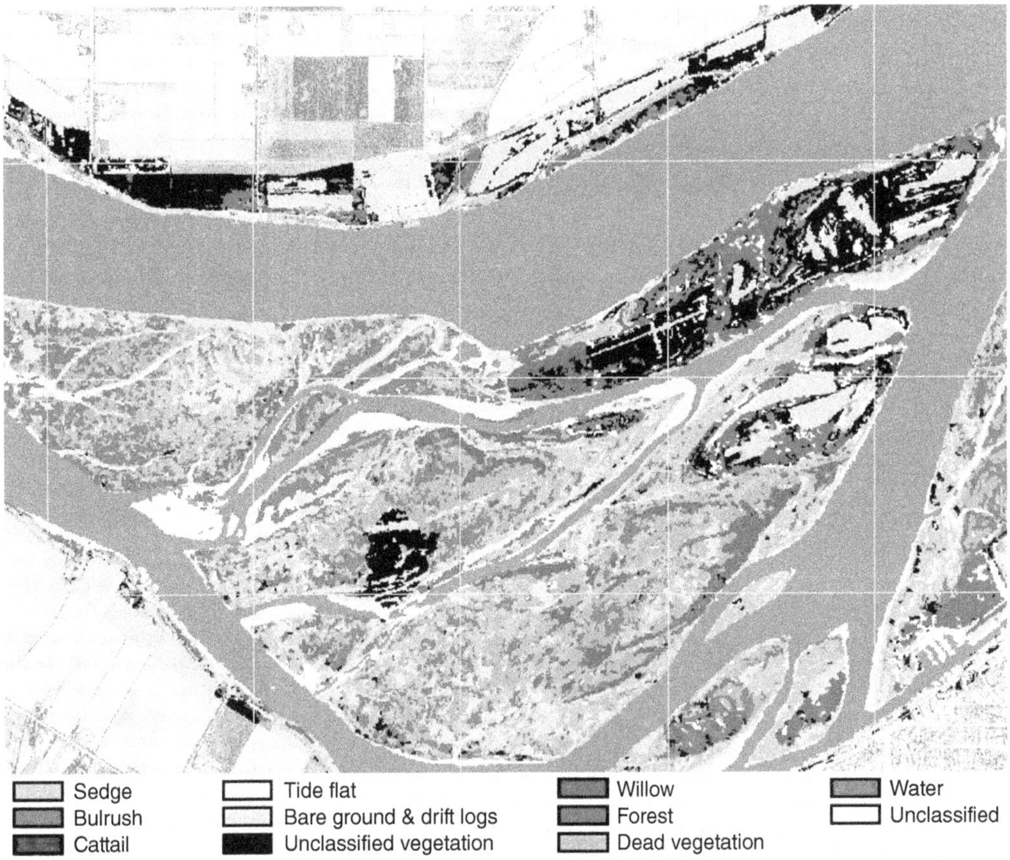

FIGURE 3.18 Classification map of wetland vegetation in the vicinity of Ladner in the Fraser River delta, British Columbia, Canada derived from 4 m × 4 m airborne CASI hyperspectral data. (Courtesy of Borstad Associates.)

which tend to cover a relatively narrow swath (20–50 km), with a high number of contiguous bands and a spatial resolution of 20–50 m.

Hyperspectral products include geocoded maps showing concentration of suspended and dissolved materials, including the location and labeling of point source pollution targets. These products are important to monitor human activity and reduce the risk of accidental or intentional spills affecting sensitive marine, coastal and inland waters. The spaceborne hyperspectral data significantly contribute to identifying the nature of the pollution event due to its increased radiometric and spatial resolution.

Ciampalini et al. (2015) reported a study using a field hyperspectral sensor and an airborne thematic mapper (ATM-2) multispectral sensor to characterize coastal environment in Pisa province coast (Tuscany, Italy). The aim of the study was to evaluate remote sensing techniques for identifying physic, morphologic and hydrodynamic features to characterize the coastal environment affected by erosive phenomena.

The study coastal area is characterized by an elevated anthropic pressure. It is about 26 km long and comprised between the Serchio River and the Scolmatore canal outlets. The coastal dune systems are mainly preserved within the park area. Three harbors in the area strongly interfere with the littoral sedimentary drift triggering erosive processes or producing sandy accumulation down or up drift of the port, respectively. The coast is characterized by sandy beaches faced a wide coastal plain formed by Late Quaternary deposits and fed, from north to south by the Magra, the Serchio, and the Arno rivers. The last two, and particularly the Arno river, feed the Pisa shoreline, even if in an insufficient way.

The field hyperspectral sensor is an Analytical Spectral Device (ASD) Fieldspec Spectroradiometer. It was used to collect samples reflectance spectra in 3–10 nm bandwidths over the 0.35–2.50 μm range. Four measurements were collected and averaged for each sample. Reference measurements on the Spectralon panel were acquired systematically every three samples. The principal component analysis (PCA) was applied to the spectral dataset to reduce the dimensionality of the original dataset.

Results obtained by the classic in-situ analyses (grain size and petrographic analysis) were used to validate those obtained using remote sensed data acquired by hyperspectral and multispectral sensors. The study concluded that the combined use of the hyperspectral sensing with the multispectral sensing is useful to retrieve spectral characteristics of coastal sediments that are related to their physical properties, such as petrographic composition and grain size distribution. The integrated approach using remote sensing and traditional techniques is useful for local authorities intervention in coastal areas. In particular, optical remote sensing technique can be profitably used in the characterization of coastal currents (longshore and rip) reducing the use of conventional analyses that are usually time and money consuming. The obtained results can be profitably used for a proper management of coastal areas allowing to plan intervention aimed at contrasting beach erosion phenomena.

3.5 FORESTRY APPLICATIONS

3.5.1 Hyperspectral Forest Products

Forestry is one of the major ecosystems of the Earth, supporting a wide diversity of plants and animals and a significant number of services to humankind. Forests store approximately 45% of terrestrial carbon and remain as a large net carbon sink by capturing one-quarter of the anthropogenic carbon dioxide (CO_2) each year (Bonan 2008, Pan et al. 2011). Monitoring current state of forest ecosystems and their changes is of prime importance for public policy and land management. The quality of forest information products currently generated by today's remote sensing technology could be improved dramatically by the use of hyperspectral sensors. Hyperspectral remote sensing offers a unique way to characterize forests as the fine spectral detail allows to characterizing the canopy chemistry and structure. New products derived from hyperspectral data that indicate forest chemistry can be used for monitoring forest health, precision forestry and environmental monitoring and assessment.

The various forestry application areas and different levels of hyperspectral forestry products are illustrated in Figure 3.19. This section describes three main application areas as follows:

- Forest inventory
- Forest chemistry
- Kyoto products

1. Forest inventory products are measurements of basic forest attributes, some of which can already be measured with multispectral remote sensing. Using hyperspectral imagery will improve these results, allowing for increased accuracy and details. These products include:

    1.1 Species Mix
    1.2 LAI and Crown Closure
    1.3 Forest Understory

2. Measurements of forest chemistry relates to element concentration and forestry health. The products include:

    2.1 Health with two subproducts of
        2.1.1 Insect
        2.1.2 Disease

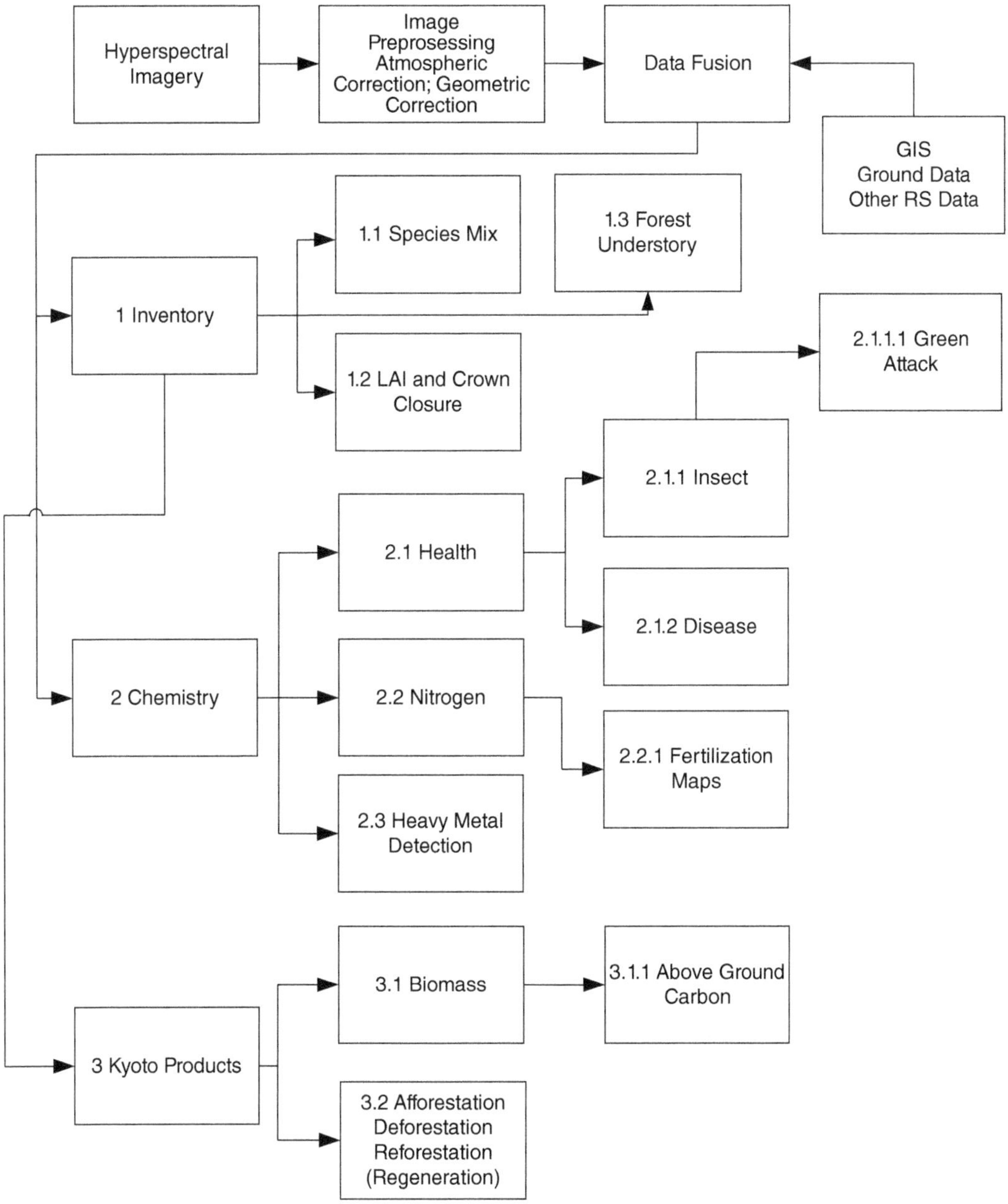

FIGURE 3.19 Block diagram of hyperspectral forestry products.

2.2 Nitrogen with subproduct of
  2.2.1 Fertilization maps
2.3 Heavy metal detection

3. The Kyoto products are a derivative of the inventory products. The products include:

3.1 Biomass with subproduct of
  3.1.1 Above-ground carbon
3.2 Afforestation, deforestation, and reforestation

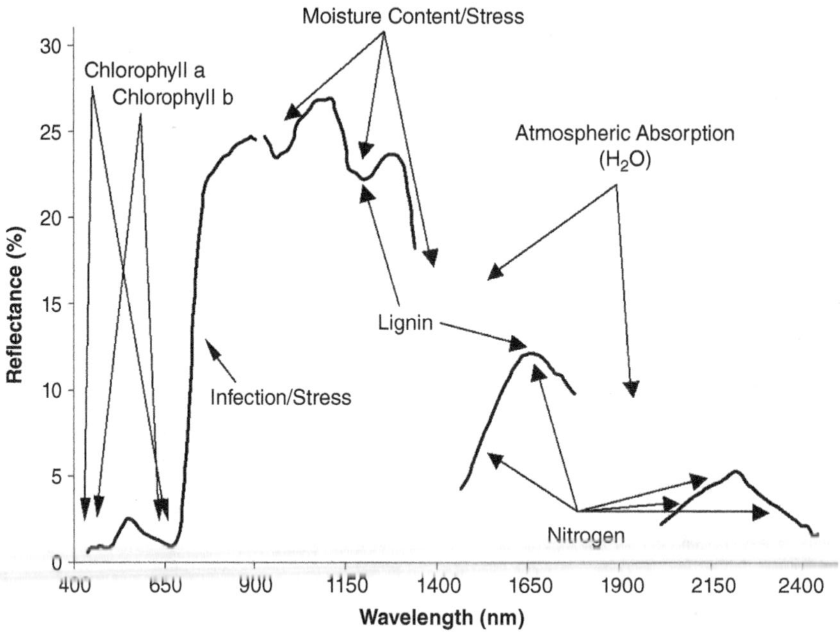

FIGURE 3.20 Spectral regions of interest for application of hyperspectral data to forest inventory and chemistry for health monitoring.

Contiguous narrow spectral bands of hyperspectral imagery data is critical for forestry applications. Figure 3.20 presents the spectral sampling regions of interest in order to identify the spectral features and signatures of canopy reflectance for a mixed deciduous forest. Selected spectral regions of interest for application of hyperspectral data to forest inventory and chemistry for health monitoring are also shown.

3.5.2 Forest Inventory

Estimating forest attributes has been the task of both governments and forest companies. Large geographic information systems (GIS) databases of forest polygons exist to provide inventories of basic tree parameters by way of species composition, quantity, density, age, height, site index, and crown closure. These parameters are then used to estimate forest species areas, timber volumes, biomass, etc. that are essential for national reporting. GIS databases vary in accuracy and detail. Hyperspectral remote sensing can provide improved, consistent, and accurate measures of forests.

The forest inventory products would consist of geocoded maps showing (1) dominant tree species, (2) estimation of LAI and crown closure, and (3) maps of forest understory with associated labels.

Forest classification using hyperspectral imagery yields higher accuracy of basic forest parameters when compared to traditional remote sensing data sources such as Landsat. Studies using airborne hyperspectral and multi-angular imagery of forest canopies have demonstrated improvements in forest-land classification results (Sandmeier and Deering 1999, Russel 2001). In these studies, the canopy optical and structural characteristics, as well as background understory variability, were shown to influence variability as well as multi-angular effects. White et al. (2002) examined the variability and extracted quantitative information on the canopy composition.

Using spaceborne hyperspectral data improved inventory maps and benefits sustainable development management practices and environmental monitoring. Goodenough et al. (2003) carried out forest classification for the Greater Victoria Watershed in British Columbia, Canada, using hyperspectral data acquired with the Hyperion sensor onboard EO-1 satellite and compared the forest classification results from Hyperion, ALI (Advanced Land Imager, a multispectral sensor also onboard EO-1 satellite), and the Enhanced Thematic Mapper Plus (ETM+) of Landsat-7. The forest classes chosen for discrimination included douglas-fir, hemlock, western redcedar, lodgepole pine, and red alder. Goodenough et al. (2003) reported that the overall classification accuracies obtained for each satellite sensor were 90.0% for Hyperion, 84.8% for ALI, and 75.0% for ETM+. Hyperspectral remote sensing provides significant advantages and greater accuracies over ETM+ for forest discrimination. A 5% difference is significant. Accordingly, Hyperion is superior in this case. Hyperion and ALI sensors provided data with excellent discrimination for Pacific Northwest forests in comparison to Landsat-7 ETM+. Figure 3.21 shows the forest classification results using the Landsat-7 ETM+ multispectral data and the Hyperion hyperspectral data.

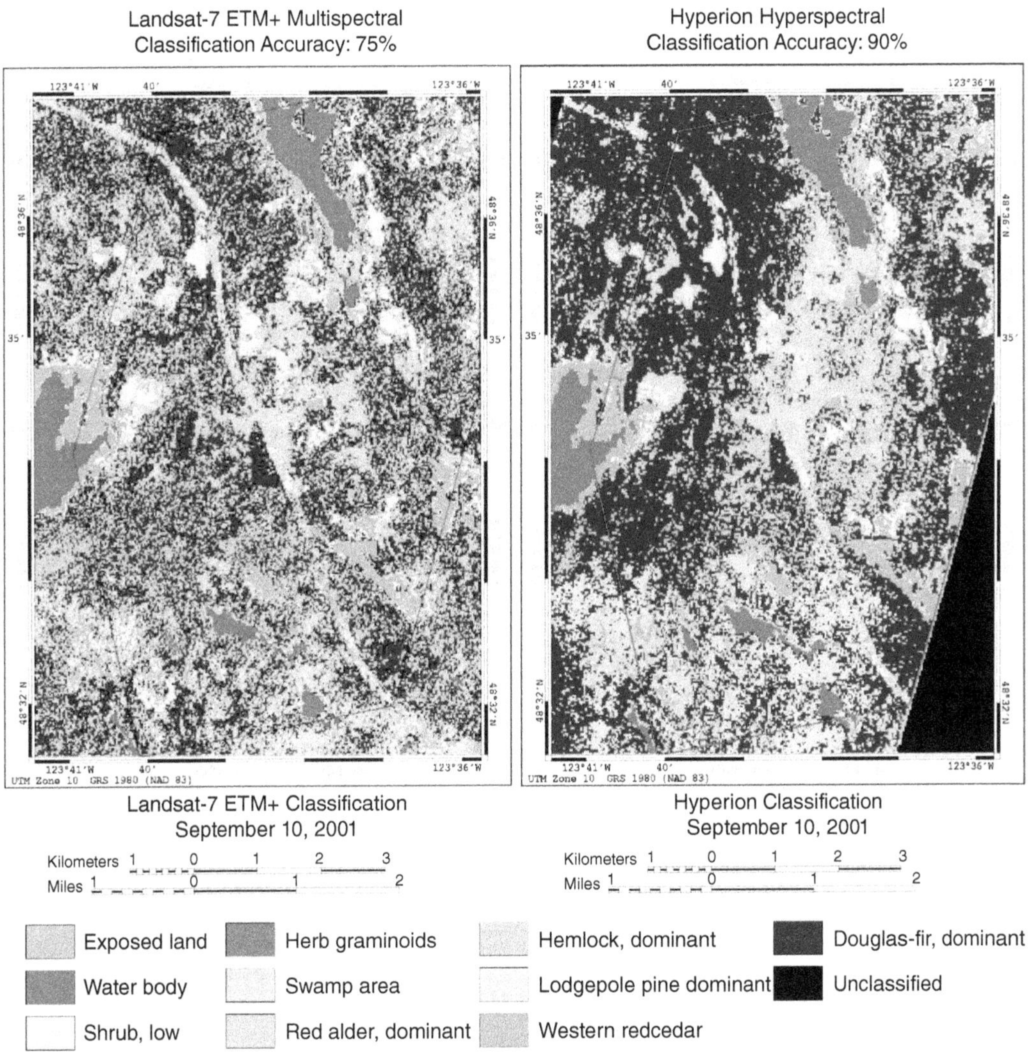

FIGURE 3.21 Forest classification of five species (red alder, hemlock, lodgepole pine, western redcedar, and douglas-fir) in the Greater Victoria Watershed, British Columbia, Canada using Landsat-7 ETM+ multispectral data and Hyperion hyperspectral data. (Courtesy of Pacific Forestry Centre, Natural Resources Canada.)

3.5.3 Forest Chemistry

Estimates of the chemical concentrations (chlorophyll *a* and *b*, nitrogen, lignin, water content) of forest canopies can be achieved using hyperspectral data. The chemistry can relate to the overall forest health or concentrations of particular chemical elements (Martin et al. 1998). Forest health is affected by disease (e.g., Dutch elm disease, root rots, mistletoes) and insect infestation (e.g., bark beetles, defoliators, weevils). Hyperspectral imagery may be able to detect stress in trees at the early stages of insect attack and, therefore, can be used to fight early insect infestations (also referred to as "green attack"). The background understory layer of a canopy can act as an early warning bio-monitor of conditions affecting canopy health, related to such variables as soil surface CO_2 effluxes (Law et al. 2001), acidification, nitrogen availability (van Dobben et al. 1999), and soil evapotranspiration. Hyperspectral imagery can allow for a more detailed forest classification related to the contribution of understory reflectance to the observed hyperspectral signal. Estimates of particular chemical elements such as nitrogen can be used for precision forestry practices. Fertilization can be applied only to areas with nitrogen deficiencies. Environmental monitoring and assessment of forested areas surrounding new mine sites, for example, can be accomplished by periodically monitoring chemical concentrations of these trees using hyperspectral imagery.

For hyperspectral imagery to estimate quantifiable concentrations of chemistry, the imagery must be calibrated and the absorption features must be well measured and understood. Hyperspectral products for forests consist of geocoded maps showing location and severity of green attack or the effects of various plant diseases. Additional products could include fertilization maps for precision forestry practices and change-detection maps showing potential impacts on forest health in areas surrounding new mine sites, and regeneration and health of harvested areas.

Hyperspectral data could provide more accurate measures of forest health and could, therefore, be used to more effectively manage the forest resource. The early detection of green attack can be incorporated into harvest planning to minimize and control green attack before it spreads. Accurate determination of tree chemistry combined with temporal monitoring could be used to assess causes of stress and could ultimately be used toward precision forestry practices.

Insects attack different parts of the tree, with defoliators feeding on leaves or needles, and bark/wood borers penetrating into the bark/wood. While some insect species are native to local ecosystems, many of the recent disturbances arise from the nonindigenous species that may pose more pernicious and unpredictable threats to forest health (Boyd et al. 2013). Forest diseases are caused by pathogens that are infectious and transmissible, such as bacteria, fungi, viruses, and helminths.

Subtle spectral discrepancies between healthy and damaged trees can be better detected by fine-spectral resolution hyperspectral images (Hatala et al. 2010). New algorithms have been developed for narrowband vegetation indices, while the others were simple modifications of the traditional vegetation indices by means of substituting narrowband for broadband reflectance, such as the red-edge NDVI. These narrowband indices have shown the potential to explain the physiological changes in forests suffering damage from insects (Oumar et al. 2013, Fassnacht et al. 2014).

The detection of early-stage forest damage offers forest managers an opportunity to perform efficient disease and insect control. The infected trees may only show a slight decline in chlorophyll levels and leaf water content. Previous efforts (e.g., Pu et al. 2008, Fassnacht et al. 2014) have confirmed the potential of applying hyperspectral remote sensing to assist with early detection of tree stress. Most of the hyperspectral data used were acquired with airborne sensors (e.g., AVIRIS, CASI, and HyMap), which makes data acquisition an expensive process. Therefore, developing Landsat-like hyperspectral sensors is a promising solution to address this challenge.

White et al. (2007) computed moisture indices derived from the Hyperion data to detect insect damages. Indices with the largest correlation ($R^2 = 0.51$) to the proportion of insect damage were those utilizing bands in the SWIR and NIR regions. Pest identification is thus a particular case of vegetation classification, but at a finer level of detection.

Oumar et al. (2013) reported a study of using field spectra resampled to the Hyperion band characteristics to predict forest damage caused by *Thaumastocoris peregrinus* (*T. peregrinus*), a sap-sucking insect that feeds on Eucalyptus leaves. Normalized indices based on NDVI ratios were calculated using the resampled VNIR bands of the Hyperion sensor to assess its utility in predicting *T. peregrinus* damage using partial least squares regression. The top 20 normalized indices between 803.3 nm and 894.9 nm were based on specific biochemical absorption features that predicted *T. peregrinus* damage with a mean bootstrapped R^2 value of 0.63 on an independent test dataset. The greedy backward variable selection model identified three indices and performed the best by predicting damage with an R^2 value of 0.74 with the lowest RMSE of 1.30% on an independent test dataset. The best three indices identified are the anthocyanin reflectance index, carotenoid reflectance index and the normalized index calculated at spectral bands between 864.4 nm and 884.7 nm.

Mapping bark beetle-induced tree mortality based on airborne HyMap hyperspectral data was reported by Fassnacht et al. (2014). They examined three scenarios of different mortality classes using a combining genetic algorithm (GA) for feature selection and a supervised support vector machine (SVM) classification. Scenario 1 included three mortality classes, while Scenario 2 and 3 each included only one general mortality-related class but with different definitions. They reported that the separation between the healthy trees and the dead trees was possible with overall accuracies between 94% and 97%.

Carter et al. (2009) carried out tamarisk invasive species mapping along the Colorado River using summer-acquired Hyperion data. Ramsey et al. (2005) developed a method to detect and monitor the establishment and spread of Chinese tallow and other invasive species by combining the Hyperion hyperspectral data with a spectral subpixel-mixing model. Characteristic spectra were significantly associated with green vegetation and senescent foliage canopy compositions and explained 78% of the percentage occurrences of red tallow as established by comparison to classifications of field images. These techniques based on the Hyperion hyperspectral data and a specialized subpixel-extraction model provided subtle spectral detection of red tallow not provided by the other satellite sensors, such as Landsat TM.

3.5.4 Kyoto Products

Hyperspectral remote sensing can be used to generate some Kyoto Protocol information products. These can be derived from parameters measured by hyperspectral remote sensing such as those described in Section 3.5.2. The temporal components of these measures provide trends for policy makers and forest managers.

The spaceborne hyperspectral products include geocoded maps of forest biomass and more specifically maps of above-ground carbon. To establish long-term trends as they relate to biomass and carbon budgets, change detection maps quantifying afforestation, deforestation, and reforestation need to be generated. Spaceborne hyperspectral data could provide more detailed and accurate forest inventory information, and the ability to provide specialized products such as above-ground carbon maps. Figure 3.22 shows a map of above-ground carbon derived from hyperspectral imagery data in Hinton, Alberta, Canada.

For example, Canada's international commitments in the Kyoto Protocol were to report on forest inventory, afforestation, deforestation, and reforestation and above-ground carbon annually starting in 2008. The Carbon Budget Model developed by the Canadian Forest Service (CBM-CFS2) was the primary system used by the Canadian Forest Service (CFS) to describe the dynamics of species groups and assess the status of carbon distributions in Canada's forests (Kurz et al. 2002). The input data for the model were from provincial forest inventories, and growth curves from temporary and permanent sample plots. The more accurate forest classifications, derived from hyperspectral data, improved the accuracies in forest inventories, which led to more accurate estimates produced by the CBM-CFS2 model. Another contribution was that forest biomass, above-ground carbon, afforestation, reforestation, and deforestation could be estimated directly from remote sensing imagery.

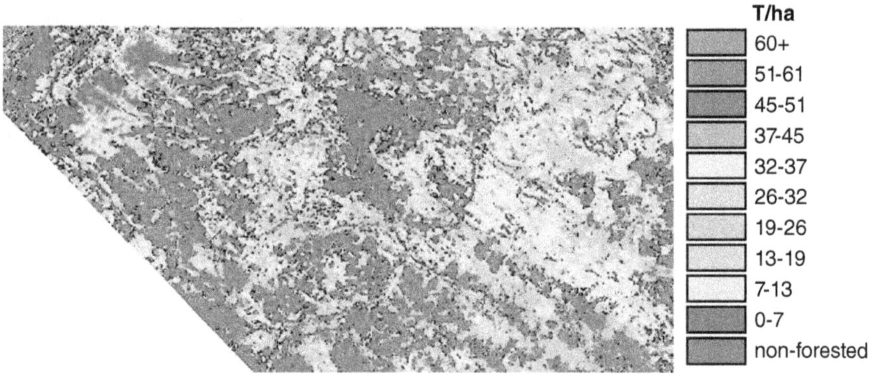

FIGURE 3.22 Map of above-ground carbon derived from hyperspectral imagery data. (Courtesy of Pacific Forestry Centre, Canadian Forest Service.)

These can be used to compare and validate the modeled biomass, carbon, and change products. The Kyoto Protocol products from hyperspectral remote sensing and the CBM-CFS2 model with more accurate forest information could help countries, such as Canada, to meet the reporting commitments on the sustainability of their forest resources.

3.6 GEOLOGY AND MINERAL EXPLORATION

Since the 1980s, mineral exploration is among the first operational applications of Earth observation data in general, and the industry is now a knowledgeable user. Typical customers include oil, gas and mining companies, as well as government authorities.

Investigations carried out over the last 30 years show that hyperspectral remote sensing can make major new contributions to the advancement of geological applications, especially in the areas of identification and mapping of minerals, and lithological mapping in arid environments. This is possible because hyperspectral sensors, in contrast to other existing broad-band multispectral sensors, are able to resolve absorption features unique to specific minerals. As a result, hyperspectral remote sensing has the potential to augment traditional surveying techniques, which can cost between $150 and $500 per square kilometer. In contrast, a hyperspectral survey costs only between $10 and $30 per square kilometer.

The following sections outline the mineral exploration applications that benefit from the availability of spaceborne hyperspectral remote sensing data. Two major application areas of hyperspectral data to mineral exploration have been selected:

1. Lithologic mapping (i.e., direct sensing of bedrock and minerals)
2. Geobotanical mapping (i.e., indirect sensing of surficial geology)

The use of hyperspectral data for lithologic mapping is more mature than geobotanical mapping, both of which represent a large potential market.

3.6.1 Lithologic Mapping

The key elements in mineral exploration are (1) to gain a preliminary understanding of a geographic area through lithological mapping, and (2) to assist in defining potential exploration targets prior to initiating intensive field exploration activities. Hyperspectral data are well suited to mapping bedrock and identifying the presence and abundance of specific diagnostic minerals at specific scales.

Products derived from hyperspectral data include categorized images of bedrock distribution with geological labels, and maps detailing the distribution of specific minerals and their abundances.

These maps provide geologists with an additional tool to decipher the overall lithologic and structural history of a region and help to define potential exploration targets.

The minerals, which have been successfully identified to date with imaging spectroscopy, can be grouped as follows: OH-bearing minerals, carbonates, sulfates, olivines, pyroxenes, iron oxides and hydroxides. The identification of minerals and the mapping of their distribution provide the necessary underpinnings for exploration purposes (precious and base metals, diamonds, etc.) and lithological (rock) mapping. The latter application is of particular interest in areas where either no maps or very generalized maps exist, such as in arctic environments. The former category also includes petroleum exploration where hyperspectral remote sensing may assist in the detection of hydrocarbon micro-seepage.

Mining companies have used several key-alteration minerals to map potential mineral prospects. Figure 3.23 shows a map of abundance for different minerals. From this map, specialists of the mining company were able to determine the signatures directly associated with a known copper (Cu) prospect. Given the similarity of signatures of the two southern features, new exploration opportunities were identified. These were later verified in the field to be legitimate prospects.

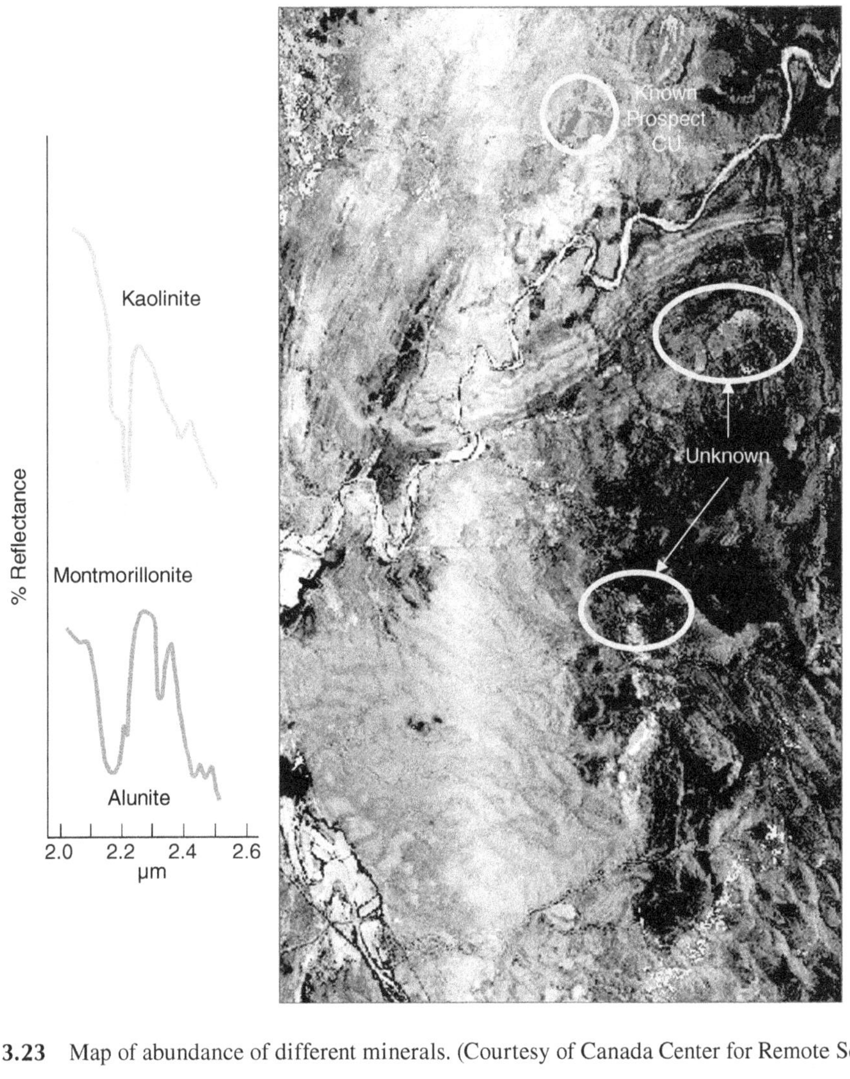

FIGURE 3.23 Map of abundance of different minerals. (Courtesy of Canada Center for Remote Sensing.)

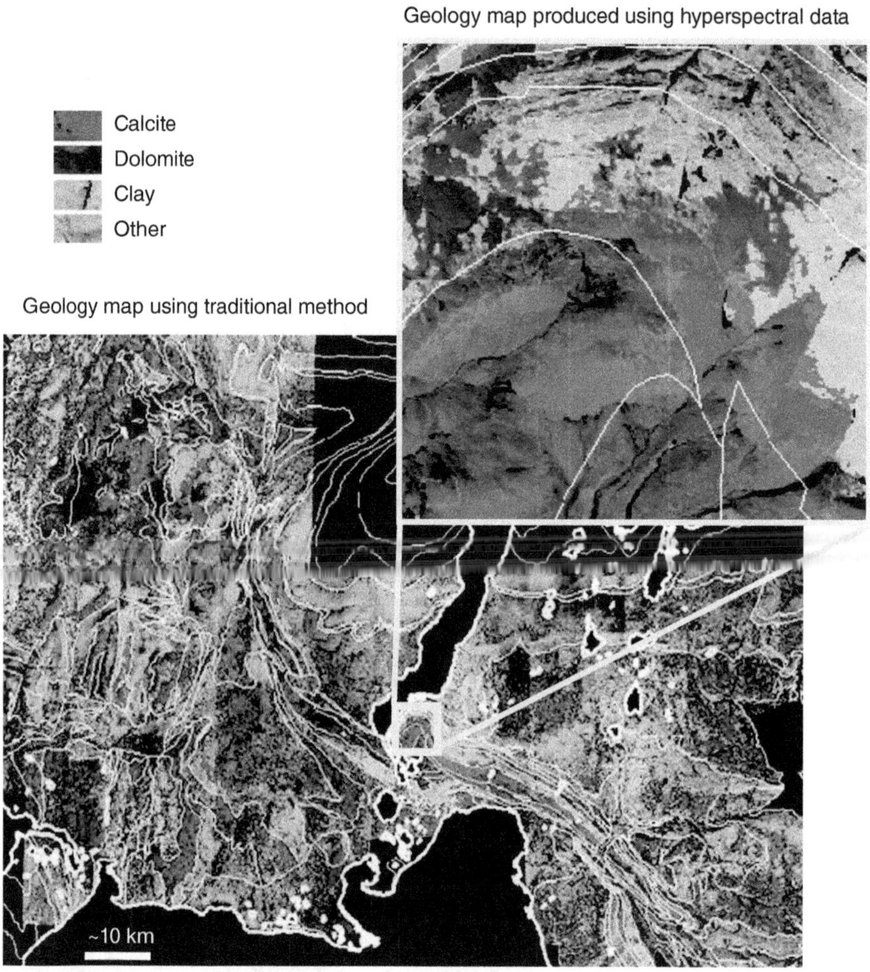

FIGURE 3.24 A geology map in the Canadian Arctic produced using the traditional method and hyperspectral data.

Figure 3.24 shows a geological map in the Canadian Arctic produced by the Geological Survey of Canada (GSC) using traditional methods draped over the image (line data). The purpose of this figure is to demonstrate the geologic mapping capability of hyperspectral data in the Canadian Arctic. The image in the detail window produced using hyperspectral data clearly illustrates much greater detail on structure, several sub-features are evident that do not appear on the regional geology map. This image demonstrates the true power of hyperspectral data in the north. Hyperspectral mapping provides an accurate map, thus reducing time and cost to produce a far superior product. Figure 3.25 shows a lithological map in the arctic in Nanisivik, Nunavut, Canada produced using Probe-1 hyperspectral data.

To be useful to the end users, products derived from hyperspectral data should be readily integratable with other data sources, possibly in a geology-specific GIS. The analysis of the products is typically conducted in conjunction with traditional geophysical data (magnetics, gamma ray). Availability of digital elevation data (DEM) at a scale of 1:20,000 is also beneficial in the analysis of hyperspectral data.

Hyperspectral products are relatively mature in arid (non-vegetated) environments. Basic research is still required to bring the product to maturity in vegetated terrains. Mining companies are focusing mainly on sparsely vegetated regions.

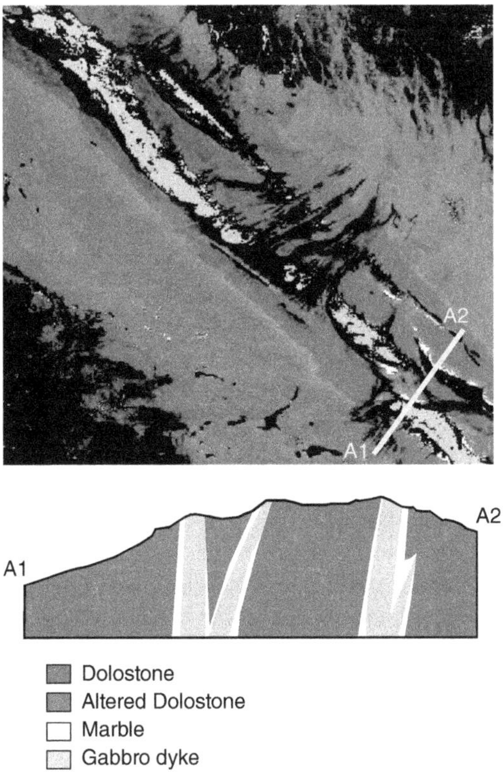

FIGURE 3.25 Lithological map in the arctic (Nanisivik, Nunavut, Canada) produced using Probe-1 hyperspectral data. (Courtesy of Canada Center for Remote Sensing, Natural Resources Canada.)

The prime benefit of hyperspectral remote sensing to the mineral user community is the ability to identify the presence and abundances of specific diagnostic minerals that will help direct mineral exploration activities to areas of high potential at considerable cost savings.

Hyperspectral data produce better discrimination of the potential geological components than multispectral data of the similar GSD. Leverington (2008, 2010) reported that hyperspectral data acquired by Hyperion were found of greater utility than data from the broadband Landsat TM to create surface cover abundance maps. They used the Hyperion data to separate several important sedimentary lithological end-members in Texas and Canada. Kruse et al. (2003) differentiated calcite from dolomite and detected differences in solid solution of micas using the Hyperion data. Gersman et al. (2008) spotted ammonium-bearing minerals with the 1558-nm band of the Hyperion data.

The capability of hyperspectral imaging for mine-waste monitoring was evaluated by Mielke et al. (2014a). They pointed out the critical revisit time of the satellite and suggested to combine hyperspectral data with multispectral data such as Sentinel-2 for repetitive area-wide mapping. The wide spectral range of hyperspectral sensors appeared to be a good advantage for geological applications.

Mielke et al. (2014b) compared the results of the Hyperion and the simulated EnMAP data with multiple multispectral sensors, including the Operational Land Imager (OLI), ALI, ASTER, ETM+, and Sentinel-2 for mine waste monitoring. They concluded that the EnMAP data provided the best overall accuracies for mapping gold-mining material, pyroxene proxies for platinum and other iron absorption feature of minerals linked to mine waste. In another study, Mielke et al. (2016) showed that the simulated EnMAP data provided better results than the Hyperion data in characterizing mineral deposits and highlighting exploration anomalies (i.e., unexpected soil properties, which are indicators of valuable soil elements). They developed an algorithm and expert system (EnGeoMAP, the EnMAP Geological Mapper) to detect metal sulfide mineral deposit sites. Despite medium spatial resolution (30 m GSD), discrimination of mafic and ultramafic rock units was feasible at a broad scale with simulated EnMAP images (Rogge et al. 2014).

3.6.2 Geobotanical Mapping

In most areas of the world, surface geology to some degree is obscured by vegetation. Geobotanical mapping (i.e., exploration in vegetated environments) has the potential of identifying the areal distribution of specific element-associated spectral changes in vegetation, which are related to soil geochemistry, lithology, or surficial materials. This approach makes use of the fact that the reflectance of vegetation is spectrally affected in the presence of heavy metals or alteration zones. For example, accumulation of heavy metals induces stress on the vegetation causing a shift of the red-edge (680–800 nm). Such a shift is only detectable with a hyperspectral imager.

The premise is that spaceborne hyperspectral data allows identification of element-specific geobotanical anomalies on the basis of absorption features or changes in the continuum of vegetation spectra. The output products consist of classified maps or images of the areal extent (and degree) of element-specific geobotanical variations.

Geobotanical anomalies associated with ore bodies may also be expressed as abrupt changes from one plant community to another as a function of the underlying surficial geology and not necessarily due to stress induced physiological changes in plants.

Although the use of geobotanical mapping is very promising, the ability of achieving element specific geobotanical products is yet to be developed and requires further basic research and development before it achieves operational status.

Spaceborne multispectral sensors are capable of providing gross-lithologic information or identification of stressed vegetation, but not detailed mineralogy or geobotany of importance for mineral exploration. This function is undertaken by field crews, which is expensive and time consuming. Hyperspectral remote sensing can potentially have a significant impact in detecting geobotanical anomalies or trends.

Distinguishing minerals talc, dolomite, chlorite, and white micas from green vegetation background using the Hyperion data was reported by Cudahy et al. (2001). Following radiative transfer correction of the Hyperion at-sensor radiance data to surface radiance (apparent reflectance), diagnostic spectral signatures (i.e., endmembers) were clearly apparent, including green vegetation, talc, dolomite, chlorite, white mica, and possibly tremolite. Even though the derived surface composition maps generated from these endmembers were noisy due to its low SNR (<40:1) of the Hyperion sensor in the SWIR, they were nonetheless spatially coherent and correlated well with the known geology.

Mapping iron oxides in vegetated shrub areas using Landsat-8 OLI and the Hyperion imagery integrated with airborne geophysical data was reported by Ducart et al. (2016). The remote detection of iron oxides was rendered difficult in heavily or sparsely vegetated areas, because vegetation obscures the reflectance characteristics of the underlying substrate. Feature extraction and spectral angle mapper (SAM) classification methods were carried out on the Hyperion hyperspectral imagery with good results for mapping high-grade iron ore, the hematite-goethite ratio, and clay minerals from regolith even in vegetated shrub areas. The Hyperion imagery is proved to be an excellent tool for fast remote mineral mapping in open-pit areas, as well as mapping waste and tailing-disposal facilities.

3.7 DEFENSE AND SECURITY

Airborne hyperspectral instruments have been available from various providers for many years and their performance keeps improving. On the other hand, the availability of spaceborne hyperspectral sensors has been very limited. Spaceborne hyperspectral sensors are described in Chapter 2, including a military demonstration hyperspectral sensor ARTEMIS.

Spaceborne hyperspectral sensors can play a vital role for the defense and security applications in monitoring territories and military operations and search and recues. Broad area surveys are required routinely, and image databases are needed to assist in the tracking of observed structures and objects over time. To improve situational awareness and decision making, intelligence, surveillance, and reconnaissance (ISR) systems often fuse information from many different sensors and sources. Remote sensing information, such as obtained from hyperspectral sensors, is one of many sources of information to be fused in ISR systems. Hyperspectral sensors on spaceborne platforms will significantly improve surveillance, target identification, terrain mapping and visualization capabilities for both land and sea operations.

3.7.1 Target Detection

Target detection and identification are one of the main applications of hyperspectral remote sensing. This includes targets, such as military vehicles, camouflages, and various man-made targets of different materials.

Hyperspectral imaging offers a range of promising civilian and military applications for target detection and material identification. The current multiplication of airborne and spaceborne collection platforms of ever finer radiometric, spectral, and spatial resolutions are allowing a wider use of this technology outside of the research and development community to clients with operation related needs. In a defense intelligence context, a capability for material identification based on spectral features is particularly interesting, potentially allowing for unattended or semi-attended search and detection of targets of interest over large areas. This is even more interesting considering that hyperspectral imaging has the capability to detect unresolved (subpixel in size) targets either too small to be imaged with the sensor's spatial resolution or partially concealed by the surrounding environment.

Among available target detection techniques, those based on unstructured statistical background modeling are widely used in the hyperspectral community, mainly because of their empirically demonstrated good detection performances and their ease of use. Methods have been proposed to increase the detection performance of those algorithms by including information on target abundance through physics-based modeling and linear spectral unmixing (Broadwater and Chellappa 2007) and by combining multiple detectors and performing detection in a multidimensional decision space (Foy et al. 2009).

3.7.1.1 Spectral Unmixing Based Target Detection

Qian and Lévesque (2009) carried out a study on target detection via hyperspectral imaging using the spectral unmixing approach. They assessed the effectiveness of a previously proposed noise reduction technology for hyperspectral imagery to examine whether it can help increase the detectability of targets after noise reduction using the unmixing technology. The hyperspectral datacube tested was acquired using an airborne SWIR Full Spectrum Imager II (SFSI-II) (Neville et al. 1995) in June 2002 at an altitude of 1800 m with a ground pixel size of 2.20 m × 1.85 m, and 240 bands covering a spectral range between 1200 nm and 2450 nm with a band interval of 5 nm. Man-made targets with different materials and sizes were deployed in the scene of the datacube that covers a mixture of sand and low-density grass. Three pieces of awnings with varying sizes ranging from 3 m × 3 m to 12 m × 12 m, two pieces of polythene, two pieces of white tarp and two pieces of white cotton with varying sizes ranging from 3 m × 3 m to 6 m × 6 m were deployed.

The acquired raw datacube was first preprocessed to remove periodic noise, dark current, slit curvature (also referred to as smile distortion) and spatial distortion (also referred to as keystone). Then a vicarious calibration was performed using calibration coefficients obtained from a previous SFSI-II survey to convert the raw data into at-sensor radiance. The radiance data were stored as 16-bit values. The size of the datacube is 140 lines by 496 pixels per line by 240 bands.

Because the targets were man-made and their spectral signatures were known, endmembers of four target materials were derived from the datacube and used to unmix the datacube into fraction images (also referred to as abundance images) corresponding to the target materials. The fraction image of a target material (e.g., awning) was then used to identify and detect the targets of that material. The spectral unmixing algorithm in the Imaging Spectrometer Data Analysis System (ISDAS) software suite (Staenz et al. 1998) was used to generate the fraction images.

The noise reduction technology for hyperspectral imagery assessed was the Hybrid Spectral-Spatial Noise Reduction (HSSNR) technique (Othman and Qian 2006). The detectability of the man-made targets derived from the original SFSI datacube and from the noise-reduced datacube with HSSNR was evaluated using both the receiver operating characteristic (ROC) approach and three predefined criteria.

The ROC curve is a graphical plot that illustrates the performance of a binary classifier system as its discrimination threshold is varied. It is created by plotting the fraction of true positives out of the positives (true positive rate, TPR) versus the fraction of false positives out of the negatives (false positive rate, FPR) at various threshold settings. TPR is also known as sensitivity, and FPR is one minus the specificity or true negative rate.

The ROC analysis provides tools to select possibly optimal models and to discard suboptimal ones independently from the cost context or the class distribution. ROC analysis is related in a direct and natural way to cost/benefit analysis of diagnostic decision making. ROC analysis has been used in medicine, radiology, biometrics, and other areas for many decades and is increasingly used in machine learning and data mining research.

The ROC curves are widely used in military target detection and the comparison of different target detection techniques (Manolakis et al. 2003). The more upward bowed the ROC curve is, the better the detectability. ROC curves of targets of awning, white tarp and cotton targets of size from 3 m × 3 m to 12 m × 12 m were created. Figure 3.26 shows some of the ROC curves of the targets derived from the original SFSI datacube and from the noise-reduced datacube using HSSNR. It can be seen from the figure that the ROC curves of the 6 m × 6 m awning target derived from both the original SFSI datacube and from the noise-reduced datacube are perfect. The two curves fully overlap the vertical line of 0 probability of false alarm (PFA) and the horizontal line of 100% probability of detection (PD). The ROC curves of the 12 m × 12 m awning target derived from both the datacubes are almost perfect and overlapped. The PD is 96% when PFA is 0, and 100% when PFA is >3%.

The ROC curves of the 6 m × 6 m white tarp and cotton targets derived from the noise-reduced datacube show better detectability than those derived from the original SFSI datacube. For the 3 m × 3 m targets, the ROC curve of the awning target derived from the noise-reduced datacube is slightly better than that derived from the original datacube, while the ROC curve of the cotton target derived from the original datacube is significantly better than that derived from the noise-reduced datacube. ROC curves derived from the noise-reduced datacube do not show constant improvement on detectability for the 3 m × 3 m targets. This is because these targets were subpixel targets due to the ground footprint of the SFSI-II instrument being 2.2 m × 1.85 m and the flying direction of the aircraft, which is not appendicular to lines of the target panels [see Section 12.5.1 in Qian (2013)]. The spectral unmixing results generated for these targets are not stable. Target detection using the spectral unmixing approach for subpixel targets has been described in detail in Section 12.5 in the book by Qian (2013).

3.7.1.2 Stochastic Detection Algorithms

Constrained energy minimization (CEM) and Kelly's spectral detection are Stochastic detection algorithms. Roy (2010) reported an easy-to-use hyperspectral target detection technique by hybridizing CEM and Kelly's spectral detection algorithms. Results show that the proposed hybrid algorithm enables the detection of subpixel targets at a false alarm rate (FAR) lower than either achieved with the CEM or Kelly's algorithms.

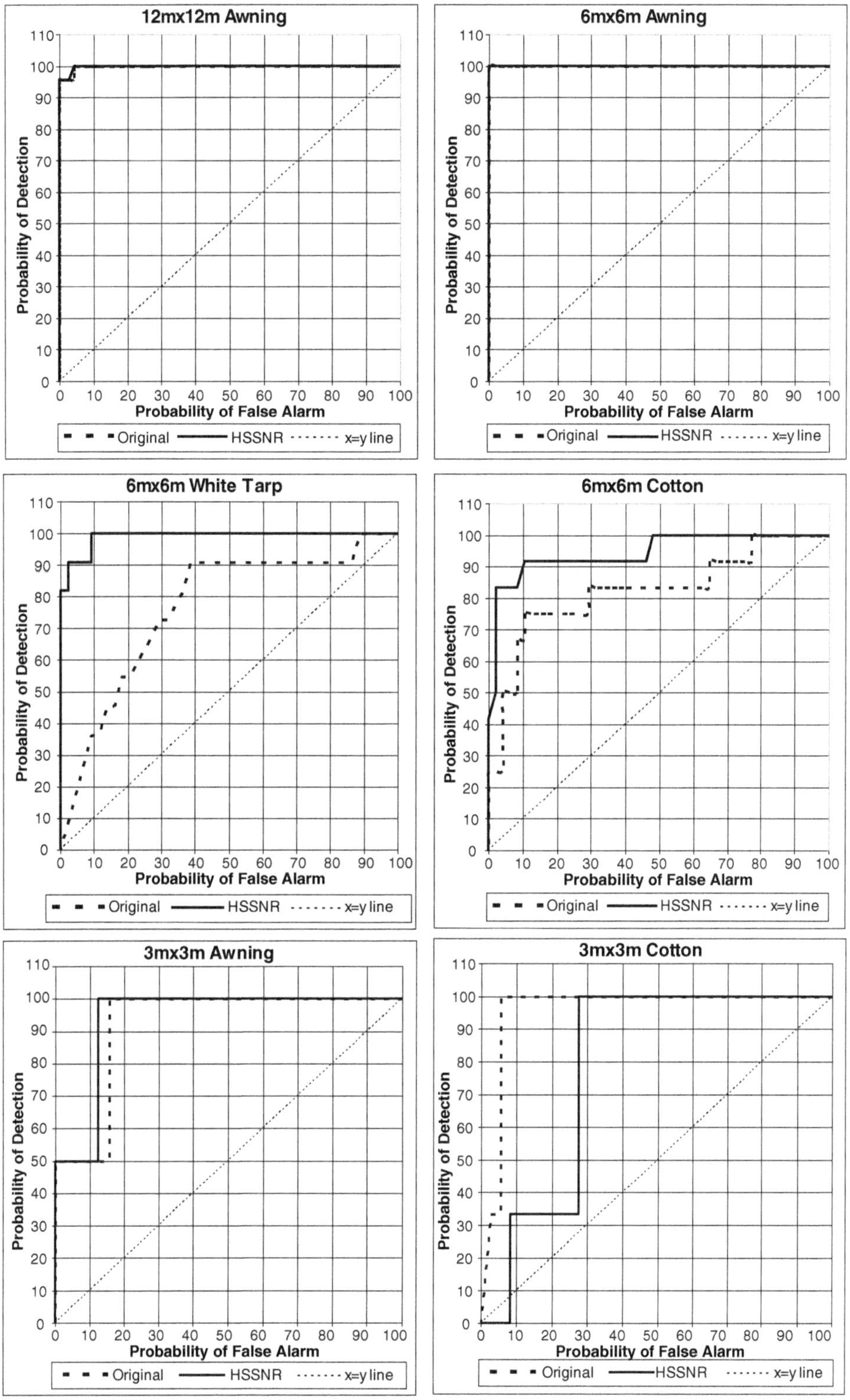

FIGURE 3.26 ROC curves of the targets derived from the original SFSI datacube and from the noise-reduced datacube using HSSNR.

The CEM algorithm is, by definition, a linear filter searching for a signature **s** under the following two constraints (Settle 2002, Du et al. 2003):

- The total energy E is a minimum and
- The filter output is constrained to a value of 1 for a pure target signature: $D(\mathbf{s}) = 1$.

The total energy E is then defined as follows:

$$E = \sum_{i=1}^{m} D_{CEM}^2 (X_i) \tag{3.1}$$

where
m is the total number of samples,
X_i is the observed signal from which the background mean signature has been subtracted, a $[N \times 1]$ vector, and
D_{CEM} is the output of the detection algorithm CEM for the observation X_i.

Under these two constraints, the CEM algorithm takes the following form:

$$D_{CEM}(X) = \frac{s^T R^{-1} X}{s^T R^{-1} s} \tag{3.2}$$

where
s is the target signal, from which the background mean signature has been subtracted, a $[N \times 1]$ vector, N is the number of spectral bands, and
R is the correlation matrix of size $[N \times N]$.

An interesting characteristic of the CEM algorithm is that its output is correlated to the target fractional abundance in signature X, assuming the mixing is linear, the target signature is well separated spectrally from the other endmembers, and the endmembers' relative abundances follow a Dirichlet distribution (Settle 2002). This is particularly useful for subpixel target detection, because the target abundance in X can then be estimated without attempting the full unmixing of signature X and the difficult determination of the complete set of pure background endmembers.

The disadvantage of the CEM algorithm is that observations X that have a large modulus in the whitened space may have a significant projection component on s even if they are not oriented in the general direction of s and, thus, be falsely flagged as candidate targets. To overcome this effect, an algorithm based on angular measurements in the whitened space can be used to confirm the abundances estimated using the CEM algorithm.

Kelly (1986) derived a detection algorithm using a generalized likelihood ratio approach, which is now known in the community as the Kelly's detection algorithm. The detection statistics for this algorithm is given by:

$$D_{KELLY}(X) = \frac{\left(s_0^T \Sigma^{-1} X_0\right)^2}{s_0^T \Sigma^{-1} s_0 \left(M + X_0^T \Sigma^{-1} X_0\right)} \tag{3.3}$$

where Σ is the background covariance, a $[N \times N]$ matrix, and
M is a parameter controlling the eccentricity of the hyperbola.

In the context of target detection using reflective hyperspectral sensors, the sign of $s_0^T \Sigma^{-1} X_0$ is important, as only positive abundances are of interest. So in practice a signed version of Kelly's algorithm is used:

$$D_{KELLY-s}(X) = sign\left(s_0^T \Sigma^{-1} X_0\right) \frac{\left(s_0^T \Sigma^{-1} X_0\right)^2}{s_0^T \Sigma^{-1} s_0 \left(M + X_0^T \Sigma^{-1} X_0\right)} \tag{3.4}$$

This algorithm describes hyperbolic decision surfaces in the whitened space, with the parameter M controlling the eccentricity of the hyperbola at a given value of $D_{KELLY-s}(X)$. Using $M = 0$ leads to degenerate hyperbolic surfaces; this version of the $D_{KELLY-s}$ algorithm is commonly known as the Adaptive Cosine Estimator in the hyperspectral target detection community, and corresponds to a pure spectral angle criterion in the whitened space. On the other hand, at very large values of M, $D_{KELLY-s}$ detection surfaces approach those of the adaptive matched filter. In order to achieve a balance between these two behaviors, a value of $M = m$ was used by Roy (2010).

A method to combine the scores of D_{CEM} and $D_{KELLY-s}$ was proposed to obtain a single score. The method is based on counting the number of pixels having scores equal to or greater than that of the test pixel, and combining this information with the $D_{KELLY-s}$ score. The proposed method to assign the empirical hybrid score to pixel X_i is as follows:

1. Count N_{Kelly} and $n_{Kelly+CEM}$; their values are defined as follows:
   - N_{Kelly} is the number of pixels in the image where their scores are equal to or greater than that of the test pixel in the Kelly algorithm

$$D_{KELLY-s}(X) \geq D_{KELLY-s}(X_i) \tag{3.5}$$

   - $n_{Kelly+CEM}$ is the number of pixels in the image where their scores are equal to or greater than that of the test pixel in both Kelly and CEM algorithms

$$D_{KELLY-s}(X) \geq D_{KELLY-s}(X_i) \quad \text{and} \quad D_{CEM}(X) \geq D_{CEM}(X_i) \tag{3.6}$$

2. Compute the ratio as follows:

$$r(X_i) = \frac{n_{Kelly+CEM}}{N_{Kelly}} \tag{3.7}$$

3. Compute the hybrid score as follows:

$$s(X_i) = r(X_i) \cdot D_{KELLY-s}(X_i) \tag{3.8}$$

In the work reported by Roy (2010), two hyperspectral datasets and scoring methodology provided by the Rochester Institute of Technology (RIT) for Target Detection Blind Test were used. The hyperspectral datasets were collected in July 2006 over Cook City, Montana, United States at a GSD of approximately 3 m using the HyMap sensor mounted on an aircraft. The datasets were provided in both at-sensor radiance and apparent ground reflectance units. Atmospheric compensation was performed by RIT using the HyCorr software. Both datasets are of the same size, with 800 × 280 spatial pixels and 126 bands. The hyperspectral images were geo-registered and geo-corrected, which implies that they were spatially resampled.

The scenes of the hyperspectral datasets contain seven (7) man-made targets of fabric (cotton and nylon with different colors) panels (F1–F7) of sizes 1 m × 1 m, 2 m × 2 m, and 3 m × 3 m and three (3) civilian vehicles (V1–V3). According to the target sizes and the GSD of the hyperspectral sensor, all targets are subpixel in size. The spectral signatures of the targets were measured in the field using an ASD FieldSpec Pro spectrometer. Samples of targets F1–F7 were also brought back to the RIT spectral laboratory and measured using a Cary 500 spectrophotometer. In the analysis, the spectral signatures measured in the laboratory for F1–F7 and spectral signatures measured in the field for V1–V3 are used as the ground references.

Two scenes are provided by the RIT. One scene intended to be used for testing the proposed algorithm: referred to as the *self-test* scene, where the ground references of all targets are known (except F5–F7) and the position of some targets is known, whereas the other scene intended to be used for evaluating the detection performance: referred to as the *blind-test* scene, where the ground references of targets F1–F4 are unknown and the position of all targets is unknown.

The performance score to detect a given target of the target detection algorithm is the number of pixels in the image that have a detection score equal to or greater than the maximum detection score for a given target. Thus, a performance score of 1 indicates perfect performance, which indicates that the target was detected without any false alarms.

Detection performances of the three detection algorithms (CEM, Kelly, and hybrid) were evaluated using both the self-test and the blind-test scenes. The ground reference provided by RIT with the self-test scene was used to compute the score for targets F1, F2, F3, F4, V1, V2, and V3 for that scene. There is no ground reference provided by RIT for targets F5, F6, and F7 in the self-test scene. Roy (2010) derived ground references for these targets based on his analysis of the detection maps. There is also no ground reference provided by RIT for targets F1, F2, F3, and F4 for the blind-test scene. Similarly the derived ground references for these targets were used.

The experimental results demonstrated a general decrease in FARs of the target detection going from the CEM to Kelly's to the hybrid algorithm. First, the number of false alarms is reduced for all targets in both scenes when using Kelly's over the CEM algorithm. Using the hybrid algorithm over Kelly's algorithm further reduces the FARs for most of the targets in both scenes. Overall, this limited set of results indicates higher performance for the hybrid algorithm than for the two other investigated algorithms.

Roy (2010) also reported the experimental results on target abundance estimated through partial spectral unmixing using the CEM algorithm. For the largest targets (3 m × 3 m fabric panels), the target abundance is between 5.7% and 10.3% of a pixel. Because those targets have a surface area similar to the ground area of a sampling pixel defined by the GSD of the sensor, in the worst-case scenario (such as when a target is split evenly between four adjacent pixels) one should observe at least one pixel over the target with an abundance close to 25%. The estimated abundances are much lower than this value, possibly indicating that one of the hypotheses for the partial spectral un-mixing is not respected. The same observation can be made for the other targets, including the vehicles.

3.7.2 Terrain Mapping

A range of military applications revolves around the identification and characterization of material on the ground for mapping purposes, terrain analysis or mission planning and support. The detection, identification, and mapping of surface types are of great value in showing materials with specific trafficability. For this purpose, ground-material is to be identified and discriminated between different ground materials to determine if the road type is concrete or asphalt, soil type, and soil moisture content. This will support military operations by analyzing trafficability and identifying lines of communications (roads, power lines, etc.) and aircraft/helicopter landing zones. Vegetation characteristics include determination of vegetation type (maple, elm, fir, oak), vegetation condition (healthy or dead) and canopy biochemistry. This information can help detect human presence, vehicles passage, contamination by chemical or biological products. Littoral studies are to collect information of land/water boundary, beach composition and vegetation presence. This information can help identification of shore landing zones.

Ardouin et al. (2007) identified the following classes of terrain features that can potentially be derived more easily from hyperspectral imagery than from multispectral imagery and that are potentially useful to produce more detailed maps to assist military operations:

- Soil characterization (soil type, moisture, disturbances, erosion, soil spectral libraries)
- Vegetation characterization (forest/plant type and species, forest upper and understory density, vegetation health, vegetation spectral libraries)
- General terrain classification: basic surface coverage (water, wetlands, vegetated/nonvegetated), lines of communications (roads and road type), hydrology, etc.

These products will identify a variety of surface materials, vegetation types, health and chemistry, all of which will aid assessing trafficability of terrain, identification of landing zones, and/or detection of human presence. The layers of generated maps will be integrated with other sources of information (i.e., topography, hydrology, roads, weather, radar) within a geographic information systems (GIS). The geomatic products will provide predictive information on the effects of terrain on military operations and will support decision-making and planning of operations.

The Defence Research and Development Canada (DRDC), which is part of the Canadian Department of National Defence, began in 2005 a project to demonstrate the military utility of space-based reflective hyperspectral imagery (0.4–2.5 μm) to the Canadian Forces (CF). The project was referred to as HYperspectral iMage EXploitation (HYMEX) and ended its activities in 2010 (Ardouin et al. 2007). Under this project four trials of trafficability and the monitoring of military training ranges to promote environmentally sustainable training were carried out. The objectives of the trials were the validation of hyperspectral processing algorithms used to derive vegetation cover information, such as type, density and height, the presence of wetlands and the determination of the soil type. These surface features are easily derived from hyperspectral imagery and can contribute to improve knowledge of the terrain for the purposes of trafficability and environmental applications. Each trial was conducted in a different vegetation background ranging from various forest biomes (deciduous, mixed, boreal) to prairie grassland. Details regarding each trial, the available ground reference and the algorithms used to analyze the various datasets were reported by Ardouin et al. (2007).

In the Canadian Forces Base (CFB) Gagetown trial, a trafficability model, which is used by the Army Meteorological Center (AMC) to plan training exercises, avoid erosion by vehicles and promote environmentally sustainable training, was improved by adding above-ground restrictions, such as the forest type (hard/softwood), density and height derived from hyperspectral remote sensing. By adding additional information of vegetation layers derived from hyperspectral remote sensing helps to produce more refined trafficability classes as illustrated in Figure 3.27. The map of off-road trafficability classes based on the current model is updated from 1D to 2D with more detail classes.

The CFB Wainwright trial was to demonstrate the portability of the trafficability model to different soil and vegetation ecosystems. The above ground trafficability was determined by a semi-arid prairie landscape with well-defined dry and wet seasons. The vegetation was composed of grassland with areas of deciduous trees (aspen, balsam poplar, and willow) and the topography was gentle. The hyperspectral vegetation products of land cover classes and LAI were used. Figure 3.28 shows the optimal route planning between point A and point B for a Wolf vehicle (an armored personnel carrier) and a LAV vehicle (an eight-wheeled amphibious armored reconnaissance vehicle) constrained by wet areas and vegetation. Figure 3.28a is RGB color-composite image of hyperspectral imagery acquired with an AISA sensor, while Figure 3.28b, c represents wet season and dry season conditions. Figure 3.28d presents the wet season situation and the requirement to move along tree lines as closely as possible (Ardouin et al. 2012).

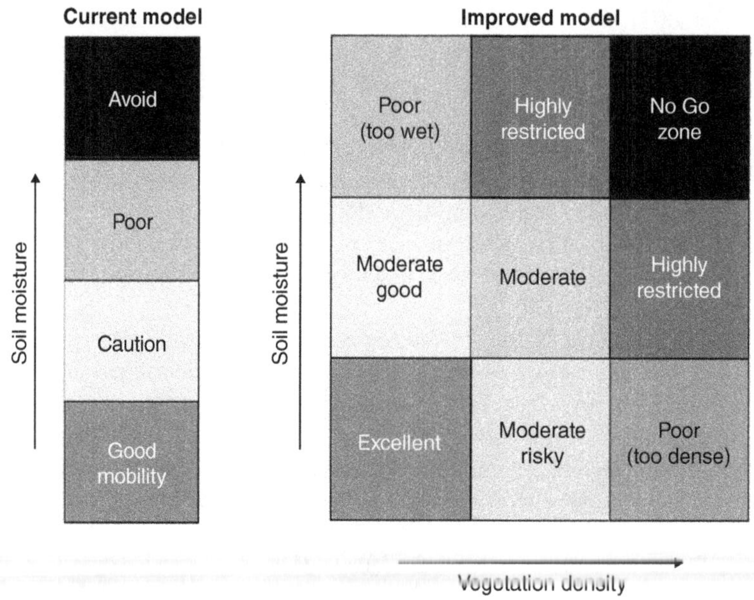

FIGURE 3.27 Off-road trafficability classes for the current model (left) and improved model using hyperspectral remote sensing (right). (Courtesy of Defence R&D Canada.)

3.7.3 Soil Characterization and Disturbance

Soil characterization is for collecting information, such as soil type, moisture, disturbances, erosion, and soil spectral libraries. Soil disturbance can be an indicator of changes, such as those produced by foot, vehicles, and buried mines. The detection of disturbed soil is based on four physical characteristics of the soil: mineralogical composition, particle size, particle coating, and lignin and cellulose content (Kelch et al. 1999a). Kelch et al. (1999b) defined 16 spectral features that can be used to detect variations in the four physical properties induced by disturbed soil. There is no universal indicator of soil disturbance. However, the authors retained four spectral features they believe to have the greatest potential to detect disturbed soil.

The strong carbonate emission peak at 2.20 μm has potential for detecting disturbed soil where carbonate minerals are brought to the surface. Moreover, the carbonate band is more defined for finer particles, which could be another indicator of soil particle size variation. The 1.7–2.5 μm spectral range is of great interest for soil rich in organic matter. Cellulose is generally found in the large-size fraction of the soil, while lignin resides in the fine size fraction. Spectral features of cellulose are near 1.7 μm and 2.1 μm, the latter being the strongest one. Lignin exhibits a "flat iron" shape feature just before 2.2 μm.

3.7.4 Marine Applications

Several potential coastal and ocean mapping products that can be derived from hyperspectral imagery were identified by Ardouin et al. (2007) during the HYMEX project definition. These products are of interest to the Canadian Forces and highly relevant to support a Rapid Environmental Assessment (REA) for maritime operations (Whitehouse et al. 2006). REA is a form of operational ocean observing that emerged in the mid-1980s when the focus of military meteorology and oceanography (MetOc) shifted from oceanic to littoral waters. REA products were initially defined for antisubmarine, mine, and amphibious warfare, but are now expanding.

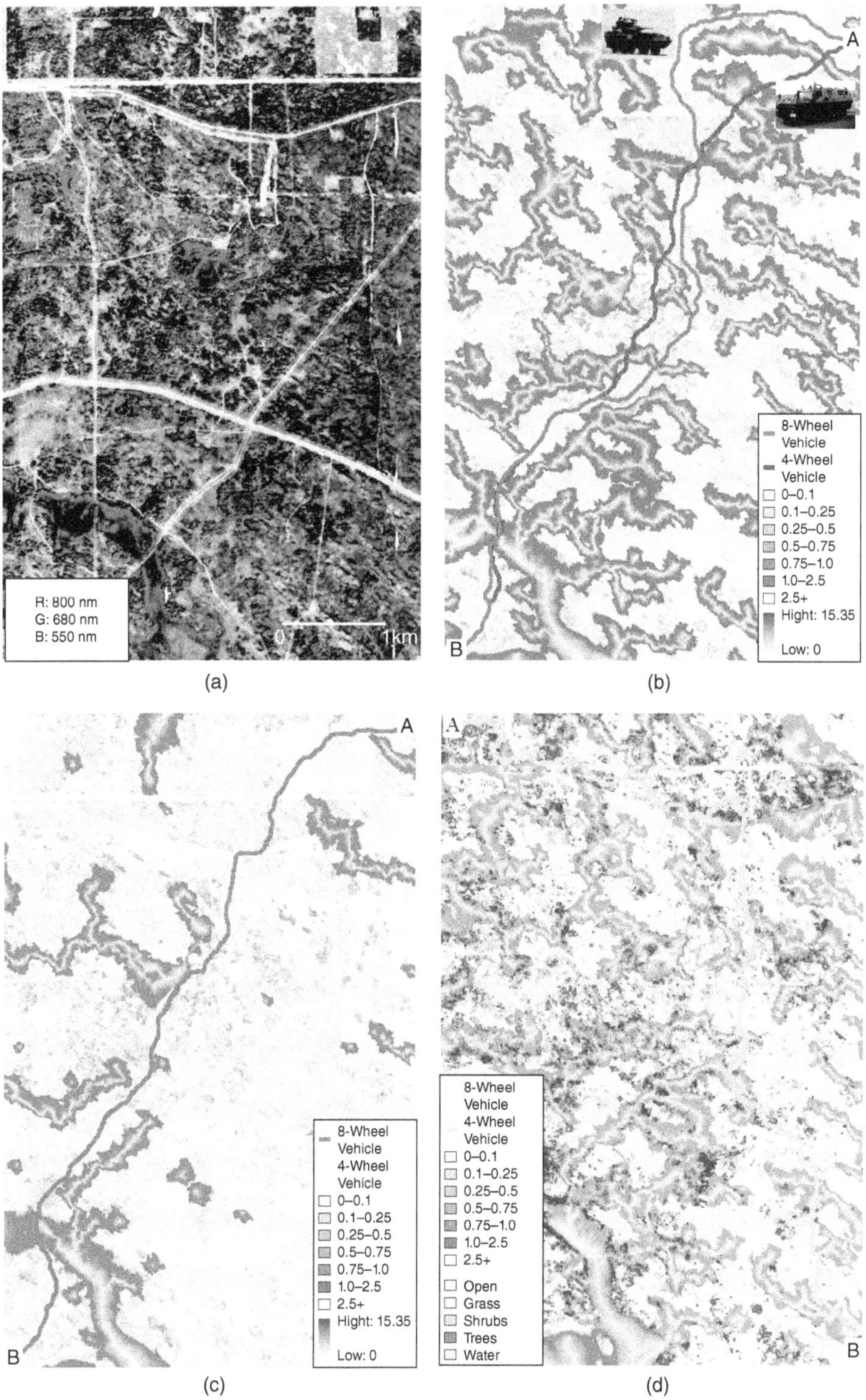

FIGURE 3.28 Optimal route trafficability planning between points A and B for a Wolf and a LAV vehicle; (a) RGB color-composite image of AISA hyperspectral data, (b) when constrained by wet season conditions, (c) when constrained by dry season conditions, and (d) when constrained by wet season conditions and the requirement to move along tree lines. (Courtesy of Defence R&D Canada.)

The HYMEX project carried out the following studies for coastal waters and ocean applications and developed several algorithms applicable to these applications:

- Near-shore bathymetry
- Beach trafficability
- Intertidal and subtidal vegetation and near-shore bottom type
- Radiative transfer model for chlorophyll, suspended material and dissolve organic matter
- Retrieval of diffuse attenuation coefficient, concentrations of chlorophyll and suspended matter

To validate the algorithms developed, two trials, one in Janvrin Island, Nova Scotia, Canada (in October 2005) and one in Georgia Strait, British Columbia, Canada (in April/August 2006), were performed.

3.7.4.1 Near-Shore Bathymetry

Knowledge of near-shore bathymetry is useful for a variety of military applications ranging from navigation safety to planning coastal operations. Mapping near-shore bathymetry must take into account variations in water depth, bottom type, illumination conditions, and scattering/absorption of the water column due to conditions, such as chlorophyll content and suspended sediment load. The best set of hyperspectral bands to extract bathymetric data will vary according to the physical properties of the water (i.e., sediment load) and the bottom characteristics. Multispectral sensors do not have this capability.

Figure 3.29 shows a bathymetry map derived from hyperspectral imagery acquired using the airborne CASI instrument in Janvrin Island trial of the HYMEX project. An empirical method similar to the one proposed by Mishra et al. (2006) was used, which is based on linear regression of reflectance at selected points of known depth (measured in the field). An R^2 fit score of 0.5 was obtained by comparing estimated depth with a digitized version of the nautical chart. This translated in an average depth error of about 2.0 m. Borstad Associates (one of the Canadian contractors of the HYMEX project) noted that they applied the same algorithm for a different project in clearer waters (in Mauritius) with greater bottom variability and obtained an average error of about 0.4 m.

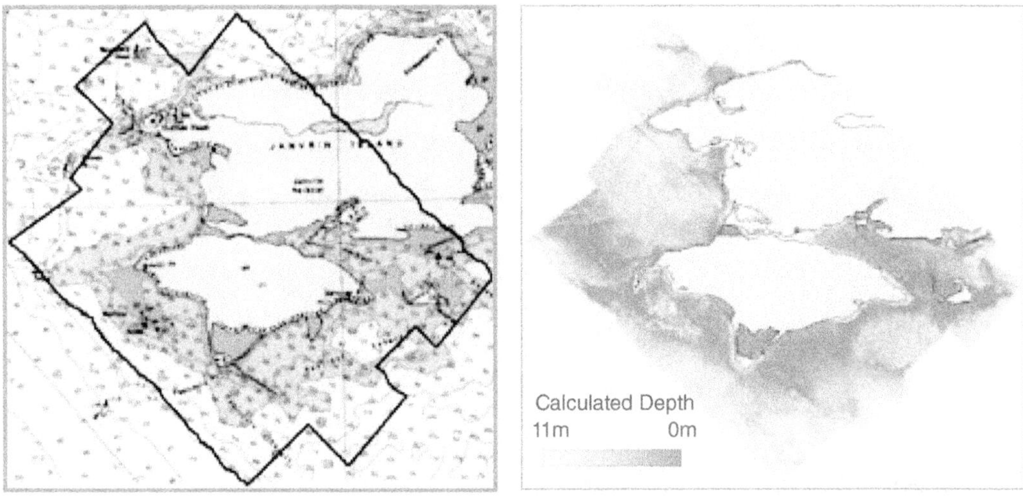

FIGURE 3.29 Bathymetry map derived from CASI hyperspectral imagery in Janvrin Island, Nova Scotia, Canada. (Courtesy of Defence R&D Canada.)

3.7.5 Detection of Landmines

The menace of landmines and unexploded ordnances is a very serious problem. According to Landmine and Cluster Munition Monitor (Landmine Monitor 2015), 61 states and areas are classified as mine-affected as of November 2015. After the end of a war, landmines remain active for a very long time. Landmines are often triggered by children and innocent civilians after the end of the war. In addition to killing and maiming innocent people, the menace of landmines also affects the socioeconomic situation in the region and prevents their development. The cost and time needed for demining are high. This motivates both the governments and the scientific community to find out demining solutions that are safer, faster, and more accurate.

DRDC is a world lead in detecting landmines using hyperspectral technology. DRDC started their research in support of the Canadian Forces on detection of landmine and unexploded ordnance in 1978 and in collaboration with ITRES Research Ltd., a Canadian company based in Calgary, using hyperspectral imaging for landmine detection in 1989.

Since its introduction in the 1980s, hyperspectral imaging technology has been applied to applications in many fields. As this technology offers the ability to distinguish different materials remotely, it has been applied to landmine detection. Every material has its special spectral signature. Therefore, knowing the mine spectral signature, by a simple comparison between the mine spectrum and the pixel spectrum observed, one can identify the presence of landmine at that specific position. It was found that spectral reflectance of each type of surface-laid mines has a constant shape between the spectral range of 500–680 nm, but varies in offset and scale according to the illuminance of the scene (Achal et al. 1995). Accordingly, the detection of this particular shape in the pixel spectrum proves the presence of the landmine.

3.7.5.1 Using VNIR Hyperspectral Imagery

Achal et al. (1995) at ITRES Research Ltd. reported identification of surface-laid mines using VNIR hyperspectral imagery acquired by a CASI sensor flown over various agricultural fields in which subpixel-size mine-like targets were laid. It was found that a linear correlation exists between the mine spectra under different incident illuminations if the spectral vector is confined between 500 nm and 680 nm. Mines were identified by classifying reflectance spectra in two ways. The first classifies individual pixels using the linear correlation coefficient as a measure of spectral similarity, while the second classifies spectra using a variant of linear spectral unmixing in which the majority of spectral members within an image are treated as background. Visually undetectable targets were determined with good results. Comparison of the classifiers revealed that the correlation method is better for high-spatial resolution data. When the targets were subpixel in size, the spectral unmixing method had a higher probability of detection than the correlation method, but had more false alarms.

McFee et al. (1996) studied the detection of buried landmines using a VNIR hyperspectral sensor. It was noticed that buried mines could not be detected by calculating the shift of the red-edge of vegetative spectra. However, by using linear correlation, some mines with low vegetative cover were detected. A further study by McFee and Ripley (1997) reported that antitank mines were more detectable than antipersonnel mines, probably due to the increased area of disturbance required to bury these relatively large mines. The probability of detection (PD), intended as the number of mines detected over all existing mines in the scene of hyperspectral imagery, obtained during the experiment varies between 33% and 100%. The FAR, measured as the number of falsely detected mines per unit area, varies between $0.10/m^2$ and $0.52/m^2$. Better PD and lower FAR could be achieved improving the classification algorithms and optimizing the spectral vectors.

The spatial resolution of the hyperspectral image data affects the performance of the mine detection. Typical antitank mines are 30 cm in diameter, whereas antipersonnel mines can have widths of 10 cm or less. As the pixel size of hyperspectral images gets smaller and closer to the size of the mines, the possibility to isolate landmines increases. Achal et al. (1999) acquired two types of VNIR

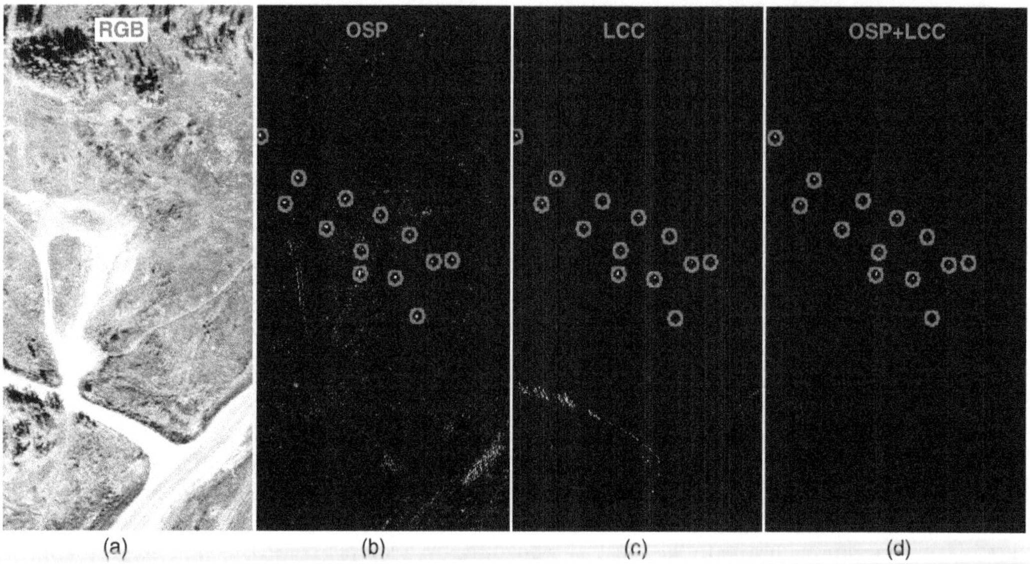

FIGURE 3.30 Detection of surface-laid mines using hyperspectral image acquired by an airborne CASI sensor; (a) is a RGB color-composite image of the test field; (b), (c), and (d) are single-band images of the same scene processed with the OSP and LCC classifiers and the product of combining the two classified images. (Courtesy of ITRES Research Ltd.)

hyperspectral images: Medium resolution images of 50 cm × 50 cm at the altitude of 300 m and high-resolution images of 15 cm × 15 cm at the altitude of 6 m in a different place. In the medium resolution experiment, they obtained a 100% PD and 0.00034/m^2 FAR. In the high-resolution experiment, all mines were detected with a FAR of 0.0043/m^2. Linear correlation coefficient (LCC) and orthogonal subspace projection (OSP) were used in classification. The best detection results were achieved when combining the two classification methods.

Ivanco et al. (2001, 2007) reported real-time detection of surface-laid mines using a VNIR imager and how software and hardware improvements can achieve real-time detection from an airborne platform.

Figure 3.30 shows an example of detection of surface-laid mines using hyperspectral imagery acquired by an airborne CASI sensor. In the field surrogates of surface-laid mines were deployed. Figure 3.30a is a RGB color-composite image of the area. Mines sizes are a few pixels wide and the mines are invisible on the color image. In Figure 3.30, images b, c, and d are single-band images of the same scene obtained by processing the 16-band CASI image with the OSP spectral classifier, LCC classifier and the product of combining the two classified images, respectively. Each circled bright blob represents a mine.

3.7.5.2 Using SWIR Hyperspectral Imagery

McFee et al. (2005) studied landmine detection using SWIR bands (1000–2500 nm). For this purpose, a SWIR hyperspectral instrument, referred to as Shortwave Airborne Spectrographic Imager (SASI) was used, which has 160 bands covering a spectral range of 850–2450 nm with a spectral sampling interval 7.5 nm, a SNR of 400:1 and f/1.8 fore-optics, and 600 pixels in cross-track direction over a 37.7° FOV. Numerous surface-laid mines, one buried mine, other surface-laid human-made items, and background materials were deployed in the scene and imaged from a horizontally scanning personnel-lift at an altitude of roughly 5 m.

With the inclusion of SWIR bands the spectral range is wider, the possibility to distinguish landmines is higher. A simple classification boundary should be able to distinguish surface-laid mines from many man-made artifacts and natural materials. However, old buried landmines are still hard to be detected using the SWIR due to buried for several years and, hence, no residual surface disturbance were recognizable.

3.7.5.3 Using TIR Hyperspectral Imagery and Combined Sensing

A project studying thermal infrared (TIR) hyperspectral imaging of landmines led to the development of a commercially available TIR hyperspectral imager at ITRES Research Ltd. suitable for airborne landmine detection (Achal et al. 2007). The instrument was used to collect imagery of surface and buried mines and improvised explosive devices over full diurnal cycles in arid, desert-like conditions and was found to provide some advantages over broadband imaging in the detection of buried threat objects (McFee et al. 2013).

Tests of landmine detection using VNIR, SWIR, and TIR imagers were reported by McFee et al. (2007). Reliable surface-laid mine detection in various weather conditions was achieved using VNIR and SWIR spectra, even if not in real time. Reliable buried landmine detection was not yet achieved even though including the TIR sensor in the trials. There is no significant difference in the VNIR region between the spectral signatures of buried landmines and background materials; however, they could be indirectly detected by observing differences in reflectance between compact soil over mines and background.

The team of DRDC started in 1997 a project testing the combination of various detection technologies called Improved Landmine Detector Project (ILDP). Because a single detection technique will not be able to detect all types of landmines in all conditions, the fusion of various techniques can be more effective (Faust et al. 2005a, 2005b, McFee et al. 2006). The team tested a small tele-operated vehicle carrying four types of instruments: a forward-looking hyperspectral infrared imager, down-looking electromagnetic induction detector, down-looking ground penetrating radar (GPR) and a thermal neutron activation detector used as confirmatory detector of suspected targets.

A Canadian research company, specialized in infrared and hyperspectral imaging named TELOPS, proved the possibility to detect buried objects using an airborne long-wave infrared (LWIR) hyperspectral imager (Gagnon et al. 2015). From an aircraft platform, they acquired thermal hyperspectral images of areas that contain man-made objects previously buried. They found that the disturbed soil right above a buried target is warmer than the undisturbed soil area next to it. By comparing the emissivity data obtained through the temperature-emissivity separation, the buried target sites show up as part of the hottest ground area within the scene. However, further classification or additional information are needed to discriminate the buried objects from other naturally hot areas.

There are some other R&D activities for detecting landmines using hyperspectral technology in the world. Makki et al. (2017) made a comprehensive review on this subject. In their review article, they highlighted the progresses attained in the detection of landmines using hyperspectral imaging and identified possible perspectives for future work to achieve a better detection in real-time operation mode.

REFERENCES

Achal, S. B., C. D. Anger, J. E. McFee, and R. W. Herring. 1999. Detection of surface-laid mine fields in VNIR hyperspectral high-spatial resolution data. *Proceedings of SPIE 3710, Conference on Detection and Remediation Technologies for Mines and Minelike Targets IV.*

Achal, S. B., J. E. McFee, and C. D. Anger. 1995. Identification of surface-laid mines by classification of compact airborne spectrographic imager (CASI) reflectance spectra. *Proceedings of SPIE 2496, Conference on Detection Technologies for Mines and Minelike Targets.*

Achal, S., J. E. McFee, T. Ivanco, and C. Anger. 2007. A thermal infrared hyperspectral imager (tasi) for buried landmine detection. *Proceedings of SPIE 6553, Conference on Detection and Remediation Technologies for Mines and Mine-like Targets XII* 6553.16.1–6553.16.11.

Adam, E. and O. Mutanga. 2009. Spectral discrimination of papyrus vegetation (*Cyperus papyrus L.*) in swamp wetlands using field spectrometry. *ISPRS Journal Photogrammetry and Remote Sensing* 64:612–620.

Ancin-Murguzur, F. J., G. Taff, C. Davids, H. Tømmervik, J. Mølmann, and M. Jørgensen. 2019. Yield estimates by a two-step approach using hyperspectral methods in grasslands at high latitudes. *Remote Sensing* 11:400.

Anderson, R. R. 1970. Spectral reflectance characteristics and automated data reduction techniques which identify wetland and water quality condition in the Chesapeake Bay. *Third Annual Earth Resources Program* 329. United States: Johnson Space Center.

Apan, A., B. Datt, and R. Kelly. September 2005. Detection of pests and diseases in vegetable crops using hyperspectral sensing: A comparison of reflectance data for different sets of symptoms. *Proceedings of SSC 2005 Spatial Intelligence, Innovation and Praxis: The national biennial Conference of the Spatial Sciences Institute.* Melbourne, Australia.

Apan, A., A. Held, S. Phinn, and J. Markley. 2004. Detecting sugarcane "Orange Rust" disease using EO-1 Hyperion hyperspectral imagery. *International Journal of Remote Sensing* 25:489–498.

Ardouin, J.-P., J. Lévesque, and T. A. Rea. 2007. A demonstration of hyperspectral image exploitation for military applications. *Proceedings of the 10th International Conference on Information Fusion (FUSION 2007)* 1–8. Quebec, Canada.

Ardouin, J.-P., J. Lévesque, V. Roy, Y. Van Chestein, and A. Faust. 2012. Demonstration of hyperspectral image exploitation for military applications. In *Remote Sensing—Applications*, ed. B. Escalante, 493–516. ISBN: 978-953-51-0651-7, InTech.

Basky, Z. and A. Fónagy. 2003. Glutenin and gliadin contents of flour derived from wheat infested with different aphid species. *Pest Management Science* 59:426–430.

Bastida, F., A. Zsolnay, T. Hernández, and C. García. 2008. Past, present and future of soil quality indices: A biological perspective. *Geoderma* 147:159–171.

Ben-Arfa, J., J. C. Bergès, G. Beltrando, K. Rim, and F. Zargouni. 2015. Mapping the land cover in coastal Gabes oases using the EO-1 Hyperion hyperspectral sensor. *Proceedings of the EGU General Assembly Conference* 17:4314. Vienna, Austria.

Ben-Dor, E., S. Chabrillat, J. A. M. Demattê, G. R. Taylor, J. Hill, M. L. Whiting, S. Sommer. 2009. Using imaging spectroscopy to study soil properties. *Remote Sensing of Environment* 113:S38–S55.

Ben-Dor, E. and J. A. M. Demattê. 2015. Remote sensing of soil in the optical domains. In *Land Resources Monitoring, Modeling, and Mapping with Remote Sensing*, ed. P. S. Thenkabail, 733–787. Boca Raton: CRC Press.

Ben-Dor, E., N. Goldshleger, O. Braun, B. Kindel, A. F. H. Goetz, D. Bonfil, N. Margalit, Y. Binaymini, A. Karnieli, and M. Agassi. 2004. Monitoring infiltration rates in semiarid soils using airborne hyperspectral technology. *International Journal of Remote Sensing* 25:2607–2624.

Best, R, M. Wehde, and R. Linder. 1981. Spectral reflectance of hydrophytes. *Remote Sensing of Environment* 11:27–35.

Böhme, L., U. Langer, F. Böhme. 2005. Microbial biomass, enzyme activities and microbial community structure in two European long-term field experiments. *Agriculture, Ecosystems & Environment* 109:141–152.

Bonan, G. B. 2008. Forests and climate change: Forcings, feedbacks, and the climate benefits of forests. *Science* 320:1444–1449.

Boyd, I. L., P. H. Freer-Smith, C. A. Gilligan, and H. C. J. Godfray. 2013. The consequence of tree pests and diseases for ecosystem services. *Science* 342(6160):1235773.

Broadwater, J. and R. Chellappa. 2007. Hybrid detectors for subpixels targets. *IEEE Transactions on Pattern Analysis and Machine Intelligence* 29(11):1891–1903.

Carter, G. A., K. L. Lucas, G. A. Blossom, C. L. L. Holiday, D. S. Mooneyhan, D. R. Fastring, T. R. Holcombe, and J. A. Griffith. 2009. Remote sensing and mapping of tamarisk along the Colorado River, USA: A comparative use of summer-acquired Hyperion, thematic mapper and quickbird data. *Remote Sensing* 1:318–329.

Chabrillat, S., H. Kaufmann, B. Merz, A. Mueller, O. Bens, and C. Lemmnitz. 2003. Development of relationships between reflectance and erosion modelling: Test site preliminary field spectral analysis. *Proceedings of the 3rd EARSeL Workshop on Imaging Spectroscopy*, 13–16. Herrsching, Germany.

Chabrillat, S., R. Milewski, T. Schmid, M. Rodriguez, P. Escribano, M. Pelayo, and A. Palacios-Orueta. 2014. Potential of hyperspectral imagery for the spatial assessment of soil erosion stages in agricultural semi-arid Spain at different scales. *Proceedings of 2014 IEEE Geoscience and Remote Sensing Symposium* 2918–2921. Quebec City, Canada.

Champagne, C., K. Staenz, A. Bannari, J.-C. Deguise, and H. McNairn. 2003. Validation of a hyperspectral curve-fitting model for the estimation of plant water content of agricultural canopies. *Remote Sensing of Environment* 87:148–160.

Ciampalini, A., I. Consoloni, T. Salvatici, F. Di Traglia, F. Fidolini, G. Sarti, and S. Moretti. 2015. Characterization of coastal environment by means of hyper- and multispectral techniques. *Applied Geography* 57:120–132.

Clark, R. N., S. Vance, and R. Green. 1998. Mineral mapping with imaging spectroscopy: The Ray Mine, AZ. *Summaries of the 7th Annual JPL Airborne Earth Science Workshop*. Pasadena, California: JPL-Publication.

Clevers, J. G. P. W., L. Kooistra, and M. E. Schaepman. 2010. Estimating canopy water content using hyperspectral remote sensing data. *International Journal of Applied Earth Observation and Geoinformation* 12:119–125.

Croft, B., R. Magarey, and P. Whittle. 2000. Disease management. In *Manual of Canegrowing*, eds. M. Hogarth and P. Allsopp, 263–289. Brisbane, Australia: Bureau of Sugar Experiment Stations.

Cudahy, T. J., R. Hewson, J. F. Huntington, M. A. Quigley, and P. S. Barry. 2001. The performance of the satellite-borne Hyperion hyperspectral VNIR-SWIR imaging system for mineral mapping at Mount Fitton, South Australia. *Proceedings of IEEE Geoscience and Remote Sensing Symposium (IGARSS 2001)* 314–316.

Curran, P. J. 1989. Remote sensing of foliar chemistry. *Remote Sensing of Environment* 30:271–278.

DeFries, R. S. and A. S. Belward. 2000. Global and regional land cover characterization from satellite data: An introduction to the special issue. *International Journal of Remote Sensing* 21(6):1083–1092.

Dehaan, R. L. and G. R. Taylor. 2002. Field-derived spectra of salinized soils and vegetation as indicators of irrigation-induced soil salinization. *Remote Sensing of Environment* 80:406–417.

Demattê, J. A. M. and D. Focht. 1999. Detection of soil erosion by spectral reflectance. *Brazilian Journal of Soil Science* 3:401–413.

Devred, E., K. R. Turpie, W. Moses, V. V. Klemas, T. Moisan, M. Babin, G. Toro-Farmer, M. H. Forget, and Y. H. Jo. 2013. Future retrievals of water column bio-optical properties using the hyperspectral infrared imager (HyspIRI). *Remote Sensing* 5:6812–6837.

Donoghue, D. N. M. 2002. Remote sensing: Environmental change. *Progress in Physical Geography* 26(1):144–151.

Du, Q., H. Ren, C.-I. Chang. 2003. A comparative study of orthogonal subspace projection and constrained energy minimization. *IEEE Transactions on Geoscience and Remote Sensing* 41(6):1525–1529.

Ducart, D. F., A. M. Silva, C. Labouré, B. Toledo, L. M. D. Assis. 2016. Mapping iron oxides with Landsat-8/OLI and EO-1/Hyperion imagery from the Serra Norte iron deposits in the Carajás Mineral Province, Brazil. *Brazilian Journal of Geology* 46:331–349.

Dutta, S., B. K. Bhattacharya, D. R. Rajak, C. Chattopadhayay, N. K. Patel, J. S. Parihar. 2006. Disease detection in mustard crop using EO-1 Hyperion satellite data. *Journal of the Indian Society of Remote Sensing* 34(3):325–330.

Eddy, P. R., A. M. Smith, B. D. Hill, D. R. Peddle, C. A. Coburn, and R. E. Blackshaw. 2013. Weed and crop discrimination using hyperspectral image data and reduced bandsets. *Canadian Journal of Remote Sensing* 39(6):481–490.

Fassnacht, F. E., H. Latifi, A. Ghosh, P. K. Joshi, and B. Koch. 2014. Assessing the potential of hyperspectral imagery to map bark beetle-induced tree mortality. *Remote Sensing of Environment* 140:533–548.

Faust, A. A., R. H. Chesney, Y. Das, J. E. McFee, and K. L. Russell. 2005b. Canadian teleoperated landmine detection systems part II: Anti-personal landmine detection. *International Journal of Systems Science* 36(9):529–543.

Faust, A. A., J. E. Mcfee, R. H. Chesney, K. L. Russell, and Y. Das. 2005a. Canadian teleoperated landmine detection systems part I. *International Journal of Systems Science* 36(9):511–528.

Fließbach, A., H.-R. Oberholzer, L. Gunst, and P. Mäder. 2007. Soil organic matter and biological soil quality indicators after 21 years of organic and conventional farming. *Agriculture, Ecosystems & Environment* 118:273–284.

Foy, B. R., J. Theiler, and A. M. Fraser. 2009. Decision boundaries in two dimensions for target detection in hyperspectral imagery. *Optics Express* 17(20):17391–17411.

Gagnon, M.-A., P. Lagueux, J.-P. Gagnon, S. Savary, P. Tremblay, V. Farley, É. Guyot, and M. Chamberland. 2015. Airborne thermal infrared hyperspectral imaging of buried objects. *Proceedings of SPIE 9649, SPIE Security+ Defence,* 96490T. International Society for Optics and Photonics.

Gersman, R., E. Ben-Dor, M. Beyth, D. Avigad, M. Abraha, and A. Kibreab. 2008. Mapping of hydrothermally altered rocks by the EO-1 Hyperion sensor, Northern Danakil Depression, Eritrea. *International Journal of Remote Sensing* 29:3911–3936.

Giardino, C., V. E. Brando, A. G. Dekker, N. Strömbeck, and G. Candiani. 2007. Assessment of water quality in Lake Garda (Italy) using Hyperion. *Remote Sensing of Environment* 109:183–195.

Goodenough, D. G. et al. 2002. Monitoring forests with Hyperion and ALI. *Proceedings of IEEE IGARSS*, Toronto, Canada.

Goodenough, D. G., A. Dyk, K. O. Niemann, J. S. Pearlman, H. Chen, T. Han, M. Murdoch, and C. West. 2003. Processing Hyperion and ALI for forest classification. *IEEE Transactions on Geoscience and Remote Sensing* 41(6):1321–1331.

Govender, M., K, Chetty, and H., Bulcock. 2006. A review of hyperspectral remote sensing and its application in vegetation and water resource studies. *Water SA* 33(2):145–151.

Hadoux, X., N. Gorretta, and G. Rabatel. 2012. Weeds-wheat discrimination using hyperspectral imagery. *Proceedings of International Conference on Agricultural Engineering* 8–12. Valence, Espagne.

Hamzeh, S., A. Ali, S. Kazem, H. Bartholomeus, and M. Herold. 2016. Assessing the accuracy of hyperspectral and multispectral satellite imagery for categorical and quantitative mapping of salinity stress in sugarcane fields. *International Journal of Applied Earth Observation and Geoinformation* 52:412–421.

Hatala, J. A., R. L. Crabtree, K. Q. Halligan, and P. R. Moorcroft. 2010. Landscape-scale patterns of forest pest and pathogen damage in the Greater Yellowstone ecosystem. *Remote Sensing of Environment* 114:375–384.

Haubrock, S., S. Chabrillat, and H. Kaufmann. 2004. Application of hyperspectral imaging and laser scanning for the monitoring and assessment of soil erosion in a recultivation mining area. In *Remote Sensing and GIS for Environmental Studies: Applications in Geography*, eds. S. Erasmi, B. Cyffka, and M. Kappas, 230–237. Göttingen: Göttinger Geographische Abhandlungen.

Haubrock, S., S. Chabrillat, and H. Kaufmann. 2005. Application of hyperspectral imaging for the quantification of surface soil moisture. *Proceedings of the 4th EARSeL Workshop on Imaging Spectroscopy* 163–171.

Havlicek, E. 2012. Soil biodiversity and bioindication: From complex thinking to simple acting. *European Journal of Soil Biology* 49:80–84.

Hbirkou, C., S. Pätzold, A. K. Mahlein, and G. Welp. 2012. Airborne hyperspectral imaging of spatial soil organic carbon heterogeneity at the field scale. *Geoderma* 175–176:21–28.

Hill, J., J. Mégier, and W. Mehl, 1995. Land degradation, soil erosion and desertification monitoring in Mediterranean ecosystems. *Remote Sensing Reviews* 12:107–130.

Hill, J., W., Mehl, and M., Altherr. 1994. Land degradation and soil erosion mapping in a Mediterranean ecosystem. In *Imaging Spectrometry—A Tool for Environmental Observations*, eds. J. Hill and J. Mégier, 237–260. Dordrecht: Kluwer Academic Publishers.

Hu, B., S.-E. Qian, D. Haboudane, J. R. Miller, A. B. Hollinger, N. Tremblay, and E. Pattey. 2004. Retrieval of crop chlorophyll content and leaf area index from decompressed hyperspectral data: The effects of data compression. *Remote Sensing of Environment* 92:139–152.

Huang, W., J. Luo, J. Zhang, J. Zhao, C. Zhao, J. Wang, G. Yang, M. Huang, L. Huang, and S. Du. 2012. Crop disease and pest monitoring by remote sensing. In *Remote Sensing—Applications*, ed. B. Escalante, 31–76. InTech, Available from: http://www.intechopen.com/books/remote-sensing-applications/crop-disease-and-pest-monitoring-by-remotesensing.

Huete, A. R., T. Miura, and X. Gao. 2003. Land cover conversion and degradation analyses through coupled soil-plant biophysical parameters derived from hyperspectral EO-1 Hyperion. *IEEE Transactions on Geoscience and Remote Sensing* 41:1268–1276.

Ivanco, T, S. B. Achal, C. D. Anger, and J. E. McFee. 2001. Casi real-time surface-laid mine detection system. *Proceedings of SPIE 4394, Detection and Remediation Technologies for Mines and Minelike Targets VI.*

Ivanco, T., S. Achal, J. E. McFee, C. Anger, and J. Young. 2007. Real-time airborne hyperspectral imaging of landmines. *Proceedings of SPIE 6553, Detection and Remediation Technologies for Mines and Minelike Targets XII.* http://dx.doi.org/10.1117/12.720442.

Johnson, D.H. and T.A. O'Neil. 2001. Wildlife-habitat relationships in Oregon and Washington. Oregon State University Press, Corvallis, OR.

Kelch, D. J., D. A. Fatora, R. K. Hill, K. Krause, P. R. Lind, E. Siler, B. L. Sink, and D. Sira. 1999a. Spectral phenomenon for detection of disturbed soil. *Proceedings of International Symposium of Spectral Sensing Research (ISSSR'99)*, Las Vegas, NV, Oct. 31–Nov. 4.

Kelch, D. J., D. A. Fatora, R. K. Hill, K. Krause, P. R. Lind, E. Siler, B. L. Sink, and D. Sira. 1999b. The physical bases for exploitable soil spectral signatures. *Proceedings of International Symposium of Spectral Sensing Research (ISSSR'99),* Las Vegas, NV, Oct. 31–Nov. 4.

Kelly, E. J. 1986. An adaptive detection algorithm. *IEEE Transactions on Aerospace and Electronic System* 22(1):115–127.

Klemas, V. K., J. E. Dobson, R. L. Ferguson and K .D. Haddad. 1993. A coastal land cover classification system for the NOAA coastal change analysis project. *Journal of Coastal Research*, 9 (3): 862–872.

Kruse, F. A., J. W. Boardman, and J. F. Huntington. 2003. Comparison of airborne hyperspectral data and EO-1 Hyperion for mineral mapping. *IEEE Transactions on Geoscience and Remote Sensing* 41:1388–1400.

Kudela, R. M., S. L. Palacios, D. C. Austerberry, E. K. Accorsi, L. S. Guild, and J. Torres-Perez. 2015. Application of hyperspectral remote sensing to cyanobacterial blooms in inland waters. *Remote Sensing of Environment* 167:196–205.

Kurz, W. A., M. Apps, E. Banfield, and G. Stinson. 2002. Forest carbon accounting at the operational scale. *The Forestry Chronicle* 78:672–679.

Landgrebe, D. 1999. On information extraction principles for hyperspectral data. *Cybernetics* 28 Part C(1):1–7.

Landmine Monitor. 2015. Landmine Monitor 2015, available at http://www.the-monitor.org/en-gb/reports/2015/landmine-monitor-2015.aspx.

Law, B. E., F. M. Kelliher, D. D. Baldocchi, P. M. Anthoni, J. Irvine, D. Moore, and S. Van Tuyl. 2001. Spatial and temporal variation in respiration in a young ponderosa pine forest during a summer drought. *Agricultural and Forest Meteorology* 110:27–43.

Leverington, D. W. 2008. Discrimination of geological end members using Hyperion imagery: Preliminary results, big bend national park, TEXAS. *Proceedings of the IEEE International Geoscience and Remote Sensing Symposium (IGARSS)* 2:40–41.

Leverington, D. W. 2010. Discrimination of sedimentary lithologies using Hyperion and Landsat Thematic Mapper Data: A case study at Melville Island, Canadian High Arctic. *International Journal of Remote Sensing* 31:233–260.

Lévesque, J., K. Staenz, J. Shang, R. A. Neville, P. Yearwood, and V. Singhroy. 1999. Temporal monitoring of mine tailings revegetation using hyperspectral data, Sudbury, Ontario. *Proceedings of the Thirteenth International Conference on Applied Geologic Remote Sensing* 2:21–28. Vancouver, Canada.

Lévesque, J., T. Szeredi, K. Staenz, V. Singhroy, and D. Bolton. 1997. Spectral unmixing for monitoring mine tailings site rehabilitation, Copper Cliff Mine, Sudbury, Ontario. *Proceedings of the Twelfth International Conference and Workshops on Applied Geological Remote Sensing* 340–347. Denver, Colorado.

Li, L., S. L. Ustin, and M. Lay. 2005. Application of multiple endmember spectral mixture analysis (MESMA) to AVIRIS imagery for coastal salt marsh mapping: A case study in China Camp, CA, USA. *International Journal of Remote Sensing* 26:5193–5207.

Lillesand, T. M. and R. W. Kiefer. 1999. *Remote Sensing and Image Interpretation*. New Jersey, NJ: John Wiley & Sons.

Lin, F., D. Zhang, Y. Huang, X. Wang, and X. Chen. 2017. Detection of corn and weed species by the combination of spectral, shape and textural features. *Sustainability* 9:1335.

Lin, C., S. L. Zhou, and S. H. Wu. 2013. Using hyperspectral reflectance to detect different soil erosion status in the subtropical hilly region of Southern China: A case study of Changting, Fujian Province. *Environmental Earth Sciences* 70:1661–1670.

Lin, C., S. Zhou, S. Wu, Q. Zhu, and Q. Dang. 2014. Spectral response of different eroded soils in subtropical china: A case study in Changting County, China. *Journal of Mountain Science* 11:697–707.

Makki, I, R. Younes, C. Francis, T. Bianchi, and M. Zucchetti. 2017. A survey of landmine detection using hyperspectral imaging. *ISPRS Journal of Photogrammetry and Remote Sensing* 124:40–53.

Manolakis, D., D. Marden, and G. A. Shaw. 2003. Hyperspectral image processing for automatic target detection applications. *Lincoln Laboratory Journal* 14(1):79–116.

Martin, M. E., S. D. Newman, J. D. Aber, and R. G. Congalton. 1998. Determining forest species composition using high resolution spectral resolution remote sensing data. *Remote Sensing of Environment* 65:249–254.

McFee, J. E. and S. Achal. 2011. Subsurface sensing. In *Infrared and Hyperspectral Systems*, eds. S. Turk, A. K. Hocaoglu, and A. A. Vertiy, 465–483. New York, NY: John Wiley and Sons, Wiley. Series in Microwave and Optical Engineering. ISBN 978-0-470-13388-0.

McFee, J. E., S. B. Achal, A. U. Diaz, and A. A. Faust. 2013. Comparison of broad-band and hyperspectral thermal infrared imaging of buried threat objects. *Proceedings of SPIE 8709, Conference on Detection and Sensing of Mines, Explosive Objects and Obscured Targets XVIII.*

McFee, J. E., S. Achal, T. Ivanco, and C. Anger. 2005. A short wave infrared hyperspectral imager for landmine detection. *Proceedings of SPIE 5794, Detection and Remediation Technologies for Mines and Minelike Targets X.*

McFee, J. E., C. Anger, S. Achal, and T. Ivanco. 2007. Landmine detection using passive hyperspectral imaging. *Proceedings of SPIE 6554, Chemical and Biological Sensing VIII, 655404,* April 26.

McFee, J. E. and T. H. Ripley. 1997. Detection of buried land mines using a CASI hyperspectral imager. *Proceedings of SPIE 3079, Detection and Remediation Technologies for Mines and Minelike Targets II.*

McFee, J. E., H. T. Ripley, R. Buxton, and A. M. Thriscutt. 1996. Preliminary study of detection of buried landmines using a programmable hyperspectral imager. *Proceedings of SPIE 2765, Detection and Remediation Technologies for Mines and Minelike Targets.* http://dx.doi.org/10.1117/12.241250.

McFee, J. E., K. L. Russell, R. H. Chesney, A. A. Faust, and Y. Das. 2006. The Canadian forces ILDS—a military fielded multi-sensor, vehicle mounted, teleoperated landmine detection system. *Proceedings of SPIE xxx, Detection and Remediation Technologies for Mines and Minelike Targets XI.*

Mielke, C., N. K. Boesche, C. Rogass, H. Kaufmann, C. Gauert, and M. de Wit. 2014b. Spaceborne mine waste mineralogy monitoring in South Africa, applications for Modern Push-Broom Missions: Hyperion OLI and EnMAP/Sentinel-2. *Remote Sensing* 6:6790–6816.

Mielke, C., N. K. Boesche, C. Rogass, K. Segl, and H. Kaufmann. 2014a. Multi- and hyperspectral satellite sensors for mineral exploration, new applications to the Sentinel-2 and EnMAP Mission. *Proceedings of 34th EARSeL Symposium* 16–20. Poland, Warsaw.

Mielke, C., C. Rogass, N. Boesche, K. Segl, and U. Altenberger. 2016. EnGeoMAP 2.0-Automated hyperspectral mineral identification for the German EnMAP space mission. *Remote Sensing* 8:127.

Mishra, D., S. Narumalani, D. Rundquist, and M. Lawson. 2006. Benthic habitat mapping in tropical marine environments using QuickBird multispectral data. *Photogrammetric Engineering & Remote Sensing* 72(9):1037–1048.

Mitsch, W. J. and J. G. Gosselink. 1993. Wetlands. New York, NY: Van Nostrand Reinhold.

Neville, R. A., N. Rowlands, R. V. Marois, and I. Powell. 1995. SFSI: Canada's first airborne SWIR imaging spectrometer. *Canadian Journal of Remote Sensing* 21(3):328–336.

Neville, R. A., K. Staenz, J. Lévesque, C. Nadeau, Q. S. Truong, and G. A. Borstad, 2001. Uranium mine detection using an airborne imaging spectrometer. *Proceedings of the Fifth International Airborne Remote Sensing Conference*, San Francisco, CA.

NRC. National Research Council. 1997. *Precision Agriculture in the 21st Century.* Washington, DC: National Academy Press.

Ogashawara, I., D. R. Mishra, S. Mishra, M. P. Curtarelli, and J. L. Stech. 2013. A performance review of reflectance based algorithms for predicting phycocyanin concentrations in Inland waters. *Remote Sensing* 5:4774–4798.

Okin, G. S., D. A. Roberts, B. Murray, and W. J. Okin. 2001. Practical limits on hyperspectral vegetation discrimination in arid and semiarid environments. *Remote Sensing of Environment* 77:212–225.

Othman, H. and S.-E. Qian. 2006. Noise reduction of hyperspectral imagery using hybrid spatial-spectral derivative-domain wavelet shrinkage. *IEEE Transactions on Geoscience and Remote Sensing* 42(2):397–408.

Oumar, Z., O. Mutanga, and R. Ismail. 2013. Predicting *Thaumastocoris peregrinus* damage using narrow band normalized indices and hyperspectral indices using field spectra resampled to the Hyperion sensor. *International Journal of Applied Earth Observation and Geoinformation* 21:113–121.

Pan, Y., et al. 2011. A large and persistent carbon sink in the world's forests. *Science* 333:988–993.

Paz-Kagan, T., E. Zaady, C. Salbach, A. Schmidt, A. Lausch, S. Zacharias, G. Notesco, E. Ben-Dor, and A. Karnieli. 2015. Mapping the Spectral Soil Quality Index (SSQI) using airborne imaging spectroscopy. *Remote Sensing* 7:15748–15781.

Penuelas, J, I. Filella, C. Biel, L. Serrano, and R. Save. 1993. The reflectance at the 950–970 nm region as an indicator of plant water status. *International Journal of Remote Sensing* 14:1887–1905.

Petropoulos, G. P., K. Arvanitis, and N. Sigrimis. 2012. Hyperion hyperspectral imagery analysis combined with machine learning classifiers for land use/cover mapping. *Expert Systems with Applications* 39:3800–3809.

Pu, R., M. Kelly, G. L. Anderson, and P. Gong. 2008. Using CASI hyperspectral imagery to detect mortality and vegetation stress associated with a new hardwood forest disease. *Photogrammetric Engineering & Remote Sensing* 74:65–75.

Pulleman, M., R. Creamer, U. Hamer, J. Helder, C. Pelosi, G. Pérès, and M. Rutgers. 2012. Soil biodiversity, biological indicators and soil ecosystem services: An overview of European approaches. *Current Opinion in Environmental Sustainability* 4:529–538.

Qian, S.-E. 2013. *Optical Satellite Signal Processing and Enhancement.* Bellingham, WA: SPIE Press.

Qian, S.-E. and J. Lévesque. 2009. Target Detection from Noise-Reduced Hyperspectral Imagery using Spectral Unmixing Approach. *Optical Engineering* 48(2):1–11.

Ramsey E., III, A. Rangoonwala, G. Nelson, and R. Ehrlich. 2005. Mapping the invasive species, Chinese Tallow, with EO-1 Satellite Hyperion hyperspectral image data and relating tallow occurrences to a classified Landsat thematic mapper land cover map. *International Journal of Remote Sensing* 26: 1637–1657.

Ritz, K., H. I. J. Black, C. D. Campbell, J. A. Harris, and C. Wood. 2009. Selecting biological indicators for monitoring soils: A framework for balancing scientific and technical opinion to assist policy development. *Ecological Indicators* 9:1212–1221.

Rogge, D., B. Rivard, K. Segl, B. Grant, and J. Feng. 2014. Mapping of NiCu-PGE ore hosting ultramafic rocks using airborne and simulated EnMAP hyperspectral imagery, Nunavik, Canada. *Remote Sensing of Environment* 152:302–317.

Rosso, P. H., S. L. Ustin, and A. Hastings, 2005. Mapping marshland vegetation of San Francisco Bay, California, using hyperspectral data. *International Journal of Remote Sensing* 26:5169–5191.

Roy, V. May 2010. Hybrid algorithm for hyperspectral target detection. *Proceedings of SPIE 7695* 1–10.

Russell, M. 2001. Monitoring regional vegetation change using reflectance measurements from multiple solar zenith angles. *Environment International* 27:211–217.

Sandmeier, S. and D. W. Deering. 1999. Structure analysis and classification of boreal forests using airborne hyperspectral BRDF data from ASAS. *Remote Sensing of Environment* 69:281–295.

Schmid, T., et al. 2012. Spectral characterization of land surface composition to determine soil erosion within semiarid rainfed cultivated areas. *Proceedings of 2012 IEEE International Geoscience and Remote Sensing Symposium* 7082–7085. Munich, Germany.

Schmid, T., et al. 2016. Characterization of soil erosion indicators using hyperspectral data from a Mediterranean rainfed cultivated region. *IEEE Journal of Selected Topics in Applied Earth Observations and Remote Sensing* 9(2):845–860.

Schmidt, K. S. and A. K. Skidmore. 2003. Spectral discrimination of vegetation types in a coastal wetland. *Remote Sensing of Environment* 85:92–108.

Settle, J. 2002. On constrained energy minimization and the partial unmixing of multispectral images. *IEEE Transactions on Geoscience and Remote Sensing* 40(3):718–721.

Shang, J., J. Lévesque, K. Staenz, P. Howarth, B. Morris, and L. Lanteigne. 2000. Investigating casi responses to different levels of tailing oxidation: Inco Copper Cliff tailings area, northern Ontario, Canada. *Proceedings of the 22nd Canadian Symposium on Remote Sensing* 565–574. Victoria, Canada.

Staenz, K., R. A. Neville, J. Lévesque, T. Szeredi, V. Singhroy, G. A. Borstad, and P. Hauff. 1999. Evaluation of casi and SFSI hyperspectral data for geological applications—two case studies. *Canadian Journal of Remote Sensing* 25(3):311–322.

Staenz, K., T. Szeredi, and J. Schwarz. 1998. ISDAS—a system for processing/analyzing hyperspectral data. *Canadian Journal of Remote Sensing* 24(2):99–113.

Steeger, C., J. Dulisse, and M.M. Machmer. 2001. Wildlife-habitat relationships in the Columbia River Basin: a British Columbia database for terrestrial vertebrate species. Report prepared for the Research and Forest Practices Branch, BC Ministry of Forests, Victoria.

Steinberg, A., S. Chabrillat, A. Stevens, K. Segl, and S. Foerster. 2016. Prediction of common surface soil properties based on Vis-NIR airborne and simulated EnMAP imaging spectroscopy data: Prediction accuracy and influence of spatial resolution. *Remote Sensing* 8:613.

Sujatha, G., R. S. Dwivedi, K. Sreenivas, and L. Venkataratnam. 2000. Mapping and monitoring of degraded lands in part of Jaunpur district of Uttar Pradesh using temporal spaceborne multispectral data. *International Journal of Remote Sensing* 21(3):519–531.

Swayze, G. A. et al. 2000. Using imaging spectroscopy to map acidic mine waste. *Environmental Science and Technology* 34:47–54.

Turpie, K. R., V. V. Klemas, K. Byrd, M. Kelly, and Y. H. Jo. 2015. Prospective HyspIRI global observations of tidal wetlands. *Remote Sensing of Environment* 167:206–217.

Ustin, S. L. and Q. F. Xiao. 2001. Mapping successional boreal forests in interior central Alaska. *International Journal of Remote Sensing* 22(6):1779–1797.

Vaiphasa, C., S. Ongsomwang, T. Vaiphasa, and A. K. Skidmore. 2005. Tropical mangrove species discrimination using hyperspectral data: A laboratory study. *Estuarine, Coastal and Shelf Science* 65:371–379.

Vaiphasa, C. K., K. A. Skidmore, W. F. de Boer, and T. Vaiphasa. 2007. A hyperspectral band selector for plant species discrimination. *ISPRS Journal Photogrammetry Remote Sensing* 62:225–235.

Van Dobben, H. F., C. J. F. Braak, and G. M. Dirkse. 1999. Undergrowth as a biomonitor for deposition of nitrogen and acidity in pine forest. *Forest Ecology and Management* 114:83–95.

Velasquez, E., P. Lavelle, and M. Andrade. 2007. GISQ, a multifunctional indicator of soil quality. *Soil Biology and Biochemistry* 39:3066–3080.

Vohland, M., M. Ludwig, S. Thiele-Bruhn, and B. Ludwig. 2017. Quantification of soil properties with hyperspectral data: Selecting spectral variables with different methods to improve accuracies and analyze prediction mechanisms. *Remote Sensing* 9(11):1103.

Wang, C., M. Menenti, M. Stoll, E. Belluco, and M. Marani. 2007. Mapping mixed vegetation communities in salt marshes using airborne spectral data. *Remote Sensing of Environment* 107:559–570.

Weng, Y. L., P. Gong, and Z. L. Zhu. 2010. A spectral index for estimating soil salinity in the yellow river delta region of china using EO-1 Hyperion data. *Pedosphere* 20:378–388.

White, J. C., N. C. Coops, T. Hilker, M. A. Wulder, and A. L. Carroll. 2007. Detecting mountain pine beetle red attack damage with EO-1 Hyperion moisture indices. *International Journal of Remote Sensing* 28:2111–2121.

White, H. P., L. Sun, C. M. Champagne, K. Staenz, and S. G. Leblanc. 2002. BRDF normalization of hyperspectral image data. *Proceedings of IEEE IGARSS 2002*. Toronto, Canada.

Whitehouse, B. et al. 2006. Rapid environmental assessment (REA) of the maritime battlespace. *Canadian Military Journal* Spring 2006:66–68.

Whiting, M. L., S. L. Ustin, P. Zarco-Tejada, A. Palacios-Orueta, and V. C. Vanderbilt. 2006. Hyperspectral mapping of crop and soils for precision agriculture. *Proceedings of SPIE 6298*, B1–15.

Xu, B. and P. Gong. 2007. Land-use/land-cover classification with multispectral and hyperspectral EO-1 data. *Photogrammetric Engineering & Remote Sensing* 73:955–965.

Yang, C., J. H. Everitt, and J. M. Bradford. 2004. Airborne hyperspectral imagery and yield monitor data for estimating grain sorghum yield variability. *Transactions of the ASAE* 47(3):915–924.

Yang, C., J. H. Everitt, and J. M. Bradford. 2007. Airborne hyperspectral imagery and linear spectral unmixing for mapping variation in crop yield. *Precision Agriculture* 8(6):279–296.

Yang, C., J. H. Everitt, and J. M. Bradford. 2008. Yield estimation from hyperspectral imagery using spectral angle mapper (SAM). *Transactions of American Society of Agricultural and Biological Engineers* 51(2):729–737.

Yang, C., J. H. Everitt, M. R. Davis, and C. Mao. 2003. A CCD camera-based hyperspectral imaging system for stationary and airborne applications. *Geocarto International Journal* 18(2):71–80.

Zádorová, T., V. Penížek, L. Šefrna, O. Drábek, M. Mihaljevic, Š. Volf, and T. Chuman. 2013. Identification of Neolithic to Modern erosion-sedimentation phases using geochemical approach in a loess covered sub-catchment of South Moravia, Czech Republic. *Geoderma* 195–196:56–69.

Zhang, N., M. Wang, and N. Wang. 2002. Precision agriculture—a worldwide overview. *Computers and Electronics in Agriculture* 36:113–132.

Žížala, D., T. Zádorová, and J. Kapicka. 2017. Assessment of soil degradation by erosion based on analysis of soil properties using aerial hyperspectral images and ancillary data, Czech Republic. *Remote Sensing* 9:28.

4 Mission Concept and Trade-Off Study

4.1 HYPERSPECTRAL SATELLITE CONFIGURATION

During the past 60 years, hundreds of satellites or spacecraft have been built in support of scientific, military, and commercial missions. Most of them can be broadly categorized as three types: (1) three-axis-stabilized spacecraft, (2) spin-stabilized spacecraft, and (3) pallets. These types are distinguished by their configurations, internal equipment, and thermal-control designs.

The most common spacecraft configuration today is three-axis-stabilized. A hyperspectral satellite belongs to this type. This type of spacecraft is characterized by a body that is roughly box-shaped and by deployable solar-array panels. The bodies of these spacecraft are usually kept inertially stable except for a slow rotation induced about one axis to keep the payload antennas or sensors continuously pointed toward Earth as the satellite orbits. The solar-array panels are then counter-rotated relative to the spacecraft body to keep them inertially fixed on the sun.

The spacecraft is commonly referred to a "payload" and a "bus," (or "platform"). The payload is the equipment that services the primary mission. For a hyperspectral satellite, the payload is the hyperspectral sensor and the panchromatic imager if deployed. The bus consists of all other spacecraft subsystems that support the payload. These subsystems typically include:

- Electrical power/distribution subsystem (EPS or EPDS): The equipment used to generate and distribute electrical power to the spacecraft, including solar arrays, batteries, solar-array controllers, power converters, electrical harnesses, battery-charge-control electronics, and other components.
- Telemetry, tracking, and command subsystem (TT&C): The electronics used to track, monitor, and communicate with the spacecraft from the ground. TT&C equipment generally includes receivers, transmitters, antennas, data recorders, and state-of-health sensors for parameters such as temperature, electrical current, voltage, propellant tank pressure, enable/disable status for various components, etc.
- Attitude/velocity control subsystem (ACS or AVCS): The devices used to sense and control the vehicle attitude and velocity. Typical components of the ACS system include sun and Earth sensors, star sensors (if high-precision pointing is required), reaction or momentum wheels, inertial measurement units (IMUs), inertial reference units (IRUs), and the electronics required to process signals from the above devices and control satellite attitude.
- Thermal-control subsystem (TCS): The hardware used to control temperatures of all vehicle components. Typical TCS elements include surface finishes, insulation blankets, heaters, and refrigerators.
- Propulsion subsystem: Liquid and solid rockets or compressed-gas jets and associated hardware used for changing satellite attitude, velocity, or spin rate. Solid rockets are usually used for placing a satellite in its final orbit after separation from the launch vehicle. The liquid engines (along with associated plumbing lines, valves, and tanks) may be used for attitude control and orbit adjustments as well as final orbit insertion after launch.

4.2 HYPERSPECTRAL SATELLITE EARTH ORBITS

A variety of orbits are used for different types of Earth observation missions. The most common orbits, in order of increasing altitude, are low Earth orbit (LEO), geosynchronous orbit (GEO), and highly elliptical orbits (HEO).

4.2.1 Low Earth Orbits

Low Earth orbits are those whose maximum altitudes are less than approximately 2000 km and have the shortest orbital periods, on the order of an hour and a half. Some of these orbits are circular, while others may be somewhat elliptical. The degree of eccentricity is limited by the fact that the orbit is not much larger than the Earth plane, whose diameter is approximately 12,760 km.

The inclination of the LEO orbits, which is the angle between the plane of the Earth equator and the plane of the orbit, typically varies from 0° to 98°. For inclinations <90°, the satellite appears to be going around its orbit in the same direction as Earth's rotation. For inclinations >90°, it appears to be going opposite Earth's rotation. In this case the orbit is known as a retrograde orbit.

Low Earth orbits are very often given high inclinations so that the satellite can pass over the entire surface of Earth from pole to pole as it orbits. This coverage is important for weather and surveillance missions.

4.2.2 Sun-Synchronous Orbits

One particular type of low Earth orbit maintains the orbit plane at a nearly fixed angle relative to the sun, as depicted in Figure 4.1. As a result of this type of orbit, the satellite passes over locations on Earth that have the same local time on every orbit, i.e., the same local sun-elevation angle. Because Earth rotates beneath the orbit, the satellite sees a different swatch of Earth's surface on each revolution and can cover the entire globe over the course of a certain period (e.g., a day), as shown in Figure 4.1. The ability to see the entire surface of Earth at the same local sun angle is important for imaging missions such as hyperspectral imaging, visual-surveillance, and weather

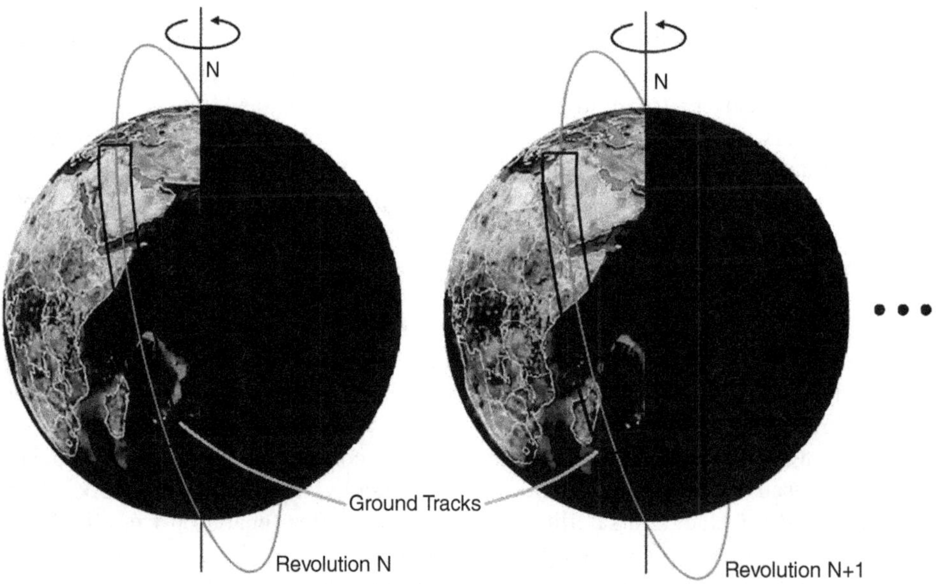

FIGURE 4.1 Illustration of sun-synchronous orbit that permits a satellite sees a different swatch of Earth's surface on each revolution due to Earth rotates beneath the orbit.

observation missions. This type of orbit is known as sun-synchronous orbit. Almost all Earth observation hyperspectral satellites use sun-synchronous orbits. Sun-synchronous orbits may be positioned so that satellites always see points on Earth at a specific time, anywhere from local sunrise/sunset to local noon. They are often known as "noon" or "morning" orbits.

4.2.3 Geosynchronous and Highly Elliptical Orbits

The next higher altitude type of common orbit is geosynchronous. These orbits are circular and have very low inclinations (<10°). They have an altitude of 35,786 km. Their distinguishing characteristic is a period matching Earth's rotation, which allows a satellite to remain over the same spot on Earth at all times. This characteristic is valuable for a wide variety of missions, including weather forecast, communication, and surveillance.

The highest common orbit type is HEO orbit, such as Molniya and Tundra orbits. For example, a Molniya orbit has an apogee of 38,900 km and perigee 550 km and is highly inclined (62°). Such extremely elongated orbits have the advantage of long dwell times at a point in the space during the approach to apogee. Bodies moving through the long apogee dwell appear to move slowly and remain at high altitude over high-latitude ground sites for long periods of time. They provide good views of the north polar region for a large portion of the orbit. Because the satellite travels very slowly near apogee, a Molniya orbit has a good view of the polar region for up to 8 h out of its 12 h period. A constellation of three satellites in Molniya orbits can provide continuous coverage of the northern hemisphere for missions such as weather forecast and communications.

One final useful observation is that most Earth-orbiting satellites travel through their orbits in a counterclockwise motion as seen from above the North Pole. They move in this direction to take advantage of the initial eastward velocity given to the satellite as a result of Earth's rotation (approximately 1500 km/h at the Kennedy).

4.3 MISSION DEVELOPMENT PHASES

Space missions are developed and executed using a phased process, as illustrated in Figure 4.2. The full life cycle includes mission definition and system design, component, subsystem and system development, system and subsystem integration, design qualification, hardware/software acceptance testing, launch, deployment, mission operations, maintenance, data handling, and disposal, if required. The mission design process occurs during conceptual design (Pre-Phase A), mission

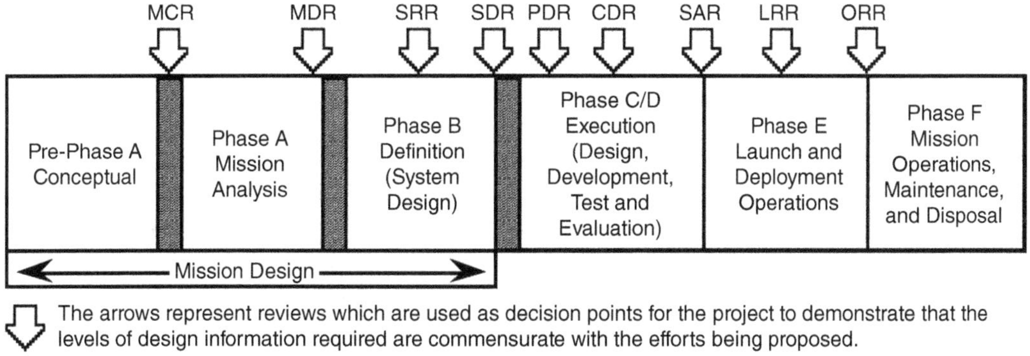

FIGURE 4.2 Space mission development phases. (Source: NASA.) (MCR: mission concept review, MDR: mission definition review, SRR: system requirements review, SDR: system design review, PDR: preliminary design review, CDR: critical design review, SAR: system acceptance review, LRR: launch readiness review, ORR: operational readiness review).

analysis (Phase A), and definition (Phase B). It is the engineering and programmatic processes. In every case, the mission design process must put a proposed mission on firm technical, cost, and schedule grounds before committing to design, development, test, and evaluation.

The mission design process generally begins with a statement of the mission objectives, mission classification, and programmatic constraints. Higher mission classification levels typically require more hardware to provide robustness, redundancy, and operational flexibility to enhance the probability of mission success.

Pre-Phase A studies are generally carried out to develop mission concepts. The mission objectives and other constraints are analyzed to define mission requirements, from which a strawman mission concept can be generated.

The mission requirements are evaluated in order to provide an overview of the available options. For each option, the mission analyst computes the information needed by the project to perform a proper trade-off between the different options and to define one or more baseline and back-up solutions for further detailed analysis, definition, and optimization.

Later on in a project, the baseline solution is studied in more detail in order to demonstrate feasibility, and further optimize performance, in addition to generating all the information needed for platform, payload, ground segment, launcher service and operations design. Pre-Phase A studies must demonstrate a viable, but not necessarily optimized, concept for pursuing a mission.

After a decision to proceed a mission is made by the sponsoring organization (e.g., a national space agency), a Phase A study formally begins to refine the overall requirements and probable costs, along with the top-level definition of the best way to accomplish the mission. The programmatic guidelines (especially overall cost and funding profile) are a major factor used in determining this. Mission requirements are further defined and refined through trade-off studies and analyses. Various system concepts are examined by means of further trade-off studies and analyses. The overall mission and system architecture, the apportionment of functionality between the various systems, and candidate system designs are evaluated. A single mission architecture and system level design is selected. Following completion of Phase A study, a formal decision to proceed into Phase B will be made.

During Phase B, further analyses and trade-off studies are conducted to establish the lower hierarchical levels required to define the system(s) down to a level such that specifications, statements of work (SOW), etc. are generated to permit entry into Phase C/D. At the completion of Phase B, a formal non-advocate review (NAR) may be initiated by the sponsoring organization. The NAR panel reviews the realism of the technical approach, the schedule, and the cost. The NAR panel may then recommend approval for proceeding into Phase C/D.

4.4 ANALYSIS OF USERS NEEDS

4.4.1 Understanding Users' Needs

Since its early development over 40 years ago, hyperspectral imaging technology has gradually expanded from its initial use as a research tool to commercial applications. Today, 20 years after the first spaceborne hyperspectral imager Hyperion launched on NASA's EO-1 satellite, hyperspectral imaging is the fastest growing research area in remote sensing.

In contrast to multispectral imagers, hyperspectral imagers generate narrowband contiguous spectra of the observed targets. With hundreds of spectral bands, a hyperspectral imager obtains a spectrum of the energy reflected or emitted by each pixel in the field-of-view. The identification of an object using a hyperspectral imager is established via its acquired spectrum, as opposed to its shape, which is used by most multispectral systems. Hyperspectral imagery is, in general, a three-dimensional data cube comprising of two spatial dimensions and one spectral dimension, where the spatial dimensions account for across track and along track directions of the satellite platform carrying the hyperspectral imager.

Hyperspectral imager applications are very diverse and generally support the generation of important products for users in all disciplines, such as industrial, agricultural, geological, and environmental, etc. As examples of the high amount of interdisciplinary effort needed, one observes that research applications are very heterogeneously covered through studies in:

- Vegetation with forestry, agriculture, e.g., detection of chlorophyll, cellulose
- Land use, e.g., detection of land cover and change
- Coastal and inland waters, e.g., detection of plankton, dissolved organic material, sediment content
- Geology and mining, e.g., detection of minerals, soil types
- Atmosphere, e.g., detection of aerosols and trace gases
- Snow and ice, e.g., snow impurity, depth, and grain size
- Defense and security

Additionally, the research fields are extremely diverse, covering technical challenges, algorithm development and application research performed by instrument manufacturers, data providers as well as the end users. The technological developments may include:

- Instrument development, for high signal-to-noise ratio (SNR), high spatial and spectral resolution, broad wavelength range, wide spatial coverage, low distortion, low self-polarization
- Processing, new parallel or grid processing schemes for large hyperspectral data volumes
- Calibration, uniformity, sensor stability, accuracies, etc.

It is essential to analyze the needs of the stakeholders and users in the early stage of a hyperspectral satellite mission. Then the needs of the stakeholders and users can be converted to the objectives and the requirements of the mission that will guide the development of the mission throughout the different development phases described in Section 4.3.

The scope of the hyperspectral applications continues to grow as users demand greater information content and more frequent and timely response. A comprehensive study of all the potential hyperspectral usages is beyond the scope of this section. This section describes user needs in following five application domains:

1. Environment
2. Coastal and inland waters
3. Agriculture
4. Natural resource, including forestry, lithology, and geology
5. Defense and security

4.4.2 Environment

Hyperspectral remote sensing has been used to address water quality and wetland issues where the data provides unique information on vegetation and pollution in specific areas where problems have already been identified.

Hyperspectral remote sensing would allow government department of environment to get ahead of the situation, by repeatedly capturing hyperspectral information over a wide range of areas. This would provide a consistent historical record showing baseline conditions, enable short-term change detection processing to detect problems as early as possible, and provide subsequent monitoring of any remediation activities. Hyperspectral remote sensing could allow many of these activities to be semiautomated by, for example, providing an "alert" process when unexpected changes occur in sensitive environmental areas. Having a readily available archive of data covering the before, during, and after conditions with sub-weekly temporal resolution would also give government environment department a much better ability to identify events which led to environmental problems. Table 4.1 tabulates a list of user needs for environment.

TABLE 4.1
Environment Domain User Needs

User Needs	Benefits	Existing Methodology	Sought Improvement Over Existing Methodology
Atmospheric quality monitoring	• Improved public safety and health	• Ground-based observation • Moderate Resolution Imaging Spectroradiometer (MODIS), Measurements of Pollution in the Troposphere (MOPITT), and other satellite	• More sensitive detection of atmospheric pollution • More frequent coverage • Larger area cover • Ability to pinpoint pollution sources
Marine and freshwater quality monitoring	• Improved public safety and health • Improved detection and prosecution of marine polluters • More rapid, expansive water quality monitoring	• Ship surveys for water quality surveillance • Demonstration spacecraft water quality monitoring from multispectral spacecraft sensors	• Increased sensitivity to water pollution from land operations • Increased resolution of land effluents, allowing sources to be identified • More sensitive and rapid detection of oil pollution • More sensitive and rapid detection of algae blooms • Expanded suite of algal bloom products (especially species discrimination)
Wetlands habitat monitoring	• More sensitive detection of wetlands water quality • More rapid detection of wetland changes		National and regional maps of wetlands showing extent, species, and health of wetland habitats
Land reclamation monitoring	• Helps enable government to ensure that corrective actions are implemented for pollution and wildlife management		• Better accuracy to assess environmental recovery • More frequent coverage to ensure timely resolution • Wider area coverage for wilderness regions which are hard to access
Sea ice monitoring	• Improved shipping safety • Improved climate change monitoring	• RADARSAT spacecraft • In situ on-ice and drifting buoys	• Improved detection of sea ice type and extent, and movement • Larger area, frequent coverage in far north
Land degradation and desertification	• Detection of local and regional changes in ecosystems to sustain wildlife, leading to corrective and/or reclamation actions		• More accurate detection of changes to vegetation and environmental conditions • More widespread and frequent observations to provide more timely detection and responses, particularly in areas with difficult access

(Continued)

TABLE 4.1 *(Continued)*
Environment Domain User Needs

User Needs	Benefits	Existing Methodology	Sought Improvement Over Existing Methodology
Land hazard mapping	• Increased public safety • More effective detection and prosecution of environmental polluters		• Increased sensitivity to man-made hazards (oil spills, leaching, etc.) • More frequent coverage of more target sites
Climate change	• More accurate reporting of carbon footprint • Public information and awareness about climate change causes and effects		• Improved monitoring of change of vegetation cover • Improved measurement of carbon intake/uptake • Improved distinction between man-made and natural change
Disturbance monitoring	• Improved wildland habitat quality • Faster detection of harmful human impacts to the environment.		• More sensitive detection of wildland disturbances • More accurate determination of disturbance causes (natural or man-made) • More accurate measurement of impact of disturbances
Environmental remediation	• Improved remediation and environmental monitoring	• Airborne and field monitoring	• Regional coverage • Spectral detection of indicators • Robust, repeatable, while complementing the different emphasis of other satellite sources
National scale indicators	• Identify cumulative impacts of climate variability and change and disturbances at regional scale	• National scale land monitoring system	• High performance hyperspectral sensors can improve calibration and validation of operational systems

4.4.2.1 Atmospheric Quality Monitoring

Monitoring atmospheric pollution is often performed using specialized space-based sensing instrumentation in space missions. Many such satellites in international missions, to observe pollution, greenhouse gases, and other atmospheric properties on a global scale, have been launched.

Hyperspectral imaging capability would provide a new tool for observing and tracking many atmospheric pollutants which are detectable in the visible and short wave infrared spectral bands. This would include observation of forest fire smoke, various industrial pollutants, volcanic ash, sulfur dioxide, and others. Hyperspectral information has a major advantage over multispectral imagers because it allows much more refined differentiation of atmospheric pollutants.

4.4.2.2 Freshwater Quality Monitoring

Water quality in lakes, rivers, and wetlands is of major public concern since this is directly related to environmental stewardship, fishing, public recreation, drinking water, and public safety. This has become a major issue on many other water bodies which are often shared by human and industrial users.

Some of the key issues include harmful algae blooms (HABs), industrial effluent run-off, suspended sediments, and marine oil pollution. These are generally detected with specific on-site testing in areas with known potential for problems, but this is not an effective method for efficient detection and monitoring of problems over large regions. Hyperspectral observations from space would allow the spectral signatures of these pollutants to be distinguished from other surface features over large areas with sufficiently frequent updates (less than weekly) to detect emerging problems and to then monitor them through subsequent remediation activities.

4.4.2.3 Wetlands Habitat Monitoring

Wetlands provide extremely important wildlife habitats and are protected by various national acts. Monitoring of such large areas on a regular basis cannot be done effectively through on-site ground sampling. Like water quality monitoring, hyperspectral remote sensing can provide a valuable tool by allowing wetlands health to be monitored through changes in vegetation conditions and extent. Hyperspectral imaging, unlike multispectral imaging, can provide much more refined mapping of wetland vegetation species, identify invasive vegetation, and detect subtle changes in vegetation stress.

The goal of a spaceborne hyperspectral sensor is to be able to establish a solid record of consistent hyperspectral remote sensing data from which change detection can be performed in a semi-automated manner where unexpected changes can be detected and alerted. Such information on a weekly basis would better allow natural or man-made events to be related to those environmental changes and to subsequently monitor any remediation progress.

4.4.2.4 Land Reclamation Monitoring

The objective of conservation and reclamation of specified land is to return the specified land to an equivalent land capability (ELC). For example, in the province Alberta, Canada, there are numerous coal, oil sands, and gravel pit mines in size ranging from a few hundred square meters to hundreds of hectares. Land reclamation auditing and inquiry assessments at well/mine sites using existing practices are extremely costly and unrealistic. Thus, the Government of Alberta has indicated strong needs for a remote sensing based system for comparing reclamation areas to surrounding natural areas and to map trends in regrowth progress as a cost-effective solution to the reclamation monitoring problem.

Among the remote sensing data, hyperspectral imagery could play a key role in generating the information products required for land reclamation monitoring.

4.4.3 Coastal and Inland Waters

The primary mandate of government department of fisheries and oceans is to maintain the health and productivity of fishing activities in nation's coastal oceans, lakes, and rivers. To do this, they regularly undertake wide-ranging science and environmental monitoring programs. Much of this activity requires in situ sampling of water, vegetation, and fish. But the effects of pollution, climate change, and overfishing occur over very large areas and can also occur on very rapid time scales (weeks) in some cases. To properly monitor such large areas on short timescales, an effective remote sensing system is needed.

To complement its in situ sampling work, the fishery and ocean government department has a strong desire to be able to acquire space-based images. Their needs for hyperspectral remote

TABLE 4.2
Coastal and Inland Waters Domain User Needs

User Needs	Benefits	Sought Improvement Over Existing Methodology
Enhanced water quality monitoring, ecological information of coastal and medium to large freshwater bodies	• Increased economic output from fisheries • Improved national management of ocean-based resources • Public information and awareness about nation's marine conditions and future	• Increased aquaculture production • Improved forecasting of ocean conditions • Improved insight into the impacts of climate change on marine ecosystems
Monitoring of hazards, discharges, effluents, and pollution events	• Improved public safety and health	• Increased sensitivity to water pollution from land operations • Increased resolution of land effluents, allowing sources to be identified • More sensitive and rapid detection of oil pollution
Assessing freshwater productivity, phytoplankton dynamics, and nuisance of potentially harmful algal blooms	• Improved public safety and health • Increased economic output from fisheries	• More frequent and rapid detection of harmful algal blooms • Reduced harmful algal blooms damage to aquaculture stocks

sensing can be broadly captured in three areas as listed in Table 4.2. A shortened version of these requirements is summarized below.

4.4.3.1 Enhanced Water Quality Monitoring

Inland waters and nearshore areas of coastal waters are often under intense pressure from the effects of nutrients loadings, contaminants, and development. But at the same time they provide areas for leisure and recreation and are important breeding grounds for fish and other wildlife as well as supporting aquaculture and fisheries. The coastal region is also home to settlements and extraction industry facilities. Regular synoptic monitoring available from Earth observation satellites including ocean color, synthetic aperture radar (SAR), scatterometer, altimeter, ocean temperature, wave height and direction, and salinity has offered the opportunity to supplement often sparse, discontinuous ground-based observations.

Current spaceborne instruments utilized for water color measurement make use of moderate resolution imaging of the MODIS/Sentinel-3 series of sensors, offering ground spatial resolution of 250–1000 m. These are excellent for open ocean assessments on moderate to large scales. There is a strong need for ground spatial resolution of 100 m less for coastal zones of oceans. Another major requirement is to have better temporal resolution of ocean color to facilitate monitoring ephemeral bloom, pollution, and ice movement events where current sampling is insufficient.

There are also developing needs to understand the movement of living material in the water as it cycles through the water column during the diurnal cycle. A constellation of satellites with the current standard of instrument would suit operational needs better than a single more advanced occasional image from a hyperspectral sensor.

4.4.3.2 Water Clarity

Water clarity is not only an important indicator of ecosystem health, but also plays a central role in both pelagic and benthic primary productivity, and through absorption affects heat transfer in the upper water column. Water clarity is directly related to the particulate and dissolved materials contained within the water, which in combination determine the degree to which light is attenuated in the upper water column.

The introduction of nonnative invasive species, point-source discharges, nutrient loading and resulting eutrophication and nuisance algal blooms have all led to notable fluctuations in the water clarity of inland waters and oceans. A range of empirical data obtained through analytical bio-geo-optical modeling methods may be adopted to reach a quantitative measure of water clarity, allowing hyperspectral imagery acquired from space to document bright-water episodes such as phytoplankton blooms, resuspension of bottom sediments, and whiting events (whereby highly scattering calcium carbonate is precipitated out of solution under specific temperature and pH conditions).

Higher spatial resolution (in the order of 100 m or less) hyperspectral imagery would allow the extension of the observations to study specific nearshore water clarity events including river plumes, nearshore blooms, and area of concern monitoring.

- The ideal temporal revisit for monitoring is 1–2 days
- Temporal revisit of 3–7 days would be useful
- Temporal revisit of >7 days is less useful for monitoring but acceptable for research and development (R&D)

A hyperspectral sensor would be valuable to enable targeting of area of concerns for optimal coverage.

4.4.3.3 Monitoring and Assessing Harmful Algal Blooms

Excess nutrients and resulting eutrophication has caused problematic algal growth (either free floating planktonic or benthic attached mats) in some lakes. Prompt information on the timing, extent, biomass, and composition of an algal bloom can provide early indication of potential socioeconomic, ecosystem or ecological effects. Many algal and cyanobacterial blooms have nuisance or harmful effects including:

- Producing toxins that may be harmful to human health
- Impairing drinking water quality
- Beach fouling (with associated impacts on recreation and tourism)
- Disrupting ecosystems and foodwebs
- Degrading benthic habitats
- Producing anoxic events from biomass degradation

Hyperspectral imagery may be used effectively to provide a simple measure of lake trophic status, to map benthic attached algae in the nearshore, to detect and track specific algal bloom events and to offer early warning of possible shoreline or water intake fouling.

Spatial and temporal distributions of algae in lakes are highly variable in response to watershed, climatic, and algal physiological variables and therefore, capturing the onset and monitoring the progress of surface blooms requires frequent revisit times. Moderate spatial resolution (in the order of 100 m or less) hyperspectral imagery may offer opportunities for observing bloom conditions in the nearshore or small water bodies currently impossible with existing Earth observation tools, while the enhanced spectral resolution offers the possibility of retrieving additional algal taxonomy information using accessory pigment signatures, allowing the discrimination of potentially harmful algal species from those more benign species.

As the availability of high spatial resolution remote sensing imagery increases, and as methods to classify spectral signatures from specific algal groups like cyanobacteria improve, so the development of effective operational and near-real-time satellite-based monitoring tools for cyanobacteria blooms may become reality.

- Ideal temporal revisit for monitoring or operational event-based targeting: 1–2 days
- Useful temporal revisit for monitoring or operational event-based targeting: 3–7 days
- Temporal revisit >7 days acceptable for benthic algal mapping, less useful for dynamic bloom monitoring but acceptable for R&D, for inadvertent capture of specific events, and in supplementing more frequent but coarser resolution remote sensing monitoring tools
- Value of targetable sensor: to enable targeting of algal blooms following alert from more frequent but coarser resolution sensors

4.4.4 Agriculture

Government departments at both national and provincial level use remote sensing information for a variety of agriculture monitoring and management activities. This information is used by scientists, practitioners, and policy makers to address the stability, competitiveness and profitability of the nation's agricultural sector. This includes, for example, business risk management, flood and drought warning, pest management, environmental assessment, market access, food security, and public health and safety.

Commercial use of remote sensing for agriculture is also rising rapidly, where the information can be used to increase yields, reduce costs, and provide more stable operations. This is predominantly led through large industrial operations in conjunction with provincial governments. Table 4.3 lists main agriculture user needs. Following subsections summarize some of the higher profile remote sensing needs.

4.4.4.1 Annual Crop Inventory Report

In Canada, the Agriculture and Agri-Food Canada (AAFC) has been continuing to develop and evolve its annual Space-Based Crop Inventory Report for Canada since 2009 (AAFC 2012). To produce this report, the AAFC has used a wide range of space-based information such as RADARSAT-2, Disaster Monitoring Constellation (DMC) satellites, Landsat, and Advanced Wide Field Sensor (AWiFS).

This report has enormous economic value. The 2011 inventory included, for example, identification of acreages that had been too wet to seed earlier in the year—these estimates fell within 3% of figures provided independently by the provinces. The inventory has also been used to validate the practices of Canadian canola producers who wish to access the foreign biofuel feedstock market, estimated to be worth $500 million annually.

However, a major limitation of the current information is the requirement to integrate >1500 satellite images from many different sources, each linked to thousands of ground data points, which are required to map Canada's entire agricultural regions annually and to validate the resulting product. Hundreds of hours of computer processing time are required to do all the calculations to produce a final high-quality classification. While the current AAFC results meet an overall target accuracy of 85% or more, the extensive effort takes many months which prevents it from being used for many operational activities. The cost of collecting and assimilating the data is also extremely high, often requiring purchase of commercial satellite data to fill-in gaps in other data sources.

Information acquired by a hyperspectral satellite would improve the inventory accuracy, allow more refined crop classifications and mapping (e.g., better detection of subspecies and invasive species), and improve the product delivery time by offering much more frequent and consistent information which could be processed by a semiautomated system.

4.4.4.2 Crop Health Monitoring

One of important activities of the government agriculture department is to monitor nation's agricultural areas for crop stress and disease. This is important to ensure that problem areas are identified quickly and that corrective actions can be taken. This information is used by provincial governments, farm operators, and insurance companies among others (McNairn et al. 2012).

TABLE 4.3
Agriculture Domain User Needs

User Needs	Benefits	Existing Methodology	Sought Improvement Over Existing Methodology
Annual crop inventory (Annual report with maps and supporting quantitative information summarizing nation's agricultural crop production and changes)	• Support sustainability • Support forestry sector • Improved market forecasts	• In situ sampling achieves 80% classification accuracy • Area coverage is limited	• New capabilities such as speciation and invasive species • Classification accuracy improved to 95% • Coverage improved
Crop health information (National and regional maps of agricultural areas showing biophysical and biochemical conditions related to crop disease and stress)	• Detection and tracking of crop stress and disease • Improved crop yield • Improved crop quality • More efficient use of fertilizers and chemicals • More efficient use of water	• Landsat, disaster monitoring constellation, RapidEye, etc. provide very basic information • Airborne hyperspectral is OK for small areas, but impractical at national scales	• New capabilities to identify and resolve crop stress and diseases • Reduced cost compared to existing methods
Precision farming (Local maps at good resolution showing crop biophysical and biochemical conditions, soil conditions, crop moisture conditions and stress)	• Detection and tracking of crop stress and disease • Improved crop yield prediction • Improved crop quality • More efficient use of fertilizers and chemicals • More efficient use of water	• In situ measurements and sampling are accurate but expensive • Airborne remote sensing is proven but is also too expensive for most operations	• More frequent information updates (weekly) • Lower cost data to allow more farmers to employ precision farming • Easier access to a nationally provided service with standardized products
Crop insurance claim verification (Regional and local maps of farmland showing the extent and nature of crop damage from various causes)	• Reduced agri-insurance claim fraud • Reduce cost of claim investigations • Faster claim processing	• Direct on-site investigation • Limited use of Landsat and other remote sensing provided limited information	• More frequent imaging to perform change detection • More detailed crop health information
Management practices (Maps showing field-level management practices of interest, such as till/no-till agriculture and crop residue cover)	• Support sustainability • Improved market forecasts • Demonstration of environmentally sustainable practices may open market access	• Direct on-site investigation • Remote sensing– based analysis mostly research, far from being implemented operationally	• Wider area coverage than small/local research studies undertaken to date • Transition to national cover/operational monitoring system
Snow cover maps (Snow cover [fraction] information [weekly] at 250 m or finer spatial resolution. May be used for extraction of snow water equivalent)	• Improved crop yield forecasting • Improved drought monitoring and forecasting • Improved hydrological characterization of agriculture lands • Improved identification of seeding dates	• Currently derived from MODIS to correspond with weekly Max-NDVI mapping	• Weekly snow maps already being delivered using MODIS data, but shortwave-infrared (SWIR) information lacking

There are two main types of crop health monitoring functions which a hyperspectral satellite would address:

1. Crop biophysical monitoring
2. Crop biochemical monitoring

Biophysical indices provide information on crop production, including such things as leaf area index (LAI), biomass, rate and extent of growth. This helps the national and provincial governments to more accurately estimate agricultural yield for market analysis and to identify areas where corrective actions may be needed. The hyperspectral information and frequent repeat coverage would provide significantly improved biophysical information, providing better yield information and more timely identification of unexpected productivity changes (e.g., due to environmental events).

Biochemical indices provide information on crop stress, including such things as moisture stress, leaf chlorophyll content, leaf nitrogen content, photosynthetic activity, and others. This allows the national and provincial governments to better assess the source of any crop production and quality issues, and coordinate responses. The detailed chemical information and timeliness afforded by hyperspectral satellite would be a dramatic improvement on the current sources of information, allow problems to be identified more quickly, determine the scale of such problems, and better assist in defining corrective actions.

4.4.4.3 Crop Insurance

The agriculture business involves managing many different risks, including the weather, disease, insects, wildlife, and market price volatility. In Canada, crop insurance is provided by a combination of federal and provincial support. While very successful, fraud and abuse by a small percentage of farmers regularly jeopardizes the performance of these programs.

In the United States, Landsat imagery is typically employed if a United States Department of Agriculture (USDA) field investigator determines that claim verification is warranted. The image investigation is performed either internally by the USDA's Risk Management Agency (RMA) or contracted out to a private remote sensing expert. Over the past a few years, the RMA has used an average of 600 Landsat scenes per year (each covering a geographic area of 7.6 million acres) for investigative purposes. Typically, about half of the Landsat image analyses support a farmer's insurance claim and half indicate fraud. Use of Landsat saves the US government on the order of $100 million per year in insurance fraud (Rocchio 2006).

Validating insurance claims starts with gathering evidence to verify the type and extent of crops planted by farmers. This can generally be performed with basic multispectral data, but needs sufficient resolution to identify field boundaries and compute total planted area. Most importantly, the data needs to be gathered at the start of the growing season and at multiple points during the growing season. During the middle of the growing season, additional data can be used to monitor the quality and expected yield of the farm—thus establishing an overall crop value. This is where frequent hyperspectral data can play a more important role since it can be used to derive crop density, crop stress, and establish whether the farm is performing comparably to other farms in the region. Following crop damage claims, remote sensing data can again be used to identify the location and extent of crop damage. Looking at the before and after data allows the insurance company to verify the cause of the damage as well as the value of crop lost.

4.4.4.4 Precision Farming

The economic impact of precision farming is potentially enormous. In the United States, precision farming techniques are estimated to have the potential of increasing corn yields by 10–20%, corresponding to an increased crop value of $7–13 billion annually in the United States alone. Other crops (wheat, oats, canola, flax, etc.) could potentially see similar improvements.

Experimental projects have used multispectral data from both airborne and spaceborne (e.g., Landsat, RapidEye, etc.) to demonstrate the associated benefits. But the high cost, infrequent coverage, and limited spectral information of these sources is impractical to use on nation-wide scales. Airborne imaging provides the highest performance data, but the cost per area is perhaps 100 times as much as equivalent satellite data. But most current satellite data, at <30 m ground sampling distance (GSD), have poor repeat coverage and limited spectral information. The ideal solution to this is hyperspectral data at 5–10 m GSD with weekly updates at national scale coverage. This is exactly the sort of information that a hyperspectral satellite need be designed to provide.

4.4.5 Natural Resource

Canada has vast natural resources. The Natural Resources Canada (NRCAN) regularly uses space-based radar (RADARSAT, RADARSAT-2, etc.) and multispectral imaging (Landsat, EO-1, SPOT satellites, etc.) for its forestry and mining applications, but has long been a proponent of developing a Canadian hyperspectral satellite and its remote sensing application capability. For the last decade, NRCAN has continued to lead much of the Canadian government's development of hyperspectral remote sensing tools such as the powerful Imaging Spectrometer Data Analysis Systems (ISDAS) software suite (Staenz et al. 1998, CCRS 2014), which is used extensively by Canadian researchers to process space-based and airborne-based hyperspectral data. NRCAN has undertaken numerous investigations, in collaboration with other government departments, to collect and process hyperspectral data for many research applications. Those applications are quickly reaching the point where they need to evolve into operational use.

The primary NRCAN uses of remote sensing data include forestry and mining. For forestry, this involves regular updates of Canada's forest inventory, monitoring the spread of insects, disease, and invasive species, and international reporting on carbon sink changes from deforestation, afforestation, and reforestation. For mining, the NRCAN needs to regularly monitor the environmental impact of mining operations, such as tailings, leaching, deforestation, changes in nearby ecosystems, etc. NRCAN also wants to collect more complete mineralogical maps of Canada's northern areas to better understand the potential resources available. The user needs of natural resource domain are listed in Table 4.4.

In this section, some higher priority needs are described in more detail below.

4.4.5.1 Forest Inventory Management

The National Forest Inventory is a collaborative effort between the federal, provincial, and territorial governments in Canada. This includes compiled information for each of Canada's forested ecozones. The Canadian Forest Service (CFS) maintains this national database and leads data analysis and reporting. Large geographic information system (GIS) databases of forest polygons exist to give inventories of basic tree parameters by way of species composition, quantity, density, age, height, site index, and crown closure. These parameters are then used to estimate forest species areas, timber volumes, biomass, etc. that are essential for national reporting. GIS databases vary in accuracy and detail.

The GIS database is currently produced through a very labor intensive ground campaign involving a grid of in situ ground sampling points across Canada covering 1% of the area. This covers most of the areas in the lower regions of the provinces. Landsat, RADARSAT, and other sources of satellite data are used to fill-in the gaps, particularly in northern Canada where there is no easy ground access. Because of the high cost of collecting this information, the data can typically only be updated every 5–10 years, with moderate accuracy. Changes which occur during this period are often not captured.

A goal of a hyperspectral satellite for this application would be to image all of Canada's forests on a regular basis throughout all of the growing seasons. Such information can provide much more consistent and accurate measures of forests across Canada by providing improved (and lower

TABLE 4.4
User Needs in the Domain of Natural Resources

User Needs	Benefits	Existing Methodology	Sought Improvement Over Existing Methodology
Forest inventory (Map of nation's forest landscapes to provide data, forecasts, health, and estimates of forest-related land use change)	• Better ability to assess sustainability and changes in nation's forests	• In situ sampling archives provides estimates	• New capabilities such as forest speciation • Improve classification accuracy
Monitoring and management of forest health	• More efficient detection and mapping of insect and disease outbreaks • More efficient harvesting planning of forests in reaction to changes in forest health	• In situ sampling • Airborne imaging	• More sensitive detection of invasive species extent and change • More sensitive detection of insect damage (e.g., pine beetles) extent and change
Environmental impact monitoring of oil sands and other oil/gas operations	• More environmentally responsible oil sands and other oil/gas extraction operations		• More sensitive detection of environmental disturbances • Faster detection of environmental impacts • Better ability to distinguish different types and sources of impacts
Treaty compliance reporting	• Improved nation's position for international climate change negotiations	• In situ sampling grid • Landsat	• More accurate estimation of carbon storage and release
Monitoring of commercial forest harvesting	• Improved monitoring of forest harvest		• More sensitive detection of logging impacts • Faster, cheaper, and more regular monitoring of logging operations
Mineral detection and lithological mapping	• Development of new mining activities	• In-sampling • Airborne hyperspectral	• Larger area coverage
Permafrost modeling	• Lasting northern infrastructure and less expensive infrastructure maintenance	• Currently uses broadband multispectral that achieves ~70% accuracy	• Better prediction of boundary surface conditions for permafrost modeling

cost) information about biodiversity. For example, researchers in NRCAN have used hyperspectral data acquired by Hyperion to map five individual tree species growing along British Columbia's coast and achieved much significantly higher classification accuracy than that using multispectral data for forest inventory (see Figure 3.21 in Chapter 3). Dominant species identification showed dramatic improvement using hyperspectral information when compared to multispectral results.

Classification results of Landsat, Advanced Line Imager (ALI), and Hyperion show how classification accuracies were improved by using hyperspectral sensors.

The researchers also used similar airborne hyperspectral imagery to map chlorophyll, water content, and nitrogen levels in west coast of Canada forest canopies' leaves, demonstrating the ability to gather key information on forest health.

A hyperspectral satellite to be developed for forestry application would be expected to provide substantially higher performance than the retired hyperspectral sensor Hyperion, with wider spatial coverage, allowing it to be used in a consistent, operational mode.

4.4.5.2 Monitoring and Management of Forest Health

Forest health is monitored using similar techniques to that used for agriculture. This includes both biophysical and biochemical properties, although some of the specific algorithms and models used to do this are obviously different.

Hyperspectral remote sensing data could be used to improve forest inventory and health information, as well as increase information about biodiversity, natural disturbances and the effects of climate change in forests. By comparing mapped leaf nitrogen, chlorophyll, or water content against known profiles of normal chemistry for species, researchers are able to pinpoint stands of highly stressed trees. Forest managers could then go and look at what is causing the stress—disease, insects, drought, or other issues—on the ground. The local government, for example, could use hyperspectral satellite imagery to identify the spread of pine beetle infestation in its forests early enough to take measures to help slow the spread and harvest forest areas before the wood is seriously damaged.

4.4.5.3 Forest Biomass Estimates

The Canada's Earth Observation for Sustainable Development (EOSD) of forests land cover map was developed primarily to support Canada's national and international reporting requirements for sustainable forest management and biomass, including the Canadian Council of Forest Ministers' Criteria and Indicators, the State of Canada's Forests annual report, and the United Nations Framework Convention on Climate Change. Meeting those commitments requires that the composition, distribution, structure, and dynamics of Canada's forests be quantified over time.

Landsat data scenes were used to produce this map. The scenes are used to provide information for biomass estimates, for climate change modeling, and for Canada's National Forest Inventory—particularly for northern forest regions where there are few ground or photo plots, and land cover classifications have been based on limited information.

However, the map can also be used to generate other information. For example, researchers are using the tiles to help them assess Canada's biodiversity in support of other national and international commitments: Canada's Species at Risk Act 2003 and the 1992 Convention on Biological Diversity. The EOSD maps help the researchers establish a baseline of habitat fragmentation for various species in Canada (NRCAN 2018).

With detailed spectral signature and high spectral resolution, satellite hyperspectral information will allow significant improvement in the EOSD process by providing much more complete information for better differentiation of forest species, more accurate quantification of biomass, and much more consistent and frequent updates throughout the year.

4.4.5.4 Environmental Impact Monitoring of Mining Operations

All mining operations have some degree of environmental impact that needs to be monitored and controlled to protect people and wildlife. Deforestation can be a direct result of mining operations. Chemicals from ore processing can propagate into local water systems and wetlands, damaging the environment over very large areas outside of the mining operation itself. Inadequate monitoring and control can result in major environmental events. The public now has a high degree of awareness and involvement in these issues due to the associated health impacts and costs. For example, in Canada some very publicly visible operations include the Alberta oil sands, gold, copper, and other metal mining in provinces British Columbia, Ontario, and Quebec, and diamond mining in northern Canada.

While on-site monitoring is generally undertaken by a mine operator itself in conjunction with national and/or provincial involvement, the very large scale of many mining operations, the even larger scale of the affected area around the mines, limited accessibility, and the need for frequent (ideally weekly) revisits often makes this very difficult and costly to perform. Identifying the contamination, tracking its source, and establishing responsibility is a complex process where remote sensing often provides key information.

With its ability to deliver detailed spectral image information on a frequent basis, satellite hyperspectral imaging could provide a significantly improved ability to properly monitor and manage the environmental impacts of mining operations. This information would be used by the mining industry, provincial governments, and the national government.

4.4.5.5 Lithological and Mineral Mapping

In a country, for example, Canada possesses vast areas that contain an enormous mineralogical wealth, these recourses continue to be revealed and exploited. Detecting minerals has traditionally been the domain of in situ sampling by mining companies that regularly perform field campaigns over promising areas. This has been enhanced in many cases, by using very high performance airborne hyperspectral imaging which can detect exposed minerals directly from their spectral signature. But this technique has limited success since most of Canada is covered in vegetation, making such minerals difficult to observe.

However, recent studies have revealed relationships between wildland vegetation and the underlying mineralogy, which has created renewed interest in this area. To take advantage of this, a spaceborne hyperspectral imaging system would be used to perform species mapping of selected types of wildland vegetation. In Canada's north, where the area is huge, and the growing season is short, the ability of a spaceborne hyperspectral imaging system to provide wide area coverage with frequent updates is key to this type of operation.

4.4.6 Defense and Security

The currently developed airborne hyperspectral remote sensing capabilities can achieve very high performance (small ground sampling distance, narrow spectral bands with extremely wide wavelength range, and high SNR). This type of performance can be used for many monitoring, research, search and rescue, and similar applications over relatively small areas of interest (e.g., about 100s of square km). Such operations typically have a high cost since they require expensive and ready-to-deploy equipment and staff, and they can be difficult to set up on short notice. For this reason, such operations are restricted to very high priority situations which can justify the costs involved.

While civilian hyperspectral satellite missions cannot hope to achieve the same level of performance as airborne hyperspectral systems, the users in defense and security domain are assessing how civilian hyperspectral satellites could address their needs for applications such as those described in Section 3.7.

4.5 A HYPERSPECTRAL MISSION CONCEPT CASE STUDY

This section uses a concept study of an imaginary hyperspectral satellite, which is referred to as Hyperspectral Mission for Environment and Resources (HyperMER) management, as an example to show the procedure of mission concept and trade-off study.

4.5.1 Mission Description and Requirements

The hyperspectral mission presented in this section is primarily for government departments at national level to monitor its environment and resources and make smart decisions to better manage its environment and resources. This hyperspectral satellite mission intends to meet most of the government stakeholders' needs described in Section 4.4 that are not covered by the existing or planned space missions.

The mission requirements of the HyperMER are tabulated in Table 4.5. These requirements were derived by reviewing and consolidating an appropriate set of the requirements which can be met

TABLE 4.5
Mission Requirements of Hyperspectral Microsatellites for Environment and Resources Management

	General requirement		
Orbit altitude and type	666±TBD km, sun synchronous (3-day repeat cycle, 14.7 orbits/day)		
Orbit local time of the ascending node (LTAN)	10:30 AM (±TBD min)		
Number of satellites	3 (spaced about 120° apart in orbit)		
Area of interest (AOI)	All of Canada land and coastal regions from 43° to 73° northern latitude Special-event tasks for medium (<500 km × 160 km) images anywhere on Earth		
Repeat coverage of AOI	≤3 days (different satellite every 3 days)		
Relook interval	≤1 day		
	Payload requirement		
	VNIR	**SWIR**	**PAN**
Imaging swath width	≥246 km, 3-day coverage for areas from 43° to 73° northern latitude with 3 satellites No gaps in the overall swath if sub-swaths must be used		
Overlap between sub-swaths	≥50 pixels		
Maximum viewing angle	≤45° from zenith at Earth surface (edge of swath, mid-point between satellite ground tracks at the equator)		
Geolocation accuracy	<±50 m 3σ, after post-processing		
Ground sampling distance (GSD)	30 m	30 m	7.5 m
Spatial resolution	<1.5 × GSD, @FWHM	<1.5 × GSD, @FWHM	<1.5 × GSD, @FWHM
Spectral range	400–1000 nm	900–2500 nm	450–750 nm
Spectral sampling interval	10 nm	10 nm	N/A
Spectral resolution	10±2 nm	10±2 nm	N/A
Band center variation across the image	<2 nm peak to peak, after post-processing	<2 nm peak to peak, after post-processing	N/A
Signal-to-noise ratio (SNR), 1σ @typical 30% albedo	Peak: 600:1 Minimum: 100:1	Peak: 400:1 Minimum: 100:1	150:1
Spectral distortion (smile)	0.1 pixel	0.1 pixel	
Spatial distortion (keystone)	0.1 pixel	0.1 pixel	
Radiometric saturation	No saturation up to 100% albedo	No saturation up to 100% albedo	No saturation up to 100% albedo
Absolute radiometric accuracy	<5% between on-orbit calibrations	<5% between on-orbit calibrations	N/A

(Continued)

TABLE 4.5 *(Continued)*
Mission Requirements of Hyperspectral Microsatellites for Environment and Resources Management

Radiometric stability	≤1% peak to peak over 30 days	≤1% peak to peak over 30 days	N/A
Radiometric quantization	≥12 bits @100% albedo	≥12 bits @100% albedo	≥12 bits @100% albedo
Stray light	<1/2 of the noise from a 30% albedo (typical) pixel in the presence of 100% albedo clouds covering 50% of the instrument FOV >1 km away from the target pixel, after post-processing		
Polarization	<3%		
Platform jitter	<3 m rad @ ≥68 Hz circular		
Optical in-band spatial co-registration calibration knowledge across each sub-swath	<0.0025 pixels, within each spectral band (for the stand-alone payload)		
Optical band-to-band spatial co-registration calibration knowledge across each sub-swath	<0.0025 pixels between spectral bands for the same spatial location (for the stand-alone payload)		
Optical stability of along-track spatial registration within each sub-swath	<0.005 pixels over 500 sec (for the stand-alone payload)		
Optical stability of co-registration between sub-swaths	<0.05 pixels over 500 sec (for the stand-alone payload)		

within the constraints of a realistic microsatellite mission while still meeting the bulk of the identified users' needs described in Section 4.4. These requirements are, of course, for the purpose of the description and demonstration of a mission concept and trade-off study.

4.5.2 Orbit and Coverage Traded-Off

This section describes some of the more important high-level trade-offs, which are required so that lower level concept design work could be completed.

4.5.2.1 Orbit Selection

Before any other concept designs can move forward, the basic interrelated issues of orbit and ground coverage need to be addressed. The design constraints are:

1. Sun-Synchronous Orbit

   The orbit must be sun synchronous to ensure repeatable good surface illumination conditions. This orbit typically has an orbit inclination of 98.2°, as shown in Figure 4.1.
2. 800 km Maximum Orbit Altitude

   The orbit altitude must be kept below about 800 km in order to keep the size and mass of the payload's optics reasonable. To achieve the same SNR with the same detector arrays, the aperture of the optical system must vary directly with the satellite altitude—so an altitude increase of 20% therefore also increases both the optics aperture and focal length by 20%. But the mass of the optics (for the same f/#) typically (depending on the design) varies by $D^{2.5}$, where D is the equivalent optics aperture diameter. So that the 20% increase in altitude leads to increase in the optical system mass by close to 60%
3. 600 km Minimum Orbit Altitude

   To prevent excessive atmospheric drag on the satellite, which would require greater propulsion fuel for a 5-year lifetime, the orbit altitude should be kept above around 600 km.

4. 3-Day Repeat Coverage

   Weekly repeat coverage is required. However, to allow for cloud cover (typically 50% average), it is normally to try to achieve semiweekly repeat coverage. In this case, a 3-day repeat coverage is chosen as the desired repeat cycle.
5. 1-Day Relook at Maximum 45° Surface Viewing Angle

   Achieving 1-day relook with a maximum surface viewing angle of 45° from zenith automatically limits the field-of-regard (FoR) from orbit and, hence, the accessible swath that the FoR represents. The 45° limit was derived from the feasibility study of the Canadian WaterSat (Qian et al. 2017).
6. Maximum Imaging Swath

   Based on the user needs, the imaging swath of the HyperMER is 246 km. The maximum imaging swath of the payload must be kept reasonable so that it is possible to achieve good optical resolution (MTF) and distortion performance. A swath of 246 km from an around 600–700 km altitude orbit yields a FOV in cross-track direction about 20°, which is deemed acceptable for 30 m GSD imaging with 8K (8192) pixels while keeping the f/# reasonably low (e.g., <4). Increasing the swath beyond this rough value also starts to incur greater slant-viewing (cosine law) effects at the edges of the swath which elongate the cross-track dimension of the pixels and degrade the image quality.

The specific swath width of 246 km was notionally chosen since this corresponds to 8192 m × 30 m pixels in the cross-track direction, which allows for a more common detector array configuration. This is only a notional value however.

4.5.2.2 Number of Satellites

In this mission concept study, the number of satellites is driven by the mission requirements, and more specifically by the following three key factors:

1. 3-Day Repeat Coverage Frequency

   A sun-synchronous low Earth orbit satellite operating in 600–800 km altitude range will typically has ground tracks that are approximately 2100 km apart at 43° northern latitude (the lower limit of Canada). The satellite will cross 43° northern latitude at an angle of roughly $[\sin^{-1}(\sin(8.2°)/\sin(47°))] = 11.2°$. To cover this longitudinal gap in 3 days at 43° northern latitude, therefore, requires a net swath width of $[2100\ \text{km}/(3\ \text{days} \times \cos(11.2°))] = 714$ km. To keep within the 246 km swath limitation, three satellites would be required, which also provides about 24 km swath margin to allow for pointing errors and overlap between adjacent swaths.

   It can be concluded that at least three satellites are required to achieve 3 day repeat coverage.

   A 246-km swath could theoretically provide full coverage down to a latitude of around 33° northern latitude without swath-to-swath overlap.
2. 1-Day Relook Frequency

   Achieving 1-day relook with a maximum surface viewing angle of 45° from zenith automatically limits the on-orbit field-of-regard (FoR) to around ±40.2° to ±39° from nadir, or a swath of about 1052–1356 km corresponding to an altitude of 600–800 km.

   Using the same approximate 2100 km ground longitudinal track gap at 43° northern latitude as described above, at least $[2100\ \text{km} \times \cos(11.2°)/1052\ \text{km}/1\ \text{day}] \approx 2$ satellites are required to achieve 1-day relook within 45° surface viewing angle. With a 20° FOV (i.e., ±10°), a satellite roll angle of around ±30° would be required to achieve this FoR, which is reasonable.

   It can be concluded that at least two satellites are required to achieve 1-day relook.
3. System Reliability

   The use of three satellites is also key for achieving good system reliability. If the HyperMER is intended to service a wide range of operational customers, it is critical that the system be tolerant of occasional equipment failures. One option here is to place a high level of redundancy on each satellite, but the cost implications are very high, particularly since there is no easy way to achieve imaging redundancy other than by using two payloads on each satellite, which drivers up the cost of the overall satellite. The other option is to use multiple satellite instead, and apply where the cost/benefit is appropriate.

While 3 HyperMER satellites can achieve 3-day repeat coverage, a loss of a single satellite can be handled by dropping the satellites' altitude to 5-day repeat cycle (~620 km) and re-phasing the satellites to 0° and 144° apart. This still provides 5-day repeat coverage and, while less tolerant of cloud cover, continues to provide full coverage to the users on a sub-weekly basis.

Based on the above factors, a constellation of 3 satellites is selected for the HyperMER system concept.

4.5.2.3 Orbit Altitude Selection

There are a number of different possible orbit scenarios which can be considered. The two most prominent options are:

1. 637-km Altitude 9-Day Repeat Cycle Orbit
   An around 637-km altitude orbit has a 9-day repeat cycle. In this case, the satellites would be phased 120° apart (i.e., at 0°, 120°, 240°) such that the satellites take turns imaging the same ground area every 3 days. This approach allows good cross-calibration between the spacecraft but might introduce more issues for short-term (<6 days) change detection applications since they would require comparing images from 3 different payloads. The lower altitude of 637 km allows the payload optics to be kept slightly smaller, but probably only saves <10% on optics mass. The lower altitude also requires more fuel to maintain orbit altitude over the mission lifetime, but could potentially avoid the need for a de-orbiting tether.
2. 666-km Altitude 3-Day Repeat Cycle Orbit
   The 666-km orbit scenario involves placing the 3 satellite 80° apart in orbit (i.e., at 0°, 80°, and 160°). In this configuration, the satellite imaging swaths will create a 3-day interlace pattern. This particular scenario has the advantage of ensuring that the same satellite re-images the same piece of ground every 3 days. The idea here is that, by using repeat images from the same satellite, this can potentially provide better sensitivity to change detection, since there is less complication from correcting the small spectral and radiometric differences that may exist between the different satellites. The altitude also provides reasonably low atmospheric drag, reducing the amount of fuel required to maintain the orbit over the mission lifetime.

The key disadvantage of this orbit is that a point on the ground (apart from small overlap regions between swaths) is only imaged by a single satellite—thereby inhibiting cross-calibration between satellites.

Both the 637-km and 666-km orbits offer different advantages and disadvantages. For the purposes of this study, the 666-km orbit altitude at roughly 10:30 AM local time of ascending node (LTAN) was selected as the baseline.

4.5.3 Hyperspectral Imaging Technology Trade-Off

Two different hyperspectral imaging approaches to the payloads have been previously studied for the Canadian Hyperspectral Mission (CHM). These were based on Dyson spectrometer and linear variable filter (LVF) techniques, respectively (Qian et al. 2013).

1. Dyson spectrometer based hyperspectral imager

   In this approach of hyperspectral imager a wide field-of-view telescope feeds radiation light on the slit of a Dyson spectrometer. Such a hyperspectral imager uses a grating to spectrally disperse a "slit" image across a 2D detector array such that the light from each ground sampling cell in a cross-track line is simultaneously observed at all wavelengths. This is the same type of spectrometer used for the MERIS onboard ENVISAT and OLCI onboard Sentinel-3 (see Chapter 2 for details of these two instruments).

   A narrow swath (38 km × 30 m GSD) dual-Dyson version of a hyperspectral imager has been previously conceptualized and studied by the Canadian Space Agency (CSA) in the concept study of its Hyperspectral Environment and Resource Observer (HERO) mission (Hollinger et al. 2006). For HyperMER mission, this would be redesigned to provide 30 m GSD sampling across 8192 cross-track pixels to achieve a wide swath (246 km × 30 m GSD) and uses two separate Dyson spectrometers to cover VNIR and SWIR regions.
2. Linear Variable Filter or On-Chip Stepped Filter

   This instrument design is based upon a traditional multispectral imaging approach using a telescope, filter assembly, and detector. In this case, the filter is a linear variable filter (LVF) with a linear gradient for spectrum, and is placed directly in front of (or built onto) a large area detector array such that different rows of the detector array image the ground scene at different wavelengths. As each of the cross-track lines in the FOV is imaged with all filter rows of the LVF while the satellite flies forward, hyperspectral imagery for the FOV is collected and a datacube can be reconstructed (see Figure 1.8). For HyperMER, up to 60 spectral bands in VNIR could be collected with 30 m GSD and 8192 pixels in cross-track direction. A LVF-based hyperspectral imager with a GDS of 500 m, referred to as HySI, has been flown onboard the Indian IMS-1 satellite (see Section 2.12 for details).

   A variant of this approach, using an on-chip stepped filter (OCSF) that has a series of spectral steps, permits time delay integration (TDI) to be combined with the imaging process, allowing the SNR to be dramatically increased—this allows the potential to reduce the GSD to as low as 10 m while still achieving sufficient SNR to meet user needs. Geelen et al. (2013) reported that OCSF in a wavelength range of VNIR (400–1000 nm) have been processed directly on top of wafer that contains silicon-based CMOS image sensor chips. The OCSF working in SWIR region (1100–1700 nm) deposited on top of InGaAs detector array has been reported (Imec 2018). This technology, while being simpler, has not yet been extensively studied or demonstrated for a hyperspectral mission and requires further work. The detailed description on LVF and OCSF can be found in Sections 1.2.2.1 and 1.2.2.2.

Table 4.6 summarizes the key trade-offs between the Dyson and OCSF payload options. Items shaded indicate where there is a significant advantage. As seen from the table, the OCSF payload has significant advantages in terms of volume, mass, and SNR performance.

4.5.4 Trade-Off of Data Collection, Compression, Storage, and Downlink

One of the key issues associated with any hyperspectral mission is the ability to transmit the very large volumes of raw data to the ground. To assess the need for ground station coverage, following assumptions are made:

1. Ground speed is approximately 6820 m/s from 666 km orbit (this speed actually varies somewhat with latitude)
2. All raw VNIR, SWIR, and PAN data will all be 12 bits/pixel
3. GSD for both VNIR and SWIR data is 30 m, GSD for PAN data is 7.5 m
4. Number of spectral bands is 60 for VNIR, 160 for SWIR, and 1 for PAN

TABLE 4.6
Trade-Off Summary of Dyson Versus OCSF Payload Options

Parameter	Dyson Spectrometer	OCSF	Comment
GSD	30 m	30 m	
Ground swath	246 km	246 km	Both concepts are based on using 8K wide detector array
f#	2.4	4.3	f/2.4 is more influential to image quality, f/4.3 is much less risky
Along track angle	IFOV	2.8°	The Dyson concept collects all spectral bands simultaneously, the OCSF approach requires at least 2.8° along-track angle to sequentially collect all spectral bands
SNR	260:1	520:1	
Volume (optics)	73.7 cm × 53.3 cm × 61.0 cm	52.7 cm × 57.7 cm × 38.3 cm	
Mass (optics)	41 kg	23 kg	

5. Both VNIR and SWIR hyperspectral data will be compressed with a ratio 10:1 using near lossless vector quantization (VQ) compression algorithm (see Section 10.3.3.6 for details)
6. PAN image data will be compressed with 3:1 using CCSDS 122 compression standard.

Based on these assumption values, the net raw, compressed data volumes of the payload, during imaging is roughly 611 Mbps, this is equivalent to approximately 364 kbit/km^2. Table 4.7 lists the calculation details.

TABLE 4.7
Payload Data Rate During Imaging

Altitude (km)	666
Ground speed (m/s)	6820
VNIR GSD (m)	30
SWIR GSD (m)	30
PAN GSD (m)	7.5
VNIR swath (pixels)	8192
SWIR swath (pixels)	8192
PAN swath (pixels)	32768
VNIR bands	60
SWIR bands	160
PAN bands	1
VNIR bits	12
SWIR bits	12
PAN bits	12
VNIR compression ratio	10:1 (near lossless)
SWIR compression ratio	10:1 (near lossless)
PAN compression ratio	3:1 (lossless)
VNIR data rate (bps)	134,086,656
SWIR data rate (bps)	357,564,416
PAN data rate (bps)	119,188,139
Total data rate (bps)	610,839,210
Net rate (bits/km^2)	364,088

TABLE 4.8
Average Daily Downlink Times for Canadian Receiving Stations

Receiving Station (Ignoring conflicts with other missions)	Average Daily Downlink Time (sec/day) From 666 km Orbit
Gatineau	1576
Prince Albert	2182
Inuvik	4168
All 3 together	6760

As an approximation, Canada, United States, and Europe each has a total land area of approximately 10 million km^2. However, due to the overlap of swaths with increasing latitude, the average coverage efficiency is only about 70%. So HyperMER will generally have to image approximately ~14 million km^2 to achieve full coverage of each of these areas of interest (AOIs). This corresponds to approximately 1.34 Tbits each.

Assuming achieving coverage of each AOI within 3 days, this corresponds to a total raw, compressed data volume of approximately 450 Gbit/day per AOI for all 3 satellites. This is equivalent to 150 Gbit/day per satellite. After allowing for roughly 10% overhead for file formatting and error correction encoding, this might be roughly 165 Gbit/day.

To transmit this amount of data to the ground, it is necessary to consider the case where 3 Canadian ground receiving stations at Inuvik, Prince Albert, and Gatineau are used. The net downlink time available for each of these stations is listed in Table 4.8. Note that these numbers ignore conflicts with other missions. In the event of conflicts, it is possible to install a specific receiving antenna at any of the existing stations.

Assuming 165 Gbit/AOI/day per satellite, with a maximum of 6760 sec/day of downlink, this requires a minimum downlink rate of about 25 Mbps for each AOI. All 3 AOIs together would then require around 75 Mbps. This is well within the capability of a microsatellite transmitter.

4.5.5 Mission Concept Overview

The HyperMER system is divided into two major portions: the space segment and the ground segment. Figure 4.3 shows the block diagram of the satellite system and its external interfaces.

The space segment collects the raw imagery data of the Earth and various calibration targets such as the moon and deep space. It also obtains, from the spacecraft bus, other information such as attitude information and orbital information needed to post-process the imagery data.

The ground segment controls the overall system, interacts with the HyperMER Users, schedules image collection activities, collects a variety of external data needed to operate the system, and processes the hyperspectral imagery data into information products which are delivered to the user community.

Figure 4.4 shows a high level HyperMER system configuration reflecting some of the major elements of the system. These are summarized below.

4.5.5.1 Space Segment

The space segment nominally includes 3 satellites in a sun-synchronous orbit. Each satellite includes a spacecraft bus supporting one hyperspectral imaging payload.

A constellation of 3 satellites was selected for a number of reasons:

1. This allows the field-of-view in cross-track direction to be kept reasonable (around 20°) for instruments which are small enough to fit a microsatellite or small spacecraft mission and can still achieve 3-day repeat coverage of AOIs (e.g., Canada, United States, of Europe) from 3 satellites.

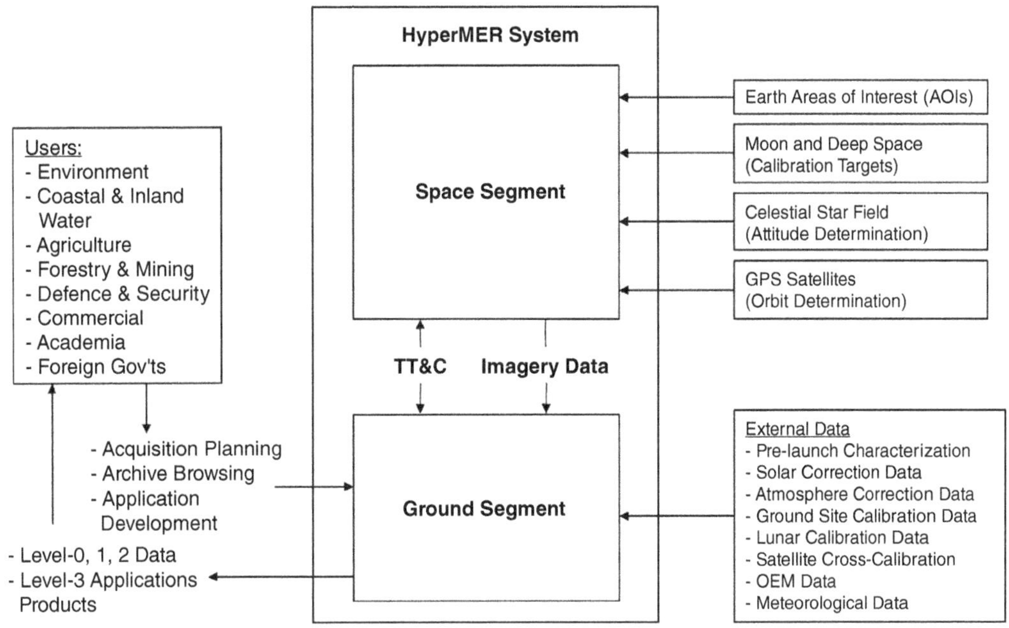

FIGURE 4.3 Block diagram of the HyperMER system and its external interfaces.

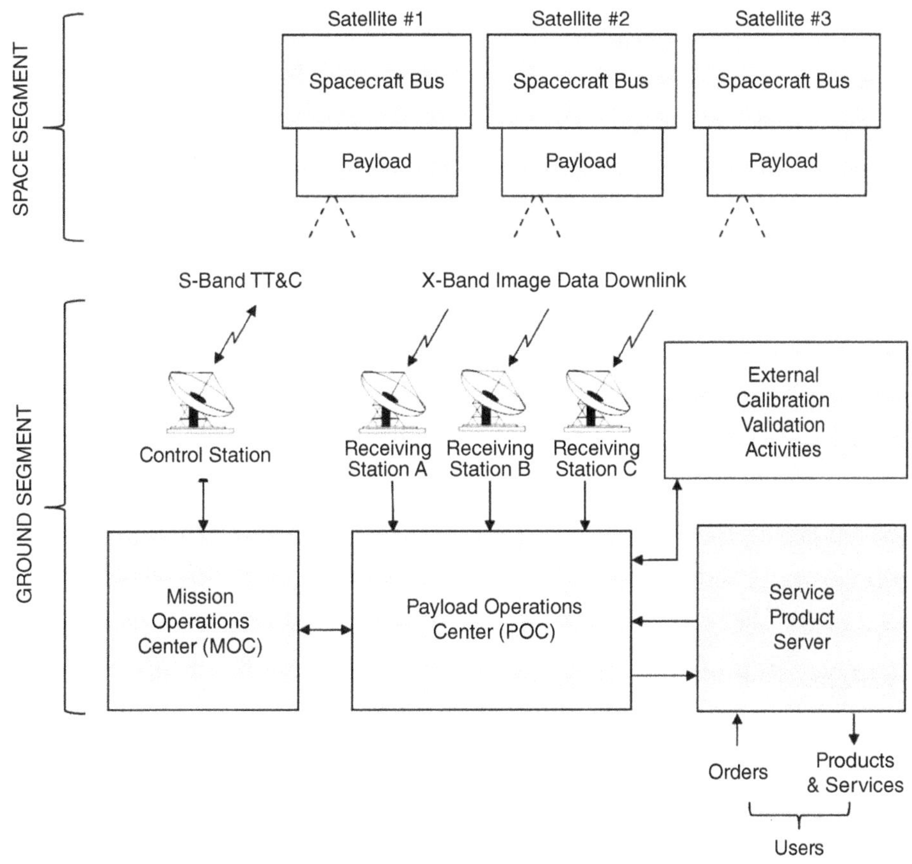

FIGURE 4.4 HyperMER mission system configuration.

2. As an operational system, having 3 satellites provides extremely important system reliability. While 3 satellites can provide 3-day repeat coverage of AOIs, a loss of one satellite would still permit 5-day repeat coverage. This is crucial for government and commercial stakeholders who demand regular delivery of information products to support their regular operations and decision making.
3. The cost of each additional satellites is relatively low compared to the cost of the development and testing required for the first satellite. Adding a third satellite therefore becomes a cost-effective means of providing enhanced coverage while simultaneously enhancing system reliability.

As described in Section 4.5.2.3, an initial orbit selection was a sun-synchronous orbit at roughly 10:30 AM local time of ascending node at 666 km altitude—this provides good surface illumination from the sun without causing problems from sun glint. In this orbit, the 3 satellites will typically be phased 80° apart such that their ground tracks are interlaced in a pattern which repeats every 3 days in such a way that each point on the ground is imaged by the same satellite every 3 days. This yields 3-day repeat coverage of all of Canada with ground tracks which are <250 km apart at Canada's lowest latitude.

Each instrument of the payload performs wide-swath (246 km) pushbroom imaging of the Earth in 3 spectral regions:

a. Visible and near-infrared (VNIR) hyperspectral bands covering 400–1000 nm in 10 nm intervals
b. Shortwave-infrared (SWIR) hyperspectral bands covering 900–2500 nm in 10 nm intervals
c. 1 panchromatic (PAN) band

The nominal ground sampling distance (pixel size) is 30 m for VNIR and SWIR imagery, and 7.5 m for PAN image.

Data from the payload is processed onboard the satellite to:

- Extract the desired areas of interest using a logical mask.
- Perform any other necessary image pre-processing such as spectral or spatial binning.
- Compress the hyperspectral imagery data using the vector quantization (VQ) near lossless compression technique, which preferentially preserves the spectral information for each image spatial sample (or CCSDS-123 compression algorithm).
- Compress the PAN data using CCSDS-122 compression, which can be configured for lossless compression, if needed.
- Store the data in a mass memory unit.
- Download the compressed files, via the spacecraft bus X-band transmitters to the ground at the specified times.

4.5.5.2 Ground Segment

The ground segment provides the means for controlling and monitoring the space segment, planning image acquisitions, receiving image data, processing the image data into higher level products, and interacting with the user community.

The Mission Operations Center (MOC) provides the primary point for controlling and monitoring the overall space segment, and communicates with the space segment via a bidirectional S-band link. The MOC concept is based upon a highly standardized set of equipment and software tools which are commonly used for a variety of space missions.

There are three receiving stations in this ground segment configuration. International receiving stations might also be utilized if necessary to support international users.

The Payload Operations Center (POC) is the hub of payload image acquisition planning, data collection, and calibration. Based on imaging requests from the HyperMER Product Sever, the POC prepares payload observation plans to collect the required data. These are then passed on to the MOC for execution. The POC also addresses regular on-orbit housekeeping to ensure that the payload is functioning properly, and plans regular calibration activities.

The Service Product Server is a crucial part of the HyperMER system. It is unique compared to the past or existing hyperspectral satellites that are mostly for users in the research and development areas. This product server enables *operational* users to obtain their desired *service products* without having to spend the efforts to develop their own facilities or to become research experts in hyperspectral image processing. The scope of such a facility will depend on the degree of available funding and user interest, but it is envisaged to include the following functionality:

- Archive and process 1–2 TB/day of HyperMER hyperspectral data.
- Provide a central web-based interactive interface point for all HyperMER users.
- Allow users to search data archives for quick-look viewing of areas of interest.
- Allow users to request *Special-Event Task* observations which are not part of HyperMER's regular routine task imaging schedule.
- Allow users to generate and download standardized "HyperMER Service Products," which utilize common application algorithms that have been previously developed and tested by experts and which are used by multiple user groups.
- Allow users to develop, test, and save their own application algorithms.
- Allow users to set up automated alerts for selected HyperMER products where public safety, defense and security, or critical events need to be detected autonomously.

4.5.5.3 System Calibration

System calibration is critical to any operational space mission, to ensure that the system performance and quality of image data products are continually maintained and validated over the life of the mission. Calibration activities typically involve a combination of:

- Prelaunch system calibration
- Use of in-flight calibration subsystem and sources
- On-orbit imaging of external targets (Earth, lunar, etc.), which can be cross-calibrated by other missions (Landsat, PRISMA, EnMAP, etc.) or from ground-based measurements
- Updating accurate models of the space segment
- Updating the image correction algorithms and associated operations used to make system spatial, spectral, and radiometric corrections

Calibration activities are planned and scheduled by the POC and its staff of experts to ensure that appropriate information is collected to achieve the required calibration and maintain product performance.

For detailed description on prelaunch calibration and in-flight calibration, readers are referred to read Chapters 12 and 9.

4.5.5.4 Concept of Operation

Users can access the HyperMER system via an interactive web-based interface, which allows users to access the mission's archive, run applications, and request special imaging tasks. Users will be able to access Level 1, 2, and 3 data products.

The HyperMER system can be designed to perform following 3 types of imaging tasks:

1. Routine tasks

   Routine tasks include repeat synoptic imaging of the AOIs such as Canada and other regions over long periods of time. The routine tasks are intended to be performed whenever there are no conflicting higher priority tasks. Such imaging will typically provide 3-day repeat coverage and is ideal for observing both short-term (days, weeks) change detection as well as long-term trends. The list of defined areas, which are to be subject to such routine imaging, is expected to be defined by a HyperMER supervisory group that periodically reviews and updates the list and responds to user requests.

2. Special-Event Tasks

   These tasks are special user-defined imaging requests which may be required in response to a disaster, public health situation, defense operation, specialized academic research, or commercial operation. Such tasks may interrupt normal synoptic imaging activities, depending on extent and timing. Very high priority special-event tasks (over smaller areas less than the imaging swath width) are expected to be executable within 24 h or less, if necessary. Repeat coverage with as little as a 1-day repeat cycle may be achievable by rolling the satellite to view the desired region from different viewing angles.

   The priority of such tasks and routine synoptic tasks can be defined by a HyperMER supervisory group, and can range from a low priority to very high priority.

3. Calibration Tasks

   Most calibration imaging is expected to involve areas already covered by regular routine tasks. There will also be a number of special calibration imaging tasks required to image various special Earth calibration targets, areas which are subject to timely cross-calibration by other spacecraft, lunar imaging, deep space imaging, and occasional diagnostic imaging.

A daily imaging schedule is prepared by a payload planning function which combines the mix of tasks into a plan which, in addition to the task priorities, also takes into account cloud forecasts. This planning schedule is sent to the MOC, which produces an overall spacecraft command file that includes spacecraft attitude maneuvers, downlink schedule, and other spacecraft control tasks. The satellite command file is then uplinked to the satellites for execution.

Data from the scheduled acquired images will generally be compressed and stored onboard by the payload until it can be downlinked via X-band to one of the ground receiving stations. The data includes both raw image data (VNIR, SWIR, PAN) as well as ancillary information (spacecraft attitude, orbital position, and various health monitoring data) needed for post-processing the data.

4.5.6 Payload Concept

This section describes the payload concept design of the Optical Unit (OU) and Instrument Control and Onboard Data Handling (ICODH) subsystem.

4.5.6.1 Optical Unit

Based on the trade-off study of hyperspectral imaging technology in Section 4.5.3, it is recommended to select the hyperspectral imaging technology that is based the on-chip stepped filter (OCSF) for the VNIR channel. As for the SWIR channel, due to the current availability of spectral coverage of the OCSF technology is only within a range of 1100–1700 nm (Imec 2018), which does not meet the required spectral range of 900–2500 nm listed in Table 4.5, a conventional dispersive element based imaging spectrometer will be used.

The OCSF technology separates the input radiation into different spectral bands at the detector chip level that omits the spectrometer and allows the instrument to be kept fairly simple. The VNIR channel could use a hybrid silicon CMOS detector array with integral Fabry-Perot filters built onto the detector array. Such a single detector array can serve the complete VNIR spectral range. As for the PAN channel, a visible region panchromatic linear detector array can also be built onto the same VNIR focal plane array, slightly offset from the VNIR array in the along-track direction. A separate PAN camera is avoid.

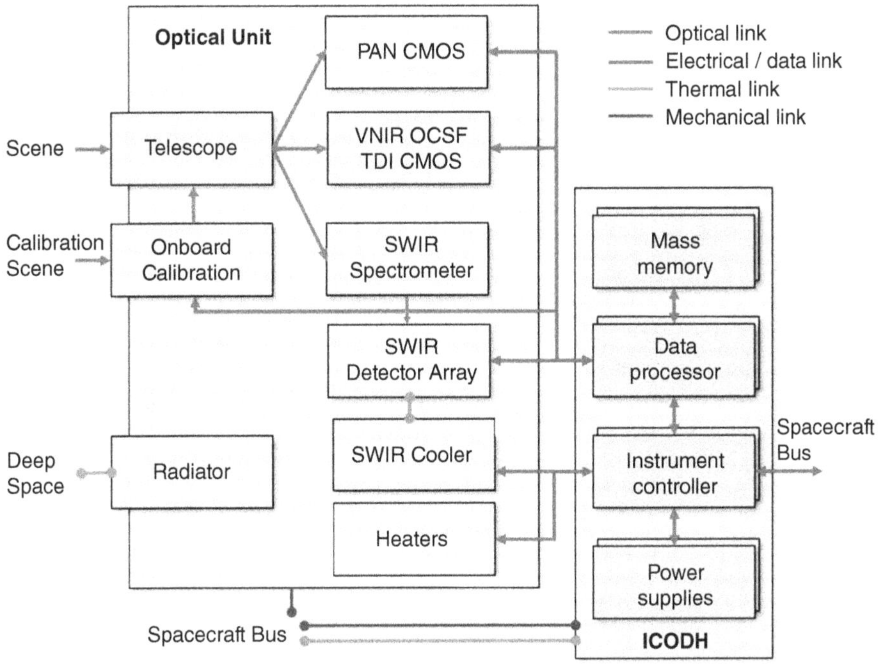

FIGURE 4.5 Black diagram of the payload concept.

The SWIR channel is a conventional imaging spectrometer, such as a Dyson form spectrometer with a SWIR detector array such as a HgCdTe detector array. The light separation between the VNIR and the SWIR can be achieved with a beam splitter if needed so that the two main spectral ranges will be closely aligned on the same optical axis.

All spectral channels are fed by a common telescope. Figure 4.5 shows the block diagram of the architecture of the payload concept. It is comprised of two main units, the Optical Unit and the Instrument Control and Onboard Data Handling (ICODH) unit. This separation simplifies both the payload accommodation on the spacecraft and the thermal management. The detailed description of optical system and design is provided in Chapter 5. Chapter 10 provides detailed description of ICODH.

The Optical Unit performs the function of hyperspectral imaging of the scene in VNIR and SWIR regions and acquires a panchromatic image, and producing raw digitized image data. The Optical Unit is also equipped with an onboard calibration facility. The main elements of the Optical Unit include:

- A telescope for all the three channels. The telescope images the scene onto the focal plane arrays of VNIR and Pan channels and the baffle that limits the out of field stray light in the instrument.
- A VNIR 2D detector array with built-in on-chip stepped filters for imaging scene in VNIR region. This includes built-in readout electronics that generate digitized VNIR region image data that is sent to the ICODH.
- A PAN linear detector array mounted on the same assembly as the VNIR detector array. This also includes built-in readout electronics that generate digitized image data that is sent to the ICODH.
- A SWIR spectrometer disperses the input radiation from the telescope on the slit into SWIR spectral bands and images them on the SWIR focal plane to be sensed by the SWIR detector arrays (multiple array may be required to cover the full swath if a single array does not have sufficient number of pixels in the spatial dimension).

- SWIR 2D detector arrays for imaging scene in SWIR region. This includes built-in readout electronics that generate digitized SWIR region image data that is sent to the ICODH.
- A passive cryocooler for the SWIR arrays to keep them at a nominal operating temperature of 150 K or lower.
- The onboard calibration facility is mainly composed of a mirror that is able to illuminate the complete optical path and a high reflectance integrating sphere. The mirror is fixed to the inner part of the main instrument cover. The integrating sphere, with lamps and LEDs mounted inside it, is located on the upper side of the optical bench (telescope side). A dedicated port is used as entrance for providing an additional calibration source from sun or moon.
- The structure that support all the elements of the Optical Unit and that interfaces with the spacecraft mounting panel.
- Thermal control including a radiator, thermal sensor and heaters. The thermal control electronics maintains the interior of the Optical Unit within its operating and survival temperature range.

4.5.6.2 Instrument Control and Onboard Data Handling

The ICODH controls the Optical Unit imaging process, collects raw image data from the Optical Unit, collects ancillary attitude and GPS information from the Spacecraft Bus, compresses the image data into files, stores the compressed data, and then retrieves the data for transmission to the ground via the Spacecraft Bus transmitters. The main elements of the ICODH include:

- A Mass memory unit (MMU). The storage capacity of the MMU can be sized to suit the mission. For HyperMER, this would typically be 1–2 TB.
- A Data processor to receive, format, and packetize the image data from the three channels into CCSDS standard data format to be downlinked to ground. Data processor may also include an onboard data compression board. The compression technique can be a vector quantization (VQ) based near lossless compression algorithm (10:1 near lossless compression) previously developed at the Canadian Space Agency (Qian et al. 2013) or the CCSDS-123 standard (CCSDS 2019). The CCSDS-122 standard can be used to compress the PAN image (typically around 2:1 lossless compression).
- Payload control is performed by the Instrument controller, which has substantial extra processing capability and external I/O interfaces, allowing it to exchange commands and other information with the Spacecraft Bus.
- An electronic power conditioner (EPC) board, which converts electrical power from the Spacecraft Bus into a variety of outputs with different voltages and characteristics required for the payload electronics.
- A support structure (chassis) to house all of the ICODH circuit cards, provide connector interfaces, and provide EMI enclosures where needed.

4.5.7 Spacecraft Concept

4.5.7.1 Selection of Spacecraft Platform

A SSTL-150 bus is selected as the baseline for HyperMER spacecraft platform. It is a microsatellite bus manufactured by Surrey Satellite Technology Ltd (SSTL) based in Guildford, Surrey, United Kingdom. The SSTL-150 bus provides versatile payload accommodation. Its straightforward interface makes it ideally suited for use with customer supplied payloads. It is capable of carrying a wide range of payloads requiring a mass and power capability of 50 kg and 50 W. SSTL-150 buses have been used by DMC+4, TopSat and RapidEye 5-satellite constellation missions. The design of SSTL's modular, flexible, and adaptable platforms allows them to accommodate a wide range of

TABLE 4.9
Key Technical Parameters of SSTL-150 Spacecraft Bus

Parameter	Value
Maximum payload mass	50 kg
Bus dry mass (without payload)	103 kg
Payload power (average)	50 W (100 W peak) @EOL
Internal payload volume	279.5 mm × 231.5 mm × 252.5 mm
External payload volume	730 mm × 455 mm × 774 mm
Types of orbits	LEO (400–1000 km) any inclination
Design mission life	7 years (Ps = 92%)
Data storage capacity	16 Gbytes, dual-redundant
Data downlink rate	80 Mbps, X-band
Pointing knowledge	25 arcsec (1σ) all 3 axes
Pointing control	36 arcsec (1σ) all 3 axes
Pointing stability (jitter)	1.5 arcsec/sec
Position knowledge	10 m
Thermal control	Passive, heaters
Attitude and orbit control	3-axis control with reaction wheels and magnetorquers
Control and data handling bus	Dual-redundant controller area network bus
Delta-V	36 m/sec
Propulsion	Hot gas xenon resistojet

mission-specific requirements. Table 4.9 lists the key technical parameters of the SSTL-150 spacecraft bus (SSTL 2019).

SSTL has a flight ready design. The SSTL-150 meets the technical needs of HyperMER. This allows a rapid and cost-effective solution with minimal risks.

The spacecraft design incorporates a standard suite of SSTL avionics, sized to meet the needs of the mission, into an existing mechanical structure. The structure is capable of accommodating the HyperMER payload in the upper payload bay. The power from the body-mounted solar panels guarantees that payload power requirements are met under all mission beta angles. The communication subsystem, with two low rate S-band Rx/Tx connected to four patch and four monopole antennas, provides quasi omnidirectional coverage for telemetry/telecommand (TTC). Two X-band transmitters (80 Mbps) provided high data rates for downlinking payload data. The attitude and orbit control system (AOCS), comprised of a series of sensors and actuators, can be configured to meet HyperMER pointing requirements. A gas propulsion system ensures that the spacecraft can be phased in the envisioned configuration. Finally, the control and data handling is done with two central onboard computers through a redundant controller area network (CAN) bus.

4.5.7.2 Spacecraft Platform Structure

The primary function of the spacecraft platform structure is to provide appropriate accommodation for the avionics subsystems, alignment for the payload and maintain acceptable environmental conditions to fulfil the mission requirements.

The structure will be subjected to inertia, vibration and acoustic loads from the launch vehicle, shock from separation, thermoelastic distortion during service and micro-vibrations from wheel operation. The SSTL-150 structure is designed to survive these loads and to only impart acceptable loads on the payload and equipment.

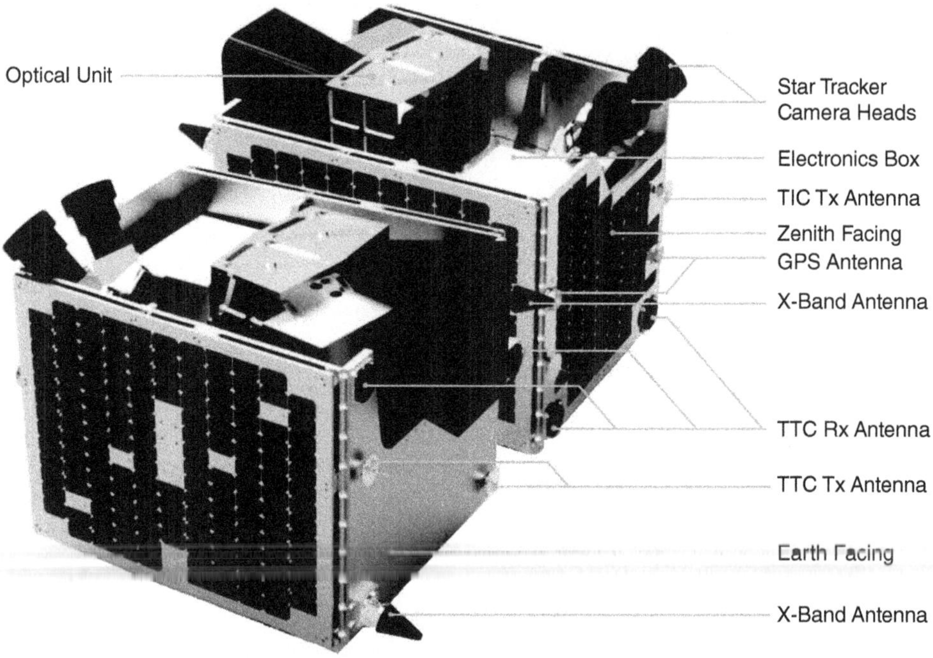

FIGURE 4.6 HyperMER exterior equipment accommodation.

The basic structure of the platform consists of a central avionics stack enclosed by honeycomb panels forming a trapezoid prism. The top panel has a negative offset with respect to the top of the side panels forming the payload bay. This configuration creates two separate compartments: the lower and upper compartments, where the lower compartment houses most avionics units and propulsion system and the upper compartment is dedicated to the payload. The static envelope of the spacecraft in stowed configuration is shown in Figure 4.6.

The platform is structured around the avionics modules, which are assembled together to create a rigid load bearing structure. Eight shear panels reinforce the avionics stack dividing the remaining space into 8 sections. The four corners are occupied by the reaction wheels, the +Y houses two X-band antennas, the Li-ion battery is located in –X and +X is occupied be stack harnessing. The propulsion system is a standalone unit that can be assembled and tested independently from the platform. During spacecraft assembly, the propulsion system is integrated in –Y side of the avionics stack.

The upper compartment of the platform houses the payload closely linked to the star tracker to minimize thermoelastic distortions. Finally, four exterior panels enclose the spacecraft. Nadir and zenith panels house TTC and GPS antennas. Velocity and anti-velocity panels are populated with triple junction solar cells.

4.5.7.3 Attitude and Orbit Control Subsystem

The Attitude and Orbit Control Subsystem (AOCS) is responsible for the attitude and orbit determination and control of the satellite. It performs the following three tasks:

1. In launch and early operations phase, it removes the initial tip-off rate after launcher separation and acquires a power and thermally safe state.
2. It controls the satellite to achieve nadir pointing attitude of the payload boresight with sufficient pointing accuracy and attitude knowledge for payload operations.
3. It leaves the satellite in a power and thermally safe condition in the event of an anomaly that causes loss of attitude control.

AOCS has five operation modes:

1. Detumbling mode (DTM)
2. Y Thomson mode (YTM)
3. Coarse pointing mode (CPM)
4. Nominal mode (NOM)
5. Stand-by mode (SBM)

After separation from launch vehicle the satellite will tumble around a random axis with some random rates. The maximum tip-off rates at separation of a SSTL spacecraft bus are generally not >10°/sec.

The first mode to be run is DTM, which starts to dump the rates around *X* and *Z* axes, leaving the rates about *Y* axis untouched. In this mode the *Y* magnetorquer is used as the actuator and the magnetometer is used as the sensor.

When the *X* and *Z* rates have been suppressed sufficiently, the mode is then changed to YTM, which continues suppressing *X* and *Z* rates but at the same time regulates the rates about *Y* axis. The actuators used in this mode are torque rods and the sensors are the magnetometers.

When the *Y* rate has been regulated, the mode is changed to CPM. In coarse pointing mode the payload boresight vector is kept at Nadir (i.e., pointing to Earth). This three axis controlled mode uses magnetometers and sun sensors. When spacecraft is in eclipse the AOCS uses magnetometers only.

Before engaging payload activities, the platform transitions to a high pointing accuracy mode, referred to as nominal mode (NOM). In nominal mode, the AOCS makes uses of star trackers to achieve the pointing accuracy and stability required to enable pushbroom imaging of the hyperspectral imager.

Transition from DTM to YTM is automatic such that the AOCS task checks if the spin rate about *Y* axis is below a given threshold and the number of consecutive cycles under this rate is above some defined number. Both of these values (limit spin rate and number of consecutive cycles) are given in the drive file. Both of these criteria need to be met before transition to YTM occurs.

The transition from YTM to CPM is semiautomatic. The AOCS task checks for spin rate and pitch angle to nadir. The operator sets a flag that authorizes AOCS task to go to CPM if the conditions are suitable. When the rate is sufficient and the spacecraft is approaching nadir; if this flag is set then the mode will automatically switch to CPM from YTM. This provides a gentle transition from YTM to CPM avoiding overloading the wheels.

HyperMER is designed to be thermally, power and payload safe in any arbitrary attitude. The Standby (STB) is the safe mode. Since there are no specific attitude requirements when the safe mode is triggered, the AOCS switches to STB mode. In this mode, AOCS will not be running any estimators or controllers; hence, it will not generate any control outputs to the actuators.

The AOCS subsystem consists of the following the units:

1. AOCS interface module (AIM)
2. Torque rods (magnetorquers)
3. Magnetometers
4. Sun sensor
5. Reaction wheels
6. GPS receivers
7. Star trackers

AIM acts as an interface for torque rods, magnetometers and sun sensors. Commands and telemetry requests via controller area network are managed by the AIM, with the appropriate actions being taken or sensor telemetry reported. AIM is cold redundant.

Three dual-wound magnetorquers are used as the torque rods in a cold-redundant manner. Each coil is able to generate 5 Am^2 magnetic momentum. The function of torque rods is to produce a strong magnetic field in the direction of its installed axis so as to interact with the surrounding Earth's field, producing a turning moment on the satellite. Each magnetorquer coil is supplied with 5 V from the AIM. The firing of the magnetorquers and the direction of the current through the coils is controlled by the AIM and commanded by the onboard controller (OBC).

Magnetometers have been chosen as coarse attitude sensors. For each axis an output of 0–5 V corresponding to a measured magnetic field of −60 μT to +60 μT is produced. The AIM provides power for the magnetometers. The magnetometer outputs are read by the AIM.

Four sun sensors are divided into two pairs connected to two separate AIMs. Both pairs are used in a cold-redundant manner. The sun sensor measures direction to the sun relative to the satellite in azimuth and elevation axis. The sun sensor produces four analogue 0–5 V outputs according to the angle measured, which are read by the AIM. The AIM also provides power for the sun sensors.

A reaction wheel is capable of providing reliable attitude control and stabilization on demanding remote sensing missions. It is controlled via a CAN interface and provides telemetry on wheel speed, current and PCB temperature via this interface.

The wheels can generate torque of up to 110 mNm and have a maximum momentum capacity of 1.5 Nms with an approximate mass of 2.6 kg.

The GPS receivers selected are SGR-07 and SGR-10 used in a cold redundant architecture. The receivers provide navigation telemetry to the satellite bus about the position, velocity, and GPS time. They also provide pulse per second (PPS) signals that are used by the OBC to ensure synchronization.

Both receivers have been designed to rapidly obtain a position fix upon start-up, and store information on the GPS constellation (ephemeris, orbital elements, etc.) in non-volatile memory.

The star tracker selected is the SSTL's Rigel-L star tracker. It has a data processing unit and a camera head unit (see Figure 4.6). The data processing unit is a LEO variant and has a 26° sun exclusion angle and a 23° Earth exclusion angle (both half cone).

4.5.7.4 Power Subsystem

The power subsystem has the following main functions:

- Generation of electrical power via solar panels
- Storage of power via batteries for use when solar power is insufficient
- Conversion of voltage and power between the solar panels and battery
- Distribution of battery voltage power to platform and payload modules

The power subsystem main elements include:

1. Solar panels
2. Battery Charge Module (BCM)
3. Battery
4. Power Condition/Distribution Module (PCDM)

Each solar panel contains one section and each section is regulated by one of the 9 battery charge regulators in the BCM. The battery charge regulator outputs are combined to form the main power bus. This main power bus is directly connected to the battery positive terminal. Any available power not demanded by the payload and platform is used to charge the battery. In the event that the battery is fully charged, the battery charge regulators scale back to only provide the required power. When the power available is insufficient, the battery discharges to support the difference. The PCDM conditions/distributes power from the bus to the payload and platform modules. During separation from launch vehicle, the activation switches close in order to connect the battery bus to the main power bus.

The solar panel configuration consists of 3 body mounted panels on zenith, velocity and anti-velocity sides, as shown in Figure 4.6, each consisting of a substrate of aluminium face skin on aluminium honeycomb. The electrical design of the velocity and anti-velocity panels consists of 6 strings of 25 solar cells in series; the zenith panel contains 4 strings of 22 solar cells. The strings on each panel are paralleled into one section. The solar cells used are Emcore ZTJM triple junction with monolithic bypass diode protection. The cells have an efficiency of 28.5% at beginning of life (BOL) at 20°C. Each section produces a BOL power of 46 W at 56.8 V at 28°C. At end of life (EOL) at a maximum operating temperature of 90°C each section will produce 36 W at 46 V.

The main function of the BCM is to convert the power from the solar arrays into a regulated output to charge the battery and supply the power bus. Power from each of the solar array sections is transferred to the corresponding independent battery charge regulator. The diodes at the output stage of each battery charge regulator isolate the regulators from each other and provide the parallel operation of each regulator. Each battery charge regulator is rated to handle an output power of 80 W and an input voltage of 80 V. Thus, the solar array sections on the solar panels cannot exceed the battery charge regulator power and voltage levels. Depending on the current charge level of the battery or the power demand of the whole system, there are two BCM operating modes: (1) Maximum Power Point Tracking mode and (2) End of Charge mode. When the battery voltage is below the threshold voltage the battery charge regulators operate in maximum power point tracking mode, acting as a current source to the battery bus to charge the battery and supply power to the modules. The maximum power point tracking on BCM is temperature compensated via thermistors placed on each solar section so that all the available power is transferred to main power bus.

The battery to be used will be from one of the two suppliers. Both suppliers have supplied batteries for SSTL spacecraft as well as numerous LEO missions. The battery from Suppler A comprises of 80 cells having a nameplate capacity of 15 ampere hours (Ah). Cells are arranged in an array of 8 cells in series and 10 parallel strings (8s10p). The parallel string topology allows for graceful degradation in the face of a battery cell failure. The battery from Suppler B uses Li-ion cells. The battery is arranged in an 8s3p configuration which means the cells are arranged in 3 strings of 8 cells. Structural integrity is achieved by clamping the cells with two horizontal plates and 4 vertical stiffeners. The battery nameplate capacity is 16.5 Ah for a nominal voltage range 21.6–32.8 V.

The PCDM contains the Power Conditioning Module (PCM) and Power Distribution Module (PDM). The PCM consists of redundant DC-DC flyback converters rated at 12.5 W, responsible for the generation of the +5 V bus within the power system. The input voltage range is 22–34 V. The +5 V bus supplies the power subsystem CAN nodes and internal logic. There are two identical hot redundant PCMs (PCM-A and PCM-B). Both are turned on when the PCDM is powered. The output of PCM-B is set to be a little higher than PCM-A so the PCM-B voltage presides. The PCM provides under-voltage and over-voltage protection for the 5-V output supply. An over-voltage condition on the 5-V line will cause PCM-B to trip off. The PCM also provides under-voltage protection for the battery by shutting down all power at a preset voltage threshold. The under-voltage protection threshold is nominally set to 25 V with the PCM restarting. The PDM consists of electronic power distribution switches and fuses, which protect the power system from failures on the subsystems and payloads as well as providing a mechanism to turn systems on and off in order to maintain the system level power budget. The unregulated battery voltage and regulated +5 V are supplied to the subsystems through the PDM. The PCM/PDM acts as a buffer between the unregulated power from the battery, BCR and the rest of the spacecraft.

4.5.7.5 Thermal Control Subsystem

The thermal control subsystem is required to maintain all the equipment and structure at acceptable temperatures throughout the mission. The process is one of design, with demonstration by analysis and tests that the temperature predictions for all of the equipment and structure fall within an acceptable temperature range. An allowance is made for modeling uncertainty in the temperature predictions. The modeling uncertainty used is ±10°C for internal parts of the spacecraft and ±15°C for external parts of the spacecraft. Where thermostatically controlled heaters are used, the thermal uncertainty is reduced to –3/10°C to account for the limiting of the lower temperature by the thermostatically controlled heater.

The general approach is one of conductive transfer of heat away from dissipating units and rejection to space via radiators. External heat inputs (Sun, Earth, and Albedo) are managed by choice of external surface properties which are covered with SSM tape which has a low solar absorptivity (α) to infrared emissivity (ε) (refer to Chapter 8 for details on thermal control design).

A design approach based on heritages is to use a dense structure consisting of machined aluminum panels and aluminum honeycomb panels with many machined aluminum inserts. This gives a stiff structure with good metal-to-metal thermal contact and good internal heat conduction.

4.5.7.6 Telemetry/Telecommand and Data Downlink

Telemetry/telecommand and data downlink includes two subsystems: telemetry and telecommand (TTC) and payload data downlink. The TTC subsystem enables ground-based operators to monitor and control the functions and behavior of the satellite throughout telecommand and telemetry. The payload data downlink subsystem transmits the data gathered by the payload through a high rate data downlink channel.

The OBCs are cross-linked to two hot redundant S-band receivers (19 kbps) and two cold redundant downlink transmitters (38 kbps). Both receivers are powered on permanently once the satellite separates from the launch vehicle. The transmitters are only powered during ground station contact and operate in cold redundant configuration.

A S-band receiver consists of three modules: the front-end band pass filter, the S-band down converter, and the intermediate frequency module. The S-band down converter consists of a low noise amplifier, band-pass filter and frequency mixer. The local oscillator port of the mixer is driven by a frequency synthesizer that allows commercial or amateur S-band frequency coverage. The intermediate frequency module is further subdivided into the radio frequency section and the digital section. The radio frequency section consists of a low-pass filter, 20-MHz band-pass filter, a low noise amplifier, narrowband-pass filter and receiver integration circuit with associated circuitry. The digital section comprises the CAN controller and down-converter synthesizer. The CAN controller is responsible for TTC decoding and interface with the CAN bus. A Miller demodulator is used to recover the 19k2 data. The S-band receiver has LVDS (low voltage differential signal) and TTL (Transistor-Transistor Logic) data interfaces as well as the ability to provide decoded data output via the CAN node.

The S-band antennas are configured to provide each receiver and transmitter with two antennas, one antenna positioned on the zenith facing facet and the second on the Earth facing facet (see Figure 4.6). This dual antenna system provides the satellite TTC subsystem with a near omnidirectional antenna pattern. An omni-directional antenna pattern guarantees communication with the satellite in any orientation. This is critical during the satellite commissioning phase when the satellite is likely to be tumbling at the start of the first link session.

The S-band transmitter features two cold redundant S-band low-rate transmitters:

- Low-rate transmitter operating frequency 2226.666 MHz
- 200 mW radio frequency output power
- Binary phase-shift keying (BPSK) modulation differential encoding as default
- Quadrature phase-shift keying (QPSK) ½ rate $k = 7$ convolution encoding configurable
- 38k4 data rate per low-rate transmitter
- Alpha roll-off = 0.6
- v.35 scrambling

The low-rate transmitter is based on a synthesized frequency source with a quartz crystal reference. Phase modulation is accomplished using an I/Q modulator, with bit shaping implement in a FPGA. Multiplexers internal to the FPGA allow interfacing with up to 8 data sources and clocks over LVDS and TTL. Good spectral purity is achieved using miniature low-loss band-pass filters and baseband analogue anti-alias filters. Baseband root-raised-cosine filtering with $\alpha = 0.6$ is implemented to provide matched filtering. Each S-band low-rate transmitter module is connected to two S-band monopole antennas through a power splitter. One monopole antenna is located on Earth facing facet and the second antenna is located on the zenith facing facet.

To downlink large volumes of the hyperspectral and PAN data the payload data downlink subsystem will use the X-band downlink. It provides high rate data downlink capability ranging from 10 Mbps to 500 Mbps depending on its configuration and antenna selection. The XTx400 utilizes 8PSK and 5/6 trellis coded modulation (TCM) in order to support the higher downlink data rates. The key parameters are following:

- 8200 MHz operational frequency
- QPSK modulation—½ rate, $k = 7$ convolutional encoding
- Data rate 80 Mbps
- v.35 scrambling
- Radio frequency output of >6 W over temperature
- Radio frequency power handling: 6 W (rms), 14 W (peak)
- Bean width: ~140° with nadir null for PFD optimization
- Axial ratio: <6 dB
- Operating temperature: –20°C to +50°C

The two X-band antennas can be seen on the side of Earth facing facet (see Figure 4.6).

4.5.7.7 Onboard Control and Data Handling

The Onboard Control and Data Handling (OBCDH) subsystem is responsible for the satellite control as well as for all data exchange and storage between the satellite electronic units. Its main tasks can be summarized in the following points:

- The control of the spacecraft platform and payload
- Provide the primary routine telecommand/telemetry interface for the satellite
- Provide file transfer capabilities between the satellite and ground
- Provide the processing capability to
  - Implement the attitude control algorithm
  - Schedule platform operations
  - Schedule payload operations
  - Maintain log files of telemetry/telecommand
  - Spacecraft safety and health monitoring
- CAN bus provides the primary bus for communicating to all other spacecraft modules
- Payload interface

The two cold redundant flight computers act as the primary OBCs and are responsible for the autonomous control of the platform and payload. Each of these OBCs is directly connected to each transmitter. The radio frequency data is modulated/demodulated by the transceiver and communicated to/from the OBC in IP packets.

The OBC is responsible for controlling the satellite during all normal modes of operation. All commands sent to the satellite are handled by the OBC. The OBC can either be commanded directly in real time or by a scheduled file. The scheduled file contains a list of time stamped commands that the OBC processes at the appropriate time. If a command is addressed to one of the OBC tasks it will be handled internally, otherwise the OBC will send the command over the CAN bus to the appropriate module.

The module power is controlled via the OBC real-time processes. In addition to turning on the appropriate power switch the safety task begins to monitor critical telemetry from the unit. If these critical telemetry points go beyond preset limits then the safety task will take action. This usually results in the unit being turned off, but depending on the device and its interaction with other units the action taken could be more complicated. The purpose of this is (a) to keep the platform safe and (b) to keep the unit safe with the priority on keeping the platform safe.

Certain telemetries from units are polled by the OBC and provide real-time telemetry to the ground station operators during a link session. The list of real-time telemetry sent by the OBC is updatable. Units that are always powered (e.g., power units and the receivers) will have a fixed set of telemetry that are always polled.

Whole orbit data files provide a similar service to real-time telemetry continuously, but store the data in files on the OBC for downloading later. This allows operators to track the health of units over the periods when the spacecraft is not in contact with the ground.

The OBC supports a file transfer protocol for moving files between the ground and the OBC. In addition the OBC has the capability to transfer files around the CAN via the CAN FTP protocol. This allows software patches for units to be uploaded to the OBC over the radio frequency link and then moved to the appropriate unit over the CAN bus.

4.5.7.8 Propulsion

The propulsion subsystem provides the ability to change the velocity of the satellite for:

- Correcting for launch injection errors
- Maintaining the mission revisit and timeliness requirement

The selected propulsion system consists of a high-pressure gas storage tank, which contains 12.0 kg of xenon at launch. The gas pressure is regulated down to the thruster inlet pressure by a bang-bang electronic pressure regulator. This uses the feedback from the downstream pressure transducer to operate two solenoid valves. The upstream valve SOL1 (or SOL3) is held open throughout the firing and the downstream valve SOL2 (or SOL4) is pulsed open to maintain the plenum chamber pressure between preset limits. The system is controlled by a propulsion controller that has two independent electrical circuits to give control redundancy.

The resistojet consists of a pair of sheathed heaters wound on a central bobbin. Helium propellant is forced to spiral through the heater assembly as there is no direct flow path through the thruster. Downstream of the heater is a conventional convergent/divergent nozzle, protected by a 10-μm filter disc. The heaters are rated at 30 W and operate at 530°C during a firing operation.

4.5.8 Launch Options

4.5.8.1 Launch Requirements

The concept of HyperMER mission is designed to be an operational mission. This imposes strict requirements on the orbit required for the constellation. The coverage and revisit requirements in particular restrict the available orbit options. These requirements essentially restrict the launch solution to be a dedicated, or primary launch. Secondary payloads are usually forced to take the most convenient ride available in their desired launch timeframe, which typically results in a compromised orbit. In addition to that, given that there are three satellites, it is unlikely all three would be accommodated on the same launch, which would complicate efforts to have all three in the same orbit.

With this launch solution in mind, available launch vehicles are reviewed to identify those that have the capability to provide the combined mass and volume of three satellites, plus some margin to accommodate them into a dispenser.

Given the selection of the SSTL-150 as the baseline spacecraft platform, it is assumed three satellites, each having a mass of 150–200 kg and a volume of 1140 mm × 827 mm × 870 mm (L × W × H).

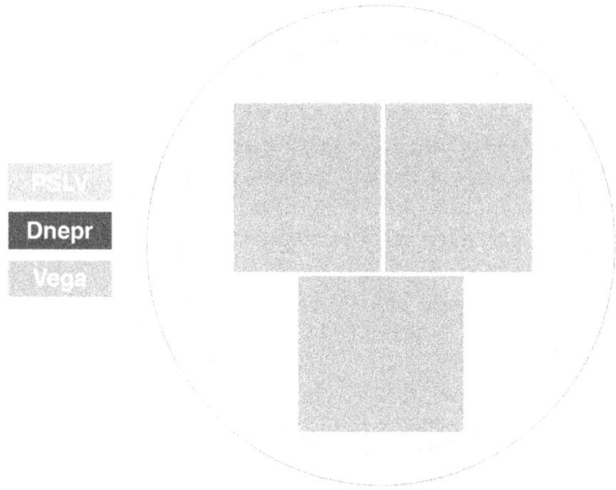

FIGURE 4.7 Layout of HyperMER satellites in various launch vehicles.

Potential launch vehicles are identified and classified as *1-level* or *2-level* and are described in next two subsections.

4.5.8.2 1-Level Launchers

Two launch vehicles are identified as potential options that would allow all 3 HyperMER satellites to be put into a 1-level stack as illustrated in Figure 4.7.

1. Dnepr Launch Vehicle
   Dnepr, a Russian launch vehicle, has a capability of roughly 900 kg to the HyperMER target orbit. This provides sufficient margin for the HyperMER mission.
2. PSLV Launch Vehicle
   The lowest capability of the PSLV rocket can bring roughly 950 kg to the HyperMER target orbit. This is an improvement to Dnepr and there is also the option to add additional solid boosters to the rocket if necessary to increase performance. There is also ample space for the HyperMER satellites, as shown in Figure 4.7.

4.5.8.3 2-Level Launchers

Three launch vehicles were identified as potential options for a 2-level HyperMER satellites stack.

1. Eurockot Launch Vehicle
   Eurockot has the capability of bringing roughly 1100 kg to the HyperMER target orbit.
2. Vega Launch Vehicle
   Vega provides the most capability to the HyperMER target orbit at roughly 1350 kg. This provides ample margin to design a 2-level dispenser. As a result Vega is seen as a viable option going forward.
3. Athena IIc Launch Vehicle
   The Athena IIc can launch roughly 1100 kg to the HyperMER target orbit. This is a modernized version of the Athena II that was flown in the 1990s before it was retired in 1999 due to lack of customers. This is a possible option going forward.

   Two strong candidates are identified for the HyperMER mission, PSLV (1-level) and Vega (2-level). There is a preference for a 1-level launch vehicle, which would result in a much simpler dispenser.

REFERENCES

AAFC. 2012. Annual Space-Based Crop Inventory for Canada (available at http://static.ow.ly/docs/AAFC%20Crop%20Inventory%20Fact%20Sheet%202012_18S5.pdf, accessed on June 9, 2019).

CCRS. 2014. Imaging Spectrometer Data Analysis System (ISDAS) (available at ftp://ftp.ccrs.nrcan.gc.ca/ad/Lixin/ISDASv2/helps/IntroISDAS.htm, accessed on April 25, 2019).

CCSDS. 2019. Low-complexity lossless and near lossless multispectral and hyperspectral image compression. Issue 2. *Recommendation for Space Data System Standards (Blue Book)*, 123.0-B-2. Washington, DC: CCSDS, February 2019 (available at https://public.ccsds.org/Pubs/123x0b2.pdf, accessed on May 23, 2019).

Geelen, B., N. Tack, and A. Lambrechts. 2013. New multi- and hyperspectral cameras cover diverse applications. *Photonics Spectra* (March) https://www.photonics.com/Articles/New_Multi-_and_Hyperspectral_Cameras_Cover/a53223.

Hollinger, A., M. Bergeron, M. Maszkiewicz, S.-E. Qian, H. Othman, K. Staenz, R. A. Neville, and D. G. Goodenough. 2006. Recent Developments in the Hyperspectral Environment and Resource Observer (HERO) Mission. *Proceedings of IGARSS'2006 IEEE International Geoscience and Remote Sensing Symposium* 1620–1623, July 2006.

Imec. 2018. Meadia release: Imec demonstrates shortwave infrared (SWIR) range hyperspectral imaging camera. SPIE Photonics West in San Francisco.

McNairn, H., A. Merzouki, A. Pacheco, J. Shang, A. Kross, X. Jiao, T. Fisette, and E. Pattey. 2012. *Space for Agriculture: Earth Observation Monitoring of Crops and Soils*. AAFC Presentation, Ottawa, Canada

NRCAN. 2018. Forest cover map (available at https://www.nrcan.gc.ca/forests/measuring-reporting/remote-sensing/13433, accessed on June 9, 2019).

Qian, S.-E. 2013. *Optical Satellite Data Compression and Implementation*. Bellingham, Washington: SPIE Press.

Qian, S.-E., R. Girard, and G. Kroupnik. 2013. Development of Canadian hyperspectral imager onboard micro-satellites. *Proceedings of IEEE International Geoscience and Remote Sensing Symposium (IGARSS'2013)* 3506–3509.

Qian, S.-E., et al. 2017. A spaceborne coastal and inland water color hyperspectral imager. *Proceedings of IEEE International Geoscience and Remote Sensing Symposium (IGARSS'2017)* 447–450.

Rocchio, L. 2006. Saving millions in government dollars: Landsat helps fight crop insurance fraud. Landsat Science (available at https://landsat.gsfc.nasa.gov/saving-millions-in-government-dollars-landsat-helps-fight-crop-insurance-fraud/, accessed on June 9, 2019).

SSTL. 2019. SSTL-150 Satellite Platform (available at https://rsdo.gsfc.nasa.gov/images/catalog/SSTL150.pdf, accessed on June 12, 2019).

Staenz, K., T. Szeredi, and J. Schwarz. 1998. ISDAS—A system for processing/analyzing hyperspectral data. *Canadian Journal of Remote Sensing* 24(2):99–113.

5 Optical System and Design

5.1 OPTICAL SYSTEM OVERVIEW

The core function of a hyperspectral imager is the identification of materials of the observing targets through their spectral signature. It extracts spectroscopic information of the observing targets in addition to the identification of their spatial features. The optical system and design are critical parts of a hyperspectral imager. Figure 5.1 shows a block diagram of typical configuration of the optical system of a hyperspectral imager. It consists of an instrument front end, fore-optics, spectrometers, focal plane arrays (FPAs), and structure thermal.

5.1.1 Instrument Front-End Functions

5.1.1.1 Input Port Cover

A cover is used to protect the optics from launch conditions, the non-acquisition phase, and known outgassing episodes of the platform (e.g., the international space station(ISS)). The cover will "seal" the instrument against contamination. To mitigate a closed failure, the cover opening function will incorporate redundancy. For example, the PRISMA hyperspectral imager is equipped with an input port cover (Meini et al. 2016). The cover may not be required for some missions.

5.1.1.2 External Baffles

External baffles are incorporated in the payload to prevent light from sources other than the scene from illuminating the pointing mirror. This includes direct illumination by the sun and moon and reflections from structures that support the payload or are otherwise external to the payload. The baffles may have contamination control features such as heaters, cold fingers, and getter.

5.1.1.3 Pointing Mirrors and Gimbals

A pointing mirror which is controlled by the instrument controller is used to transfer the scene on Earth or a planet to the fore-optics. The pointing mirror should have negligible wavefront error and should minimize in-field stray light (low scatter).

If a hyperspectral imager is deployed on a dedicated spacecraft, a payload level pointing mechanism may not be needed if the spacecraft has provided pointing capability in cross-track direction. In this case, a pointing mechanism along the satellite flight track is required only if it is for the purpose of performing ground motion compensation (GMC) to increase the signal-to-noise ratio (SNR).

If a hyperspectral imager is deployed as a secondary payload, the spacecraft flight track is not suitable for scanning of the image acquisition. In this case, at least a pointing capability in the along-track direction, or in both along-track and cross-track directions is required.

5.1.1.4 Spectral Calibration Source

Narrowband spectral calibration sources with well-defined wavelengths are used to determine the absolute spectral registration across the visible and near-infrared (VNIR) and shortwave infrared (SWIR) detector arrays. The spectral calibration sources may be used to determine in-flight variation in band spacing and bandwidth, across the VNIR and SWIR detector arrays. The source(s) shall illuminate the instrument aperture uniformly. The source shall illuminate selected portions of the instrument field of view (FOV).

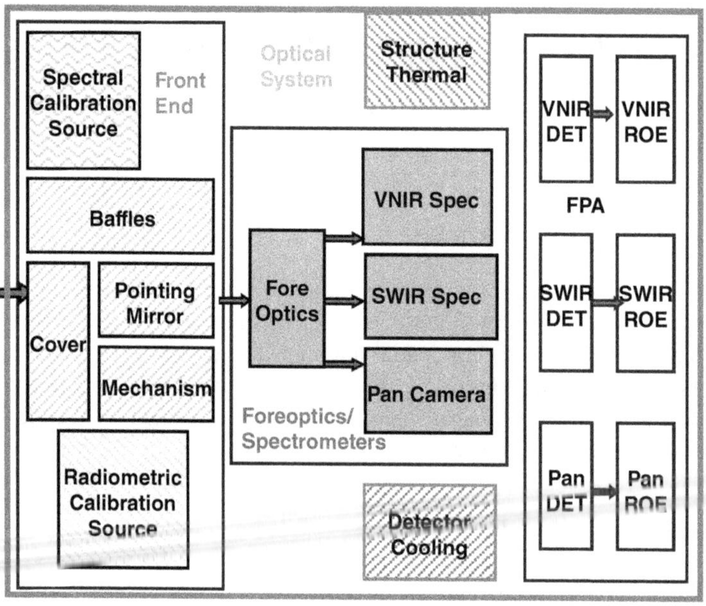

FIGURE 5.1 Typical configuration of the optical system of a hyperspectral imager.

5.1.1.5 Radiometric Calibration Source

A broadband (continuum) radiometric source is used to monitor the absolute radiometric response of the instrument (VNIR, SWIR spectrometers). Absolute radiometric calibration to a level of approximately 1–5% (depending on a particular mission) in the final data products is the requirement. This may be implemented as periodic views of an integrating sphere with solar and lamp inputs; separate monitoring of the lamp radiance would be required.

5.1.2 Fore-Optics

5.1.2.1 Telescope

The function of the fore-optics is to use a telescope to image the ground scene onto the slit assembly of each spectrometer (i.e., VNIR and SWIR, if the wavelength range needs to be covered by two separate spectrometers). The primary image from the telescope should also be imaged or relayed to a panchromatic (Pan) camera if applicable. For each detector array, the telescope should meet the requirements for the SNR, the FOV, the spatial resolution, and the wavelength range. Other design requirements are that it meets the modulation transfer function (MTF) and panchromatic camera requirements, and that the in-field stray light is minimized.

5.1.2.2 Slit Assembly

A slit assembly provides stable co-registered FOVs for the VNIR and SWIR spectrometers. The direction along the slit is nominally referred to as the cross-track direction (or the spatial direction). Curved slits may be used to correct for field distortion or field curvature in the spectrometers or in the telescope (if off-axis). The slit assembly should facilitate packaging of the spectrometers and the panchromatic camera. The slit assembly may have multiple slits for each spectrometer (if spectrometers are modular).

The slit assembly is co-registered with the FOV of the panchromatic camera.

The slit assembly should incorporate covers or coatings that prevent light falling outside the FOV from reflecting onto optical or structural elements that can be directly viewed by the spectrometers.

5.1.2.3 Internal Baffles

The baffles in the fore-optics attenuate the out-of-field stray light and prevent optical cross talk within and between the spectrometers and the panchromatic camera.

5.1.3 Spectrometers and Associated Detector Arrays

5.1.3.1 VNIR Spectrometer

The VNIR spectrometer typically covers the spectral range from 400 nm to 1000 nm. It should have sufficient spectral dispersion, oversampling, spatial and spectral resolution, and the MTF to meet the requirements. The design should provide sufficient throughput to meet the overall SNR requirement. Multiple spectrometers (with corresponding slits) may be used to meet the FOV requirements.

The spectrometer design should minimize the spatial distortion (keystone) and the curvature of spectral line images (spectral distortion, or smile), when the optical system is illuminated with monochromatic light. The spectrometer design should minimize stray light that induces cross talk between nominally distinct spectral bands.

5.1.3.2 VNIR Detector Array(s)

The elements of the detector array convert the incident photons into electrical signals efficiently at the spectral range of interest. The detector array maps the spatial and spectral distributions in the FOV. Hence, the elements represent well-defined spatial and spectral samples at well-defined locations in the VNIR spectrometer focal plane.

The VNIR detector array should efficiently detect incident light within a typical wavelength range of 400–1000 nm and perform complete spectral-spatial array readout consistent with the spatial resolution. Multiple arrays may be required to achieve the spatial resolution and frame rate. It should have a noise performance (from all sources including the readout noise, shot noise, 1/f noise, and cross talk), which is consistent with the overall SNR requirement.

5.1.3.3 SWIR Spectrometer

The SWIR spectrometer is typically a stigmatic spectrometer covering a typical wavelength range from 1000 nm to 2500 nm with sufficient spectral dispersion, oversampling, spatial and spectral resolution, and the MTF to meet the requirements. The spectrometer design should provide sufficient throughput to meet the overall SNR requirement. Multiple spectrometers (with corresponding slits) may be used to meet the FOV requirements.

The spectrometer design should minimize the keystone and smile when the optical system is illuminated with monochromatic light. The spectrometer design should minimize stray light that induces cross talk between nominally distinct spectral bands.

5.1.3.4 SWIR Detector Array(s)

The detector array converts the incident photons into electrical signals efficiently at the spectral range of interest. The detector array maps the spatial and spectral distributions in the field of view. Hence, the elements represent well-defined samples at well-defined locations in the SWIR spectrometer focal plane.

The SWIR detector array should efficiently detect incident light within the typical wavelength range of 1000–2500 nm and perform complete spectral-spatial array readout consistent with the along-track spatial resolution. Multiple arrays may be required to achieve the spatial resolution and frame rate. It shall have a noise performance (from all sources including the readout noise, shot noise, 1/f noise, and cross talk), which is consistent with the overall SNR requirement.

5.1.3.5 Overlapping Spatial and Spectral Sampling

There should be spatial and spectral overlap of the VNIR and SWIR spectrometer(s) and detector(s) to facilitate alignment and co-registration.

5.1.3.6 Preamplifiers

The preamplifiers are used to amplify the low-level signal from the VNIR and SWIR detector arrays. Their location and low-noise design require that they be located as close to the detector arrays as possible to provide high-fidelity signals.

5.1.3.7 Focal Plane Cooling

The dark current from the detectors should not impact the noise performance. Sufficient cooling should be provided to the VNIR and SWIR detector arrays in order to minimize and stabilize dark current. The detector housing's primary function is to facilitate detector cooling, but the housing may have a secondary function of protecting the detector array from molecular contamination via the use of a getter and/or cold finger(s).

5.1.3.8 Panchromatic Camera

The function of the panchromatic camera is to provide a spatially co-registered high-resolution image of the scene to sharpen the imagery acquired (help locate and identify the spatial features seen) by the VNIR and SWIR spectrometers. The panchromatic camera should be spatially registered to the VNIR and SWIR spectrometers within a particular number of pixels over the entire field. The panchromatic camera provides a pushbroom type image of the scene centered on the scene imaged by the VNIR and SWIR spectrometers. It should provide a spatial resolution that is typically 2–4 times greater than that of the VNIR and SWIR spectrometers. It should provide a swath width that is typically ~50% greater than that of the VNIR and SWIR spectrometers.

The panchromatic camera optics should meet the requirements for the SNR, swath, and spatial resolution. Where necessary, optical elements should provide correction of the field distortion or curvature to achieve the required spatial co-registration.

5.1.3.9 Panchromatic Linear Detector

The panchromatic linear detector should efficiently detect incident light within a typical wavelength range of 500–900 nm and perform complete spatial readout consistent with the spatial resolution. It should have a noise performance that is consistent with the overall SNR requirement.

5.1.4 Readout Electronics

The Readout Electronics (ROE) is used to convert the detector output signals into a digital form for transmission to the data-handling system. One of the most critical parameters in the electronics and the detectors is the clock out speed of the detector elements. The detector array and electronics should be capable of clocking out at a frame rate of 250 Hz. The noise of the circuit is minimized to ensure that SNR requirement is achieved. The ROE should have functions to allow row and/or column selection, detector area selection, and element binning. The control of these functions should reside in the data processor.

5.1.4.1 VNIR Array Readout

The ROE for the VNIR array is expected to have a fixed logic command interface and interpreter to select the fixed logic clock generation. The ROE should be capable of on-chip spatial and spectral binning. It should provide DC bias levels for low-noise voltage regulation. It should have a correlated double sampling, at least 12-bit A-to-D conversion and a fast serial data output to the data preprocessor. It is possible that up to 4 channels are required to clock out all data at the speed desired. It should be capable of synchronizing to a master clock.

5.1.4.2 SWIR Array Readout

The SWIR array ROE shall have all general functions as the VNIR array. Row and column selection is required. Note that due to the nature of the detectors, correlated double sampling will be implemented digitally. It should be capable of synchronizing with a master clock.

5.1.4.3 Panchromatic Readout

The panchromatic camera linear array ROE shall have the same general functions as the VNIR. It requires only a single channel high-data rate output to the data preprocessor. It should be capable of synchronizing with a master clock.

5.1.5 Instrument Structure

The structure of the payload provides suitable attenuation/control of launch loads so that on-orbit performance is unchanged from that measured in ground testing and characterization. It provides mechanical stability on-orbit, such that the optical performance (MTF, spatial, and spectral resolution, etc.) as well as the pointing and spatial and spectral co-registration requirements are maintained in all operational thermal environments. Cooling of the focal plane arrays is described above.

5.2 OPTICAL DESIGN CONSIDERATIONS

The optical design of an imaging spectrometer is recommended to focus on the achievement of the following three goals: (1) high SNR, (2) high-calibration accuracy, and (3) high-response uniformity (Mouroulis and Green 2018).

The design should maximize the system throughput to achieve high SNR. This implies an optically fast system with high transmittance. In general, an optical design should seek to minimize the number of optical surfaces as well as provide as small F-number as possible.

Calibration accuracy depends largely on the selected calibration methodology, facility, and procedures. However, at the design stage, it is important to ensure that accurate calibration is enabled and facilitated by the instrument characteristics.

The third goal is essentially a restatement of the following self-evident requirements: that the spectrum of a sample in the scene should be independent of where it falls within the FOV of the spectrometer and that all wavelengths in a spectrum should arise from the same area in the scene. These requirements have far-reaching consequences for the design of imaging spectrometers.

As described in Chapter 2, all dispersive element based spaceborne hyperspectral imagers use a slit-scan spectrometer with 2D detector arrays. The imaging spectrometers to be designed in this chapter are assumed being slit-scan imaging spectrometers. A slit-scan imaging spectrometer images an entire line of ground typically (but not necessarily) in the cross-track direction while the spacecraft provides the forward (along-track) scan. The image of the cross-track line, formed on the spectrometer slit, is then dispersed onto a 2D detector array, which provides spatial information of each pixel in the line along one axis and spectral information along the other. This architecture effectively integrates as many individual spectrometers as there are pixels in cross-track direction into a single instrument. It, therefore, permits substantial mass and volume savings, an often-definitive advantage especially in resource-constrained space missions. Other potentially significant advantages are an increase in the integration time and hence SNR, avoidance of moving parts for whiskbroom scanning, and perhaps reduced polarization sensitivity. In addition, the large increase in integration time permits the use of smaller detector pitch size to reduce the mass and volume of the optics. A disadvantage is nonuniformity, calibration complexity, readout artifacts, and striping due to the use of a 2D detector array.

5.2.1 Telescope Design Considerations

In the design of the telescope, it is important to consider the characteristics such as field curvature, telecentricity, channel separation, stray light performance, manufacturability, alignment, and thermal stability.

For several reasons, off-axis telescopes appear to be more suitable for hyperspectral imagers than on-axis telescopes. For instance, they provide a folded system with easier access to the slit/focal plane and they can have stray light advantages, particularly since mounting spiders and obscurations of the on-axis telescopes can be stray light scatters.

Reflecting or catadioptric telescopes are best able to perform over the wide wavelength range while minimizing weight and volume. Normally, an optical designer primarily considers reflective designs and considers refractive optical elements only for field flatteners and corrector plates, where needed.

The simplest interface from the telescope to the spectrometer is for the telescope to produce a flat image plane. Nevertheless, a curved slit on a non-flat focal surface can be an advantage in some instrument designs. Some of designs produce flat image planes, while others produce curved image planes.

Two-mirror telescopes generally do not provide sufficient degrees of freedom to control the location of the exit pupil and field curvature. Three-mirror telescopes provide sufficient degrees of freedom to correct for aberrations. A larger number of mirrors generally make the mirrors easier to manufacture, but at the expense of overall system complexity and usually mass and volume. Manufacturing spherical mirrors is easier than manufacturing aspheric or conic mirrors.

Mouroulis and Green (2018) recommend three fundamental requirements that the telescope design must satisfy:

1. Maximum transmission
2. Essentially zero transverse chromatic aberration (TCA) or less than a small fraction (such as 1%) of a pixel
3. Minimum (ideally zero) variation of response with wavelength

Maximum transmission is a goal so that the available photons will not be wasted on their way to the detector, where they would contribute to a higher SNR.

Zero TCA is a geometric uniformity requirement. It implies that the telescope will be typically all-reflective so as to have identically zero TCA. Some catadioptric solutions using a refractive field corrector can also be designed with sufficiently small amounts of TCA with great care in the design. Fully refractive solutions are extremely limited and probably impractical, certainly over a wide spectral range. A few examples have been given in the literature (Fisher and Welch 2006, Mouroulis 2007), but it can be argued that they fail requirement 1, since the large number of elements reduces transmission due to the difficulty of producing efficient antireflection coatings over a broad range.

The third requirement implies that the diffraction spread should be well contained within the pixel or slit width even for the longest wavelength used. This means that the telescope should have a large relative aperture or low F-number. For a broadband system, the diffraction spread is more important than the geometric spot size.

All these requirements must be satisfied for a potentially very wide spectral range, extending over 2.5 octaves or more. In attempting to satisfy these requirements, a designer can make use of the following degrees of freedom:

1. The pupil matching between the telescope and the spectrometer need not be very close, provided the telescope aperture is oversized to account for the mismatch.
2. The telescope need not have a real aperture stop (when not in the thermal infrared range), provided the mirrors are sufficiently oversized to permit the stop inside the spectrometer to act as the limiting aperture.
3. The telescope does not normally need to be diffraction limited, provided the energy is reasonably contained within the pixel. In other words, the maximum spatial resolution is not usually a goal, since the spectrometer detects objects through their spectrum form rather than their spatial form.

5.2.2 Spectrometer Design Considerations

For a successful design of the spectrometer, Mouroulis and Green (2018) found that it is beneficial to adhere to the following principles:

1. Geometric distortions should be controlled to ~1% of a pixel at the design stage or ~3% after tolerancing.
2. More than ~75% of diffraction energy should be contained within the pixel at all wavelengths and fields.
3. Degraded spot sizes are acceptable and desirable (subject to rule 2 above), provided they improve uniformity.
4. The grating or dispersive element is the preferred stop location for the full system, although if this is not an option, this condition may be violated subject to careful assessment.

In addition, two practice procedures, which cannot be properly called principles, have been found beneficial:

1. Integrating the deterministic part of stray light assessment (predictable ghosts and reflections) into the first design pass.
2. Assessing uniformity and image quality in terms of cross-track spatial response function, along-track spatial response function, and spectral response function, while using the more common measures, such as wavefront, spot size, and MTF, only as a temporary shortcut if they are much faster to compute, and assuming they can be shown to correlate with the response functions.

The implications of these principles are far reaching. Unless a designer incorporates the geometric distortion minimization into the merit function, the tendency of the optical design software will be to correct the point-imaging aberrations often well beyond what is necessary. The inverse of this condition is that a design with well-controlled point-imaging but large geometric nonuniformity may have unacceptable spot sizes once the distortion is corrected. Response uniformity must be enforced at the beginning (Mouroulis et al. 2000). It is for this reason that designs with minimum aberrations but uncontrolled uniformity may not be good starting points for a final uniform design. Such may be the case if one starts from a design form for which the distortion is inherently large and difficult to correct.

5.3 TELESCOPE DESIGN

An optical telescope focuses and collects light, mainly from the visible to shortwave infrared of the electromagnetic spectrum to create a magnified image for direct viewing or on the silt of a spectrometer to be dispersed spectrally. The capability of a telescope to gather light and show small details is dependent on the diameter (i.e., the aperture) of the telescope's primary mirror that gathers and focuses the light of the objective. The larger the aperture, the greater the ability of the telescope to gather and focus light, and the greater its ability to show small details in the FOV. Based on the method that is used to focus the image into the viewing device, a telescope falls into one of three classes: refracting telescopes, reflecting telescopes, or catadioptric telescopes.

5.3.1 Refracting Telescopes

A refracting telescope uses lenses to gather and focus light. The refracting telescope gathers a greater amount of light into the lens than is possible to gather with the naked eye. This presents the observer with a brighter, clearer, and magnified image of the object being observed. This is accomplished by focusing the parallel light onto a focal point while the peripheral light is focused onto a focal plane.

The refracting telescope was the first telescope to be invented and used since early 17th century. Hans Lippershey, a Dutch lens maker, invented a telescope and filed a patent in 1608. He intended to use the telescope for military purposes. Galileo was the first person credited with applying a refracting telescope to the study of the sky. With the telescope, he was able to see the craters on the moon, the four major moons of Jupiter, and the rings of Saturn.

There are several inherent problems with a refracting telescope. Refracting lenses lend themselves to what are called chromatic and spherical aberrations. Chromatic aberration occurs when a lens fails to focus all the colors to the same focal point, as shown in Figure 5.2. This defect shows as a fringe of color along the boundaries that separate dark and bright parts of the image. This was originally dealt with by increasing the focal length of the lens, which led to an extremely long

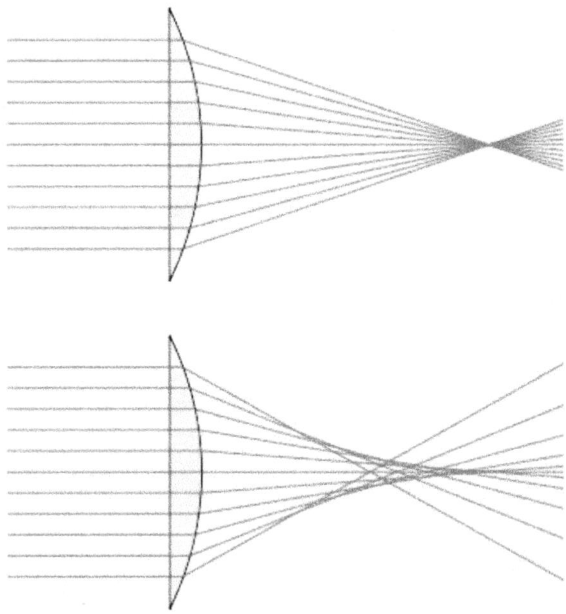

FIGURE 5.2 Illustration of chromatic and spherical aberrations of a refracting telescope.

telescope. Spherical aberration occurs due to the increased refraction of light rays when they strike a lens near its edge. This causes the outer rays of light to be focused more tightly away from the focal point, which causes the image to be imperfect.

There is also a problem with the lenses themselves. Lenses are flawed with small air bubbles trapped within the glass, which is also opaque to certain wavelengths of light. Even visible light is dimmed by the reflection and absorption when the light passes through the glass.

Due to their inherent chromatic aberrations, refracting telescopes are not popular for use in hyperspectral imagers. If a refracting telescope is selected for use, a correction mechanism is required. An example is the HyperSpectral Imager (HySI) onboard the Indian Mini Satellite-1 (IMS-1) that used a refracting telescope. HySI was a linear variable filter (LVF) based hyperspectral imager. As described in Sections 1.2.2 and 2.12, unlike the conventional dispersive element based imaging spectrometers, the optical system of the HySI consisted of only a telescope without a spectrometer assembly. The telescope collected the light from the scene and focused onto a focal plane. A LVF was placed in front of the detector array to disperse the spectrum.

The refracting telescope of HySI was a seven-element telecentric lens assembly with f/# = 4, as shown in Figure 2.10. A thermal filter was placed at the front to protect the telescope. The telecentric design (i.e., a chief ray at each image field is parallel to the optical axis) was adopted to minimize the spectral shift with the look angle. All the lens elements had spherical surface profiles and were placed very close to each other to make the telescope compact. The design optimized the correction of chromatic aberrations and achieved the performance up to spatial frequency of 20 lp/mm over a FOV of ±13°. The thickness of each of the lens elements was minimized to the extent possible to reduce the mass. The lens elements were assembled into a titanium barrel. The barrel design took into account various parameters such as the stresses in the lens due temperature changes, dynamic loads, material mismatches, the constraints on the flange focal distance with temperature, mounting requirements onto structure, stiffness, and miniaturization requirements.

5.3.2 Reflecting Telescopes

A reflecting telescope uses mirrors to gather and focus light. It was inspired by the problems with refracting telescopes. The ideas for developing a telescope that used curved mirrors instead of lenses began to circulate in the 17th century. This idea had been introduced in the 11th century by Alhazen in his widely read work, *Book of Optics*. Although the idea had been present for several centuries, no practical application of the theory occurred until 1673 when Robert Hooke created the first working reflecting telescope. Isaac Newton has been credited with the creation of the first practical reflecting telescope using a spherically metal primary mirror and a small diagonal mirror. This design is now known as the Newtonian telescope, as shown in Figure 5.3, and it is still popular for amateur telescopes. Although the theoretical advantages of the reflecting telescope compensated for many of the disadvantages of the refracting telescope, it took over 100 years for the reflecting telescope to become popular due to the poor performance of speculum metal that

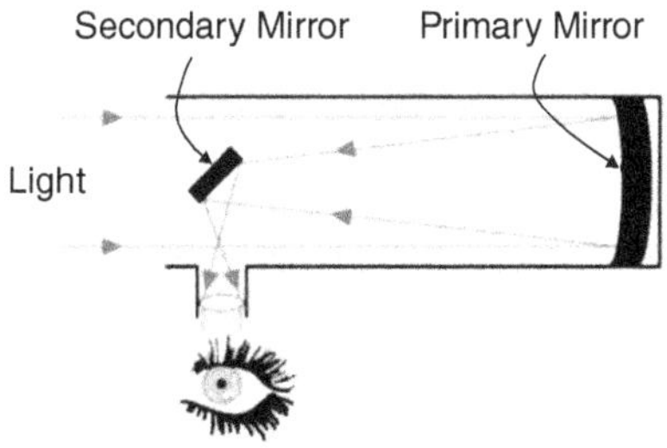

FIGURE 5.3 An illustration of a Newtonian reflecting telescope.

was being used as the reflective surface at the time. With the perfection of parabolic mirror fabrication in the 18th century, silver-coated mirrors in the 19th century, and long-lasting aluminum-coated mirrors in the 20th century, the reflecting mirror has become the telescope of choice for astronomers worldwide.

A reflecting telescope uses a combination of curved mirrors that reflect light and form an image into a viewing device. A curved primary mirror is the basic optical element and creates an image at the focal plane. A viewing device, such as an eyepiece, might be present for viewing the image, or a detector array may be located at the focal plane to record the image.

Mirrors within a reflecting telescope eliminate the problem of chromatic aberration, but this type of telescope may produce other types of aberrations, namely spherical. This was originally the design flaw within the mirrors of the Hubble Space Telescope. There are other types of aberrations as well, which can be corrected with more advance telescope design.

Reflection-type telescopes are widely used in remote sensing instruments. The most common design of telescopes for hyperspectral imagers is a three-mirror anastigmat (TMA) and its several variations. A TMA telescope has excellent features of correcting aberrations, wide spectral range, and shorter physical (i.e., smaller volume) requirement. As a result, TMA telescopes have become more and more attractive and have been adopted in space missions.

According to Cook (1992), there are three design versions for TMAs:

- TMAs with virtual entrance pupils (Class I)
- TMAs with real entrance pupils (Class II)
- TMAs with intermediate images between the secondary and tertiary mirrors, a relay TMA's (Class III)

The virtual entrance pupil TMA is telecentric. It does not have an intermediate image, thus its exit pupil can be at infinity. An advantage is that it can be matched to a standard spectrograph designed for a distant entrance pupil. Cook noted that this type telescopes tend to have higher performance than TMAs with real pupils, since the virtual entrance pupil retains annular field symmetry. However, in the analyses carried out by the author of the book and his team, a significant performance difference between the virtual and real aperture TMA designs was not found in terms of resolution.

A TMA telescope with a real entrance pupil has the advantage of locating a pointing mirror at the real entrance pupil of the fore-optics. The real entrance pupil would also likely be used for calibration.

The relay TMA is non-telecentric. An intermediate image is formed between the secondary and the tertiary mirrors; this image is then relayed by the third mirror. Thus, its focus is not telecentric. The exit pupil is not distant. These kinds of telescopes have stray light advantages and are very useful in thermal wave bands where cold shielding is required.

The two most important TMA variants recommended by Mouroulis and Green (2018) are a telecentric version with the stop on the secondary and a non-telecentric version, which provides a cold stop location before the slit of the spectrometer.

Two-mirror telescopes have some advantages over three-mirrors telescopes in that they are more stable and easier to adjust. Because there are only two mirrors, it is easier to prevent tilt and decentration errors. However, two-mirror telescopes do not offer enough degrees of freedom to control the location of the exit pupil and field curvature. A common two-mirror telescope design is the Ritchey-Chrétien design, which is a specialized Cassegrain telescope that utilizes two hyperbolic mirrors instead of a primary parabolic one that is common in the standard Cassegrain design.

In practice, the following three telescope designs have been found most useful for hyperspectral imagers in satisfying the requirements described in Section 5.2: (1) a classic TMA, (2) a two-mirror modified Schwarzschild (TMS), and (3) a Cassegrain, typically with a field corrector. Each has a distinct range of applicability.

Specifically, the TMS telescope is an unobscured design that can provide the widest possible FOV and lowest f/#, in addition to having the minimum number of mirrors, and thus the highest transmittance. These advantages are achieved at the expense of a virtual aperture stop and large geometric distortion, which does not seem to be an issue given the fact that these concerns are normally not relevant in imaging spectrometers. A more significant concern is that the telescope form is essentially an inverted telephoto and therefore largely relative to its focal length. For this reason, its application is limited to wide-field systems and from low to medium spatial resolution.

A TMA telescope offers flat fields with comparably good aberration correction and comparably good telephoto advantage and provides a good accessibility of the focal plane. It can be designed with dimensions roughly equal to its focal length and therefore can be more compact. It can also achieve relatively wide FOV, though not readily as wide as that of a TMS telescope. Since the focal length and FOV are inversely related given a fixed detector size and pixel count, this is not so much a disadvantage as the delineation of its range of applicability. However, it cannot normally reach f/# well below 2 because the mirrors are pushed to large off-axis angles to avoid vignetting.

The compactness of a Cassegrain telescope design makes it the only choice for high-resolution (long focal length) applications. However, the basic two-mirror front must be typically followed by either a refractive or preferably a reflective corrector to widen the field and achieve the final required f/#. The obscuration and the light loss of this form as well as the larger number of elements are inevitable compromises. A Schmidt corrector plate has also been incorporated; however, its chromatic effect over a broad spectral range would require careful assessment.

5.3.3 Catadioptric Telescopes

The word "catadioptric" contains within it an explanation of what it is. "Catoptric" is a term applied to those telescopes that use mirrors, i.e., reflecting telescopes. "Dioptric" is a term that refers to those telescopes that use lenses, i.e., refracting telescopes. Hence, the fusion term "catadioptric" indicates a system that combines both lenses and mirrors, which is exactly what a catadioptric telescope is.

In short, a catadioptric telescope is a combination of a refracting telescope and a reflecting telescope. It has a compound or lens/mirror combination that offers a combination of the benefits of the refracting and reflecting telescopes.

The catadioptric telescope uses a combination of specifically shaped lenses and mirrors to form an image and transmit it to the eyepiece or a focal plane.

Because of its combination of mirrors and lenses, the catadioptric telescope allows the folding of the light within the barrel of the telescope. The combination of lenses and mirrors also compensates for the weaknesses of each element. That is, the use of mirrors with additional correcting lenses eliminates the spherical aberrations encountered in the reflecting telescopes.

Figure 5.4 illustrates the principle of a catadioptric telescope and how it works. Light enters the telescope through a correcting lens, which corrects the aberrations often encountered as a result of later steps of the process. The light is then reflected off the primary mirror. Light strikes the secondary mirror in the center of the correcting lens. Light is reflected back toward a hole in the primary mirror and into the eyepiece.

A catadioptric telescope has the following advantages:

1. Folding of the light within the barrel of the telescope allows for a shorter telescope that results in a small volume without reducing the focal length.

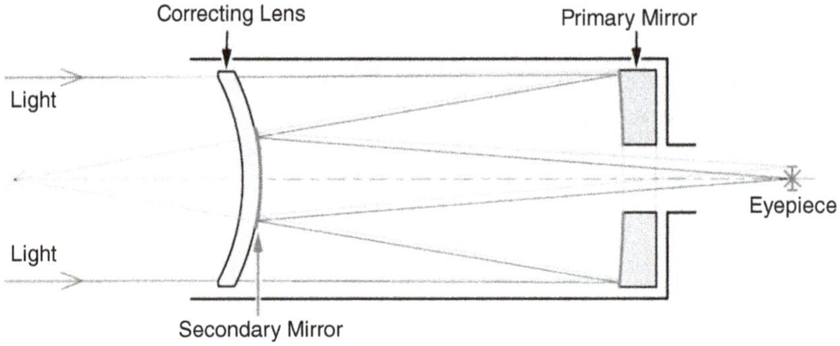

FIGURE 5.4 Principle of a catadioptric telescope.

2. Combination of lenses and mirrors compensates for the weaknesses of each element individually.
3. Use of shaped mirrors creates a true beam of light that creates more detailed images of the FOV.

A catadioptric telescope has disadvantages of possibly slight loss of light due to the position of the secondary mirror in the middle of the primary light-gathering mirror.

Catadioptric telescopes have been used in spaceborne imaging spectrometers. The examples are medium-resolution imaging spectrometer (MERIS) and ocean and land color imager (OLCI) described in Sections 2.7 and 2.16. In order to cover a wide FOV of 68.6°, the MERIS instrument had five identical imaging spectrometers, each of which had a FOV of 14° with 0.4° overlap between adjacent cameras. The OLCI instrument onboard the European Space Agency (ESA)'s Sentienl-3A satellite is the successor of the MERIS and has the same imaging spectrometers based on the heritage design from the MERIS.

The telescope of the MERIS is an off-axis catadioptric form design with a 67.3-mm focal length made up of three large lenses in fused silica, a concave primary mirror, a convex secondary mirror, and a field lens in fused silica located in the image plane, as shown in Figure 5.5. The telescope is corrected for a 50-mm diameter entrance pupil area. The lenses have antireflection coating optimized for the spectral range 390–1040 nm. The entrance pupil area has a 50-mm diameter, of which an approximately rectangular area of 20 mm × 32 mm is used as the entrance pupil.

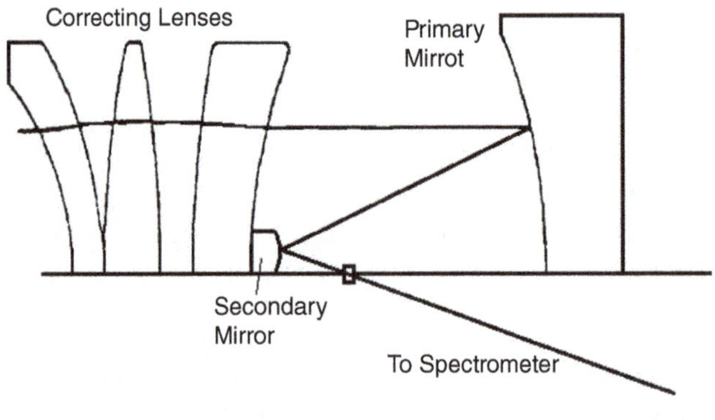

FIGURE 5.5 An off-axis catadioptric telescope of the MERIS and OLCI hyperspectral imagers.

5.3.4 Telescope Design Examples

5.3.4.1 Telescope A—A Non-Coaxial TMA

The first example is a non-coaxial TMA design, as shown in Figure 5.6. Its design is based on the requirements listed in Table 5.1.

The designed telescope has a focal length of 700 mm and an aperture of 240 mm corresponding to a f/# of 2.92. Its FOV is 2.46°. The shape of the three mirrors is aspherical with only conic constants. This solution is very compact and without obstruction and assures excellent optical quality with a minimum number of optical elements. The secondary mirror is on-axis. The off-axis values of the primary mirror and of the tertiary mirror are very small in order to facilitate the mirror manufacturing. The position of the aperture stop lies on surface of the secondary mirror. The extraction of the telescope focal plane is quite large in order to facilitate the insertion of a folding mirror and the overall mechanical accommodation.

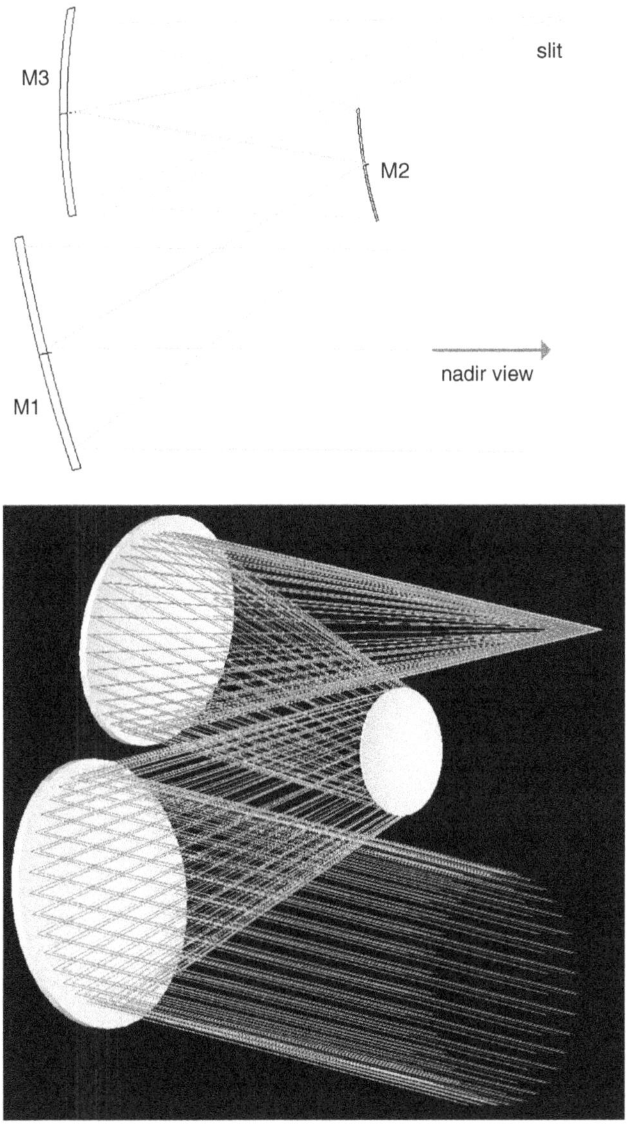

FIGURE 5.6 The optical layout of a non-coaxial TMA telescope.

TABLE 5.1
Requirements for Telescope Design Example One

Parameter	Value	Note
Satellite altitude	700 km	
Swath width	30 km	For both hyperspectral and panchromatic channels
Ground sampling distance (GSD)	30 m × 30 m	Hyperspectral channel
	5 m × 5 m	Panchromatic channel
Wavelength range	0.4–2.5 μm	Hyperspectral channel (VNIR + SWIR)
	0.4–0.7 μm	Panchromatic channel
Field of view (FOV)	2.46°	
Focal length	700 mm	
Aperture	240 mm	
f/#	2.92	
Slit length	30 mm	Corresponding to a swath 30 km wide with 1000 pixels of 30 m for the hyperspectral channel or 6000 pixels of 5 m for the Pan channel
Slit width	30 μm	Corresponding to 30-m GSD in the along-track direction of the hyperspectral channel
Slit width	5 μm	Corresponding to 5 m GSD in the along-track direction of the Pan channel

The image quality of the telescope is excellent in terms of spot diagrams computed for the axial, the zonal, and the marginal FOV, as shown in Figure 5.7. The spots are well enclosed in a 30-μm box. The spot diagrams are relevant to points located on the spectrometer slit.

The TMA distortion is about 30 μm (1 pixel) at full field. This implies that the on-ground projection is not straight but slightly curved. The maximum sag of this curve is 30 m.

The polychromatic MTF curves computed up to the Nyquist frequencies of the hyperspectral imager (16.7 cl/mm) and of the Pan imager (76.9 cl/mm) are shown in Figures 5.8 and 5.9 and compared with the diffraction-limited curve. The polychromatic point spread function (PSF) for the axial, marginal, and zonal FOVs is shown in Figure 5.10.

All these curves indicate that the selected design example is almost diffraction limited. The convolution of the slit profile and the telescope PSF determines the slit illumination profile. Due

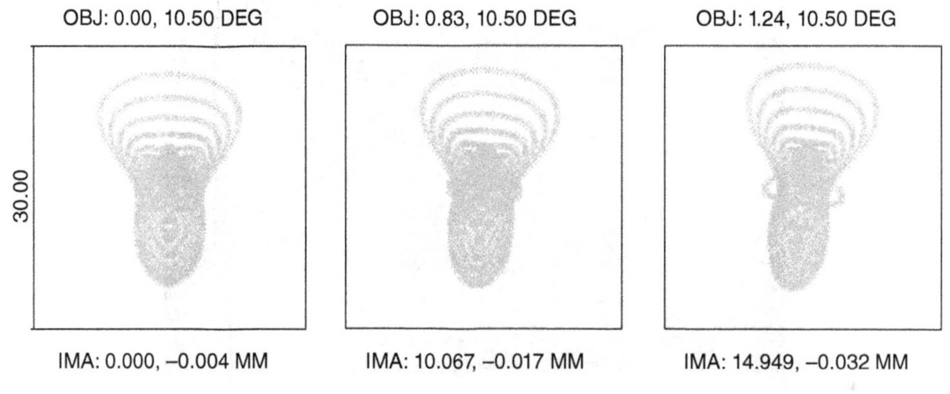

FIGURE 5.7 A spot diagram of the non-coaxial TMA telescope.

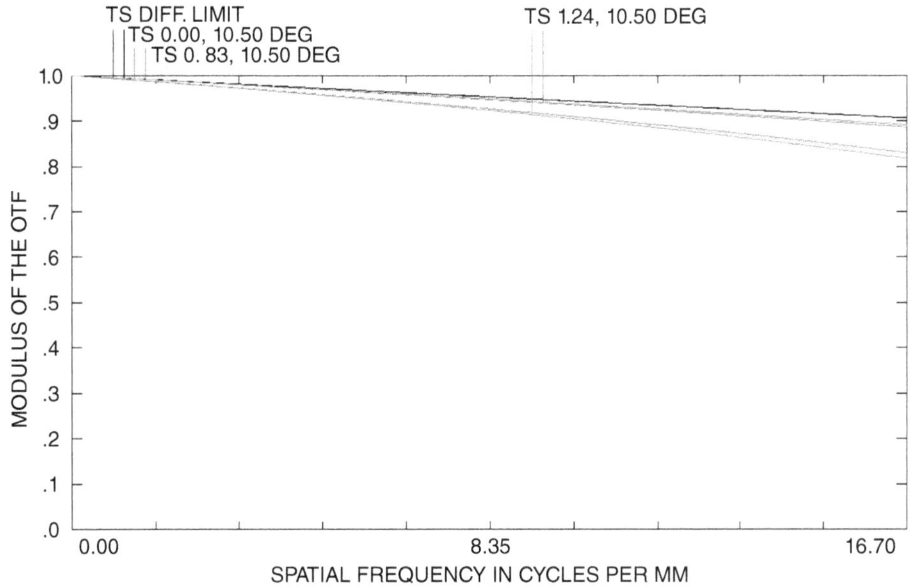

FIGURE 5.8 The TMA polychromatic MTF up to Nyquist frequency of the hyperspectral imager (16.7 cl/mm).

to the nearly diffraction-limited performances of the telescope the slit illumination profile is almost ideal.

5.3.4.2 Telescope B—An On-Axis TMA

The second example of the telescope design is an on-axis Class III TMA as defined by Cook (1992). The requirements for the telescope are similar to those of the first telescope. The three mirrors of

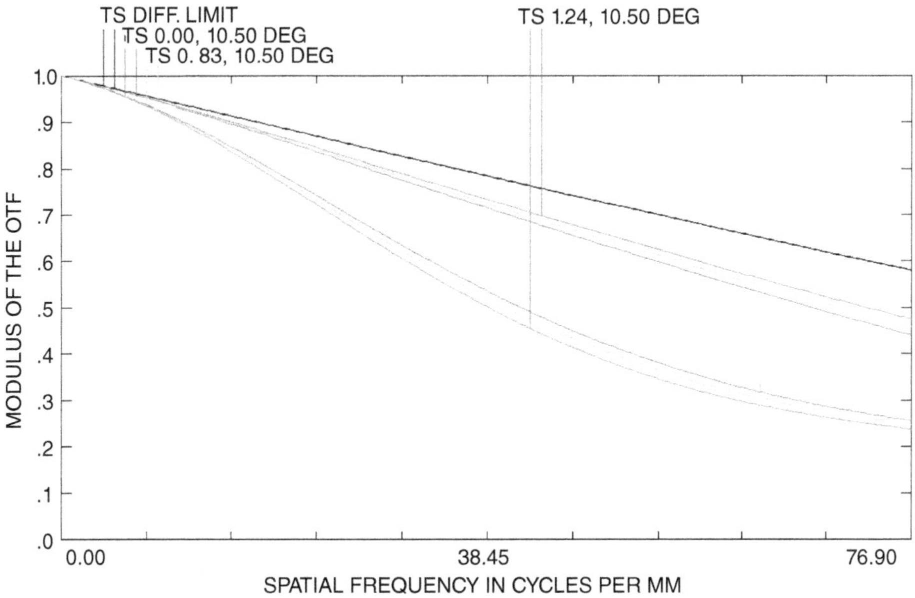

FIGURE 5.9 The TMA polychromatic MTF up to Nyquist frequency of the Panchromatic imager (76.9 cl/mm).

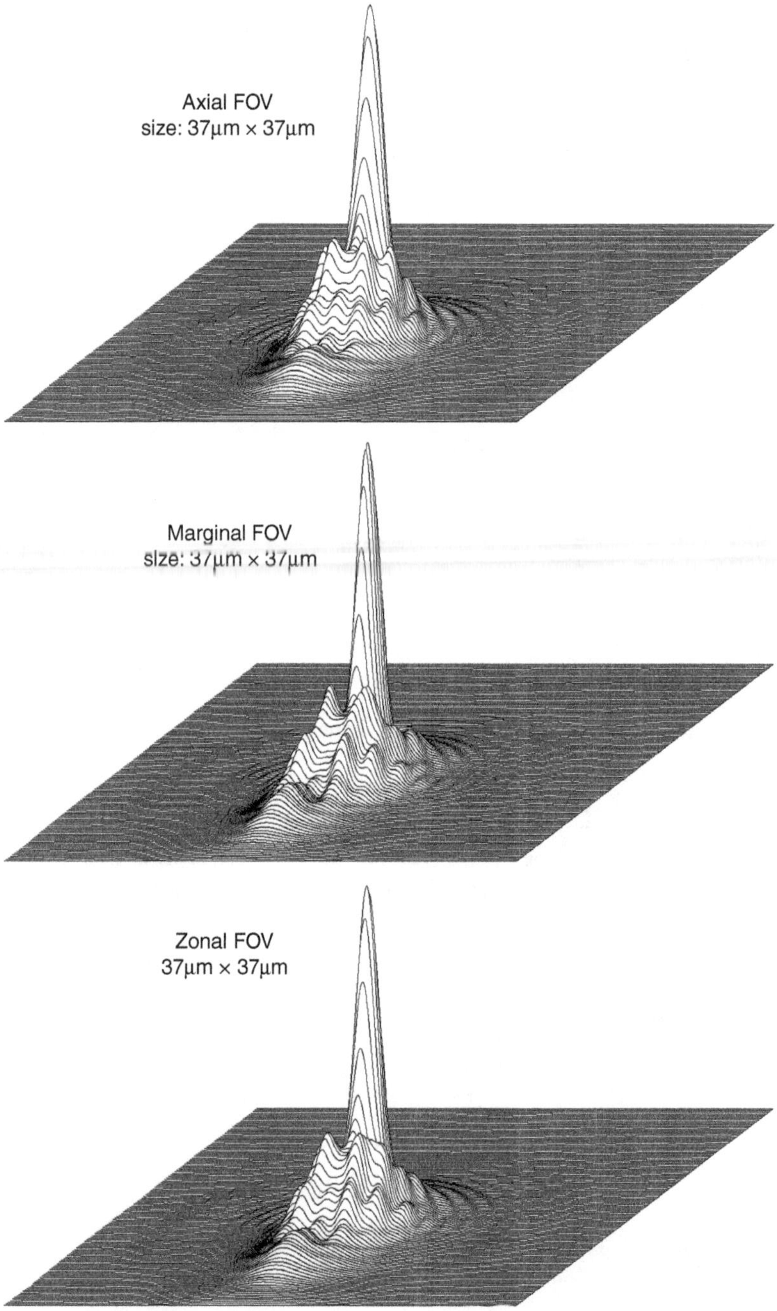

FIGURE 5.10 Polychromatic PSF for the axial, marginal, and zonal FOVs.

the telescope are on-axis, as shown in Figure 5.11 for XZ projection, and in Figure 5.12 for YZ projection. The primary mirror, which is the largest, is a simple conic mirror and is at the entrance pupil, thus it can be circular with a 40.5-cm aperture. It has an intermediate focus between the secondary mirror and the tertiary mirror. There is also a real exit pupil after the tertiary mirror at about the same location as the intermediate focus, but the rays are travelling in the opposite direction.

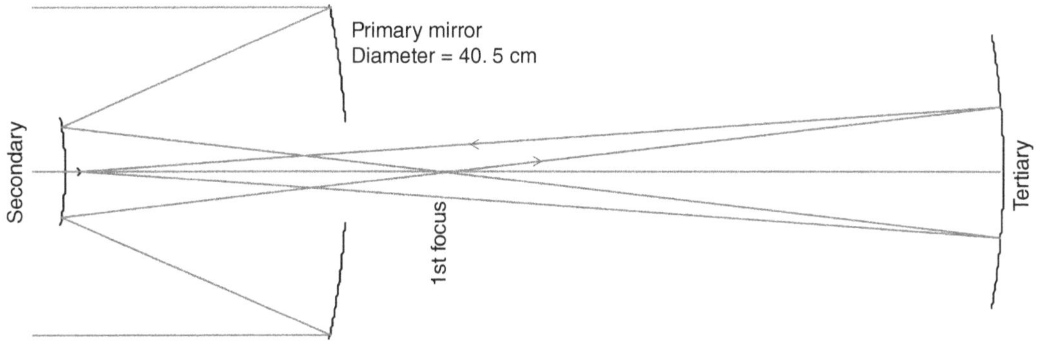

FIGURE 5.11 The optical layout of Telescope B in XZ projection.

There, the field and the pupil have different dimensions, thus a baffle located there cannot be of the exact shape of either an aperture stop or a field stop, but it must always be outside the extreme rays. However, it does help to reduce stray light.

TMA telescopes usually work in off-axis because of obstruction caused either by the back of the detector array or by the back of a flat folding mirror inserted to direct the light to the detector array. This telescope design is unusual in that it is an on-axis TMA but has a full, unobstructed FOV. Although the telescope has central obstruction around the secondary mirror and from spiders holding the secondary mirror, it does not suffer from any obstruction of the FOV even though it is an on-axis TMA. The reason is that it does not have a folding mirror between the secondary and tertiary mirrors, which obstructs the central part of the FOV. The focus of the telescope is internal, where the slit mirrors are located, close to the secondary mirror, within the shadow of the vane that supports the secondary mirror, and thus the slit mirrors do not obstruct any light, even though they are centered on the axis. This can be seen in Figures 5.11 and 5.12.

A major reduction of stray light is achieved by a baffle at the secondary mirror, and by a baffle extending up from the primary mirror around the beam from the tertiary mirror, as indicated in Figure 5.12. Baffles in these spaces are standard in two-mirror Cassegrain telescopes, but in this case the baffle is not circular. It is oval-shaped in the direction of the slit mirrors. The baffle at the secondary is 175-mm long in the direction of the slit Y, and 127-mm wide in X. The corners are rounded at a radius of 63.5 mm.

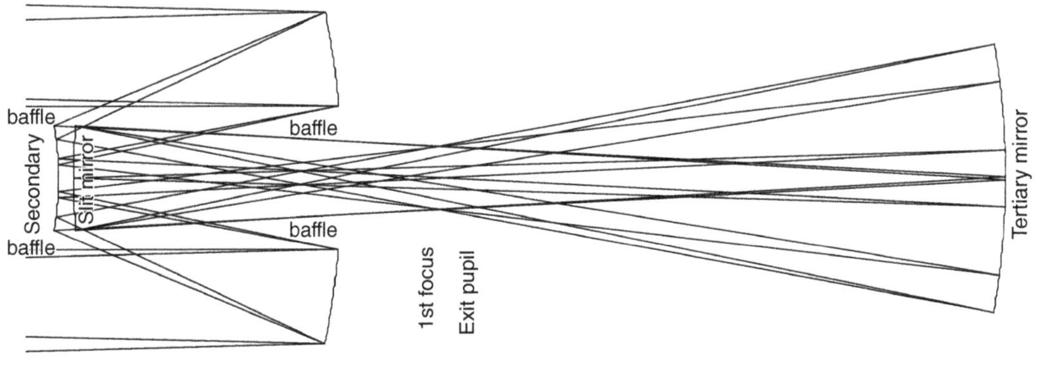

FIGURE 5.12 The optical layout of Telescope B in YZ projection.

There is also obstruction of the incident light by the oval baffle and by spider arms holding the secondary mirror. The main thicker arm is parallel to the slit, thus its width shades the slit mirrors from the incident light from the primary mirror. Ideally, the slit mirrors should see light only from the tertiary mirror. The main thicker spider should extend down to the primary mirror. This sharpens the shadow on the slit mirrors. This deep arm should taper in thickness where it is in the converging beam from the primary mirror. The converging rays are shown in Figure 5.12. The surface of this deep vane should have sharp grooves to reduce the reflection of light at grazing incidence (which is substantial even on blackest black but smooth surfaces). This deep spider also has a structural advantage in supporting the secondary mirror. Also, the conical baffle (Figure 5.12), extending part way up from the primary mirror, can be supported by this vane. It is because the FOV of the telescope is slit-shaped that such a deep vane can be used without obstructing wanted light. The thinner spider arms should have much less depth because they do obstruct wanted light.

It is because the focal surface is in the form of a narrow slit that this TMA telescope can be on-axis. Two cylindrical slit mirrors are located at the internal focus, located 20 mm in front of the secondary mirror, as shown in Figures 5.11 and 5.12. The illuminated part of each (side-by-side) slit mirror is 127-μm wide and 127-mm long, each slit projects to 30 m × 30 km on the ground.

The radius of curvature of the cylindrical slit mirrors is 1298 mm concave. However, because the slit mirrors are tilted ±45° to direct the light to the VNIR and SWIR spectrographs on opposite sides of the telescope, the telescope sees a different curvature of the focal surface. Thus, if the telescope is tested by itself, with an untilted focal surface, the radius of curvature of the focal surface is 918 mm, concave to the incident light.

The image quality of this fore-optic telescope is very high. The geometrical root mean square (RMS) error on the ground ranges from 0.1 m to 0.7 m, which is smaller than the Airy disc that is 1.4 m at 680 nm and 0.8 m at 400 nm. Thus, the telescope is diffraction limited, as shown by the MTF in Figure 5.13. Because the telescope has central obstruction, the diffraction limit (black dashed lines) is not a straight line, having a temporary decrease in slope at about modulation 0.3. This is a

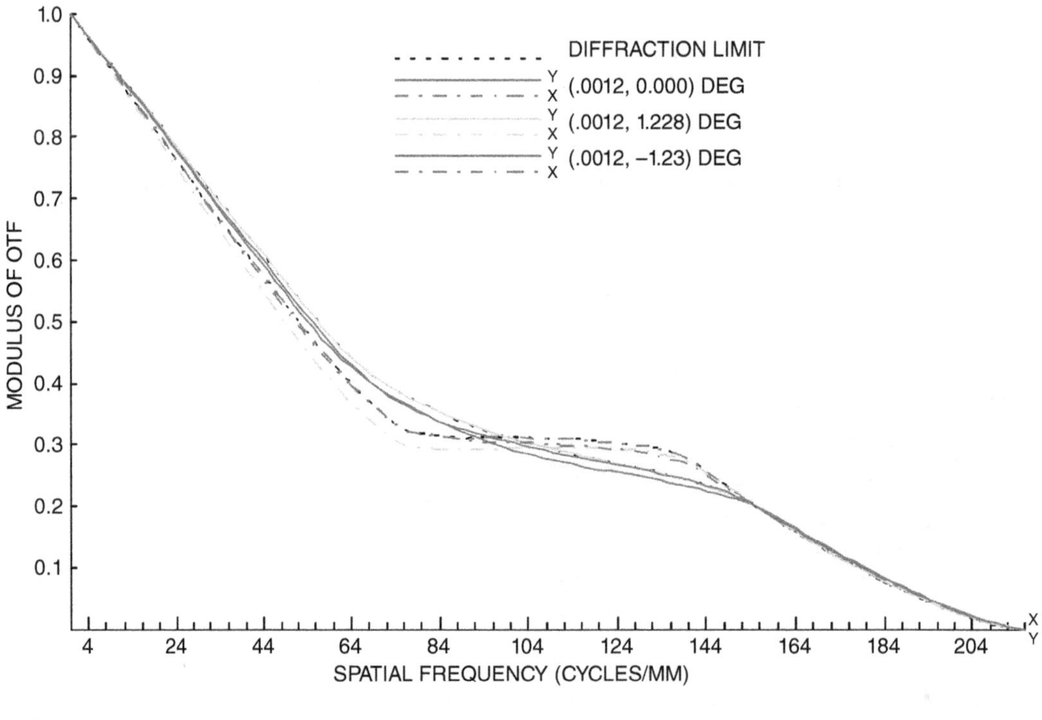

FIGURE 5.13 The MTF of Telescope B.

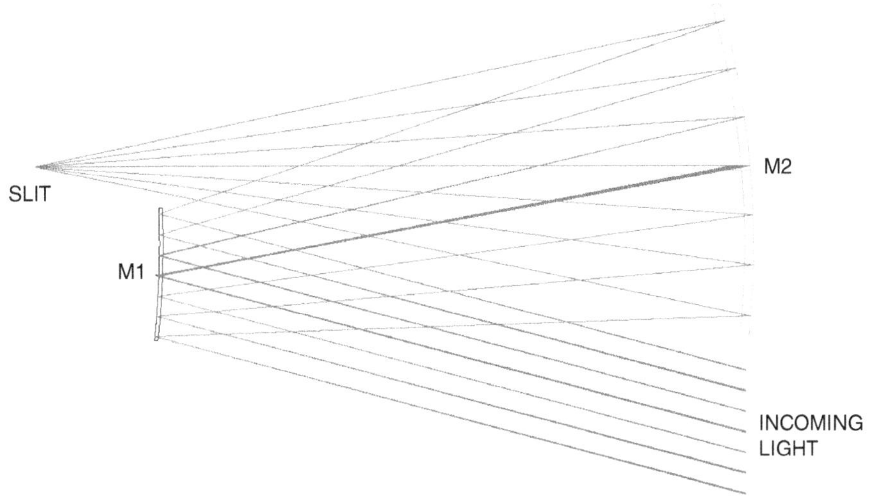

FIGURE 5.14 The optical layout of a two-mirror telescope.

characteristic of all telescopes with central obstructions. The telescope is unusual in having a central obstruction that is longer in one dimension (Y, the slit direction) than in the other.

For packaging, the distance from the secondary mirror to the tertiary mirror is constrained to be no more than 1.2 m.

5.3.4.3 Telescope C—A Two-Mirror Compact Telescope

The third example is a two-mirror compact telescope design depicted in Figure 5.14. It was designed for a hyperspectral imager to be deployed onboard a microsatellite on an orbit of altitude 700 km covering a swath of 150 km with a ground sampling distance (GSD) of 100 m. It is a diffraction-limited unobstructed telescope with a focal length of 14 cm and an aperture of 5.6 cm, resulting in a fast optical throughput of f/# = 2.5. Its effective FOV is 12° with an instantaneous FOV of 0.14 mrad that results in a 100-m GSD from 700-km altitude and an image swath of 150 km. The image swath is slightly curved (2.8%) and has a maximum distortion of 3.2% along the swath.

The telescope design achieved excellent performance. Polychromatic spot diagrams are shown in Figure 5.15 for three wavelengths from the shortest (0.45 μm) to the longest (0.95 μm).

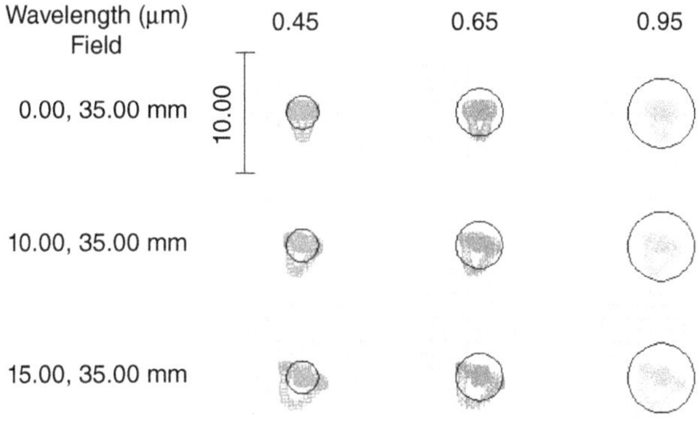

FIGURE 5.15 Polychromatic spot diagrams of Telescope C (airy radius 3.2 μm).

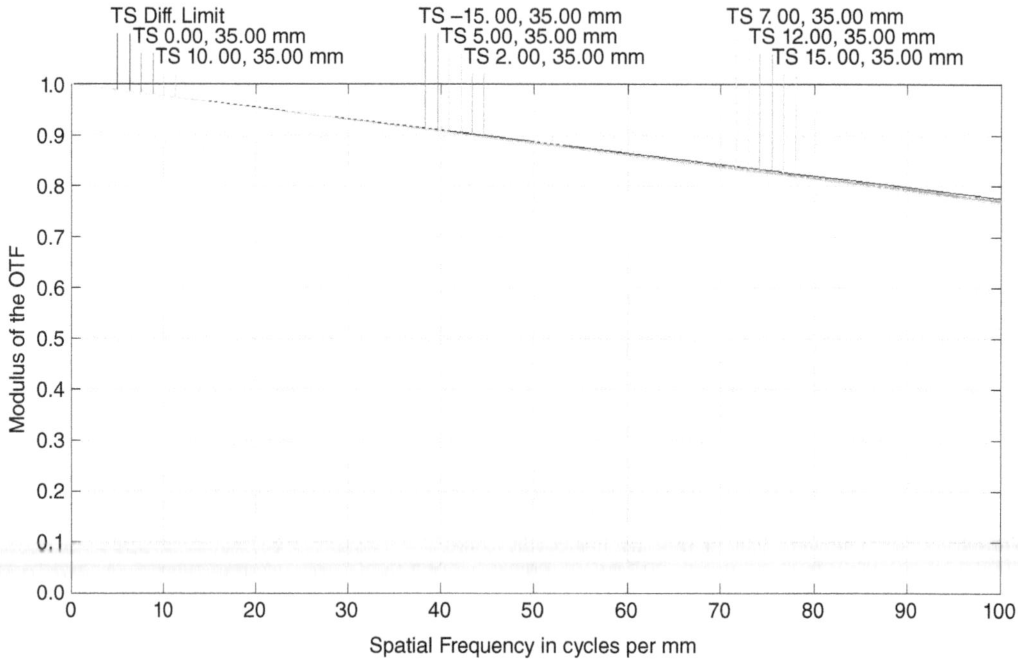

FIGURE 5.16 Polychromatic diffraction MTF of Telescope C.

The aberrations are well within the 10-μm boxes. Polychromatic diffraction MTF of the telescope is shown in Figure 5.16.

5.4 SLIT SCAN SPECTROMETERS

As described in Chapter 2, all spaceborne hyperspectral imagers used or use 2D detector arrays and operated or operate in the pushbroom mode, except for the moderate-resolution imaging spectroradiometer (MODIS) that uses 1D linear detector arrays and operates in the whiskbroom mode. The spectrometers of these hyperspectral imagers are all slit-scan spectrometers when the spectrum dispersion is implemented by gratings or prisms. For this reason, this chapter focuses on slit-scan spectrometers.

5.4.1 Principle of Dispersive Spectrometers

A dispersive spectrometer is an optical device containing two lenses/mirrors that produces an image of the input slit on the detector array. A dispersive element is placed between the lenses/mirrors to disperse different wavelengths in different angles. This causes different wavelengths of light entering the input slit to be imaged at different positions on the detector array.

Figure 5.17 shows a transmitting setup of a grating spectrometer. In reality, reflecting grating systems are primarily used instead of transmitting systems. It is chosen for easier description. As shown in the figure a dispersive element based spectrometer consists of five components: (1) an entrance aperture, (2) collimating optics, (3) a dispersing element, (4) focusing optics, and (5) an exit aperture.

The entrance aperture is usually a slit, a pinhole, or the end of a fiber. It is to limit the spatial distribution of the entering light and help in defining the light cone required to illuminate the optical system.

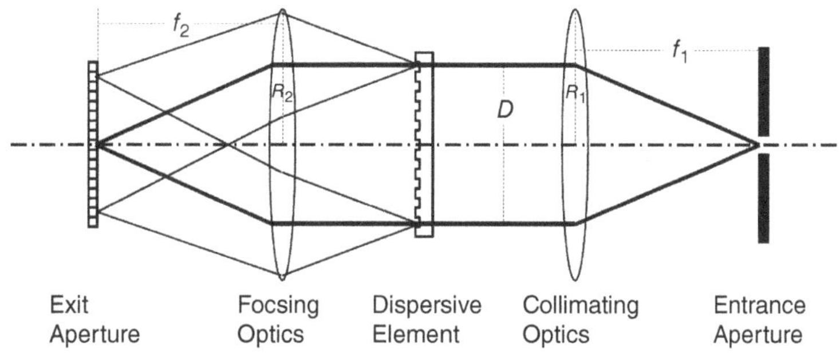

FIGURE 5.17 An illustration of a dispersive spectrometer.

The collimating optics can be either a mirror or a lens. It ensures that the light hits the dispersing element in a parallel, collimated fashion.

The dispersing element can be either a grating or a prism or a grism. A grating or a prism only works correctly when it is illuminated with collimated light. The dispersed light leaves the disperser in parallel bundles of light. Each wavelength leaves under a separate angle. All the light of the same wavelength and spectral order leaves under the same angle.

The focusing optics can be either a mirror or a lens. It collects the dispersed light (i.e., spectrum) and refocuses the spectrum in the output plane.

The exit aperture can be a detector array, a pinhole, or a fiber to collect the parallel arriving wavelengths. In the optimal case, the entrance aperture is reimaged at the output with wavelength being dispersed.

If the optical components are perfectly spherical, foci of the parallel output light arrives on the output radius of the focusing element, i.e., the light arrives in a curved field.

In Figure 5.17, R_1 and R_2 are radii of the mirrors. The larger the radii R_1 and R_2, the better the dispersion, resolution, and quality of spectral and image transfer, but the lower the light flux. D (diameter) is the used size of the mirrors. The bigger the diameter of the mirrors, the higher the possible light flux. The light-collecting power called f-number $= f/D$, where f is the focal length, and D is used diameter. A larger f-number produces lower transfer quality. The wider the slit, the more the entry of light, but the lower the resolution. The longer the slit, the better the light flux at the expense of transfer quality and therefore the resolution and image quality.

5.4.2 Grating Spectrometers

Gratings are common optical elements used to produce dispersion in imaging spectrometers. Gratings are optical surfaces or volumes that contain a periodic structure with features on the order of an optical wavelength that diffracts incident radiation into possibly multiple diffracted orders, each of which is spectrally dispersed. By controlling the dimensions and shape of a periodic structure, one can control the amount of dispersion and the amount of energy diffracted into a particular order. Due to imperfections in the production of fine grating structures, gratings tend to exhibit more scattering than prisms, reducing the spectral purity of grating spectrometers. Gratings also exhibit other limitations due to multiple orders and a wavelength-dependent grating efficiency.

While transmission gratings are sometimes used for imaging spectrometers, reflection gratings are more commonly used. The primary reason for this is that gratings made from metallic materials can achieve higher diffraction efficiency across broad spectral ranges, since they avoid surface reflection losses that occur with transmission gratings. Such gratings have traditionally been manufactured using ruling machines but more recently have been produced by precision diamond turning or a combination of electron-beam lithography and an etching process.

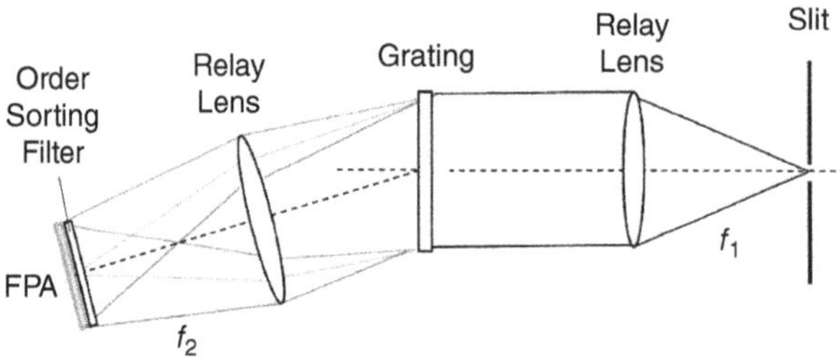

FIGURE 5.18 A traditional transmission grating spectrometer.

A traditional transmission grating spectrometer design is depicted in Figure 5.18. A conventional flat grating is placed in the collimated space of a one-to-one, or unit magnification refractive relay. The relay shown in the figure has unity magnification because the lens focal lengths f_1 and f_2 are identical. In general, focal lengths f_1 and f_2 can be different and the spectrometer relay can have a magnification of f_2/f_1. An order-sorting filter is presented in front of the FPA to compensate for the spectral order overlap caused by the multi order diffraction of the grating.

Multi-order diffraction of the grating is a unique design issue that must be contended with grating spectrometers. If the spectral range of a system falls within the limits imposed by Equation 5.1, it can be easily accommodated by filtering all of the optical energy within these limits somewhere in the system before the FPA.

$$\lambda_{max} - \lambda_{min} < \frac{\lambda_{min}}{m} \tag{5.1}$$

where $[\lambda_{min}, \lambda_{max}]$ is the interval of an optical filter, and m is an integer defining the diffraction order.

However, it is often desirable to achieve a larger spectral range; this can be obtained by employing order-sorting filters. For example, an order-sorting filter for a VNIR spectrometer covering the 0.4–1.0 μm spectral range, without any such filtering, the second order corresponding to the 0.4–0.50 μm spectral region overlaps the first order of the 0.8–1.0 μm region. However, the second order can be rejected over this part of the FPA by placing a high-pass filter that begins to transmit radiation at 0.7 μm over the FPA region associated with the 0.75–1.0 μm band, as depicted in Figure 5.18.

The order-sorting concept can be extended to relatively larger spectral ranges, but this can involve more filters with tighter tolerances, since multiple diffracted orders can be involved. It is common to design an order-sorting filter so that the edges between two different filter sections fall within spectral regions of less importance to the intended remote sensing applications, such as water absorption bands.

In some cases, the multi-order nature of gratings can be a benefit to spectrometer design. For example, consider a case where an imaging spectrometer is to be designed across the entire VNIR/SWIR range from 0.4 μm to 2.5 μm. The detector array for the VNIR region is silicon, while detector array for the SWIR region requires another material such as mercury-cadmium-telluride (HgCdTe). Due to multi-order diffraction of the grating the first order will fall off considerably across this extensive spectral range. However, the second order, associated with the 0.4–1.25 μm bands, aligns perfectly with the first order of the 0.8–2.5 μm bands and achieves similar diffraction efficiency and optical aberrations. Therefore, it is possible to use a single spectrometer for this entire spectral range with two FPAs by using both orders and separating VNIR and SWIR energy between the spectrometer and FPAs with a low-pass spectral filter that transmits VNIR and reflects SWIR. This filter is known as a dichroic filter (Eismann 2012). The dichroic filter is described in Section 5.5.2.

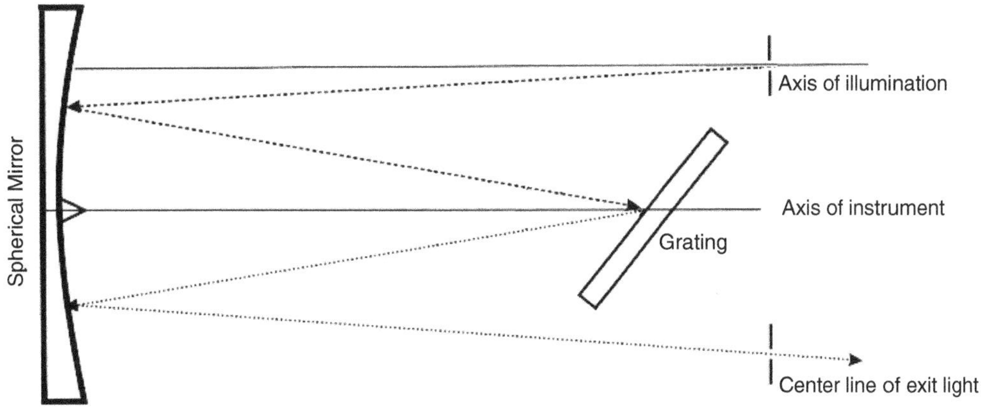

FIGURE 5.19 Configuration of an Ebert-Fastie spectrometer.

5.4.2.1 Ebert-Fastie Spectrometer

The main problem with a Littrow spectrometer is the close vicinity of the entrance and exit.

The Ebert-Fastie spectrometer avoids the problem by placing the entrance and exit on two sides of the grating, far enough that their beams do not touch the grating edge, as depicted in Figure 5.19. This configuration was invented by Hermann Ebert in the 1890s and improved by William Fastie in the 1950s. However, a much wider mirror is required, and the entrance and exit show a small tilt angle versus the perpendicular of the front plate of the instrument. This allows the two distances, mirror-grating and mirror-entrance, to be of equal length.

An Ebert-Fastie system is symmetric by design and thus ideal for the application of curved slits, which compensates for the differences in ray travel. The Ebert-Fastie spectrometer has only three degrees of freedom in adjustment and is very well suited for attachments. It also makes for superb multiple-stage systems. All types of gratings and prisms can be used as the dispersive element. As summarized by Neumann (2014) an Ebert-Fastie spectrometer has the following advantages:

- Only one mirror
- Symmetric design resulting in very good stigmatic transfer
- High degree of flexibility
- Large curved slits to offer high resolution combined with high light flux
- Ability to work with all kinds of dispersers

An Ebert-Fastie spectrometer has the following disadvantages:

- Risk of internal reflection and multiple dispersions at small grating angles
- Problems can appear in image reproduction with detectors >6 mm in height if no extra correction is applied.

5.4.2.2 Czerny-Turner Spectrometer

The concept of a Czerny-Turner spectrometer is based on the Ebert design. The mirror in Ebert design is separated into a collimator and a focusing element, as depicted in Figure 5.20. The Czerny-Turner design offers much more variety for construction. Both mirrors are spherical or parabolic. They can be made in different sizes, which has advantages when illuminating wide areas. Furthermore, they can be made toroidal to allow stigmatic transfer with straight slits. The beam

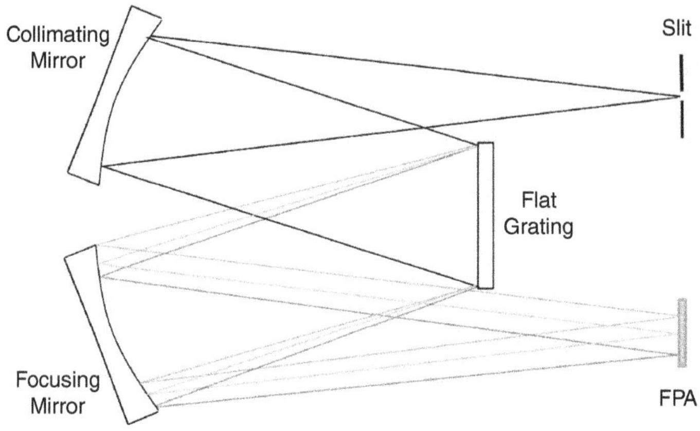

FIGURE 5.20 Configuration of a Czerny-Turner spectrometer.

configuration can be designed such that all entrances and exits are plane parallel to the housing. The grating might be off the symmetry axis of the instrument. However, Czerny-Turner has nine degrees of adjustment freedom, which can create some problems. A clear advantage is that the freedom of building the instrument is much wider than an Ebert-Fastie spectrometer with the same focal length and f-number. This design minimizes internal reflections and multiple dispersions, and eventually produces less stray light.

5.4.2.3 Rowland Spectrometer and Paschen-Runge Spectrometer

A Rowland spectrometer has two spherical mirrors and a plane grating, as depicted in Figure 5.21. It focuses the separated wavelengths at the sector of a circle known as the Rowland circle. All wavelengths directed at the circle are well organized and focused. There are two different detector arrangements for recording the spectrum: (1) an array of detectors is fixed at the circle to receive a predefined wavelength, (2) a single detector travels on the rail and can be moved to the wavelength of interest.

In the Rowland spectrometer configuration, if the grating is concave, the second mirror becomes redundant, and the performance is similar. This variant version is named the Paschen-Runge

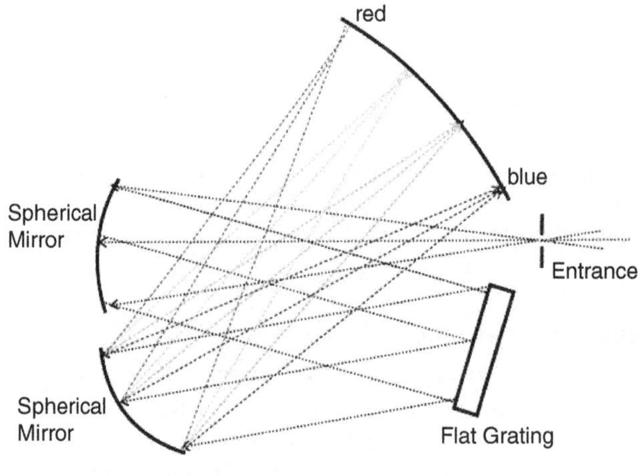

FIGURE 5.21 Configuration of a Rowland spectrometer.

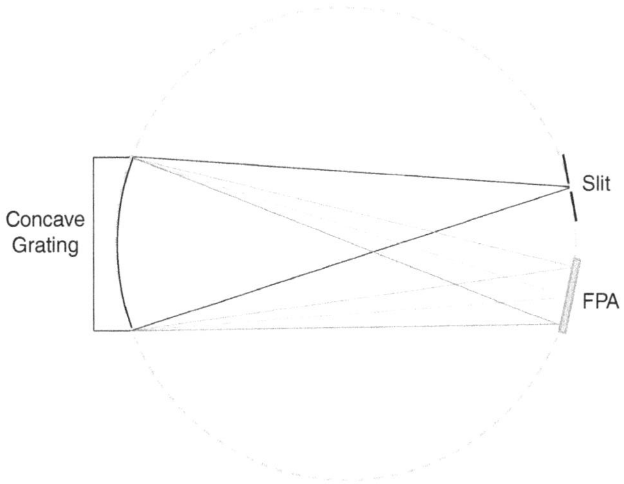

FIGURE 5.22 Configuration of a Paschen-Runge spectrometer.

spectrometer. Figure 5.22 illustrates a configuration of a Paschen-Runge spectrometer with a concave grating.

5.4.3 Prism Spectrometers

Prisms produce angular dispersion of light due to a combination of material dispersion and nonparallel optical surfaces. Since the refractive index (RI) of the material decreases with wavelength in regions of low absorption, prisms refract short-wavelength radiation at greater deviations than they refract long-wavelength radiation. The amount of achievable dispersion is limited by the material characteristics available and optical aberrations introduced by the prism elements. These limitations are addressed by using curved prism surfaces in the spectrometer design. The use of prisms as dispersive elements can be advantageous due to the excellent spectral purity that prism spectrometers can achieve. Prism spectrometers are well suited for hyperspectral remote sensing applications requiring moderate spectral resolution.

A prism-based spectrometer inherently suffers from geometrical distortion along the spectral axis (nonlinear dispersion) caused by the nonlinearity of the change in the refractive index with wavelength. The spectral sampling interval (SSI), which denotes the difference in central wavelength for neighboring bands, is not constant over the spectrum as in a grating spectrometer. Combining different prism materials, leading to less SSI variation at the cost of increased system complexity, can alleviate the problem.

A grating spectrometer can be converted to a prism version if the grating is replaced with a prism and rotated around the pivoting point. However, the progress of the turn is not a sine or cosine. The output wavelength and dispersion of a prism spectrometer depend mainly on three parameters: the refractive index at the given wavelength λ, the prism's internal angle α, and the illumination angle θ_1. The wavelength function will follow a polynomial curve. The resulting dispersion depends on the focal length.

Figure 5.23 depicts the optical layout of a typical prism spectrometer design. In the figure, a Littrow prism is used and the pivoting point (dot) is shown. It has a cut end at apex angle α. Shortening the prism end has no disadvantage because it is not used, but the "nose" can interfere with the beam. The cut offers the advantage of increasing the base length, which is important for the resolution of the prism, without increasing the other sides.

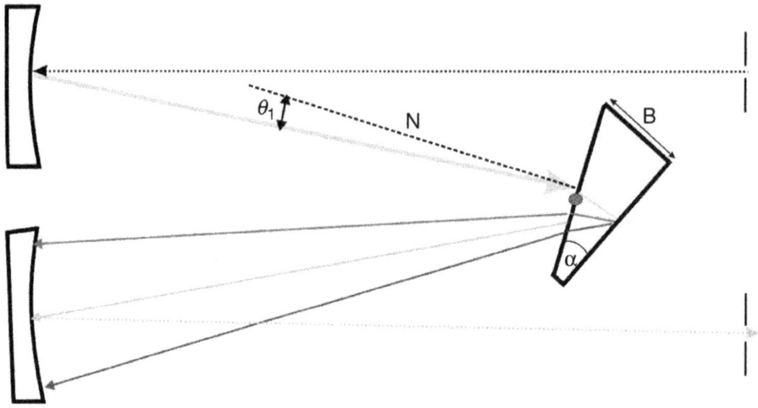

FIGURE 5.23 Configuration of a typical prism spectrometer.

Both the resolving power and spectral range of a prism spectrometer are determined by the optical properties of the prism material; specifically, resolving power is defined by material dispersion, and spectral range is defined by the optical transmission spectrum.

For prism-based spectrometers, the classical spectrometer design with a collimator-prism-imager configuration (Ulbrich et al. 2004) or an alternative approach using curved prisms in a non-collimated beam, as initially suggested by Féry (1911), are possible.

Figure 5.24 depicts a common prism spectrometer design form for an imaging spectrometer (Sidran et al. 1966). This design with a Littrow arrangement exhibits some basic features of a practical imaging spectrometer. The parabolic mirror eliminates the limited optical transmission, chromatic aberrations, and spherical aberrations due to refractive relay lenses. The prism is located in a collimated region and operates in a double-pass way, doubling the angular dispersion for a given prism thickness. The parabolic mirror, however, introduces significant off-axis aberrations that are compounded by the prism, limiting the achievable FOV of the instrument (Eismann 2012).

Replacing the parabolic mirror in the Littrow design with a different mirror form can reduce off-axis aberrations at the expense of on-axis performance. To further reduce aberrations, a Schmidt

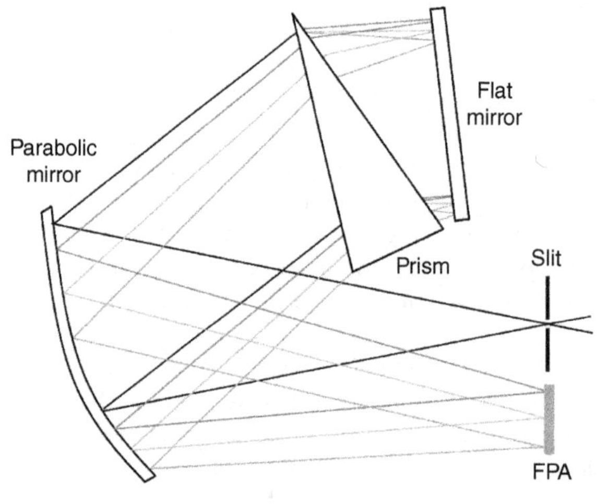

FIGURE 5.24 Configuration of a Littrow prism spectrometer.

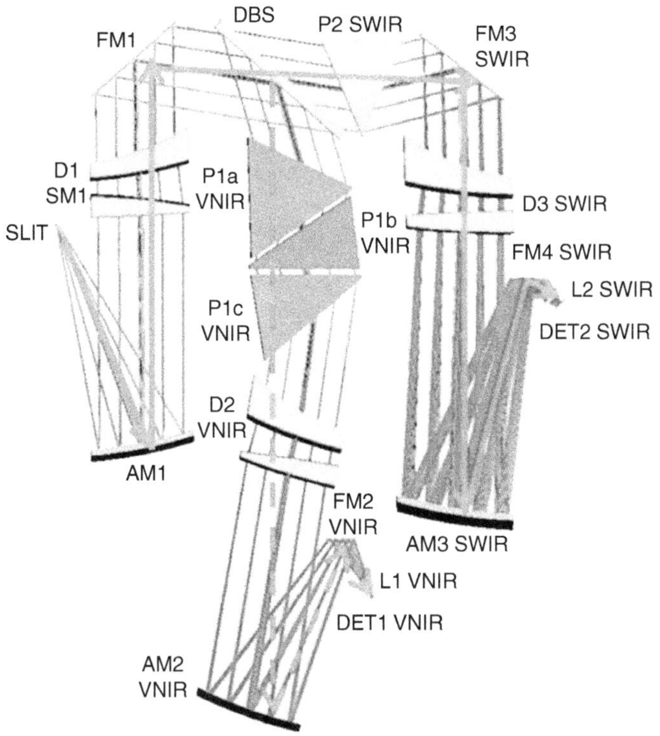

FIGURE 5.25 Optical configuration of the prism-based imaging spectrometers. (Courtesy of SELEX ES.)

corrector can be introduced near the prism, either as an additional refractive element or by introducing curvature into the flat mirror. One such reflecting Schmidt imaging spectrometer design uses an elliptical relay mirror and an aspheric mirror to provide the Schmidt aberration correction (Breckinridge et al. 1983). This approach supports FOV on the order of 15° and maintains good spectral resolution across a broad 0.4–2.5 μm spectral range with 50-μm detectors and f/4 optics.

The traditional prism spectrometer design concept is generally not suitable for hyperspectral imaging applications due to significant aberrations introduced by the thick, flat-sided prism element over the imaging field. To minimize aberrations over the field, prism-based imaging spectrometers tend to employ some combination of prisms with curved surfaces, wavefront correction optics, and double-pass spectrometer arrangements.

An example of spaceborne prism-based imaging spectrometers is the VNIR and SWIR spectrometer of the Italian hyperspectral mission PRISMA depicted in Figure 5.25. The light coming from the folding mirror 1 (FM1) transmits through the dichroic beam splitter (DBS) for the SWIR channel and reflects for the VNIR channel. The prism of the VNIR spectrometer is a combination of a set of three prisms together with a set of correction mirrors (D2 VNIR), whereas the prism of the SWIR spectrometer is a single prism with a set of correction mirrors (D3 SWIR). There is another set of correction mirrors (D1 SM1) before FM1.

5.4.4 Two Typical Spectrometers Used in Spaceborne Hyperspectral Imagers

The designs of most spaceborne pushbroom imaging spectrometers generally fall into two categories. The first category uses different and separated optics for collimation and reimaging. The second category uses the same optics to perform both collimation and reimaging. In either category, the spectral dispersion can be accomplished by a diffraction grating or a prism.

Imaging spectrometer designs for hyperspectral applications must meet very demanding requirements with respect to image quality and distortion. One successful approach to achieving excellent spectrometer performance in a compact optical system has been to employ concentric imaging design forms (Lobb 1994). Concentric imaging systems have a rotational symmetry about any line through a common center of curvature. This results in perfect imaging for any radial line from the optical axis, with no axial aberrations, coma, or sagittal field curvature. The designs known to be capable of meeting the required performance are the Offner- and Dyson-type spectrometers utilizing a grating disperser (Mouroulis 1999). Both Offner and Dyson spectrometers are concentric.

5.4.4.1 Offner Spectrometers

An Offner spectrometer is an adaptation of classical Offner imaging relay in the form of a spectrometer, as shown in Figure 5.26. It consists of two spherical mirrors and a spherical convex diffraction grating. It is an example of the first category of imaging spectrometer designs. It is simple as it does not require conversion of spherical wavefront to plane wavefront and vice versa. A concentric imaging relay has an excellent image quality and is compact and achromatic as it comprises only mirrors.

As a result of using only reflective elements, there are no chromatic aberrations in an Offner implementation. There is tangential field curvature (i.e., curvature of the best focal surface for tangential lines) that can be controlled at the expense of other aberrations by deviation from perfect concentricity. The convex diffraction grating can be produced by ruling or diamond turning but has most successfully been produced by electron-beam lithography.

Because of the asymmetry introduced by grating diffraction, a classical Offner spectrometer usually suffers from spectral-spatial distortions (keystone and smile). To cope with this, a split-Offner design is employed, where orientation of the two mirrors is slightly asymmetric. The use of this design has resulted in the imaging spectrometers with extremely low values of spatial-spectral distortion. On the other hand, with advancement in fabrication techniques and implementation of volume phase holographic (VPH) gratings, various advanced designs, like half-Offner and spherical transmission grating spectrometer (STGS), are now available and

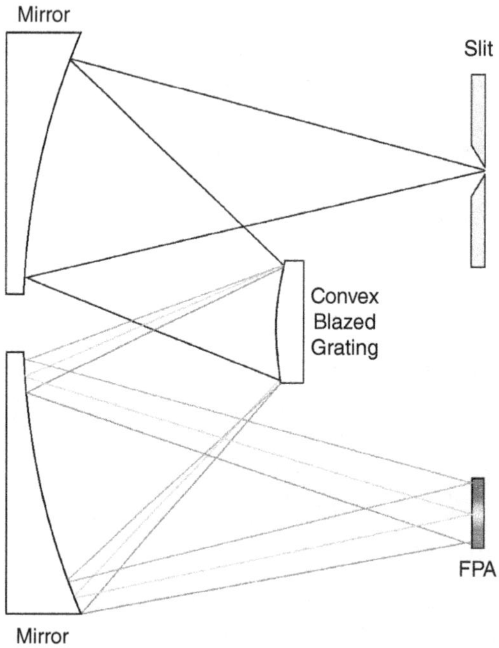

FIGURE 5.26 The layout of an Offner spectrometer.

adopted, which has resulted in superior performance compared to that of the classical Offner spectrometers (O'Donoghuea and Clemens 2014).

The Offner-type imaging spectrometers have no refractive components and find utility in very broad spectrum range applications. They have been used on a number of space missions, such as Cassini (Brown et at. 2004), VIRTIS for three planetary missions: Rosetta (Coradini et al. 2007), Venus-Express (Piccioni et al. 2008), and NASA-Dawn mission, M3 (Green et al. 2011) and HICO (Kappus et al. 2016). These imaging spectrometers have been described in detail in Chapter 2 except the Cassini.

However, because the Offner-type imaging spectrometers possess steep internal incidence angles (>20°), this makes them polarization sensitive. Due to larger incidence angles, the polarization sensitivity increases with decreasing f-number. Offner-type imaging spectrometers have a large amount of air spaces between the optical elements. This tends to make the system more bulky than the second category of imaging spectrometer types.

The optical elements in high-performance Offner-type imaging spectrometers generally do not share the same optical axis. Thus, their manufacture demands tight tolerances (i.e., spacings and tilts).

5.4.4.2 Dyson Spectrometers

Another concentric imaging design is the Dyson variety based on a spherical mirror, a thick double-pass plano-convex lens (hemispherical lens, often referred to as a Dyson block) and optional corrector optics. The curved lens surface and mirror are concentric and the image plane is located near the flat surface of the lens. The Dyson concept can be translated into a grating spectrometer by replacing the spherical mirror by a grating on the concave surface, as depicted in Figure 5.27 (Mouroulis et al. 2008). Dyson-type systems are superb examples of the second-category imaging spectrometers designs. They have much smaller volumes than Offner-types for the equivalent performance and throughput. A Dyson-form spectrometer is comparatively compact compared with the conventional prism and grating spectrometers. To minimize the spherical and chromatic aberrations, a meniscus corrector lens is frequently used.

Dyson-type systems have small incidence angles (typically <5°). Thus, they have less polarization sensitivity than Offner-type systems. Also, the Dyson lens and grating share a common optical axis. This allows for relaxed manufacturing tolerances. Diverse examples of Dyson-type imaging spectrometers are found in space missions, such as MERIS (Bezy et al. 2016), OLCI/Sentinel-3A (Nieke et al. 2016), HyspIRI (JPL 2018), HERO (Bergeron et al. 2008), and WaterSat (Qian et al. 2017). Since 2015, NASA/JPL scientists and engineers have developed a new hyperspectral imager

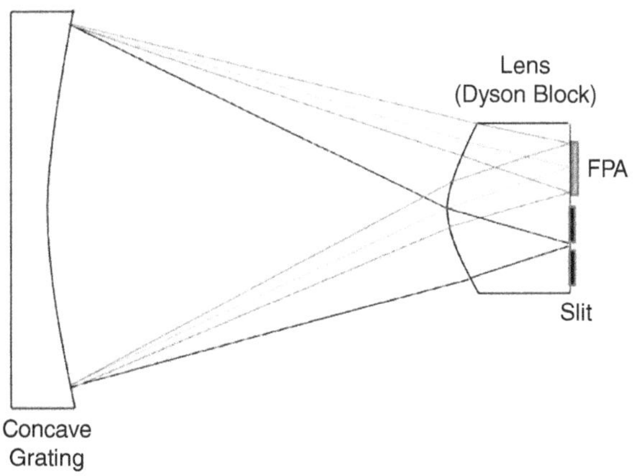

FIGURE 5.27 The layout of a Dyson spectrometer.

based on Dyson design aiming at evolving the Landsat satellites to hyperspectral capability by replacing the current multispectral payload Operational Land Imager (OLI) with a hyperspectral imager (Mouroulis et al. 2016).

A Dyson spectrometer requires the manufacture of a blazed reflection grating on a spherical surface, which is concave in this case. Compared to the prism and grating spectrometers, a Dyson spectrometer is more sensitive to fabrication tolerance and misalignment of components (Warren et al. 2008).

5.4.4.3 Dyson Spectrometer Versus Offner Spectrometer

Table 5.2 compares Dyson-type and Offner-type spectrometers for hyperspectral imagers by summarizing their advantages and disadvantages.

TABLE 5.2
Comparison of Dyson Spectrometers With Offner Spectrometers

	Dyson		Offner	
Parameter	**Advantage**	**Disadvantage**	**Advantage**	**Disadvantage**
Grating complexity		Blazed gratings more difficult to procure	Blazed gratings generally available (E-beam)	
Optical throughput	Comfortably handles f/# down to f/1			f/2 limit achieved with difficulty, size increases rapidly with aperture
Distortion	Can achieve very low levels of distortion		Can achieve very low levels of distortion	
Spectral and spatial uniformity	Can achieve high levels of spectral and spatial uniformity		Can achieve high levels of spectral and spatial uniformity	
Compactness	Very compact, smaller than Offner for the equivalent performance and throughput		Very compact	
Polarization Sensitivity	Less sensitive due to small incidence angles			Relatively more sensitive than Dyson, increase with decrease of f/#
Product of spectral resolution and spectral range	Higher than Offner			Relatively limited
Stray light		Refractive interfaces create ghosts and scatter, ghost reflections from detector must be considered	All reflective, typically no significant ghost problems	
Package		Proximity of object and image can cause severe packaging problems	Typically amenable to a variety of packaging options	

It can be seen from Table 5.2 that despite the Dyson-type spectrometers have highly desirable characteristics they still have two key shortcomings from the perspectives of implementation and stray light control. The first shortcoming is that there is very little back focal length to align and focus the slit, Dyson block, diffraction grating and FPA with respect to each other. Greater back focal length allows for a more forgiving, robust, and precise focusing and alignment process. The second shortcoming is that the FPA is prone to stray light from the full-spectrum light exiting the entrance slit and reflecting off the Dyson's lens' convex surface. The degree of stray light is dependent on the slit-FPA separation, f-number, Dyson lens material (index of refraction and glass quality—bubbles and inclusions, striae, stress birefringence, etc.), and the efficiency of the Dyson lens' antireflection coating. Even with the theoretically best available ultra-broadband antireflection coatings (>99.5% transmission across 360–1050 nm), the FPA will experience measurable stray light.

Achal (2013) invented a modified-Dyson spectrometer that addresses these shortcomings. The modified-Dyson spectrometer maintains the strong attributes of the Dyson format. It efficiently blocks and baffles stray light resulting from the reflection of the full-spectrum light off the spectrometer's curved refractive surfaces. It also has much greater back focal length to facilitate the precise focusing and alignment of all the spectrometer's components. In addition, the modified-Dyson design utilizes two thin lenses, instead of one very thick Dyson lens. The thinner lenses have better thermalization characteristics and are lighter than the Dyson lens. This modified-Dyson design has been adopted and built in the airborne demonstrator of the WaterSat, the Canadian coastal and inland water color hyperspectral microsatellite (Achal 2019). ITRES Research Ltd. plans to use this modified-Dyson design in its future ultraviolet (UV), VNIR, SWIR, visible-shortwave infrared (vis-SWIR), mid-wave infrared (MWIR), and very longwave infrared (VLWIR) airborne hyperspectral systems.

5.4.5 Spectrometer Design Examples

This section presents design examples of six spectrometers for hyperspectral imagers:

1. A Dyson spectrometer for VNIR
2. A Dyson spectrometer for SWIR
3. A grating spectrometer for VNIR
4. A grating spectrometer for SWIR
5. A prism spectrometer for SWIR
6. A Dyson spectrometer for VNIR with a larger entrance pupil

Table 5.3 lists the basic requirements of the spectrometers. Smile and keystone are distortions of particular importance to hyperspectral imagers. A hyperspectral imager operating in the pushbroom mode exhibits spectral line curvature, sometimes referred to as slit curvature, and commonly in the remote sensing community as spectral "smile." Generally, these terms refer to the misregistration of

TABLE 5.3
Optical Requirements of Spectrometers

Parameter	Requirement	Description
Spectral distortion (Smile)	0.1 pixel	Departure from straightness of the swath image
Spatial distortion (Keystone)	0.1 pixel	Swath length error
Spectral linearity	0.5 nm	Departure from linearity of the swath position on detector versus wavelength
MTF	>0.7	At the Nyquist limit

the detector elements across the FOV of the imager in the spectral dimension (Qian 2013). The smile can also be viewed as the deviation of the swath image from the straight line that passes through and is tangential to the swath image at its center. Variation of smile with wavelength is especially important, because it creates numerical problems resolving the spatial images into discrete spectral wave bands.

In addition to the spectral distortion described above, there is also spatial distortion in hyperspectral imagers. This spatial distortion exhibits inter-band spatial misregistration along the spectral direction. This spatial distortion is often referred to as "keystone" in the hyperspectral remote sensing community. Keystone is the variation of slit magnification with wavelength, but it is measured as an absolute length value or fraction of pixel patch size of a detector array. The keystone distortions may be caused by geometric distortion, as can be seen in camera lenses, or by chromatic aberration, or a combination of both. In general terms, it refers to the cross-track spatial misregistration of the imaged pixels of the ground samples of the various spectral bands of the spectrograph (Qian 2013).

The spectral linearity is defined as the variation of spectral sampling interval (SSI) with respect to wavelength. It is calculated from the derivative of swath-image position with respect to wavelength. The derivative is estimated from a five-number forward- or backward-difference formula for wavelengths at either end of the spectral range and from a four-number central-difference formula for other wavelengths. The SSI is then the reciprocal of the derivative, scaled to units of nanometer per pixel.

In general, prism spectrometers have a relatively good throughput but poor spectral linearity. Grating spectrometers have a relatively poor throughput for the required spectral bandwidths but good spectral linearity.

MTF is a measure of image sharpness and is as important for hyperspectral imagers as for any other imaging system. The MTF data at the Nyquist limit will be provided for each spectrometer design example.

At the end of the section, a comparison of the designs is given.

5.4.5.1 A Dyson Spectrometer for VNIR

This design assumed an unobstructed off-axis TMA telescope as the fore-optics, which images a cross-track line on the ground and focuses it on the slit of the spectrometer. The designed Dyson spectrometer is depicted in Figure 5.28.

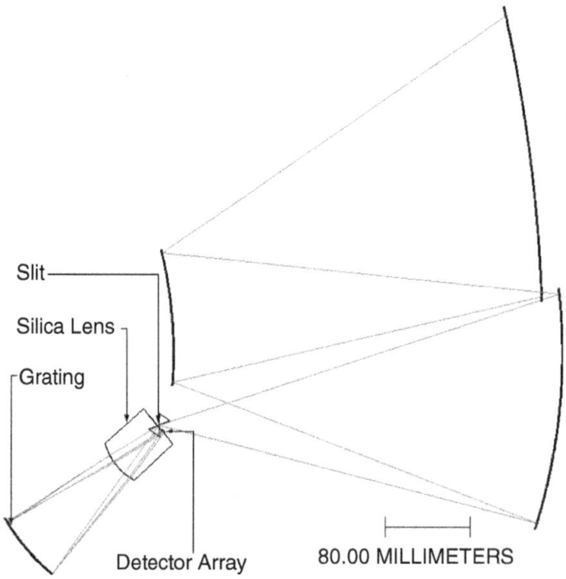

FIGURE 5.28 The optical layout of a Dyson spectrometer for VNIR.

The designed spectrometer comprises a concave reflective grating and a thick, fused-silica field-flattener (i.e., a Dyson block). This arrangement retro-reflects the spectrum adjacent to the spectrometer slit.

The Dyson spectrometer form has some interesting trade-offs. The design offers excellent correction of distortions together with a very simple design. However, packaging the slit and detector is a challenge. The design has a total-internal reflection scheme to facilitate packaging the slit and the detector. There is also a ghost image that needs control. It is produced from the curved surface of the field-flattener.

This spectrometer has the following paraxial characteristics:

- Spectral range: 0.4–1.0 μm
- SSI: 7.55 nm
- FOV: 2.80°
- Entrance pupil diameter: 250 mm
- Focal plane array format: $1000 \times B_n$ with a detector pitch size of 22 μm × 22 μm, where B_n is the number of spectral bands desired.

The design analysis indicates that the maximum smile is <0.01 pixel for all the ground samples in the cross-track line within the spectral range. The keystone is <0.025 pixel in the spectral range 0.4–1.0 μm. The maximum spectral nonlinearity is 0.04 nm in the spectral range.

The MTF Nyquist limit of the spectrometer is 23 cl/mm for the detector pitch size of 22 μm × 22 μm. The MTF values at the Nyquist limit are better than 0.91, 0.95, and 0.92 at $\lambda = 0.4$, 0.7, and 1.0 μm, respectively.

This design meets all the requirements listed in Table 5.3.

5.4.5.2 A Dyson Spectrometer for SWIR

This design also assumed an unobstructed off-axis TMA telescope as the fore-optics, which images a cross-track line on the ground and focuses it on the slit of the spectrometer. The designed Dyson spectrometer is depicted in Figure 5.29.

This spectrometer has the following paraxial characteristics:

- Spectral range: 1.0–2.5 μm
- SSI: 7.55 nm
- FOV: 2.80°
- Entrance pupil diameter: 250 mm
- Focal plane array format: $1000 \times B_n$ with a detector pitch size of 22 μm × 22 μm, where B_n is the number of spectral bands desired.

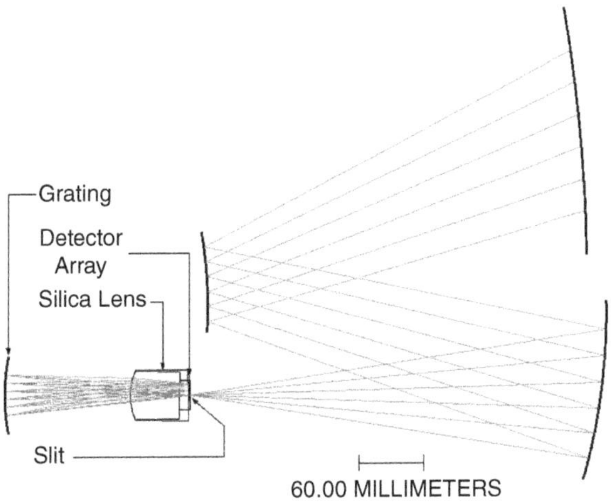

FIGURE 5.29 The optical layout of a Dyson spectrometer for SWIR.

The spectrometer design is very similar to the scheme used in Section 5.4.5.1. The design is very simple and achieves excellent correction of distortion. However, packaging the slit and detector is a challenge. These issues restrict the fore-optics to systems with a large clearance around the image.

The design analysis indicates that the maximum smile is <0.03 pixel for all the ground samples in the cross-track line within the spectral range of 1.0–2.5 μm. The keystone is <0.015 pixel in the spectral range 1.0–2.5 μm. The maximum spectral nonlinearity is 0.16 nm in the spectral range.

The MTF Nyquist limit of the spectrometer is 23 cl/mm. The MTF values at the Nyquist limit are better than 0.92, 0.88, and 0.83 at $\lambda = 1.0$, 1.75, and 2.5 μm.

This design meets all the requirements listed in Table 5.3.

5.4.5.3 A Grating Spectrometer for VNIR

This design assumed a centered TMA of 35 cm as the fore-optics. It comprises three axially centered mirrors, producing an image of a cross-track line on the ground on a curved slit to couple light into the spectrometer. The curved view of the slit profile also serves to null the distortion of the fore-optics.

The designed VNIR grating spectrometer is depicted in Figure 5.30. It comprises a flat grating within a highly decentered reflecting optical system. The figure also shows the unused parts of the mirrors, to make the geometry clear. In real implementation, the unused parts of the mirrors are removed to avoid blocking light and save the mass.

This spectrometer has the following paraxial characteristics:

- Spectral range: 0.4–1.0 μm
- SSI: 7.50 nm
- FOV: 2.64°
- Entrance pupil diameter: 350 mm
- Focal plane array format: $1000 \times B_n$ with a detector pitch size of 30 μm × 30 μm, where B_n is the number of spectral bands desired.

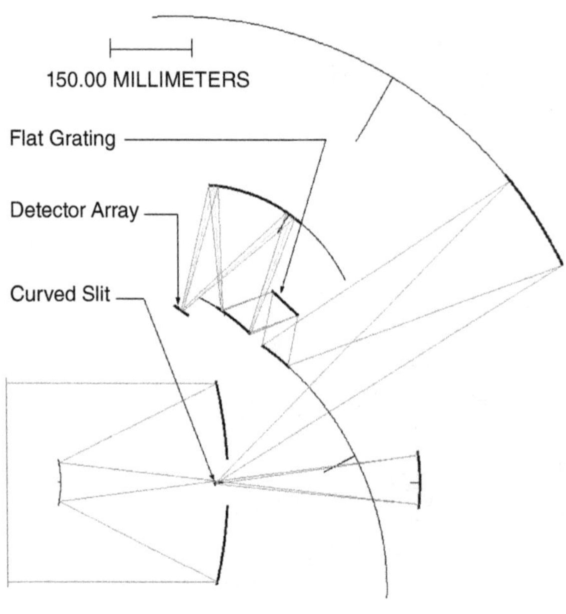

FIGURE 5.30 The optical layout of a grating spectrometer for VNIR.

The design analysis indicates that the maximum smile is <0.03 pixel for all the ground samples in the cross-track line within the spectral range. The keystone is <0.07 pixel in the spectral range 0.4–1.0 μm. The maximum spectral nonlinearity is 0.05 nm in the spectral range.

The MTF Nyquist limit of the spectrometer is 17 cl/mm for the detector pitch size of 30 μm × 30 μm. The MTF values at the Nyquist limit are better than 0.96, 0.95, and 0.92 at $\lambda = 0.4$, 0.7, and 1.0 μm.

These numbers show the design of the VNIR grating spectrometer is compliant with the requirements in Table 5.3.

5.4.5.4 A Grating Spectrometer for SWIR

This design assumed the same centered TMA of 35 cm, as discussed in Section 5.4.5.3, as the fore-optics. It comprises three axially centered mirrors, producing an image of a cross-track line on the ground on a curved slit to couple light into the spectrometer. The slit is slightly displaced from the position occupied by the VNIR slit.

The designed SWIR grating spectrometer is depicted in Figure 5.31. It comprises a flat grating of 40 g/mm within a highly decentered reflecting optical system.

This spectrometer has the following paraxial characteristics:

- Spectral range: 1.0–2.5 μm
- SSI: 7.48 nm
- FOV: 2.64°
- Entrance pupil diameter: 350 mm
- Focal plane array format: $1000 \times B_n$ with a detector pitch size of 30 μm × 30 μm, where B_n is the number of spectral bands desired.

The design analysis indicates that the maximum smile is <0.04 pixel for all the ground samples in the cross-track line within the spectral range. The keystone is <0.12 pixel in the spectral range 1.0–2.5 μm. The maximum spectral nonlinearity is 0.04 nm in the spectral range.

The MTF Nyquist limit of the spectrometer is 17 cl/mm. The MTF values at the Nyquist limit are better than 0.92, 0.90, and 0.85 at $\lambda = 1.0$, 1.75, and 2.5 μm.

These numbers show that the design of the SWIR grating spectrometer is compliant with the requirements in Table 5.3 except for the keystone that is 0.12 pixel, slightly larger than the required 0.1 pixel.

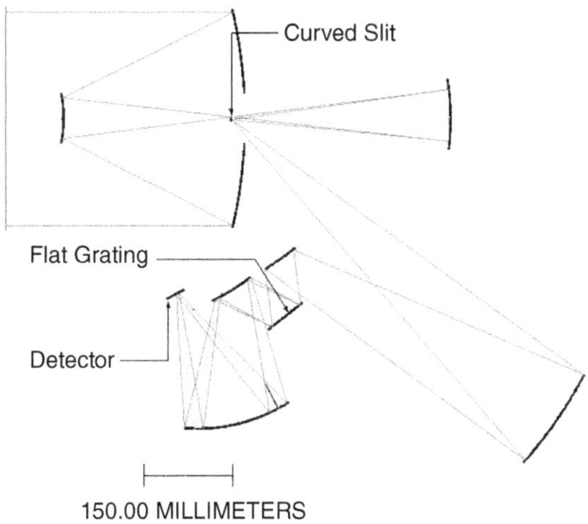

FIGURE 5.31 The optical layout of a grating spectrometer for SWIR.

5.4.5.5 A Prism Spectrometer for SWIR

This design assumed a centered TMA of 25 cm as the fore-optics. It comprises three axially centered mirrors and produces an image of a cross-track line on the ground on a curved slit to couple light into the spectrometer. The curved view of the slit profile also serves to null the distortion of the fore-optics.

The designed SWIR grating spectrometer is depicted in Figure 5.32. It consists of a prism with cylindrical faces enclosed in a highly decentered reflective optical system. The figure shows the complete parent surface of each decentered mirror to indicate the geometry. In a real instrument only the used portions of the mirrors will be presented.

This spectrometer has the following paraxial characteristics:

- Spectral range: 1.0 to 2.5 μm
- SSI: 15 nm
- FOV: 2.64°
- Entrance pupil diameter: 250 mm
- Focal plane array format: $1000 \times B_n$ with a detector pitch size of 30 μm × 30 μm, where B_n is the number of spectral bands desired.

The design analysis indicates that the maximum smile is <0.23 pixel for all the ground samples in the cross-track line within the spectral range. The keystone is <0.24 pixel in the spectral range 1.0–2.5 μm. Both the smile and keystone do not satisfy the requirement of 0.1 pixel.

The maximum spectral nonlinearity is 4.0 nm in the spectral range, which is two orders of magnitude larger than that of the VNIR spectrometers described in Sections 5.4.5.1, 5.4.5.3 and Sections 5.4.5.4, 25 times larger than that of the SWIR spectrometer described in Sections 5.4.5.2. It has been discussed in Section 5.4.3 that a prism-based spectrometer has an inherent disadvantage of nonlinear spectral dispersion.

The MTF Nyquist limit of the spectrometer is 17 cl/mm for the detector pitch size of 30 μm × 30 μm. The MTF values at the Nyquist limit are better than 0.83, 0.86, and 0.72 at $\lambda = 1.0$, 1.75, and 2.5 μm.

This design does not meet the requirements listed in Table 5.3.

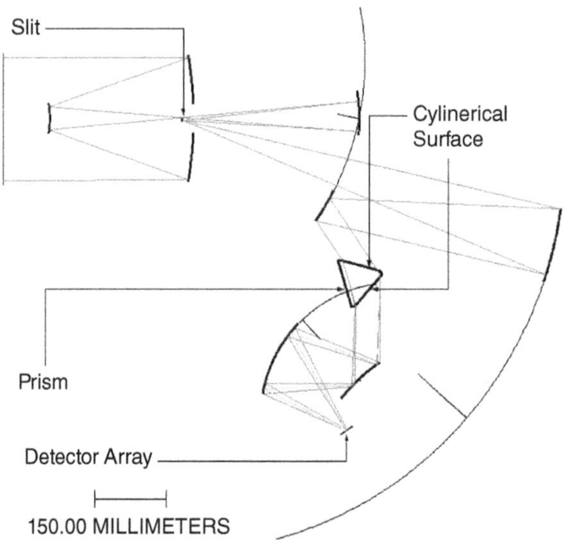

FIGURE 5.32 The optical layout of a prism spectrometer for SWIR.

5.4.5.6 A Dyson Spectrometer for VNIR With a Larger Entrance Pupil

This design assumed an unobstructed off-axis TMA telescope as the fore-optics, which images a cross-track line on the ground and focuses it on the slit of the spectrometer. The designed Dyson spectrometer is depicted in Figure 5.33.

The spectrometer design is similar to the Dyson spectrometers described in Sections 5.4.5.1 and 5.4.5.2 but with a larger entrance pupil diameter and for a larger detector pitch size of 30 μm × 30 μm.

This spectrometer has the following paraxial characteristics:

- Spectral range: 0.4–1.0 μm
- SSI: 7.51 nm
- FOV: 2.64°
- Entrance pupil diameter: 363 mm
- Focal plane array format: $1000 \times B_n$ with a detector pitch size of 30 μm × 30 μm, where B_n is the number of spectral bands desired.

The design analysis indicates that the maximum smile is <0.01 pixel for all the ground samples in the cross-track line within the spectral range. The keystone is <0.047 pixel in the spectral range 0.4–1.0 μm. The maximum spectral nonlinearity is 0.04 nm in the spectral range.

The MTF Nyquist limit of the spectrometer is 17 cl/mm. The MTF values at the Nyquist limit are better than 0.93, 0.96, and 0.93 at $\lambda = 0.4$, 0.7, and 1.0 μm.

This design meets all the requirements listed in Table 5.3.

5.4.5.7 Comparison of the Design Examples

In this section, the spectrometer designs are compared in terms of the following categories:

1. Throughput
2. Optical correction and distortions
3. Manufacturability

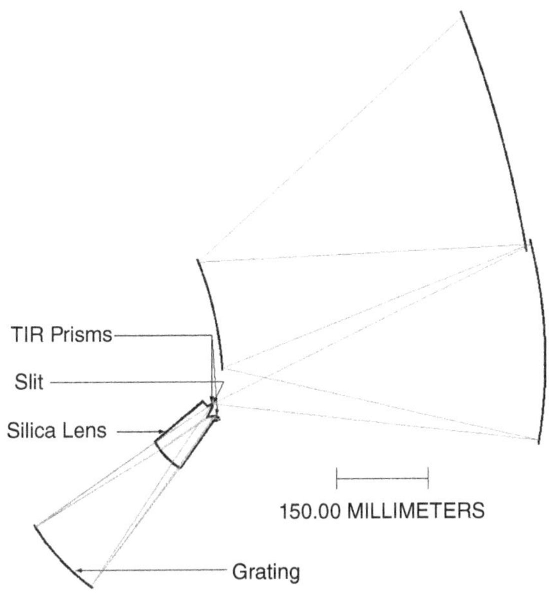

FIGURE 5.33 The optical layout of a Dyson spectrometer for VNIR with a larger entrance pupil.

TABLE 5.4
Comparison of Throughput-Related Characteristics of the Spectrometers

Spectrometer Design	Pupil Area (cm^2)	Reflective Surfaces	Transmissive Surfaces	Dispersion	Transmission
Dyson spectrometer for VNIR	491	3	4	Grating	55%
Dyson spectrometer for SWIR	491	3	4	Grating	55%
Grating spectrometer for VNIR	922	8	0	Grating	51%
Grating spectrometer for SWIR	922	8	0	Grating	51%
A prism spectrometer for SWIR	466	8	2	Prism	84%
Dyson spectrometer for VNIR with larger entrance pupil	1035	3	4	Grating	55%

1. **Instrument Throughput**
   In the calculation of the throughput, following assumptions are used:
   - The mirrors have a reflectance of 98%
   - Total internal reflection has a reflectance of 100%
   - Glass/air surfaces have a transmission of 99.5%
   - Gratings have a mean reflectance of 60% over the spectral range
   - Material absorption is negligible

   These assumptions are speculative and may not match the real cases. They are only for facilitating the comparisons. Table 5.4 lists the throughput related characteristics of the designs.
2. **Optical Correction**
   Table 5.5 lists items related to the optical correction of the spectrometer designs. The comparisons in the table require some judgment and the classifications there have the following meaning:
   - Excellent: significantly exceeds the requirement
   - Good: meets the requirement
   - Poor: does not meet the requirement
3. **Spectrometer Manufacturability**
   Prediction of manufacturability is a matter of judgment. Following are the assumptions used:
   - Aspheric mirrors are readily produced in metal on commercial diamond-turning lathes.
   - The world manufacture markets have commercial diamond-turning lathes that handle parts up to 700 mm in diameter.

TABLE 5.5
Optical Correction of the Spectrometers

Spectrometer Design	Smile	Keystone	Spectral Linearity	MTF (Shortwave)	MTF (Mid-Wave)	MTF (Long-Wave)
Dyson spectrometer for VNIR	Excellent	Excellent	Excellent	Excellent	Excellent	Excellent
Dyson spectrometer for SWIR	Excellent	Excellent	Excellent	Excellent	Excellent	Excellent
Grating spectrometer for VNIR	Excellent	Excellent	Excellent	Excellent	Excellent	Excellent
Grating spectrometer for SWIR	Excellent	Poor	Excellent	Excellent	Excellent	Excellent
Prism spectrometer for SWIR	Poor	Poor	Poor	Good	Good	Good
Dyson Spectrometer for VNIR with larger entrance pupil	Excellent	Excellent	Excellent	Excellent	Excellent	Excellent

TABLE 5.6
Manufacturability of the Powered-Optics of the Spectrometers

Spectrometer Design	Aspheric Surfaces	Only Spherical Polishing	Can Be Made on a Commercial Diamond-lathe	Can Be Custom-made
Dyson spectrometer for VNIR	No	Yes	N/A	Yes
Dyson spectrometer for SWIR	No	Yes	N/A	Yes
Grating spectrometer for VNIR	Yes	No	No	Yes
Grating spectrometer for SWIR	Yes	No	No	Yes
Prism spectrometer for SWIR	Yes	No	No	Yes
Dyson spectrometer for VNIR with larger entrance pupil	No	Yes	N/A	Yes

In Table 5.6, which lists some manufacturing issues with the powered optics, a "No" in the commercial diamond-lathe column indicates that some of the required mirrors have apertures too large from the rotational axis for the capacity of the largest commercial diamond-turning lathes.

5.5 SLIT ASSEMBLY AND BEAM SPLITTER

As described in Chapter 2, all the grating- or prism-based hyperspectral imagers consist of slit-scan spectrometers. In this kind of hyperspectral imagers, the telescope images the ground scene and focuses the image onto a slit assembly, which limits the ground scene to a cross-track line. There are generally two approaches to distributing the light to the VNIR and SWIR spectrometers: (1) directly by the slit assembly, or (2) after the light passes through the slit and gets separated internally by a beam splitter (i.e., a dichroic filter).

For example, in the optical design of the PRISMA hyperspectral imager depicted in Figure 5.25, the light focused on the slit by the telescope has travelled via mirror AM1 and correction lenses D1 SM1 before being separated to the VNIR and SWIR spectrometers inside the instrument. The light coming from the folding mirror 1 (FM1) transmits through the dichroic beam splitter (DBS) for the SWIR channel and reflects for the VNIR channel (Meini et al. 2016).

The hyperspectral image of the HISUI mission is an example of light distribution using a slit assembly. As depicted in Figure 2.20, the telescope feeds the ground scene image on the slit assembly, which contains two parallel slits, one transmitting to the VNIR spectrometer, another reflecting to the SWIR spectrometer (Matsunaga et al. 2016).

5.5.1 Distributing Light Using a Slit Assembly

As an example, the hyperspectral imager of the EnMAP satellite uses a slit assembly of dual aperture to distribute light to VNIR and SWIR spectrometers (Kaufmann et al. 2016). A transmissive aperture design for the slits is adopted to separate the telescope compartment to the spectrometers for good control of the stray light and polarization sensitivity. Figure 5.34 illustrates the principle design of the slit assembly. There are two parallel slits, each of which is 24-μm wide and 24-mm long. Light from the telescope passes the slit assembly. After passing through the two slits, the radiation for the SWIR spectrometer is redirected by a gold-coated micro-mirror to the SWIR spectrometer. The VNIR is passed to the spectrometer without further reflection. In this way, the incoming light is separated. The distance between the two slits must be kept as small as possible in order to keep the separation between the spectrometer FOVs as low as possible for good data

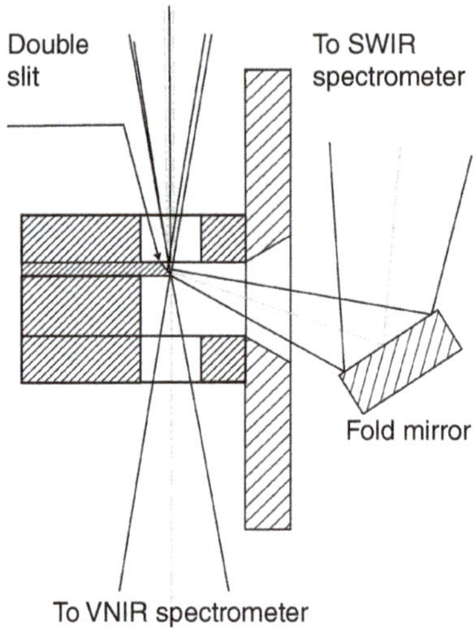

FIGURE 5.34 An example of distribution of light to VNIR and SWIR spectrometers using a slit assembly. (Courtesy of German Aerospace Center Space Agency.)

co-registration. Technological constraints limit the minimum distance between them being 480 μm, because the micro-mirror becomes increasingly fragile to manufacture and coat.

Figure 5.35 shows a design of another arrangement of the slit assembly of a hyperspectral imager for distributing light to the VNIR and SWIR spectrometers using two parallel slits of size 45-μm wide and 30-mm long. The slit assembly comprises two adjacent slits: a reflective slit for the SWIR spectrometer and a transmissive slit for the VNIR spectrometer. Both slits lie in the image plane of the fore-optics, perpendicular to the rotational axis. The specular, black-coated portions form part of the arrangement to suppress stray light. The width of the slits is 45 μm, which is 1.5 times larger than the width of the pitch size of the detector. This value comes from a scheme to reduce spectral aliasing.

5.5.2 Distributing Light Using a Beam Splitter

A beam splitter is an optical device for dividing a beam into two or more separate beams. A simple beam splitter may be a very thin sheet of glass inserted in the beam at an angle to divert a portion

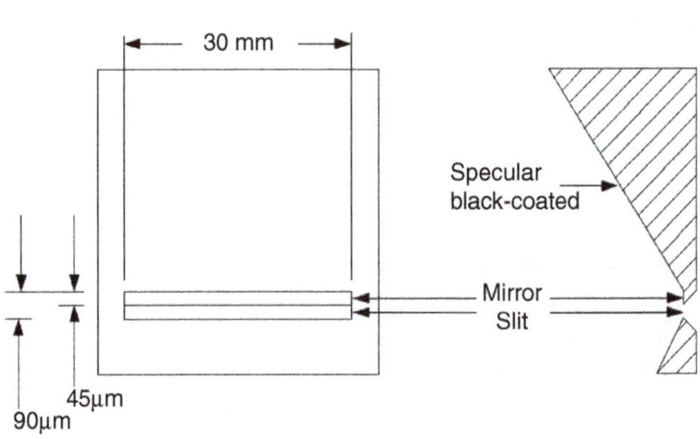

FIGURE 5.35 Arrangement of a slit assembly, two adjacent slits: a reflective slit for the SWIR spectrometer, and a transmissive slit for the VNIR spectrometer.

of the beam in a different direction. A more sophisticated type consists of two right-angled prisms cemented together at their hypotenuse faces. A dichroic mirror, also referred to as a dichroic filter, is the optical device commonly used as a beam splitter in imaging spectrometers to divert the beam to VNIR and SWIR spectrometers. It spectrally separates light by transmitting and reflecting light as a function of wavelength. A long-pass dichroic mirror is highly reflective below the cutoff wavelength and highly transmissive above it, while a short-pass dichroic mirror is highly transmissive below the cutoff wavelength and highly reflective above it.

Figure 5.36a depicts a design of a dichroic filter for a hyperspectral imager containing a VNIR spectrometer covering a spectral range of 380–1000 nm and a SWIR spectrometer covering a spectral range of 900–2540 nm. The dichroic filter reflects the light from 380 nm to 900 nm with a reflectivity R >95% and transmit the required SWIR spectral bands, 900–2540 nm, with a transition region near 900 nm. The transmission curve of the dichroic filter is shown in Figure 5.36b. The dichroic filter is a customized Commercial off-the-shelf (COTS) solution. It is made with established multilayer thin-film coating process based on alternating SiO_2 and ZrO_2 thin-film layers.

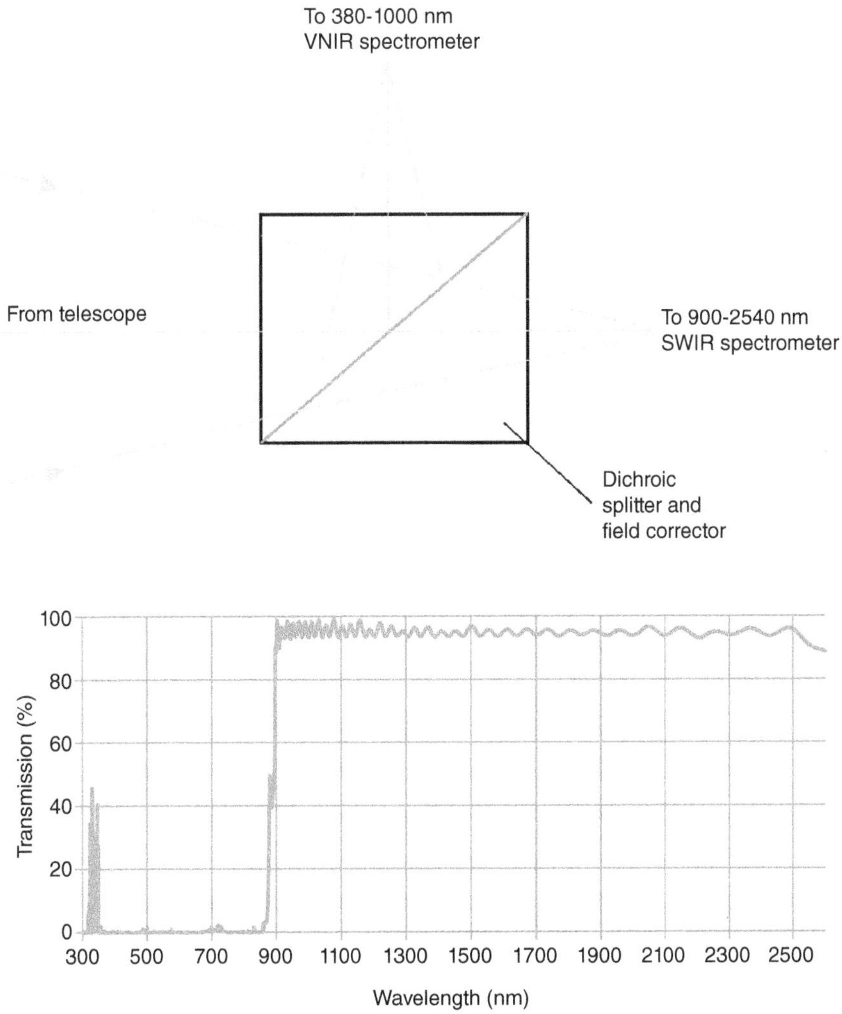

FIGURE 5.36 A dichroic beam-splitter and its transmission curve.

5.6 DEPOLARIZER

5.6.1 Polarization Sensitivity

Sensitivity to polarization is one of the major optical design drivers of dispersive imaging spectrometers, because it impacts the accuracy of the radiometric measurement, especially for ocean and water color applications given the extremely low albedo of the observing targets. Polarization is a major source of radiometric error for spaceborne spectrometers. For example, a 5% polarization sensitivity could add additional 4% error to the total radiometric error in water color applications. Very low overall polarization sensitivity is critical for successful high-quality data retrieval.

While the measured Earth radiance observed from space in the UV, visible, and near-IR regions has a strong and highly variable linearly polarized component, most essential components that are inherently sensitive to polarization are scan mirrors, gratings, and dichroics in the imaging spectrometers. Polarization sensitivity can also come from coatings on the optical elements, and other sources.

Minimization of the resulting radiometric errors is a challenge and cannot be achieved only with careful optical designs. Depolarization by using passive optical components, such as birefringent polarization scramblers, has been demonstrated in the last generation of spaceborne hyperspectral imagers, such as the MERIS onboard ENVISAT (Bezy et al. 2016), and atmosphere-monitoring instruments, such as the ozone-monitoring instrument (OMI) onboard the Aura satellite for ozone observation (Levelt et al. 2006).

Polarization sensitivity of spectrographs tends to be more significant at the blue and red ends. The typical polarization sensitivity of water color sensors and multispectral sensors is around 2–5%. The MODIS and the visible infrared imaging radiometer suite (VIIRS) do not use depolarizer and are therefore sensitive to polarization. MODIS radiance measurements vary by up to ±5.4% for totally linearly polarized radiance, depending on the orientation of the sensor relative to the plane of polarization.

The polarization sensitivity of the Hyperspectral Imager for the Coastal Ocean (HICO) was 2% between 450 nm and 650 nm, 4% at 900 nm, and 15% at 350 nm (Kappus et al. 2016). For the MERIS instrument, the polarization sensitivity was about 3% L_{typ} in the absence of the depolarizer. Its polarization sensitivity was reduced to 0.25% L_{typ} when the depolarization scrambler was applied. MERIS's special instrument design for polarization allowed the user to make the assumption that the instrument is insensitive to polarization states, which simplifies and makes the retrieval of ocean color products much more robust (Bezy et al. 2016).

5.6.2 MERIS Type Depolarizer

Figure 5.37 depicts the configuration of the MERIS depolarizer (also called the scrambler). It was installed right behind the UV filter and in front of the entrance pupil at about 80 mm in front of the telescope, as shown in Figure 2.5. The MERIS depolarizer is made of three cemented crystal wedges, two wedges in quartz for depolarization and one wedge in fused silica for chromatic correction. Quartz has the property to alter the polarization state of the transmitted light. The change in polarization state depends on the thickness of quartz. The shape of the quartz wedge imposes polarization-dependent phase delays that vary over the pupil so that the exiting polarization state depends on the position of the pupil. The depolarizing effect is a function of wedge angles. Larger angles produce more cycles in change of polarization state across an optical aperture of given dimensions. Each quartz wedge produces a split of the ground image, which has the effect of reducing the resolution of the ground image. The two quartz wedges together produce four images in a square array. In MERIS scrambler design, the wedge angles have been selected to provide acceptable degradation on resolution with acceptable effective depolarization. The MERIS depolarizer made the instrument insensitive to the Earth's spectral radiance polarization status with a radiometric error lower than 0.25%.

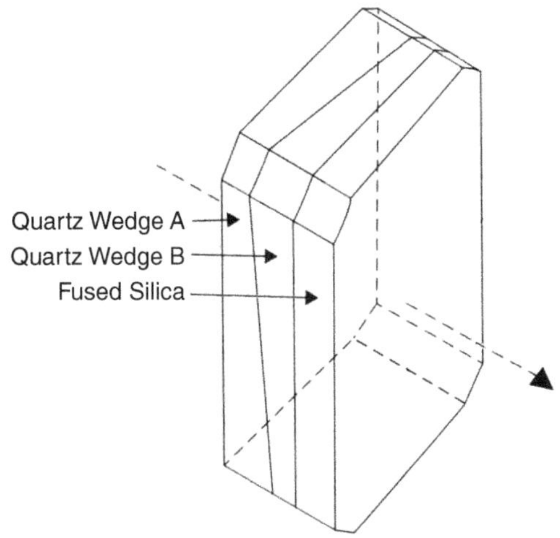

FIGURE 5.37 A MERIS depolarization scrambler.

A MERIS scrambler is very efficient in depolarizing the input illumination, but it shows rapid spectral oscillations in its residual polarization sensitivity, which prevent its use in atmospheric measurements. This is because in the MERIS scrambler, for each wedge, it was assumed that the thickness t depends on the pupil coordinates (x,y)

$$t(x,y) = t_0 + \Delta t(x, y) \tag{5.2}$$

with t_0 being the thickness at the pupil center; the phase difference between an ordinary ray and an extraordinary ray has the following form:

$$\Delta\varphi_{MERIS}(x,y) = \frac{2\pi}{\lambda}(n_e - n_o)\left[t_0 + \Delta t(x,y)\right] \tag{5.3}$$

The phase shift changes very rapidly with the wavelength due to the large thickness t_0 divided by λ. This is the origin of the fast oscillations.

5.6.3 Dual Babinet Depolarizer and Variant

A Dual Babinet depolarization scrambler is proposed, as shown on the right in Figure 5.38 (Caron et al. 2012). In the Dual Babinet scrambler, the wedges A and B in the MERIS design are replaced by two pairs of wedges W1 + W2 and W3 + W4 with orthogonally oriented crystal axes. Both pairs are rotated against each other by a specific angle (e.g., 30°). The ordinary ray in W1 becomes the extraordinary ray in W2, similarly for the pair W3 + W4. In addition, W1 and W2 (similarly for W3 and W4) have opposite wedge angles so that the assemblies W1 + W2 and W3 + W4 have parallel external sides. In this way, there is no need for a fused silica for chromatic correction. In the Dual Babinet scrambler, the phase difference between an ordinary ray and an extraordinary ray in the first pair of wedges becomes:

$$\Delta\varphi_{DB}(x,y) = \frac{2\pi}{\lambda}(n_e - n_o)2\Delta t(x,y) \tag{5.4}$$

Due to the much smaller thickness involved in this expression, the fast spectral oscillations that occurred in the MERIS scrambler now disappear.

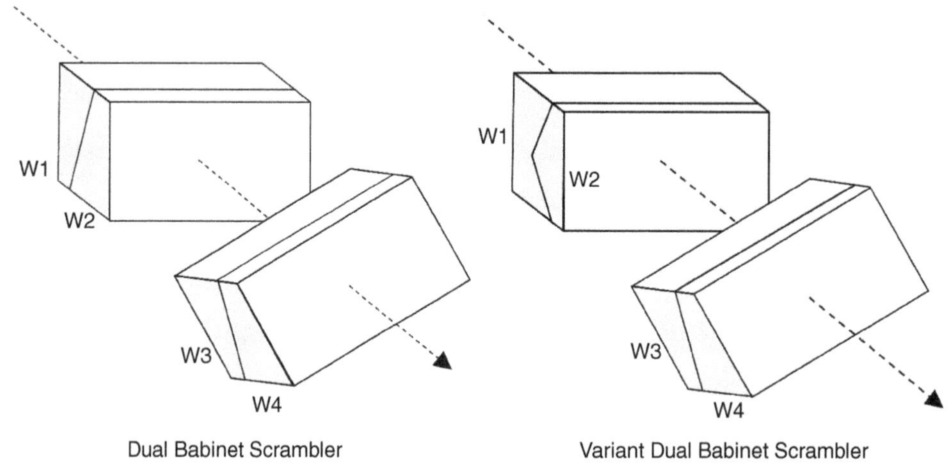

FIGURE 5.38 A Dual Babinet scrambler and a variant version.

The Dual Babinet scrambler splits the incident light in 4 mutually diverging linearly polarized beams with a far-field distribution corresponding to a split of each image point into a pattern of four (Windpassinger et al. 2014). The barycenter of the point pattern intensity profile shifts with the polarization angle. To avoid the shift of the image points and the co-registration error generated by the Dual Babinet scrambler, a variant of the Dual Babinet scrambler has been proposed (Windpassinger et al. 2014). The variant replaces the first pair wedges in the Dual Babinet scrambler with a block having a roof-shaped internal surface, as depicted in Figure 5.38, effectively representing the lateral combination of the two pairs with opposing wedge angles. The variant of the Dual Babinet scrambler eliminates the shift of the image points by superimposing two mutually mirror-inverted patterns. A similar design has also been reported by Kruizinga et al. (2013).

5.6.4 Examples of Depolarizers in Space Missions

The location of the depolarizer in the optical system is important, because the polarization introduced by the preceding optical components of the depolarizer will not be reduced. Therefore, the desired location of the scrambler is the foremost position of the optical system, if possible. Table 5.7 lists the spaceborne spectrometers that use depolarization scramblers. It can be seen from the table that both MERIS and its successor OLCI, onboard Sentinel-3A, have the depolarizer installed in front of the entrance pupil of the fore-optics, the foremost position of the optical system, to effectively reduce the polarization sensitivity of the subsequent elements of the optical system. The OMI used a Dual Babinet scrambler, which was installed inside the telescope.

The ultraviolet/visible/near-Infrared (UVN) instrument onboard Sentinel-4 is to provide continuous monitoring of the atmospheric composition at high temporal and spatial resolutions with respect to air quality, focusing on tropospheric gases in particular.

It is a hyperspectral spectrometer operating in UV-visible (305–500 nm) and near-infrared regions (750–775 nm) in the solar reflectance spectrum with a spatial sampling of 8 km × 8 km. It uses a variant of Dual Babinet scrambler, which is placed inside the telescope.

The imaging spectrometer UVNS (ultraviolet visible near-infrared shortwave infrared) to be deployed onboard Sentinel-5 uses a variant version of Dual Babinet scrambler, as depicted in Figure 5.38. The scrambler is placed inside the telescope, subsequent to mirror M4 right before the focused beam to the slit of the spectrometer. The scrambler, placed in the telecentric beam path, does not introduce any chromatic lateral errors. Sentinel-5 is an atmospheric monitoring mission within the European Copernicus programme. The UVNS is intended to provide a spatial resolution

TABLE 5.7
Spaceborne Imaging Spectrometers that Use a Depolarizer

	MERIS	OLCI	OMI	UVN	UVNS
Platform	ENVISAT	Sentinel-3A	AURA	Sentinel-4	Sentinel-5
Application	Land, ocean, atmosphere	Land, ocean, atmosphere	Ozone	Atmosphere	Atmosphere
Instrument overview	Medium spatial resolution, pushbroom, 5 cameras in fan-shape	Medium spatial resolution, pushbroom, 5 cameras in fan-shape	High spatial resolution, pushbroom	High spatial resolution, east-west scan mirror	High spatial resolution, pushbroom
Field of view	68.5° (14° per camera)	68.6° (14° per camera)	115°	4.2°	108.4°
Depolarizer	MERIS scrambler	MERIS scrambler	Dual Babinet scrambler	Variant of Dual Babinet scrambler	Variant of Dual Babinet scrambler
Depolarizer position	Before instrument pupil	Before instrument pupil	Inside telescope	Inside telescope	Inside telescope

of 7 km × 7 km on Earth (at Nadir) at a FOV of 108.4° in the cross-track direction corresponding to a ground swath width of about 2670 km. The instrument is implemented as a pushbroom spectrometer, scanning the Earth by measuring one cross-track of ground pixels at a time, and providing the respective spectrum for each pixel. The image dimension in the along-track direction is acquired by the satellite motion.

5.7 IMAGING SPECTROMETER OPTICAL DESIGN EXAMPLES

This section presents design examples of three imaging spectrometer systems. Their requirements are listed in Table 5.8. In the table, B_n is the number of spectral bands desired or to be defined, which have an impact on the spectral dispersion of the spectrometers and on the number of pixels of the

TABLE 5.8
Requirements of Three Imaging Spectrometer Systems

	System 1	System 2	System 3
Orbit altitude (km)	625	632	632
Swath width (km)	30	20	30
Ground sampling distance (GSD) (m)	30	20	20
Spectral sampling interval (SSI) (nm)	10	5	10
Focal plane format (pixel)	1000 × 62 (VNIR) 1000 × 153 (SWIR)	1000 × 120 (VNIR) 1000 × 256 (SWIR)	1500 × 60 (VNIR) 1500 × 160 (SWIR)
Detector pitch size (μm)	30	30	30
Spectral range (μm)	0.4–2.5	0.4–2.5	0.4–2.5
Spectral distortion (smile)	<0.1 SSI	<0.1 SSI	<0.1 SSI
Spatial distortion (keystone)	<0.1 GSD	<0.1 GSD	<0.1 GSD
Spectral sampling nonlinearity	<0.5 nm	<0.5 nm	<0.5 nm
Modulation transfer function	0.70	0.70	0.70
Spatial registration error	<0.1 of GSD	<0.1 of GSD	<0.1 of GSD
Spectral registration error	<0.1 of SSI	<0.1 of SSI	<0.1 of SSI

selected detector array in the spectral dimension. Each system uses two separate VNIR and SWIR spectrometers to cover the required spectral range.

The requirement of spatial registration error is defined as that images of a point-like object at any two wavelengths in the spectral range shall be co-registered within 10% of the GSD. This means that the peak-to-valley error should not exceed 10% GSD. The requirement of spectral registration error refers to that the absolute registration of each spectral band in the spectrometer shall be within 10% of the SSI.

The telescope images a cross-track line on the ground on the slit(s) of the SWIR and VNIR spectrometers. The three systems have a similar four-mirror telescope design. It comprises four, rotationally symmetric, aspheric mirrors, mounted in a coaxial fashion. The symmetry of the mirrors and the axis of rotation are depicted in Figure 5.46. However, the apertures are decentered and the used portions of the telescope mirrors are indicated by the heavier lines. The telescope is reflective and thus completely achromatic over the spectral range of 0.4–2.5 μm.

The length of the slit of the spectrometers corresponding to the cross-track line of images is 30 mm, the same as the width of the detector array in spatial dimension. Thus, the spectrometers operate at 1:1 conjugates. The distortion of the telescope is very small. A straight swath on the ground produces a straight image on the slit. The image surface is accurately plane. This characteristic interacts with the slit geometry to maintain a sharp focus at both sides of the slit and over the entire length.

5.7.1 Imaging Spectrometer System 1

5.7.1.1 System 1 Overview

Imaging Spectrometer System 1 is composed of a four-mirror axially symmetric telescope, a VNIR grating spectrometer, and a SWIR prism spectrometer. Table 5.9 lists the characteristics of the system. This design has a separated entrance pupil and can be used with an external steering scan mirror. The optical layout of System 1 is shown in Figure 5.39. The spectrometers for System 1 are very fast, operating at approximately *f*/1.5. Field lenses are necessary to make them function.

The optical paths are split between the two spectrometers by the adjacent reflective and transmissive slits. The reflective slit for the SWIR spectrometer, whereas the transmissive slit for the VNIR spectrometer. Figure 5.40 shows the arrangement of the slit assembly. The reflective slit comprises a mirror coating on one face of a small glass prism. One side of the slit is formed by the edge of the coating. The other side comes from the edge of the prism mount, bonded to the glass.

TABLE 5.9
Imaging Spectrometer System 1 Characteristics

Orbit altitude	625 km
Swath width	30 km
Ground sampling distance	30 m
Aperture diameter	380 mm
Central obstruction	108 mm × 96 mm (elliptical)
Stop position	800 mm from the primary mirror
Slit length	30 mm
Slit width	30 μm
Focal plane format (pixel)	1000 × 62 (VNIR) 1000 × 153 (SWIR)
Detector pitch size	30 μm × 30 μm

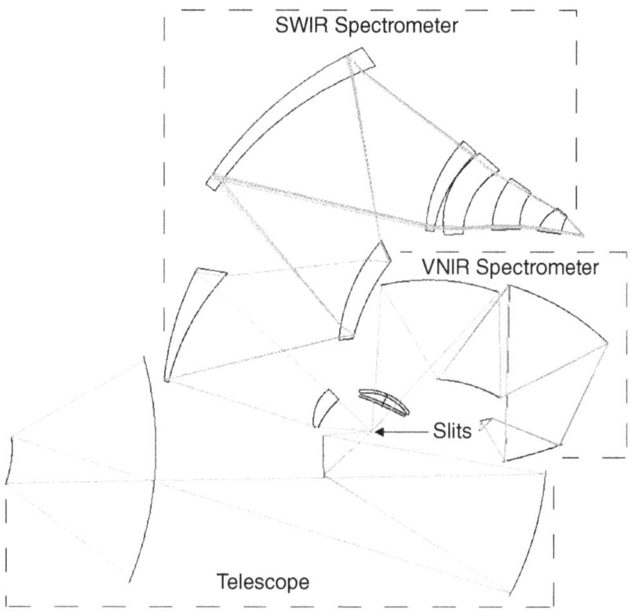

FIGURE 5.39 Configuration of imaging spectrometer System 1.

The transmissive slit is formed by the edge of the mirror coating and the edge of the prism. Both the reflective and transmissive paths are indicated by the rays.

5.7.1.2 System 1 VNIR Grating Spectrometer

Table 5.10 lists the parameters of the VNIR spectrometer of System 1. It is a grating-based spectrometer. Figure 5.41 shows its optical layout. The fore-optics is a four-mirror axially symmetric telescope. The symmetry of the mirrors and the axis of rotation are shown in Figure 5.46. The spectrometer configuration is primarily a rotationally symmetric all-reflecting design.

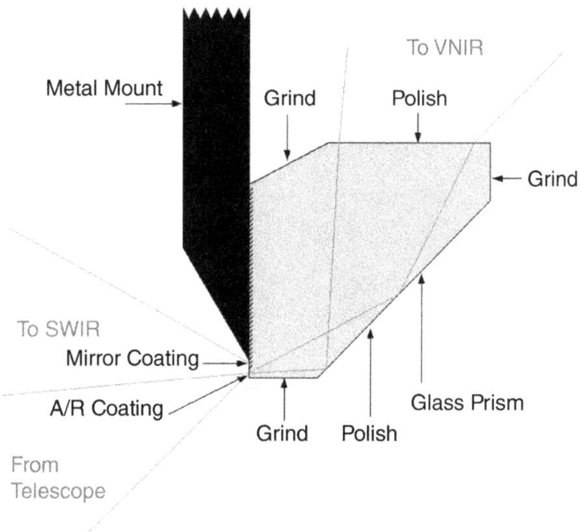

FIGURE 5.40 The slit assembly of Imaging Spectrometer System 1.

TABLE 5.10
System 1 VNIR Spectrometer Parameters

Dispersive Element	Diffracting Grating
Spectral range	0.4–1.025 μm
Spectral sampling interval	9.8–10.12 nm
Number of spectral bands	62

The rotational axis is indicated in the figure and the used portions of the mirror surfaces are shown by the heavier lines.

One departure from this design concept is the use of a refracting field lens. This is because the spectrometers for System 1 are very fast, operating at approximately *f*/1.5. Field lenses are necessary to make them function properly.

The dispersive element for this spectrometer is a diffraction grating, formed on one of the mirrors. The refractive dispersion from the field lens contributes a small (about 0.4-nm) departure from constant spectral sampling shown in Table 5.10.

5.7.1.2.1 Smile

For a single wavelength, the cross-track line on the ground should form a narrow and straight spectral band on the detector array, parallel to the spatial direction. In reality, spectral smile occurs. The spectral band deviates from straightness and forms distortion perpendicular to the swath image, in the spectral direction, as a function of the field (i.e., swath). Figure 5.42 shows the spectral distortion (smile) of the VNIR spectrometer at five spectral bands: 0.400, 0.556, 0.712, 0.868, and 1.025 μm for the semi-field (i.e., half swath of 15 km). The positions are calculated by ray tracing, as a geometric centroid. The data for each plot in the figure is the mean swath-image position, in the spectral direction, for the corresponding wavelength. It can be seen from the figure that the spectral distortion at the two ends of the spectral range (i.e., 0.400 and 1.025 μm) is relatively smaller. The peak-to-valley values are 0.033 and 0.023 pixel (i.e., SSI), respectively. The spectral distortion at the middle spectral bands (0.556, 0.712, and 0.868 μm) is relatively larger. The peak-to-valley values are 0.047, 0.044, and 0.035 pixel. The maximum smile of the VNIR spectrometer is 0.047 pixel, which is smaller than the requirement of 0.10 pixel listed in Table 5.8.

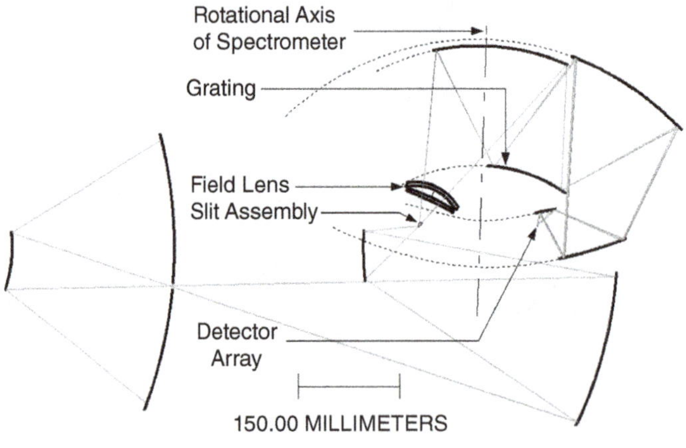

FIGURE 5.41 The optical layout of the VNIR grating spectrometer of System 1.

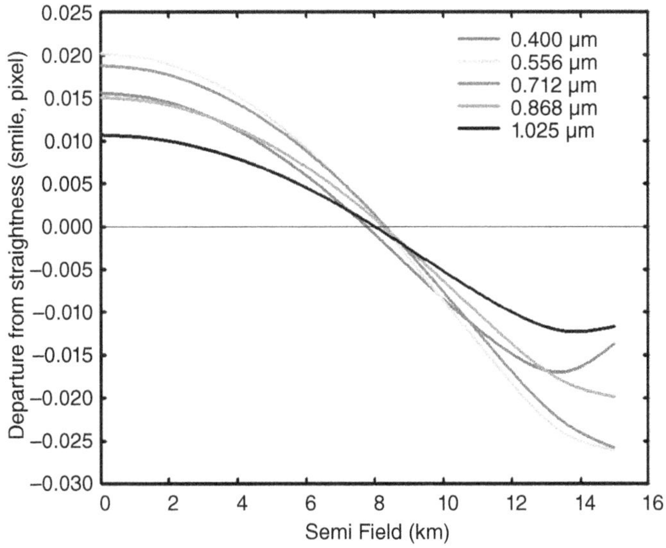

FIGURE 5.42 Spectral distortion (smile) of the VNIR spectrometer of System 1.

5.7.1.2.2 Keystone

For a single point target in the cross-track line (i.e., swath), the image for all wavelengths should form a straight line on the detector array in the spectral direction, perpendicular to the swath image. In reality, keystone distortion occurs. Deviations from this line produce distortions in a direction parallel to the swath image. The keystone calculation is based on a geometric centroid and uses the mean position for the corresponding field as a data.

Figure 5.43 shows the spatial distortion (keystone) at 4 spatial locations in the cross-track line. It can be seen from the figure that the keystone values increase with the spatial locations moving toward

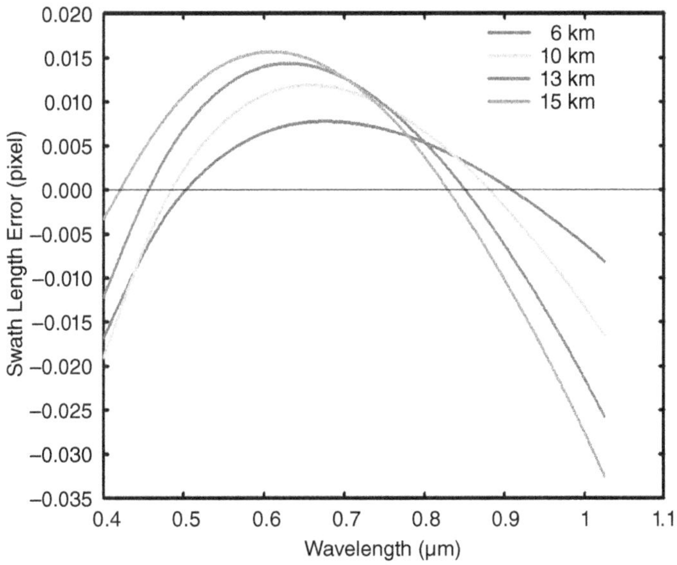

FIGURE 5.43 Spatial distortion (keystone) of the VNIR spectrometer of System 1.

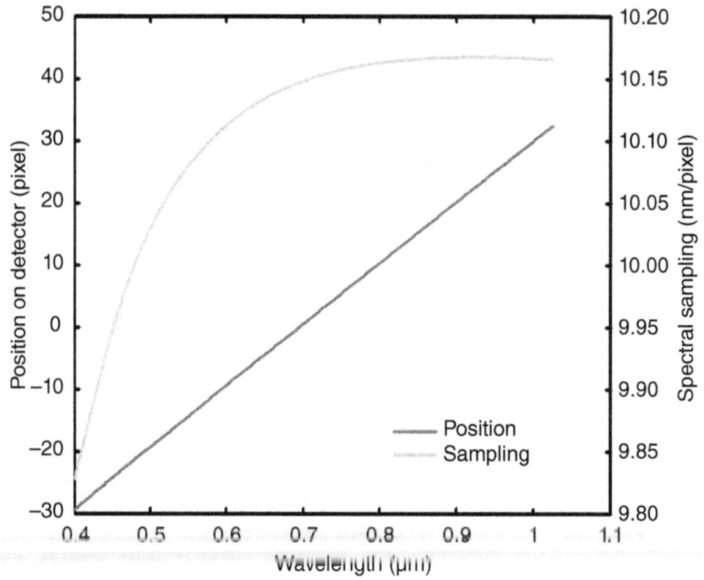

FIGURE 5.44 Spectral sampling interval (SSI) versus wavelength and spectral sampling nonlinearity of the VNIR spectrometer of System 1.

the edges of the swath. For the spatial locations of 6, 10, 13, and 15 km, the peak-to-valley keystone values are 0.032, 0.032, 0.040, and 0.048 pixel. The maximum keystone value is 0.048 pixel, which is smaller than the requirement of 0.10 pixel listed in Table 5.8.

5.7.1.2.3 Spectral Nonlinearity

Figure 5.44 shows the plot of SSI versus wavelength and the spectral sampling nonlinearity of the VNIR spectrometer in the spectral range. It can be seen that the SSI linearly varies from 9.80 nm to 10.12 nm. The maximum nonlinearity of spectral sampling is 0.2 nm, which meets the requirement of 0.5 nm listed in Table 5.8.

5.7.1.2.4 MTF

Figure 5.45 shows plots of the MTF of System 1 VNIR spectrometer, at three wavelengths: $\lambda = 0.4000$, 0.7125, and 1.0250 μm spanning the operating spectral range. The range of spatial frequencies is from 0 to 17 cycles per mm—approximately the Nyquist limit for a detector array with a 30-μm pitch. The plots within each figure correspond to various fields and azimuths. For all cases, the MTF is within 0.80 of the diffraction limit, except the worst MTF at wavelength of 0.4 μm, which is 0.77. All the MTF values are within the MTF requirement of 0.7.

5.7.1.2.5 Spatial Registration

The error of cross-track co-registration of the VNIR spectrometer is computed for 11 selected wavelengths (0.4000, 0.4625, 0.5250, 0.5875, 0.6500, 0.7125, 0.7750, 0.8375, 0.9000, 0.9625, and 1.0250 μm) versus 4 positions (6.0, 10.0, 13.0, and 15.0 km) in the FOV (i.e., swath). For each position and wavelength, the computed error is the difference from the mean swath position across the wavelengths and scaled to fractional nominal GSD. The peak errors are the maximum departures from the data. The maximum peak-to-valley error is 0.048 GSD, which is well within the requirement of 0.1 GSD listed in Table 5.8.

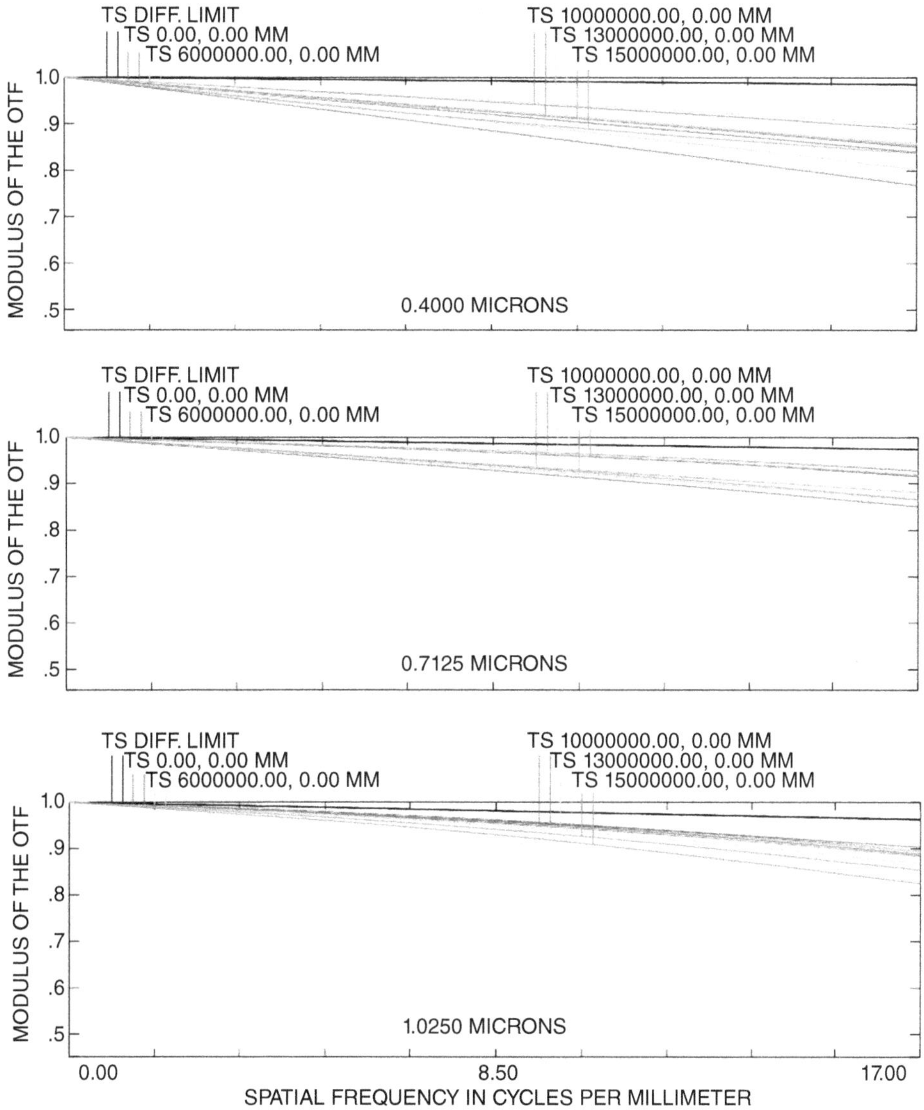

FIGURE 5.45 Plots of the MTF of VNIR spectrometer of System 1.

5.7.1.2.6 Spectral Registration

Five wavelengths 0.4000, 0.5563, 0.7125, 0.8513, and 1.0250 μm are selected as the mean wavelengths. Eleven positions in the swath from the center of the swath to one end with a 1.5-km gap are selected. The spectral registration is computed as the departure from the mean wavelength for the selected wavelengths against the selected cross-track positions. The errors are scaled as fractions of both the nominal pixel size and, by applying the dispersion for the wavelength, the SSI. The errors increase with cross-track positions moving the extreme field position. The maximum peak-to-valley error is 0.046 SSI, which comfortably satisfies the requirement of 0.1 SSI listed in Table 5.8.

5.7.1.3 System 1 SWIR Prism Spectrometer

Table 5.11 lists the parameters of the SWIR spectrometer of System 1. It is a prism-based spectrometer. Figure 5.46 shows its optical layout. There is an overlap of spectral coverage from

TABLE 5.11
System 1 SWIR Spectrometer Parameters

Dispersive Element	Combined Prism
Spectral range	0.975–2.5 μm
Spectral sampling interval	7.2–11.0 nm
Number of spectral bands	153

0.975 μm to 1.025 μm between the VNIR and SWIR for better calibration of the data generated by the two spectrometers.

The fore-optics is the same as the VNIR spectrometer. For information purpose, Figure 5.46 depicts the symmetry of the mirrors and the axis of rotation of the telescope. The apertures in the telescope are decentered and the used portions of the telescope mirrors are indicated by the heavier lines.

The SWIR spectrometer comprises three rear surface mirrors and a combined prims set, which consists of a series of wedged lenses. Each of these optical elements has spherical surfaces. The wedged nature of the elements provide the spectrometer's dispersion. The spectrometer is not axially symmetric, but it does have a plane of mirror symmetry.

5.7.1.3.1 Smile

Figure 5.47 shows the smile distortion of the SWIR spectrometer at five spectral bands: 0.975, 1.356, 1.737, 2.118, and 2.500 μm for the semi-field (i.e., half swath). The positions are calculated by ray tracing, as a geometric centroid. The data for each plot in the figure is the mean swath-image position, in the spectral direction, for the corresponding wavelength. It can be seen from the figure that smile distortion values at the five spectral bands show the same trend as the VNIR spectrometer depicted in Figure 5.42. That is, the smile distortion peak-to-valley values at the two SWIR ends (i.e., 0.975 and 2.500 μm) are relatively smaller. They are 0.039 and 0.045 pixel, respectively. The smile distortion peak-to-valley values at spectral bands 1.356, 1.737, and 2.118 μm are 0.049, 0.048,

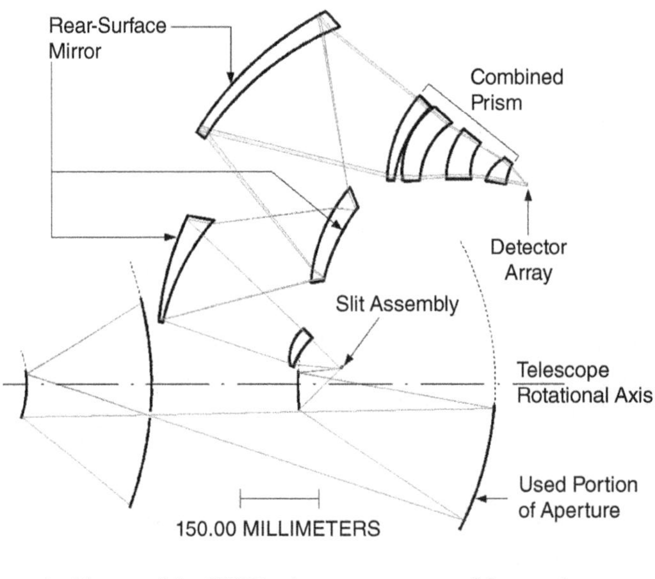

FIGURE 5.46 The optical layout of the SWIR prism spectrometer of System 1.

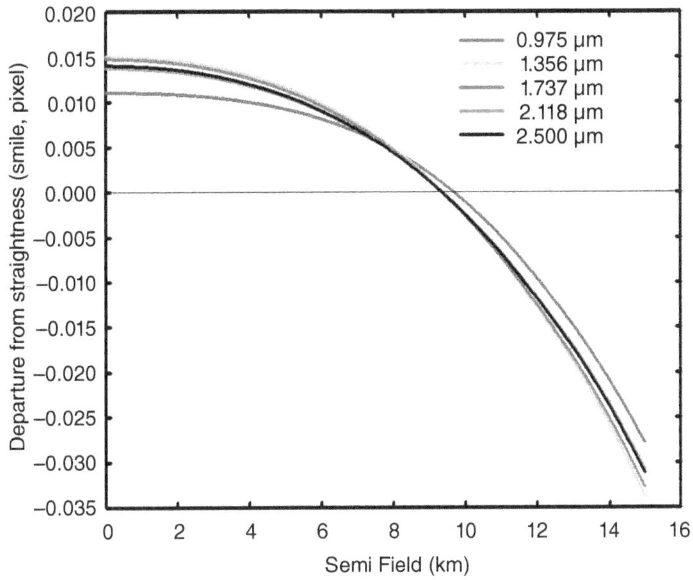

FIGURE 5.47 Spectral distortion (smile) of the SWIR spectrometer of System 1.

and 0.045 pixel. The maximum smile of the SWIR spectrometer is 0.049 pixel, which is smaller than the requirement of 0.10 pixel listed in Table 5.8.

5.7.1.3.2 Keystone

Figure 5.48 shows plots of keystone distortion of the SWIR spectrometer at 4 spatial locations in the cross-track line. For the spatial locations of 6, 10, 13, and 15 km, the peak-to-valley keystone values are 0.046, 0.057, 0.041, and 0.024 pixel. The maximum keystone value is 0.057 pixel, which is smaller than the requirement of 0.10 pixel listed in Table 5.8.

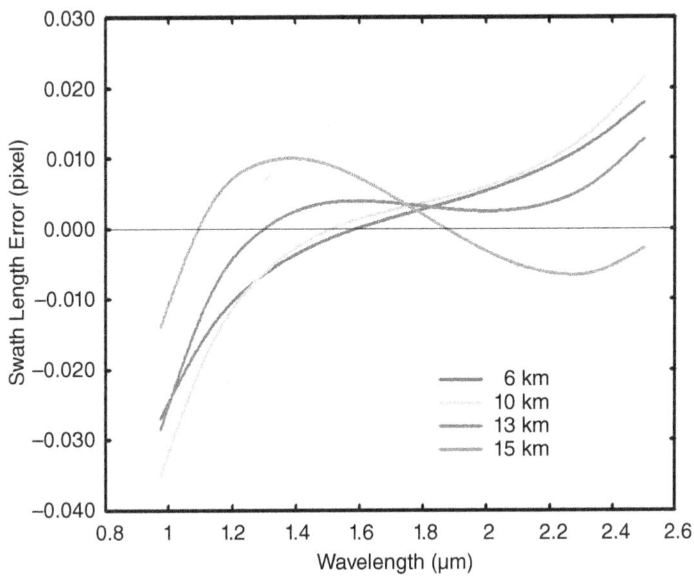

FIGURE 5.48 Spatial distortion (keystone) of the SWIR spectrometer of System 1.

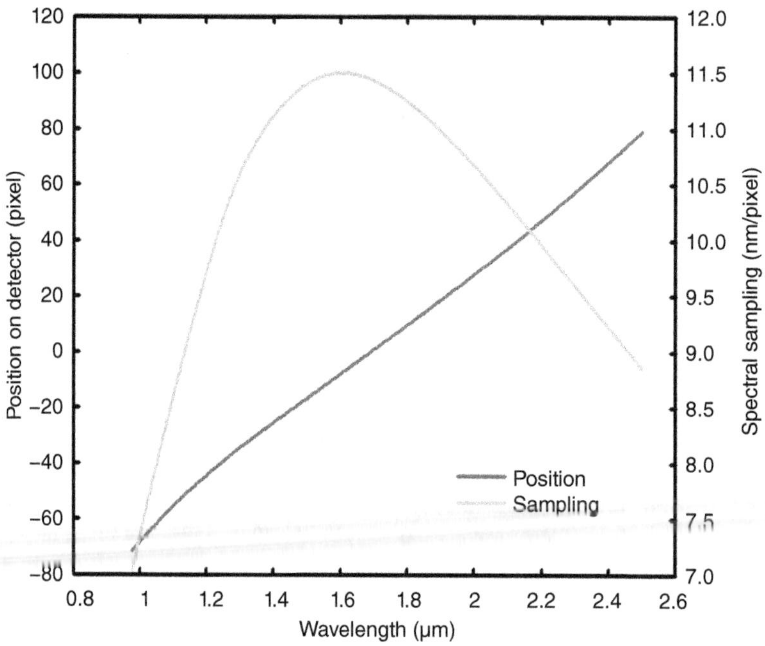

FIGURE 5.49 Spectral sampling interval (SSI) versus wavelength and spectral sampling nonlinearity of the SWIR spectrometer of System 1.

5.7.1.3.3 Spectral Nonlinearity

Figure 5.49 shows the plot of SSI versus wavelength and the spectral sampling nonlinearity of the SWIR spectrometer in the spectral range. It can be seen that the SSI linearly varies from 7.2 nm to 11.0 nm. The maximum nonlinearity of spectral sampling is 2.8 nm, which does not meet the requirement of 0.5 nm listed in Table 5.8.

It has been discussed in Section 5.4.3 that a prism-based spectrometer has an inherent disadvantage of nonlinear spectral dispersion. Recall the optical design of the prism-based SWIR spectrometer described in Section 5.4.5.5. In that case, the maximum spectral nonlinearity is 4.0 nm, which is even worse than the 2.8 nm in the current case. The reason for the improvement in the spectral nonlinearity in the current case is the use of the combined prism set that consists of a series of wedged lenses.

The VNIR and SWIR spectrometers of the EnMAP satellite also use prism as the disperser. To alleviate the inherent problem of spectral nonlinearity, a pair of two curved prisms is used in each spectrometer, and different materials were chosen for them. The SSI varies from 4.8 nm to 8.2 nm with an average of 6.5 nm over the full VNIR spectral range. For the SWIR spectrometer, an average SSI is 10 nm with variations of +20% and −25% over the SWIR spectral range (refer to Section 2.24 for detailed description of the EnMAP hyperspectral imager).

5.7.1.3.4 MTF

Figure 5.50 shows plots of the MTF of System 1 SWIR spectrometer, at three wavelengths: λ = 0.9750, 1.7375, and 2.5000 μm, for spatial frequencies up to the Nyquist limit (17 cl/mm) of the detector array. For each field, azimuth and wavelength combination, the MTF exceeds 80% of the corresponding diffraction-limited value. All the MTF values are within the MTF requirement of 0.7.

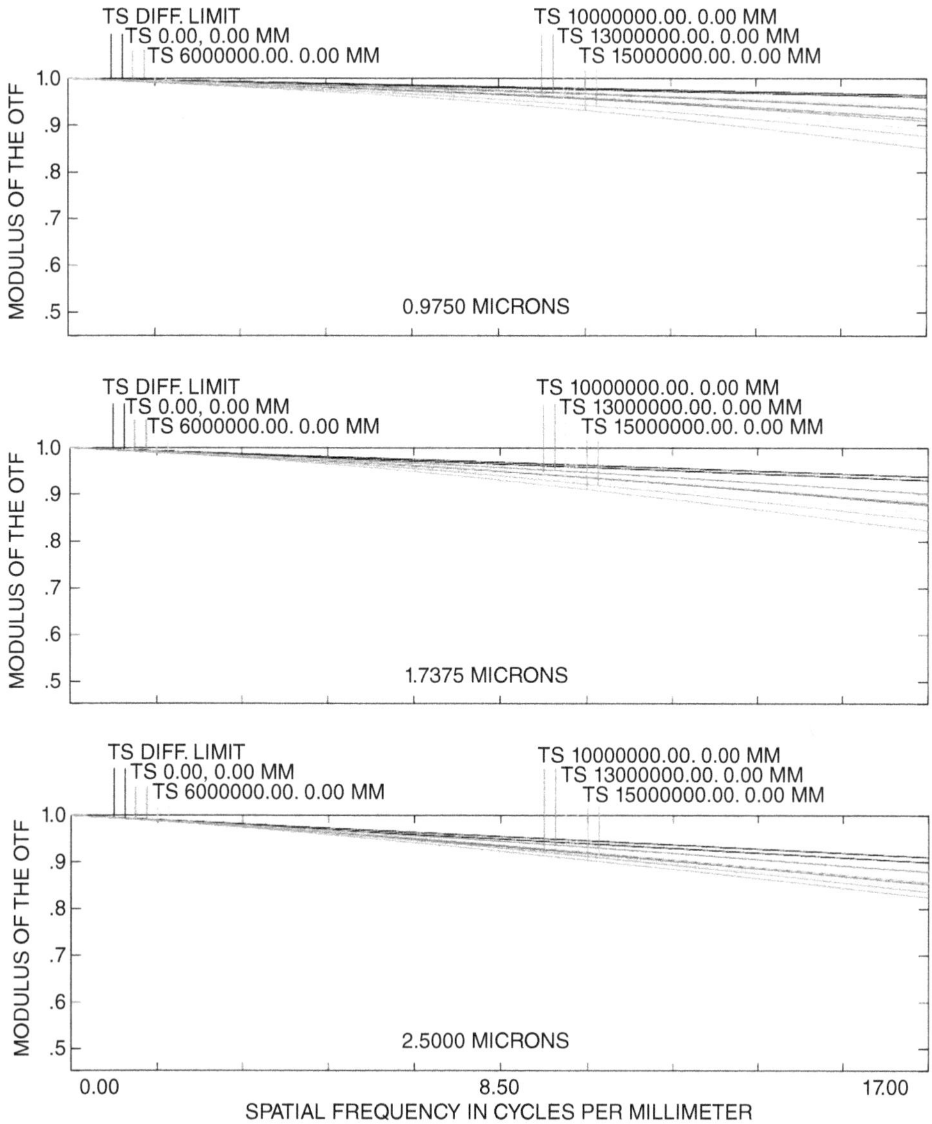

FIGURE 5.50 Plots of the MTF of SWIR spectrometer of System 1.

5.7.1.3.5 Spatial Registration

The error of cross-track co-registration of the SWIR spectrometer is computed for 11 selected wavelengths (0.90, 1.06, 1.22, 1.38, 1.54, 1.70, 1.86, 2.02, 2.18, 2.43, and 2.50 μm) versus 4 same positions in the FOV (i.e., swath) as selected for the VNIR spectrometer. For each position and wavelength, the computed error is the difference from the mean swath position across the wavelengths and scaled to fractional nominal GSD. The peak errors are the maximum departures from the data. The maximum peak-to-valley error is 0.056 GSD, which is well within the requirement of 0.1 GSD listed in Table 5.8.

5.7.1.3.6 Spectral Registration

Five wavelengths 0.9750, 1.3563, 1.7375, 2.1188, and 2.500 μm are selected as the mean wavelengths. The same eleven positions in the swath from the center of the swath to one end with a 1.5-km gap are selected. The spectral registration is computed as the departure from the mean wavelength for the selected wavelengths against the selected cross-track positions. The errors are scaled as fractions

TABLE 5.12
Imaging Spectrometer System 2 Characteristics

Orbit altitude	632 km
Swath width	20 km
Ground sampling distance	20 m
Aperture diameter	350 mm
Central obstruction	88 mm × 80 mm (elliptical)
Stop position	On the primary mirror
Slit length	30 mm
Slit width	30 μm
Focal plane Format (pixel)	1000 × 120 (VNIR) 1000 × 256 (SWIR)
Detector pitch size	30 μm × 30 μm

of both the nominal pixel size and by applying the dispersion for the SSI. The errors increase with cross-track positions moving the extreme field position. The maximum peak-to-valley error is 0.054 SSI, which comfortably satisfies the requirement of 0.1 SSI listed in Table 5.8.

5.7.2 Imaging Spectrometer System 2

5.7.2.1 System 2 Overview

Imaging Spectrometer System 2 is composed of a four-mirror axially symmetric telescope, a VNIR grating spectrometer, and a SWIR prism spectrometer. Table 5.12 lists the overall characteristics for System 2. Unlike System 1, the entrance pupil of System 2 is on the primary mirror. This arrangement is more compact but does not lend itself to operation with an external steering mirror. The design is suitable for an agile satellite configuration. System 2 relaxes the SWIR SSI to a nominal 6.25 nm to allow the use of the 256 rows of the detector array.

Figure 5.51 shows the layout of System 2. It comprises a four-mirror telescope, which images two adjacent swaths on the Earth's surface onto an adjacent pair of slits, a grating spectrometer for VNIR and a prism spectrometer for SWIR.

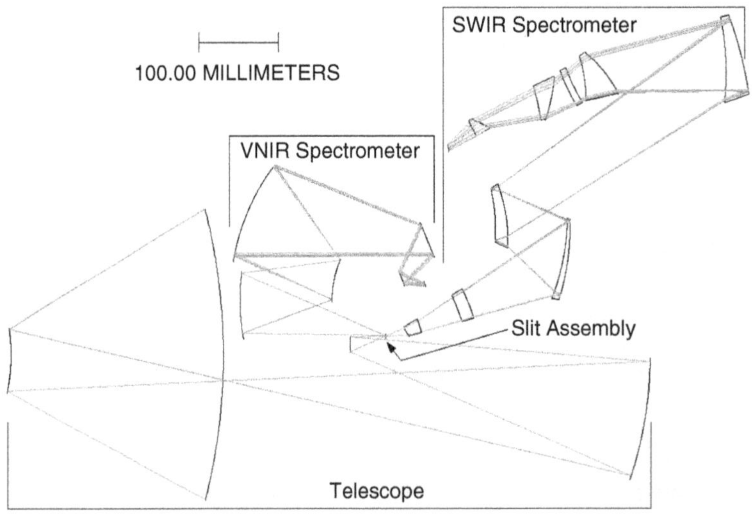

FIGURE 5.51 Configuration of imaging spectrometer System 2.

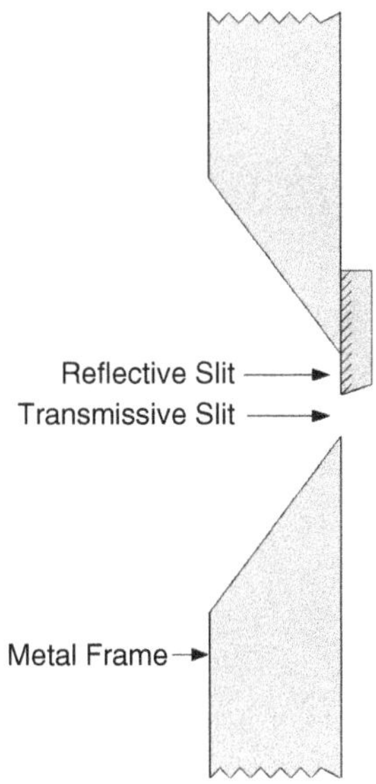

FIGURE 5.52 The slit assembly of imaging spectrometer System 2.

Figure 5.52 depicts the arrangement of the slit assembly. It comprises two, adjacent slits—one reflective and one transmissive. These two slits are arranged to intercept the adjacent swath images and pass them to the appropriate spectrometers. The reflective slit is for the VNIR spectrometer and the transmissive slit is for the SWIR spectrometer.

5.7.2.2 System 2 VNIR Grating Spectrometer

Table 5.13 shows the spectral characteristics of the System 2 VNIR spectrometer. It is a grating-based spectrometer. The design provides linear spectral sampling together with the required sampling value. Figure 5.53 shows the optical layout of the VNIR spectrometer, which comprises five front-surface mirrors in a rotationally symmetric arrangement. The diffraction grating is formed on one of the mirrors.

5.7.2.2.1 Smile

Figure 5.54 shows the smile distortion of the VNIR spectrometer at five spectral bands: 0.40, 0.55, 0.70, 0.85, and 1.00 μm for the semi-FOV (0–10 km). The positions are calculated by ray tracing, as a geometric centroid. The data for each plot in the figure is the mean swath-image position, in the spectral direction, for the corresponding wavelength. It can be seen from the figure that the peak-to-valley smile values are 0.044, 0.035, 0.023, 0.014, and 0.032 pixel at the five spectral bands. The maximum smile of the VNIR spectrometer is 0.044 pixel, which is smaller than the requirement of 0.10 pixel listed in Table 5.8.

TABLE 5.13
System 2 VNIR Spectral Characteristics

Dispersive Element	Diffracting Grating
Spectral range	0.4–1.0 μm
Spectral sampling interval	5.0–5.01 nm
Number of spectral bands	120

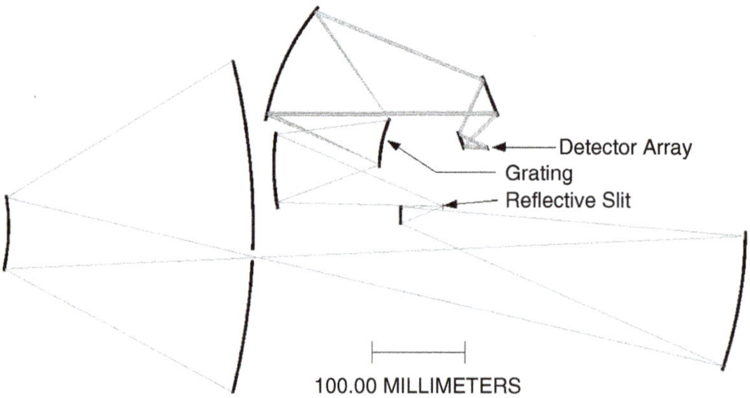

FIGURE 5.53 The optical layout of the VNIR grating spectrometer of System 2.

5.7.2.2.2 Keystone

Figure 5.55 shows the keystone distortion at 4 spatial locations in the in the half FOV (swath). It can be seen from the figure that the keystone distortion values are very small and close. For the spatial locations of 5, 7, 9, and 10 km in the half swath, the peak-to-valley keystone values are 0.007, 0.008, 0.008, and 0.007 pixel. The maximum keystone value is 0.008 pixel, which is over one order of magnitude smaller than the requirement of 0.10 pixel listed in Table 5.8.

5.7.2.2.3 Spectral Nonlinearity

Figure 5.56 shows the plot of SSI versus wavelength and the spectral sampling nonlinearity of the VNIR spectrometer of System 2 in the spectral range. It can be seen that the SSI linearly varies from

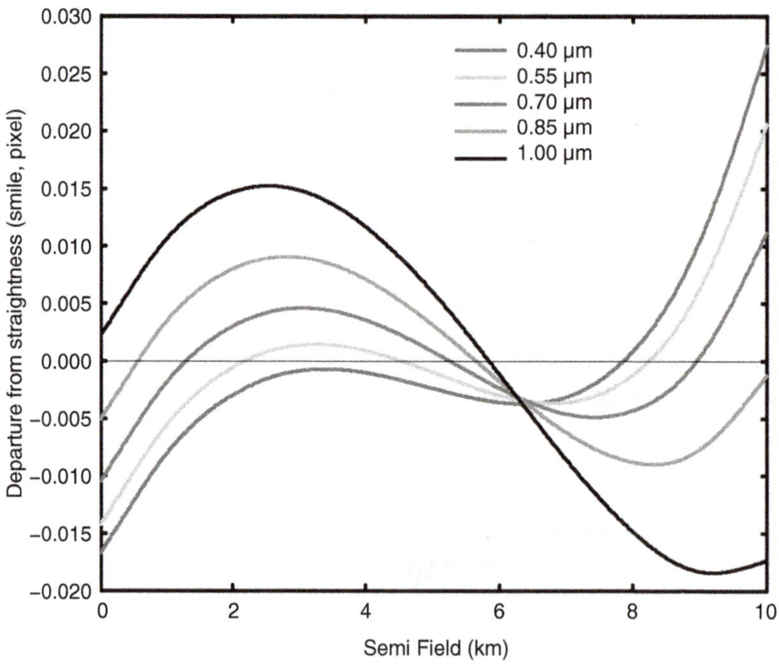

FIGURE 5.54 Spectral distortion (smile) of the VNIR spectrometer of System 2.

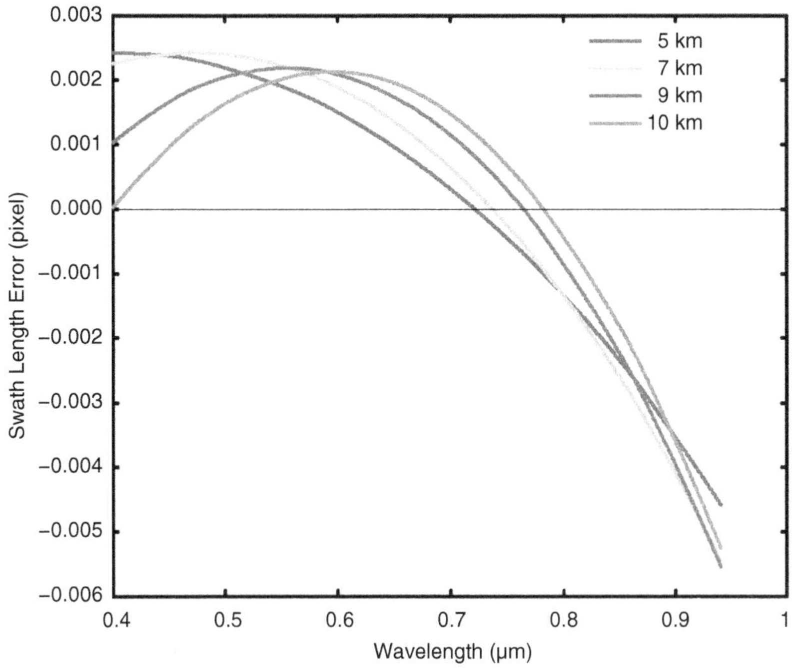

FIGURE 5.55 Spatial distortion (keystone) of the VNIR spectrometer of System 2.

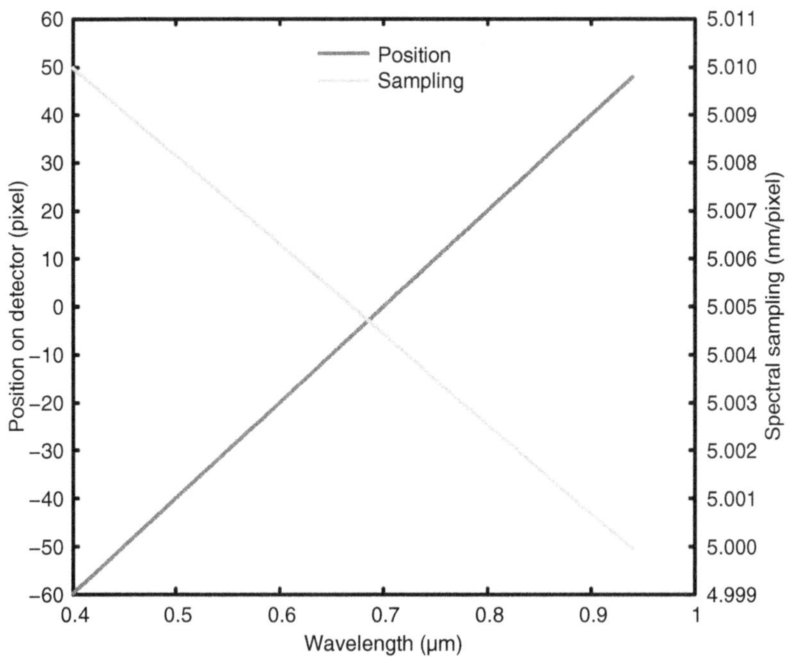

FIGURE 5.56 Spectral sampling interval (SSI) versus wavelength and spectral sampling nonlinearity of the VNIR spectrometer of System 2.

5.00 nm to 5.01 nm, which is very small. The maximum nonlinearity of spectral sampling is 0.01 nm, which is 50 times better than the requirement of 0.5 nm listed in Table 5.8.

5.7.2.2.4 MTF

Figure 5.57 shows plots of the MTF of the System 2 VNIR spectrometer, at three wavelengths: λ = 0.40, 0.70, and 1.00 μm spanning the operating spectral range. The range of spatial frequencies is from 0 to 17 cycles per mm—approximately the Nyquist limit for a detector array with a 30-μm pitch. The plots within each figure correspond to various fields and azimuths. For all cases, the MTF is within 0.90 of the diffraction limit. All the MTF values are within the MTF requirement of 0.7.

5.7.2.2.5 Spatial Registration

The error of cross-track co-registration of the VNIR spectrometer is computed for 10 selected wavelengths (0.40, 0.46, 0.52, 0.58, 0.64, 0.70, 0.76, 0.82, 0.88, and 0.94 μm) versus 4 positions

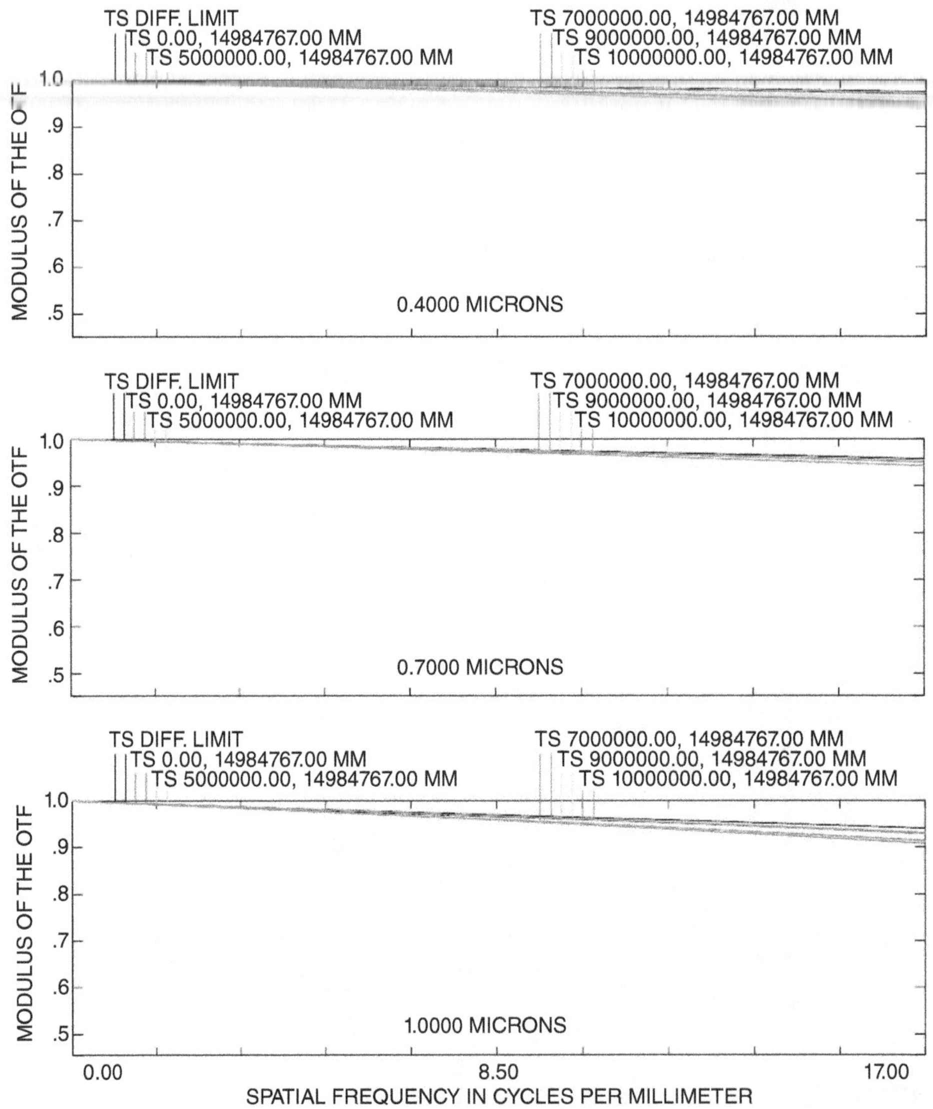

FIGURE 5.57 Plots of the MTF of the VNIR spectrometer of System 2.

TABLE 5.14
System 2 SWIR Spectral Spectrometer Parameters

Dispersive Element	Combined Prism
Spectral range	0.9–2.5 μm
Spectral sampling interval	5.05–7.10 nm
Number of spectral bands	256

(7.51, 10.51, 13.51, and 15.00 km) in the FOV (i.e., swath). For each position and wavelength, the computed error is the difference from the mean swath position across the wavelengths and scaled to fractional nominal GSD. The peak errors are the maximum departures from the data. The maximum peak-to-valley error is 0.008 GSD, which is over one order of magnitude smaller than the requirement of 0.1 GSD listed in Table 5.8.

5.7.2.2.6 Spectral Registration

Five wavelengths 0.40, 0.55, 0.70, 0.85, and 1.00 μm are selected as the mean wavelengths. Eleven positions in the swath from the center of the swath to one end with a 1-km gap are selected. The spectral registration is computed as the departure from the mean wavelength for the selected wavelengths versus the selected cross-track positions. The errors are scaled as fractions of both the nominal pixel size and, by applying the dispersion for the wavelength, the SSI. The errors increase with cross-track positions moving the extreme field position. The maximum peak-to-valley error is 0.043 SSI, which comfortably satisfies the requirement of 0.1 SSI listed in Table 5.8.

5.7.2.3 System 2 SWIR Spectrometer

Table 5.14 lists the parameters of the SWIR spectrometer of System 2. It is a prism-based spectrometer. There is an overlap of spectral coverage from 0.90 μm to 1.0 μm between the VNIR and SWIR for better calibration of the data generated by the two spectrometers. Figure 5.58 shows its optical layout. It comprises three rear-surface mirrors and a combined prism set. The spectral dispersion comes from the combined prism set that consists of a high degree of wedge in a series of lenses.

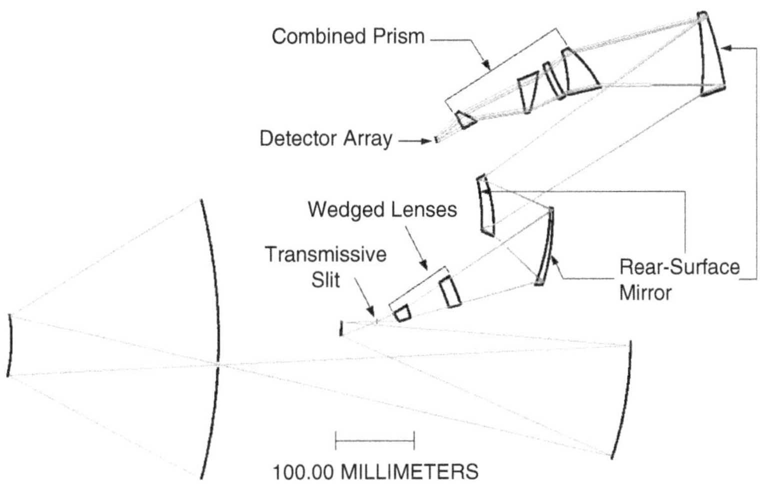

FIGURE 5.58 The optical layout of the SWIR prism spectrometer of System 2.

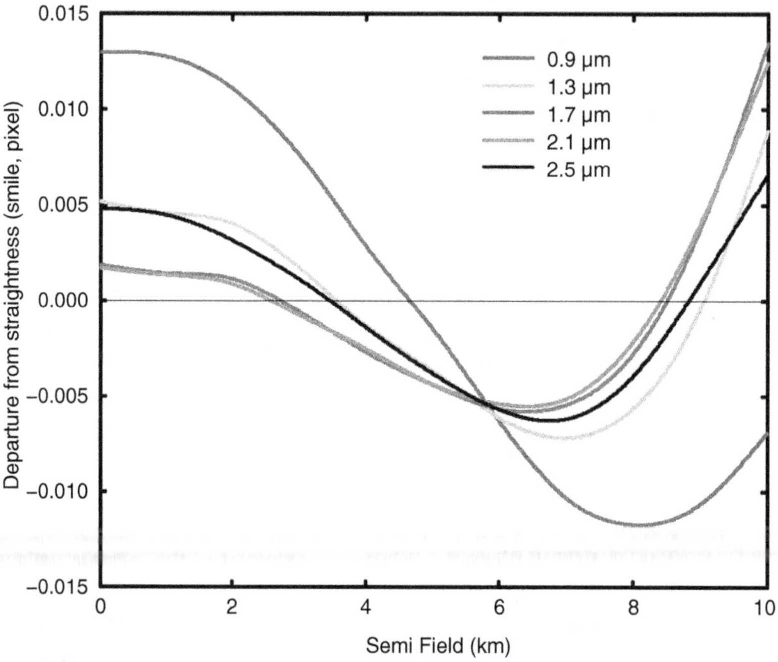

FIGURE 5.59 Spectral distortion (smile) of the SWIR spectrometer of System 2.

This design does not meet the SSI requirement of 5 nm listed in Table 5.8, but it achieves a varying SSI in a range of 5.05–7.10 nm, as shown in the table by dispersing the wavelength range over the entire 256-pixel width of the detector array. For this wavelength range, a 5-nm SSI requires 300 pixels in the spectral direction of the detector array.

5.7.2.3.1 Smile

Figure 5.59 shows the smile distortion of the SWIR spectrometer at five spectral bands: 0.9, 1.3, 1.7, 2.1, and 2.5 μm for the semi-FOV (0–10 km). The positions are calculated by ray tracing, as a geometric centroid. The data for each plot in the figure is the mean swath-image position, in the spectral direction, for the corresponding wavelength. It can be seen from the figure that the peak-to-valley smile values are 0.025, 0.016, 0.019, 0.019, and 0.014 pixel at the five spectral bands. The maximum smile of the SWIR spectrometer is 0.025 pixel, which is 4 times smaller than the requirement of 0.10 pixel listed in Table 5.8.

5.7.2.3.2 Keystone

Figure 5.60 shows the keystone distortion at 4 spatial locations in the half FOV (i.e., swath). It can be seen from the figure that the keystone distortion values are slightly larger than those of the VNIR spectrometer. For the spatial locations of 5, 7, 9, and 10 km in the half swath, the peak-to-valley keystone values are 0.014, 0.018, 0.026, and 0.036 pixel. The maximum keystone value is 0.036 pixel, which is smaller than the requirement of 0.10 pixel listed in Table 5.8.

5.7.2.3.3 Spectral Nonlinearity

Figure 5.61 shows the plots of SSI versus wavelength and the spectral sampling nonlinearity of the SWIR spectrometer of System 2 in the spectral range. It can be seen that the SSI linearly varies

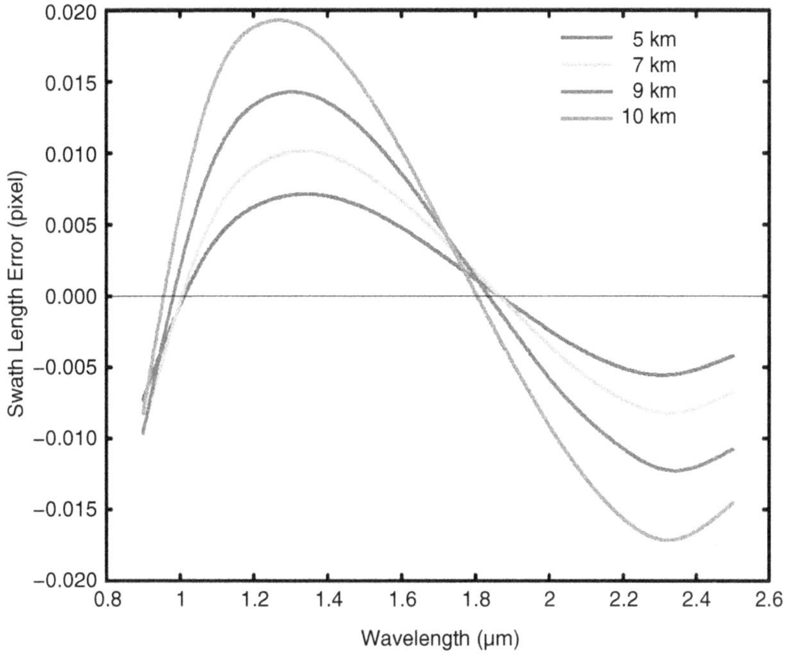

FIGURE 5.60 Spatial distortion (keystone) of the SWIR spectrometer of System 2.

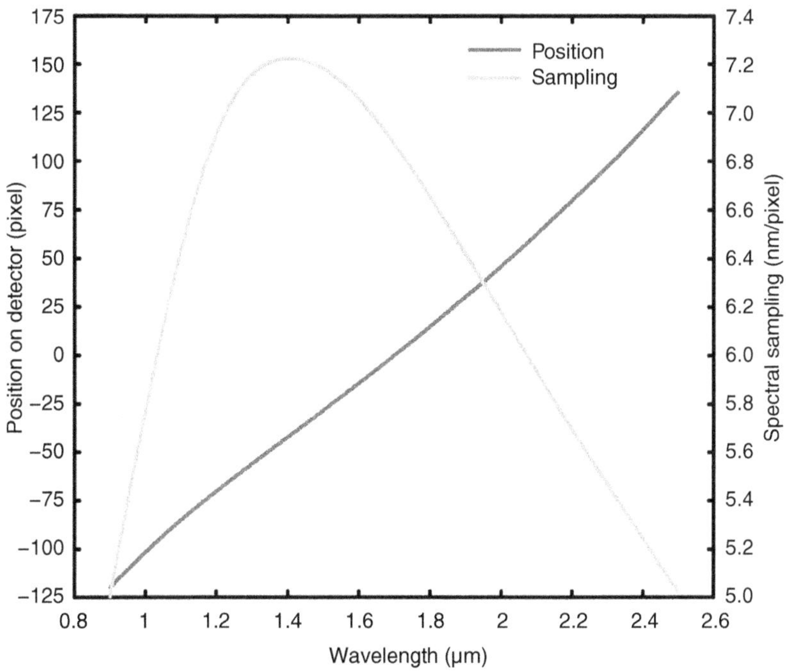

FIGURE 5.61 Spectral sampling interval (SSI) versus wavelength and spectral sampling nonlinearity of the SWIR spectrometer of System 2.

from 5.05 nm to 7.10 nm. The maximum nonlinearity of spectral sampling is 2.1 nm, which is slightly smaller than 2.8 nm of the SWIR prism spectrometer of System 1 described in Section 5.7.1.3.3, where the reason of spectral sampling nonlinearity of prism spectrometers is discussed. The maximum spectral sampling nonlinearity of the SWIR spectrometer is over 4 times larger than the requirement of 0.5 nm listed in Table 5.8.

5.7.2.3.4 MTF

Figure 5.62 shows plots of the MTF of the System 2 SWIR spectrometer at three wavelengths spanning the operating spectral range. The range of spatial frequencies is from 0 to 17 cycles per mm—approximately the Nyquist limit for a detector array with a 30-μm pitch. The plots within each figure correspond to various fields and azimuths. At λ = 0.9, 1.7, and

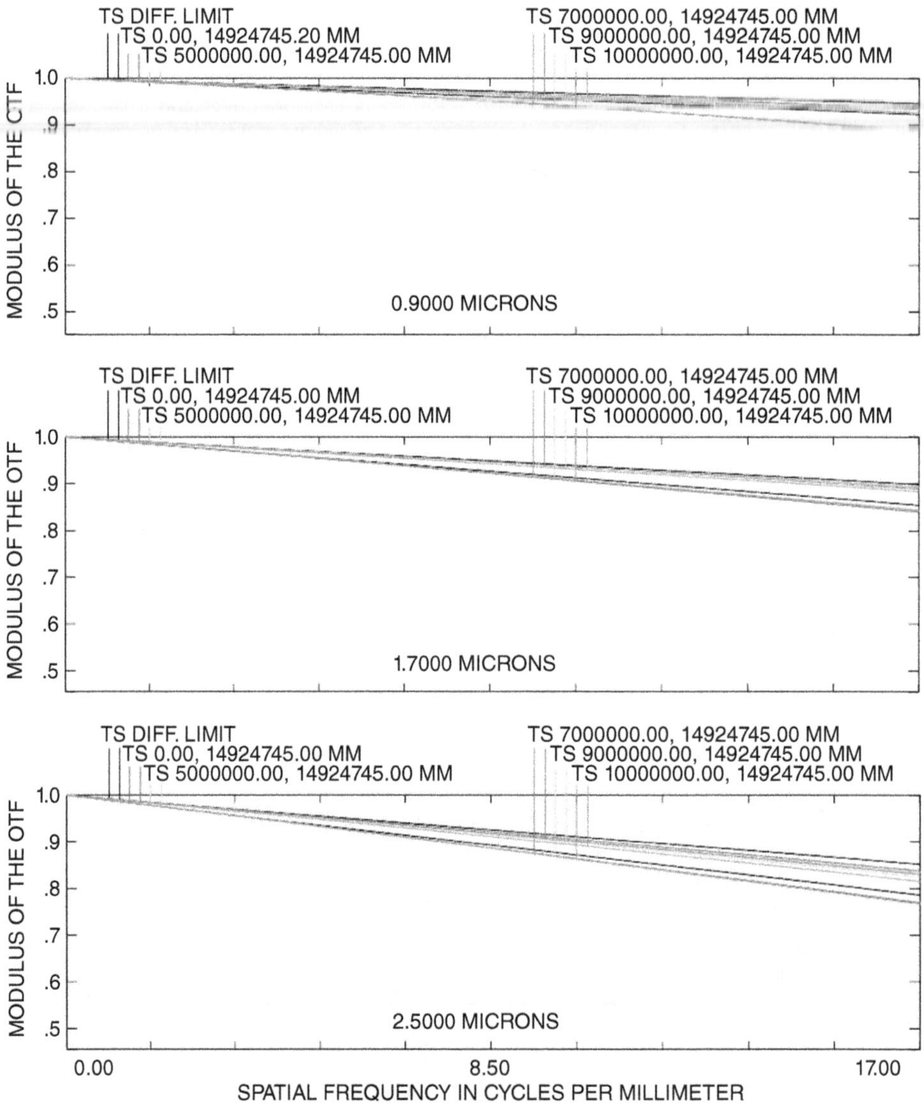

FIGURE 5.62 Plots of the MTF of the SWIR spectrometer of System 2.

2.5 μm, the worst case MTF is 0.89, 0.84, and 0.77 of the diffraction limit. All the MTF values are within the MTF requirement of 0.7.

5.7.2.3.5 Spatial Registration

The error of cross-track co-registration of the SWIR spectrometer is computed for 11 selected wavelengths (0.90, 1.06, 1.22, 1.38, 1.54, 1.70, 1.86, 2.02, 2.18, 2.43, and 2.50 μm) versus 4 positions (7.51, 10.51, 13.51, and 15.00 km) in the FOV (i.e., swath). For each position and wavelength, the computed error is the difference from the mean swath position across the wavelengths and scaled to fractional nominal GSD. The peak errors are the maximum departures from the data. The maximum peak-to-valley error is 0.036 GSD, which is well within the requirement of 0.1 GSD listed in Table 5.8.

5.7.2.3.6 Spectral Registration

The five wavelengths 0.90, 1.30, 1.70, 2.10, and 2.50 μm and the same 11 spatial positions as in the VNIR spectrometer are selected. The spectral registration is computed as the departure from the mean wavelength for the selected wavelengths against the selected cross-track positions. The errors are scaled as fractions of both the nominal pixel size and, by applying the dispersion for the wavelength, the SSI. The errors increase with cross-track positions moving the extreme field position. The maximum peak-to-valley error is 0.027 SSI, which comfortably satisfies the requirement of 0.1 SSI listed in Table 5.8.

5.7.3 Imaging Spectrometer System 3

5.7.3.1 System 3 Overview

Imaging Spectrometer System 3 is composed of a four-mirror telescope, a VNIR grating spectrometer, and a SWIR prism spectrometer. Table 5.15 lists the overall characteristics for System 3. It is similar in concept to System 2, but has a 30-km swath, which is wider than 20 km in System 2. This requires 50% more pixels of the detector array in the spatial direction. Another difference from System 2 is the coarser SSI, which is 10 nm instead of 5 nm in System 2.

Figure 5.63 shows the layout of System 3. It comprises a telescope, which images two adjacent swaths on the Earth's surface onto an adjacent pair of slits, a grating spectrometer for VNIR and a prism spectrometer for SWIR. The arrangement of the slit assembly is similar to System 2 except that the length of the two slits is 50% longer to cover a 50% wider swath.

TABLE 5.15
Imaging Spectrometer System 3 Characteristics

Orbit altitude	632 km
Swath width	30 km
Ground sampling distance	20 m
Aperture diameter	350 mm
Central obstruction	93 mm × 80 mm (elliptical)
Stop position	On primary mirror
Slit length	45 mm
Slit width	30 μm
Focal plane format (pixel)	1500 × 60 (VNIR) 1500 × 160 (SWIR)
Detector pitch Size	30 μm × 30 μm

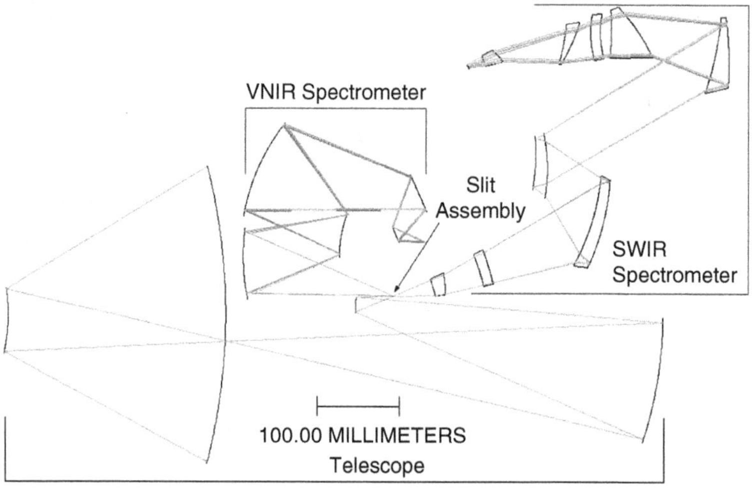

FIGURE 5.63 Configuration of the imaging spectrometer of System 3.

5.7.3.2 System 3 VNIR Spectrometer

Table 5.16 shows the spectral characteristics of the System 3 VNIR spectrometer. It is a grating-based spectrometer. The design provides linear spectral sampling together with the required sampling value of 10 nm. Figure 5.64 shows the optical layout of the VNIR spectrometer. It comprises five front-surface mirrors in a rotationally symmetric arrangement. The diffraction grating is formed on one of the mirrors.

5.7.3.2.1 Smile

Figure 5.65 shows the smile distortion of the VNIR spectrometer at five spectral bands: 0.40, 0.55, 0.70, 0.85, and 1.00 μm for the semi-FOV (0–15 km). The positions are calculated by ray tracing, as a geometric centroid. The data for each plot in the figure is the mean swath-image position, in the spectral direction, for the corresponding wavelength. It can be seen from the figure that the peak-to-valley smile values are 0.069, 0.059, 0.045, 0.028, and 0.036 pixel at the five spectral bands. The maximum smile of the VNIR spectrometer is 0.069 pixel, which is smaller than the requirement of 0.10 pixel listed in Table 5.8.

5.7.3.2.2 Keystone

Figure 5.66 shows the keystone distortion at 4 spatial locations in the in the half FOV (swath). It can be seen from the figure that the keystone distortion values are very small and close. For the spatial locations of 6, 9, 12, and 15 km in the half swath, the peak-to-valley keystone values are 0.004, 0.005, 0.006, and 0.006 pixel. The maximum keystone value is 0.006 pixel, which is over one order of magnitude smaller than the requirement of 0.10 pixel listed in Table 5.8.

TABLE 5.16
System 3 VNIR Spectral Characteristics

Dispersive Element	Diffracting Grating
Spectral range	0.4–1.0 μm
Spectral sampling interval	10.0 nm
Number of spectral bands	60

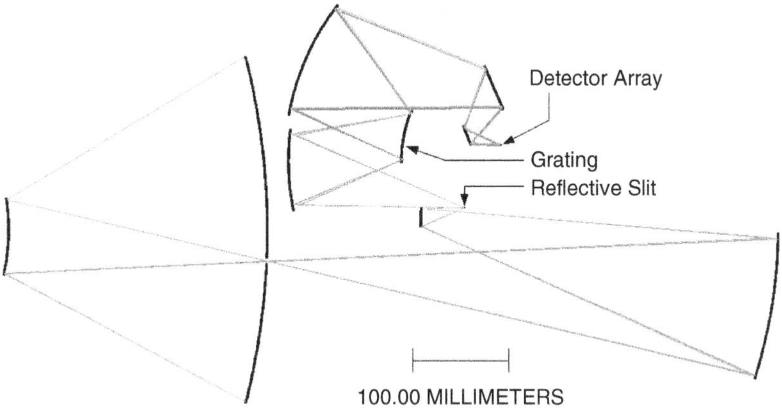

FIGURE 5.64 The optical layout of the VNIR grating spectrometer of System 3.

5.7.3.2.3 Spectral Nonlinearity

Figure 5.67 shows the plot of SSI versus wavelength and the spectral sampling nonlinearity of the VNIR spectrometer of System 3 in the spectral range. It can be seen that the SSI linearly varies from 10.0050 nm to 10.0105 nm, which can be comfortably rounded to 10.0 nm. The maximum nonlinearity of spectral sampling is 0.015 nm, which is 33 times better than the requirement of 0.5 nm listed in Table 5.8.

5.7.3.2.4 MTF

Figure 5.68 shows plots of the MTF of the System 3 VNIR spectrometer, at three wavelengths spanning the operating spectral range. The range of spatial frequencies is from 0 to 17 cycles per mm—approximately the Nyquist limit for a detector array with a 30-μm pitch. The plots within

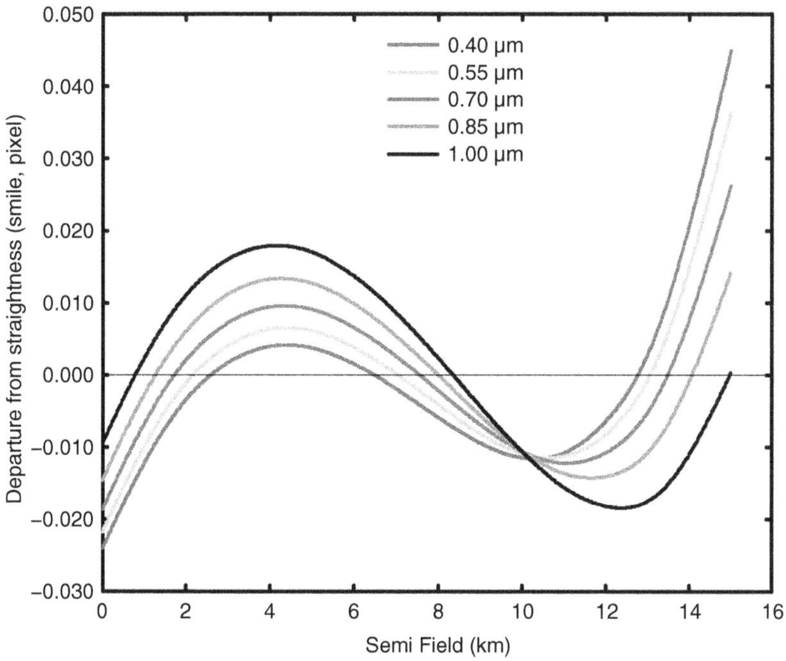

FIGURE 5.65 Spectral distortion (smile) of the VNIR spectrometer of System 3.

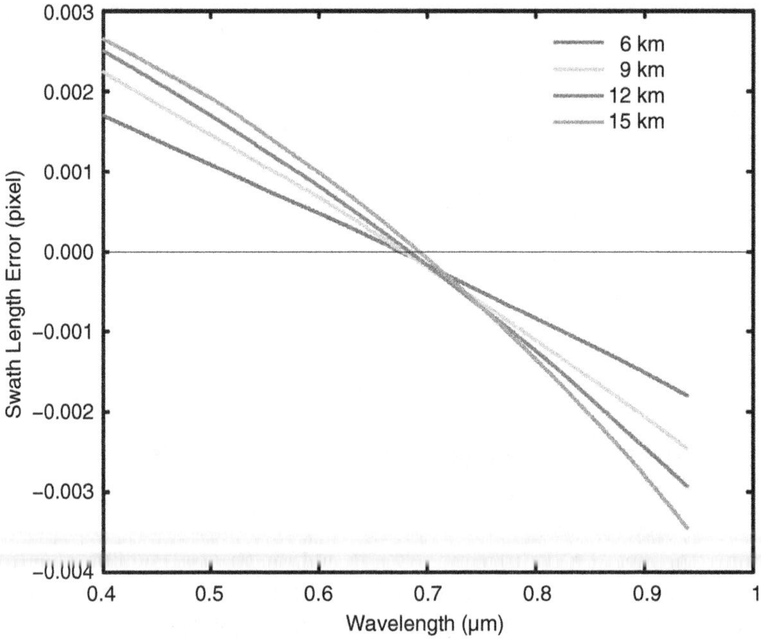

FIGURE 5.66 Spatial distortion (keystone) of the VNIR spectrometer of System 3.

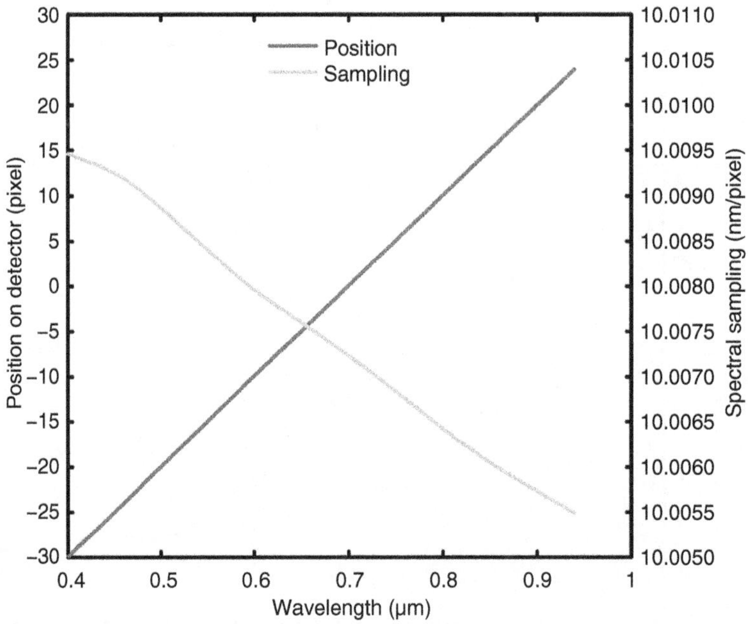

FIGURE 5.67 Spectral sampling interval (SSI) versus wavelength and spectral sampling nonlinearity of the VNIR spectrometer of System 3.

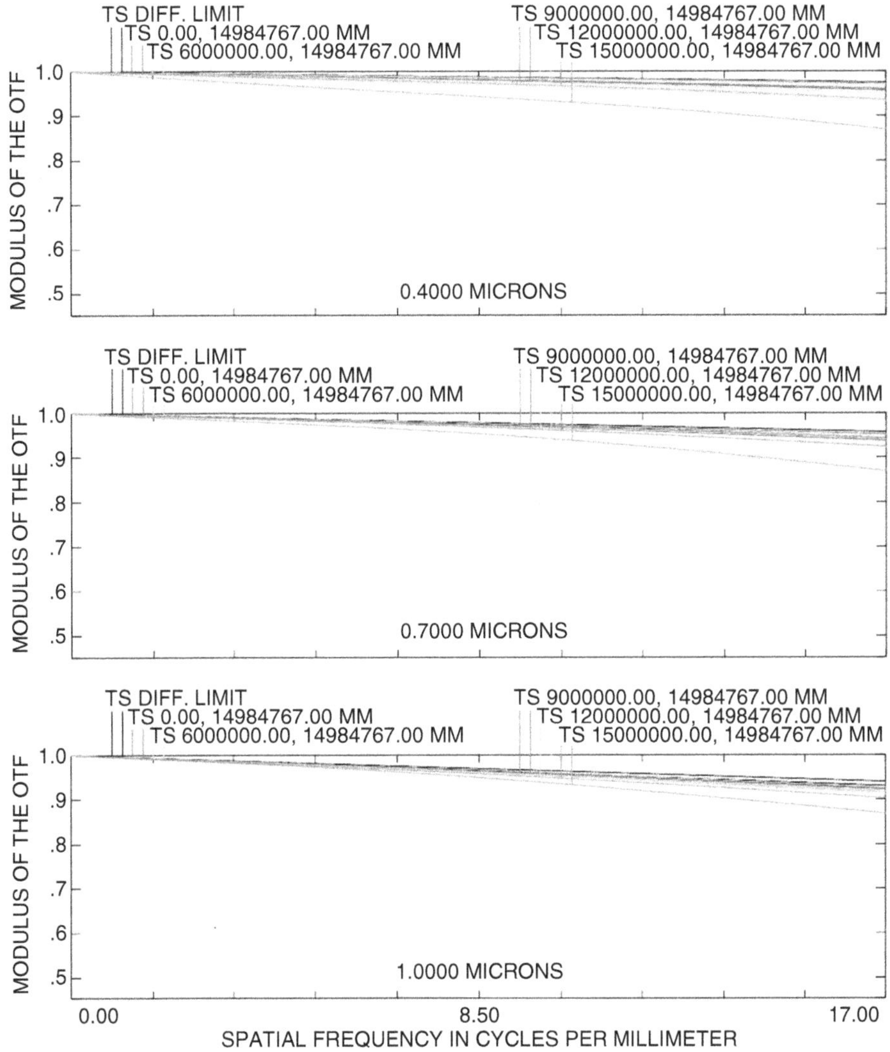

FIGURE 5.68 Plots of the MTF of the VNIR spectrometer of System 3.

each figure correspond to various fields and azimuths. At $\lambda = 0.40$, 0.70, and 1.00 μm, the worst MTF is 0.87, 0.87, and 0.87 of the diffraction limit. All the MTF values are within the MTF requirement of 0.7.

5.7.3.2.5 Spatial Registration

The error of cross-track co-registration of the VNIR spectrometer is computed for 10 selected wavelengths (0.40, 0.46, 0.52, 0.58, 0.64, 0.70, 0.76, 0.82, 0.88, and 0.94 μm) versus 4 positions (9.03, 13.53, 18.02, and 22.50 km) in the FOV (i.e., swath). For each position and wavelength, the computed error is the difference from the mean swath position across the wavelengths and scaled to fractional nominal GSD. The peak errors are the maximum departures from the data. The maximum peak-to-valley error is 0.006 GSD, which is over one order of magnitude smaller than the requirement of 0.1 GSD listed in Table 5.8.

TABLE 5.17
System 3 SWIR Spectral Spectrometer Parameters

Dispersive Element	Combined Prism
Spectral range	0.9–2.5 μm
Spectral sampling interval	7.5–11.5 nm
Number of spectral bands	160

5.7.3.2.6 Spectral Registration

Five wavelengths 0.40, 0.55, 0.70, 0.85, and 1.00 μm are selected as the mean wavelengths. Eleven positions in the swath from the center of the swath to one end with a 1.5-km gap are selected. The spectral registration is computed as the departure from the mean wavelength for the selected wavelengths versus the selected cross-track positions. The errors are scaled as fractions of both the nominal pixel size and, by applying the dispersion for the wavelength, the SSI. The errors increase with cross-track positions moving the extreme field position. The maximum peak-to-valley error is 0.068 SSI, which satisfies the requirement of 0.1 SSI listed in Table 5.8.

5.7.3.3 System 3 SWIR Spectrometer

Table 5.17 lists the parameters of the SWIR spectrometer of System 3. It is a prism-based spectrometer. There is an overlap of spectral coverage from 0.90 μm to 1.0 μm between the VNIR and SWIR for better calibration of the data generated by the two spectrometers. Figure 5.69 shows its optical layout. It comprises three rear-surface mirrors and a combined prism set. The spectral dispersion comes from the combined prism set that consists of a high degree of wedge in a series of lenses.

5.7.3.3.1 Smile

Figure 5.70 shows the smile distortion of the SWIR spectrometer at five spectral bands: 0.9, 1.3, 1.7, 2.1, and 2.5 μm for the approximate semi-FOV (0–13.5 km). The positions are calculated by ray tracing, as a geometric centroid. The data for each plot in the figure is the mean swath-image position, in the spectral direction, for the corresponding wavelength. It can be seen from the figure that the peak-to-valley smile values are 0.014, 0.021, 0.023, 0.014, and 0.023 pixel at the five spectral bands. The maximum smile of the VNIR spectrometer is 0.023 pixel, which is over 4 times smaller than the requirement of 0.10 pixel listed in Table 5.8.

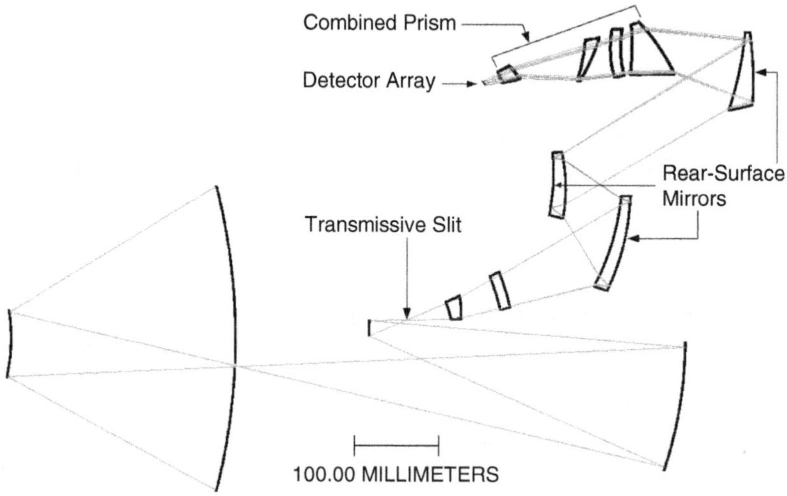

FIGURE 5.69 The optical layout of the SWIR prism spectrometer of System 3.

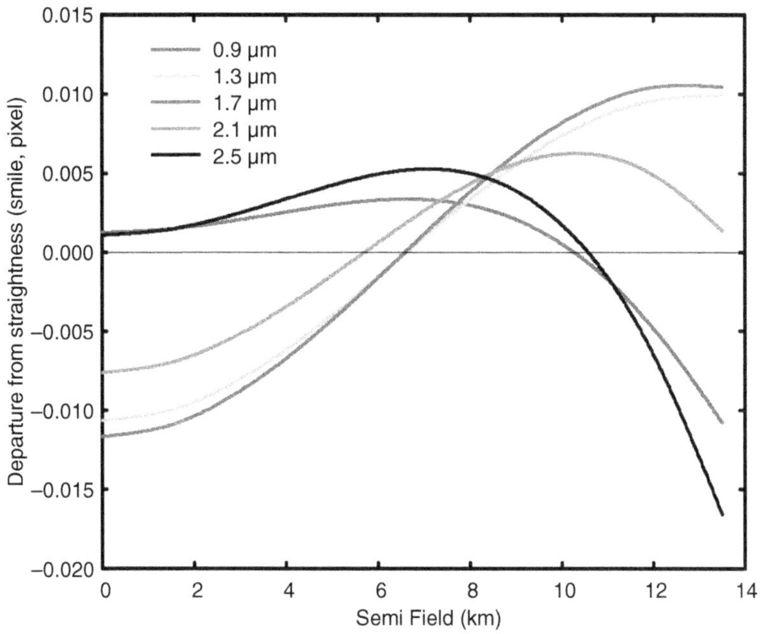

FIGURE 5.70 Spectral distortion (smile) of the SWIR spectrometer of System 3.

5.7.3.3.2 Keystone

Figure 5.71 shows the keystone distortion at 4 spatial locations in the half FOV (swath). It can be seen from the figure that the keystone distortion values are significantly larger than those of the VNIR spectrometer. For the spatial locations of 5, 10, 13, and 15 km in the half swath, the peak-to-valley keystone values are 0.090, 0.108, 0.052, and 0.039 pixel. The maximum keystone value is

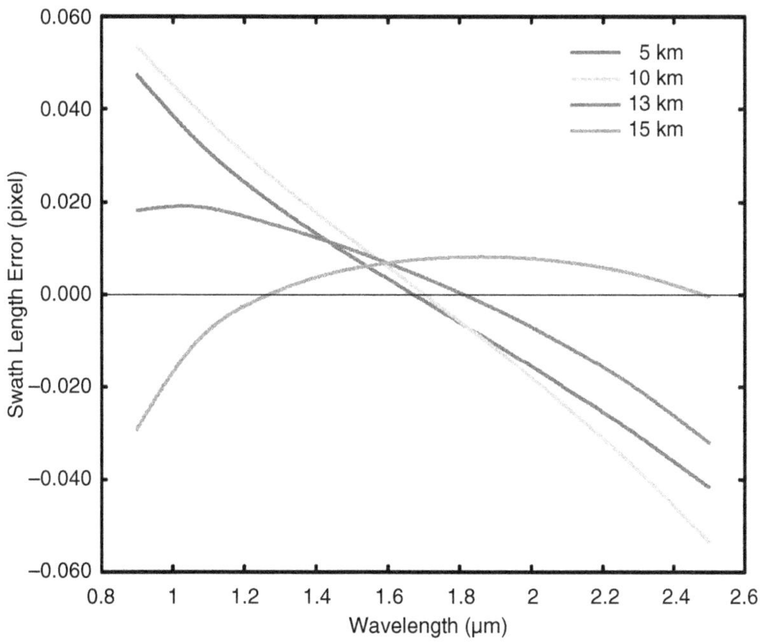

FIGURE 5.71 Spatial distortion (keystone) of the SWIR spectrometer of System 3.

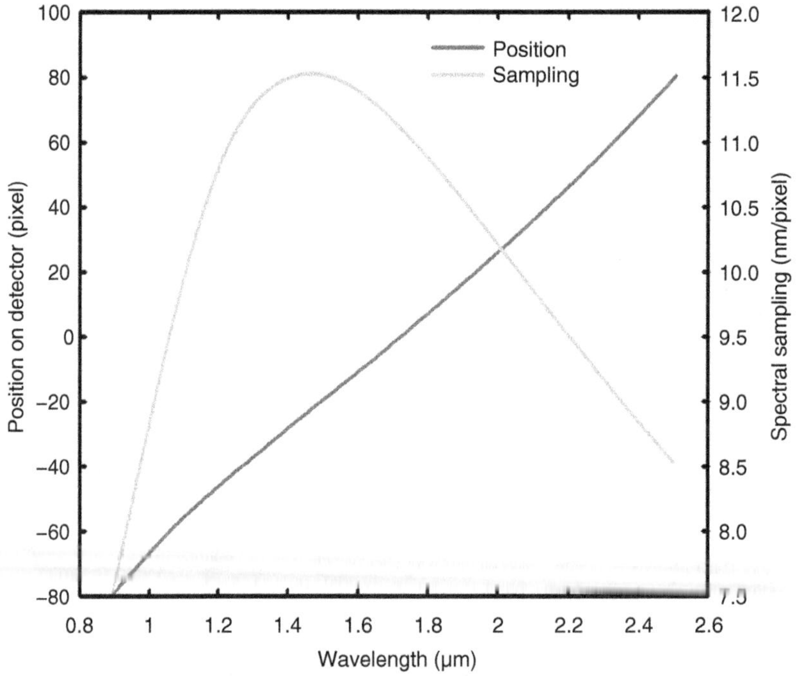

FIGURE 5.72 Spectral sampling interval (SSI) versus wavelength and spectral sampling nonlinearity of the SWIR spectrometer of System 3.

0.108 pixel at the location of 10 km in the swath. The keystone value at the location of 5 km in the swath is also larger than those in other two locations. The keystone of the SWIR spectrometer (0.11) is slightly larger than the requirement of 0.10 pixel listed in Table 5.8.

5.7.3.3.3 Spectral Nonlinearity

Figure 5.72 shows the plots of SSI versus wavelength and the spectral sampling nonlinearity of the SWIR spectrometer of System 3 in the spectral range. It can be seen that the SSI linearly varies from 7.5 nm to 11.5 nm. The maximum nonlinearity of spectral sampling is 3.0 nm, which is worse than 2.8 nm and 2.1 nm of the SWIR prism spectrometers of System 1 and System 2 described in Sections 5.7.1.3.3 and 5.7.2.3.3. The reason of spectral sampling nonlinearity of prism spectrometers is discussed in Section 5.7.1.3.3. The maximum spectral sampling nonlinearity of the SWIR spectrometer is 6 times larger than the requirement of 0.5 nm listed in Table 5.8.

5.7.3.3.4 MTF

Figure 5.73 shows plots of the MTF of System 3 SWIR spectrometer at three wavelengths spanning the operating spectral range. The range of spatial frequencies is from 0 to 17 cycles per mm—approximately the Nyquist limit for a detector array with a 30-μm pitch. The plots within each figure correspond to various fields and azimuths. At $\lambda = 0.9$, 1.7, and 2.5 μm, the worst case MTF is 0.82, 0.84, and 0.72 of the diffraction limit. All the MTF values are within the MTF requirement of 0.7.

5.7.3.3.5 Spatial Registration

The error of cross-track co-registration of the SWIR spectrometer is computed for 11 selected wavelengths (0.90, 1.06, 1.22, 1.38, 1.54, 1.70, 1.86, 2.02, 2.18, 2.34, and 2.50 μm) versus 4 positions (7.59, 15.12, 19.57, and 22.50 km) in the FOV (i.e., swath). For each position and wavelength, the computed error is the difference from the mean swath position across the wavelengths and scaled

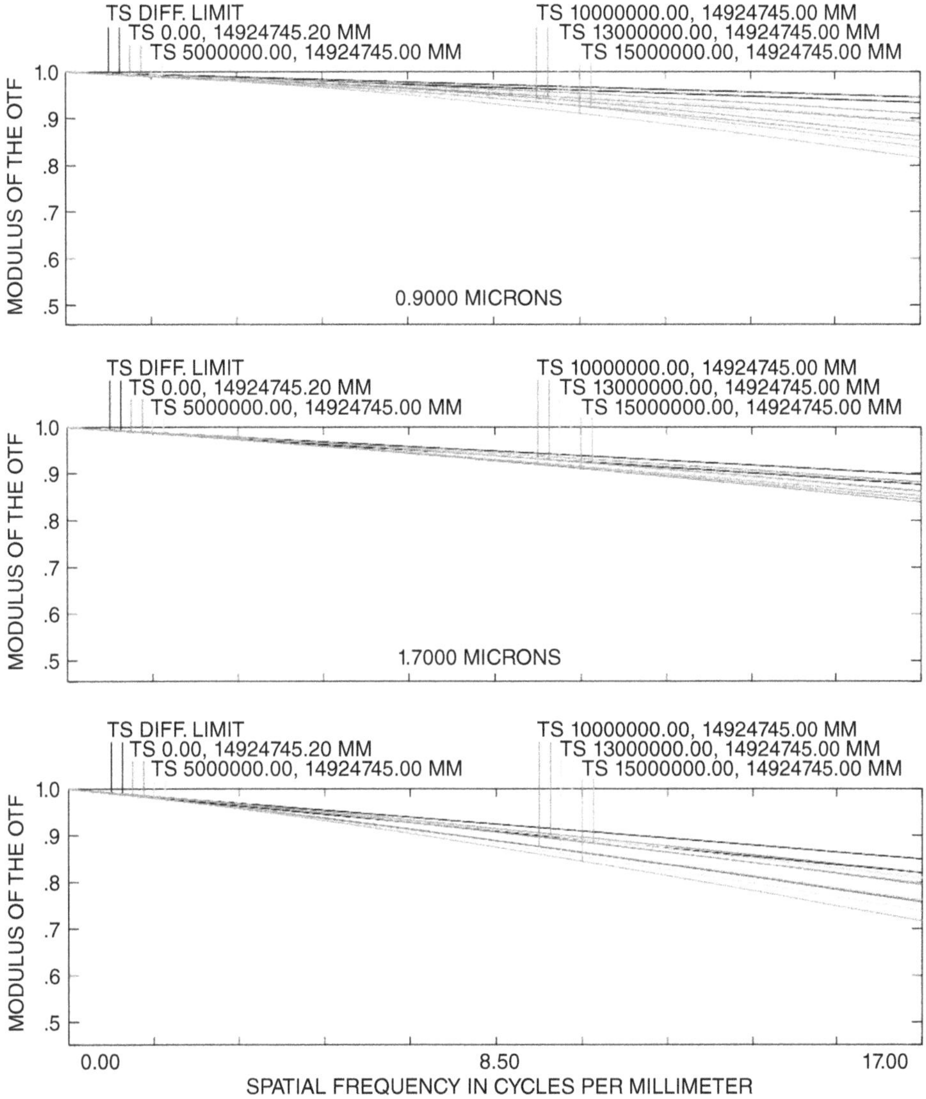

FIGURE 5.73 Plots of the MTF of the SWIR spectrometer of System 3.

to fractional nominal GSD. The peak errors are the maximum departures from the data. The peak-to-valley errors of cross-track co-registration are: 8.9%, 10.7%, 5.1%, and 3.7% corresponding to the 4 positions. The maximum peak-to-valley error is 10.7% GSD, which is slightly larger than the requirement of 0.1 (10%) GSD listed in Table 5.8.

5.7.3.3.6 Spectral Registration

The five wavelengths 0.90, 1.30, 1.70, 2.10, and 2.50 μm and the same 11 spatial positions as in VNIR spectrometer are selected. The spectral registration is computed as the departure from the mean wavelength for the selected wavelengths against the selected cross-track positions. The errors are scaled as fractions of both the nominal pixel size and, by applying the dispersion for the wavelength, the SSI. The errors increase with cross-track positions moving the extreme field position. The maximum peak-to-valley error is 0.064 SSI, which meets the requirement of 0.1 SSI listed in Table 5.8.

TABLE 5.18
Summary of Achieved Design Performance of the Three Imaging Spectrometer Systems

	System 1		System 2		System 3	
Requirement	**VNIR**	**SWIR**	**VNIR**	**SWIR**	**VNIR**	**SWIR**
Swath width ≥20 or 30 km	30	30	20	20	30	30
Ground sampling distance ≤20 or 30 m	30	30	20	20	20	20
Spectral range: 0.4–2.5 μm	0.4–1.025	0.975–2.5	0.4–1.0	0.9–2.5	0.4–1.0	0.9–2.5
Spectral sampling interval (nm)	9.8–10.12	7.2–11.0	5.00–5.01	5.05–7.10	10	7.5–11.5
Smile distortion <0.1 pixel	0.047	0.049	0.044	0.025	0.069	0.023
Keystone distortion <0.1 pixel	0.05	0.057	0.008	0.036	0.006	0.108
Spectral sampling nonlinearity <0.5 nm	0.02	2.8	0.01	2.1	0.015	3.0
Modulation transfer function <0.7	0.77	0.80	0.90	0.77	0.87	0.72
Spatial registration error <0.1 GSD	0.048	0.056	0.008	0.036	0.006	0.107
Spectral registration error <0.1 SSI	0.046	0.054	0.043	0.027	0.068	0.064

5.7.4 Summary of Three Designed Imaging Spectrometer Systems

The achieved design performances of the three imaging spectrometer systems are summarized in Table 5.18. It can be seen that the three designed systems achieved the requirements listed in Table 5.8 except five items (shaded in the table). The prism-based SWIR spectrometer of System 3 has three incompliances to the requirements: (1) the keystone distortion slightly exceeds the requirement, (2) the spatial registration error slightly exceeds the requirement, (3) spectral sampling non-linearity is larger than the requirement. The three prism-based SWIR spectrometers all have the same disadvantage of spectral sampling non-linearity. Their relatively high values of spectral sampling non-linearity compared to those of the grating-based spectrometers are understandable and may be acceptable depending on different applications. In general, both Systems 1 and 2 are good candidates in terms of optical performance.

REFERENCES

Achal, S. 2013. Compact, light-transfer system for use in image relay devices, hyperspectral imagers and spectrographs. *US Patent Application US 2013/0148195 A1.*

Achal, S., S.-E. Qian and M. Bergeron. 2019. WaterSat imaging spectrometer experiment (WISE) for Canadian microsatellite mission. *Proceedings of EARSel* 2019.

Bergeron, M., A. Hollinger, K. Staenz, M. Maszkiewicz and S.-E. Qian. 2008. Hyperspectral environment and resource observer (HERO) mission. *Canadian Journal of Remote Sensing* 34(S1):1–11.

Bezy, J.-L., J.-P. Huot, S. M. Delwart, L. Bourg, R. Bessudo, and Y. Delclaud. 2016. Medium resolution imaging spectrometer for ocean colour onboard ENVISAT. In *Optical Payloads for Space Missions*, ed. S.-E. Qian, 91–120. West Sussex UK: John Wiley & Sons.

Breckinridge, J. B., N. A. Page, R. R. Shannon, and J. M. Ridgers. 1983. Reflecting Schmidt imaging spectrometers. *Applied Optics* 22:1175–1180.

Brown, R. H., et al. 2004. The CASSINI visual and infrared mapping spectrometer (VIMS) investigation. *Space Science Reviews* 115:111–168.

Caron, J., J.-L. Bezy, G. Bazalgette, M. Richert, and D. Loiseaux. 2012. Polarization scramblers in earth observing spectrometers: Lessons learned from Sentinel-4 and 5 phases A/B1. *Porceedings of SPIE* 10564: 105642B-1–105642B-9.

Cook, L. G. 1992. The last three-mirror anastigmatic (TMA). In *Lens Design: A Critical Review*, ed. W. J. Smith, 310–324, CR41. Bellingham: SPIE Optical Engineering Press.

Coradini, A., F. Capaccioni, P. Drossart, et al. 2007. VIRTIS: An imaging spectrometer for the ROSETTA mission. *Space Science Reviews* 128(1–4):529–559.

Eismann, M. T. 2012. *Hyperspectral Remote Sensing.* Bellingham: SPIE Press.

Féry, C. 1911. A prism with curved faces for spectrograph or spectroscope. *Astrophysical Journal* 34:79–87.

Fisher, J. and W. C. Welch. 2006. Survey and analysis of fore-optics for hyperspectral imaging systems. *Proceedings of SPIE* 6206: 62062R-1–62062R-11.

Green, R. O., et al. 2011. The Moon mineralogy mapper (M3) imaging spectrometer for lunar science: Instrument description, calibration, on-orbit measurements, science data calibration and on-orbit validation. *Journal of Geophysical Research* 116(E00G19):1–31.

JPL. 2018. HyspIRI Mission Concept Report. https://hyspiri.jpl.nasa.gov/downloads/reports_whitepapers/HyspIRI_FINAL_Report_1October2018_20181005a.pdf

Kappus, M., et al. 2016. Hyperspectral imager for the coastal ocean on the international space station. In *Optical Payloads for Space Missions*, ed. S.-E. Qian, 29–52. West Sussex UK: John Wiley & Sons.

Kaufmann, H., et al. 2016. Environmental mapping and analysis program—A german hyperspectral Mission. In *Optical Payloads for Space Missions*, ed. S.-E. Qian, 161–181. West Sussex UK: John Wiley & Sons.

Kruizinga, B., H. Visser, and M. D. Nijkerk. 2013. Depolarizer, telescope and remote sensing device and method. *EP 2602646 (A1)*, 12 June 2013.

Levelt, P. F., G. H. J. van den Oord, M. R. Dobber, A. Mälkki, H. Visser, J. de Vries, P. Stammes, J. Lundell, and H. Saari. 2006. The ozone monitoring instrument, *IEEE Transactions on Geoscience and Remote Sensing* 44:1093–1101.

Lobb, D. R. 1994. Theory of concentric designs for grating spectrometers. *Appl. Optics* 33:2648–2658.

Matsunaga, T., et al. 2016. Hyperspectral imager suite (HISUI): Japanese spaceborne hyperspectral imager for resource and environmental mapping. In *Optical Payloads for Space Missions*, ed. S.-E. Qian, 215–222. West Sussex UK: John Wiley & Sons.

Meini, M., A. Bini, L. Giunti, E. Fossati, and R. Formaro. 2016. Hyperspectral payload for Italian PRISMA programme. In *Optical Payloads for Space Missions*, ed. S.-E. Qian, 183–213. West Sussex UK: John Wiley & Sons.

Mouroulis, P. 1999. Spectral and spatial uniformity in pushbroom imaging spectrometers. *Proceedings of SPIE* 3753:133–141.

Mouroulis, P. 2007. Rear landscape on steroids. *Proceedings of SPIE* 6667: 666705-1–666705-8.

Mouroulis, P. and R. O. Green. 2018. Review of high fidelity imaging spectrometer design for remote sensing. *Optical Engineering* 57(4):040901.

Mouroulis, P., R. O. Green, and T. G. Chrien. 2000. Design of pushbroom imaging spectrometers for optimum recovery of spectroscopic and spatial information. *Applied Optics* 39:2210–2220.

Mouroulis, P., R. O. Green, B. V. Gorp, L. B. Moore, D. W. Wilson and H. A. Bender. 2016. Landsat swath imaging spectrometer design. *Optical Engineering* 55(1):015104.

Mouroulis, P., R. O. Green, and D. W. Wilson. 2008. Optical design of a coastal ocean imaging spectrometer. *Optics Express* 16:1087–1096.

Neumann, W. 2014. *Fundamentals of Dispersive Optical Spectroscopy Systems.* Bellingham: SPIE Press.

Nieke, J., C. Mavrocordatos, C. Donlon, B. Berruti, T. Garnier, J.-B. Riti, and Y. Delclaud. 2016. Ocean and land color imager on sentinel-3. In *Optical Payloads for Space Missions*, ed. S.-E. Qian, 223–245. West Sussex UK: John Wiley & Sons.

O'Donoghuea, D. and J. C. Clemens. 2014. Spherical grating spectrometers. *Proceedings of SPIE* 9147:9147–9148.

Piccioni, G., et al. 2008. VIRTIS: The visible and infrared thermal imaging spectrometer. *ESA-SP-1295, ESA Publications Division SP-1295*, Noordwijk, The Netherlands: 1–27.

Qian, S.-E. 2013. *Optical Satellite Signal Processing and Enhancement.* Bellingham: SPIE Press.

Qian, S.-E., et al. (other 18 coauthors). 2017. A spaceborne coastal and inland water color hyperspectral imager. *Proc. of IEEE International Geoscience and Remote Sensing Symposium (IGARSS'2017)* 447–450.

Sidran, M., H. J. Stalzer, and M. H. Hauptmann. 1966. Drum dispersion equation for Littrow-type prism spectrometers. *Applied Optics* 5:1133–1138.

Ulbrich, G. J., R. Meynart, and J. Nieke. 2004. APEX-airborne prism experiment: The realization phase of an airborne hyperspectral imager. *Proceedings of SPIE* 5570:453–459.

Warren, D. W., D. J. Gutierrez, J. L. Hall, and E. R. Keim. 2008. Dyson spectrometers for infrared earth remote sensing. *Proceedings of SPIE* 7082:70820R-1–70820R-8.

Windpassinger, R., J. Schubert, and D. Kampf. 2014. Proposed concept and preliminary design for the Sentinel-5 UVNS spectrometer. *Proceedings of SPIE* 1056343-1–1056343-9.

6 Focal Plane Arrays

6.1 FROM INSTRUMENT REQUIREMENTS TO FOCAL PLANE ARRAY SPECIFICATION

Space missions of today and tomorrow rely heavily on electro-optical sensors to fulfil their operational goals and science objectives. The focal plane array (FPA) is the key of an optical payload, as it dominates the sensitivity, resolution, and data quality of the payload. Hyperspectral satellite missions require sensitive and large format detector arrays. Modern innovative sensor technologies offer optical system designers a broad range of choices. While in many optical systems the FPA is the performance-limiting component, the optical design, cryogenics, and signal processing must satisfy the conditions of the FPA in order to achieve the performance.

In selecting the FPA technologies in each unique hyperspectral space mission, the performance requirements must flow down from the mission level to the instrument, from the instrument to the system, from the system to its subsystems, and from subsystems to components. The challenges associated with the applications of the state-of-the-art detector arrays include optimization of array format, device architecture, operating temperature, quantum efficiency (QE), readout noise, and dark current for each technology. In addition, space applications require highly reliable, radiation-tolerant devices incorporated into unique space-qualified flight packages. Detailed and comprehensive studies of the physics, architecture, operation, and characterization of FPAs operating in the UV, visible, and IR spectral regions including their optimizations for demanding applications have been reported in literature (Dereniak and Boreman 1996, Janesick 2001, Rogalski 2010, Holst and Lomheim 2011, Theuwissen 2013, Djazovski 2016).

The selection process begins with an understanding of the overall mission requirements and science-driven constraints on the performance of the electro-optical system. For example, operational goals of an Earth observation hyperspectral satellite can be translated into five basic requirements: signal-to-noise ratio (SNR) (i.e., detectability), spatial resolution, spectral resolution, wavelength coverage, and temporal resolution (i.e., revisit period). Since these requirements are often conflicting, it is essential to establish the observing requirements for each component such as source flux, type of instrument, band-pass, and SNR as close as possible to the one that the completed payload is expected to execute (Djazovski 2016).

The next step involves an analysis of the system requirements document (SRD) of the mission, which specifies the high-level functional requirements and overall system configuration parameters. Each of the subsystems has its own detailed requirements, which flow from the SRD. For example, the key driving requirements for the optical configuration may include the spectrometer type, effective aperture, field of view (FOV), etendue, spectral coverage, system image quality, etc. The electro-optical sensor design process starts from this point as it is necessary to relate spectral radiance $L(\lambda, T)$ to electrons in a detector pixel, n_{pe}. The spectrometer formula equation commonly used to compute the minimum and maximum signal in electrons per pixel is as follows (Holst and Lomheim 2011):

$$n_{pe} = \int_{\lambda 1}^{\lambda 2} \eta(\lambda)\lambda \frac{\pi}{4} \frac{L(\lambda,T) t_{int} A_d}{(f/\#)^2 (1+M_{opt})^2} T_{opt} T_{atm} d\lambda \tag{6.1}$$

where $\eta(\lambda)$ is the QE of the detector array, t_{int} is the integration time, A_d is the detector pixel area, T_{opt} is the optical system transmittance, T_{atm} is the intervening atmospheric transmittance, M_{opt} is the optical magnification, and f/# is the optical system focal ratio, i.e., the ratio of the focal length (*f*) to the lens diameter (*D*):

$$f/\# = \frac{f}{D} \tag{6.2}$$

The minimum and maximum values of the parameter $L(\lambda, T)$ in Equation 6.1 are usually calculated by specialized software tools (e.g., MODTRAN).

The initial selection of the FPA technology can be made at this stage, because the wavelength range and pitch size of detector array are used in the calculations of the minimum aperture diameter, and effective focal length and spectral band requirements are defined. From the optical system design point of view, smaller detector pixels are preferable. Thus, for approximate Nyquist sampling the optimal focal ratio at the detector is given by:

$$\frac{f}{D} = \frac{2p}{\lambda} \tag{6.3}$$

where *p* is the pitch size of detector array and λ is the operating wavelength. Since at a given wavelength, the focal ratio is proportional to the pitch size, a smaller pitch size results in more compact optics. However, this smaller pitch size is in conflict with large detector pixel area, A_d, defined in Equation 6.1 for collecting more electrons for high SNR.

The throughput requirements address losses when light travels through the optical system and the ability of a detector array to utilize that light. The output of throughput analysis would yield refined requirements for QE, detector flatness, operating temperature, and temperature stability. The noise budget analysis addresses the shot noise from detector photon response, output amplifier noise, noise from the front end electronics (or the ASIC chip), contributions from the back end electronics boards, and contributions from other sources of system noise. Calculating signal and noise in electrons per pixel allows determination of a fundamental noise source associated with the photon-to-electron conversion process.

The full well capacity needed to meet the SNR requirements in the integration time, t_{int}, can be estimated. Steps of a more detailed analysis include detector thermal noise; 1/*f* noise; the noise associated with all of the photon-related noise sources, including the pixel readout and multiplexing; video-analog signal processing; and quantization of pixel digitization. Holst and Lomheim (2011) discussed the detailed equations associated with this process along with dynamic range requirements and analog-to-digital converter (ADC) selection for specific applications. The noise-equivalent figures of merit, such as Noise Equivalent Angle (NEA), Noise Equivalent Reflectance Difference (NE$\Delta\rho$), or Noise Equivalent Temperature Difference (NETD), are defined in terms of focal plane Noise Equivalent Irradiance (NEI) to facilitate the further iterations of the flow-down process. This is the preferred method of analyzing focal plane SNR. Also, implementers of the FPA prefer this metrics because SNR is amenable to direct measurement.

In addition to SNR metrics for systems that require high image quality, other metrics of the systems also need be computed, such as the Modulation Transfer Function (MTF) and the relative edge response. These two metrics characterize how diffraction and other inherent limitations of the optical system blur sharp features of a scene, such as coastlines or edges of objects. For example, during the design synthesis of a hyperspectral sensor, the designer might simultaneously vary ground sample distance, relative edge response, SNR, etc., accounting for variations in detector response, band-to-band misregistration, and mis-calibration.

Quantitative analyses of the requirements and expected FPA performance are usually conducted using simulations and engineering analysis tools. By applying various models, such as MTF, noise levels, and uniformity to a higher quality image, the expected FPA performance output can be produced. The results of the simulation can be fed into algorithms used to extract data from the image to determine actual error levels and then the parameters can be modified to meet the instrument-level specification. The number of trials required to find an optimal design varies with the complexity of the relationship between the metrics and the instrument parameters upon which they depend. The output of the quantitative analysis is the preliminary FPA specification that includes spectral range, array format, pixel pitch, operating temperature, integration time, frame rate, noise equivalent metrics, saturation level, minimum usable illumination level, dynamic range, and resolution of ADC.

6.2 CCD FPAs

Charge-coupled devices (CCDs) were invented in 1969. Since their invention, they have become the detectors of choice for high-quality imaging in a wide range of fields. They have the advantages of high resolution, 100% fill factor, greater than 90% peak QE, excellent charge transfer efficiency, and very low dark current when with sufficient cooling (Janesick 2001). CCD detectors are arrays of individual light sensitive elements (or pixels) formed on the silicon wafer by a series of closely spaced electrodes separated from the underlying substrate by a thin insulating layer. The detectors are referred to as a linear array (one-dimensional) if individual pixels are arranged in a single row, while they are called a two-dimensional (2D) array when the pixels are arranged in rows and columns. Photons induced on the CCD surface generate charges in each pixel. The quantity of charges that are collected in the well is linearly related to the intensity of the photon flux and to the time over which the light falls on the pixel (integration time). After an integration period, the array readout involves simultaneously clocking all rows of charge packets, which transfers the packet sequentially from the site of collection to a charge output amplifier. The resulting data stream is a pixel-by-pixel, row-by-row representation of the scene being imaged.

The state-of-the-art scientific-grade CCDs are available in a number of distinct technologies (Janesick 2001, Holst and Lomheim 2011, Theuwissen 2013), such as frontside-illuminated (FSI) CCDs, thinned backside-illuminated (BSI) CCDs, and BSI deep-depletion CCDs, as shown in Figure 6.1.

In a traditional FSI CCD depicted in Figure 6.1a, light passes through the polysilicon gate that defines each pixel and generates electric charge in the collecting well when the pixel is electrically biased. Due to reflection and absorption losses in the poly gate structure QE of a FSI device

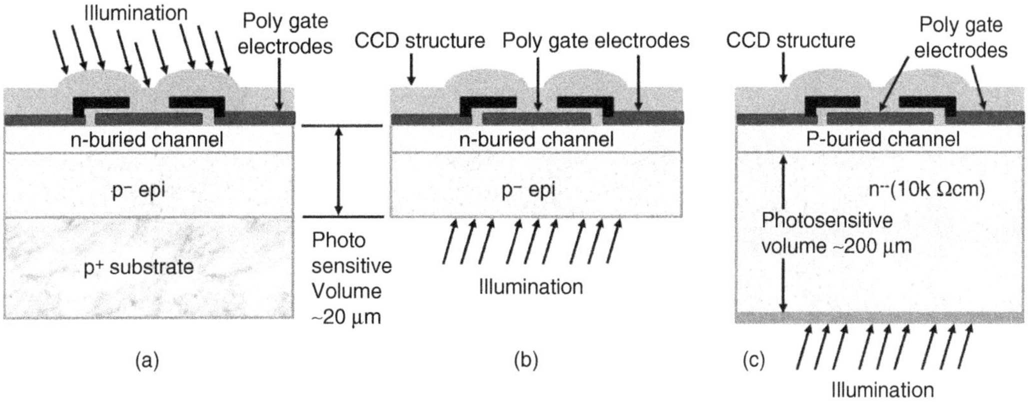

FIGURE 6.1 CCD technologies. (a) Frontside-illuminated CCD, (b) thin backside-illuminated, and (c) backside-illuminated deep depletion CCD.

is only about 50%. To improve QE, the Si substrate material can be uniformly removed to attain approximately 10–15 μm thickness so that light can be illuminated from backside directly onto the photosensitive area of the CCD without absorption losses in the gate structure as depicted in Figure 6.1b. Antireflection coatings can be used to modify, to some extent, the wavelength uniformity and coverage. Compared to front-illuminated CCDs, the thinned back-illuminated devices have a higher QE across the visible spectrum with peak QE > 85%.

To further improve QE, especially for near-infrared (NIR) imaging applications, the CCDs are manufactured from high-resistivity silicon materials with a thickness ranging from 50 μm to 300 μm as depicted in Figure 6.1c to produce a "deeper" depletion region (i.e., larger active photosensitive volume). This architecture allows longer-wavelength photons to interact within the depletion region layer as opposed to merely penetrating it, ultimately helping to increase QE in the NIR spectral region (Holland et al. 2007). Figure 6.2 shows examples of QE curves for the three CCD manufacture technologies.

Linear (i.e., 1D) CCD arrays are used in line-scan cameras and in satellite multispectral imagers operating in the pushbroom mode, where the spacecraft motion provides the frame scan (Holst and Lomheim 2011). A series of scans is taken continuously with the pixel rates synchronized with the speed of the motion forming a 2D image. Since only a single row of data needs to be transferred out before the next exposure, the advantages of line scan imaging include very high speed (up to thousands of lines per second) and possibility to produce continuous imaging with wide swath of area coverage. Although large-format linear silicon detector arrays have been fabricated since 1990s, for some applications a single linear array CCD may not be adequate to cover an entire swath when the swath is very wide. Therefore, many earth observation instruments utilize mosaic focal planes that are composed of individual linear CCDs mounted on a single base plate in order to form very wide detector arrays. For example, the panchromatic focal plane unit of the IKONOS satellite consists of three overlapping 4648-pixel wide linear arrays, providing a total of 13,500 pixels per line. The detector subsystem of the QuickBird satellite consists of 6 linear CCD arrays, providing 27,568 pixels in the cross-track direction for the panchromatic band (Yuan et al. 2012).

The development of large-format area (i.e., 2D) CCDs has been primarily driven by the scientific requirements of the ground-based astronomical community. Most of the recent astronomy observations in the visible region of the spectrum have been made with 2D CCDs. The most popular scientific-grade 2D CCDs are the 2048 × 4096 (2K × 4K) and 4096 × 4096 (4K × 4K) arrays with 15-μm pitch size, which are used as building blocks of large mosaic FPAs. These 2D detector arrays

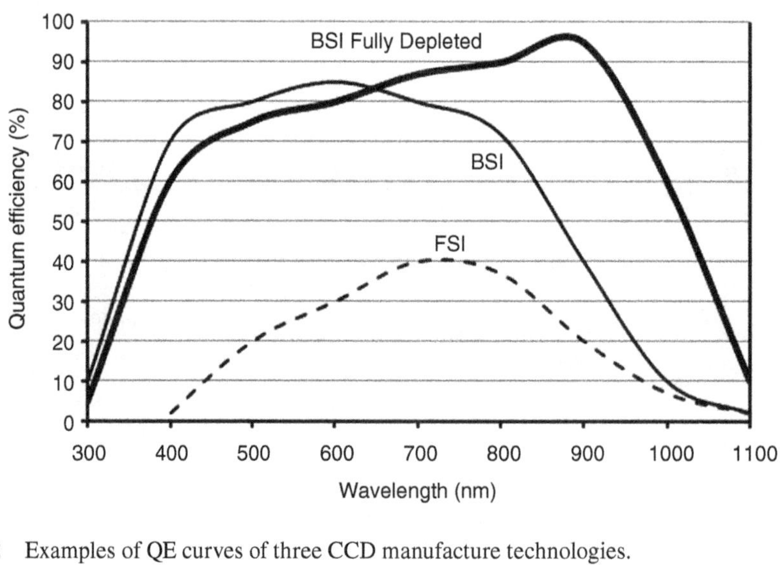

FIGURE 6.2 Examples of QE curves of three CCD manufacture technologies.

have been selected as a reasonable compromise between manufacturing yield that favors smaller devices and integration complexity of mosaic FPAs that favors larger blocks. The current state of the art for astronomical CCDs are large format detector arrays and commercially available as front- and back-illuminated devices with remarkably low readout noise and dark current at an operating temperature in a range of 170–220 K.

Some larger-format 2D CCDs are being fabricated, such as 9216 × 9216 pixels arrays of 10-μm pixel pitch and 92 mm × 92 mm physical dimensions manufactured by e2V. Even the largest CCD STA-1600 with a format of 10,560 × 10,560, a pixel pitch of 9 μm and 95 mm × 95 mm physical dimensions was designed, developed, and tested at Semiconductor Technology Associates (Zacharias et al. 2007). Each of these devices fills the 6-in. silicon wafer. One highest resolution CCD is a custom 252 megapixel (17,216 × 14,656 pixels with 5.6 μm pixel pitch) image sensor, which has been designed by DALSA exclusively for Intergraph (Jacobsen 2011). The STA 1600 and 252 megapixel CCDs were fabricated at Teledyne-DALSA facility in Bromont, Quebec, Canada. The largest focal plane assembly ever flown in space is the mosaic of 106 large area CCDs manufactured by e2V for the ESA's astrometry satellite Gaia launched on December 19, 2013. Each CCD is 4.5 cm × 5.9 cm, providing a 4500 × 1966 pixel array. The entire assembly represents a remarkable 937 megapixel camera system. Each pixel contains a supplementary buried channel (SBC) and anti-blooming drain (ABD) that are unique in astronomical CCDs making the Gaia pixel one of the most complicated ever manufactured (Bruijne 2012).

Time delay integration (TDI) CCD is a special type CCD technology that is used for imaging of earth observation, mapping, or surveillance. The TDI operation is based on the concept of accumulating multiple exposures of the same object by the detector consisting of *N* detector rows (*N*-stage TDI). Figure 6.3 depicts the concept of a linear CCD array with TDI operating in pushbroom mode. Photo-generated electrons are transferred from one stage to another in the charge domain. The pixel clock rate must match to the imaging velocity. The TDI integration time of a cross-track line image on ground will be *N* times longer after *N* stages. The signal charge increases by *N*-fold. A typical TDI CCD might have $N = 64$ rows, enabling 64 times improvement in signal and 8 times improvement in SNR. This is because shot noise also increases as the square root of *N*, resulting in an SNR improvement of the square root of *N*. The practical limit on the number of TDI stages is determined by the accuracy of synchronization between the CCD array's vertical-shift frequency and the velocity of the moving object. Examples of spaceborne instruments with TDI implementation include the SeaWiFS instrument of Sea-Star, the Infrared Spectral Imaging Radiometer (ISIR) on NASA's Shuttle flight STS-85, MODIS (Moderate Resolution Imaging Spectroradiometers) on Terra and Aqua satellites, the OSA instrument of the IKONOS-2 satellite, BGIS of QuickBird, PIC-2 of the EROS-B satellite program, and MSRS (Multi-Spectral Resolution System) of the Diamant satellite (Kramer 2001).

Another example of TDI implementations is a 24,000-line multispectral TDI imager for KOMPSAT-3, an earth observation satellite developed by the Korean Aerospace Research

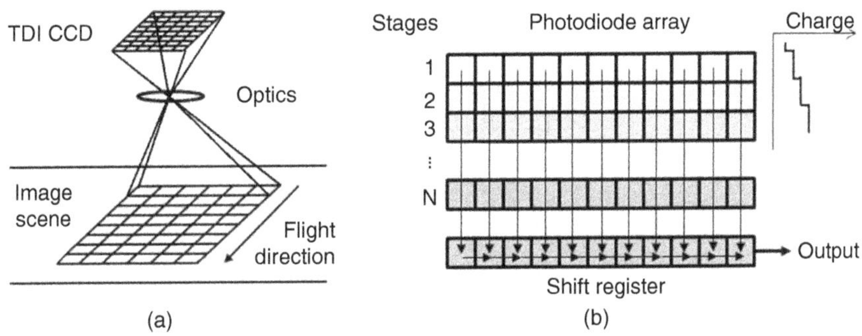

FIGURE 6.3 Concept of a linear CCD array with TDI operating in pushbroom mode.

Institute (KARI). Since the size of the device exceeded the width of the silicon wafer needed to produce the sensor, two 12,000-pixel (12K) devices were butted together to extend the length to over 8 in. of imaging area. This very large TDI imager supports KARI in gathering geographical, agricultural, and oceanographic image data for environmental monitoring applications (Kong et al. 2009). A variety of TDI CCD arrays have been developed by e2V, Fairchild Imaging, and Teledyne-DALSA.

6.3 CMOS FPAs

The complementary metal oxide semiconductor (CMOS) detector arrays were developed at the same period of time as CCD devices in the late 1960s. The first generation of CMOS detector arrays were passive-pixel sensors (PPS). They were designed in a 2D structure, with access wire shared by pixels in the same row, and output wire shared by column. Each pixel converts photons into electric charges that are then carried off the sensor and amplified by an amplifier connected at the end of each column. These sensors were small and suffered from a slow readout due to the large capacitive load and high level of noise that appears as a background pattern in the image. To cancel out this noise, PPS sensors often use additional processing steps. However, CCDs became dominant in the imaging sensors market until the 1990s because they produced better image quality with the available manufacturing technology.

6.3.1 Monolithic CMOS FPAs

In a monolithic CMOS FPA, the photodetector array and accompanying readout circuits are fabricated on the same substrate. These devices were developed and commercialized in the mid-1990s (Ohta 2008). They significantly reduce the noise that occurred with PPS by incorporating active circuitry in each pixel that determines what its noise level is and cancels it out. The general architecture of a CMOS active-pixel sensor (APS) is shown in Figure 6.4. The CMOS FPA consists of an array of pixels and each pixel contains a photodetector and an amplifier connected to signal processing and interface electronics formed on the same die. The monolithic integration of the photodetector and readout electronics in a single silicon die simplifies fabrication and scale-up manufacturing. The size of the pixel array determines the spatial resolution of the CMOS FPA. Within each pixel, there are a photo-sensitive area and three or more transistors to convert the integrated charges into a voltage, buffer and enhance the signal, and perform noise rejection function.

The 3- and 4-transistor (3T and 4T) per active pixel are the most popular designs. A typical 3T pixel design consists of a photodiode, reset gate, row selector, and source follower transistors (Holst and Lomheim 2011). An active pixel is first reset to the voltage level, V_{DD}, using the reset gate. Then the pixel integrates light, decreasing the voltage across the photodiode while photons are converted to charges. The final voltage level is stored on the common column line. Next, the pixel is reset a

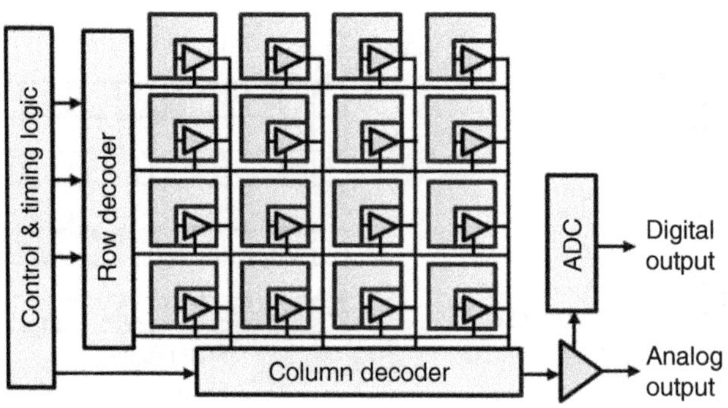

FIGURE 6.4 Architecture of CMOS active-pixel sensor.

second time. The reset voltage is measured and also stored on the column line. The column circuit samples the signal and then the reset value, and subtracts these two values to obtain the difference between the voltage levels at reset and after light integration. This non-correlated double sampling method is sufficient to reduce the pixel-to-pixel offsets but does not eliminate the temporal noise, because each time the photodiode is reset, it resets to a slightly different voltage. A small difference in the reset levels before and after the integrations results in random noise, which is referred to as reset noise or sometimes as kTC noise given its association with the gate capacitor and temperature. In addition, since the reset level is measured after the pixel has been integrated, the reset and pixel values are not truly correlated. In addition to kTC noise, residual charge on the photodiode results in image lag in fast-changing dark to light settings, which can cause ghost images.

The 4-transistor (4T) pixel design adds an additional transistor as the transfer gate and a floating diffusion. The photodiode is a pinned diode instead of a normal photodiode in the standard 3T CMOS pixel (Holst and Lomheim 2011). A pixel is reset when the reset gate and transfer gate are turned on simultaneously, setting both the floating diffusion and the photodiode to the V_{DD} voltage level. Next, the transfer gate is turned off (disconnecting the photodiode and floating diffusion) and the photodiode is left to integrate light. After integration, the signal measurement occurs. First, the reset transistor is turned on and off to reset the floating diffusion. Immediately after this, the reset level is sampled from the floating diffusion and stored on the column circuit. Next, the transfer gate is turned on and off, which allows charges on the photodiode to transfer to the floating diffusion. Once charge transfer is complete, the charges (the photodiode signal level plus the floating diffusion reset level) are measured and stored on the column circuit as well. These two stored voltages are then differenced to determine the photodiode signal level. The 4T design allows for true correlated double sampling (CDS) operation to occur, as the reset level used to determine the absolute pixel level is now measured before the signal level and the same reset level is referenced throughout the measurement. By mimicking the operation of a CCD device, this 4T pixel design significantly improves the performance of the standard 3T architecture, reducing both read noise and image lag (Djazovski 2016).

While some designs employ address decoders to enable random pixel access, a photodiode array is usually *x-y* addressed by shift registers. By enabling all row selector transistors within a single row of pixels, the array is accessed one row at a time. The individual pixels within the row are selected and read out column-by-column at the bottom of the array. The resulting video signal may be fed through a multiplexer to an analog output amplifier, a serial ADC, or an array of column-parallel ADCs. The rolling shutter mode used in 4T CMOS FPAs for exposure control results in offset between adjacent row exposures, which gives rise to motion artifacts when imaging fast moving objects. Distortion would be more apparent in cases where larger objects are moving at a rate that the image readout could not match, while distortion is less likely when relatively small objects are moving at a rate that is being temporally oversampled by the frame rate.

More advanced 5T and 6T CMOS pixel designs enable global shutter, which can work by abruptly exposing and then obstructing all pixels at once, allowing CMOS imagers to freeze motion like electronic shutter in CCD imagers. The state-of-the-art CMOS pixel architectures utilize 8T pixel designs by incorporating more electronics in the pixel-architecture at the cost of increased noise and reduced fill factor (Ohta 2008).

In order to overcome the sensitivity reduction due to a fill factor <100% in COMS FPAs, micro-lenses have been introduced to the designs to focus the incoming light onto the photosensitive area of the pixel. While micro-lenses improve the pixel sensitivity, there are many design challenges associated with the application of microscale lens arrays. One key issue is cross talk that is an erroneous signal from neighboring pixels, which degrades spatial resolution, color separation, and overall sensitivity. Several components contribute to cross talk in a sensor, including optical (due to micro-lenses and color filter), spatial, and electrical (due to pixel design). Reducing cross talk in small pixels has thus become one of the most difficult and time-consuming tasks in CMOS sensor design. At pixel dimensions below 1.45 μm, it becomes more and more difficult to guide enough light to the photosensitive element through the frontside metallization of a CMOS sensor. Advanced light guiding structures need to be used to reduce cross talk. Many of these issues have been solved when CMOS FPAs manufacturers introduced back-thinning process.

The majority of CMOS FPAs are frontside illuminated, i.e., incoming light falls on the processed side of the semiconductor die illuminating photosensitive area, transistors, and processing electronics. For backside-illuminated CMOS FPAs, the wafer fabrication process is mostly the same as frontside-illuminated sensors. Photodiodes are laid out first and then the metal layer. After that, the silicon wafer is flipped over and the excess bulk silicon is ground down until the wafer is only about a 10 μm thick (Ohta 2008). This process has been used for many years for manufacturing high-performance CCDs. Grinding the silicon down allows the wafer to only be as thick as it needs to be about enough to expose the photodiodes. After grinding, the color filters and micro-lenses are laid out as with frontside-illuminated sensors. This architecture improves fill factor to nearly 100%. By contrast, typical fill factors of frontside-illuminated sensors range from 30% to 70%, i.e., up to 70% of the photodiode area may be shielded by transistors and stacked or interleaved metallic bus lines, which are optically opaque and absorb or reflect a majority of the incident photons colliding with the structures. Due to wavelength-dependent nature of photon absorption, the fill factor is a critical parameter in determining the QE of a CMOS FPA, especially at shorter wavelengths.

Disadvantages associated with backside-illuminated CMOS FPAs include lower production yields and higher costs, because the thinner silicon wafer makes dies more fragile to produce and more complex to package. In some designs the silicon membrane is very thin to minimize the spreading of charge in the field-free region outside the photodiode that eliminates NIR sensitivity at wavelengths beyond 650 nm which is often desired for scientific applications.

It is important to note that performance metrics of commercial CMOS FPAs is based on a statistical model of the human visual response to light which is restricted to the wavelength range from about 390 nm to 720 nm. In contrast, scientific CMOS FPAs tend to require response anywhere and/or everywhere from ultraviolet to 1.0 μm.

Shrinking pixel size of CMOS FPAs poses significant challenges to sensor design. As pixel size gets smaller, the well capacity shrinks, resulting in smaller dynamic range. The photon collection area also shrinks, causing less sensitivity. Noise, coupling, and nonuniformity become more severe and visible. The pixel cross talk also inevitably gets higher.

Details on monolithic CMOS FPAs have been provided by Djazovski (2016).

6.3.2 Hybrid CMOS FPAs

Hybrid CMOS technology combines the best quality detector material with the superior performance achieved by the state-of-the-art CMOS integrated circuit technologies. The detector array and CMOS read out integrated circuit (ROIC) are optimized independently, which allows to fine-tune key performance parameters, such as spectral response, QE, full well capacity, noise floor, MTF, etc., for each specific application. The CMOS ROICs are usually fabricated in the same silicon foundries that produce computer chips.

A hybrid CMOS detector array is shown schematically in Figure 6.5. The fabrication of hybrid CMOS arrays involves three main process steps (Bai et al. 2012):

1. Growth and processing of the detector layer.
2. Design and fabrication of the CMOS ROIC.
3. Flip-chip bonding (hybridization) of the detector layer to the CMOS ROIC.

Various flip-chip technologies have been used for manufacturing hybrid CMOS arrays (Kozlowski and Kosonocky 2010). A process of indium bump-to-bump interconnection (hybridization) is adopted, which is achieved by aligning the bumps and applying high normal forces to cause the bumps on opposing surfaces to fuse (cold-weld) together. This process is widely used in volume

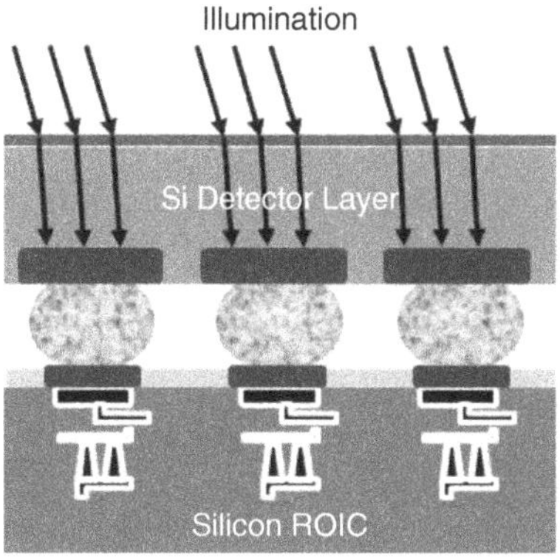

FIGURE 6.5 Cross-section view of a hybrid CMOS FPA.

manufacture of infrared (IR) arrays. However, large FPAs with fine-pitched bumps may require substantial force to achieve reliable interconnection, which creates the potential for damaging delicate semiconductor detector layers. This limits the thickness of active layer to 75–100 μm, which is beneficial for improving NIR performance but may be a problem for blue- and UV-enhanced devices and MTF. Also, there is always a risk of misalignment associated with the joining process and "hybrid slip." In addition, even small variation in camber between the mounted surfaces can cause the indium bumps to separate after removal of the normal forces applied during the hybridization process that makes it a challenge to achieve 100% interconnectivity in large area arrays.

Some of these issues can be resolved by using a fusion bonding (or direct bond) process. This process usually involves a room temperature alignment of wafers via van der Waals interactions followed by annealing that creates permanent covalent interfacial bonds between upper and lower wafers. This eliminates misalignment errors (as fusion bonding results in consistent submicron alignment accuracies) and many potential problems associated with compression bonder (hybrid slip, layer damage, bump separation, etc.). Also, fusion bonding makes it feasible to produce very large arrays up to 10 K × 12 K format and thin active layers (<10 μm), which is beneficial for devices with enhanced blue and UV performance and MTF (Kean et al. 2012). The drawback of this method is its extreme sensitivity to surface flatness, roughness, and particles.

An alternative technology is an adhesive bonding between the active detector layer and support wafer. This provides low temperature bonding and alignment accuracy in a 1–2 μm range. This technology eliminates concerns associated with wafer flatness, roughness, and particles. However, the adhesive has to be carefully selected and tested for mechanical rigidity, temperature stability, and outgassing. Inter-connections must be made after bonding and thinning. The active layer can be as thin as 4 μm, which is very attractive for UV-enhanced devices and MTF (Jerram et al. 2010).

Scientific CMOS FPAs with hybrid architecture are available in formats as large as 5 K × 5 K four-side buttable, which is essential for making large FPA mosaics, and radiation tolerant by design. Hybrid CMOS FPAs achieve the same performance as the best deep depletion CCDs. (Bai et al. 2012, Kean et al. 2012). Figure 6.6 shows examples of QE curves of monolithic and hybrid CMOS arrays. It can be seen that hybrid CMOS FPAs achieved significant QE enhancement. Hybrid CMOS FPAs also have very low power consumption and allow nondestructive random access to the pixels and on-chip integration of analog and digital circuitry, which significantly reduces the imaging system mass, volume, power, and thermal output and overall complexity for space use.

During the operation of a hybrid CMOS FPA, the silicon ROIC converts the photo-generated charges to voltage with an amplifier in each pixel, and transfers the signal to the edge of the array. One of the key advantages of hybrid CMOS arrays over monolithic devices for hyperspectral imagers is the possibility to use the same ROICs for both detector arrays dedicated to visible and near-infrared (VNIR) and shortwave infrared (SWIR) regions. This provides a common electrical

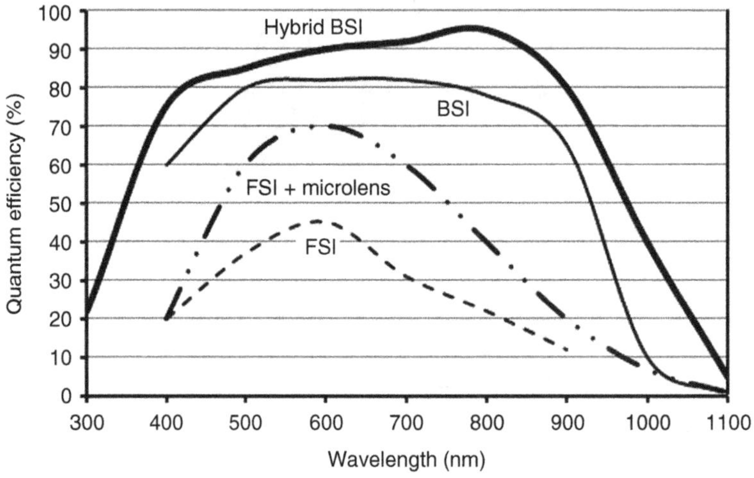

FIGURE 6.6 Examples of quantum efficiencies of monolithic and hybrid CMOS FPAs.

interface to both types of detector arrays and greatly simplifies design of hyperspectral instruments. Detailed description of ROICs for hybrid detector arrays is provided in Djazovski (2016).

6.3.3 Comparison CCD Versus CMOS FPAs

CCD and CMOS detector arrays are all manufactured from silicon. Both technologies can support two types of photo sensitive elements: the photogate and the photodiode. Both of them convert incident light (i.e., photons) into electronic charges (i.e., electrons) by the same physical process known as photoelectric effect. Both technologies have fundamentally similar properties and offer excellent sensitivity and imaging performance over the visible and near IR spectrum. Therefore, the criteria for the decision on which technology to use thus depend on the applications.

CCD technology is currently the mainstream technology used for space imaging. Almost all of the more recent astronomy observations in the visible region of the spectrum have been made with CCDs. The main advantages of CCD arrays are their high QE over a wide spectral range, high dynamic range, high linearity, low read noise, and low dark current. Scientific CCD arrays still represent important imaging sensor technology in high-end applications (such as large area FPAs for telescopes). Improvements are still being made to CCD technology. Current Earth observation satellites and planetary observation (e.g., Hubble, XMM, HiRISE, ENVISAT, Pleiades, Spot, Helios, GCOM, etc.) all rely on CCD technology (Bogaerts 2014).

The main disadvantages of CCD arrays are complex power supply and support electronics, high power consumption, and performance degradation in radiation environment. CCD arrays tend to be inherently power hungry because they are essentially capacitive devices that need external control signals and large clock swings (5–15 V) to achieve acceptable charge transfer efficiencies. The clock amplitude and shape are critical to successful operation of a CCD detector array, resulting in the need for specialized clock drivers that are used to generate correctly sized and shaped clocks. CCD arrays often require multiple non-standard supply voltages and off-chip support circuitry that dissipates significant power. In addition, frame transfer CCD arrays generate smear effect during the transfer of collected charges in the imaging zone to the storage zone. Smear is an unwanted signal that appears as a bright vertical strip that needs to be corrected sufficiently on ground in order to obtain real signal. Table 6.1 summarizes the main advantages and disadvantages of CCD FPAs.

Driven by the huge amount of investment in CMOS FPAs for mobile phone cameras and consumer electronics, CMOS technology is catching up with CCD technology in terms of image quality.

TABLE 6.1
Advantages and Disadvantages of CCD FPAs

Advantages	Disadvantages
Large format devices with good spatial resolution	Long readout times with only a few readout ports
Very low readout noise (a few electrons rms)	Noise increases with increased readout rate
Very low dark current (from <1 [typical] to ~0.001 e^{-}/sec/pixel at an appropriate low temperature)	Destructive readout
Common amplifier for each readout segment	Serial readout (pixels cannot be randomly accessed)
Correlated double sampling each clock cycle	Numerous off-chip power suppliers and voltage regulators and support circuitry for operation
Photo response nonuniformity <1%	High power consumption (2–5 W) due to large amplitude clock voltages (5–15 V)
Low fixed pattern noise, typically <1%.	Complex clock amplitude and phasing requirements; requiring individual adjustments for each CCD
Linear response and a large dynamic range (≥90 dB)	Blooming and smear effects
Mature technology with significant flight heritage	Charge transfer efficiency degradation in radiation environment

Advantages of CMOS detector arrays over CCD arrays are their relatively simple operation without the need to fine-tune clocking voltages to the characteristics of individual devices. CMOS detector arrays, like CMOS integrated circuits, operate at considerably lower voltages than CCD arrays, typically between 1.8 V and 5 V depending on process form and architecture. This enables easy integration of CMOS arrays with other CMOS circuits such as digital signal processors, digital bus interfaces, etc., which allow for the reduction of overall system complexity and component count. In addition, CMOS FPAs offer more functionality such as timing and control electronics, signal processing electronics, ADC, and digital interface integrated on one chip. This allows achieving a cost-effective, highly integrated and compact imaging system, known as a camera-on-a-chip which has evolved dramatically in recent years in consumer electronics and industrial applications. When a scientific CMOS FPA is coupled with a newly developed application-specific integrated circuit (ASIC) readout chips (Loose et al. 2006), very compact system is possible that are highly scalable. For instruments requiring low detector operating temperatures, for example, combined visible and IR instruments, CMOS FPAs offer the advantage of operating down to about 30 K without problems (Bai et al. 2012). Some silicon CMOS devices can be designed to operate at temperature below 10 K. The CCD devices begin to stop working at liquid nitrogen temperature (77 K or −196°C) because of a physical limit known as carrier freeze out (Janesick 2001). For space applications, the better radiation hardness of CMOS devices is another important advantage. CMOS FPAs offer a great flexibility in readout modes without smear effect, from fast reading to low-noise slow reading, the possibility to guide on brighter stars in the field, and the ability to achieve, in up-the-ramp sampling, very high dynamic range, and the rejection of cosmic ray events during the integration. Table 6.2 summarizes the main advantages and disadvantages of hybrid CMOS FPAs.

CMOS detector arrays were mainly introduced in space as FPAs for star trackers because the flexibility and ease of implementation provided by (multiple) windowing, nondestructive readout, and on-chip ADC make CMOS-based sensors very appealing for such applications. Hybrid silicon detectors have been used for various Earth observation instruments, such as MODIS (see Section 2.4 for details), EO-1, and Landsat now on orbit. Although for future space missions, CCD technology is still considered as the most mature technology, many government organizations and space agencies support the development of CMOS FPAs as a promising alternative (Bogaerts 2014).

TABLE 6.2
Advantages and Disadvantages of Hybrid CMOS FPAs

Advantages	Disadvantages
Independent optimization of the detector and ROIC designs and processing, free of smear effect	Fixed pattern noise due to transistor offsets (correctable)
Nondestructive readout allows multiple sampling during long integration	Pixel-to-pixel gain differences due to independent amplifier per pixel
Addressable readout of selected image areas (for guider readout); random access to pixels (windowing)	Correlated double sampling (CDS) possible only over longer time intervals
Low voltage (3.3 V); low power (1–200 mW)	Higher read noise than in CCDs
Large dynamic range by nondestructive readout "up the ramp"	Capacitive (deterministic) cross talk to adjacent pixels
System-level functionality integrated on-chip (timing, control and signal processing electronics, ADC, and digital interface)	Relatively new technology (some arrays not yet proven on space telescopes)

6.4 SWIR FPAs

The main motivation for the research and development of SWIR FPAs is due to the properties of solar reflectance region extending from 0.4 μm to 2.5 μm, which provides the best opportunity to identify the materials in a scene observed from airborne or spaceborne platform by matching the reflectance spectra of the scene to a library of known spectra (Rogalski 2010). The scientific interest in the SWIR spectral region arises because it plays a main role in the Earth's radiation budget and also because it is employed to validate models used in solar physics (Loose et al. 2006). This wavelength region is of key interest in astronomy because it would enable investigations of the distance redshift relationship and the evolution of cosmic structures by measuring shapes and redshifts of galaxies and clusters and facilitate the search for planets in our galaxy (Zombeck 2006).

6.4.1 HgCdTe FPAs

Mercury-cadmium-telluride (MCT or HgCdTe) is one of the most important semiconductor materials used for the manufacture of high-performance IR FPAs. It is the highest performance IR material currently available and has the largest manufacture and investment base for IR materials. It is a unique IR detector material because its bandgap can be adjusted by varying the relative proportion x of Hg and Cd constituting elements ($Hg_{1-x}Cd_xTe$) to cover all wavelengths between 0.7 μm and 20 μm. Figure 6.7 shows the curves of cutoff wavelength of HgCdTe material as a function of the composition parameter x for temperatures of 77 K and 200 K.

In spite of the variable bandgap, there is very little variation in the value of the lattice constant across all range of $Hg_{1-x}Cd_xTe$ compositions, which allows for simplification of many of the enabling techniques used in FPAs, such as substrate selection, passivation, ROIC optimization, etc. In principle, with the development of HgCdTe technology, it has become possible to select a wavelength band in the IR region to optimize the system performance and to build the FPA to meet the need.

HgCdTe is a direct bandgap semiconductor material with high absorption coefficient for the IR radiation that results in high QE even in a relatively thin detector structure (approximately 10 μm thick). The direct bandgap allows operation of the IR detector arrays at a relatively higher temperature than other mid- and long-wavelength IR detector arrays such as InSb and SiAs. These fundamental advantages have led to a considerable investment in HgCdTe technology over the years as the material of choice for high-performance SWIR (1–3 μm), MWIR (3–5 μm), and long-wave

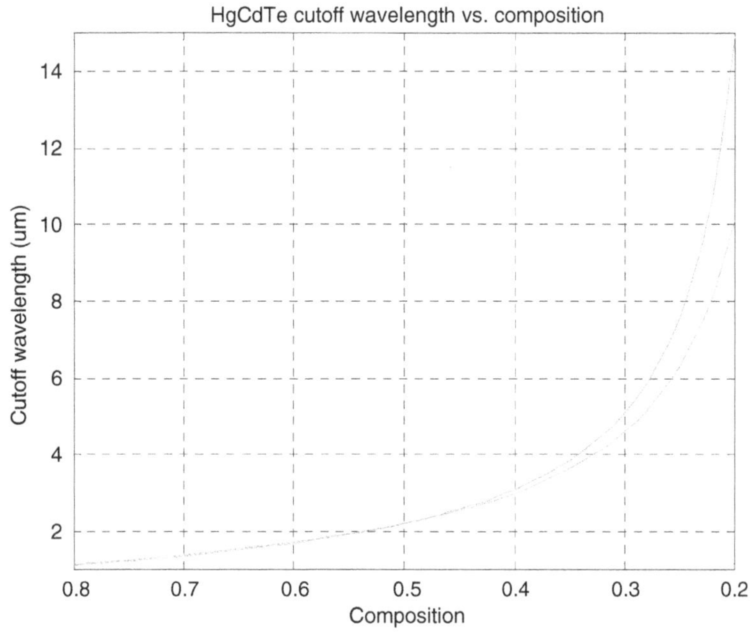

FIGURE 6.7 Plot of cutoff wavelength of $Hg_{1-x}Cd_xTe$ as a function of the composition parameter x for temperatures of 77 K (blue) and 200 K (red).

infrared (LWIR; 8–12 μm) FPAs. Scientific applications of HgCdTe FPAs include 1D and 2D detector arrays in thermal imaging systems, ground and space astronomical telescopes, and surveillance imaging systems, spectrometers, and radiometers (Dereniak and Boreman 1996, Norton 2003, Rogalski 2010).

The HgCdTe FPAs have the following common features even though their architecture and manufacturing processes vary significantly from one manufacturer to another. First, they are all based on epitaxial growth of the HgCdTe material. Second, arrays of photodiodes are fabricated using a sequence of doping, layering, and etching steps. Third, all detector dies are connected to a silicon ROIC via a flip chip bonding.

Various processes have been utilized to grow HgCdTe material, including liquid phase epitaxy (LPE), metal-organic chemical vapor deposition (MOCVD), molecular beam epitaxy (MBE; Capper 2006). LPE has been the fabrication process most favored for volume manufacture of small- to mid-sized single-color p-n heterojunction device structures with excellent structural quality. The need to address the particular mission requirements that can only be satisfied with the development of more sophisticated device structures has motivated an increased level of research investment in the MOCVD and MBE growth of HgCdTe FPAs. MOCVD allows material growth on low-cost substrates. The composition, thickness and doping levels of individual layers can be programmed to achieve any desired device structure. Over the course of the past decade significant progress has been achieved with MBE growth technology that has enabled the synthesis of complicated multilayer IR detector structures at the lowest growth temperature under an ultrahigh vacuum environment. The development and integration of in situ sensors for monitoring and control of the sensitive HgCdTe MBE growth process have significantly advanced the capabilities and consistency of the technology, resulting in demonstrations of excellent device performance in a wide variety of IR device architectures such as multispectral detectors, high-performance MWIR detectors, and megapixel arrays on Si substrates. Companies such as Teledyne, DRS, Raytheon, BAE, Sofradir, and many others have published numerous papers on their progress with commercial products.

The predominant interest in SWIR FPAs is driven by the astronomy community, which requires large format and extremely low dark current due to the low backgrounds involved the observations. An example is the Wide Field Camera 3 (WFC3) program that is an instrument upgrade for the Hubble Space Telescope (HST). During the course of this program Teledyne developed a new type of HgCdTe 1024 × 1024 (1K × 1K) detector array, called WFC3-1R, with cutoff wavelength of 1.7 μm and 150 K operating temperature. It made possible to significantly improve QE, dark current, and read noise due to material developments required for WFC3. Many of the performance increases are associated with the development of substrate removal technology. The substrate-removed FPAs have attained QEs very close to ideal (>90%) within cutoff wavelength of 1.7 μm, readout noise of 30 e^- rms with double correlated sampling, and dark current <0.2 e^-/sec/pixel at 150 K (Piquette et al. 2008).

To meet the rapidly rising astronomical instrumentation market demand for large-format SWIR arrays Teledyne developed HAWAII-2RG arrays of 2048 × 2048 (2K × 2K) pixels with 2.5 μm cutoff wavelength. Several NIR instruments at European Southern Observatory (ESO) are equipped with these arrays that delivered excellent performance (Beletic et al. 2008). Raytheon Vision Systems and Teledyne have developed the 2048 × 2048 format, 20-μm pixel pitch FPAs for the United Kingdom's Visible and Infrared Survey Telescope for Astronomy (VISTA) program (Hodapp and Hall 2006). Large format (1K × 1K and 2K × 2K) NIR detectors manufactured by Teledyne and Raytheon Vision Systems were used by the Super-Nova/Acceleration Probe. The ESA has funded Selex ES to develop large format NIR detector arrays for its future space and ground-based programs (Atkinson et al. 2012).

Future flagship space missions such as NASA's Wide Field Infrared Survey Telescope (WFIRST), ESA's Euclid mission, as well as next-generation ground-based surveys require large mosaic FPAs sensitive in both visible and NIR wavelengths. HgCdTe is the key material for the development of a new generation of large-format SWIR detector arrays. Teledyne was fabricating the SWIR arrays for Euclid (Maciaszek et al. 2014) and developing the new H4RG-10 FPAs for WFIRST space telescope instrument (Piquette et al. 2014). The 4096 × 4096 arrays with 10-μm pixel pitch are fabricated in a fully substrate removed configuration to provide high QE, response to visible wavelengths, and cosmic ray and fringing mitigation. These hybrid sensors can be manufactured as either a silicon p-i-n detector array (e.g., HyVISI) or a HgCdTe photodiode array with IR cutoff wavelength of 1.75, 2.5, or 5.3 μm. Initial test results have demonstrated a low mean dark current <0.01 e^-/sec/pixel at ~100 K, low read noise ~11 e^-, and array operability >99% pixels (Piquette et al. 2014).

6.4.2 InGaAs FPAs

The indium-gallium-arsenide (InGaAs) FPAs can be made from either photodiodes or optoelectronic devices. They exhibit high responsivity in spectral range from 0.9 μm to 1.7 μm at room temperature. The detector materials can be prepared using MBE, MOCVD, LPE, hydride-transport vapor phase epitaxy (VPE), and atomic layer epitaxy (ALE) (Capper et al. 2006). $In_xGa_{1-x}As$ layers are typically grown lattice-matched to the InP substrates using an alloy composition of $x = 0.53$. In this, material planar diodes are formed by diffusion. The detector array is attached to a silicon ROIC via indium bump interconnection to form a hybrid FPA.

By increasing the In proportion $x = 0.82$, the spectral response of $In_xGa_{1-x}As$ can be extended to 2.6 μm. However, crystal defects originating from the lattice mismatch with the InP substrate degrade the diode performance at the longer cutoff wavelengths. The test results reported for back-side-illuminated FPAs indicate that the standard $x = 0.53$ $In_xGa_{1-x}As$ photodiodes have the best room temperature detectivity, D^*. With increasing cutoff wavelength, there is a marked decrease in D^*. It is decreased by approximately one order of magnitude when cutoff wavelength is extended to 1.95 μm and decreased by approximately another order of magnitude when cutoff wavelength is extended to 2.25 μm. With the standard $x = 0.53$ $In_xGa_{1-x}As$ photodiodes detector, the QE measured across the 1.2–1.6 μm bands is about 80% over a temperature range of 65–40°C. The spectral response of InGaAs devices can be extended to visible wavelength region by removing the InP substrate that results in QE about 40% over the entire visible band (Nelson et al. 2008).

SWIR imaging for surveillance, military, and aerospace applications is driven primarily by potential phenomenological advantages in detecting objects using target signatures that are dominated by reflection of external sources of illumination as opposed to thermal emission of radiation that occurs in the longer IR wavelengths. The main advantage of SWIR imaging is that it produces significantly better quality images in challenging atmospheric conditions such as haze, mist, fog, and rain compared to visible imagers. In addition, SWIR imaging excels in low-light conditions that make SWIR imaging very attractive for object or individual identification applications. Unlike HgCdTe or indium antimonide (InSb) PFAs, InGaAs SWIR FPAs operate at 20°C, require no cooling or mechanical shutters, and never need field nonuniformity corrections (NUCs). As a result, InGaAs FPAs have been designed with the goal to restrain the size, weight, and power consumption (SWaP) of cameras on small platforms for persistent large area surveillance under low light conditions.

The dark current of InGaAs FPAs can be reduced through designing semiconductor structures with tailored electronic and optical properties and optimized fabrication processes. For example, sidewall-related dark current can be minimized to negligible levels with appropriate surface passivation of the exposed p-n junction. By focusing on the growth conditions for the InGaAs absorption layer, hetero-interfaces, and the passivation layer, a dark current density <0.5 nA/cm^2 has been demonstrated for a 1024 × 1280 InGaAs PIN FPAs with 12.5-μm pitch size at 15°C (Yuan et al. 2014). InGaAs arrays with formats of 640 × 512 and 1280 × 1024 pixels and pitch size ranging from 12.5 μm to 17 μm are commercially available from AIM Infrarot-Module GmbH, Sofradir, Spectrolab, UTC Aerospace Systems, and Xenics.

InGaAs FPAs are not widely used for space scientific instrumentation, despite their potential advantages and technological advances. Linear InGaAs arrays have been used in the Descent Imager/Spectral Radiometer (DISR) aboard the Cassini-Huygens mission to Titan (Tomasko et al. 2003) and in the SCIAMACHY instrument aboard of ESA's ENVISAT (Kleipool et al. 2007). SCIAMACHY employed one lattice-matched InGaAs detector and two wavelength-extended InGaAs with 2.4-μm cutoff wavelength. Since wavelength-extended InGaAs FPAs are used for the first time in a spaceborne instrument, the performance of the detector arrays has been monitored since launch. The lessons learned will help for their future space utilizations. The results show only minor degradation of the silicon-based UV-Vis detector arrays and the lattice-matched InGaAs NIR detector array. In contrast, the degradation of wavelength-extended InGaAs detector arrays is unexpectedly large. The degradation of individual detector pixels is abrupt and mostly leads to excessively high dark current. The results reported by Kleipool et al. (2007) have shown that nonionizing radiation due to trapped protons leads to displacement damage to the InGaAs detector material. The current degradation rate leads to a loss of approximately 50 out of 1024 pixels per channel per year.

Another application of InGaAs FPAs is in the multispectral instrument of the Proba-V satellite (Bentell et al. 2010). Three large linear 1024 pixels arrays with 25-μm pitch size are used to form a nearly continuous line of 3072 pixels and installed in the multispectral instrument.

6.4.3 HgCdTe Versus InGaAs for SWIR FPAs

High-quality epitaxial layers of HgCdTe and of InGaAs FPAs can be grown by the state-of-the-art thin film growth techniques such as MBE and MOCVD with precise control of composition, doping, and thickness of semiconductor structures. Both materials have high QE and low thermally excited carrier density relative to extrinsic photoconductors. In terms of performance, large area InGaAs detectors with high and flat QE (above 80%) over the spectral range from 0.9 μm to 1.7 μm, low dark current and noise, and ability to operate at room temperature are commercially available. In addition, technology for HgCdTe, such as substrate removal, which is used to increase short-wavelength performance and minimize nonuniformities at cutoff wavelength and cosmic ray effects, also exists for InGaAs devices. These advantages offer imagers based on InGaAs technology, the potential to be a lower cost, higher reliability alternative to those based on HgCdTe.

During the course of the mission concept development of SuperNova/Acceleration Probe (SNAP) program led by Lawrence Berkeley National Laboratory, the performance of a lattice-matched InGaAs array was compared with a HgCdTe array hybridized to a Teledyne's multiplexers (Seshadri et al. 2007). The results showed that although InGaAs has a cutoff wavelength of 1.7 μm at room temperature, its bandgap is temperature dependent, and the cutoff shifts to 1.57 μm at the required operating temperature of 140 K. As discussed earlier, it is possible to fabricate InGaAs layers with cutoff wavelengths longer than 1.7 μm, though the devices exhibit degraded performance due to increasing lattice mismatch with the InP substrate. In contrast, HgCdTe maintains a nearly constant lattice parameter over the entire range of alloy composition and can be easily optimized for 140 K without any performance degradation.

InGaAs FPA may be the detector of choice, when cost is taken into account and when the longest detection wavelength is 1.7 μm.

It should be noted that detection of SWIR can also be achieved with FPAs based on InSb. InSb is a traditional semiconductor material used for photodiode arrays with response up to 5.5 μm. The bandgap of InSb cannot be tuned. InSb FPAs have dark current performance which is consistent with the long cutoff wavelength SWIR FPAs. These devices have been used in hyperspectral instrumentation (e.g., the HYDICE airborne instrument). However, InSb must be cooled to <77 K, which is a significant drawback for the space application, while InGaAs and HgCdTe can be used at room temperature. This makes InSb less competitive for SWIR applications.

6.5 HYPERSPECTRAL FPAs

As described in Section 5.4, all spaceborne hyperspectral imagers use 2D detector arrays and operated in pushbroom mode. The 2D FPA of a slit scan hyperspectral imager can be viewed as an assembly of many linear detector arrays (rows) combined into a 2D array. In the hyperspectral imager, an optical dispersing element such as a grating, prism, or a wedged filter splits the incoming light into many narrow adjacent wavelength bands illuminating on each detector row with different wavelength. Therefore, a distinguishing feature is that in hyperspectral FPA the entire spectrum is distributed across the device and may be divided into bands as required as depicted in Figure 1.2, whereas in multispectral FPA, discrete broad spectral bands are filtered and imaged on separate linear arrays as depicted in Figure 2.2. The satellite flight provides the scan on the ground scene and the imaging sensor obtains the scene image with the spectral content of incoming light being split across the FPA. The column data from the FPA then provide spatial information and the row data provide spectral information. The combination of spectral data and spatial detail enables analysis of color, chemical content, uniformity, quality, and a host of other spectrophotometric sensing applications.

Hyperspectral FPAs can obtain scene image in spectral bands from the UV to LWIR spectral regions. A majority of hyperspectral imagers have been designed to operate over the VNIR and SWIR regions to exploit solar illumination to detect and identify materials on the basis of their reflectance spectra. Hyperspectral FPAs require innovative design considerations on sensitivity, efficiency, and dynamic range to overcome challenges associated with a large variation in signal intensity between the different spectral bands and to avoid optical cross-talk of the high-intensity spectral bands with low-intensity bands. Most instrument designs of hyperspectral imagers use two separate FPAs for the VNIR and SWIR regions (for example, as described in Section 5.7). Key attributes for developing leading-edge hyperspectral FPAs include ultra-high QE, backside thinning, high-speed read rate, low defects, and advanced packaging techniques (Jerram et al. 2011).

CCD FPAs have been successfully used for many years for VNIR hyperspectral imaging applications (Jerram et al. 2011, Kullar et al. 2011). Examples include the Medium Resolution Imaging

Spectrometer (MERIS) on ESA's ENVISAT (Bézy et al. 2016), the Compact High Resolution Imaging Spectrometer (CHRIS) on PROBA (Cutter 2000), and Land and Ocean Colour Imager (LOCI) onboard Sentinel-3A (Nieke et al. 2016). CCD FPAs are being considered for a number of upcoming missions such as Sentinel 4 and 5 (Jerram et al. 2011).

Recent advances in CMOS FPAs technology result in their wider acceptance for VNIR hyperspectral imaging applications. For example, the technology trade-off studies for the EnMAP mission (Kaufmann et al. 2016) have highlighted the performance advantages of CMOS FPAs over CCD FPAs for hyperspectral imaging applications. The main advantage is that CMOS technology eliminates the frame-shift time and spectral smear. The smear of CCD must be removed in data post-processing. This processing step is known to have a detrimental effect on data quality by reducing the SNR depending on the spectral signature of the recorded data. In CMOS FPAs, the intense spectral lines can be read multiple times within a frame resulting in a significant increase in dynamic range. In addition, CMOS technology eliminates issues associated with CTE degradation in CCD devices. The FPA selected for EnMAP is a back thinned silicon CMOS of format 1024 × 146 with 24-μm pixel pitch manufactured by Fairchild Imaging. The FPA is based on a pinned photodiode with a 5-transistor pixel architecture to allow global electronic shuttering in a stare-while-scan read out mode. The full well capacity is 1 million electrons (Me^-) with 2% nonlinearity. The linearity of the sensor data will be corrected to 0.2% by implementing a look-up table (LUT) based algorithm with individual correction LUTs for each pixel at the ground post-processing stages.

SWIR hyperspectral FPAs have been developed and fabricated. Sofradir in France manufactures 1000 × 256 and 500 × 256 HgCdTe FPAs with 30-μm pixel pitch for SWIR (0.8–2.5 μm) and visible-infrared (VIS-IR; 0.4 up to 2.5 μm) spectral regions (Fièque et al. 2010). These FPAs have been extensively tested for space applications, including radiation tests, thermal cycling, and ageing. Various packaging designs have been validated for shocks and vibration resistance. For Sentinel 2 and PRISMA programs (Meini et al. 2016), Sofradir has developed FPA designs compatible with passive cooling.

AIM in Germany has fabricated 1024 × 256 SWIR hyperspectral FPAs covering a wavelength range 0.90–2.45 μm (Sang et al. 2008) with a rectangular pixel pitch of 24 μm × 32 μm. While 24-μm pixel pitch along the spatial dimension of the FPA is compatible with standard silicon foundry stepper rules and fits into AIM's standard Dewars, the 32-μm pixel pitch along the spectral dimension of the FPA provides a sufficient additional area to fit the CTIA circuitry and in-cell logic (Kaufmann et al. 2016). The FPA operates with frame rates up to 245 Hz in the full frame mode and up to 25 kHz in the 8 × 8 pixels subframe mode.

Teledyne manufactured the Configurable Hyperspectral Readout for Multiple Applications (CHROMA) FPA designed for VNIR to SWIR imaging spectrometers. The FPAs based on Teledyne's HyViSi hybrid silicon p-i-n VNIR offer a spectral coverage of 0.33–1.05 μm, whereas the FPAs based on Teledyne's latest substrate removed HgCdTe technology cover a wavelength range 0.30–2.50 μm (VNIR+SWIR) or 3.80–5.50 μm. The FPAs are available in formats ranging from 320 × 480 to 1600 × 480 pixels with a pixel pitch of 30 μm and various full well capacities, all of which are optimized for low noise performance (Demers et al. 2013).

Raytheon has designed the high dynamic range Multispectral Adaptive Networked Tactical Imaging System (MANTIS) sensor chip assembly (SCA) and demonstrated a variety of FPAs with formats ranging from 640 × 512 to 2048 × 2048, and pixel pitch ranging from 10 μm to 30 μm (Acton et al. 2009). The baseline SCA was designed for high-resolution SWIR hyperspectral imaging with 20-μm square pixels in a 1280 × 1024 format and flat QE of 85% from 0.9 μm to 2.5 μm at 80 K operating temperature. Other formats are also available, including 640 × 512, 1920 × 512 pixels, as well as FPAs with Visible-SWIR detector layer, SWIR InGaAs, and LWIR HgCdTe detector materials.

Performance specifications of VNIR and SWIR hyperspectral FPAs (Sang et al. 2008, Fièque et al. 2010, Jerram et al. 2011, Kullar et al. 2011) are listed in Table 6.3.

TABLE 6.3
Characteristics of the State-of-the-Art VNIR and SWIR FPAs for Hyperspectral Imaging

Parameter	Fairchild VNIR CMOS (EnMAP)	e2V VNIR CMOS	AIM SWIR (EnMAP)	Sofradir VIS-IR	Teledyne CHROMA	Raytheon MANTIS
Spectral range (μm)	0.42–1.00	0.40–1.00	0.90–2.45	0.80–2.50	0.33–1.05 0.35–2.50 0.38–5.50	0.40–0.90 0.50–1.70 0.90–2.50
Array format	1024 × 146	1024 × 256	1024 × 256	1000 × 256	From 320 × 480 to 1600 × 480	1280 × 1024 1920 × 512 640 × 512
Pixel size (μm)	24 × 24	24 × 24	24 × 32	30 × 30	30 × 30	10 × 10 15 × 15 20 × 20 30 × 30
Full well capacity (Me⁻)	1.0	0.1–0.3	1.2 (low gain) 0.3 (high gain)	1.6 (low gain) 0.4 (high gain)	0.7–5.0	0.07–2.2
Noise (e⁻)	30 (high gain) 200 (low gain)	15	140 (high gain) 290 (low gain)	450 (low gain) 150 (high gain)	80–600 (depending on full well)	15
Number of outputs	8	8	8	8	2–10	4 or 8

6.6 CONSIDERATIONS OF FPA SELECTION

In the design of spaceborne hyperspectral imagers, the selection of the FPAs is critical as they directly impact the design of the instrument. The FPAs operate at high frame rate and can generate up to 60 Mpixel/sec of data that must subsequently be digitized with low noise and high dynamic range. Integration time of the FPAs must be well controlled to make optimum use of the available light and the pixel full well capacity to achieve high SNR as well as to enable all onboard calibration procedures. In the previous sections, many aspects of FPAs, such as array format, device architecture, operating temperature, QE, readout noise, and dark current, have been described. In the selection of FPAs of a hyperspectral imager, the following six parameters of the FPAs are worth to consider.

6.6.1 Detector Pixel Size

This section uses an example to show the determination of detector pixel size. In the design of a hyperspectral imager, one of the driving detector requirements comes from a combination of the SNR requirement and the manufacturability of the optics. In an effort to determine the detector pixel size, two combinations of ground sampling distance and orbital altitude were chosen in a radiometric model to determine instrument apertures that produce similar SNRs. A SNR of 250:1 at wavelength 2.0 μm under typical observing and target conditions is used for the modeling.

Table 6.4 lists the common parameters used in the radiometric model. An assumption was made that the overall instrument throughput was fixed at 60%, including detector QE. The modeling resulting apertures are (1) 38 cm for the case of 20-m Ground Sampling Distance (GSD) at 650-km orbital altitude, and (2) 22 cm for the case of 30 m GSD at 650-km orbital altitude.

For a range of detector pixel sizes, the GSDs listed in Table 6.4 are used to obtain the corresponding range of focal lengths. The combination of apertures (as per Table 6.4) and focal lengths gives the overall optical system *f*-numbers shown in Figure 6.8.

TABLE 6.4
Input Parameters for Modeling Signal-to-Noise Ratio at 2.0 μm

Parameter	Value
Sensor view	Nadir
Solar illumination angle	30°
Orbital altitude	650 km
Orbital velocity	6.9 km/sec
Ground Sampling Distance (GSD)	20 m, 30 m
Integration time	2.9 msec, 4.3 msec
Spectral Sampling Interval (SSI)	10 nm
Readout noise	100 e^-
Dark current	3–5 e^-/sec/pixel
Residual dark current nonuniformity	1%
Residual photo-response nonuniformity	0.1%

Given the requirements on aperture reimaging and relay, an imaging spectrometer with significantly faster than f/2 optical system is challenging to fabricate, align, and keep stable. In the design, it is recommended to avoid f/# being significantly <2.0. As shown in Figure 6.8, the detector pixel sizes greater than 20 or 25 μm are required. When detector pixel size is 20 μm, f/# is 1.7 for the 20-m GSD case and f/# is approximately 2.0 for the 30-m GSD case. When detector pixel size increases to 25 μm, f/# increases to 2.2 for the 20-m GSD case and to approximately 2.5 for the 30-m GSD case.

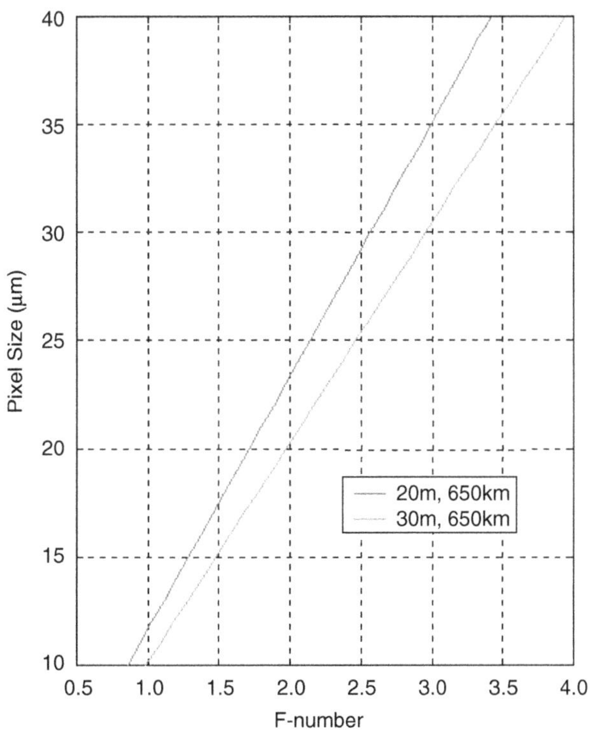

FIGURE 6.8 Hyperspectral instrument optics f/# verses detector pixel size.

The detector pixel size of 30 μm will further loose the demand on f/#. However, as defined by Equation 6.3 in Section 6.1, larger detector pixel size will increase the size of the optical system.

6.6.2 Quantum Efficiency

Obviously a detector array with the possible highest QE is desired for an imaging spectrometer, but in practice the true performance metrics are the resulting SNR, full well capacities, and readout rates. Silicon is the detector of choice for the 0.4–1.0 μm wavelength region because the technology is mature and high quantum efficiencies can be obtained. IR detector arrays generally cut on in the 1.0-μm region and the cutoff can be tuned for long wavelength. Recent work with thinned IR detector arrays (HgCdTe and InSb for example) has allowed these detector materials to have cut on wavelength down to 0.4 μm with excellent QE. Thus, it must be kept in mind in the selection of FPAs that the traditional wavelength divisions may not be a design requirement for the instrument.

6.6.3 Dark Current

A key detector parameter that is desirable to minimize is the dark current produced in the detector cells during the integration period. The dark current can vary markedly for different kinds of detector arrays. It depends heavily on the detector technology and operating temperature, but can also depend on the quality of the devices, the process control during fabrication, and the details of how the detector arrays are biased and clocked. The dark current produces both a potentially variable background and a contribution to the shot noise in the collected signal in the detector cells. Two different regimes of dark current can be considered: (1) the dark current shot noise can be allowed to dominate, or (2) the accuracy of the dark current nonuniformity (DCNU) can be allowed to dominate. The first regime would allow significantly higher dark currents than the second regime, assuming that only 1% correction accuracy can be achieved. The second regime would require a dark current <3–5 e^-/sec/pixel, while the first regime would require a dark current <40 e^-/sec/pixel.

6.6.4 Dark Current Nonuniformity

Because of the maturity of the technology, silicon-based FPAs typically have relatively small pixel-to-pixel variations in the dark current, or dark current nonuniformity. It is on the order of <5–10% of the average dark current. However, manufacturers usually only specify <20%.

IR detectors typically have much larger pixel-to-pixel variations than that of silicon-based FPAs due to the nature of the less controlled growth processes in IR materials. These variations can be as much as few times the typical dark current. At some dark current level an operability limit is usually defined in terms of an absolute maximum dark current. Obtaining a moderate DCNU (excluding the outliers, with dark current greater than the operability limits) of <100% of the average dark current is possible, it is simply a question of yield at the manufacturer as to how many detector arrays with this limit can be reliably produced.

Since subtraction (i.e., correction) of dark reference images is performed in the application, the issue is how accurately can the DCNU be corrected. Note that in the application the very fast integration time limits the dark current accumulation and much of the variation in the dark image will not be DCNU due to electronic offset variation. As long as the origin of the electronic offset (biases, clocking overlap, etc.) is constant, the dark frame correction will take care of both variations at the same time.

6.6.5 Photo-Response Nonuniformity

Photo-response nonuniformity (PRNU) is the pixel-to-pixel variation in the effective QE. Again because of the maturity of the technology, silicon-based detector arrays can be expected to have much better PRNU than IR detector arrays. Silicon CCD FPAs routinely achieve <3% PRNU. While the variations in QE can be much greater in IR detector arrays, in part for the same reasons that the dark current variations are larger. The root cause of much greater pixel-to-pixel variability in IR detector arrays is the variation in the material properties across a detector array. A variation in cutoff wavelength will show up as both a larger PNRU and a larger DCNU. DCNU is much more sensitive than PRNU to cutoff wavelength variations. IR detector arrays can have fairly small PRNU levels, on the order of 5–10%. At this level of PRNU, the instrument radiometric calibration can correct for these variations. The variation is not typically great enough to affect the dynamic range of individual pixels.

6.6.6 Readout Noise

The lowest possible readout noise is generally to be preferred in the selection of FPAs. In the case of the hyperspectral application once the readout noise is below the 100 e^- range, the PRNU correction precision becomes the limiting factor to the SNR in VNIR region, and the DCNU correction precision becomes the limiting factor to the SNR in SWIR region.

At the high readout rates being contemplated for a hyperspectral mission, it will become difficult for the analog chain between the detector array and the A/D converter to achieve a noise performance below the 100 electron level (<0.3 mV, given a 3 $\mu V/e^-$ detector gain). Thus, as long as the detector array is capable of a readout noise <100 electrons, it will likely satisfy the hyperspectral requirements.

REFERENCES

Acton, D., M. Jack, and T. Sessler. 2009. Large format short wave infrared (SWIR) focal plane array (FPA) with extremely low noise and high dynamic range. *Proceedings of SPIE* 7298:72983E-1–72983E-13.

Atkinson, D., et al. 2012. Operation and performance of new NIR detectors from SELEX. *Proceedings of SPIE* 8453:84530U1–84530U10.

Bai, Y., et al. 2012. 4K × 4K format, 10 μm pixel pitch H4RG-10 hybrid CMOS silicon visible focal plane array for space astronomy. *Proceedings of SPIE* 8453:84530M-1–84530M-18.

Beletic, J. W., et al. 2008. Teledyne imaging sensors: Infrared imaging technologies for astronomy & civil space. *Proceedings of SPIE* 7021:70210H-1–70210H-14.

Bentell, J., et al. 2010. 3000 pixel linear InGaAs sensor for the Proba-V satellite. *Procedings of SPIE* 7862:786206-1–786206-10.

Bézy, J.-L., J.-P. M. Huot, S. M. Delwart, L. Bourg, R. Bessudo, and Y. Delclaud. 2016. Medium resolution imaging spectrometer for ocean colour onboard ENVISAT. In *Optical Payloads for Space Missions*, ed. S.-E. Qian, 91–120. Sussex, UK: John Wiley & Sons.

Bogaerts, J. 2014. Complementary metal-oxide-semiconductor (CMOS) image sensors for use in space. In *High Performance Silicon Imaging: Fundamentals and Applications of CMOS and CCD Image Sensors*, ed. D. Durini, 250–280. Cambridge: Woodland Publishing.

Bruijne, J. 2012. Science performance of Gaia, ESA's space-astrometry mission. *Astrophysics and Space Science* 341(1):31–41.

Capper, P. 2006. Narrow band gap II-VI semiconductors growth. In *Springer Handbook of Electronic and Photonic Materials*, eds. S. Kasap and P. Capper, 303–324. New York: Springer.

Capper, P., S. Irvine and T. Joyce. 2006. Epitaxial crystal growth: Methods and materials. In *Springer Handbook of Electronic and Photonic Materials*, eds. S. Kasap and P. Capper, 271–301. New York: Springer.

Cutter, A. M. 2000. Compact high resolution imaging spectrometer (CHRIS). Proceedings of RTO MP-61 Symposium on Space-Based Observation Technology, 16–18 October 2000, Island of Samos, Greece.

Demers, R., et al. 2013. The CHROMA focal plane array: A large-format, low-noise detector optimized for imaging spectroscopy. *Proceedings of SPIE* 8870:88700J-1–88700J-7.

Dereniak, E. and G. Boreman. 1996. *Infrared Detectors and Systems*. Chichester: John Wiley and Sons.

Djazovski, O. 2016. Focal plane arrays for optical payloads. In *Optical Payloads for Space Missions*, ed. S.-E. Qian, 793–837. Chichester: John Wiley and Sons.

Fièque, B., P. Chorier, and B. Terrier. 2010. Sofradir detectors for hyperspectral applications from visible up to VLWIR. *Proceedings of SPIE* 7826:78261I-1–78261I-12.

Hodapp, K. and D. Hall. 2006. Introduction to detectors: Possible status in 2010-2020. In *The Scientific Requirements for Extremely Large Telescopes*, eds. P. A. Whitelock, M. Dennefeld and B. Leibundgut, 40–51. Cambridge: Cambridge University Press.

Holland, S., et al. 2007.Technology development for 4K × 4K, back-illuminated, fully depleted scientific CCD imagers. *Proceedings of IEEE Nuclear Science Symposium, NSS '07* 3:2220–2225.

Holst, G. and T. S. Lomheim. 2011. *CMOS/CCD Sensors and Camera Systems*, 2nd ed. Bellingham: SPIE Press.

Jacobsen, K. 2011. Recent developments of digital cameras and space imagery. *Proceedings of GIS Ostrava 2011*, 8S-1–8S-8.

Janesick, J. 2001. *Scientific Charge Coupled Devices*. Bellingham: SPIE Press.

Jerram, P., et al. 2010. Back-thinned CMOS sensor optimisation. *Proceedings of SPIE* 7598:759813-1–759813-12.

Jerram, P., et al. 2011. Hyperspectral CMOS imager. *Signatures, Newsletter of the ISRS–AC* 23(1 & 2):39–43.

Kaufmann, H. et al. 2016. Environmental Mapping and Analysis Program – A German Hyperspectral Mission. In *Optical Payloads for Space Missions*, ed. S.-E. Qian, 161–181. West Sussex UK: John Wiley & Sons.

Kean, B., et al. 2012. Advancements in large-format SiPIN hybrid focal plane technology. *Proceedings of SPIE* 8511:851111-1–851111-10.

Kleipool, Q. L., et al. 2007. In-flight proton-induced radiation damage to SCIAMACHY's extended-wavelength InGaAs near-infrared detectors. *Infrared Physics & Technology* 50(1):30–37.

Kong, J.-P., A. Eckardt, and B.-G. Kim. 2009. The design of the focal plane assembly for AEISS. *Proceedings of 60th IAC International Astronautical Congress IAC-09*, 12–16 October, 2009, B1.I.6-1–B1.I.6-10, Daejeon, Korea.

Kozlowski L. J. and W. F. Kosonocky. 2010. Infrared detector arrays. In *Handbook of Optics, Volume II: Design, Fabrication and Testing, Sources and Detectors, Radiometry and Photometry*, 3rd ed., eds. M. Bass, C. DeCusatis, J. M. Enoch, V. Lakshminarayanan, G. Li., C. MacDonald, V. N. Mahajan, and E. V. Stryland. 33.1–33.33. New York: McGraw-Hill.

Kramer, H. 2001. *Observation of the Earth and Its Environment: Survey of Missions and Sensors*, 4th ed. Berlin: Springer-Verlag.

Kullar, S., et al. 2011. Multispectral and hyperspectral sensors for remote sensing. *Signatures, Newsletter of the ISRS–AC* 23(1 & 2):34–36.

Loose, M., J. Beletic, J. Garnett, and N. Muradian. 2006. Space qualification and performance results of the SIDECAR ASIC. *Proceedings of SPIE* 6265:62652J-1–62652J-11.

Maciaszek, T., et al. 2014. Euclid near infrared spectrophotometer instrument concept and first test results at the end of phase B. *Proceedings of SPIE* 9143:91430K-1–91430K-14.

Meini, M., A. Bini, L. Giunti, E. Fossati, and R. Formaro. 2016. Hyperspectral payload for Italian PRISMA programme. In *Optical Payloads for Space Missions*, ed. S.-E. Qian, 183–213. West Sussex, UK: John Wiley & Sons.

Nelson, M. J., et al. 2008. High sensitivity infrared extended wavelength response InGaAs detectors. *Proceedings of SPIE* 7021:702122-1–702122-9.

Nieke, J., C. Mavrocordatos, C. Donlon, B. Berruti, T. Garnier, J.-B. Riti, and Y. Delclaud. 2016. Ocean and land color imager on sentinel-3. In *Optical Payloads for Space Missions*, ed. S.-E. Qian, 223–245. West Sussex, UK: John Wiley & Sons.

Norton, P. 2003. Detector focal plane technologies. In *Encyclopedia of Optical Engineering*, ed. R. Driggers, 320–348. New York: Marcel Dekker.

Ohta, J. 2008. *Smart CMOS Sensors and Applications*. Boca Raton, FL: CRC Press.

Piquette, E. C., et al. 2008. Substrate-removed HgCdTe-based focal-plane arrays for short-wavelength infrared astronomy. *Journal of Electronic Materials* 37(9):1396–1400.

Piquette, E. C., et al. 2014. Progress in development of H4RG-10 infrared focal plane arrays for WFIRST-AFTA. *Proceedings of SPIE* 9154:91542H-1–91542H-8.

Rogalski, A. 2010 *Infrared Detectors*, 2nd ed. Boca Raton, FL: CRC Press.

Sang, B., et al. 2008. The EnMAP hyperspectral imaging spectrometer: Instrument concept, calibration, and technologies. *Proceedings of SPIE* 7086:708605-1–708605-15.

Seshadri, S., et al. 2007. Comparing the low-temperature performance of megapixel NIR InGaAs and HgCdTe imager arrays. *Proceedings of SPIE* 6690:669006-1–669006-14.

Theuwissen, A. J. 2013. *Solid-State Imaging With Charge-Coupled Devices*. Dordrecht: Kluwer.

Tomasko, M. G., et al. 2003. The descent imager/spectral radiometer (DISR) experiment on the Huygens entry probe of Titan. In *The Cassini-Huygens Mission: Overview, Objectives and Huygens Instrumentarium*, ed. C. T. Russel, 469–551. Doerdrecht: Kluwer Academic Publishers.

Yuan P., J. Chang, J. C. Boisvert, and N. Karam. 2014. Low-dark current 1024 × 1280 InGaAs PIN arrays. *Proceedings of SPIE* 9070:907007-1–907007-6.

Yuan, X., S. Ji, J. Cao, and X. Yu. 2012. Geometric processing and positioning techniques. In *Advanced Remote Sensing: Terrestrial Information Extraction and Applications*, eds. S. Liang, X. Li, and J. Wang, 40–42. Oxford: Elsevier.

Zacharias, N., et al. 2007. Realization and application of a 111 million pixel backside-illuminated detector and camera. *Proceedings of SPIE* 6690:669008-1–669008-8.

Zombeck, M. V. 2006. *Handbook of Space Astronomy and Astrophysics*, 3rd ed. Cambridge: Cambridge University Press.

7 Hyperspectral Imager System Performance Modeling

This chapter describes a collection of performance models of three exemplary hyperspectral imagers with a series of input parameters and assumptions. These models simulate the performance of different designs of the hyperspectral imager systems. Based on these outcomes, evaluation and analysis as well as trade-off can be performed on the different designs. This process helps optimize the system design before a real hyperspectral imager is actually built.

For each hyperspectral imager design, the models include the signal-to-noise ratio (SNR), modulation transfer function (MTF), spectral response, and spatial response.

7.1 CONFIGURATION OF EXEMPLARY HYPERSPECTRAL IMAGING SYSTEMS

It is assumed that the three hyperspectral imager systems have a ground swath of 30 km with a ground sampling distance (GSD) of 30 m. From the instrument point of view, they all have an entrance slit 30-mm long in the cross-track direction and 30-μm wide in the along-track direction. A ground pixel of size 30 m × 30 m is imaged to a 30 μm × 30 μm point on the focal plane, i.e., the detector pitch size is 30 μm × 30 μm.

7.1.1 Telescope

A detailed design of the telescope is described in Section 5.3, where three telescope designs are presented as examples. For the simplicity of the modeling, the telescopes of the three exemplary hyperspectral imager systems are the same. A three-mirror anastigmat (TMA) telescope is used for all the three systems (Cook 1992).

7.1.2 Spectrometers

A detailed design of the spectrometers is also described in Chapter 5. In system modeling, the three hyperspectral imager systems all use separate spectrometers for the visible and near-infrared (VNIR) (0.4–1.0 μm) and shortwave infrared (SWIR) (1.0–2.5 μm) spectral regions. Their spectrometer design options are evaluated and modeled.

The three hyperspectral imager systems being modeled are configured as follows:

1. The dual Dyson system: TMA + VNIR Dyson spectrometer + SWIR Dyson spectrometer (Mouroulis et al. 2008), as depicted in Figure 7.1.
2. The dual Offner system: TMA + VNIR Offner spectrometer + SWIR Offner spectrometer (Prieto-Blanco et al. 2006), as depicted in Figure 7.2.
3. The dual prism system: TMA + VNIR prism spectrometer + SWIR prism spectrometer, as depicted in Figure 7.3.

7.1.3 Detector Arrays

Two sets of detector array options are considered in system modeling. They are all hybrid complementary metal oxide semiconductor (CMOS) devices as described in Chapter 6. A detector array comprises an array of pixels of photon-sensitive layer of indium bump bonded to a CMOS readout

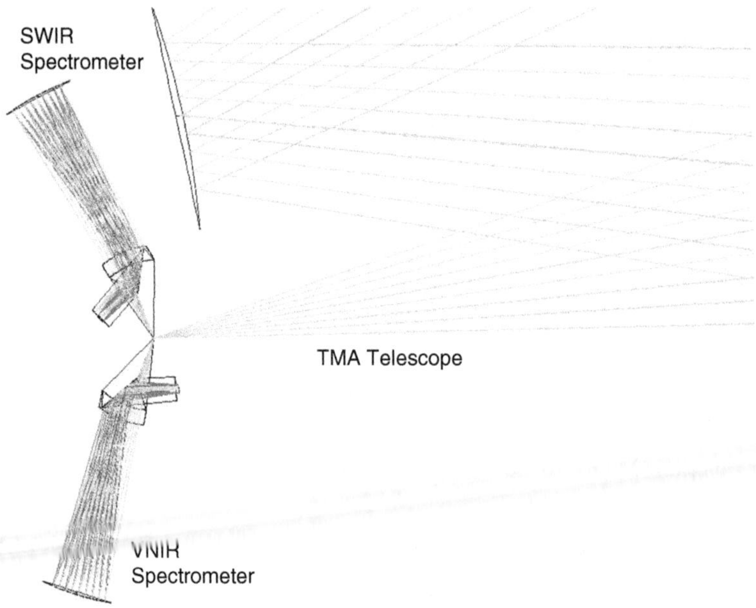

FIGURE 7.1 The optical layout of a hyperspectral imager with dual Dyson spectrometers.

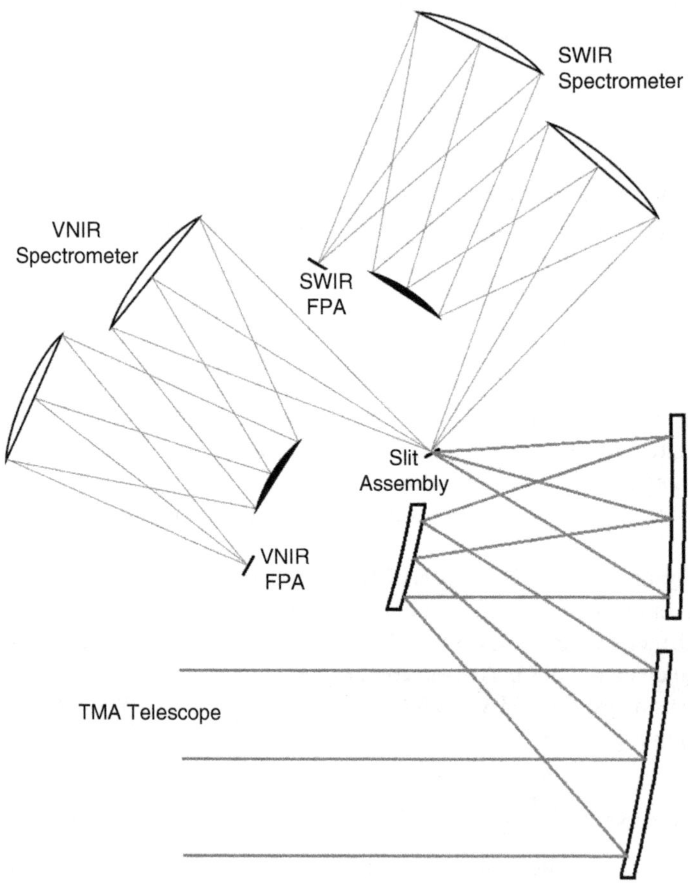

FIGURE 7.2 The optical layout of a hyperspectral imager with dual Offner spectrometers.

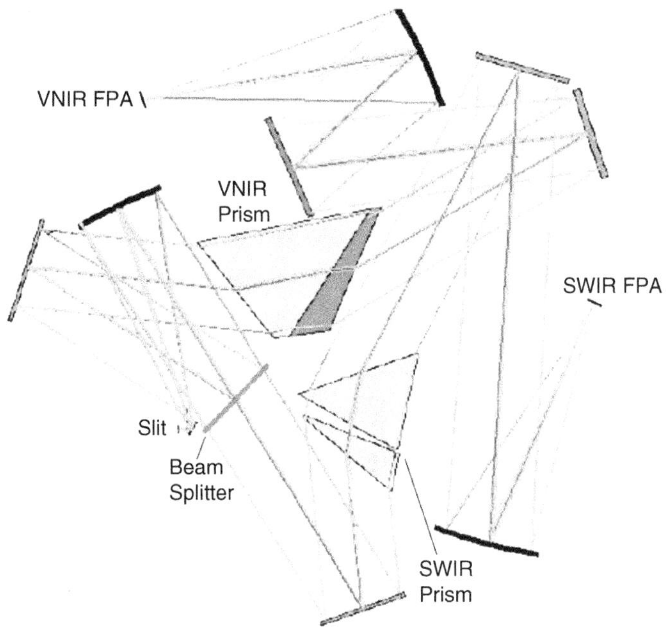

FIGURE 7.3 The optical layout of a hyperspectral imager with dual prism spectrometers.

integrated circuit (ROIC; Djazovski 2016). The ROIC is assumed to have an ability to read 1024 rows (corresponding to the spatial dimension in the cross-track direction) and 256 columns (corresponding to the spectral dimension) of detector pixels via 8 readout ports, each at a rate of up to 8 Mpixel/sec.

In Option 1, the detector material is assumed to be silicon for the VNIR spectrometer and mercury-cadmium-telluride (MCT or HgCdTe) for the SWIR spectrometer. In the modeling, the Teledyne's HyViSi silicon-based hybrid detector array is used for the VNIR spectrometer as an example (Demers et al. 2013). To achieve the desired balance between the full well capacity (FWC) and readout noise, the parameters are assumed to be different for the VNIR and SWIR detector arrays. This would require two different versions of the ROIC, thus increasing the development complexity and possibly the cost.

To avoid the need for different ROIC for VNIR and SWIR spectrometers, focal plane array (FPA) manufacturers (e.g., Teledyne and Lyndred) have developed substrate-removed HgCdTe detector arrays to extend the cut-on wavelength to 0.4 μm. This allows the same detector material (e.g., HgCdTe) and ROIC combination to be used for both VNIR and SWIR and has the further advantage of having two different integration capacitors available to each pixel that allows the FWC and readout noise to be optimized band by band. Option 2 uses substrate-removed HgCdTe detector arrays for both VNIR and SWIR spectrometers. In the modeling, the Lyndred (Sofradir)'s substrate-removed HgCdTe detector arrays are used as an example (Fièque et al. 2010).

7.2 SIGNAL-TO-NOISE RATIO SIMULATION

This section simulates SNR of the three exemplary hyperspectral imager systems after the definition of the SNR model and discussion on the assumptions for the SNR simulations.

7.2.1 Signal-to-Noise Model

SNR is a key parameter of a hyperspectral imager system. It is defined as the ratio of signal power to the noise power to quantify the sensitivity or detectability of an instrument. Satellite data users always prefer and/or require data/images with high SNR in order to accurately derive information and better serve their applications. The following subsections define the equation of SNR to be used in the simulation of SNR of the three exemplary hyperspectral imager systems.

7.2.1.1 Signal Model

Because the noise is random, it is not possible to predict the exact value of a noisy signal at a given time. However, it is possible to predict the mean of the signal and its standard deviation. The mean value of the signal is often referred to simply as *the signal* and its standard deviation is often referred to as *the noise*. Despite the simplified nomenclature, one should never forget that these are statistical values only.

It is common practice to make the approximation that the scene spectral radiance and the instrumental response function are spectrally flat over the relatively narrow spectral sampling intervals (SSI) of hyperspectral sensors. This significantly simplifies the calculations. So, approximating the spectral radiance as constant at the center wavelength over a sub-band and the spectral response function (SRF) as the trapezoidal convolution of the spectral slit width and the detector width, the number of signal electrons S from a spectral band can be given by:

$$S = \sum_{n} n\Delta\lambda L(\lambda_n)E\tau\eta(\lambda_n)\Gamma_o(\lambda_n), \tag{7.1}$$

where

n is the number of samples summed (sub-bands summed to create a spectral band and binned pixels samples to create an effective spatial sample),
$\Delta\lambda$ is the spectral width of a sub-band,
$L(\lambda_n)$ is the spectral radiance,
λ_n is the wavelength at the centre of the sub-band,
τ is the integration time,
$\eta(\lambda_n)$ is the quantum efficiency (QE) of the detector,
$\Gamma_o(\lambda_n)$ is the total efficiency of the optics, and
E is the etendue that can be approximated by:

$$E = \frac{A_o A_d}{f^2}, \tag{7.2}$$

where

A_o is the collecting area of the optics,
A_d is area of the detector pixel, and
f is the effective focal length.

7.2.1.2 Noise Model

The noise can be given by the root sum squared of all independent noise contributions:

$$N = \sqrt{n\left(N_S^2 + N_D^2 + N_T^2 + N_R^2 + N_E^2 + N_Q^2 + N_F^2 + N_J^2\right)}, \tag{7.3}$$

1. N_S is the signal shot noise. It is due to the discrete nature of photons and electrons and occurs when photoelectrons are generated within the detector. Assuming the Poisson distribution, the root-mean-square (rms) value of the signal shot noise is equal to the signal S:

$$N_S = \sqrt{S} \tag{7.4}$$

This noise cannot be avoided. An instrument can be considered optimum if N_S is the dominating noise term in Equation 7.3.

2. N_D is the dark current shot noise. It is due to the discrete nature of the thermally generated electrons within the detector. Assuming the Poisson distribution, the rms value of this noise source is given by mean value of dark electrons D:

$$N_D = \sqrt{D} \tag{7.5}$$

3. N_T is the thermal background shot noise. It is proportional to the square root of the electrons T_e generated by the thermal emission from the internal surfaces of the instrument.

$$N_T = \sqrt{T_e} \tag{7.6}$$

It is very difficult to estimate T_e without using detailed ray tracing. As a first approximation, T_e can be given by:

$$T_e = \tau \int_{\lambda_1}^{\lambda_2} \frac{2c\eta(\lambda)\epsilon(\lambda)\Gamma_{cf}(\lambda)}{\lambda^4 \left(e^{\frac{hc}{kT\lambda}} - 1\right)} d\lambda, \tag{7.7}$$

where

c is the speed of light,
h is Planck's constant,
$\Gamma_{cf}(\lambda)$ is the transmittance of the cold filter (if used),
T is the temperature of the emitting surfaces, and
ε is the effective emissivity of the emitting surfaces.

For short wavelengths, this contribution is usually negligible. In the event when it is not negligible, it will be necessary to make a more refined estimation and consider the cascade of thermal emission of the optical components in the optical path (especially after dispersion) and of other elements of the instrument that can be seen by the detector.

4. N_R is the noise associated with the detector readout circuits, sense node, and the on-chip amplifier. This noise source is independent of the signal magnitude and may be sufficient to treat it as a single quantity. It is given in terms of number of electrons and depends on the FWC of the detector array. The larger the FWC, the higher the readout noise. Normally, the manufacturer of the detector arrays provides this value as one of the key performances of the detector array.
5. N_E is the noise associated with the electronics between the detector readout and the analog-to-digital converter.
6. N_Q is the quantization noise given by

$$N_Q = \frac{C}{2^M \sqrt{12}}, \tag{7.8}$$

where C is the FWC of the detector array, and M is the number of quantization steps.

7. N_F is the $1/f$ or flicker noise. It has a dependence from the total detector current (photocurrent + dark current) generated from the photodiodes that charge the equivalent capacitance of the detector. The rms value for the $1/f$ noise is given in the following formula:

$$N_F = K_1 (S + D)^{0.76} \left(\frac{\tau}{\tau_{\max}}\right)^{0.24} \sqrt{K_2 - \ln\left(\frac{\tau}{\tau_{\max}}\right)}, \tag{7.9}$$

where K_1 and K_2 are constants.

8. N_J is the time jitters noise. If the integration time is not completely stable, small fluctuations of the signal will be generated. This produces a noise that is directly proportional to the signal:

$$N_J = (S + D)\left(\frac{\delta_\tau}{\tau}\right) \tag{7.10}$$

where δ_τ is the standard deviation of the integration time.

The SNR is simply the average signal, S, in Equation 7.1 divided by the standard deviation of the noise in the same units and in the same domain, N, in Equation 7.3:

$$SNR = \frac{S}{N} \tag{7.11}$$

7.2.1.3 Saturation Signal Model

The saturation signal is given by:

$$S_{Sat} = \sum_n n\Delta\lambda L_{Sat}(\lambda_n)E\tau\eta(\lambda_n)\Gamma_o(\lambda_n)(1+3PRNU) + S_D, \tag{7.12}$$

where S_{Sat} is the saturation spectral radiance, $PRNU$ is the photo response nonuniformity, and S_D is the maximum dark charge, and it is given by:

$$S_D = T_e + 3N_T + D(1 + DCNU) + 3N_D + 3N_R + 3N_E, \tag{7.13}$$

where $DCNU$ is the dark signal nonuniformity. The above equation adds a margin for noise equal to three standard deviations. This does not guarantee that the noisy signal will not be saturated, but this is often considered a sufficient margin for estimation of saturation.

The ratio of the saturation signal to the FWC of the detector array is defined as the saturation fraction:

$$F_{sat} = \frac{S_{Sat}}{FWC} \tag{7.14}$$

7.2.2 Assumptions for SNR Simulation

7.2.2.1 Input Radiance

In the SNR simulation, two scenes with different levels of radiance are used: a typical level and a saturation level. The scene with a typical level radiance is assumed to have an albedo of 0.3 and illuminated at a solar zenith angle of 30°, whereas the scene with a saturation level radiance is assumed to have an albedo of 1.0 and illuminated at a solar zenith angle of 0°.

Figure 7.4 depicts the spectral radiance of a typical scene and a saturation scene in the VNIR region. Figure 7.5 depicts the spectral radiance of a typical scene and a saturation scene in the SWIR region.

7.2.2.2 Quantum Efficiency

As described in Section 7.1.3, there are two options in selection of detector arrays for VNIR and SWIR spectrometers. Option 1 uses a silicon-based hybrid CMOS detector array for the VNIR and

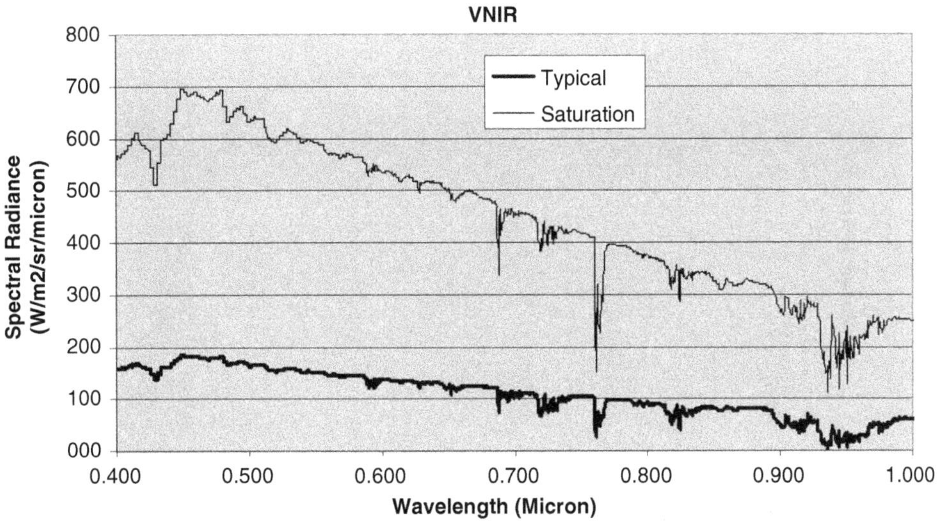

FIGURE 7.4 Spectral radiance of a typical scene and a saturation scene in the VNIR region.

a hybrid HgCdTe-based detector array for the SWIR. Option 2 uses substrate-removed HgCdTe detector arrays for both VNIR and SWIR.

1. **Detector Selection Option 1**
   The QE of the VNIR region for Option 1 detector arrays is derived from the data supplied by the Teledyne for the HyViSi device and from the data of 2.5-μm cutoff wavelength HgCdTe device for the SWIR (Demers et al. 2013).
   Figure 7.6 depicts the QE curve in the VNIR region of a silicon-based hybrid CMOS FPA. Figure 7.7 depicts the QE curve in the SWIR region of a HgCdTe FPA with a 2.5-μm cutoff wavelength.

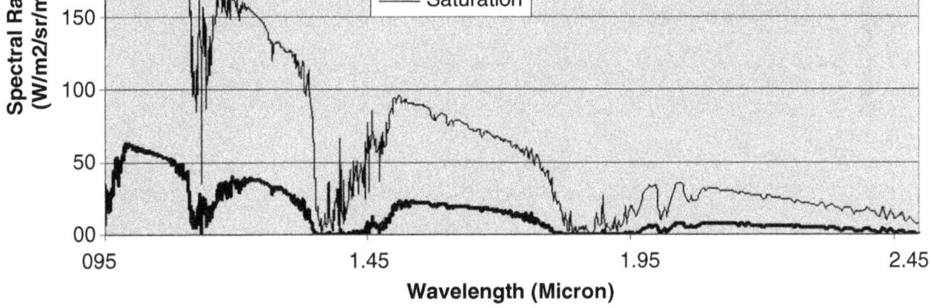

FIGURE 7.5 Spectral radiance of a typical scene and a saturation scene in the SWIR region.

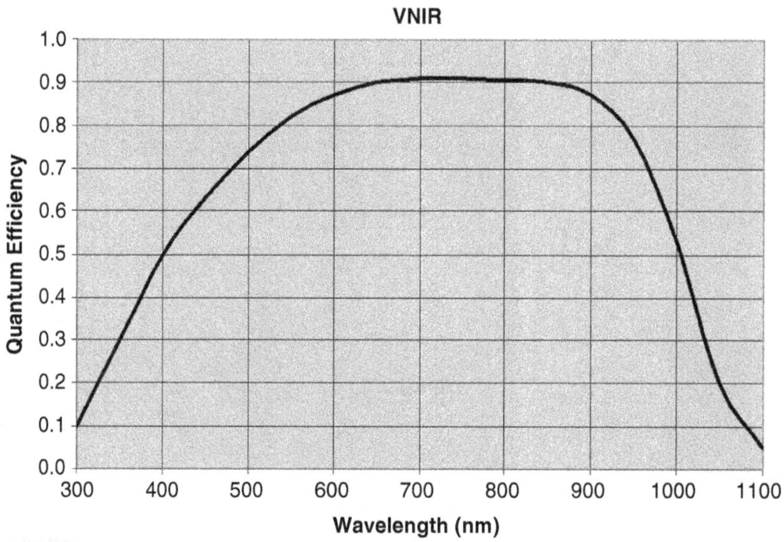

FIGURE 7.6 Quantum efficiency curve in the VNIR region of a silicon-based hybrid CMOS detector array.

2. **Detector Selection Option 2**

   The QE for the Sofradir HgCdTe detector arrays is assumed being similar to that of the detector arrays in Option 1. The substrate-removed HgCdTe FPA is assumed to extend the same QE into the VNIR as indicated by the Sofradir data, showing the responsivity to vary linearly with wavelength over the entire range.

   Figure 7.8 depicts the QE curve in the VNIR region of the substrate-removed HgCdTe FPA. Figure 7.9 depicts the QE curve in the SWIR region of the HgCdTe FPA with a cutoff wavelength of 2.5 μm.

7.2.2.3 Detector Parameters

The FWC, pitch size, readout noise, and dark current are key parameters of the detector arrays. Their values used in the SNR modeling of the two detector selection options are listed Tables 7.1 and 7.2.

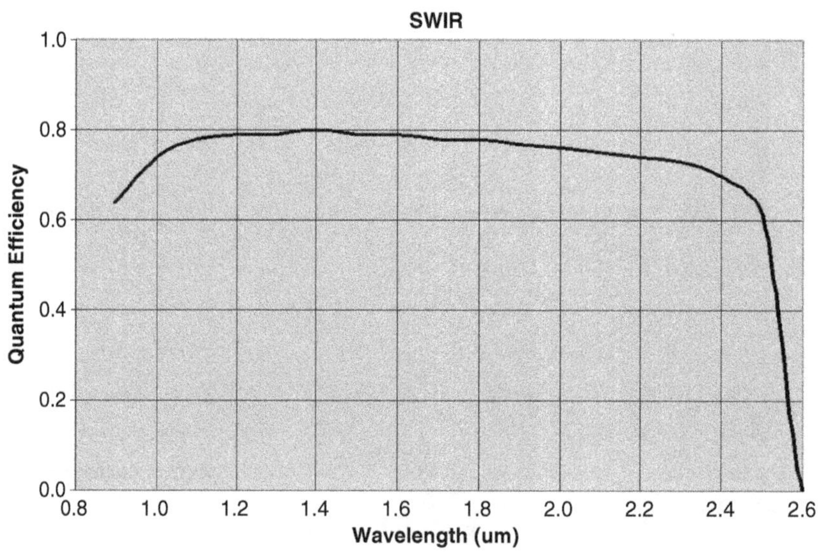

FIGURE 7.7 A quantum efficiency curve in the SWIR region of a hybrid HgCdTe detector array.

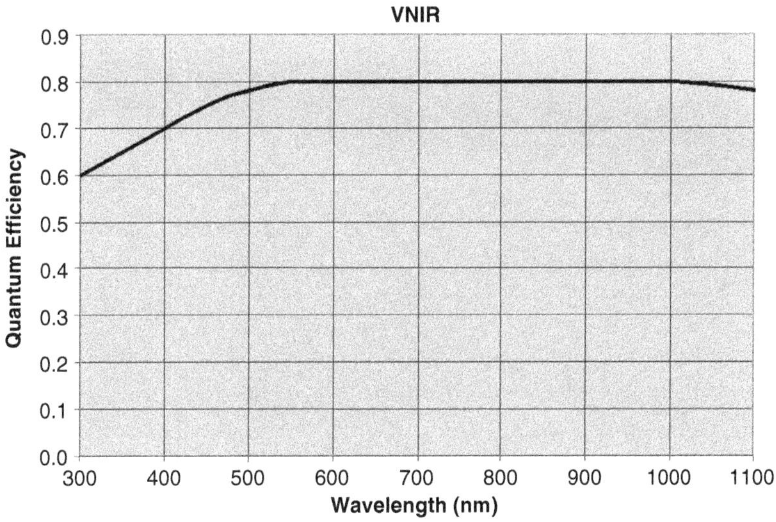

FIGURE 7.8 A QE curve in the VNIR region of the substrate-removed HgCdTe FPA

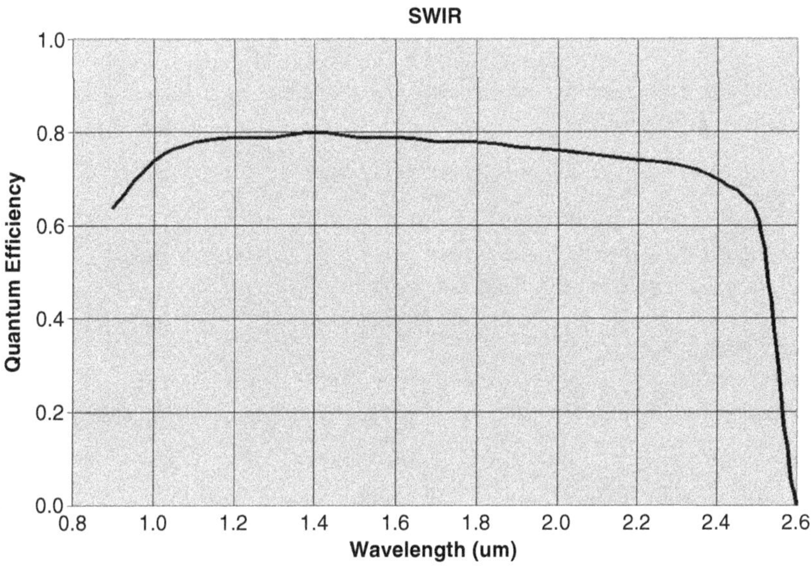

FIGURE 7.9 A QE curve in the SWIR region of the HgCdTe FPA.

TABLE 7.1
Parameters of the Option 1 Detector Arrays Used in the SNR Modeling

Parameter	VNIR	SWIR
Full well capacity (e–)	2.0×10^6	1.0×10^6
Pixel pitch size (μm)	30	30
Readout noise (e–)	300	200
Dark current (nA/cm^2)	0.2	1.0

TABLE 7.2
Parameters of the Option 2 Detector Arrays Used in SNR Modeling

Parameter	VNIR	SWIR1	SWIR2
Wavelength range (μm)	0.40–1.00	1.00–1.78	1.78–2.50
Full well capacity (e–)	2.6×10^6	2.6×10^6	5.8×10^5
Pixel pitch size (μm)	30	30	30
Readout noise (e–)	339	339	145
Dark current (nA/cm²)	0.2	1.0	1.0

7.2.2.4 Grating Efficiency

The grating efficiency assumed for the dual Dyson system and the dual Offner system is derived from the data for flat ruled, reflective gratings, which have similar groove density and blaze angle to those selected from the Newport Corporation catalogue. The blaze angles are adjusted slightly to balance the overall instrument throughput at each end of the wavelength range. Figures 7.10 and 7.11 depict the grating efficiency in the VNIR and SWIR regions.

7.2.2.5 Optical Throughput

The measurement of the optical throughput of a hyperspectral imager system must consider the efficiency of all optical components in the optical path. In the exemplary modeling, it takes into account the following four aspects:

1. Mirror reflectivity
2. Transmittance of the antireflection (AR) coating
3. Transmittance of the order-sorting filter
4. Transmittance or reflectance of the dichroic beam splitter

1. **Mirror Reflectivity**
   The mirror reflectivity of a hyperspectral imager system is measured by first counting the total number of reflective surfaces for the combination of the TMA telescope and

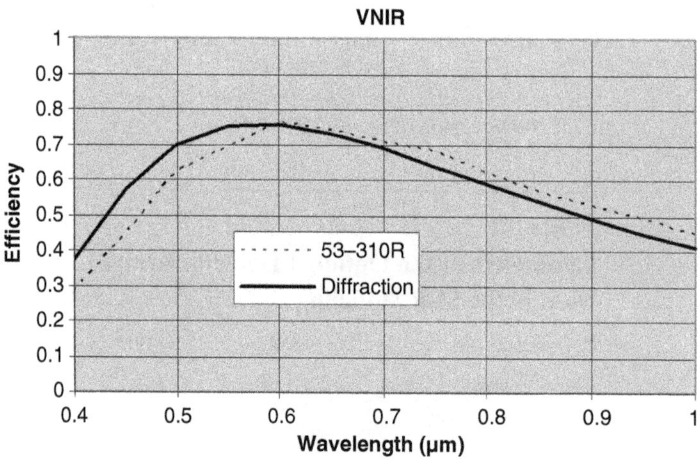

FIGURE 7.10 Grating efficiency in the VNIR region.

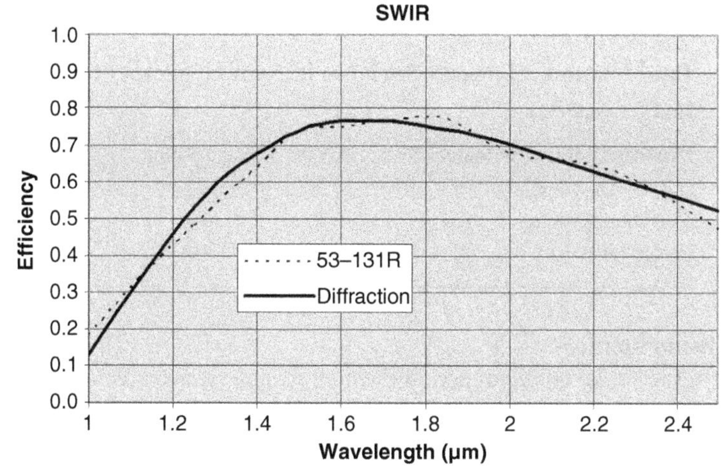

FIGURE 7.11 Grating efficiency in the SWIR region.

the spectrometers and then by multiplying a typical reflectivity curve, as depicted in Figure 7.12. Table 7.3 lists the total number of reflective surfaces of the three exemplary hyperspectral imager systems.

2. **Transmittance of Antireflection Coating**
   The transmittance of each AR-coated surface is assumed as a constant 0.98 across the VNIR region and a constant 0.99 across the SWIR region.

   The total number of AR-coated surfaces (excluding the order-sorting filter) for each spectrometer in each hyperspectral imager system is listed in Table 7.4.
3. **Order-Sorting Filter**
   The order-sorting filter required for the dual Dyson system and the dual Offner system is assumed to have a transmittance of 0.97 in the passband based on the measurement of a prototype order-sorting filter used previously.

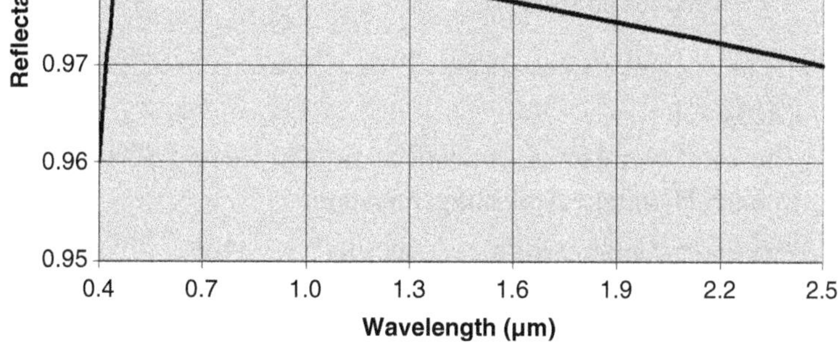

FIGURE 7.12 Typical mirror surface reflectivity.

TABLE 7.3
Total Number of Mirror Surfaces in Each Hyperspectral Imager System

Hyperspectral Imager System	VNIR	SWIR	Total
Dual Dyson system	3	3	6
Dual Offner system	5	5	10
Dual prism system	9	8	17

4. **Dichroic Beam Splitter**
   As shown in Figure 7.3 configuration of the dual prism system, a dichroic beam splitter is used to distribute the input radiative light for VNIR and SWIR spectrometers in the dual prism system. Figure 7.13 depicts the reflectance and transmittance of the beam splitter used in the SNR modeling.

7.2.2.6 Spectral Dispersion

The spectral dispersion for the dual Dyson system and dual Offner system is assumed to be a constant 7.5 nm per pixel (i.e., SSI is 7.5 nm), because both of them use grating as the spectral disperser. It has been known from the description in Chapter 5 that the spectral dispersion of gratings is quite linear.

In the dual prism system, two combined prism sets are used for VNIR and SWIR spectrometers, as shown in Figure 7.3. The spectral dispersion of prisms is nonlinear. The dispersions of the two combined prism sets assumed in the SNR modeling are depicted in Figures 7.14 and 7.15.

7.2.2.7 1/*f* Noise and Electronics Noise

The VNIR spectrometers are assumed to have no significant 1/*f* noise contribution. For the SWIR spectrometers, K_1 and K_2 in Equation 7.9 are taken as $K_1 = 3.79 \times 10^{-3}$ and $K_2 = 3.215$.

The electronics noise level is taken as 20 electrons for both VNIR and SWIR regions.

7.2.2.8 Ageing Degradation

It is assumed that the resulting instrument performances are modeled at the beginning of life (BOL) and at the end of life (EOL) with overall ageing degradation of 4% per year for a mission life of 5 years.

7.2.3 Results of SNR Simulation

This section shows SNR simulation results of five hyperspectral imager systems:

1. Dual Dyson hyperspectral imagers
2. Dual Offner hyperspectral imagers

TABLE 7.4
The Total Number of Antireflection (AR) Coated Surfaces in Each Hyperspectral Imager System

Hyperspectral Imager System	VNIR	SWIR	Total
Dual Dyson system	4	4	8
Dual Offner system	0	0	0
Dual prism system	6	6	12

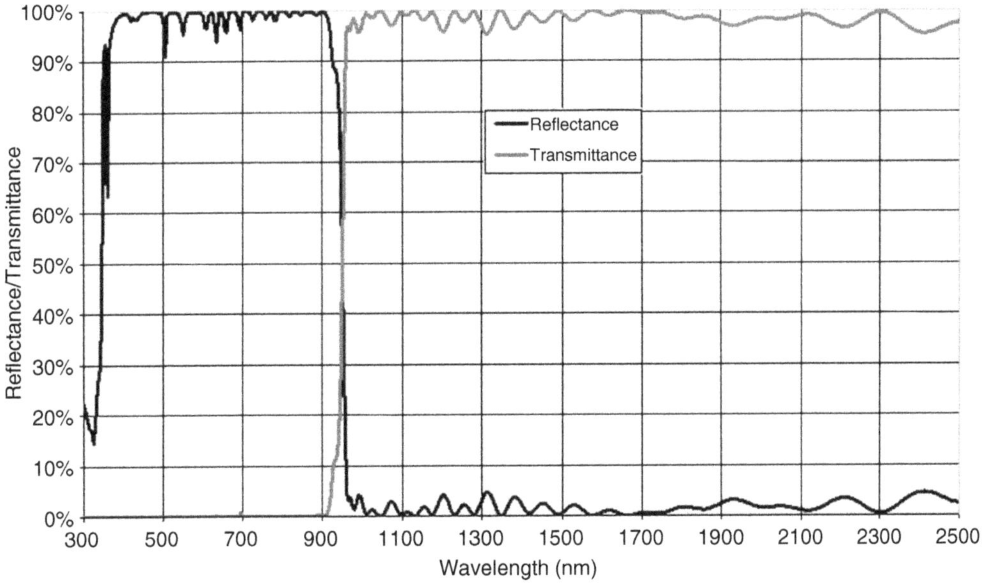

FIGURE 7.13 Reflectance and transmittance of the beam splitter used in the dual prism system.

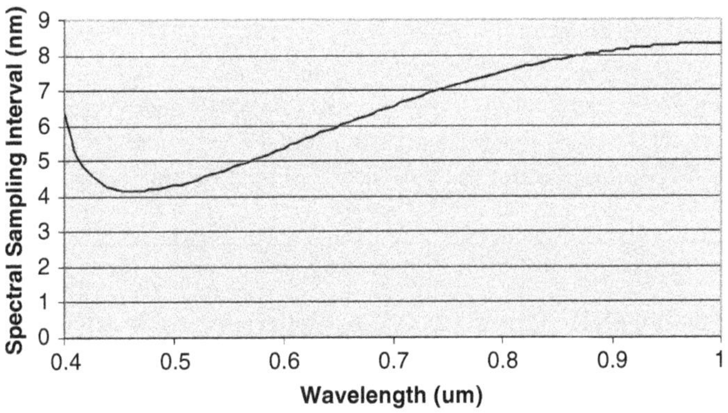

FIGURE 7.14 Spectral dispersion of the VNIR spectrometer of the dual prism system.

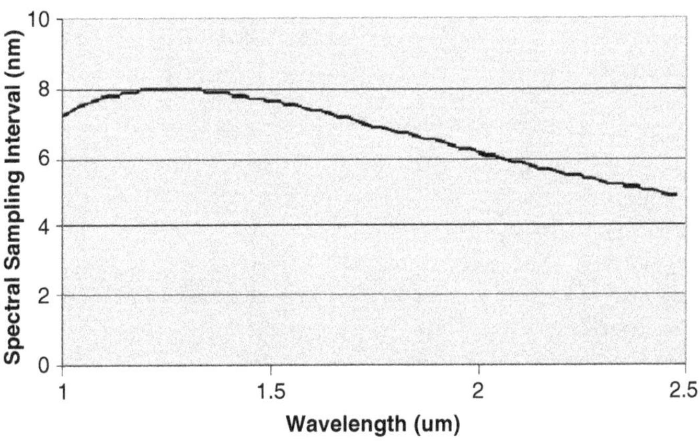

FIGURE 7.15 Spectral dispersion of the SWIR spectrometer of the dual prism system.

3. Dual prism hyperspectral imagers—Case 1
4. Dual prism hyperspectral imagers—Case 2
5. Dual prism hyperspectral imagers—Case 3

7.2.3.1 Dual Dyson Hyperspectral Imagers

The aperture of the designed dual Dyson imaging spectrometer system is 25.5 cm with a f/# of 2.7, which corresponds to a focal length of 68.85 cm. The FPAs used are Option 1 detector arrays, i.e., a silicon-based hybrid CMOS detector array for the VNIR spectrometer and a HgCdTe detector array for the SWIR spectrometer.

All the assumptions discussed in Section 7.2.2 are used in the SNR modeling. Table 7.5 summarizes the parameters used as the inputs for the SNR modeling.

Under these conditions, the SNR of both the dual Dyson system and dual Offner system are simulated. Because the throughput difference between the two systems is very small, the results are very similar. For the sake of brevity only the results related to the dual Dyson system are presented in this section. These results are assumed representative of the dual Offner system.

For the dual Dyson system, the simulated SNR values of nine representative wavelengths at the BOL and EOL are listed in Table 7.6. In the table, the SNR margin with respect to each required value of nine representative wavelengths is also listed. It can be seen from the table that the SNRs of the dual Dyson system meet all the requirements with some margin at both the BOL and EOL except the one at wavelength 1950 nm, where the SNR has a deficit of 54.8% and 63.1% at the BOL and EOL (shaded in Table 7.6).

The saturation fraction is the ratio of the saturation signal to the FWC of the detector array, which is defined by Equation 7.14. The maximum saturation fractions at the BOL and EOL in VNIR, SWIR1, and SWIR2 regions are listed in Table 7.7. VNIR and SWIR1 are saturated by 39% and 29% at the BOL. This means that the saturation signal is 39% and 29% more than the FWC of the detector arrays. VNIR and SWIR1 are still saturated by 25% and 3% at the EOL.

TABLE 7.5
Inputs for Modeling SNR of the Dual Dyson Hyperspectral Imager System

Parameter	VNIR	SWIR1	SWIR2
Aperture (cm)	25.5	25.5	25.5
f/#	2.7	2.7	2.7
Detector pixel size (μm)	30	30	30
Detector full well capacity (e−)	2×10^6	1×10^6	1×10^6
Detector readout noise (e−)	300	200	200
Detector dark current (nA/cm^2)	0.2	1.0	1.0
Wavelength range (μm)	0.40–1.00	1.00–1.78	1.78–2.50
Spectral sampling interval (nm)	7.5	7.5	7.5
Quantization step (e−/DN)	488.3	244.1	244.1
Max integration time (ms)	4.43	4.43	4.43
Integration factor to max integration time at BOL (%)	89	100	100
Integration factor to max integration time at EOL (%)	100	100	100
Readout electronics noise (e−)	20	20	20
Photo response nonuniformity (%)	3	3	3
Detector dark signal nonuniformity (%)	10	10	10
Blaze wavelength (nm)	580	1665	1665
Annual degradation (%)	4.0	4.0	4.0
EOL degradation (%)	20	20	20

TABLE 7.6
SNR at the BOL and EOL as well as the Margin of the Dual Dyson System

Wavelength (nm)		Signal-to-Noise Ratio			SNR Margin	
		Required	BOL	EOL	BOL	EOL
VNIR	400	200	280	259	39.9%	29.4%
	650	600	734	690	22.3%	15.0%
	1000	200	287	267	44.4%	33.7%
SWIR	1000	200	235	202	17.6%	1.0%
	1550	400	482	424	20.4%	6.1%
	1750	200	381	333	90.6%	66.7%
	1950	100	45	37	−54.8%	−63.1%
	2100	200	264	228	32.0%	13.9%
	2350	100	123	103	22.8%	2.8%

The SNR values at the EOL for each spectral band in the VNIR, SWIR1, and SWIR2 regions covering the entire wavelength range (400–2500 nm) are depicted in Figure 7.16. The figure also shows the curve of the saturation fraction values (see the vertical axis on the right of the figure for the scale).

7.2.3.2 Dual Offner Hyperspectral Imagers

The aperture of the designed dual Offner imaging spectrometer system is 26.5 cm with a f/# of 2.6, which corresponds to a focal length of 68.9 cm. The FPAs used are Option 2 detector arrays, i.e., a substrate-removed Sofradir HgCdTe detector array for the VNIR spectrometer and a Sofradir HgCdTe detector array with a 2.5-μm cutoff wavelength for the SWIR spectrometer.

All the assumptions discussed in Section 7.2.2 are used in the SNR modeling. Table 7.8 summarizes the parameters used as the inputs for the SNR modeling.

Under these conditions, the SNR of both the dual Dyson system and the dual Offner system are simulated. Because the throughput difference between the two systems is very small, the results are very similar. For the sake of brevity only the results of the dual Offner system are presented in this section. These results are assumed representative of the dual Dyson system.

For the dual Offner system, the simulated SNR values of nine representative wavelengths at the BOL and EOL are listed in Table 7.9. In the table, the SNR margin with respect to each required value of nine representative wavelengths is also listed. It can be seen from the table that the SNRs of the dual Offner system meet almost all the requirements with some margin at both the BOL and EOL except at wavelengths of 1550 nm and 1950 nm, where the SNR has a deficit of 3.6% at both the BOL and EOL for 1550 nm, and a deficit of 51.6% at both the BOL and EOL for 1750 nm (shaded in Table 7.9).

TABLE 7.7
Maximum Saturation Fraction at BOL and EOL of the Dual Dyson System

Parameter	VNIR	SWIR1	SWIR2
Max saturation fraction at BOL	139%	129%	70%
Max saturation fraction at EOL	125%	103%	56%

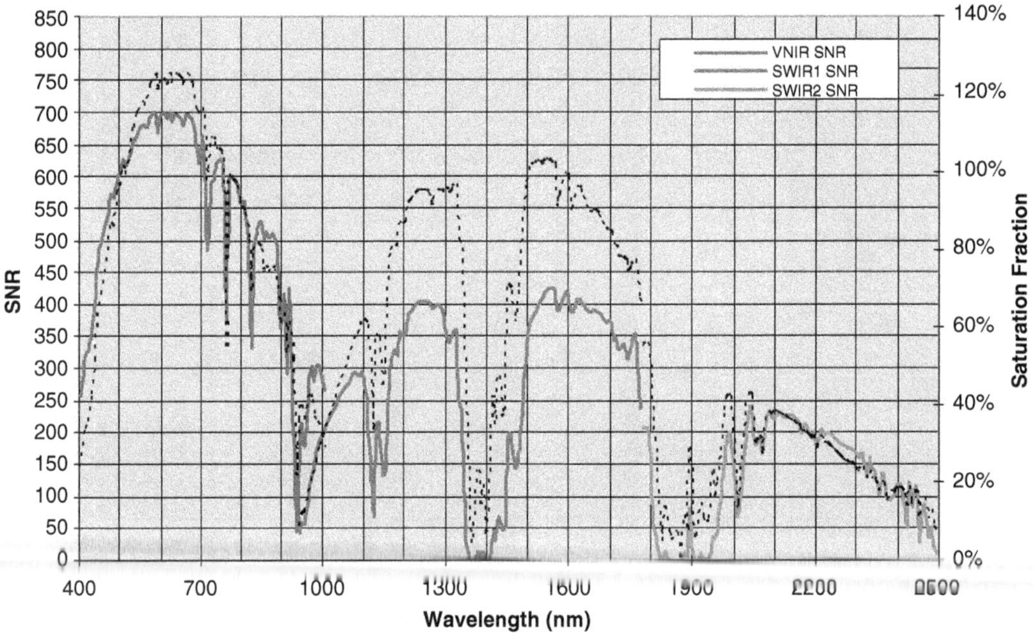

FIGURE 7.16 SNR and saturation fraction (thin dotted line) of the dual Dyson hyperspectral imager system at the EOL as a function of the wavelength.

TABLE 7.8
Inputs for Modeling SNR of the Dual Offner Hyperspectral Imager System

Parameter	VNIR	SWIR1	SWIR2
Aperture (cm)	26.5	26.5	26.5
f/#	2.6	2.6	2.6
Detector pixel size (μm)	30	30	30
Detector full well capacity (e–)	2.6×10^6	2.6×10^6	5.8×10^5
Detector readout noise (e–)	339	339	145
Detector dark current (nA/cm²)	0.2	1.0	1.0
Wavelength range (μm)	0.40–1.00	1.00–1.78	1.78–2.50
Spectral sampling interval (nm)	7.5	7.5	7.5
Quantization step (e–/DN)	634.8	634.8	634.8
Max integration time (ms)	4.43	4.43	4.43
Integration factor to max integration time at BOL (%)	82	80	80
Integration factor to max integration time at EOL (%)	100	100	100
Readout electronics noise (e–)	20	20	20
Photo response nonuniformity (%)	3	3	3
Detector dark signal nonuniformity (%)	10	10	10
Blaze wavelength (nm)	580	1550	1550
Annual degradation (%)	4.0	4.0	4.0
EOL degradation (%)	20	20	20

TABLE 7.9
SNR at the BOL and EOL as well as the Margin of the Dual Offner System

Wavelength (nm)		Signal-to-Noise Ratio			SNR Margin	
		Required	BOL	EOL	BOL	EOL
VNIR	400	200	330	323	65.2%	61.1%
	650	600	666	654	11.1%	9.0%
	1000	200	375	367	87.7%	83.7%
SWIR	1000	200	256	256	28.1%	28.1%
	1550	400	386	386	−3.6%	−3.6%
	1750	200	278	278	39.1%	39.0%
	1950	100	48	48	−51.6%	−51.6%
	2100	200	249	249	24.4%	24.4%
	2350	100	120	120	20.2%	20.2%

The maximum saturation fractions at the BOL and EOL in VNIR, SWIR1, and SWIR2 regions are listed in Table 7.10. Compared to those described in Section 7.2.3.1, there is no saturation in the dual Offner system. This is encouraged. This is because the FWC of the detector arrays chosen in this section is larger than that of the detector arrays chosen in Section 7.2.3.1.

The SNR values at the EOL for each spectral band in the VNIR, SWIR1, and SWIR2 regions covering the entire wavelength range (400–2500 nm) are depicted in Figure 7.17. The figure also shows the curve of the saturation fraction values. The vertical axis on the right of the figure is for the scale of saturation fraction. Note that the maximum saturation fraction at the EOL is 100% (i.e., no saturation) in this figure instead of 125% (i.e., 25% more) in Figure 7.16.

7.2.3.3 Dual Prism Hyperspectral Imagers—Case 1

The aperture of the designed dual prism imaging spectrometer system is 20.0 cm with a f/# of 3.5, which corresponds to a focal length of 70.0 cm. The FPAs used are Option 1 detector arrays, i.e., a silicon-based hybrid CMOS detector array for the VNIR spectrometer and HgCdTe detector array for the SWIR spectrometer.

All the assumptions discussed in Section 7.2.2 are used in SNR modeling. Table 7.11 summarizes the parameters used as the inputs for the SNR modeling.

For the dual prism system Case 1, the simulated SNR values of nine representative wavelengths at the BOL and EOL are listed in Table 7.12. In the table, the SNR margin with respect to each required value of nine representative wavelengths is also listed. It can be seen from the table that the SNRs of the dual prism system—Case 1 do not meet the requirements for majority of the representative wavelengths at both the BOL and EOL. The SNRs that do not meet the requirement are shaded in the table.

TABLE 7.10
Maximum Saturation Fraction at BOL and EOL of the Dual Offner System

Parameter	VNIR	SWIR1	SWIR2
Max saturation fraction at BOL	100%	48%	100%
Max saturation fraction at EOL	97%	48%	100%

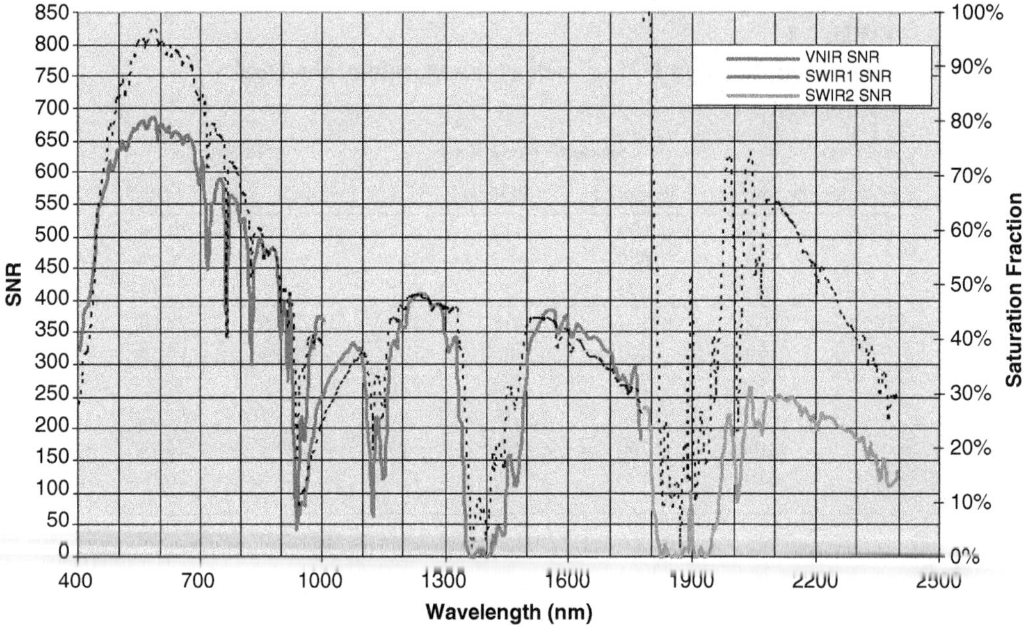

FIGURE 7.17 SNR and saturation fraction (thin dotted line) of the dual Offner hyperspectral imager system at the EOL as a function of the wavelength.

TABLE 7.11
Inputs for Modeling SNR of the Dual Prism Hyperspectral Imager—Case 1

Parameter	VNIR	SWIR1	SWIR2
Aperture (cm)	20.0	20.0	20.0
f/#	3.5	3.5	3.5
Detector pixel size (μm)	30	30	30
Detector full well capacity (e–)	2×10^6	1×10^6	1×10^6
Detector readout noise (e–)	300	200	200
Detector dark current (nA/cm^2)	0.2	1.0	1.0
Wavelength range (μm)	0.40–1.00	1.00–1.74	1.74–2.50
SSI Max (nm)	8.3	8.0	6.6
SSI Min (nm)	4.0	6.6	4.6
Quantization step (e–/DN)	488	244	244
Max integration time (ms)	4.43	4.43	4.43
Integration factor to max integration time at BOL (%)	100	100	100
Integration factor to max integration time at EOL (%)	100	100	100
Readout electronics noise (e–)	20	20	20
Photo response nonuniformity (%)	3	3	3
Detector dark signal nonuniformity (%)	10	10	10
Background temperature (K)	150	150	150
Annual degradation (%)	4.0	4.0	4.0
EOL degradation (%)	20	20	20

TABLE 7.12
SNR at the BOL and EOL as well as the Margin of the Dual Prism System—Case 1

Wavelength (nm)		Signal-to-Noise Ratio			SRN Margin		
		Required	BOL	EOL	BOL	EOL	SSI (nm)
VNIR	400	200	302	257	51.0%	28.5%	6.3
	650	600	554	482	−7.7%	−19.7%	5.9
	1000	200	230	194	15.0%	−3.1%	8.3
SWIR	1000	200	370	323	85.1%	61.7%	7.3
	1550	400	390	341	−2.5%	−14.7%	7.5
	1750	200	284	246	41.9%	22.8%	6.9
	1950	100	39	32	−60.9%	−68.1%	6.3
	2100	200	183	155	−8.6%	−22.4%	5.8
	2350	100	93	77	−7.1%	−22.9%	5.1

The performance of saturation fraction defined by Equation 7.14 is also provided. The maximum saturation fractions at the BOL and EOL in VNIR, SWIR1, and SWIR2 regions are listed in Table 7.13. VNIR and SWIR1 are saturated by 3% and 51% at the BOL. This means that the saturation signal is 3% and 51% more than the FWC of the detector arrays. The SWIR1 is still saturated by 21% at the EOL.

The results of simulated SNR and saturation fraction indicate that the dual prism system—Case 1 is not a good design, and therefore it is not of much interest to the users and scientists.

The SNR values at the EOL for each spectral band in the VNIR, SWIR1, and SWIR2 regions covering the entire wavelength range (400–2500 nm) are depicted in Figure 7.18. The figure also shows the curve of the saturation fraction values, whose scale is shown by the vertical axis on the right of the figure.

7.2.3.4 Dual Prism Hyperspectral Imagers—Case 2

In the dual prism hyperspectral imagers—Case 1, the simulated SNRs in many spectral bands do not meet the required values. In addition, there are signal saturations in VNIR and SWIR1 regions at the BOL as well as in the SWIR1 region at EOL.

In the design in Case 2, the aperture of the dual prism imaging spectrometer system is increased to 24.0 cm from 20.0 cm in Case 1. The optical system is designed faster than in Case 1 by reducing the f/# to 2.92 from f/# 3.5 in Case 1. This results in a focal length of 70.08 cm in Case 2, almost the same as in Case 1 (70.0 cm). The same FPAs as in Case 1 are used, i.e., a silicon-based hybrid CMOS detector array for the VNIR spectrometer and a HgCdTe detector array for the SWIR spectrometer.

All the assumptions discussed in Section 7.2.2 are used in the SNR modeling. The parameters used as the inputs for SNR modeling in Case 2 are the same as listed in Table 7.11, except the aperture of the system being 24.0 cm and f/# being 2.92.

TABLE 7.13
Maximum Saturation Fraction at BOL and EOL of the Dual Prism System—Case 1

Parameter	VNIR	SWIR1	SWIR2
Max saturation fraction at BOL	103%	151%	33%
Max saturation fraction at EOL	82%	121%	26%

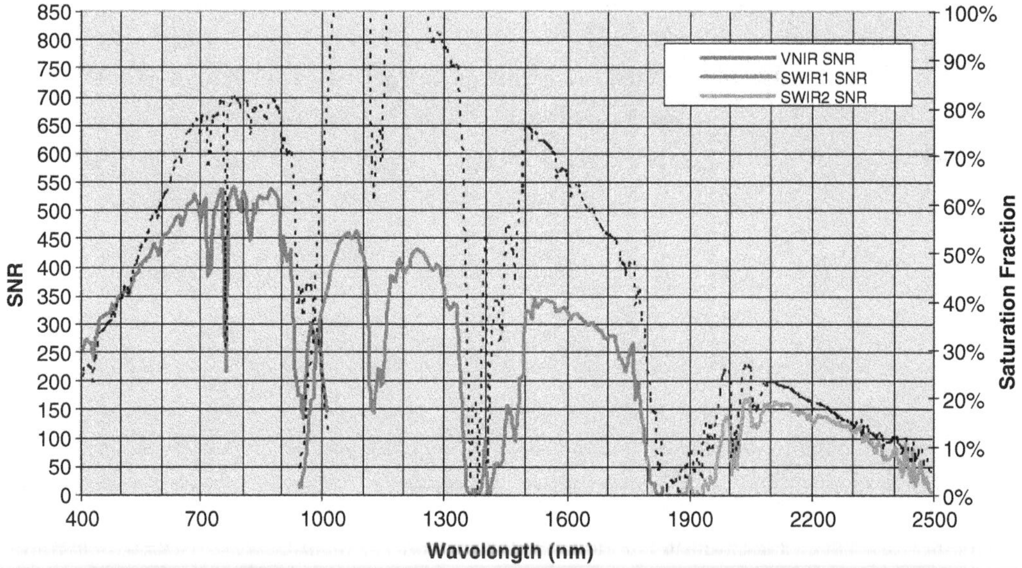

FIGURE 7.18 SNR and saturation fraction (thin dotted line) of the dual prism hyperspectral imager system—Case 1 at the EOL as a function of the wavelength.

For the dual prism system Case 2, the simulated SNR values in nine representative wavelengths at the BOL and EOL are listed in Table 7.14. In the table, the SNR margin with respect to each required value of nine representative wavelengths is also listed. It can be seen from the table that the SNRs of the dual prism system—Case 2 meet all the requirements with some margin at both the BOL and EOL except the one at wavelengths of 1950 nm, where the SNR has a deficit of 45.6% at BOL and 55.5% at EOL (shaded in Table 7.14).

The performance of saturation fraction defined by Equation 7.14 is also provided. The maximum saturation fractions at the BOL and EOL in VNIR, SWIR1, and SWIR2 regions are listed in Table 7.15. VNIR and SWIR1 are saturated by 48% and 118% at the BOL. This means that the saturation signal is 48% and 118% more than the FWC of the detector arrays. The VNIR and SWIR1 are still saturated by 18% and 74% at the EOL.

TABLE 7.14
SNR at the BOL and EOL as well as the Margin of the Dual Prism System—Case 2

		Signal-to-Noise Ratio			SNR Margin		
Wavelength (nm)		Required	BOL	EOL	BOL	EOL	SSI (nm)
VNIR	400	200	388	333	93.8%	66.7%	6.3
	650	600	688	603	14.6%	0.5%	5.9
	1000	200	300	256	50.1%	27.8%	8.3
SWIR	1000	200	458	403	128.8%	101.3%	7.3
	1550	400	481	424	20.3%	5.9%	7.5
	1750	200	356	310	77.9%	55.2%	6.9
	1950	100	54	45	–45.6%	–55.5%	6.3
	2100	200	235	202	17.7%	1.0%	5.8
	2350	100	124	104	24.5%	4.3%	5.1

TABLE 7.15
Maximum Saturation Fraction at the BOL and EOL of the Dual Prism System—Case 2

Parameter	VNIR	SWIR1	SWIR2
Max saturation fraction at BOL	148%	218%	47%
Max saturation fraction at EOL	118%	174%	38%

From the signal saturation perspective, the dual prism system—Case 2 is not an optimal design. A new design is needed to improve this performance.

The SNR values at the EOL for each spectral band in the VNIR, SWIR1, and SWIR2 regions covering the entire wavelength range (400–2500 nm) are depicted in Figure 7.19. The figure also shows the curve of the saturation fraction values, whose scale is shown by the vertical axis on the right of the figure. Note that the maximum scale value of the axis is 200% in this figure instead of 100% in Figure 7.18 due to the large saturation fraction values.

7.2.3.5 Dual Prism Hyperspectral Imagers—Case 3

It has been known in modeling the SNR for the dual prism hyperspectral imager system—Case 2 that the simulated SNRs meet almost all the required values except at one spectral band. However, there are signal saturations in VNIR and SWIR1 regions at both BOL and EOL.

In Case 3, a substrate-removed Sofradir HgCdTe detector array is chosen for the VNIR spectrometer, and a Sofradir HgCdTe detector array with a 2.5-μm cutoff wavelength is chosen for the SWIR spectrometer. These detector arrays have larger value of the FWC of charges in both VNIR and SWIR1 regions (2.6×10^6 rather than 1.0×10^6). Because the saturation is caused by the signal-induced charges more than the FWC of the detector array, an increase in the FWC will help reduce the chances of saturation.

In the design in Case 3, the aperture of the dual prism imaging spectrometer system is further increased to 26.0 cm from 24 cm in Case 2 and 20.0 cm in Case 1. The optical system is designed

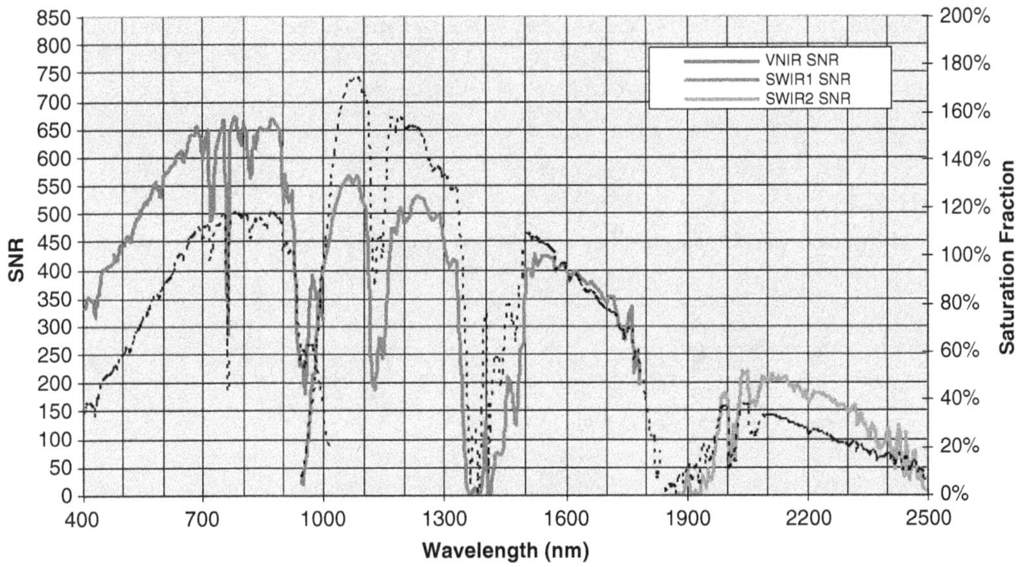

FIGURE 7.19 SNR and saturation fraction (thin dotted line) of the dual prism hyperspectral imager system—Case 2 at the EOL as a function of the wavelength.

TABLE 7.16
Inputs for Modeling SNR of the Dual Prism Hyperspectral Imager System—Case 3

Parameter	VNIR	SWIR1	SWIR2
Aperture (cm)	26.0	26.0	26.0
f/#	2.69	2.69	2.69
Detector pixel size (μm)	30	30	30
Detector full well capacity (e–)	2.6×10^6	2.6×10^6	5.8×10^5
Detector readout noise (e–)	339	339	145
Detector dark current (nA/cm^2)	0.2	1.0	1.0
Wavelength range (μm)	0.40–1.00	1.00–1.74	1.74–2.50
SSI Max (nm)	8.3	8.0	6.6
SSI Min (nm)	4.0	6.6	4.6
Quantization step (e–/DN)	635	635	142
Max integration time (ms)	4.43	4.43	4.43
Integration factor to max integration time at BOL (%)	84	100	100
Integration factor to max integration time at EOL (%)	100	100	100
Readout electronics noise (e–)	20	20	20
Photo response nonuniformity (%)	3	3	3
Detector dark signal nonuniformity (%)	10	10	10
Background temperature (K)	150	150	150
Annual degradation (%)	4.0	4.0	4.0
EOL degradation (%)	20	20	20

further faster than in Case 2 and Case 1 by reducing the f/# to 2.69 from 2.92 in Case 2 and 3.50 in Case 1. This results in a focal length of 69.94 cm in Case 3, close to 70 cm in both Case 2 and Case 1.

All the assumptions discussed in Section 7.2.2 are used in the SNR modeling. Table 7.16 summarizes the parameters used as the inputs for SNR modeling.

For the dual prism system Case 3, the simulated SNR values in nine representative spectral bands at the BOL and EOL are listed in Table 7.17. In the table, the SNR margin with respect to each required value of nine representative spectral bands is also listed. It can be seen from the table that the SNRs of the dual prism system—Case 3 meet all the requirements with some margin at both the BOL and EOL except the one at wavelengths of 1950 nm, where the SNR has a deficit of 21.1% at BOL and 34.2% at EOL (shaded in Table 7.17).

TABLE 7.17
SNR at the BOL and EOL as well as the Margin of the Dual Prism System—Case 3

		Signal-to-Noise Ratio			SNR Margin		
Wavelength (nm)		Required	BOL	EOL	BOL	EOL	SSI (nm)
VNIR	400	200	453	440	126.5%	120.0%	6.3
	650	600	616	600	2.7%	0.0%	5.9
	1000	200	391	380	95.7%	89.8%	8.3
SWIR	1000	200	441	380	120.6%	90.0%	7.3
	1550	400	467	403	16.8%	0.8%	7.5
	1750	200	328	279	64.2%	39.7%	6.9
	1950	100	79	66	–21.1%	–34.2%	6.3
	2100	200	285	249	42.6%	24.7%	5.8
	2350	100	164	140	63.5%	40.1%	5.1

TABLE 7.18
Maximum Saturation Fraction at BOL and EOL of the Dual Prism System—Case 3

Parameter	VNIR	SWIR1	SWIR2
Max saturation fraction at BOL	100%	99%	96%
Max saturation fraction at EOL	96%	79%	77%

The performance of saturation fraction defined by Equation 7.14 is also provided. The maximum saturation fractions at the BOL and EOL in VNIR, SWIR1, and SWIR2 regions are listed in Table 7.18. From the table, it can be seen that there is no saturation at both the BOL and EOL in this system design. This is an excellent design in terms of signal saturation.

The SNR values at the EOL for each spectral band in the VNIR, SWIR1, and SWIR2 regions covering the entire wavelength range (400–2500 nm) are depicted in Figure 7.20. The figure also shows the curve of the saturation fraction values, whose scale is shown by the vertical axis on the right of the figure.

7.2.3.6 Summary of SNR Simulation

In this chapter, a total of seven hyperspectral imager systems have been evaluated for their SNR and saturation. The results for the five of them are presented. In the evaluations, all the assumptions and parameters discussed in Section 7.2.2 are used as the common inputs in the SNR modeling of the seven systems. The variables of the evaluations are as follows:

1. Spectrometer configuration form, such as Dyson, Offner, and prism based
2. Aperture size of the optical system, when the spectrometer configuration form is the same
3. f/#, when the spectrometer configuration form is the same
4. FPAs used, when the spectrometer configuration form is the same

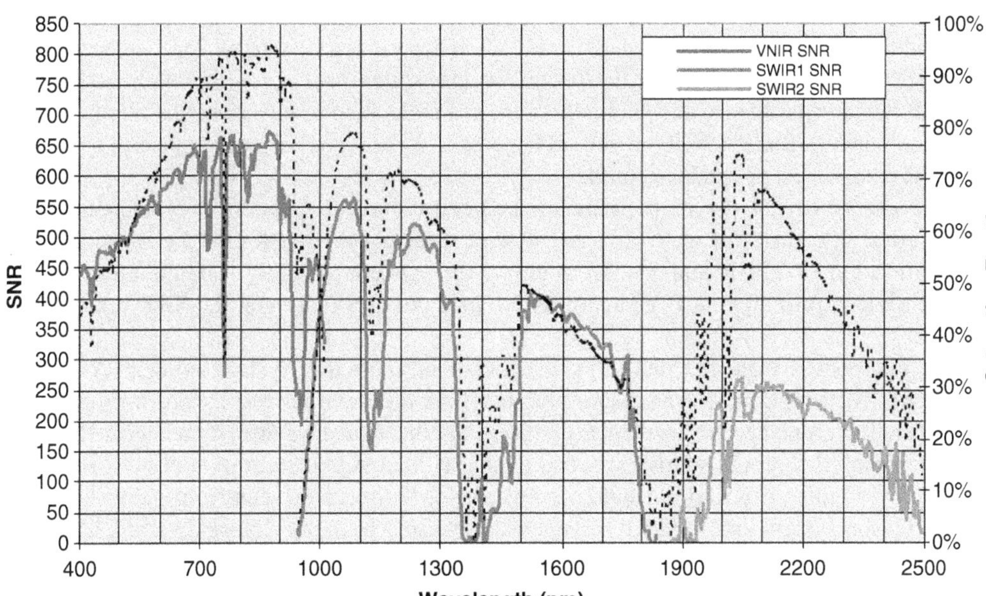

FIGURE 7.20 SNR and saturation fraction (thin dotted line) of the dual prism hyperspectral imager—Case 3 at the EOL as a function of the wavelength.

TABLE 7.19
Summary of SNR Modeling Results and Signal Saturation of Five Designed Hyperspectral Imager Systems

Hyperspectral Imager System	Optical Aperture	f/#	FPA Used	# of Bands Whose SNRs Meet the Requirement		Signal Saturation at BOL		
				BOL	EOL	VNIR	SWIR1	SWIR2
Dual Dyson	25.5 cm	2.7	Opt.1	8/9	8/9	139%	129%	70%
Dual Offner	26.5 cm	2.6	Opt.2	7/9	7/9	100%	48%	100%
Dual prism Case 1	20.0 cm	3.5	Opt.1	4/9	3/9	103%	151%	33%
Dual prism Case 2	24.0 cm	2.92	Opt.1	8/9	8/9	148%	218%	47%
Dual prism Case 3	26.0 cm	2.69	Opt.2	8/9	8/9	100%	99%	96%

There are two options for the selection of the FPAs: Option 1 and Option 2.

The evaluation excludes the telescope of the hyperspectral imager systems. A common TMA telescope is assumed for all the seven systems in the modeling.

Table 7.19 summarizes the evaluation results of the SNR in nine representative spectral bands that reached requirement at BOL and EOL, as well as the signal saturation at BOL for VNIR, SWIR1, and SWIR 2 regions of the seven systems. In the table, the bad results are shaded.

It can be seen from the table that the best system is the dual prism hyperspectral imager system Case 3, which meets the SNR requirement in 8 representative spectral bands out of 9 bands without signal saturation.

The second best system is the dual Dyson hyperspectral imager. It meets the SNR requirement in 8 representative spectral bands out of 9 bands, but with signal saturation 139% and 129% in VNIR and SWIR1 regions, respectively.

The third best system is the Dual prism hyperspectral imager system Case 2. It also meets the SNR requirement in 8 representative spectral bands out of 9 bands, but with signal saturation in VNIR and SWIR1 regions higher than the dual Dyson hyperspectral system.

The design of the dual Offner hyperspectral imager system that uses FPAs with larger FWC (Option 2) reduced the chance of signal saturation. The simulation results show that no signal saturation occurs in VNIR and SWIR1 regions at the cost of the reduced number of spectral bands that meet SNR requirement (7 bands out of 9).

The worst system is the dual prism hyperspectral imager Case 1. This system design used smallest aperture (20 cm) with the slowest optical speed (f/# of 3.5). In addition to signal saturation in VNIR and SWIR1 regions, this system has only 4 bands out of 9 that meet the SNR requirement at BOL and only 3 bands out of 9 that meet the SNR requirement at EOL.

For a fair comparison, the dual Dyson hyperspectral imager system using FPAs with a larger FWC (i.e., Option 2) was also simulated (but the detailed results are not presented here). With the enhanced performance of the FPAs the dual Dyson system meets the SNR requirement in all 9 representative spectral bands without signal saturation at both BOL and EOL. This system configuration would be ranked No. 1. As can be seen from Table 7.19, both the dual Dyson system and the dual prism system Case 3 have very close optical parameters (aperture 25.5 cm versus 26.0 cm; f/2.70 versus f/2.69). Using the same FPAs for both the systems is a fair comparison. This fair comparison indicates that a Dyson spectrometer based hyperspectral imager outperforms a conventional spectrometer based hyperspectral imager. This result is consistent with the discussion of the optical system and design described in Chapter 5.

7.3 MODELING MODULATION TRANSFER FUNCTION

7.3.1 MTF Model

In the cross-track direction, the MTF of a hyperspectral instrument is given by:

$$MTF_{Cross} = MTF_{Optics} MTF_{Detector} MTF_{Diffusion} \tag{7.15}$$

In the along-track direction, the MTF of a hyperspectral instrument is given by:

$$MTF_{Along} = MTF_{Optics} MTF_{Slit} MTF_{Smear} MTF_{Diffusion} \tag{7.16}$$

The MTF of the optics is to be determined by an optical modeling tool such as Zemax or by test. In the absence of such a tool or a design, for example, during early conceptualization phases, the MTF of the optics can be approximated as:

$$MTF_{Optics} = MTF_{Optics\_Diffration} MTF_{Optics\_Geometric} \tag{7.17}$$

The MTF of the optics due to diffraction is given by:

$$MTF_{Optics\_Diffraction} = \frac{1}{\pi}\left\{2arccos\left(\frac{\lambda\eta}{D}\right) - \sin\left[2arccos\left(\frac{\lambda\eta}{D}\right)\right]\right\}, \tag{7.18}$$

where λ is the wavelength, η is the angular spatial frequency (cycles per unit angular subtense), and D is the diameter of the aperture.

The MTF of the optics due to geometrical aberrations is modeled as a Gaussian distribution given by:

$$MTF_{Optics\_Geometric} = exp\left(-2\pi^2 b^2 \gamma^2\right), \tag{7.19}$$

where b is the diameter of the blur spot in the focal plane, and γ is the spatial frequency in the focal plane (cycles per unit distance).

The MTF of the detector (or of the slit in the along-track direction) is a geometric term arising from sampling with discrete element of finite size. The MTF of the detector is given by:

$$MTF_{Detector} = \frac{\sin(\pi d\eta)}{\pi d\eta}, \tag{7.20}$$

where d is the pixel pitch, and η is the spatial frequency (in the focal plane domain).

The MTF of the slit of the instrument is modeled as a uniform aperture given by:

$$MTF_{Slit} = \frac{\sin(\pi\theta\eta)}{\pi\theta\eta}, \tag{7.21}$$

where θ is the projected slit width at the detector plane, and η is the angular spatial frequency (in the focal plane domain).

The MTF due to diffusion and cross talk of charges between pixels, $MTF_{diffusion}$, is a term that must be given by the detector vendor, taking care that their value does not include the geometric term due to finite sampling (Equation 7.20).

In optics, smear is used to refer to deterministic image motion that has a low temporal frequency relative to the integration time. This typically results from a relative rate of the image with respect to the detector. Due to the finite integration time of image collection, the footprints on the ground are moving linearly with the satellite velocity, thus blurring in the direction of flight (Johnson 1993).

Smear is typically differentiated from jitter, which has a higher frequency relative to the integration time. Smear refers to a relatively constant rate during the integration time, while jitter refers to a relatively sinusoidal motion during the integration time.

The equation for the optical MTF associated with smear is the standard sinc function:

$$MTF_{Smear} = \frac{\sin\left(\pi\theta_\tau \frac{\tau}{\tau_{\max}} \upsilon\right)}{\pi\theta_\tau \frac{\tau}{\tau_{\max}} \upsilon}, \tag{7.22}$$

where υ is the spatial frequency, τ is the integration time, $\tau_{\max}$ is the maximum integration time (frame time), and θ_τ is the angular aperture given by:

$$\theta_\tau = v\tau_{max}, \tag{7.23}$$

where v is the along-track ground velocity.

Spatial jitters can also degrade the spatial resolution and this is another effect that can impact the MTF and should be added into Equations 7.15 and 7.16 if it is important. It can be modeled as a Gaussian distribution given by:

$$MTF_{Jitters} = exp\left(\frac{-2\pi^2\tau^2\sigma^2\nu^2 d^2}{X^2}\right), \tag{7.24}$$

where σ is the RMS position jitters amplitude (in the scene domain and for a timescale comparable to the integration time, τ), and X is the projected pixel size in the scene domain.

7.3.2 Assumptions for MTF Modeling

MTF values used for the optics in Equations 7.15 to 7.17 are calculated by adjusting the blur spot diameter to match the overall MTF derived for the five hyperspectral imager systems modeled in Section 7.2.

The detector MTF used in the modeling is derived from data provided by Teledyne for a HyViSi silicon-based CMOS detector array and 2.5-μm cutoff HgCdTe detector array. Figure 7.21 depicts the MTF of these detector arrays.

7.3.3 MTF Results

The MTFs of the five hyperspectral imager systems are modeled based on Equations 7.15 to 7.24. Figure 7.22 shows the MTF curves of the dual Dyson hyperspectral imager as a function of spatial frequency in the range between 0 cycles/mm and 20 cycles/mm. The vertical red line indicates the Nyquist frequency of 16.7 cycle/mm. The requirement of the MTF of the overall instrument at the Nyquist frequency is marked as a diamond (allocation). The figure depicts the MTF values of the instrument in both the cross-track direction (i.e., MTF_{Cross}) and the along-track direction (i.e., MTF_{Along}) at three representative wavelengths 400 nm, 1000 nm, and 2500 nm. These MTF values are the overall values after taking into account all the components of the instrument. For example, for the worst case of MTF_{Along} at wavelength 2500 nm, the MTF value is equal to 0.33 at the Nyquist frequency. It is the product of three terms defined by Equation 7.16 after ignoring $MTF_{diffusion}$:

$$MTF_{Along} = MTF_{Optics} MTF_{Slit} MTF_{Smear} = 0.815 \times 0.637 \times 0.637 = 0.331.$$

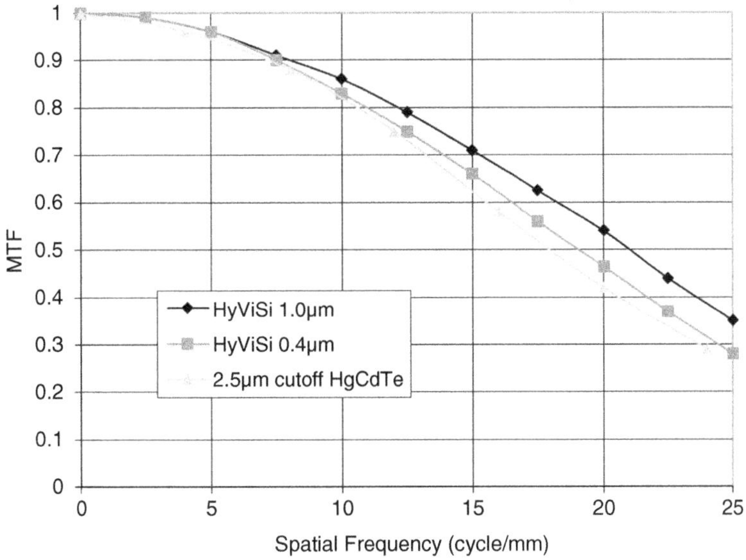

FIGURE 7.21 MTF of a HyViSi silicon-based CMOS detector array and a 2.5-μm cutoff HgCdTe detector array.

The $MTF_{Optics} = 0.815$ at the Nyquist frequency is the total MTF of the optical system including the telescope, beam splitter, and spectrometer.

For a hyperspectral instrument, the requirement for the overall instrument MTF is typically set to 0.3 at the Nyquist frequency. The worst case overall instrument MTF of the dual Dyson hyperspectral imager is 0.33 and larger than 0.3. The design of this hyperspectral imager meets the MTF requirement.

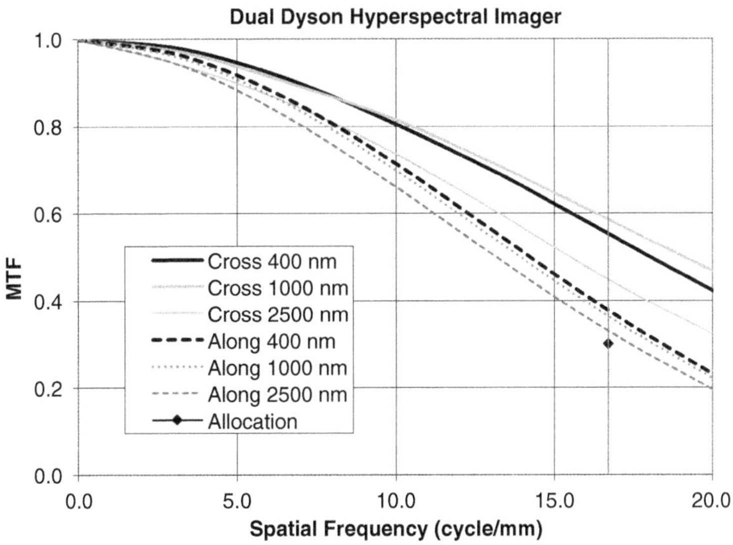

FIGURE 7.22 MTF as a function of spatial frequency in cross-track and along-track directions at three representative wavelengths (400 nm, 1000 nm, and 2500 nm) of the dual Dyson hyperspectral imager. The vertical red line indicates the Nyquist frequency (16.7 cycle/mm).

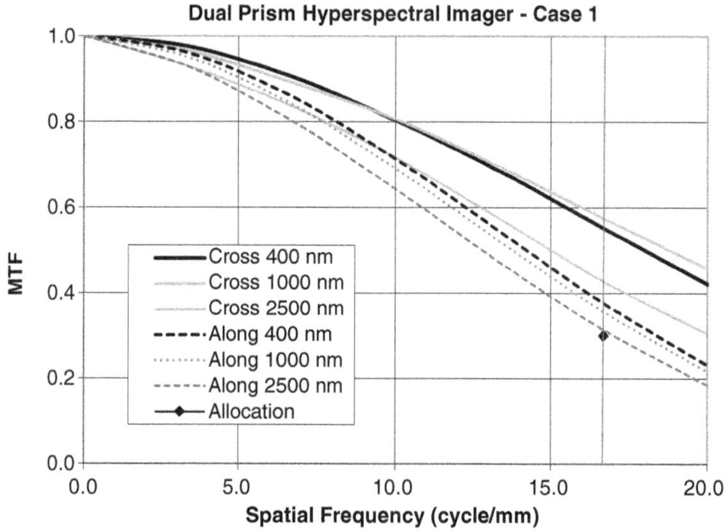

FIGURE 7.23 MTF as a function of spatial frequency in cross-track and along-track directions at three representative wavelengths (400 nm, 1000 nm, and 2500 nm) of the dual prism hyperspectral imager—Case 1.

Figure 7.23 shows the MTF curves of the dual prism hyperspectral imager—Case 1 as a function of spatial frequency in the range between 0 and 20 cycles/mm. The worst case of MTF_{Along} at wavelength 2500 nm is equal to 0.315 at the Nyquist frequency, which is smaller than that of the dual Dyson hyperspectral imager, but still meets the MTF requirement of 0.30. This MTF loss is caused by the relatively lower optical system $MTF_{Optics} = 0.777$ at the Nyquist frequency. This MTF result is consistent with its poor SNR result modeled in Section 7.2 and summarized in Table 7.19.

7.4 MODELING SPECTRAL RESPONSE

7.4.1 Spectral Response Model

It has been recognized that a hyperspectral imager operating in the pushbroom mode must satisfy stringent spectral and spatial uniformity requirements in order to provide accurate spectroscopic and spatial information. Ideally, a hyperspectral imager disperses the spectral and spatial information in two orthogonal directions preferably aligned with the columns and rows of the detector array. Due to the imperfections in the fore-optics and in particular the optical components of the spectrometers, wavelength-dependent spectral and spatial distortions along with spatial and spectral blur are introduced in the acquired image (Mouroulis 1999).

Extraction of accurate spectroscopic information depends on how well the spectral response function (SRF) of the various pixels is known. If, for the sake of simplicity, we accept that the SRF has a simple form that can be characterized by its peak location and half-width, then these two parameters must be known to within a small fraction of the nominal spectral bandwidth. Since, in a pushbroom hyperspectral imager, there are tens to hundreds of spectral bands, it is preferable to ensure that those two parameters remain approximately constant within the spectral bands, thus reducing dramatically the need for calibration of individual bands.

The modeling of SRF of a hyperspectral instrument is given by:

$$SRF_{Inst} = SRF_{Slit} * SRF_{Detector\_Geometric} * SRF_{Detector\_Diffusion} * SRF_{Opical\_Along}, \tag{7.25}$$

where $*$ denotes convolution operation.

In modeling, the worst-case blur is used for all the components. The spatial response functions for the slit and detector geometry were taken as uniform apertures. The detector diffusion was derived by matching a Gaussian distribution to the detector MTF data as same as described in Section 7.2.

7.4.2 Spectral Response Results

This section shows the results of spectral response modeling for the dual Dyson hyperspectral imager and the dual prism hyperspectral imager Case 2.

Figure 7.24 depicts the spectral response of the dual Dyson hyperspectral imager in spectral band 1000 nm. This system has an aperture of 25.5 cm with a constant SSI of 7.5 nm. The figure shows the spectral response of each component including slit, detector geometric, detector diffusion, and optics as well as the overall instrument. The spectral resolution of the hyperspectral imager is 8.5 nm at the full width at half maximum (FWHM). The spectral resolution is about 1.13 times of the SSI, which is superior. Normally, a factor of not larger than 1.2 is acceptable.

Figure 7.25 depicts the spectral response of the dual Dyson hyperspectral imager in spectral band 2500 nm. Compared to the spectral response curves for the 1000-nm band in shown Figure 7.24, the spectral responses of detector geometric and slit are almost the same, but the spectral responses of detector diffusion and optics in the along-track direction become wider. This results the spectral response of the overall instrument becoming wider. The spectral resolution at FWHM is 9.2 nm, which is 1.23 time of the SSI.

Figures 7.26 and 7.27 depict the spectral response of the dual prism hyperspectral imager Case 2 in spectral bands 1000 nm and 1300 nm, respectively. Due to the prisms used as the spectral disperses, the SSI of the hyperspectral imager is not linear. The SSI in spectral band 1000 nm is 8.3 nm, as shown in Figure 7.14, whereas the SSI in spectral band 1300 nm is 8.0 nm, as shown in Figure 7.15. The spectral response curves of the components are very similar corresponding to bands 1000 nm and band 1300 nm, except the detector diffusion, which is the main contributor to the difference of the two spectral response curves of the instrument. The spectral resolution at FWHM of the instrument is 9.4 nm in spectral band 1000 nm. It is 1.13 times of the SSI, which is superior. The spectral resolution at FWHM of the instrument is 9.7 nm in spectral band 1300 nm. It is 1.21 times of the SSI, which is very close to 1.2 nominal value.

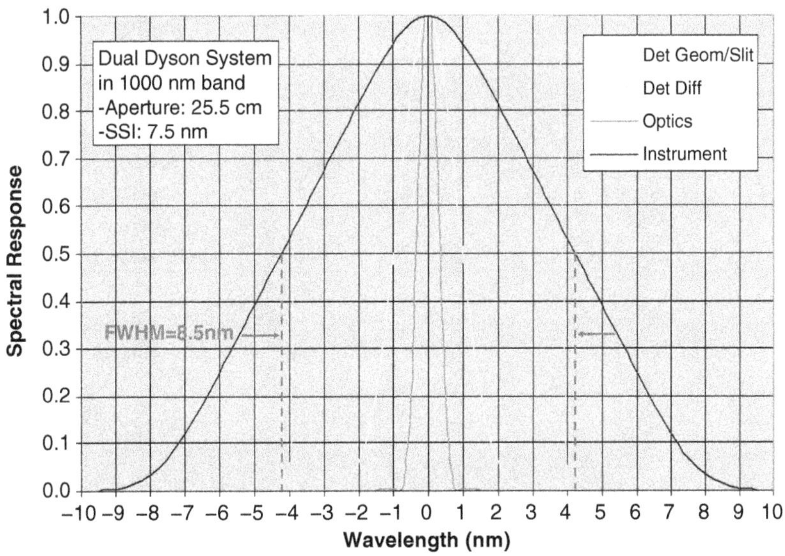

FIGURE 7.24 Spectral response of the dual Dyson hyperspectral imager in spectral band 1000 nm.

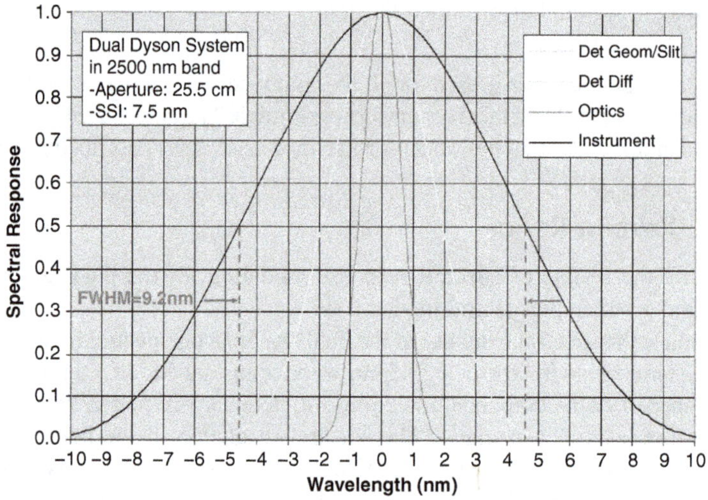

FIGURE 7.25 Spectral response of the dual Dyson hyperspectral imager in spectral band 2500 nm.

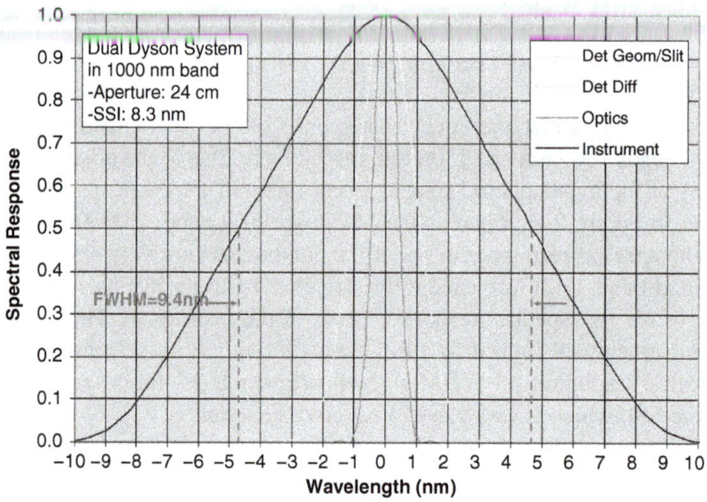

FIGURE 7.26 Spectral response of a dual prism hyperspectral imager in spectral band 1000 nm.

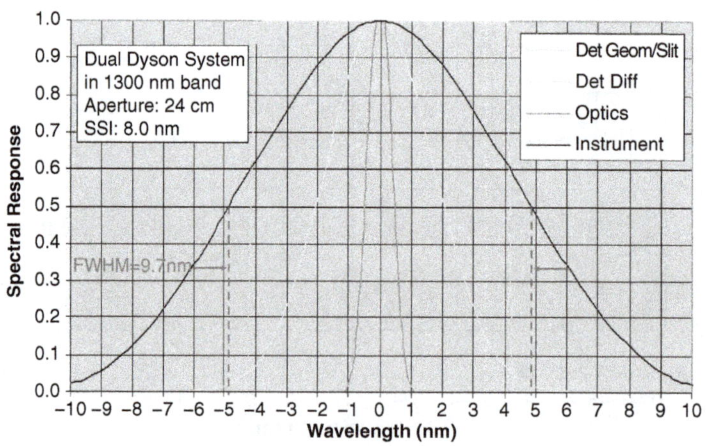

FIGURE 7.27 Spectral response of a dual prism hyperspectral imager in spectral band 1300 nm.

7.5 POINT SPREAD FUNCTION ANALYSIS

7.5.1 Point Spread Function Model

Point spread function (PSF) is the spatial domain version of the optical transfer function of an imaging system. It describes the response of an imaging system to a point source in spatial dimensions. A more general term for the PSF is a system's impulse response, the PSF being the impulse response of a focused optical system. The PSF is the Fourier transform of the MTF and is another description or quantification of the spatial resolution.

Pixels acquired by a hyperspectral imager may contain mixed spectra for two reasons, either because there are materials with different spectral signatures in the same pixel, or because the optical PSF has a finite extent, and thus causes light from one point on the ground to strike two or more pixels on the FPA simultaneously. The second case is of concern because steps can be taken during optical design to facilitate spectral unmixing of the acquired hyperspectral imagery. Specifically, if all PSFs were of the same form, then algorithms could be developed that would take the PSF into account. At the very least, an attempt should be made to minimize the PSF variation, or equivalently, minimize the variation of the spatial response function of the spectrometer.

Since it is impossible to produce identical PSFs throughout wavelength and the field of view (FOV) of a hyperspectral imager, the PSF variation is constrained in the spectral and spatial directions individually. The ideal spectrum of a point source generated by a hyperspectral imager should have the following characteristics: the image point is within the square (or rectangle) pixel of the focal plane and aligned with the photodetector element. The width of the PSF for any one wavelength is constant, independent of the FOV. Also, the height of the PSF is constant, independent of wavelength and also independent of the FOV.

Let $o(u,w)$ be an image of a reference object acquired by a hyperspectral imager and let $g(u,w)$ be the corresponding ideal undistorted image defined by the properties of the reference object. The main goal of the PSF characterization is finding a response function $h(u,w)$ that describes the transformation between the ideal image and the corresponding acquired image of the reference object that contains distortions introduced by the hyperspectral imager (Kosec et al. 2013).

Adopting a polynomial baseline model $b(u,w)$, the transformation between the reference image and corresponding acquired image can be described within error $\varepsilon(u,w)$ as:

$$o(u,w) = g(u,w) * h(u,w) + b(u,w) + \varepsilon(u,w), \tag{7.26}$$

where $*$ denotes convolution. To solve Equation 7.26, the response function $h_{i,j}(u,w)$ in each point (i, j) of the imaging plane is modeled by an appropriate basis function:

$$h_{i,j}(u,w) = f\big(u,w,\lambda_0(i,j),\lambda_1(i,j),\ldots,\lambda_n(i,j)\big), \tag{7.27}$$

where λ_k $(k = 0,1,\ldots n)$ are parameters of the basis function f and are expected to be smooth within the domain of the imaging plane. Therefore, a bivariate cubic spline approximation l parameterized by $\boldsymbol{v}_k(\boldsymbol{c})$ defined at knots $\boldsymbol{c}$ is used to describe the variation:

$$\lambda_k(u,w) = l\big(u,w,\boldsymbol{c},\boldsymbol{v}_k(\boldsymbol{c})\big) \tag{7.28}$$

Likewise, the smooth baseline of the hyperspectral imager is modeled by a polynomial surface:

$$b(u,w) = \sum_{i=0,\ k=0}^{J,K} \beta_{j,k} \cdot u^j \cdot w^k \tag{7.29}$$

Let $s(u,w)$ represent an estimate of the acquired reference object image that can be expressed as:

$$s(u,w) = g(u,w) * h(u,w) + b(u,w) \tag{7.30}$$

then the problem of finding the spatial response function $h(u,w)$ is translated to matching the modeled $s(u,w)$ and acquired $o(u,w)$ images of the reference object.

7.5.2 Point Spread Function Modeling Results

The PSFs of the VNIR and SWIR regions have been computed for three wavelengths (one central spectral band and two extreme bands) and for two FOVs (at a marginal and the center of the slit), including both the spectrometer and telescope contribution for the five hyperspectral imager systems described in Sections 7.1 and 7.2.

Figure 7.28 depicts the six PSFs of the dual prism hyperspectral imager Case 3 in the VNIR region. Three spectral bands in 0.40, 0.70, and 1.0 μm are selected. For each spectral band, two spatial locations at one end (0, 0) and at the center (15, 0) of the slit are selected. The figure shows the spectral band, location, and size of the point spread. It can be seen from the figure that the PSFs in spectral bands 0.40 μm and 0.70 μm are all enclosed inside the pixel pitch size of 30 μm × 30 μm on the focal plane. The size of PSFs in spectral band 1.0 μm is 32.6 and 32.3 μm at one end and at the center. They are slightly outside the detector pitch dimension. In the same spectral band, comparing the performance of PSFs for spatial location it is found that the PSF at the center of the slit is slightly better.

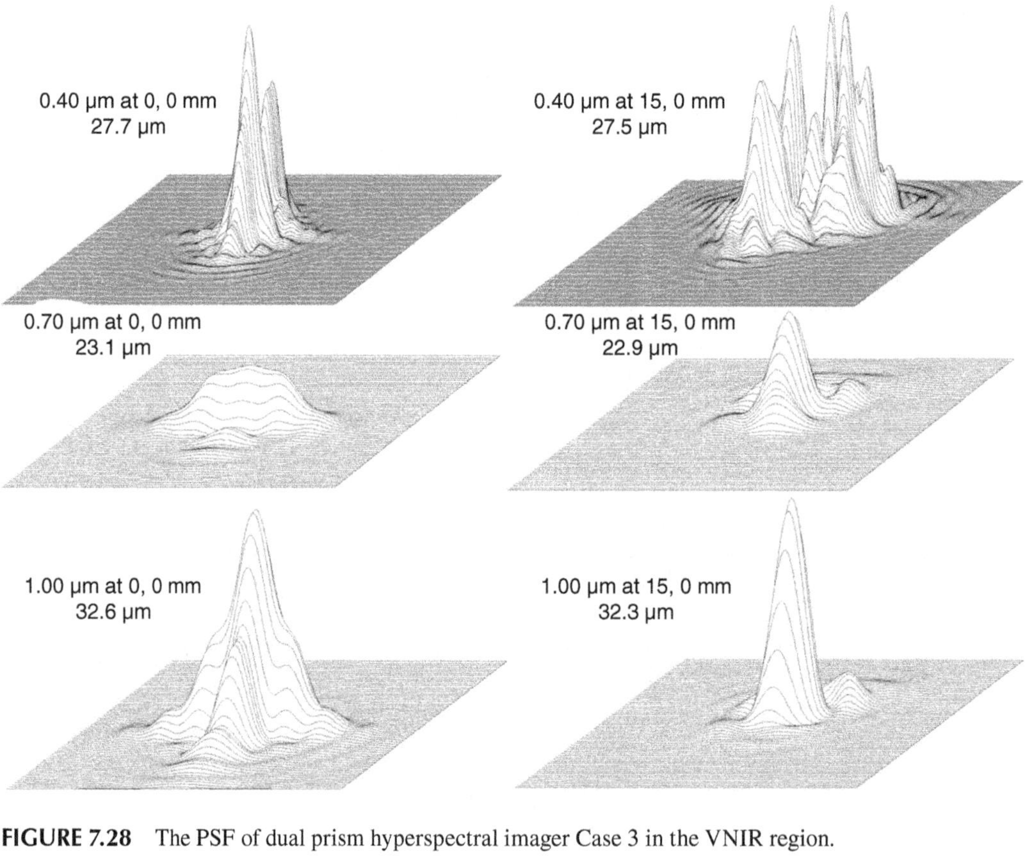

FIGURE 7.28 The PSF of dual prism hyperspectral imager Case 3 in the VNIR region.

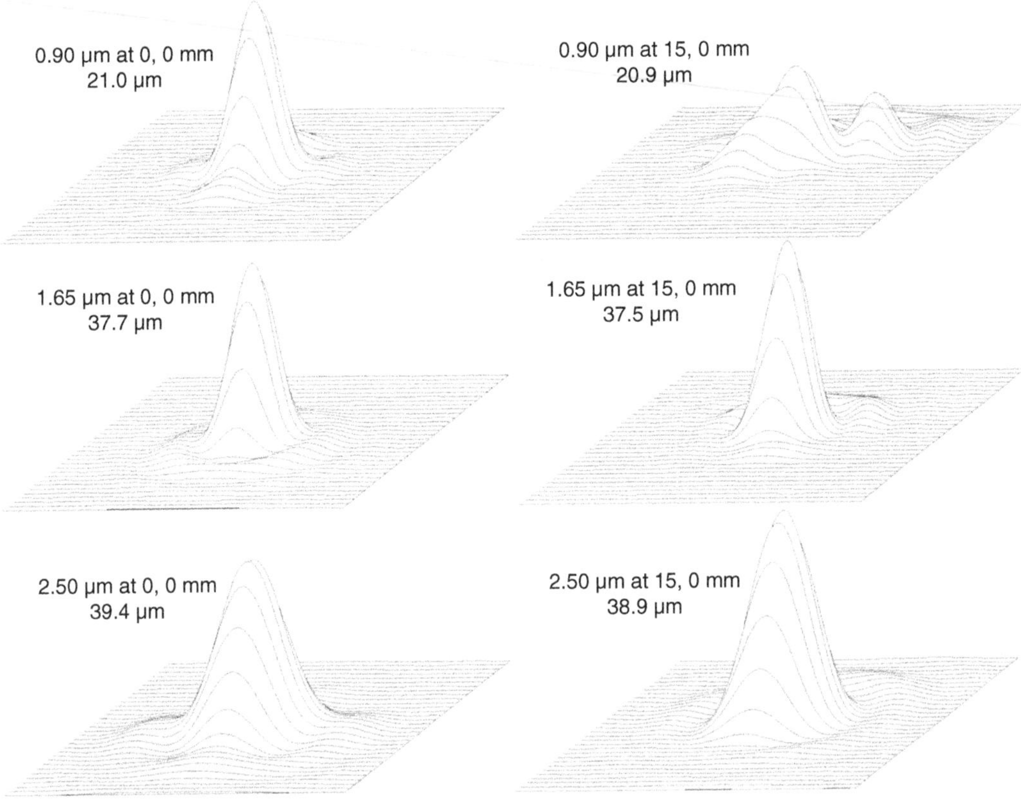

FIGURE 7.29 The PSF of dual prism hyperspectral imager Case 3 in the SWIR region.

Figure 7.29 depicts the six PSFs of the dual prism hyperspectral imager Case 3 in the SWIR region. It can be seen from the figure that the two PSFs in spectral band 0.90 μm are enclosed inside the detector pitch size of 30 μm × 30 μm on the focal plane. The size of all PSFs in spectral bands 1.65 μm and 2.50 μm is larger than the detector pitch size. The longer the wavelength, the larger the size of the PSF. The size of PSFs in spectral band 1.65 μm is 37.7 and 37.5 μm at one end and at the center. The worst size of PSFs in spectral band 2.50 μm is 39.4 and 38.9 μm at one end and at the center.

In general, it is difficult to design and build a hyperspectral imager system with its PSF of all pixels being enclosed inside the detector pixel on the focal plane in the FOV in all spectral bands. The PSF modeling results of the dual prism hyperspectral imager Case 3 shown in Figures 7.28 and 7.29 indicate that the optical quality of the system is reasonably good.

REFERENCES

Cook, L. G. 1992. The last three-mirror anastigmatic (TMA). In *Lens Design*, ed. W. J. Smith, 310–324, CR41. Bellingham: SPIE Optical Engineering Press.

Demers, R., et al. 2013. The CHROMA focal plane array: A large-format, low-noise detector optimized for imaging spectroscopy. *Proceedings of SPIE* 8870:88700J-1–88700J-7.

Djazovski, O. 2016. Focal plane arrays for optical payloads. In *Optical Payloads for Space Missions*, ed. S.-E. Qian, 793–837. Chichester: John Wiley & Sons.

Fièque, B., P. Chorier, and B. Terrier. 2010. Sofradir detectors for hyperspectral applications from visible up to VLWIR. *Proceedings of SPIE* 7826:78261I-1–78261I-12.

Johnson, J. F. 1993. Modeling imager deterministic and statistical modulation transfer functions. *Applied Optics* 32(32):6503.

Kosec, M., M. Burmen, D. Tomazevic, F. Pernus, and B. Likar. 2013. Characterization of a spectrograph based hyperspectral imaging system. *Optics Express* 21:12085–12099.

Mouroulis, P. 1999. Spectral and spatial uniformity in pushbroom imaging spectrometers. *Proceedings of SPIE* 3753:133–141.

Mouroulis, P., R. O. Green, and D. W. Wilson. 2008. Optical design of a coastal ocean imaging spectrometer. *Optics Express* 16:1087–1096.

Prieto-Blanco, X., C. Montero-Orille, B. Couce, and R. De la Fuente. 2006. Analytical design of an Offner imaging spectrometer. *Optics Express* 14:9156–9168.

8 Thermal and Mechanical Design

8.1 BASICS OF THERMAL AND MECHANICAL DESIGN

Thermal and mechanical design is crucial within the overall system design of a hyperspectral satellite. The study of heat transfer, which is a subset of mechanical engineering, encompasses and is often concerned with predicting and controlling temperatures in engineering systems. It is important to maintain temperatures of both mechanical and electrical components within certain ranges because most material properties vary with temperature. If the temperature is allowed to fluctuate too widely, components may fail or not function as intended. Additionally, many of the failure mechanisms in electrical components involve thermally activated processes, and elevated temperatures can result in thermal expansion mismatch, corrosion, and other chemical reactions that amount to component aging and ultimately decreased reliability.

The purpose of thermal control is to maintain all the components and subsystems of the satellite and its payloads within their required temperature limits throughout all phases of the mission. Temperature limits include a cold temperature limit and a hot temperature limit to define a range. A component must not go below the cold temperature and not exceed the hot temperature. Two thermally related ranges are often used: operational limit range and survival limit range. The components and subsystems must remain within the operational limits during the period of operation. The instrument performance is at risk of being out of tolerance when operational limits are exceeded. The components and subsystems must remain within the survival limit at all times, even when not powered. Exceeding the survival limits can result in permanent damage of the components and subsystems. A large temperature gradient could cause structural exceedances including deformation such that pointing could be impacted. Thermal controls are applied to ensure that components remain with the specified ranges as well as to limit temperature gradients, where the spatial variation of temperature could be detrimental. Temperature gradients are driven by the magnitude of the heat load and the thermal resistance within components or contacts between components.

Thermal control is critical to ensure the performance and survival of satellite and its payloads. The techniques used are generally divided into two categories: passive thermal control and active thermal control. Passive thermal control can be achieved by control of conductive and radiative heat paths through selection of the proper thermal coatings, surface finishes (such as insulation blankets, or second surface mirrors), radiator, sun shields, surface thermos-optical properties, heat sinks, geometrical configurations, and phase-change materials (Jewell 1996). A passive system does not involve moving parts or fluids (Lam et al. 2002).

The spacecraft component temperatures are maintained within the desired range by appropriate control of the dissipated energy between all spacecraft elements through the conductive and radiative heat paths. However, to execute a design in which the passive thermal control cannot deal with environmental extremes or to accommodate equipment dissipating high power, use of active thermal control may be more efficient. In such cases, designs can be executed by employing heaters, louvers, heat pipes, thermoelectric coolers, cryogenic coolers, and pumped fluid loops. Obviously active thermal control is generally more complex and expensive than passive thermal control, though necessary to achieve high levels of temperature stability.

Mechanical design deals with the structural and mechanism subsystems, which supports all other satellite subsystems and attaches the satellite to the launch vehicle. The mechanical design must satisfy all strength and stiffness requirements of the satellite and its interface to the launch vehicle. The primary structure must carry loads associated with joining together all major subsystems of a satellite, for both on-ground and in-orbit cases. The secondary structure supports precise positioning of certain devices (instruments, reaction wheels, sensors, antennas) and movable mechanisms such as for unfolding antennas, solar generators.

The mechanical design needs to take into account the constraints related to the payload accommodation (the orbit, mission phases and maneuvers, position on top of the satellite) and the environmental conditions (such as launch, deployment, and in-orbit phase). Mechanical design is closely connected to and often driven by the thermal requirements, considering that the use of spectrometers implies a tight control of temperature variations to guarantee the stability of all instrument features once in orbit. In particular, material selection of the mechanical mounts can be thermally driven to thermal expansion properties for positioning, or for heat transfer. In particular, thermal design effort must be devoted to the detector cooling. The use of active mechanical coolers, due to intrinsic micro-vibrations of reciprocating elements, may be a potential cause of a modulation transfer function (MTF) degradation, especially for very narrow instantaneous field of view instruments (IFOVs), and moreover is an additional risk in terms of life-cycle reliability. This is sometimes referred to as "jitter." However, sophisticated vibration cancellation systems have been developed to actively cancel such vibrations to very low levels, such as have been demonstrated for Stirling and pulse tube mechanical cooler systems (Ross 2003).

8.2 THERMAL ENVIRONMENT

Environmental heating plays a major role in satellite thermal control. Over the course of the development and operational life cycle, a satellite will be exposed to different environments, including from ground testing, transportation to launch site, launch, orbital transfer to operational orbit with nominal and safe-hold attitudes. During the ground testing cycle, convection with ambient air and conduction to thermal interfaces (and to a lesser degree, radiant heat exchange with surrounding objects except for high temperature differences) represent the main thermal environment related factors. During launch ascent, radiant heating from the inside surface of the booster fairing and free molecular heating due to friction with the atmosphere are the main environmental influences. Once above the upper reaches of the atmosphere, direct sunlight, sunlight reflected by the Earth (or other planetary bodies), and infrared energy emitted from the planet's atmosphere or surface are the dominant sources of environment heat. In most cases, the thermal control subsystem is designed to maintain all satellite components within allowable temperature limits in the environments encountered on-orbit, while compatibility with ground operations and launch ascent conditions is assured by controlling the environment of limiting the satellite's exposure to it (Gilmore et al. 2006).

Satellite thermal control is a process of energy management. Since there is no atmospheric convection in space, the overall thermal control of a satellite on orbit is usually achieved by balancing the heat emitted by the satellite as infrared radiation against the heat dissipated by its internal components and the heat absorbed from the environment. It is generally more practical and cost-effective to custom-tailor a thermal control design to each satellite. This is because a generic thermal control system that is capable of managing all conceivable scenarios for satellites in different environments would be impractical and likely inefficient. The design analysis of a satellite thermal control must consider the variable environmental heat loads on the satellite together with the worst case hot and cold combinations of waste heat generated by the satellite components in various operation modes. The design must also account for anticipated degradation aspects, including material degradation due to radiation, component degradation (such as electrical component efficiency), and interfaces including thermal contacts and surface thermo-optical properties over the mission duration.

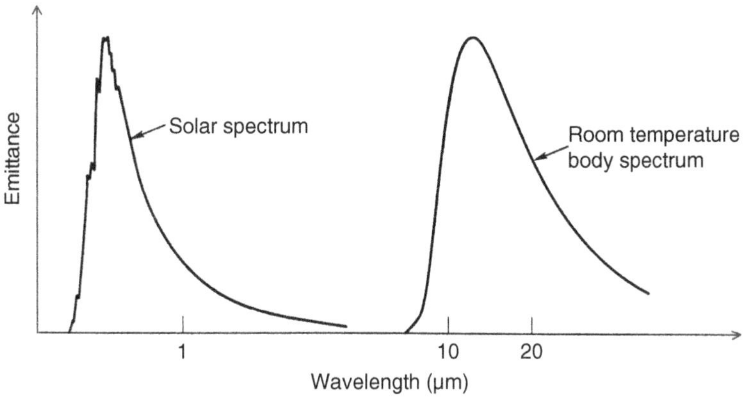

FIGURE 8.1 Spectral distribution of emitted energy by solar infrared radiation and a room-temperature body.

8.2.1 Sunlight

Sunlight is the main source of environmental heating on most satellites. The sun is a very stable energy source with a constant to a variation within a fraction of a percent. The intensity of sunlight reaching Earth varies with the distance to the sun. The intensity of sunlight at Earth's mean distance to sun (1 AU) is equal to 1367 W/m^2, which is known as the solar constant. When Earth is farthest from the sun, the intensity is at a minimum value of 1322 W/m^2, which is approximately 3.5% lower than the solar constant. When Earth is closest to the sun, the intensity is at a maximum value of 1414 W/m^2, which is approximately 3.5% higher than the solar constant (Gilmore 2002).

The intensity of sunlight also varies as a function of wavelength. The energy distribution is approximately 7% in ultraviolet, 46% in visible, and 47% in near-infrared (NIR). It is worth noting that the wavelength of infrared energy emitted by sun is much shorter than that of the emitted energy by a body at near room temperature as shown in Figure 8.1. This wavelength distinction makes it possible to select effective thermal control finishes. By using thermal control finishes that are very reflective in the solar infrared spectrum but highly emissive at room temperature infrared, the solar heat loads can be minimized while maximizing a satellite's ability to dissipate waste heat.

8.2.2 Albedo

Sunlight reflected from the Earth, a planet, or moon is known as albedo. Albedo is defined as the diffuse reflection of solar radiation out of the total solar radiation received by Earth or a planet. It is highly variable, measured on a scale from 0 (corresponding to a black body that absorbs all incident radiation) to 1 (corresponding to a body that reflects all incident radiation). Reflectivity is greater over land than over oceans. Albedo increases with decreasing local solar elevation angles and increasing cloud coverage. Albedo increases at high latitude due to great snow and ice coverage, decreasing solar elevation angle and increasing cloud coverage.

Table 8.1 lists the representative albedo values for Earth orbits of different inclinations and angle of sun out of orbit plane. It can be seen that, for example, on an orbit with 60–90° inclination, the minimum and maximum albedo is 23% and 30%, respectively, when the angle of sun out of the orbit plane is 0°. The minimum and maximum albedo increases to 50% and 57%, respectively, when the angle of sun out of the orbit plane increases to 90°. Large variations of albedo make selection of the more accurate albedo value for thermal analysis rather uncertain.

TABLE 8.1
Typical Albedo Values for Various Earth Orbits (Data Courtesy NASA)

Orbit Inclination (°)	Angle of Sun out of Orbit Plane (°)	Albedo (%)	
		Min	Max
0–30	0	18	28
	90	45	55
30–60	0	23	30
	90	50	57
60–90	0	23	30
	90	50	57

8.2.3 Earth Infrared

All incident sunlight that is not reflected as albedo is absorbed by Earth and eventually re-emitted as infrared energy. The intensity of infrared energy emitted by Earth varies with times and location on Earth. Even at a same given time from a same location, the intensity can vary considerably depending on the factors such as the local temperature of the Earth's surface and amount of cloud coverage. A warmer surface region will emit more radiation than a colder region. Tropical and desert regions receive the maximum solar heating, thus emit the highest infrared radiation compared with other regions. The intensity of infrared energy decreases with increasing latitude. Cloud cover reduces Earth's emitted infrared energy. This is because cloud tops are cold and clouds effectively block up-welling radiation from the Earth's warm surface below. Although these variations in Earth-emitted infrared radiation are significant, they are much less severe than the variations in albedo.

Table 8.2 shows the representative orbit-average values for Earth infrared radiation. It can be seen that the maximum value is 275 W/m^2 when the orbit inclination is 0–30°, whereas the minimum value is 218 W/m^2 when the orbit inclination is 30–90°. The ratio of the maximum variation is 1.26 compared to the ratio of the maximum variation in albedo of 3.17.

The wavelength of Earth's infrared energy emission is much longer than that of the solar infrared energy emission described in Section 8.2.1. It is of approximately the same wavelength as that emitted by the satellite itself, since satellites normally operate at a nominal temperature that is not far from room temperature. Earth's infrared emission incident on a satellite cannot be designed to be reflected away from radiator surfaces with special thermal control coatings, because the same

TABLE 8.2
Typical Orbit-Average Earth Infrared Radiation Values for Various Orbits (Data Courtesy NASA)

Orbit Inclination (°)	Angle of Sun out of Orbit Plane (°)	Emitted Radiation (W/m^2)	
		Min	Max
0–30	0	228	275
	90	228	275
30–60	0	218	257
	90	218	257
60–90	0	218	244
	90	218	244

coatings would prevent the radiation of waste heat emitting away from the satellite. Because of this fact, Earth's infrared energy emission can present a particularly heavy heat load on satellite radiators in low Earth orbits (LEOs). Earth has an effective average temperature around −18°C. A satellite is usually warmer than the effective Earth temperature. The net heat transfer is from satellite to Earth, but is minimal when the temperature difference is small.

8.2.4 Interplanetary Environments

Most of spaceborne hyperspectral imagers are developed for Earth observation. At the time of this book being written, at least one imaging spectrometer had flown on lunar orbit—the Moon Mineralogy Mapper (M3)—and at least one imaging spectrometers had flown on Mars orbit—the Compact Reconnaissance Imaging Spectrometer for Mars (CRISM). The Visible Infrared Thermal Imaging Spectrometer (VIRTIS) had been built for three copies with slight variant, which had flown three planetary missions: (1) ESA's Rosetta mission (where ESA is European Space Agency) on a comet orbit (Coradini et al. 2007, Auster et al. 2010, Pätzold et al. 2011), (2) ESA's Venus Express on a Venus orbit (Piccioni et al. 2007), and (3) NASA-Dawn mission on orbits of proto-planets Vesta and Ceres (Russell et al. 2007). Detailed description of these spaceborne imaging spectrometers is provided in Chapter 2. Another planetary imaging spectrometer called Moons And Jupiter Imaging Spectrometer (MAJIS) for ESA's JUpiter ICy moons Explorer (JUICE) mission is under development and will be deployed onboard JUICE spacecraft to be sent to visit the Jovian system for studying three of Jupiter's Galilean moons: Ganymede, Callisto, and Europa (Plaut et al. 2014).

Because of the increasing number of interplanetary hyperspectral imagers, interplanetary environments are also needed to be considered in thermal and mechanical design of spaceborne hyperspectral imagers.

Interplanetary trajectories can expose spacecraft to a range of thermal environments much more severe than those encountered on Earth orbits. During an interplanetary cruise, the only environmental heat that comes is from direct sunlight, which falls off as the square of the distance to the sun. During a flyby, or after a spacecraft enters into an orbit around a planet, it is also exposed to infrared radiation and albedo heat loads from the planet.

Planetary and lunar infrared emission and albedo vary widely due to various planet's distances to the sun and different atmosphere and surface characteristics. A detailed characterization of the thermal environments of the planets is provided by Clawson et al. (2002).

8.3 PASSIVE THERMAL CONTROL

Passive thermal control is a desired approach to maintain the limits of temperatures. It makes use of surface finishes, insulation, radiators, and specific materials chosen for properties such as their thermal conductivity and heat capacity. It is a more reliable and cost-effective means.

8.3.1 Surface Finish

The surfaces of a spacecraft are radiatively exposed to external sources of energy, such as sunlight and Earth-emitted infrared radiation, so their radiative properties must be selected to achieve an energy balance at the desired temperature between spacecraft internal dissipation, external sources of heat, and reradiation to space. In most cases, the desired temperature range can be driven by specific equipment, such as electronics.

Thermal control surface finishes are classified into four basic categories: solar reflector, solar absorber, flat reflector, and flat absorber. The solar reflector reflects incident solar energy while absorbing and emitting infrared energy. Solar reflectors are characterized by a very low α/ε ratio, where α is the solar absorptivity and ε is the infrared emissivity. Solar absorbers absorb solar energy while emitting a significantly lower percentage of the infrared energy. Flat reflectors reflect energy in both the solar and infrared regions, while flat absorbers absorb in both the solar and infrared regions.

As described in Section 8.2.1 and as shown in Figure 8.1, the wavelength of the sunlight emitted energy is much shorter than that of the emitted energy by a body at near room temperature. In thermal designs this fact can be beneficial if thermal control finishes are effectively selected. By using thermal control finishes that are very reflective in the solar infrared spectrum but highly emissive to room temperature infrared, the solar heat loads can be minimized while maximizing a satellite's ability to dissipate waste heat.

A wide variety of wavelength-dependent thermal control coatings have been exploited. The problems of in-space stability, outgassing, and mechanical adhesion to the substrate have all been resolved for most coatings. Many fully qualified coatings are available (but the lifetime needs to be verified according to the mission duration). Solar reflectors, such as second-surface mirrors and white paints or sliver or aluminum backed Teflon, are used to minimize absorbed solar energy and at the same time to emit energy almost like a blackbody.

To minimize both the absorbed solar energy and infrared emission, polished metal finishes such as aluminum foil or gold plating is utilized. The principal external surface finishes seen on most spacecraft are the outer layer of insulation blankets, radiator coatings, and paints. Electronics boxes located inside the spacecraft, and the structural panels to which they are attached, are usually painted to achieve a high emittance. While most paints have the required high emittance regardless of color, black paint is commonly used on the interior of a spacecraft to facilitate radiant heat transfer among internal components, to distribute heat from hotter components to cooler regions, reducing gradients. Internal temperature sensitive components that do not dissipate much heat often utilize a low-emittance finish of aluminum or gold (including minimizing parasitic heating in cryogenic instruments).

The performance of the thermal control finishes is degraded by charged particles, ultraviolet radiation, high vacuum, micrometeoroid impacts, and the contaminant films that deposit out on almost all spacecraft surfaces. To compensate the degradation, spacecraft radiators must be oversized to account for the substantial increase in absorbed solar energy over the mission. These oversized radiators may cause the spacecraft to run much cooler in the early years of the mission, sometimes necessitating the use of heaters to avoid under temperatures of electronic components. The stability of coating properties is important in order to limit maximum temperatures and minimize heater-power requirements.

A detailed description of the thermal control surface finishes has been reported by Gilmore et al. (2002).

8.3.2 Insulation

Thermal insulation is the reduction in the transfer of thermal energy between objects of differing temperature in thermal contact or in range of radiative influence. It provides a region of insulation in which thermal conduction is reduced or thermal radiation is reflected rather than absorbed by the lower-temperature body. Multilayer insulation (MLI) and single-layer radiation shields are among the most common thermal control insulation applications in spacecraft. MLI blankets are used either to prevent excessive heat loss from a component or excessive heating from environmental fluxes, rocket plumes, and other sources (Donabedian et al. 2002).

Most spacecraft flown today are covered with MLI blankets, with cutouts provided for areas where radiators reject internally generated waste heat. MLI blankets are also typically used to protect internal propellant tanks, propellant lines, and cryogenic dewars. Single-layer radiation shields are sometimes used in replace of MLI where less thermal isolation is required, since they are lighter and cheaper to manufacture.

MLI is composed of multiple layers of low-emittance films with low conductivity spacer material between layers. The simplest MLI construction is a layered blanket assembled from thin (1/4 mil thick, 1 mil = 0.001 in) embossed mylar sheets, each with a vacuum-deposited aluminum finish on one side. As a result of the embossing, the sheets touch at only a few points, and conductive heat paths between layers are thus minimized. The layers are aluminized on one side only so that the mylar can act somewhat as a low-conductivity spacer. Higher-performance construction is composed of mylar film metalized with aluminum or gold on both surfaces with silk or dacron netting as the low-conductance spacers.

Heat transfer through MLI is a combination of radiation, solid conduction, and gaseous conduction under atmospheric conditions. Imperfections are unavoidable and include stitched seams where conductances are increased by the practicality of having too close joints. These forms of heat transfer are minimized in different ways. Radiative heat transfer is minimized by interposing as many enclosing reflective surfaces (metalized sheets) as is practical between the object being insulated and its surroundings. Solid-conduction heat transfer is minimized by keeping the density of the low-conductance spacers between the reflective surfaces as low as possible and making the blanket fluffy to minimize contact between layers. Gaseous-conduction heat transfer is minimized by allowing the insulation to vent gas to space after the spacecraft is launched.

The emissivity ε for a multilayer blanket theoretically varies directly with the number of layers. However, in practice, simply increasing the number of layers will not improve performance past a certain value. As the number of layers increases, radiative heat transfer becomes small compared with conductive shorts between layers and other losses. Thus, a point of diminishing returns is reached. Taking all of these factors into account, it is found that about 25 layers usually suffice to obtain a minimum overall conductance value.

Experience has shown that when a blanket is configured for spacecraft application, an ε more like 0.015–0.030 is representative of current design, manufacturing, and installation methods for medium-area applications. As the size of the blanket increases, the performance generally increases. This relationship results from the smaller relative influences of heat leaks that result from edge effects, seams, cable penetrations, etc. For very large area blankets with minimal penetrations, the laboratory performance approaches 0.005 at 30 layers.

Blanket design must accommodate requirements for durability, flammability, contamination control, launch loads, pressure decay, spacecraft venting, glint minimization, and restrictions on magnetic materials. Because most launch sites are near beaches (or even in the middle of the ocean, as in the case of Sea-Launch), exposure to salt spray and other corrosive agents is possible, so blanket design must take that exposure into account. All of these functions and design requirements must be addressed by blanket developers, who are also striving to minimize mass, cost, risk, and development time, while maintaining relatively durable blanks to survive assembly operations, testing and vibration associated with transportation and launch.

8.3.3 Radiator

Radiators are key passive thermal control components. They dissipate waste heat of a satellite to space by infrared radiation from their surfaces. Radiators are configured in several different forms, such as satellite structural panels, flat-plate radiators mounted to the side of the satellite, and panels deployed after the satellite is on orbit. The radiating power depends on the surface's emissivity, area, and temperature. Radiators must dissipate both the satellite waste heat plus any radiant-heat loads from the environment or other satellite surfaces that are absorbed. Most radiators are therefore given surface finishes with high infrared emissivity ($\varepsilon > 0.8$) to maximize heat leaving and low solar absorptivity ($\alpha < 0.2$) to limit heat loads from the sun. These fall under the solar reflector category described in 8.3.1. Typical finishes include quartz mirrors, silvered or aluminized Teflon, and white paint (Gilmore 2002).

The radiating power of a radiator is a strong function of temperature. The total heat dissipation of a radiator surface is given by the equation below:

$$Q = A\varepsilon\sigma T^4 \tag{8.1}$$

where A is surface area, ε is emissivity, σ is the Stefan-Boltzmann constant (5.669×10^{-8} W/m²·K⁴), and T is absolute temperature (K).

Because Q is proportional to T^4, temperature increase results in a large increase in radiating capability. For example, the radiating power at 50°C is about twice that at 0°C. At cryogenic temperatures the effect is even more pronounced. A radiator at 70 K has only 1/300th the heat-dissipating capability as at a room-temperature. This characteristic makes cryogenic radiators extremely sensitive to environmental heating and heat leaks through insulation and supports, or by small view factors to warmer equipment, leading to special design considerations.

Satellite radiators can dissipate between 100 W and 350 W of internally generated waste heat per square meter. The upper end of this range is typical of a radiator that runs at a fairly high temperature and experiences a relatively modest heat backload from the environment or satellite surfaces. The lower end of the range might represent a radiator running below room temperature on a LEO, where environmental backloads can be substantial. The actual radiator size is determined by a thermal analysis that considers the desired operating temperature, worst-case satellite waste heat, environmental heating, and radiative and conductive interactions with other satellite surfaces. Mass for radiators typically vary from almost nothing, if an existing structural panel is used as a radiator, to around 12 kg/m² for a heavy deployable radiator and its support/deployment structure.

Assuming that the total satellite waste heat is ΣQ_{int} and environment heat loads are $A\alpha$(solar + albedo) + $A\varepsilon$(IR), the steady state of the radiators of a satellite should be:

$$A\varepsilon\sigma T^4 = A\alpha(solar + albedo) + A\varepsilon(IR) + \sum Q_{int} \tag{8.2}$$

A structural radiator is the simplest and most common radiator. For example, an existing aluminum honeycomb panel wall of a satellite serves both as part of the structure and as a radiator. In this case, there is no additional mass added in as it has been charged to the structures subsystem. The panel face sheets spread out waste heat from the electronics boxes with an area on the outside face acting as the radiating surface. Heat is conducted fairly well from the inner to outer face sheets through the aluminum honeycomb core. Lateral heat conduction, however, occurs mainly in the face sheets, since the internal epoxy core normally has low conductivity.

If high-power electronics boxes mount to a honeycomb-panel radiator, then the lateral conduction in the face sheets may not be sufficient to spread the heat out over an area large enough to radiate it to space. This situation would result in very large temperature gradients in the panel and cause the high-power boxes to exceed their upper temperature limits. Doublers or increased face-sheet thickness may be used to overcome this problem. However, at a certain point these techniques will result in an unacceptably large mass increase. To avoid this mass penalty, heat pipes can be used and mounted on the panel to spread the heat.

Sometimes structural panels may not be convenient for use as a radiator, or the radiator may need to run at a temperature different from that of the rest of the satellite. In such situations a body-mounted radiator may be used. The radiator itself may be a honeycomb panel or a stiffened aluminum plate. Heat is transported from the heat-dissipating components to the radiator using fixed or variable conductance heat pipes, loop heat pipes, or capillary pumped loops, and additional heat pipes may be used to spread the heat out in the radiator panel itself.

Most satellites can dissipate internal waste heat using structural panels or body-mounted radiators. However, as satellite size is reduced or as satellite power levels increase and thus waste heat increase, the satellite at some point simply lacks enough area to dissipate the internally generated waste heat. In such a situation, deployable radiators are sometimes required to increase the available radiating area.

8.4 ACTIVE THERMAL CONTROL

8.4.1 Heaters

When temperature variations in a passive design are out of the range that components can tolerate, active thermal control needs to be used. In a thermal design, heaters are used to protect components under cold case environmental conditions or to make up for heat that is not dissipated when an electronics box is turned off. Another common use for heaters is to warm up components to their minimum operating temperatures before the components are turned on, such as to prepare motors to operate the pour point of a liquid lubricant. Heaters may be used with thermostats or solid-state controllers to provide precise temperature control of a particular component (Gilmore 2002).

The patch heater is the most common type of heater used on board satellite. It consists of an electrical resistance element sandwiched between two sheets of flexible electrically insulating material, such as Kapton™. Depending on whether or not redundancy is required, a patch heater may contain either a single circuit or multiple circuits within it. Redundancy is generally used to increase the reliability, because heater circuits can fail. The redundancy can be provided either within the patch heater by using multiple circuits within it or externally through the use of two separate patches. The shape of heaters in most instances is a simple rectangle of standard dimensions, where the area increases according to peak heat generation required.

A cartridge heater is another type of heater. It consists of a wound resistor enclosed in a cylindrical metallic case. This kind of heater is often used to heat blocks of material or high-temperature components. Typically, a hole is drilled in the component to be heated and the cartridge is inserted into the hole. Another attachment technique is the use of a clamp or small bracket to hold the heater. Cartridge heaters are usually a quarter inch or less in diameter and up to a few inches long.

Almost all heaters have some sort of switch or control. This typically involves a relay to enable or disable power being supplied to the heater, a fuse to protect the spacecraft from a short circuit, and a thermostat or solid-state controller to turn the heater on and off. Over their operation on satellites, most heaters require some automatic heater control to keep a component at a desired temperature and to minimize the amount of time the heater is on so as to reduce power consumption. Onboard computers normally take the task to monitor temperatures at specific locations and turn heaters on and off at appropriate times using relays.

8.4.2 Thermoelectric Cooler

Thermoelectric coolers (TEC) are miniature solid-state heat pumps. Their operating principle is based on the Peltier phenomenon that exists when electric current is applied to the junction of two different conducting materials (typically semiconductors). One material, *p*-type, contains positive charge carriers (holes) and the other, *n*-type, contains negative charge carriers (electrons) as depicted in Figure 8.2. Heat is pumped from the cold junction to the hot junction. The net cooling is diminished by the effects of Joule losses generated by the current, and heat conduction through semiconductor material from the hot to the cold junction. Semiconductors, principally bismuth-telluride (Bi_2Te_3), have made these devices practical. Prior to the advent of such semiconductors, parasitic conduction through metal elements largely negated any useful cooling (Chuchra and Stevenson 2002).

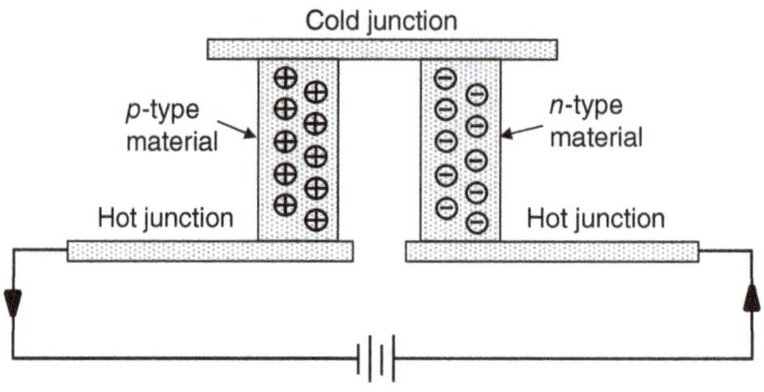

FIGURE 8.2 Peltier thermoelectric couple.

In general, thermoelectric materials can be categorized as uniform bulk materials or thin films with periodic variation in structure and composition (Rowe 1995, 2006). Bulk materials can pump heat fluxes on the order of 10 W/cm^2, while thin films can reach an order of magnitude higher heat fluxes with direct sensor cooling capabilities (Venkatasubramanian et al. 2001). Efforts to develop thermoelectric materials for cooling at cryogenic temperatures were reported by Bentien et al. (2007).

Thermoelectric coolers can be used as single-stage or multistage assemblies. One stage generally contains many elements electrically joined in series, with all cold junctions soldered to metallization on one ceramic plate and all hot junctions soldered to the opposite ceramic plate.

The coefficient of performance (COP) is defined as the useful cooling divided by the input power to a TEC to measure the efficiency of the TEC. TECs with low COPs consume more power and produce excessive waste heat. A designer can select the number of stages, the number of couples, and the geometry of the *p* or *n* post to optimize the COP. Space applications have favored multistage TECs where the desired temperature difference exceeds the 25–40°C range. The shape of a multistage TEC looks like a multitiered wedding cake, with the upper layer representing the cold stage and the base layer representing the hot stage. Each stage needs to be larger than the one above it to handle progressively larger amounts of waste heat. This geometry also stems from the need to minimize the parasitic thermal coupling between stages.

Compared to common heat pumps, thermoelectric coolers possess great advantages for space applications because they are solid-state devices having no moving parts (noiseless, vibrationless) and are miniature, low mass, highly reliable, and easy to integrate into the system. However, their use in spacecraft is limited by their relatively low COP, particularly with large temperature differences. Because of their limited efficiency, they are best suited to situations with modest heat loads, cold temperatures not below 150 K, and hot-to-cold side differences not exceeding 100°C. Other cooling methodologies are generally better suited (more efficient) for applications with greater heat loads or larger temperature-difference requirements. TECs are not recommended for use below 130 K because of their prohibitively low efficiencies.

A major issue of a TEC is the structural integrity of bismuth-telluride and soldered joints when it is subjected to differential thermal expansion stresses. Externally imposed stresses are commonly resolved through the use of a compliant conductive strap on the cold side of the device. Thanks to TECs as fairly high reliability redundant TECs are generally not used. This should be assessed per system and would be greatly dependent on lifetime, where electronics suffer reliability limitations with longer life missions.

It must be noted that the reliability of such systems is strongly dependent on the way they are installed and clamped, as high stress situations can arise from gradients in the "sandwich" assemblies that are often developed. In other cases, pressure changes will affect contact resistances and even gaps may open. Low volatility grease can help but will not be a complete substitute for good thermal contact. There is no substitute for early testing and characterization of such systems to establish the reliability of the design in a range of operating and environmental conditions. Further, safety limits must be actively monitored to prevent inadvertent damage to critical solder joints during start-up or extremes of operation.

8.4.3 Stirling Cryocooler and Pulse Tube Cryocooler

Cryocoolers play an important role in satisfying the need for cryogenic cooling of space-based infrared detectors. They can be generally classified as either recuperative or regenerative cryocoolers (Radebaugh 2000). Recuperative cryocoolers use only recuperative heat exchangers. Their compressor operates with a fixed inlet pressure and a fixed outlet pressure with a steady flow of working fluid in the system. If the compressor is a reciprocating type, it must have inlet and outlet valves to provide steady flow. The most common recuperative cryocooler is the Joule-Thompson refrigerator, which is used in domestic applications.

The regenerative cryocoolers use at least one heat exchanger (i.e., regenerator) and operate with oscillating flow and pressure. In a regenerator, incoming hot gas transfers heat to the matrix of the regenerator, where the heat is stored for a half cycle in the heat capacity of the matrix. In the second half of the cycle, the returning cold gas, flowing in the opposite direction through the same channel, picks up heat from the matrix and returns the matrix to its original temperature before the cycle is repeated. At equilibrium, one end of the regenerator is at room temperature while the other end is at the cold temperature. Enhanced heat transfer is obtained in regenerators through the use of stacked fine-mesh screens or packed spheres because of their large effective surface areas. Stirling cryocoolers and pulse tube cryocoolers (PTCs) are the most common regenerative cryocoolers.

Stirling cryocoolers are based on the principle of forcing a working gas to undergo a Stirling cycle, which ideally includes 2 constant volume processes and 2 isothermal processes. A Stirling cryocooler is most commonly configured with a reciprocator and a cold-head (i.e., cold finger). The reciprocator is a compressor that generates pressure oscillations. The reciprocator is connected to the cold finger by a thin pipe, called a transfer tube. The cold finger consists of a regenerator, helical spring, and a moving regenerator, which is made of cascaded sheets (discs) of very fine mesh or packed spheres. Stirling cryocoolers were the first active coolers to be used successfully in space and have proved to be reliable and efficient. Recent years have seen the development of two-stage devices for space, which extend the lower temperature range from 50–80 K to 10–30 K. Larger capacity devices have also been developed by scaling units or operating at higher frequencies to remove larger heat loads, as higher performing detector systems increase demands. A key disadvantage introduced is vibration due the moving parts (Gibson et al. 2008).

Most Stirling cryocoolers are configured using linear drive motors with a dual-opposed arrangement to cancel first-order harmonics of vibration. The linear drive reduces side forces between the piston and the cylinder and increases mean time to failure, though it does not eliminate them as the pistons are normally mounted on diaphragms that involve some rotation. The working gas system is sealed within the overall assembly, so that any leakage of gas past the piston is contained and happens in reverse on the counter stroke.

The Stirling cryocooler was first used in space applications after flexure bearings were developed for supporting the piston and displacers in their respective cylinders with little or no contact in a clearance gap of about 15 μm. The technology originated from another optical device referred to as a pressure modulated cell, developed at Oxford University, for the purpose of correlation spectroscopy (Werrett et al. 1985). The first instruments to fly Stirling cryocoolers were launched in 1991 on ATSR (Along Track Scanning Radiometer) on ERS-1 (ESA Remote-sensing Satellite-1) and soon after on ISAMS (Improved Stratospheric and Mesospheric Sounder) on UARS (Upper Atmospheric Research Satellite).

PTCs are similar to Stirling cryocoolers except that a pulse tube replaces the moving displacer. The thermodynamic processes of the PTCs are somewhat different from those of Stirling cryocoolers. A PTC is mechanically simpler than Stirling cooler because it has no moving parts at the cold end. This can increase cooler reliability of a system and make integration with the payload much simpler if the thermal margins are comparable. Stirling cryocoolers have fundamentally higher efficiency where optimized, requiring lower input power to achieve the same cold temperature, which is emphasized at lower cold-tip temperatures. Stirling cryocoolers also offers some functionality to shift heat loads and adjust performance by controlling the phase of the displacer, which could be advantageous if ever a leak affected the resonance of the unit. Stirling cryocoolers can use the displacer (cold-head mechanism of a Stirling) and can help to achieve lower vibration at the cold-tip itself, where this has been problematic in the case of some PTC designs. One cannot assume that the vibration at a pulse tube cold-tip is null. The compressors are balanced, but the gas slug moving the cold-head can be a significant source of micro-vibration.

The development of pulse tubes was started by Gifford and Longsworth (1964). They observed that when a metallic tube's one end was connected to a pressure wave generator and the other end was closed, the temperature at the closed end increased. This tube is known as the pulse tube (Popescu 2001) and is shown in Figure 8.3a. Further research proved that placing a thermal regenerator between the reciprocator and the pulse tube could increase temperature at the closed end (hot zone) in addition to cooling at the other end (cold zone). This concept was named the basic pulse tube refrigerator as shown in Figure 8.3b and achieved a minimum temperature of 124 K. However, a basic pulse tube refrigerator was impractical due to its very low efficiency.

Mikulin et al. (1984) introduced the concept of adding an orifice and reservoir at the hot zone to the basic pulse tube refrigerator, which increases the cooling efficiency of the PTC and reached 105 K as shown in Figure 8.3c. Radebaugh et al. (1986) changed the location of the orifice and reached 60 K.

Zhu et al. (1990) increased the efficiency of orifice PTCs by adding a second orifice between the generator and the hot zone allowing the gas to travel directly between the compressor and the warm zone of the pulse tube, thereby bypassing the regenerator. This configuration is called the double inlet orifice pulse tube refrigerator. The bypass flow reduces the flow through the regenerator, thereby reducing the regenerator loss. The secondary orifice has been used on a majority of small PTCs built since then (Popescu 2001).

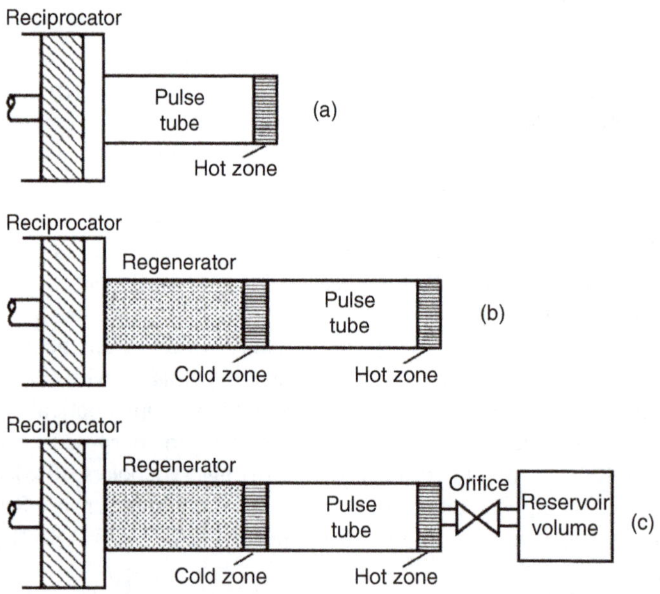

FIGURE 8.3 Evolution of the development of pulse table cryocoolers. (a) a pulse tube, (b) a basic pulse tube refrigerator, (c) a pulse tube refrigerator with an orifice and a reservoir.

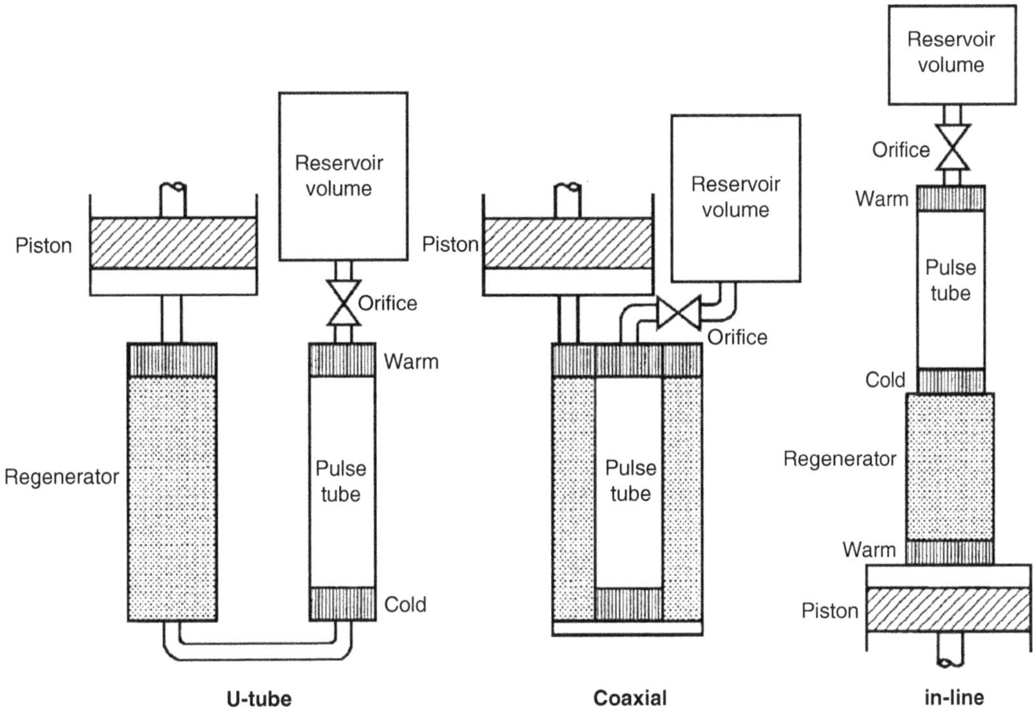

FIGURE 8.4 Different geometric arrangements of PTCs. (Source: National Institute of Standards and Technology.)

Zhu et al. (1997) reported that the performance of the PTCs can be improved by replacing the single orifice by a long thin tube (called inertance tube). The gas in the tube is subjected to inertia, and the resulting cooler has become known as the inertance PTC (De Boer 2002). As a result of these improvements, PTCs have become competitive with Stirling cryocoolers for many space and terrestrial applications (Radebaugh 2000).

There are three different geometric arrangements that have been used with PTCs as shown in Figure 8.4. The in-line arrangement is the most efficient because it requires no void space at the cold end to reverse the flow direction, nor does it introduce turbulence into the pulse tube from the flow reversal. The disadvantage is having the cold plate located between the two warm ends. The coaxial arrangement is the most compact and the most like that of the Stirling cryocooler. The disadvantage is having the potential problem of a mismatch of temperature profiles in the regenerator and in the pulse tube that would lead to steady heat flow between the two components and a reduced efficiency. For many applications it is preferable that the cooling is produced at the end of the cooler. In a U-tube arrangement, both hot ends can be mounted on the flange of the vacuum chamber at room temperature. Coaxial versions were a natural extension of the U-tube configuration and have now become the most common shape of PTCs, primarily for packaging reasons, with the heat dissipation region co-located with the heat source of the compressor.

The absence of a moving displacer in PTCs gives them potential advantages over Stirling cryocoolers for the cooling of infrared sensors. Early PTCs were not nearly as efficient as Stirling cryocoolers, but advances have greatly improved their efficiency, in particular in the range of 60–80 K. A great deal of interest in PTCs has led to further improvements, including three-stage units to reach temperatures below 15 K, but these lower temperature units require significantly more power to operate versus an optimized Stirling cryocooler. For this reason, interest still remains in Stirling cryocoolers for applications in astronomical instruments in the far infrared and particularly as a precooler for other very low-temperature cooling methods.

8.4.4 Heat Pipe

A heat pipe is a heat-transfer device. It uses a closed two-phase liquid-flow cycle with an evaporator and a condenser. It is used to transport heat from one location to another without electrical power. As a thermal control device, it can change the flux density of a heat flow and create isothermal surfaces as a thermal "transformer." The heat driving mechanism in a heat pipe is a capillary pumping action induced by the temperature difference between the evaporator wall and the condenser wall.

Figure 8.5 illustrates the working principle of a heat pipe. In the figure, a heat pipe is in an equilibrium and isothermal environment. The vapor in the vapor space and the liquid in the wick are at saturation. If heat is applied to the evaporator to raise its temperature, liquid in the wick evaporates to remove some of the added heat, which depresses the meniscus in the evaporator because less liquid remains there. This process also raises the local vapor pressure, because the pressure must be in saturation with the heated liquid in the wick. The difference between the increased curvature of the meniscus in the evaporator wick and the unchanged meniscus in the condenser wick causes a difference in capillary pressure sufficient to pull liquid from the condenser wick toward the evaporator wick. This action refills the liquid in the evaporator wick. At the same time, heated vapor flows from the evaporator to the condenser, which is at a lower pressure. When this vapor comes in contact with the cooler surfaces of the condenser, it condenses. This cycle of evaporation and condensation is repeated. Only small amounts of fluid need to flow to transport significant quantities of heat because of the high latent heat of vaporization of most heat-pipe working fluids. The required driving mechanism is also small.

Because the driving mechanism in a heat pipe is capillary pumping, a relatively weak force that is provided by a wick, the pipe may be susceptible to severe performance degradation when operating in a gravitational field. Planning and specific design effort is therefore needed to facilitate the ground testing of systems that include heat pipes.

A heat pipe is used to move heat from one location to another (possibly changing the flux density in the process) or to isothermalize a surface. The most obvious application of a heat pipe is for physical separation of the heat source and sink. If a heat pipe is used, it is not necessary to mount all hardware to be cooled directly on radiator panels, or to use relatively inefficient conductive coupling.

There are three types of heat pipes: (1) constant-conductance heat pipes, (2) diode heat pipes, and (3) variable-conductance heat pipes.

Constant-conductance heat pipe are the most basic type of heat pipes. A constant-conductance heat pipe consists of a working fluid, a wick structure, and an envelope. Different working fluids can be used depending on the temperature range required. In most spacecraft applications, ammonia is the working fluid. The most common wick envelope design consists of axial grooves in the wall of extruded aluminum tubing. This type of wick is very consistent and has relatively low cost to manufacture, but is very susceptible to gravity effects during ground testing. Grooves are typically rectangular or trapezoidal in shape, but more complex shapes such as the teardrop or keyhole have been extruded (Prager et al. 2002).

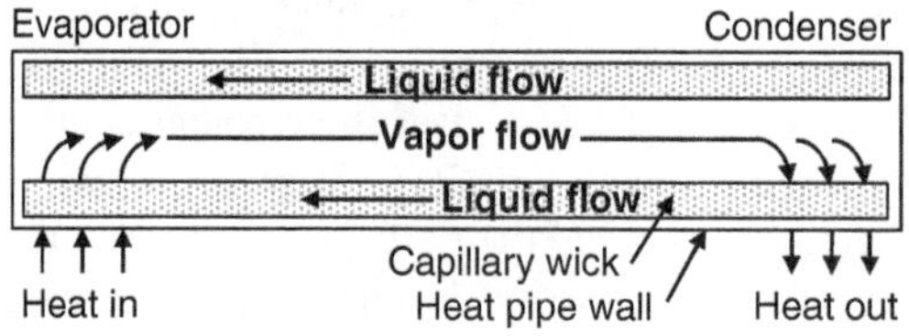

FIGURE 8.5 Heat pipe operating principle.

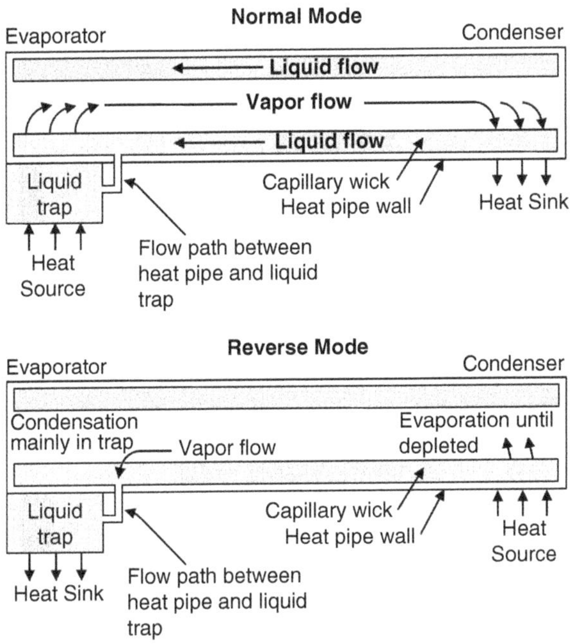

FIGURE 8.6 Liquid trap diode heat pipe operating principle.

Diode heat pipes transfer heat in one direction, but cease when attempt is made to transfer in the other direction. Common diode heat pipes are the liquid-trap, liquid-blockage, and gas-blockage diodes.

The principle of a liquid-trap diode heat pipe is depicted in Figure 8.6. In the normal mode of operation, which is the case for a constant-conductance cryocooler, the liquid trap is dry and heat is transferred from the heat source to the heat sink. In reverse mode operation, which is the case when a cryocooler turns off, the liquid trap becomes the cold end of the heat pipe. Condensation begins to occur within the liquid trap and evaporator. As liquid accumulates in the trap, the heat pipe wick becomes under-filled, which causes a reduction in transport capability. The reduction in transport capability eventually leads to complete shutdown as long as the trap is slightly colder than the evaporator and large enough to accommodate the fluid inventory of the heat pipe. The design shown in figure ensures the trap is always colder than the evaporator. Once shutdown occurs, reverse heat leak occurs only by conduction in the tube wall and the wick.

Variable-conductance heat pipes use a gas reservoir connected to the end of the condenser. The reservoir is filled with a non-condensable gas to control the operating area of the condenser based on the evaporator temperature.

In spacecraft design, a heat pipe can be used when a heat source and the sink are required to be separated. If a heat pipe is used, it is not necessary to mount all hardware to be cooled directly on radiator panels, and relatively inefficient conductive couplings need not be used. Further, heaters need not be mounted directly on hardware to be heated if a heat pipe is employed.

Heat pipes have also been used to reduce temperature gradients in structures to minimize thermal distortion. An example is the telescope tube of the NASA Orbiting Astronomical Observatory (OAO). Three ring-shaped heat pipes were used to minimize circumferential temperature gradients. The ammonia heat pipes worked throughout the 8 years of mission life.

If a spacecraft has at least one of the radiators free of any direct solar load at all times during the orbit, a diode heat pipe would couple the device to the cold radiator, while preventing heat from leaking back into the system from the radiator on the sun side.

Variable-conductance heat pipes can control the amount of active radiator area, providing reasonably good temperature control without the use of heaters. This capability is particularly attractive if electrical power is limited, and this type of design has been flown on a number of satellite experiments.

8.5 THERMAL DESIGN ANALYSIS

8.5.1 Thermal Design Process Overview

The thermal design process is a combination of design selection and supporting analysis. The selection of a viable thermal-design approach may become almost intuitive for an experienced thermal engineer. Detailed thermal analyses are required to verify and refine the design. Experience can help minimize the number of time-consuming analysis iterations required to close in on a final design (Gilmore et al. 2002).

Establishing a thermal design for a spacecraft is usually a two part process. The first part is to select a thermal design for the body of the spacecraft that will serve as a thermal sink for all internal components. The second part is to select thermal designs for various components located both within and outside the spacecraft body.

As described in Sections 8.3 and 8.4, many existing thermal control hardware and techniques are available, from simple surface finishes to complex cryocooler systems. The basic design philosophy is to keep the thermal design as simple as possible and avoid the use of active components in order to minimize weight, reduce the cost, and simplify test complexity of the spacecraft. A design that relies only on passive thermal control components, such as surface finishes and insulation blankets, will be lighter, far less expensive to build, more reliable, and easier to test than a design involving active thermal control components, such as heat pipes and cryocoolers. Therefore, although active or semi-active components will sometimes be required, they should be avoided wherever possible.

The goal of the thermal design is to provide a reliable thermal control system at minimum cost for the satellite. Before starting the thermal design analysis process, the overall effort must be planned. All of the analysis, design, and testing activities are only tools to be used to reach that goal, and any unnecessary expenditure of time or money should be avoided. Thus, the design should be no more complex than is required to do the job and the thermal mathematical models should have just slightly more than the minimum number of nodes needed to verify the design. If any difficulties are encountered in the analysis, design, or testing, it should be asked whether a simpler means is available to cross-check, including via a simple test. It is easy to make errors in an overly complex design, and some experience is required to know the optimum trade between complexity and practicality.

The first step in the thermal design process is to make an assessment of the mission requirements and hardware that will drive the design. This process begins with reviewing the top-level requirements and understanding the objective and any ground rules or constraints. The project ground rules and constraints must also be considered, since these will affect the thermal design effort. These factors will play a major role in structuring the effort.

Once the objectives and ground rules are understood, more specific requirements that will be useful in developing the thermal design must be derived. An approach to the solution of the problem must be selected. The approach could be an analysis, tests, hand calculations, adaptation of the thermal design from a thermally similar device, or a combination of these activities. Each potential approach must be evaluated to determine whether its elements are meaningful to the solution of the problem. To identify the approach, one must also consider schedule, budget, any risk, available expertise, and reliance on new or unproven technologies or analysis software.

If the spacecraft packaging space permits, the simplest thermal control scenario would be to mount dissipative components directly on the inner surface of the spacecraft external structure panels. Basically, minimizing the path length between the source and the exit point of the heat will be the most efficient method. Some or the entire external surface of the panel is used to radiate heat loads to space. Local heat loads into the panel are spread over the panel area via conduction through the panel. In higher heat flux situations, thicker face sheets or double plates may be used.

High-power components are usually mounted on the walls of the spacecraft, which provides them with a direct conduction path to the radiating surfaces on the outside. Conductive doublets or heat pipes may be used to spread heat out over a larger area if component powers are high enough. Heaters are sometimes used to heat components to the required operating temperature or to protect components from cold-case environment conditions.

An analysis of the thermal design in the principal mission thermal environment must be performed to determine whether all temperature control requirements can be met. This is done by constructing two analytical models: geometric math model and thermal math model. The geometric math model is a mathematical representation of the physical surfaces of the satellite or component. It is used for calculating radiation interchange factors between all the surfaces in the model, as well as heating rates to each surface from external flux sources such as sunlight, Earth infrared, and albedo radiation. The thermal math model is a lumped-parameter representation of the thermal capacitance of each node and thermal conduction terms between the nodes. These nodes are constructed using a combination of computer-aided design technology and hand calculations. A number of runs of the thermal model may be required to determine the worst-case combinations. Once a satisfactory design is obtained, estimates of the thermal control subsystem weight and power can be made (Gilmore et al. 2006).

8.5.2 Hyperspectral Satellites Thermal Design Considerations

Hyperspectral satellites for Earth observation missions are typically launched into a sun-synchronous LEO with an altitude between 400 km and 800 km. For this kind of orbits, the infrared radiation from the Earth and albedo sunlight reflected from Earth's surface are also significant sources of heat loads in addition to the major heat loads from direct sunlight. Sun-synchronous orbits have fairly constant eclipse period and a small range of sun angles to contend with and have a cold side of the spacecraft that is never in direct sunlight. This cold side can be used to benefit the thermal control.

In interplanetary missions, hyperspectral satellites can have considerable variation in solar flux, which varies inversely with the square of the distance from the sun. For example, the VISTIS onboard spacecraft of ESA's Venus Express mission orbiting Venus, the solar constant at Venus is reduced to 15 W/m^2 due to far away from the sun compared to 1367 W/m^2 at Earth. Decreasing solar energy intensity presents a huge challenge for thermal control of the interplanetary missions.

Hyperspectral satellites are three-axis stabilized spacecraft. The same basic approach to thermal control for this kind spacecraft can be used to hyperspectral satellites, though radioactive heating may be necessary. For instance, the spacecraft is insulated from the space environment with MLI blankets, and radiator areas with low solar absorption and high infrared emittance are provided to dissipate waste heat. Wavelength-dependent surface finishes or coatings are adopted for various thermal control purposes. Solar reflector materials, including white paints, or silver- or aluminum-backed Teflon are used to minimize absorbed solar energy. Inside a spacecraft black paints are commonly used if it is desired to exchange energy with the compartment or other devices. Components mounted inside the spacecraft on the shelves, panels, and structures radiate their waste heat to the outside walls of the spacecraft, where the heat is dissipated to space.

The payload of a hyperspectral satellite consists of optical subsystem, onboard control electronics, and thermal and structure subsystem. The optical subsystem sometimes is also referred to as Optical Unit, or Optical Module, or Optical Head. It typically consists of the fore-optics and spectrometers, optical bench, forward baffle, calibration equipment, moveable aperture cover, detector arrays and pre-amps, cooling equipment for detector arrays, external radiator surfaces, a light-tight enclosure, etc. The Optical Unit is a complicated optical system. There are numerous optical surfaces and mounts distributed over an optical bench. Optical elements do not like large cold temperature excursions due to coefficient of temperature expansion mismatches between the optical element glass and the mounting substrate as well as variation of the index of refraction with temperature. Cold conditions may overstress the mounting system due to coefficient of temperature expansion mismatch between the optics and mounting surface, leading to cracked optics. Bulk temperature control of the optical bench may be necessary to maintain end-to-end alignment of the optical path, which can be facilitated with the use of trim heaters. The kinematic constraints and material matching must be considered to eliminate distortion. Also note that the bench material itself can affect the distance between optical components, changing focal properties, which may drive the design to use lower expansion materials (such as AlBeMet™ or SiC composites), though these can drive up cost considerably.

With regard to thermal design, the following should be considered:

- The temperature of the Optical Unit and the spatial temperature distribution within the Optical Unit should be controlled within the specifications.
- The temperature of the shortwave infrared (SWIR) detector array should be maintained at its required low temperature.
- The temperature of the visible and near-infrared (VNIR) detector array should be maintained at its required low temperature.
- The optical bench would be thermally isolated from the rest of the satellite.
- The Optical Unit would be "cold-biased," i.e., if left in equilibrium with the Earth-space-satellite boundary conditions, the Optical Unit would get slightly colder than desired. Therefore, local heaters and temperature sensors and an autonomous controller would be used to bring it to the desired temperature, and to balance out the spatial distribution of temperatures within the Optical Unit if necessary.
- Optical Unit temperatures would be controlled throughout the entire orbit, i.e., dayside and the nightside. Because of the heat capacity of the optics and concerns for thermal cycling of sensitive components, it may be preferred to maintain a steady temperature during the nightside of the orbit (not to allow them to cool during the nightside of the orbit and reheat prior to the next imaging opportunity time).
- The optical bench interface of the Optical Unit would be isolated from the platform of the spacecraft and kinematically mounted to manage stress. Clearly, it would be strongly coupled thermally to the heat removal path (active or passive).
- The exterior surfaces of the Optical Unit will have coatings with the appropriate α/ε values (solar reflectors) to maintain the Optical Unit at a constant nominal temperature in line with electronics operability. Interior heaters would be used to maintain (trim) the temperature throughout the orbit, and to minimize spatial temperature gradients.

Detector arrays are the key components in hyperspectral imagers. These need to be cooled down to the required operational temperature to optimize performance. The VNIR detector arrays are typically CCD (charge-coupled device) devices or silicon hybrid CMOS devices. The temperature of these detector arrays is normally maintained at room temperature (e.g., 20°C) or lower (e.g., 10°C) to reduce the dark current.

The following requirements typically drive the need to maintain the SWIR detector arrays and their surroundings at adequately cold temperatures:

- The need to minimize detector noise—dark current, shot noise, on-chip read-out noise, residual photo response nonuniformity (PRNU)—in order to achieve the required high signal-to-noise ratio (SNR).
- The need to maximize the number of SWIR detector pixels that operate according to their specified performance values.

- The need to minimize thermal emission sources (i.e., "warm" surfaces) that can thermally be "seen" by the SWIR detector array particularly at or around the 2.5 μm nominal cutoff wavelength of its responsivity range.
- The need to have a sharp cutoff for the SWIR detector array to assure it has negligible responsivity to background thermal radiation that is above 2.5 μm.

Most SWIR detector arrays in hyperspectral imagers operate at cryogenic temperatures, typically 125 K or lower to minimize dark current, shot noise, and other noise components, and to maximize detector pixels that operate within specs. To achieve this goal two types of cryogenic cooling methods are normally used in hyperspectral satellites: (1) passive radiators and (2) active coolers.

In method 1, passive radiators are based on the normal principle of radiative heat dissipation to space. This method is limited to reach lower cryogenic temperature range. The radiators are located on the +Z side of the satellite, that is always oriented toward deep space during the orbit. The radiators are often composed of two stages in order to more facilitate reaching lower temperatures and maintaining the detector arrays inside a limited range of temperatures. The cold zone is surrounded by a boundary area having an intermediate temperature in order to reduce the parasitic heat fluxes coming from the external environment and directed inside the SWIR. Material selection is critical to prevent parasitic heat leaks via conduction. Multiple stage radiators remove these parasitic heat inputs. The final coldest stage provides the actual sink for the detector array cooling. A dedicated Earth shield can be used to reduce the heat flux from the Earth on the cold radiator. Intermediate stages may also be used to pre-cool baffles/shields or lead wires to reduce parasitic losses.

In method 2, active coolers may include Stirling cryocoolers, PTCs, and Peltier thermoelectric coolers. They use electrical power to achieve a heat pump action in order to remove heat of the detector arrays and reject it at a higher temperature. Thermoelectric coolers possess great advantages for cooling of space-based infrared detector arrays because they are solid-state devices having no moving parts and are miniature, highly reliable, and relatively straightforward to integrate into the system. However, these relatively low-cost devices become impractical to reach temperatures below 130 K. Efficiency of these devices drops off dramatically at lower temperatures making them less attractive.

A concept design of a passive radiator for cooling a SWIR detector array is showed in Figure 8.7. It consists of a cold radiator, a warm radiator, insulating trusses, a thermal strap, and a cold shield. The cold radiator is a rectangular plate made of aluminium alloy, painted white on its external

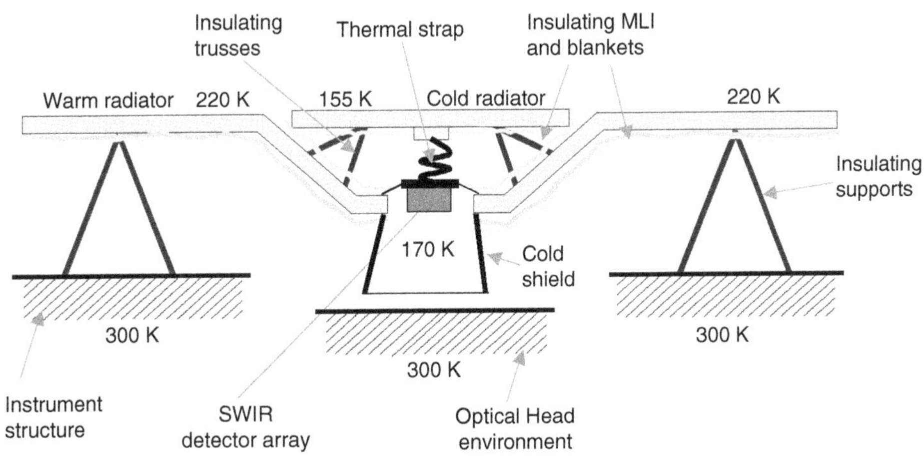

FIGURE 8.7 Schematic view of concept design of a SWIR detector cooler using passive radiators.

surface. It is structurally connected to the warm radiator by insulating trusses in order to reduce the parasitic heat flux coming from the warm radiator. At steady state, the total heat flux exchanged between the cold radiator and space includes the dissipation of the detector array plus the parasitic heat fluxes radiatively and conductively coming from the warmer surrounding structure and from the cabling conductive contribution.

By shielding the Earth view with a properly sized Earth shield, the cold radiator is always protected from the associated heat loads during the instrument imaging and calibration phases of the orbit. This allows reaching and maintaining the required detector array low operational temperatures.

The warm radiator is a rectangular panel, painted white on its external surface. It is a sandwich panel made of aluminium to maintain low mass and high stiffness. The warm radiator is mechanically connected to the main structure by thermal insulating supports. It is thermally insulated from the cold radiator and from the main structure by MLI blankets. The function of the warm radiator is to generate an intermediate temperature zone where the cold radiator is connected for reducing the temperature gap and so the parasitic heat fluxes.

The SWIR detector plus housing is connected with the cold radiator by a flexible thermal strap to guarantee a good thermal connection. The SWIR detector is surrounded by a cold shield, thermally coupled to the warm radiator.

The Optics Head has only parasitic heat exchanges with spacecraft panel. An appropriate MLI cover should be located on the optics head to protect it from the external environment.

Heat sources, such as electronic boxes, that may introduce parasitic heat fluxes are positioned far from the cold zones compatibly with other requirements and preferably outside the Optics Head.

About the proximity electronics more cases are possible:

a. The proximity electronics is allocated inside the spacecraft.
b. The proximity electronics is external to the spacecraft, conductively coupled to it, and protected by MLI.
c. The proximity electronics is external to the spacecraft, insulated from the spacecraft with one or more walls white painted to work as radiators. Survival heater is necessary in this configuration when the unit is OFF (cold case).

A cold shield around the detector arrays can prevent and limit the thermal background of the detector arrays, in particular in the spectral bands close to 2.5 μm.

The cold zones and the warm enclosure are externally covered with MLI to radiatively separate these elements from the outer environment. An inner MLI blanket, sustained by the thermal shield, is used also to separate, from radiative point of view, the insulated part and the cold parts from the warm enclosure.

A low conductive heat load between cold and warm enclosures is achieved by mounting the cold box on low conductivity supports (e.g., nonmetallic such as fiberglass material, Torlon™, or thin-section titanium spacers) whose number and attachment positions are selected to avoid distortion upon cooling from room temperature.

8.5.3 Hyperspectral Sensor Thermal Design Examples

Different hyperspectral satellites have their unique thermal control designs due to different orbits, system design, and applications. This section provides thermal design examples of two well-known hyperspectral satellites for Earth observation: Hyperion and EnMAP, and one planetary exploration mission: CRISM. Thermal design examples of other three spaceborne hyperspectral sensors for Earth observation mission (PRISMAT) and planetary exploration missions (VISTIS and MAJIS) are described together with the structure design in Section 8.8. Detailed description of these hyperspectral satellites or sensors is provided in Chapter 2.

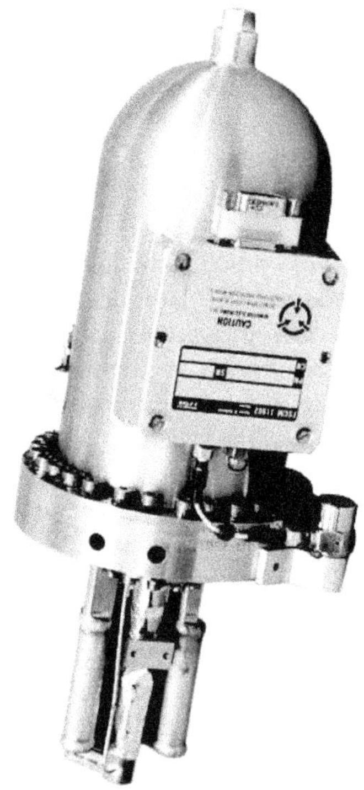

FIGURE 8.8 Miniature pulse tube cryocooler used in Hyperion. (Source: NASA.)

8.5.3.1 Hyperion Thermal Design

Hyperion is well known in the remote sensing community and often regarded as the first spaceborne hyperspectral sensor although at least two other hyperspectral sensors had been launched before it (see Section 2.2 to Section 2.5 for more details). It was onboard NASA's New Millennium Earth Observing-1 (EO-1) launched in November 2000 as a technology demonstrator to support evaluation of hyperspectral technology for future Earth observing missions.

The thermal design of Hyperion hyperspectral payload was straightforward. The Hyperion sensor assembly (HSA) consisted of a single telescope and two Offner form spectrometers to span the VNIR and SWIR channels. The HSA enclosure controlled the optics thermal environment. The HSA housing was maintained at 293 ± 2 K for precision imaging and alignment. The focal plane array of the VNIR spectrometer was passively cooled by a radiator and operated at 283 K. The focal plane array of the SWIR spectrometer was actively cooled to 110 K by a Thompson Ramo Wooldridge Inc. (TRW) mini PTC as shown in Figure 8.8 (Tward et al. 1999, Ross and Boyle 2007). The Hyperion operated for >16 years and was decommissioned in March of 2017.

8.5.3.2 CRISM Thermal Design

The Compact Reconnaissance Imaging Spectrometer for Mars (CRISM) was designed to identify minerals on the surface of Mars, especially those likely to be formed in the presence of water. CRISM is one of six major science instruments on NASA's Mars Reconnaissance Orbiter (MRO) spacecraft launched in August 2005.

CRISM's thermal design was driven by the following five requirements (Murchie et al. 2007):

1. Maintaining the spectrometer housing at or below –75°C (198K) to minimize infrared thermal background radiation.
2. Maintaining the temperature of the VNIR detector array at –60°C (213K) to retain quantum efficiency at spectrum >900 nm.

3. Maintaining the temperature of the infrared detector array at <–153°C (120 K) to minimize dark current.
4. Maintaining the temperature of the optics near –60°C to control their thermal emission at the longest infrared wavelengths.
5. Maintaining the temperature of the body of the mechanical cryocoolers above –40°C (233K) for cooling efficiency and to prevent leakage of their helium refrigerant.

These requirements may sometimes be conflicting, but were met through a combination of passive and active cooling to establish the required thermal plateaus.

Three radiators were used to provide passive cooling. Heaters were used for the components where a minimum temperature was required to be maintained. The anti-sunward radiator is the largest of the three radiators. It faces toward the terminator from MRO's 3 PM local solar time sun-synchronous orbit. It was designed to cool the spectrometer housing and was constructed from an oriented-strand composite that provides more thermal conductivity than alternative metallic materials for radiator construction. Its duck-foot shape increases radiator area while not interfering with the spacecraft deck when the optical sensor unit (OSU) scans. The anti-sunward radiator was connected to the spectrometer housing by a flexible link. Where the radiator stem passed through the gimbal base, a polymer spacer provided thermal isolation. A trim heater on the spectrometer housing prevented temperatures from falling low enough to cause icing of the optics or to overwhelm the heater on the VNIR detector.

The optical bench was exposed on the exterior of the instrument and there is one half of the planet-facing radiator that is oriented toward Mars during the primary science phase. It cooled the telescope optics to minimize thermal emission at the longest infrared wavelengths. The other half of the planet-facing radiator is the plate to which the bodies of the three mechanical cryocoolers were thermally sunk. The function of this second half of the planet-facing radiator was to dissipate heat from the cryocoolers. A trim heater, together with a spacecraft survival heater kept the bodies of the cryocoolers above –25°C during cold excursions, such as during cruise to Mars. A narrow gap between the two halves of the planet-facing radiator maintained their distinct temperatures.

The infrared detector array was cooled to 120 K or below by an active Stirling cryocooling system that consisted of three redundant helium-charged mechanical cryocoolers to either reduce mission risk or enable mission feasibility. The fundamental requirement for a redundant cryocooler system was for "on" cryocooler to be able to handle the parasitic heat leak from the "off" (i.e., redundant) cryocooler(s).

For payloads with sufficient cooling capacity, a thermal switching system was unnecessary and flexible conductive links were used to make the multiple cryocooler-to-cooled component thermal connections. But for payloads with limited cooling capacity, a thermal switching system was needed to reduce the parasitic heat leak from the off cryocooler(s). Due to its low input power and flight heritage, the CRISM program selected a small commercial cryocooler (Ricor K508) with a 1-year lifetime and limited cooling capacity. To provide the required 2-year mission life with margin, three cryocoolers were therefore needed. To minimize parasitic heating from the two "off" cryocoolers, and thus provide sufficient cooling capacity, a thermal switching system was a necessity to integrate the cryocoolers with the infrared detector array (Bugby et al. 2005).

To meet the requirement of thermal switching system, the CRISM program examined three thermal switching options: (1) a cryogenic loop heat pipe, (2) a differential thermal expansion cryogenic thermal switch, and (3) a cryogenic diode heat pipe (CDHP). Since neither of the first two options had prior space flight experience, the CDHP was selected. CDHP systems with prior space flight experience include the oxygen and methane CDHPs flown on the CRYOFD (Thienel et al. 1998) flight experiment, the methane CDHP flown on the Three-Color (Misselhorn et al. 1987) flight experiment, and the oxygen CDHP flown on the Cryo-System (Sugimura et al. 1995) flight experiment. The CDHPs used were liquid trap diode heat pipes that are described in Section 8.4.4.

Using a methane (CH_4) charged CDHP, each of the three Stirling cryocoolers was linked to the infrared detector array. The CDHPs provided thermal conductivity from the infrared detector array to a cryocooler that is on, and thermal isolation to a cryocooler that was "off" to switch between the three cryocoolers. When the cryocooler was powered off and the detector array cold, the liquefied methane was in the liquid trap and not circulating; with the wick design providing low thermal conductivity via effectively long, narrower passages. When the cryocooler was powered on and the condenser end reached approximately 138 K, the methane began to liquefy there and wicking commenced to the evaporator for heat to be rejected, while vapor returned to the condenser end to continue the cycle. After cryocooler turned on, the system reached the operating temperature in approximately 12 h.

With the cryocooler body temperature near −20°C, approximately 700 mW of heat lift was available. The cryocooler cold tip is connected to a heat pipe with a copper braid. Each cryocooler draws 450–550 mA of current at 15 V to cool the detector to around 104 K. During the mission life of the CRISM, the cryocoolers were run one at a time using the heat pipes to switch in another cryocooler when the proceeding one reaches wear out (Ross and Boyle 2007).

The heat pipes are enclosed in a shoebox-shaped cold shroud that envelops the assembly and minimizes radiative coupling to the rest of the instrument. The shroud was thermally coupled by a flexible link to the optical bench and suspended from the warm side of the instrument housing by a set of Kevlar cables. The heat pipes were supported inside the shroud by an another set of small-diameter Kevlar cables, creating a dual-nested Kevlar cable thermal isolation/structural support system. The shroud was gold-plated for low emissivity to minimize radiative coupling.

The OSU had been baked out in vacuum during cruise to Mars for approximately 60 days to dissipate water adsorbed during prelaunch conditions. Water adsorbed on surfaces inside the cold shroud would increase emissivity, increase radiative coupling of the heat pipes to their surroundings, and increase the amount of heat that each cryocooler has to remove. Ultimately some of the water desorbed from other parts of the instrument interior would condense on the 105 K surfaces of the infrared detector array and the active heat pipe, increasing heat load on the cryocooler and perturbing radiometric calibration especially near 3000 nm. It turns out that the bakeouts had been successful: approximately 10% less power is drawn by each cryocooler to maintain the infrared detector array at the operating cryogenic temperature than was required during ground testing in vacuum after 4 days of bakeout. This effect has been seen in other cryocooler systems as well, where successive bakeout remove trapped vapor and the performance improves significantly versus the performance immediately after commissioning (Caldwell et al. 2005).

8.5.3.3 EnMAP Thermal Design

The Environmental Mapping and Analysis Program (EnMAP) is a German hyperspectral satellite that aims to characterize the Earth's environment on a global scale. It is capable of measuring the solar radiance reflected from the Earth's surface as a continuous spectrum in the spectral range of 420–2450 nm, with a spectral sampling interval (SSI) of 6.5 nm in VNIR region and 10 nm in SWIR region. EnMAP's swath is 30 km and its ground sampling distance is 30 m (Kaufmann et al. 2016).

The spectrometers of EnMAP use glass prisms as the spectral dispersers and NiP-coated aluminum mirrors. The optical design of the instrument is not athermal. A finite element analysis of the mirrors shows that the surfaces of the large mirror optically distort in a significant manner when the temperature differs from the manufacturing environment by more than ±1°C. This is caused by the mismatch of coefficient of thermal expansion (CTE) between the NiP coating and the aluminum structure. The focal shift of the image location induced by the combination of glass and aluminum for this temperature range is $<20\ \mu m/°C$. As a consequence, the first requirement for the thermal design of the instrument is that the absolute temperature change of the instrument in space must be kept within ±1°C of the operational temperature throughout the mission.

A second requirement for the thermal design of the instrument is that the temperature of the Optical Unit must be kept stable to ±0.5°C between the spectral calibration cycles to be conducted every 2 weeks to meet the specifications for spectral calibration and line of sight stability. Because the change caused by temperature fluctuations in the refractive index of the prisms, a lateral shift is induced in the image location along the spectral axis of the detectors.

A third requirement for the thermal design is that the spectrometer housing must be kept within ±0.2°C for the offset calibration of the infrared background signal in the SWIR region. Because data for offset correction are recorded prior to every data capture, this temperature stability requirement relates only to the maximum duration of a data capture period of approximately 2 min.

The FPA of the VNIR spectrometer is a high-speed silicon CMOS 1024 × 108 pixel detector array. It is optimized for a large full well capacity (1 Me^-) featuring on-chip correlated double sampling to provide low noise in snapshot mode, including stare-and-scan capabilities, high quantum efficiency, high resolution, high-speed operation, low power consumption, and low noise. The VNIR detector array can be operated at room temperature 21°C, thanks to its small dark current due to the short integration time. However, to retain the accuracy of calibration of the sensor throughout the operation, its temperature is to be thermally controlled to stabilization at 21°C.

The FPA of the SWIR spectrometer is a mercury-cadmium-telluride (HgCdTe) 1024 × 256 pixel photovoltaic array. With indium bumps the photovoltaic array is attached to the silicon read out integrated circuit (ROIC) chip forming the so-called IR-hybrid. The hybrid is optimized with respect to quantum efficiency and sensitivity. This SWIR array must operate at a nominal temperature of 150 K or lower to reduce thermal noise and dark current.

The thermal power dissipated within the EnMAP hyperspectral imager increases by about 70% from the standby to data-capture operation. In addition, the external heat load to the radiators temporarily increases from an average of about 50 W/m² up to 185 W/m² during data-capture due to off nadir pointing in combination with Earth's infrared radiation and albedo. The instrument thermal control system has to compensate for these internal and external fluctuations.

The thermal architecture of the EnMAP hyperspectral imager consists of a mixture of passive and active thermal control means combined with design rules for isolating different thermal control zones by conductive and radiative decoupling.

In order to achieve the thermal requirements, the Optical Unit is mounted to the platform inside the instrument envelope structure using its own thermally isolating suspension. Hence, the instrument enclosure structure and the platform act as a first thermal barrier to the space environment. Whenever possible, sources of power dissipation and heat load, and the baffles of the full aperture diffuser and the telescope are mounted to the instrument envelope structure. Heat sources, which must be mounted to the Optical Unit such as the SWIR cryocooler and camera electronics, transfer their dissipated heat directly to external radiators via dedicated thermal links.

In order to cool the SWIR detector array, an active Stirling cryocooler with a split pulse tube cold-finger has been implemented. An extremely compact integrated detector cryocooler assembly was developed, which houses the SWIR detector array, the ROIC, the cryocooler, and a cold shield for background-signal reduction. The SWIR detector array is cooled to 150 K. The pulse tube uses helium as a working medium and is driven by a symmetric flexure-bearing compressor, of which two pistons are opposed to achieve vibration cancellation, minimizing the micro-vibration output of the cryocooler. Using a moving magnet concept and low out gassing materials, the cooler is designed and proven to have a mean time to failure of >60,000 h. The dissipation of the cooling system is <30 W for a total heat input at the cold tip of 1 W and a heat sink temperature of 21°C.

In order to control the temperature of the VNIR detector array to retain the accuracy of calibration of the sensor throughout the operation, a 10 W thermoelectric cooler is used to provide thermal stabilization during operation to a base temperature of 21 ± 0.05°C.

Besides employing standard methods of thermal control such as thermal coatings and finishes, MLI, heaters, and radiators, the thermal design also contains 12 two phase flow heat transport devices: mini loop heat pipes. The thermal design has the following main features:

1. The material of the Optical Unit structure exhibits high thermal conductivity and possesses a high thermal inertia.
2. Once launched and in operation, the Optical Unit is maintained at 21 ± 1°C with a temporal stability of better than ±0.3°C during data captures. The spectrometers and telescope assembly structure are cold biased and equipped with trim heaters to adjust the temperature level within the required range.
3. Heat fluxes between the instrument and the spacecraft platform and the inner side of the instrument enclosure structure are minimized to reduce the instrument's internal temperature gradients.
4. The Optical Unit structure holds the heat dissipating units. In order to minimize any unacceptable spatial gradients within the Optical Unit structure, these units are connected to the radiators by miniature loop heat pipes. These heat pipes provide an adjustable thermal resistance that enables the stabilization of temperature to manage effect of the heat sources with an accuracy in the 1°C range under varying dissipation levels and varying external radiator loads. The temperature for the corresponding control loops is set to 21°C. This minimizes the temperature gradients of the Optical Unit structure and it minimizes the heat fluxes between Optical Unit structure and the dissipating units. In order to ensure proper functioning of the miniature loop heat pipes control loops, the dissipating units are radiatively and conductively decoupled from their environment.
5. Three electronic units, power unit, instrument control and processing unit, and cooler control electronics, are mounted on the top panel of the instrument enclosure structure in the vicinity of the Optical Unit. Their housings may get warm. In order to reduce the heat radiation to the Optical Unit, the external sides of the housings are taped with vacuum-deposited aluminum tape (which provides a lower emissivity than conventionally machined aluminum, closer to that of polished aluminum).
6. The instrument enclosure structure is wrapped externally with MLI. Dedicated cutouts provide radiator areas for the electronic units mounted underneath to the inner face sheets. The inner instrument enclosure structure surfaces are taped with vacuum-deposited aluminum tape to minimize the radiative heat exchange with the Optical Unit.

The use of loop heat pipes for thermal control in optical instruments results in increased freedom of design and testability as well as facilitating the minimization of residual forces into the optical and optoelectronic system from the thermal control system. In addition, micro-vibration sensitive systems are designed with a low stiffness link between evaporator and condenser, allowing for simple reduction of micro-vibrations by passive means at spacecraft level (Lang et al. 2013).

8.6 STRUCTURE ANALYSIS AND CONSIDERATION

Structures must endure exposure to all environment from satellite manufacturing to the end of the mission. The structural analysis of a hyperspectral satellite needs to consider structural requirements (static and dynamic loads coming from launcher then spacecraft interfaces) and involves verification of the response of the instruments to stand the proto-qualification campaign.

The following considerations may drive the selection of payload's interface with the spacecraft:

- Availability of suitable positions for the payload.
- Reduction of the launch mechanical loads.
- Maximization of the available free surfaces for the radiator in order to avoid the active cooling.

The structure of the payload is based on an architecture that hold all the subsystems and components. With the hyperspectral satellite, the Optics Head is the key subsystem. The structure of Optics Head is constituted by two main structures: an internal structure containing the optical system and an external structure supporting the radiator, the entrance port, and the boxes of the proximity electronics. The Optics Head thermomechanical main functions are to provide a mechanical support for the optical subsystem, while maintaining thermal isolation from the structure and keeping some parts at colder temperatures. For thermal reasons, the Optics Head is located at a specific distance from the mounting plate.

The main considerations that may driver architecture of Optics Head are as follows:

- Positioning of the cold zones (e.g., SWIR detector array) as closely as possible to the radiator for the most efficient heat rejection, facilitating passive cooling for the SWIR detector array.
- Positioning of the radiator surfaces oriented toward cold space.
- Necessity to concentrate as much as possible the mass of the payload toward the spacecraft mechanical interface to reduce the mechanical loads.

For these purposes, the Optics Head is normally divided into two principal functional parts: a warm enclosure and a relatively cold enclosure. The cold enclosure is sometimes referred to as Cold Box (see Section 8.8.1 for VIRTIS hyperspectral instrument). It includes the optics bench that houses the cold detector arrays and contains the cold shield. The Cold Box includes the optical subsystem, mounted on a cold ledge that also supports the cover mechanisms, the baffle, and the passive cooling hardware, such as the cold radiator.

The warm enclosure includes the Optics Head baseplate, the telescope, in some cases it can contain the Proximity Electronics box (depending on the maximum allowable heat flux), the thermal shield, and the structural truss, which sustains the upper Cold Box while thermally insulating from the baseplate. The warm enclosure is generally conductively coupled to the spacecraft.

The mounting feet are positioned on the baseplate external perimeter and their conductivity to the spacecraft is reduced in order not to have any thermal load from the spacecraft.

In Section 8.8.1 for VIRTIS and Section 8.8.3 for MAJIS hyperspectral instrument, examples of instruments that use the similar structural architecture philosophy are described.

The structural analysis should consider the spacecraft dynamics loads, such as sine and random vibration levels, and verify that the instrument can stand without any degradation and damage.

Other important considerations are to have the main instrument resonance frequency above the main modes of the spacecraft, to separate the spacecraft's dynamic loads from the instrument dynamic loads, and to reduce the stresses at instrument and their component level.

Typical spacecraft and launcher main frequencies, those ones that might affect the main structure of the instrument, are typically maintained below 100 Hz. It is expected that the selected mechanical configuration should stand the typical space mission dynamic loads and allow to have the first Eigen frequency above 100 Hz.

8.7 MECHANISMS

A spaceborne hyperspectral instrument typically has three port cover mechanisms: a port cover mechanism, a solar port cover mechanism, and a shutter mechanism (SM).

The main port cover mechanism (MPM) integrated with calibration mirror is to open and close main scene entrance cover of Optical Head for protecting optical components during launch phase and non-image-acquisition phase, and the mirror mounted on the inner side used for calibration.

The solar port mechanism (SPM) opens and closes the solar entrance window to permit instrument calibration with the sun.

The SM opens and closes the spectrometer entrance slit so as to permit dark acquisition for calibration purposes.

Open position of the MPM allows observation of the scene, while the closed position protects the instrument internal optics from particulate and molecular contaminants. During ground transportation and launch phases, this cover must be in the closed position.

The MPM allows to perform important tasks of this kind of instrument: the calibration and sun protection. The spectral and solar calibrations are performed when the cover is closed. The light beams coming from the internal calibration sources or sun light entering from a solar port, while the SPM is open, illuminate the entire optical path until the focal planes; thanks to the reflection on the inner surface of the cover that is treated as a light diffuser. To achieve accurate calibration, the position of the diffuser must be located with very high precision and high repeatability with respect to the Optical Head during both flight and ground operation.

Sunlight protection is necessary during satellite maneuvers that could result in sunlight entering directly into the optical path used for observation of the scene with detrimental consequences to the optical system and/or the detector arrays. This port cover is to be closed before starting maneuvers to avoid the occurrence of such a situation.

The design of the MPM should consider the requirements to allow on-ground testing of the operational states: open and closed. To avoid the motor being oversized to overcome the force of gravity during ground testing, a removable counterweight may be adopted for ground testing. Torque margin must be considered to provide adequate torque to allow for degradation over the lifetime of the mechanism and minor anomalies/build variations, following the relevant standards (such as ECSS or NASA mechanism standard Factors of safety).

When in the closed position, the MPM cover must be secured for unwanted movement or hammering against the seat (end-stop) where this could affect the following positioning precision. It is good practice (as noted in the European ECSS Mechanisms specification) to use well-defined contacts that are repeatable by design and resilient to small changes in alignment, which could occur due to vibration or thermal effects (avoiding flat-on-flat configurations in particular). Contact materials must be selected to avoid adhesion and significant yielding. During ground and launch phases, due to relevant inertia of the cover, the motor detent torque may not be adequate to prevent such occurrences, therefore a launch-locking device would likely be implemented. Even if this subassembly is only a one-shot mechanism in flight, it must be designed for reversible operation, in particular for ground testing (including retesting) and transport.

Considering the extremely high reliability requirement for port cover opening systems, whose failure might result in loss of payload functionality, even where it is based on a well-proven and reliable concept with high-reliability components, the MPM is required to be designed to fail open or be equipped with an emergency actuator device (EAD) to force the cover into an open position in case of failure of the main drive system. The EAD is also required to be able to force the cover in closed position in order to ensure its closing for sun protection and calibration. The EAD must be a high-reliability mechanism, designed to open and close the cover starting from any position (including intermediate positions) for the specified number of actuations (Meini et al. 2016).

8.8 THERMAL AND STRUCTURE DESIGN EXAMPLES

This section presents three examples of hyperspectral instrument's thermal and structural design. This includes two hyperspectral payloads for planetary missions: VIRTIS and MAJIS, and one hyperspectral payload for Earth observation: PRISMA (PRecursore IperSpettrale della Missione Applicativa).

8.8.1 VIRTIS for Three Planetary Missions

The Visible and Infrared Thermal Imaging Spectrometer (VIRTIS) is a spaceborne hyperspectral imager with three focal planes in two channels: (1) hyperspectral imaging channel, and (2) spectroscopic channel. The hyperspectral imaging channel, referred to as VIRTIS-M, has two 2D focal planes covering the visible region (0.28–1.1 μm) and the infrared region (1.05–5.13 μm). The spectroscopic channel, referred to as VIRTIS-H, has a single aperture covering a wavelength range 1.84–4.99 μm with high SSI. Three slightly variant VIRTIS payloads had flown in three planetary missions: (1) ESA's Rosetta cometary mission (Coradini et al. 2007), (2) ESA's Venus Express mission (Piccioni et al. 2007), and (3) NASA-Dawn mission (Russell et al. 2007). Detailed description of VIRTIS is provided in Section 2.8.

8.8.1.1 VIRTIS Thermal Design

VIRTIS consists of four modules: (1) the Optics Module (OM) containing the VIRTIS-M and VIRTIS-H optical heads, (2) the VIRTIS-M proximity electronics module (PEM-M), (3) the VIRTIS-H proximity electronics module (PEM-H), and (4) the main electronics module (ME). The Optics Module is shown in Figure 2.6. It can be seen from the figure that the VIRTIS-M is mounted on the cold side (upper portion) of the support structure at right and the VIRTIS-H is at left upper portion.

The two infrared focal plane arrays of the VIRTIS-M and VIRTIS-H are of HgCdTe material on sapphire substrates. These infrared detector arrays have the same 5.1-μm cutoff wavelength and were manufactured from the same wafers. The infrared detector arrays require cooling to minimize the detector dark current owing to the thermally generated Johnson noise. To reduce the dark current, these infrared detector arrays must be cooled to temperatures below 100 K. This has been achieved by means of active Stirling cryocoolers, which guarantees operability for >10,000 h of operation in the planetary environment. Passive cooling has been excluded because of the low emissive power of passive radiators and the risk of planetary contamination.

The VNIR focal plane array of the VIRTIS-M is a CCD device. It is a frame transfer device and need to be cooled to lower than room temperature to reduce the dark current. The challenge for VIRTIS-M is to keep the CCD device temperature above 155 K to prevent degradation of the charge transfer efficiency. Another challenge is that the CCD housing is only 9-mm away from the infrared FPA housing that is cooled down to below 100 K, so both detector arrays have adjacent edges free of bond wires.

The main thermal and mechanical function of the Optics Module is to provide mechanical support for the two optical heads of VIRTIS-M and VIRTIS-H while maintaining them at a specific operating temperature. It provides cooling to the two infrared focal plane arrays to maintain their operation temperature below 100 K. In order to minimize the thermal background radiation seen by the two infrared detector arrays, their surrounding temperature is maintained below 150 K. To achieve this goal, the Optics Module is divided into two thermal zones as shown in Figure 8.9: a Warm Enclosure and a Cold Box.

1. The Warm Enclosure is at the bottom of the Optics Module. The pallet baseplate directly interfaces with the warm spacecraft (i.e., it is thermally controlled by the spacecraft). The two active cryocoolers are mounted on the pallet baseplate.
2. The Cold Box is at the upper portion of the Optics Module. Its structure is rigidly mounted to the much warmer pallet baseplate but thermally insulated from it. The Cold Box houses the two Optical Heads and the cooled detector arrays. The VIRTIS-M Optical Head is mounted on the support ledge at the right, whereas the VIRTIS-H Optical Head is mounted on the support ledge at the left. The Cold Box is thermally disconnected from the spacecraft and passively cooled to below 150 K by radiating one of its surfaces toward cold space.

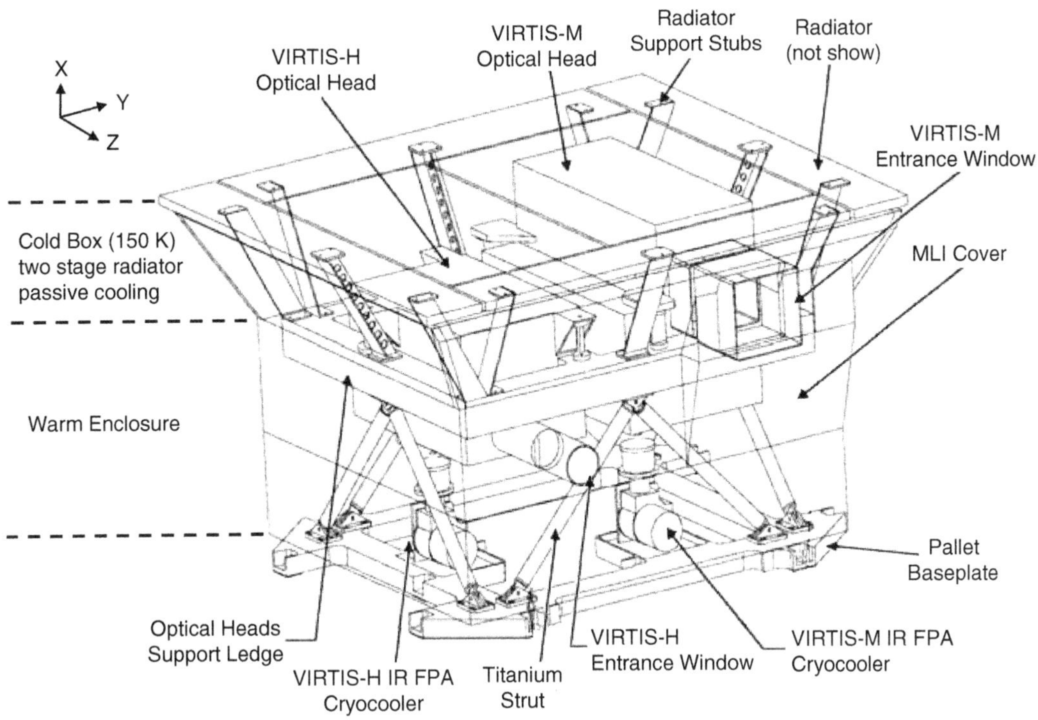

FIGURE 8.9 VIRTIS Optical Module structure and mounting of the two optical heads of VIRTIS-M and VIRTIS-H. (ESA courtesy.)

Two active cryocoolers in the Warm Enclosure are connected to their corresponding infrared detector arrays in the Cold Box through cold fingers, which maximize the thermal pathway from the cryocoolers to the infrared detector arrays while remaining mechanically pliant. In contrast, the stand-off insulators (eight titanium hollow struts connecting the Cold Box to the baseplate of the pallet) minimize the thermal pathway between the warm spacecraft and the cold optical subsystems, while remaining mechanically rigid. In this way, the structure and the delicate subsystems that it supports were not only predicted to survive launch vibrations, but the structure would also help to minimize the usual thermal gradients that adversely affect the alignment of low-temperature optical systems.

There are differences between the Rosetta and Venus Express instruments owing to the very different targets: the cold nucleus of a comet and the hot planetary atmosphere of Venus. The Venus Express design improves thermal dissipation of the VIRTIS payload.

8.8.1.2 VIRTIS Structure Design

The Optics Module is mounted externally on the X panel of the spacecraft with the two Optical Heads co-aligned and viewing the scenes in the Z direction. Both Optical Heads have their slits parallel to the Y axis. VIRTIS-M can point and scan by rotating its primary mirror around the Y-axis. The structure and inside view of the Optics Module are shown in Figures 2.6 and 8.9.

The mechanical interface of the Optics Module with the spacecraft is the pallet baseplate, placed at the bottom of the Optics Module as shown in Figure 8.9. The baseplate is connected to a special supporting structure provided by the spacecraft. The size of the baseplate is 450 mm × 510 mm. The pallet is a hollow box 130-mm high with the two cryocoolers mounted on it. The pallet must dissipate 15 W heat, of which 13 W heat is conducted to the spacecraft at 300 K for radiative dissipation. The remaining 2 W of heat is radiated directly by the Optics Module at lower temperatures.

The Cold Box is mounted on the pallet baseplate via eight titanium hollow struts whose number and attachment positions were selected to avoid distortion upon cooling from room temperature. This structure design has achieved a low conductive heat load. The Cold Box's surface adjacent to the pallet is covered with MLI (see Figure 2.6). A conservative assumption was made that much of the thermal FOV of the surfaces inside the Cold Box is of high emissivity parts. That is why MLI also covers at least three of the remaining surfaces. The empty space inside the Cold Box is intentional to allow for sufficient radiating surface area without requiring a separate and complex radiator. The result is a low-density module with sufficient stiffness structures only where necessary. This reduces mass while maintaining the surface area. Irregular shapes were eliminated to facilitate efficient MLI attachment.

The Warm Enclosure includes the pallet baseplate, the brackets for interunit harnesses, two of the active cryocoolers, and the structural truss that thermally insulates the upper Cold Box from the baseplate. The Cold Box includes the two optical heads, mounted on a cold ledge that also supports the cover mechanisms, the two baffles, and the passive cooling hardware. A set of four thermal pipes in the Warm Enclosure is dedicated to the thermal power dissipation of the two active cryocoolers. In this way, the power generated on the baseplate, mostly from the cryocoolers, is conductively dissipated in an efficient manner to deep space via a dedicated radiator.

The passive cooling to keep the Cold-Box below 150 K consists of a primary radiator and two secondary radiators each for VIRTIS-M and VIRTIS-H. The primary radiator conductively connects to the optical head support ledge. The three radiators are placed on the spacecraft's X-axis, looking at cold space. Their overall envelope is about 590 mm × 650 mm. The primary radiator cools the support ledge by dissipating the heat from the truss and from part of the internal cabling. The two secondary radiators are each mounted directly on top of the VIRTIS-M and VIRTIS-H heads.

The Cold Box and the Warm Enclosure are externally covered with MLI. From a radiative point of view, the MLI blanket separates the Cold Box from the Warm Enclosure. The low-conducting heat load between the Cold Box from the Warm Enclosure is achieved by mounting the Cold Box on eight titanium hollow struts. All of the struts are blanketed with MLI to reduce the thermal load and avoid differential heating of the cold ledge.

The heat flow through the Cold Box is small enough, thus the VIRTIS optical heads are inside an essentially isothermal cavity. To maintain this isothermal cavity, the covers and actuators for the VIRTIS-M and VIRTIS-H entrance ports are located inside the Cold Box between the optical heads and the external MLI blanket.

Both VIRTIS-M and VIRTIS-H heads have a cover to protect the instrument from direct sun illumination and to preserve the cold environment inside the Cold Boxes when the sun is within the FOV and they are not operational. Each cover has an emergency actuator to force it open in case of malfunction. The covers are machined as retro-reflectors for in-flight calibration. The measurement of the CCD detector's dark current plus background radiation requires calibration shutters to be placed at the entrance light-limiting slit of each spectrometer.

8.8.2 PRISMA Hyperspectral Imager

PRISMA is an Italian hyperspectral satellite aiming to qualify the technology, contribute to develop applications, and provide products to institutional and scientific users for environmental observation and risk management. It was launched on March 22, 2019, on a Vega launch vehicle in Kourou into a sun synchronous orbit.

PRISMA payload is composed of a hyperspectral imager and panchromatic (PAN) camera. The hyperspectral imager is made up of a VNIR prism based spectrometer and a SWIR prism based spectrometer to cover spectral bands ranging from 400 nm to 1010 nm and from 920 nm to 2505 nm. It provides hyperspectral images of the Earth with 30-m ground sampling distance (GSD), 30-km swath width, and spectral bands at a SSI of 12 nm.

The thermal and mechanical design of the satellite was driven by thermal requirements. The use of prism-based spectrometers requires a tight control of temperature variations to ensure the stability of all instrument features on orbit. Particular thermal design efforts were devoted to the cooling of detector arrays, which is achieved without the use of mechanical coolers. The use of a mechanical cooler, due to intrinsic micro-vibrations, is a potential cause of modulation transfer function degradation, especially for very narrow IFOVs in addition the risk of life cycle reliability (Meini et al. 2016).

The mechanical design was guided by the constraints related to the satellite accommodation and the environmental conditions. The satellite accommodation includes the positions on the top of the spacecraft by taking into account the orbit, the mission phases, and maneuvers. The environmental conditions include launch, deployment, and in-orbit phase.

To deal with the constraints, the mechanical design was based on a closely integrated optical bench structural for providing support to the mirrors of the three mirrors anastigmat telescope on the upper side and support to the spectrometers optical elements on the lower side.

The thermal design of the payload has the following features:

- The SWIR and VNIR detector arrays are cooled in a range between 140 K and 185 K using two-stage passive radiators facing cold space.
- The cold chain of each detector array uses a thermal strap and a high performance ethane heat pipe.
- Specific parts of the spectrometers, such as the detector arrays and their cold shield, are positioned close to the radiator using a short thermal strap to improve the system efficiency.
- The heat fluxes from the warm electronic part to the cold zones of the instruments are very low.
- The prisms subsystems are maintained thermally insulated from the supporting structure.
- The optical elements are mounted in an insulated enclosure on the optical bench and are maintained in a relatively small range of temperature by a dedicated thermal control so to reach a high thermomechanical stability.

The opto-mechanical design of the Optical Head utilizes reflective optical components and structural elements all use aluminum alloy to obtain a quasi-athermalized design. The optical bench is closed by the upper and lower covers equipped with patch heaters to stabilize the entire instrument at 20°C. The insulation of the Optical Head from the spacecraft conductive and radiative heat loads is achieved by means of insulating, mounting its feet (titanium alloy) positioned on the baseplate of the payload interface and wrapping with MLI blankets.

The SWIR and VNIR detector arrays are mounted on the lower side of the Optical Head. They are cooled between 140 K and 185 K by a cold chain connected to two-stage passive radiators as shown in Figure 8.10. The main element of the cold chain is a high efficiency ethane axial grooved heat pipe connected with a thermal strap that allows the structural decoupling of the detector arrays from the radiator. The passive cooling system consists of two-stage radiators, detector housings, thermal straps, heat pipes, heat pipe jackets, insulating supports, and mechanical support.

The heat sources inside the Optical Head are the proximity electronics boxes; the detector arrays of SWIR, VNIR, and PAN spectrometers; and the motors and actuators for the covers and the heaters. The proximity electronics boxes are located far from the cold zones. They are mounted on the baseplate outside (externally to optical bench) and thermally sunk to the baseplate to minimize parasitic heat flux.

There are three kinds of heaters that are used distributed on various parts of the optical bench structure for different purposes. The operational heaters enable accurate thermal control during operation. The survival heaters are for thermal conditioning before operation and in-between operations. The third kind heaters are for decontamination purposes.

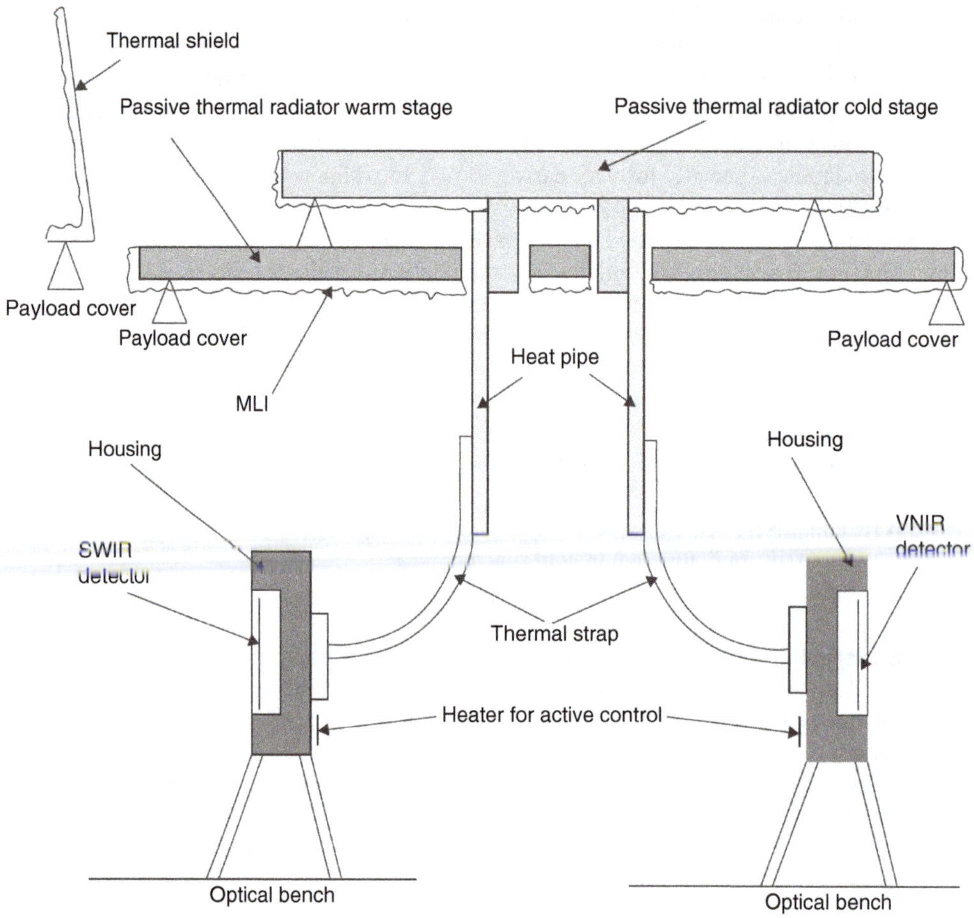

FIGURE 8.10 Schematic layout of the passive cooling system for VNIR and SWIR detector arrays. (Courtesy of SELEX ES.)

The Optical Head is accommodated in the spacecraft and has three thermal interfaces with the spacecraft platform at the top panel, with solar array panel on the backside, and with external environment. Thermal shields are installed to thermally decouple the Optical Head from outer thermal environment. The thermal shields are aluminum sandwich panels bolted together and covered externally by MLI. The optical bench is protected from the outer environment by the double-level MLI: one inner MLI between the optical bench covers and the payload panels, and an external MLI mounted on the lateral panels.

8.8.3 MAJIS for Exploring Galilean Moons of Jupiter

The MAJIS is a hyperspectral imager covering VNIR and IR regions. It was selected by ESA in February 2013 for JUICE mission to survey the Jovian system with a special focus on the three Galilean Moons: Callisto, Ganymede, and Europa. The JUICE mission plans to flyby Callisto, Ganymede, and Europa, then a 1-year orbital fly around Ganymede, the largest moon of Jupiter. It is scheduled to be launched in 2022 (Langevin et al. 2014).

MAJIS consists of two grating-based spectrometers with the VNIR spectrometer covering wavelengths from 0.5 μm to 2.35 μm and the IR spectrometer covering wavelength from 2.25 μm to 5.54 μm. The two spectrometers use two HgCdTe detector arrays of size 400 × 508 pixels.

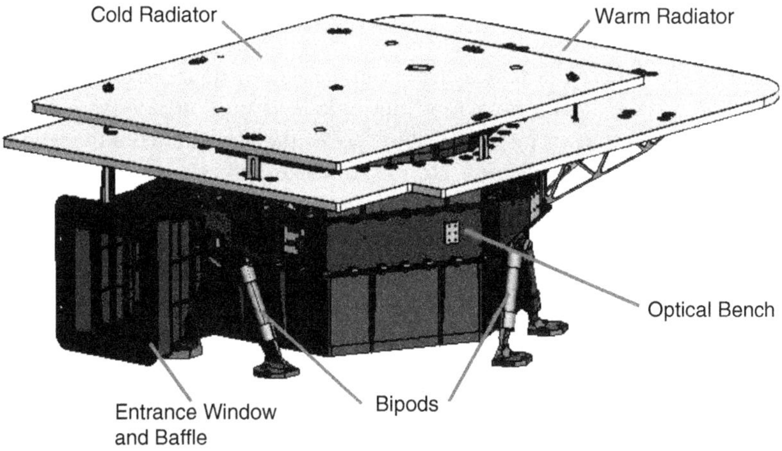

FIGURE 8.11 MAJIS Optical Head structure. (Courtesy of Leonardo SpA, Airborne & Space Systems.)

The MAJIS will produce 2 × 508 spectral bands from 0.5 μm to 5.5 μm at SSI 3.6 nm and 6.4 nm with 400 spatial pixels. The SNR will exceed 100 over most of the spectral range. The instantaneous FOV of 150 μrad of the instrument corresponds to a spatial sampling distance of 75 m from a 500-km circular orbit over Ganymede and to 150 km for observations of the atmosphere of Jupiter.

The MAJIS is made of many different subsystems. The main subsystem is the Optical Head, which includes a scanning mechanism, a telescope, two spectrometers, and a calibration unit. The thermal design of MAJIS is based completely on passive radiators. It provides operational temperatures <90 K for the infrared detector array, and <140 K for VNIR detector array, and the optical bench. These low temperatures are required to reduce detector noise and background radiation disturbances on the measured signals. Figure 8.11 shows the structure of the Optical Head. It can be seen that two-stage radiators are mounted on the top of the Optical Head.

8.8.3.1 MAJIS Thermal Design

In the Optical Head there are two different thermal requirements:

1. The optical bench, optical elements of the VNIR and IR spectrometers as well as the VNIR detector array need to be maintained at around 140 K or lower, which is achieved by thermally connected to the relatively warm radiator.
2. The IR detector array needs to be cooled below 90 K, which is achieved by thermally connecting to a second passive cold radiator. This cold radiator is surrounded by the relatively warm radiator, which acts as a shielding element against external sources of heat as shown in Figure 8.11.

MAJIS payload will be mounted on the external panel of the JUICE mission spacecraft. The temperature of the spacecraft during operation is between 263 K and 293 K (–10°C to 20°C). It is required to maintain the temperature of the Optical Head below 140 K. The payload supporting structure that connects the Optical Head and the spacecraft has to withstand a maximum temperature difference of 153 K, in addition to providing the strength and stiffness required to hold the Optical Head and radiators. Therefore, one of the thermo-structural design drivers is to minimize the thermal conductance between the payload supporting structure and the spacecraft. It is critical to select the supporting structural material that has sufficient thermal insulation between the instrument and the spacecraft. In order to provide the requested thermal insulation, a configuration based on composite material three bipods that connect to titanium heads and feet has been designed and adopted as shown in Figure 8.11. Selection of bipods material and geometry has been done following accurate trade-off between their thermal and structural properties.

In order to reduce the radiative heat load on the Optical Head, a double thermal barrier is designed for it by applying MLI to the outer surfaces and an single-layer insulation to the inner surfaces. This barrier has also been applied to enclose the volume below the warm radiator to avoid sun trapping.

Survival and anticontamination heaters are implemented and are controlled by the spacecraft. The survival heaters are aimed at maintaining the temperature of the Optical Head within the nonoperative temperature range of each subsystem, while the anticontamination heaters are used during the first part of the cruise after launch and in case of need to prevent water ice formation on the optics and the detectors (Tommasi et al. 2018).

8.8.3.2 MAJIS Structure Design

Thermomechanical design of the MAJIS is driven by the following requirements (Saggin et al. 2014):

- Allocated mass for the instrument support interface is 2.7 kg and the mass for the radiators is 2.9 kg.
- A quasi-static design load is 500 m/s^2.
- First natural frequency is above 130 Hz to avoid coupling with lower frequency satellite modes.
- Sine vibration amplitude is 240 m/s^2 between 20 and 100 Hz.
- Random vibration is between 20 and 2000 Hz with amplitude of power spectral density up to 0.102 g^2/Hz.

Optical elements of the Optical Head are mounted and aligned on the optical bench that constitutes the main structure of the instrument and provides support to all the other subsystems. The optical bench surface is stiffened by a series of ribs that provide the necessary bending stiffness with a limited cost of mass. It is highly beneficial to machine such components from a single piece of material (or to minimize the number). The entire external border of the optical bench would typically be used, providing reinforcement by a robust continuous rib that increases the stiffening of the structure, designed to provide adequate strength and stiffness for the connection of supports.

The strict thermal requirements and the passive cooling strategy have a strong impact on the structural design. The Optical Head supports should provide the required resistance under mechanical loading and a proper dynamic behavior, but at the same time, insulating the Optical Head from the warm spacecraft. As described in Section 8.8.3.1, three bipods of composite material connected to titanium heads and feet are adopted. These bipods are placed on three reinforced zones of the optical bench located at about 120° among them, which are in turn connected to the network of inner ribs. This configuration of the bipods ensures that the direction of minimal flexural stiffness of each bipod is directed toward a common point located in the proximity of the center of mass of the instrument, resulting in an isostatic design. The thermal elongation of the optical bench is thus compensated by the bipod transversal flexibility minimizing the induced stress in the optical bench. Preliminary structural analysis focused at bipods design verification showed that the first Eigenfrequency is well above the 100 Hz requirement.

The two-stage passive radiators are on top of the Optical Head. To meet the allocation of the radiators requires an accurate control of all the thermal exchanges in the minimization of the dimensions and mass of the radiators.

The Optical Head includes two mechanisms, a scan unit and a shutter:

1. The scan unit is equipped with a lightweight mirror supported by two frictionless flexural pivots. The rotation of the mirror is pre-programmable in order to synchronize its position with scientific data acquisitions. Moreover, the scan unit can direct the telescope beam

toward the line of sight of the internal calibration unit for calibration purpose. The scan mirror and other a few components in the scan unit are made of beryllium. The combination of the coefficient of thermal expansion (CTE) of beryllium and the CTE of the scan axis ensure that the mirror remains blocked at room temperature and is free to rotate at cryogenic temperatures. This temperature difference driven CTE ensures maximum stability of the mechanism during the launch.

2. The shutter mechanism is placed at the entrance slit of instrument. The closure of the shutter allows for acquisition of dark current and thermal background data from the spectrometers, which are particularly relevant for an IR spectrometer.

Since the Optical Head has to operate in the harsh radiation environment of the Jovian system, radiation hardiness has been taken into account for the design of the structure in order to protect most sensitive items, especially detector arrays and optical coatings, from ionizing and nonionizing radiation. Dedicated analyses were iteratively performed during the design phase to assess the implemented architecture against radiation doses and dedicated radiation shielding has been added when required. This was coupled with a qualification plan that includes radiation testing to the foreseen levels of the most sensitive parts, especially optical coatings and calibration sources. Whenever possible, testing will be carried on at cryogenic temperature with an electron source, for the best representative radiation environment experienced by the JUICE mission. A maximum of 50 Mrad total ionizing dose is foreseen on the scan mirror, which is the optical element most exposed to spare radiation.

REFERENCES

Auster, H. U., I. Richter, K. H. Glassmeier, G. Berghofer, C. M. Carr, and U. Motschmann. 2010. Magnetic field investigations during Rosetta's 2867 Šteins flyby. *Planetary and Space Science* 58(9):1124–1128.

Bentien, A., S. Johnsen, G. K. H. Madsen, B. B. Iversen, and F. Stegkich. 2007. Colossal Seebeck coefficient in strongly correlated semiconductor $FeSb_2$. *Europhysics Letters* 80(17008):1–5.

Bugby, D., J. Garzon, M. Beres, C. Stouffer, D. Mehoke, and M. Wirzburger. 2005. Cryogenic diode heat pipe system for cryocooler redundancy. In *Cryogenic Optical Systems and Instruments XI*, ed. J. B. Heaney and L. G. Burriesci, Proceedings of SPIE 5904:321–329.

Caldwell, D., J. Hackett, A. S. Gibson, J. R. Drummond, and F. Nichitiu. 2005. The design and flight performance of the MOPITT instrument mechanisms. *Proceedings of the 11th European Space Mechanisms and Tribology Symposium, ESMATS 2005*:99–106.

Chuchra, A. and T. Stevenson. 2002. Thermoelectric coolers. In *Spacecraft Thermal Control Handbook Vol. I: Fundamental Technologies*, 2nd ed., ed. D. G. Gilmore, 473–487. El Segundo: The Aerospace Press.

Clawson, J. F., G. T. Tsuyuki, B. J. Anderson, C. G. Justus, W. Batts, D. Ferguson, and D. G. Gilmore. 2002. Spacecraft thermal environments. In *Spacecraft Thermal Control Handbook Vol. I: Fundamental Technologies*, 2nd ed., ed. D. G. Gilmore, 21–70. El Segundo, CA: The Aerospace Press.

Coradini, A., et al. 2007. VIRTIS: An imaging spectrometer for the Rosetta mission. *Space Science Review* 128(1–4):529–559.

De Boer, P. C. T. 2002. Performance of the inertance pulse tube, *Cryogenics* 42:209–221.

Donabedian, M, D. G. Gilmore, J. W. Stultz, G. T. Tsuyuki, and E. I. Lin. 2002. Insulation. In *Spacecraft Thermal Control Handbook Vol. I: Fundamental Technologies*, 2nd ed., ed. D. G. Gilmore, 161–206. El Segundo, CA: The Aerospace Press.

Gibson, A. S., J. Reed, T. W. Bradshaw, and M. Linder. 2008. Heritage overview: 20 years of commercial production of cryocoolers for space. *AIP Conference Proceedings 985*:493–505.

Gifford, W. E. and R. C. Longsworth. 1964. Pulse tube refrigeration. *Transactions of the ASME, Journal of Engineering for Industry* 86(3):264–268.

Gilmore, D. G. (ed.). 2002. *Spacecraft Thermal Control Handbook Vol. I: Fundamental Technologies*, 2nd ed. El Segundo, CA: The Aerospace Press.

Gilmore, D. G., B. E. Hardt, R. C. Prager, E. W. Grob, and W. Ousley. 2006. Thermal. In *Space Mission Analysis and Design*, eds. W. J. Larson and J. R. Wertz, 428–458. El Segundo, CA: Microcosm Press.

Gilmore, D. G., W. K. Stuckey, and M. Fong. 2002. Thermal Surface Finishes. In *Spacecraft Thermal Control Handbook Vol. I: Fundamental Technologies*, 2nd ed., ed. D. G. Gilmore, 139–160. El Segundo, CA: The Aerospace Press.

Jewell, C. I. 1996. Cryogenic cooling systems in space (invited). Submillimetre and far-infrared space instrumentation. *Proceedings of the 30th ESLAB Symposium held in Noordwijk*, eds. E. J. Rolfe and G. Pilbratt. ESA SP-388, 24–26 September 1996, 267–276. Paris: European Space Agency.

Kaufmann, H. et al. 2016. Environmental mapping and analysis program – A German hyperspectral Mission. In *Optical Payloads for Space Missions*, ed. S.-E. Qian, 161–181. West Sussex, UK: John Wiley & Sons.

Lam, T. T., G. C. Birur, and P. Bhandari. 2002. Pumped Fluid Loops. In *Spacecraft Thermal Control Handbook Vol. I: Fundamental Technologies*, 2nd ed., ed. D. G. Gilmore, 405–472. El Segundo, CA: The Aerospace Press.

Lang, D., L. Morgenroth, F. Harms, B. Sang, L. Dandaleix, T. Coquard, and O. Berder. 2013. EnMAP hyper spectral instrument thermal design and test verification. *Proceedings 43rd International Conference on Environmental Systems.* Vail, CO.

Langevin, Y., et al. 2014. The MAJIS VIS-NIR imaging spectrometer for the JUICE mission. *Proceedings of 45th Lunar and Planetary Science Conference*: #2493-1–2.

Meini, M., A. Bini, L. Giunti, E. Fossati, and R. Formaro. 2016. Hyperspectral payload for Italian PRISMA programme. In *Optical Payloads for Space Missions*, ed. S.-E. Qian, 183–213. West Sussex, UK: John Wiley & Sons.

Mikulin, E. I., A. A. Tarasov, and M. P. Shkrebyonock. 1984. Low temperature expansion pulse tubes. *Advances in Cryogenic Engineering* 29:629–637. New York: Plenum Press.

Misselhorn, J., P. Brennan, L. Thienel, and G. Fleischman. 1987. Application of heat pipes to a cryogenic focal plane space experiment. *Proc. AIAA 22nd Thermophysics Conference*, Honolulu, HI.

Murchie, S., et al. 2007. Compact reconnaissance imaging spectrometer for Mars (CRISM) on Mars reconnaissance Orbiter (MRO). *Journal of Geophysics Research* 112(E05S03):1–57.

Pätzold, M., T. P. Andert, S. W. Asmar, J. D. Anderson, J.-P. Barriot, et al. 2011. Asteroid 21 Lutetia: Low mass, high density. *Science* 334(6055):491–492.

Piccioni, G., et al. 2007. VIRTIS: The visible and infrared thermal imaging spectrometer. *ESA-SP-1295, ESA Publications Division* SP-1295. Noordwijk, The Netherlands.

Plaut, J. J., et al. 2014. Jupiter Icy Moons Explorer (JUICE): Science Objectives, Mission and Instruments. *Proceedings of 45th Lunar and Planetary Science Conference*: #2717-1–2.

Popescu, G, et al. 2001. A critical review of pulse tube cryogenerator research. *International Journal of Refrigeration* 24:230–237.

Prager, R. C., M. Nikitkin, and B. Cullimore. 2002. Heat Pipes. In *Spacecraft Thermal Control Handbook Vol. I: Fundamental Technologies*, 2nd ed., ed. D. G. Gilmore, 489–522. El Segundo, CA: The Aerospace Press.

Radebaugh, R. 2000. Pulse tube cryocoolers for cooling infrared sensors. *Proceedings. of SPIE* 4130:363–379.

Radebaugh, R., J. Zimmerman, D. R. Smith, and B. Louie. 1986. A comparison of three types of pulse tube refrigerators: New methods for reaching 60 K. *Advances in Cryogenic Engineering* 31:779–789, New York: Plenum Press.

Ross, R. G., Jr. 2003. Vibration suppression of advanced space cryocoolers: An overview. *Proceedings of the SPIE 5052 Smart Structures and Materials*: 1–12.

Ross, R. G., Jr. and R. F. Boyle. 2007. An overview of NASA apace cryocooler programs—2006. *Proceedings of the International Cryocooler Conference*, Boulder, CO.

Rowe, D. M. 1995. *CRC Handbook of Thermoelectric.* Boca Raton, FL: CRC Press.

Rowe, D. M. 2006. *CRC Handbook of Thermoelectric: Macro to Nano.* Boca Raton, FL: CRC Press.

Russell, C. T., et al. 2007. Dawn mission to Vesta and Ceres. *Earth Moon Planet* 101:65–91.

Saggin, B., D. Scaccabarozzi, M. Tarabini and G. Piccioni. 2014. Thermo-mechanical design feasibility study of an imaging spectrometer for the Jovian system. *Proceedings of IEEE International Workshop on Metrology for Aerospace*, May 2014: 450–455.

Sugimura, R., S. Russo, and D. Gilman. 1995. Lessons learned during the integration phase of the NASA in-step cryo-system experiment. *Cryocoolers* 8:869–882.

Thienel, L., P. Brennan, M. Stoyanof, and C. Gerhart. 1998. Design and performance of the cryogenic flexible diode heat pipe (CRYOFD) flight experiment. *Proceedings of the 28th International Conference on Environmental Systems (ICES)*, Danvers, MA.

Tommasi, L. et al. 2018. The advanced optical and thermo-mechanical design of the JUICE/MAJIS spectrometer. *Proceedings of the 5th IEEE International Workshop on Metrology for Aerospace*: 247–251.

Tward, E., et al. 1999. Miniature space pulse tube cryocoolers. *Cryogenics* 39(8):717–720.

Venkatasubramanian, R., E. Siivola, T. Colpitts, and B. O'Quinn. 2001. Thin-film thermoelectric devices with high room-temperature figures of merit. *Nature* 413:597–602.

Werrett, S. T., et al. 1985. Development of a small stirling cycle cooler for spaceflight applications. *Advances in Cryogenic Engineering* 31:791–799.

Zhu, S., P. Wu, and Z. Chen. 1990. Double inlet pulse tube refrigerators: An important improvement. *Cryogenics* 30:514–520.

Zhu, W., S. L. Zhou, N. Yoshimura, and Y. Matsubara. 1997. Phase shift effect of the long neck tube for the pulse tube refrigerator. *Cryocoolers* 9:269–278, New York: Plenum Press, 1997.

9 In-Flight Calibration Design

9.1 IMPORTANCE OF ONBOARD CALIBRATION

Hyperspectral satellites or sensors acquire signals of the reflection or emission of radiation by the Earth's or a planet's surface. These signals enable the retrieval of respective physical quantities (essential variables). The physical basis for the retrieval is the interaction of the radiation with the materials of the Earth's or a planet's surface. This interaction is defined by radiative transfer, which favors the use of radiances and their respective units within retrieval methods.

The signals acquired by hyperspectral satellites or sensors are electrical signals that are then converted into digital numbers (DNs). Hence, calibration has to be applied in order to relate the digital numbers to the incoming radiances, consequently, to the physical units of interest. The relation between the digital numbers and the radiances is given by the calibration coefficients, which can be derived by comparison of the signals acquired by the sensor with an absolute standard reference prior to launch. This process is referred to as on-ground calibration prior to launch or prelaunch calibration, which is the subject of and described in Chapter 12. The calibration coefficients can also be derived by comparison to the signals acquired when looking at onboard references during flight.

9.1.1 In-Flight Calibration for Monitoring Instrument Behavior

Hyperspectral satellites or spaceborne hyperspectral sensors are all well designed and calibrated prior to launch. Unfortunately, no matter how well and sophisticated the instruments are built and calibrated prelaunch, once on orbit they are subject to degrade with time, for example, due to exposure to space radiation, or thermal, mechanical, or electrical effects. The radiometric and spectral performances of hyperspectral sensors are affected by deviation during their mission life. Such fluctuations are mostly due to vibrations/shocks, temperature and pressure changes at the moment of launch or due to contamination and other aging-driven degradation of the instrument during flight. For example, the center wavelength of a hyperspectral sensor may shift due to the distortion in the spectrometer, the optics, and the detector assembly. The deviation of the center wavelengths from the nominal band position and radiometric degradation has an impact on precise data utilization. Therefore, it is necessary to monitor and track on-orbit radiometric and spectral stability. For the data product retrievals for science and operational uses, the instrument calibration is essential. Calibration of hyperspectral satellites or sensors is, therefore, the basis of reliable data retrieval that ensures the quality of the derived data and products.

In-flight calibration is devoted to track the instrument behavior on orbit. It is very important to monitor the development of the instrument performance and especially the radiometric and spectral properties over the entire mission life to ensure a high data quality for the scientific interpretation and operational uses. In all cases, the objective of in-flight calibration is the traceability of the data calibration accuracies to the International System of Units (SI) for science users and data products with consistent quality for the broader user community.

However, in-flight calibration with reference standards of known accuracy is difficult to manage in terms of metrology. Onboard calibration systems being interpreted as the reference standard are subject to degradation processes as well, especially in the visible wavelength region. Moreover, many satellite sensors (especially in the visible spectral region) are not adjusted by onboard calibration. Hence, in any case, in-flight calibration is quite a challenging task, but is essential for physically defined retrieval of essential variables.

Many different sources contribute to the error in the calibrated data. With effort, the resultant error can be reduced. The degree to which it can be reduced is a function of the level of effort applied to the design of both the hyperspectral instrument itself and the calibration methodology and associated system, both in the prelaunch calibration on ground and the in-flight calibration on-orbit. In general, the more accurate the calibration, the higher the cost in terms of not just money but also volume, mass, and operational sophistication of the satellite. For a hyperspectral satellite mission with a fixed budget, as are all missions, the challenge confronting the mission and payload designers will be to optimize the use of the project resources in terms of producing the most accurate system outputs. By defining an end-to-end calibration error budget to develop an understanding of the relative size of the various error sources for various candidate systems, the trade-offs involved in optimizing the use of the mission resources can be evaluated. The exact nature of the error sources must be understood as must the level to which they are theoretically correctable with the various techniques.

9.1.2 Instrument Calibration and Validation

It is also of importance to define the terms "calibration" and "validation" and their relationship. As defined by the Committee on Earth Observation Satellites (CEOS 1995).

- Calibration is the process of quantitatively defining the system response to known, controlled inputs.
- Validation is the process of assessing by independent means the quality of the data products derived from the system outputs.

Calibration of satellite data for Earth observation or planetary exploration can normally be considered as one of three types of activities: prelaunch calibration, in-flight calibration using standard light sources, and in-flight vicarious calibration. The prelaunch calibration of the sensor provides accurate measurements of the system parameters required to accurately convert the acquired data to the physical units of interest, normally the at-sensor or top-of-atmosphere (TOA) spectral radiance. These data are then monitored in orbit by onboard calibration devices using secondary or tertiary standard light sources (lamps, sun, or others), or using vicarious methods with images of well-characterized specific ground targets or the moon.

Dinguirard and Slater (1999) provided five general guidelines for spaceborne multispectral sensors, which are worthwhile repeating here:

1. To the extent possible, the same geometry and spectral radiance distribution and levels should be used in the calibration as occur in the operational image-acquisition mode of the sensor. This minimizes differences between measurement and use, due to stray light, detector nonlinearity, out-of-band rejection, etc. To meet this condition, the calibration should be full aperture, full field, and full dynamic range and should use an appropriate source spectral distribution.
2. Several different and independent techniques should be used preflight and in-flight to determine if systematic errors exist in one or more techniques and, to the extent possible, identify, remove, or account for them in the calibration results.
3. The characterization of the sensor should be as detailed as possible. This requires measuring such parameters as MTF, stray light and ghosting, out-of-band spectral rejection, linearity, polarization, etc. Most of these measurements can be made more precisely in the laboratory than in flight. In some cases, their precise determination can allow corrective algorithms to be written and applied to improve the radiometric accuracy of measurements in the vicinity, for example, of cloud edges.

4. Related to the last point, the user needs to be informed about the limits of applicability of absolute calibration values associated with the scene data. Pixels near a cloud edge or pixels in a complex scene are unlikely to correctly represent the actual surface radiance for the reasons mentioned in item 3 above. Calibration data are the result of measurements made with an extended spatially uniform source such as an integrating-sphere source or a solar diffuser (SD) that will fill the entrance pupil of the sensor. The result of calibrations under these conditions, where there is no target detail in the image, will not apply accurately, without correction, to pixels in the image of a complex scene.
5. The user also needs to know uncertainties in the published calibration data in detail. Typically, calibration data are presented as a single, 1σ value for a sensor over a wide spectral range, for example, all the sensor's bands in the visible and near-infrared, from 400 nm to 1000 nm. If the calibration was done with a standard lamp, a single value is inappropriate because the lamp itself varies in uncertainty of calibration over this range. Furthermore, other quantities, for example, the signal-to-noise ratio, also vary spectrally, so band-by-band calibration uncertainties should be published.

It is common for many hyperspectral instruments to have an onboard calibration (OBC) system to provide the primary method of monitoring the changes in the calibration coefficients and to use vicarious techniques as the method of providing independent validation as suggested in item 2 above. Although this combination of calibration and validation method is a desirable approach, it was predicted by Teillet et al. (2001) and Thome et al. (2001) that weight and cost factors of including an onboard calibration system may make use of vicarious calibration techniques as the primary method of calibrating and validating the sensor once in-orbit. Such an approach has become more promising due to a couple of key developments. First, the accuracy of undertaking vicarious calibration has improved significantly as the techniques employed have matured over the past decades. Secondly, as the number of satellite instruments increase and the desire for cross calibration of the sensor systems increase, the level of effort behind the optimization and automation related to the measurements of the vicarious calibration sites also increases.

The importance of traceability to SI was discussed by Fox (2001). It is essential that each spaceborne instrument be calibrated traceably against a standard which can be shown to be stable over the timescale of not only the instrument itself, but also considerably longer so that the measurements made by an instrument can be compared with those made by other instruments. However, it was stated that while prelaunch activities help in evaluating the extent to which the instruments meet specifications, it is the post-launch environment that the issue of traceability to SI units becomes critical. In the visible and near-infrared (VNIR) and shortwave-infrared (SWIR) spectral regions, this may be achievable through the use of vicarious calibration techniques whose accuracy is becoming to rival that of OBC techniques using standard light sources. At longer wavelengths, onboard calibration sources are simpler and vicarious calibration is more complicated and less accurate, so that the trade-off between onboard calibration devices and vicarious calibration is often in favor of the former.

Radiometric calibration can be seen as the application of a transfer function that converts the more or less arbitrary raw sensor data into physical radiometric units. Determining the coefficients of the transfer function so that can be applied to the uncalibrated data is the goal of the radiometric calibration process.

For narrow spectral band instruments such as hyperspectral sensors, the variation of the spectral response of the instrument is often considered spectrally flat over the narrow spectral interval between two adjacent spectral points. That hypothesis considerably simplifies the calibration transfer function as it makes it independent of the observed scene. In comparison, the radiometric calibration equation of broadband radiometric instruments depends on the spectral profile of the observed scene and some a priori assumption about the target must be made.

If, over the dynamic range of interest for the system and for the required accuracy, the instrument can be considered linear, then the radiometric transfer function can be as simple as a linear function. In that case, only two coefficients have to be determined per function: the slope and the ordinate (sometimes referred as the gain and the offset). The determination of these two unknowns requires the measurement of two different radiometric references. If the instrument is not linear, then the calibration equation must incorporate higher-order terms and more coefficients must be determined, which requires more references. In Section 9.2, we will see what some selected instruments use as radiometric references.

As the calibration coefficients are different for every pixel of the detector array, a set of coefficients must be determined for every single pixel in the instrument swath and for every spectral pixel along the spectral axis.

Spectral calibration is a different process. It consists in determining a mathematical relationships and its coefficient, between the pixel coordinates along the spectral axis and wavelength. This is commonly achieved by making measurements of targets containing sharp spectral features of known wavelengths. A more or less arbitrary fit is then made between the pixel coordinate number and the wavelengths of the observed spectral features. The spectral calibration coefficients have to be determined for every line of spectral pixels in the instantaneous instrument swath.

Although both calibration processes are independent, spectral errors can be a source of radiometric error. This is discussed in the next section.

9.1.3 Relationship Between Spectral and Radiometric Calibration

A discussion is required with respect to the relationship between the spectral and radiometric calibration. An accuracy can be specified for the knowledge of a channel's spectral characteristics, including center wavelength, full width at half maximum (FWHM) and shape of its passband, based upon the application of the measured spectrum to a particular application of hyperspectral data analysis. However, the accuracy required to perform atmospheric correction of the data far surpasses those requirements. Small errors in the spectral knowledge can lead to large radiometric errors near sharp spectral absorption features.

Green (1998) evaluated this relationship between spectral and radiometric errors in great detail. For example, for spectral bands with 10 nm FWHM bandwidth, the analysis predicted that for a misalignment of 1.0 nm, radiometric errors of up to 25% can be expected. For 0.5 nm and 0.1 nm misalignments, the radiometric errors of up to 15% and 3%, respectively, are expected. Similar sensitivity analysis on the effect of 1.0, 0.5, and 0.1 nm errors in the FWHM bandwidth shows errors of up to 18%, 9%, and 2% can be expected. It was concluded that to measure a radiance spectrum without significant spectrally distinct errors, a 1% or better knowledge is required for both the spectral calibration accuracy of a band position and the band shape. Fortunately, the spectral locations of those sharp spectral features are often well known and they can therefore be used to undertake a spectral calibration and/or validation that will be most accurate at those precise spectral locations. If the accuracy of the calibration degrades somewhat between these spectral features, the radiometric impact will not be as severe as if that degraded accuracies were applied to the location of the features themselves.

In order to quantify the radiometric impact on the accuracies of the spectral parameters, one would need to undertake a sensitivity analysis given a specific measured spectrum. An activity such as this has not yet been seen being reported. At this time, when considering the radiometric calibration, the spectral knowledge is always considered accurate. The radiometric impact on the spectral inaccuracies can be considered similar to that determined by Green and mentioned in the above paragraph.

9.2 REVIEW OF ONBOARD CALIBRATION SYSTEMS

9.2.1 MERIS

MEdium Resolution Imaging Spectrometer (MERIS) onboard calibration was based on the measurements of sunlight. It was performed at the orbital South Pole, where diffuser plates were deployed by rotating the selection disk, as shown in Figure 9.1, into any of the five positions described below:

1. Shutter, used for dark calibration as well as for protecting the instrument from contaminants during launch
2. Earth observation, a diaphragm was introduced in the field of view
3. Radiometric calibration, the sun illuminated a *white* diffusers plate inserted in the field of view
4. Diffuser degradation monitoring, the reference white diffuser plate was deployed, on the consecutive orbit following a radiometric calibration, every 3 months to monitor the degradation of the frequently used plate, by deploying it
5. Spectral calibration, an erbium-doped *pink* diffuser was deployed with MERIS configured to sample its absorption features. Spectral absorption features introduced in the reflectance spectrum by erbium allows to accurately monitoring MERIS spectral behavior.

MERIS radiometric calibration was performed with the use of the two onboard sun-lit radiometric calibration white diffusers described above. The diffusers, made of Spectralon, provided a uniform illumination signal over the large FOV of the instrument. The radiometric calibration sequence was executed every 2 weeks. It started with a dark calibration with the calibration mechanism in shutter position, averaging 1024 frames, followed by the radiometric calibration with the frequently

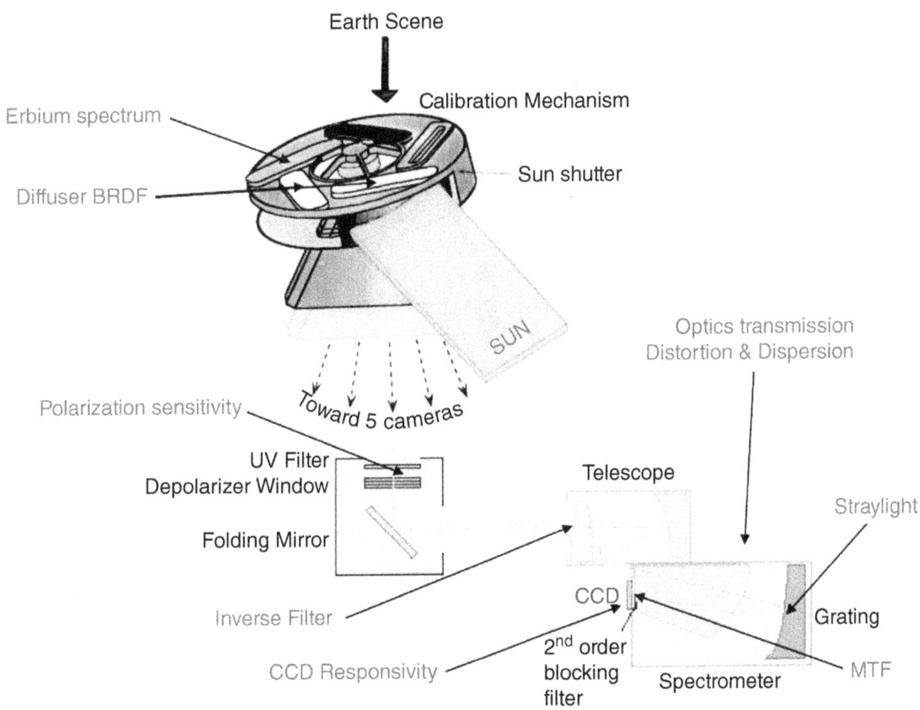

FIGURE 9.1 MERIS onboard calibration system. (Courtesy of ESA.)

illuminated diffuser deployed, averaging 512 frames, centered when the solar illumination angle on the diffuser was at 65.5°.

The spectral calibration sequence was carried out every 3 months on two consecutive orbits. On both orbits, similar sequences to that of the radiometric calibration were executed but with the instrument spectral band setting centered around an erbium spectral features. The first orbit was used for radiometric calibration of the contiguous fine bands (1.25 nm), while the second orbit measures the erbium absorption spectrum centered near 520 nm. Every 6 months, this sequence was repeated twice over four consecutive orbits, with the instrument spectral band setting centered on two different erbium spectral features (410 nm and 520 nm).

The diffuser ageing calibration sequence was carried out every three months on two consecutive orbits. On both orbits, similar sequences to that of the radiometric calibration are executed. On the first orbit, the radiometric diffuser is deployed, while on the second orbit the reference diffuser is deployed.

The dark offset measures were regularly acquired during dedicated calibration sequences: 1024 frames were acquired as part of every radiometric calibration, then averaged temporally and transmitted to ground. This allowed a regular monitoring of the dark offset as well as the generation of dark offset correction coefficients.

The accuracy of the radiometric calibration relied on the sun diffusers characterization and onboard measurements. Its processing used the same radiometric model as Earth observation processing. It can be summarized in four main steps:

1. A radiometric calibration measurement generates instrument digital numbers $X_{cal}(\lambda, k)$, where λ stands for the spectral band number and k for the spatial position.
2. Instrumental correction (including nonlinearity, dark offset, smear) produces corrected digital numbers, $X'_{cal}(\lambda,k)$, considered as proportional to the radiance at instrument entrance.
3. Instrument gain coefficients $G(\lambda,k)$ are then computed such as:

$$X'_{cal}(\lambda,k) = G(\lambda,k) \cdot L_{cal}(\lambda,k) \tag{9.1}$$

4. $L_{cal}(\lambda,k)$ is estimated from $E_0(\lambda)$, the sun extraterrestrial irradiance at the MERIS wavelength λ, illumination and viewing geometry and the diffuser bidirectional reflectance distribution function (BRDF).
   a. The diffuser BRDF had been characterized on ground.
   b. E_0 is derived from a model in Thuillier et al. (2003), the seasonal variation of the Sun-Earth distance, and the instrument spectral characterization (channels central wavelengths and spectral response curves).
   c. Geometry was derived from the satellite position and attitude computations and instrument pointing characterization.

MERIS onboard spectral calibration mechanism included an erbium-doped *pink* diffuser dedicated to spectral characterization, which offers a number of spectral absorption features in the spectral range of MERIS for spectral calibration. In addition, a number of observation campaigns had been dedicated to spectral calibration, thanks to the instrument flexibility in programming the position and bandwidth of the spectral bands. These measurements include:

- Earth observation of the atmospheric oxygen (O_2-A) absorption feature at 760 nm using a number of independent wavelength derivation methods
- Observation on the radiometric calibration diffuser of the sun Fraunhofer lines circa 393, 485, 588, 655, 855, and 867 nm.

The erbium-doped Spectralon has a number of spectral features found within the MERIS's spectral range (390–1040 nm). Features centered at 408 nm (Peak 1) and 522 nm (Peak 3) were selected for spectral calibration. The spectral calibration using the onboard erbium doped diffuser was performed as follows. First, the instrument was configured to have 15 adjacent bands—at the highest spectral bandwidth—centered on the spectral feature (either Peak 1 or Peak 3), then the pixels were calibrated using the radiometric calibration white diffuser, finally on the following orbit the pink diffuser was deployed and the signal from this spectral feature acquired. Processing on ground determined the position on the detector array of the peak of the erbium absorption feature. The results, as applied to the 522-nm erbium spectral feature, showed a stability of better than 0.02 nm on average and 0.1 nm peak residual errors.

The spectral calibration using the absorption feature of oxygen O_2-A made use of a pressure minimization technique (Delwart et al. 2007) and an artificial neural network method. In general terms, the TOA radiance depends mainly on the O_2 absorption and therefore on the surface elevation or pressure. The O_2 absorption can be determined from the ratio of two reflectance spectra, one measured within the oxygen A-band and one measured in a close nonabsorbing band. MERIS data was acquired to support this by programming it to acquire a set of 10 contiguous, single row bands (FWHM = 1.25 nm) that span the oxygen A-band feature. An apparent surface pressure can then be derived for each of the bands for assuming a wavelength shift in 0.1 nm steps. By searching the results for minimum pressure dispersion, the actual spectral shift can be identified. The technique had been applied to a MERIS data set and shown to have an accuracy of ±0.02 nm (Delwart et al. 2003, Ramon et al. 2003).

The spectral calibration using the solar Fraunhofer absorption features was performed using raw data of SD observations—calibration would remove the spectral signature inherent to the solar illumination. The algorithm is based on a shape matching technique comparing measurements to subsampled versions of a high-resolution solar spectrum, accounting for the lack of radiometric calibration in the considered instrument configuration through a linear normalization process. The six observed Fraunhofer lines were acquired during three successive orbits with band settings focused on two lines each time.

In the MERIS instrument, the needs to adequately characterize the spectral parameters of the pixels across the entire FOV, to effectively characterize the spectral smile, took on added meaning as the MERIS instrument was composed of five identical hyperspectral cameras whose imagery was designed to be mosaicked into a single wide-swath image. If not adequately performed the result would be a discontinuity in the resulting data product between adjacent cameras. The smile correction was applied by adjusting TOA radiances to a nominal wavelength by linear interpolation of the adjacent wavelengths (Delwart and Huot 2002a). Examples of the impact on Level 2 data products were shown by Delwart and Huot (2002a, 2002b).

9.2.2 Hyperion

The approach to on-orbit calibration/validation of the Hyperion instrument was a combination of a simple onboard calibration scheme with the analysis of atmospheric and solar absorption features (Barry et al. 2002). The Hyperion sensor assembly was equipped with a motorized aperture cover, which was an integral part of the instrument calibration process. The cover has three positions: open, closed, and calibration. The backside of the cover was coated with a Spectralon diffuse reflecting white and silicone thermal control paint. With the cover in calibration position (partially open at 37°), solar illumination reflecting off the diffuse painted surface was used as a source for the on-orbit radiometric calibration. With the cover closed, the internal calibration source (lamps) reflected off the painted cover. For these two calibration techniques, the paint was characterized for reflectivity as a function of angle (Pearlman et al. 2003).

An absolute determination of the solar irradiance with Hyperion measurements was made as a cross calibration of absolute responsivity. Additional cross-calibration efforts using the Landsat data and a ground truth site at Lake Frome, Australia which do not depend on the cover reflectance were also used.

The internal calibration source (ICS) used four quartz tungsten halogen (QTH) lamps. It illuminated the back of the telescope cover painted with diffuse, reflecting, white, silicone, thermal control paint when the cover was in the closed position. The lamps were powered in two pairs making a primary and a secondary set. Two lamps per set were required to achieve an adequate level of illumination.

The sunlight was incident on the back of the telescope cover at a 53° angle of incidence for radiometric calibration. The solar calibration data were collected at 53° with the FOV of Hyperion 0.43°. A special solar data collection was made by scanning the spacecraft so the sun was occulted by the solar baffle in a way to ensure the scatter angle was correct. The comparison of the solar irradiance in the three models (Kurucz et al. 1984, Wehrli 1986, Thuillier et al. 1998) with the Hyperion measurement showed an agreement about ±2% in the VNIR (500–850 nm) spectral region.

The Hyperion diffuse reflectance panel used for the solar calibration was cross-checked against the results of the ground truth campaign carried out at Lake Frome, Australia on January 5 and 21, 2001. The TOA radiances were calculated using an atmospheric effects model and measurements of the spectral reflectance of the ground at the several locations. This replaced the uncertainty of the Hyperion cover panel diffuse reflectance with the ground truth characterization uncertainty in an entirely independent way. Another cross-check of on-orbit calibration was performed by comparison with Landsat 7 measurements made of the same ground site at the same time. Hyperion spectral bands were aggregated radiometrically to give the same in-band radiance as the Landsat spectral bands.

Frames of the reflected solar spectrum from the SD were acquired as the sun was rising through the Earth's limb. A 12-sec window existed for Hyperion to acquired such data as it was crossing the pole. The viewed spectrum consisted of not only features contributed by the reflectance spectrum of the paint used on the SD, but also of spectral features from the sun and the atmosphere (Barry et al. 2002).

Nineteen useable features were identified in the SWIR region that included the three spectral features introduced by the diffuser paint (1190, 1690, and 1740 nm). The only useful features in the VNIR were the atmospheric 762.5 nm oxygen absorption line and the 520-nm solar Fraunhofer line. As with any other spectral calibration technique, the observational data must contain clearly defined spectral features that can be identified and traced to a reference spectrum. For each spectral feature, a cubic spline was used to identify its exact location. The use of the spline was expected to introduced a ±1.1 nm error. As this was performed for each spatial pixel in the field-of-view, a two dimensional surface of calculated wavelength shift was plotted for the SWIR detector array, with the x-axis representing the pixel location of the spectral feature based upon original spectral calibration and the y-axis being the field-of-view location. A low-order polynomial was then applied to reduce statistical noise in the data and produced a complete band to wavelength map of the focal plane. For the VNIR detector array, as there were only two points, a simple offset and tilt was determined for each spatial pixel.

The experimental results showed that the on-orbit measurements were within 5 nm, or half a bandwidth of the nominal Hyperion FWHM, of the prelaunch calibration in the SWIR. The VNIR results were slightly better, although with less confidence. The accuracy of this technique was limited by the accuracy of the reference spectra (feature dependent) and the error in the spline determination of the feature locations (±1.1 nm).

It was concluded that although this procedure should be applied periodically throughout the mission, the spectral calibration of the sensor was expected to be constant since the spectrometer was temperature controlled. The methodologies in the determination of the center wavelength positions have been improved. Gao et al. (2002), Neville et al. (2003), Green and Pavri (2002), and Green et al. (2002) have all presented results of improved spectral matching techniques applied to Hyperion data.

The techniques reported by Gao et al. and Neville et al. make use of superior spectral feature matching techniques to that described above. The data sets upon which the analyses are applied in both cases are that of, unlike the Barry procedure, ground scenes. Gao et al. suggest that for instruments with 10 nm spectral resolution (FWHM), seven spectral features should be suitable for application of the described analysis (H_2O: 820 nm, 940 nm, 1140 nm; O_2: 760 nm; CO_2: 1580, 2010, 2060 nm). For instruments with spectral resolution of 5 nm (FWHM) or better, the 430-nm Fraunhofer line becomes useable and for 2.5 nm spectral resolution instruments, the additional Fraunhofer lines at 516, 656, 854, and 863 nm could be used. Use of any of these spectral features is, however, dependent on adequate signal-to-noise ratio of the data sets. This includes adequate removal of pixel-to-pixel nonuniformities. Neville et al. apply an iterative procedure to precondition the data to eliminate the impact of any scene variations on the measurement results. The results of applying both techniques, however, were found to be comparable. Neville et al. infer from analysis of the AVIRIS data set that the procedure is capable of making single pixel measurements down to ±0.05 nm.

In the work reported by Barry et al. (2002) and Pearlman et al. (2003), validation of the bandwidths determined preflight was not discussed. The techniques developed by Gao et al. (2002) and Neville et al. (2003) were also applied to the determination of the bandwidths. However, the techniques will function only if the band response function is known. It will not work if the band response function is to change in an unknown manner although this is not expected to be of concern. For example, Neville et al. reported that when applied to AVIRIS data, the bandwidth in the vicinity of the 942 nm water vapor band, was found to increase 2.06% ± 0.05%.

As with the technique reported by Barry et al. (2002), the results must be interpolated or extrapolated from the analyzed spectral features to each of the bands for each of pixels across the field-of-view. The suitability of undertaking this process is evaluated by applying an atmospheric correction to the data set based upon the new sensor wavelength calibration. When this is done, the "spikes" in the corrected scene spectra that occur when the atmospheric correction algorithm is spectrally misaligned with the data set, for the most part, disappear. The only remaining spikes are found at the CO_2 lines and may be attributable in part to incorrect offset correction.

The techniques reported by Gao et al. (2002) as well as Green and Pavri (2002) are applied to data sets acquired in a comprehensive field campaign undertaken in February of 2001 in Argentina to validate the radiometric and spectral calibration of the Hyperion instrument. The spectrally flat reflectance of the surfaces of the scene found at the remote Salar de Arizaro site made this scene an ideal location for the acquisition of ground imagery for application to these spectral feature-matching procedures. The campaign, which is reported by Green et al. (2002), included under flights of AVIRIS instrument and in situ measurements that, in terms of spectral calibration, were used to ensure accuracy of the input parameters used in executing the MODTRAN simulation used in the spectral matching procedures.

The process applied to derive the optimal on-orbit spectral shift with respect to the laboratory spectral calibration for the bands in the vicinity of a given spectral feature is to (1) normalize the MODTRAN radiance to that of the Hyperion data; (2) calculate the sum of the radiance differences for those bands while shifting the Hyperion data in 0.1 nm steps. The shift where that radiance difference is minimized should indicate the optimal shift for that cross-track pixel and wavelength region. This can be redone for each cross-track pixel in order to determine the spectral smile of the instrument. A third-order polynomial is fitted to this curve in order to remove instrument noise.

For the data set of Salar de Arizaro acquired on February 7, 2001, 100 lines of data over the calibration site were summed together. Application of the spectral matching technique described above to the 760 nm O_2 absorption feature indicated a +1.3 nm shift to a pixel near the center of the image. Over the complete field-of-view, the shift varied between 0.38 nm and 1.39 nm or 3.8–13.9%, respectively, of the nominal 10-nm instrument bandwidth.

The spectral fitting technique is also applied in a similar manner to the determination of the optimal FWHM. Although it is not specifically described, it is assumed that the center wavelength optimization is first implemented, and then the minimization of the radiance difference by adjustment of the FWHM is applied. Deltas in the FWHM derived prior to flight and this on-orbit determination vary from −0.28 nm to +0.71 nm across the field-of-view at 760 nm. It should be noted that the only suitable absorption feature available for this process was O_2 feature near 760 nm. It is believed that at least part of the reason was the low Hyperion SNR estimated in the paper as near 140:1 for the test condition found at this sight. Since the VNIR spectrometer is a self-contained subsystem, it was assumed that these results, identified for only the single spectral area of each spatial pixel, could be applied to the entire VNIR spectral region.

In the SWIR region two spectral features are identified as suitable for application of this technique, the 1140-nm water vapor and the 2010-nm carbon dioxide atmospheric absorption features. As previously mentioned, the in situ measurements were used to determine the atmospheric water vapor in order to allow the water vapor band to be utilized. Application of this technique to these SWIR features resulted in shifts with respect to the laboratory calibration of +2.2 nm to +2.4 nm at the 1140 feature and +3.0 nm to +3.5 nm for the 2020-nm feature. A statement in the paper suggested that the similarity of the shifts at the two SWIR spectral features is similar enough to justify using them to improve the spectral calibration across the full spectral range.

Four supplementary data sets were acquired and analyzed in order to determine the changes in the spectral calibration parameters with respect to the ground calibration. Data was acquired on February 10, March 30, April 25, and May 1 of 2002. Analysis of this time series for each of the three spectral features shows stable spectral registration within ±0.5 nm with the exception of the two SWIR results for May 1 that shows a shift of 1.5 nm.

9.2.3 MODIS

MODIS was designed and built with state-of-the-art onboard calibrators (OBCs), which include a SD, a solar diffuser stability monitor (SDSM), a blackbody, and a spectro-radiometric calibration assembly (SRCA). In addition, a space view port is part of the OBCs that enables measurements of instrument background and detector offsets. The MODIS SD panel was made of space grade Spectralon materials and is used primarily for in-flight calibration of the 25 reflective solar bands. Its on-orbit degradation is tracked by the SDSM, which is a separate ratioing radiometer that makes measurements of the sunlight reflected from the SD panel and direct sunlight through a pinhole attenuation screen. The SDSM has nine filtered detectors inside a 2-in. diameter solar integration sphere. These filters are similar to those used for the MODIS VIS/NIR spectral bands with center wavelengths ranging from 0.41 μm to 0.94 μm. A rotating mirror within the SDSM system enables the detectors inside the integration sphere to alternately collect data from its sun view, SD view, as well as a dark view. The SDSM is operated during each scheduled SD calibration event. Multiple scans of data from the sun view, SD view, and dark view are collected by the SDSM using the same optics, detectors, and electronics. The SD degradation at each SDSM detector's wavelength can be determined from on-orbit changes of its SD view responses relative to its sun view responses (Xiong et al. 2016).

The SRCA, as shown in Figure 9.2, operates in three different modes to perform sensor radiometric and spectral calibration for the reflective solar bands (RSB) and thermal emissive bands (TEB). It is essentially a grating monochromator with internal light sources and spectral calibration capability. Replacing the grating by a flat mirror, the SRCA can be configured into a collimator that, together with a set of reticles, can also be used for spatial characterization. The spatial mode is used to monitor the band-to-band registration of all 36 spectral bands, including FPA to FPA relative shifts.

The SRCA is a lamp illuminated integrating sphere whose output passes through order filters and an entrance slit, is then dispersed by a grating, and focused through an exit slit and onto the scan mirror, the first element of the MODIS optical system. The key to SRCA on-orbit spectral calibration is to accurately calibrate its source wavelength. For this purpose, the SRCA was designed

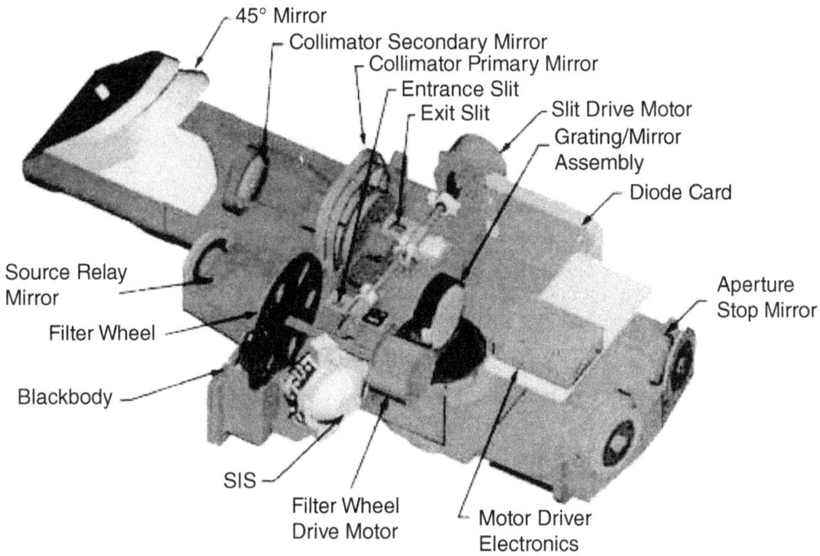

FIGURE 9.2 MODIS spectro-radiometric calibration assembly. (Courtesy of NASA.)

with internal wavelength calibration capability. It is self-calibrating by viewing a signal through a didymium glass filter located behind an auxiliary slit located adjacent to the exit slit. The centroid of each band can be determined to within 1.0 nm uncertainty with 0.5 nm precision.

This calibration approach is very expensive and has a great deal of complexity, including significant mass, volume, and power budgets. MODIS represents the first time that a monochromator has been put in orbit to monitor the spectral characteristics of an on-orbit radiometer (Montgomery et al. 2000) and it is not likely that this method (or a similar one) will be replicated in future missions (Teillet 1997).

Despite the fact that this approach to on-orbit spectral calibrations is unlikely to be undertaken again as well as the fact that MODIS is a filter-based multispectral system as opposed to a dispersive element based hyperspectral instrument, it is useful to review the results provided by this two of a kind (MODIS Terra, MODIS Aqua), onboard spectral calibration system. After 15 years on-orbit for MODIS Terra and over 12 years on-orbit for MODIS Aqua with spectral calibration checks occurring every 3 months and late reduced to every 4 months, the MODIS Terra and Aqua report that for most of the solar reflective bands, the center wavelength have been found to vary by less than 1.0 nm from their prelaunch determined values (Xiong et al. 2016). The change in bandwidths over the course of each mission is also small. The bands showing the greatest change (bands 1 and 19) have a 50-nm bandwidths that is the largest among the measured VIS/NIR bands. Variations in the relative spectral response of the MODIS passbands were simply specified as having an effect of less than 0.5%. This expensive and complicated system provides the MODIS mission with spectral characterization sufficient to ensure that its mission requirements are met.

9.2.4 SCIAMACHY

The SCIAMACHY (Scanning Imaging Absorption spectroMeter for Atmospheric CHartographY) was one of ten instruments aboard of ESA's Environmental satellite (ENVISAT). It was an extremely complicated instrument and different from the hyperspectral instrument described in this book. Despite this, there are a number of interesting aspects of the SCIAMACHY spectral calibration that are worth being described here.

The SCIAMACHY is an imaging spectrometer, however, it acquired one spectrum data at a time, with the image being built up through scanning of an elevation and azimuth mirrors (Frerick et al. 2002). The input light was broken up into eight "channels" and dispersed by gratings across a 1-dimensional 1024 element detector array (Si or InGaAs) within each channel covering a total spectral range from 240 nm to 2380 nm. The spectral sampling of each element ranges from 0.2 nm in the ultraviolet (UV) region to 1.5 nm in the SWIR region.

Its spectral calibration was performed on the ground using an external PtCrNe hollow cathode emission spectral line source (SLS) to determine both the center wavelength of each element and the slit function (spectral response function). The SLS provided an adequate number of spectral lines (>100) to undertake the spectral calibration of channels 1 through 6 (240–1772 nm) (Bovensmann et al. 2002). However, there were inadequate lines available in channel 7 (2 lines) (1932–2043 nm) and although there were more lines available in channel 8 (2255–2385 nm), they were all located at the lower end of the channel's spectrum. Data acquired in the laboratory using trace gas absorption cells were used to supplement the spectral calibration data in these channels. A baseline fourth-order polynomial was determined for channels 1 through 6 and second-order polynomial for channels 7 and 8 from these ground-based measurements.

A similar SLS was installed internally in order to undertake in-flight calibrations (Bovensmann et al. 2002, Frerick et al. 2002). The light from the internal SLS was introduced into the optical path. The internal lamp was tested on the ground prior and found to provide results nearly identical to those obtained with the external SLS in terms of the center wavelength. For the slit function characterization a discrepancy was found in the results acquired from the two SLS lamps (Bovensmann et al. 2002). The slit function was found to be approximately 0.1 of a pixel smaller for channel 1 through 6 when determined with the internal SLS than that with the external SLS. This was determined to be due to a partial blockage of the light path of the internal SLS. In channels 7 and 8 it proved to be very difficult to accurately determine the slit function due to the lack of spectral lines as well as higher noise levels.

Once in orbit, the internal SLS had been used to monitor the spectral calibration of the instrument. Updated spectral calibrations were maintained as deltas with respect to the baselined ground calibration. For channels 1 through 6 the differences were found to be < ±0.05 nm. For reasons identified in the previous paragraph, channels 7 and 8 had deltas of up to 0.3 nm. Spectral information from atmospheric absorptions features was used to supplement the SLS information in channels 7 and 8 in a manner similar to that used with the gas absorption cells on the ground. Stability of the in-flight spectral calibration over two calibrations periods roughly 100 days apart showed a variability within ±0.02 nm.

Since it was possible to acquire SLS data at any location throughout the orbit, on-orbit variations in the spectral calibration parameters were investigated and found to be small but measurable (Frerick et al. 2002). Spectral calibration data was acquired in each of 12 different "orbital regions." Extraction of spectral calibration information from this data showed variations close to or <0.01 pixels over an orbit.

Checks for difference in spectral calibration data acquired with *s* and *p* polarized measurements were undertaken. Although small, the differences were found to not be negligible.

A wavelength shift in the measured spectra was observed in limb radiance measurements as a function of tangential height (Skupin et al. 2002). This shift has been attributed to a combination of under sampling and inhomogeneous illumination of the slit, effects that change as the scan angle of the elevation mirror varies. This effect was found to cause spectral shift of 0.01 nm or 0.1 pixels.

9.2.5 VIRTIS

VIRTIS was equipped with several in-flight calibration devices in order to check the spectroscopic and radiometric parameters from time to time during the mission. An intense calibration campaign was performed prelaunch at the subsystem level for VIRTIS-M and VIRTIS-H (Drossart et al. 2004) and then at the system level.

VIRTIS-M is the first imaging spectrometer that used a new generation of lamps for spectral calibrations. The two custom lamps were optimized for the respective visible and infrared spectrometers (Melchiorri et al. 2003). They are highly stable. An intense qualification campaign was carried out on the qualification of the lamps while checking the limited degradation with time. The lamps were characterized in a wide spectral range with a blackbody-like emission with an effective temperature of about (2400–2600 K), thereby covering the whole VIRTIS-M's spectral range (0.2–5.0 μm). The lamps offer a wider spectral calibration in comparison with the quasi-monochromatic LED emission. A precise spectral calibration was achieved by adding two special filters in front of the window. These filters contain many narrow absorption lines. The Holmium filter is for the visible region and the polystyrene filter for the IR region (Piccioni et al. 2007).

The lamps can be seen by both visible and IR spectrometers, providing redundancy. The lamp illuminates the inner part of the VIRTIS-M cover, designed to reflect the radiance diffusely back toward the instrument. The cover has two areas of different materials chosen to have higher spectral reflectance for the lamp and the related channel. The areas correspond roughly to the visible and IR part of the instrument pupil. This choice comes from a trade-off between the mechanical constraints and the need for the best calibration. The lamp's spot size is equal to the pupil dimension and also spatially homogeneous at the distance of the cover.

Although the spectral response has demonstrated to be very stable during the entire mission without any relevant variation, it was detected that there was a constant spectral shift in the infrared spectrometer of about 7–9 nm (about one full band) for all the bands in the infrared region, which affects all observations since the start of the nominal mission (Bézard et al. 2009). The reason for this spectral shift was believed caused by thermo-mechanical stabilization of the instrument during the Venus orbit insertion.

Another spectral accuracy related issue detected by the in-flight calibration was the spectral distortion, often referred to as "smile." In the case of VIRTIS-M-IR, the dedicated prelaunch calibration studies have shown that the smile induced spectral variations on the order of 10% of the actual spectral width for each band (Bézard et al. 2009). However, the measured spectral variation was about 17 nm, which is larger than what was measured during ground calibrations (Moinelo et al. 2010).

For VIRTIS-H spectrometer, one radiometric and two spectral in-flight calibration modules were equipped. The radiometric calibration module used a tungsten lamp through the entrance cover mirror. The spectral calibration modules used a tungsten lamp and a temperature-calibrated Fabry-Pérot spectrometer that gives 20–10 lines depending on the grating order. One of the modules was used through the entrance cover mirror that returns the flux to the telescope and slit. The other was used through the back of the slit that returns the flux to the collimator.

The Fabry-Pérot system allows the spectral registration of the instrument to be confirmed after a shift is suspected, for example, following the vibrations induced by launch. The absolute position of each known spectral line on the detector was Gaussian-fitted for each order and used to reconstruct the spectral registration of the instrument to a 2-order polynomial function. Those functions were then used to give the correspondence between the pixel number and the wavelength for each pixel.

9.2.6 PRISMA

The PRISMA hyperspectral instrument is equipped with an in-flight calibration unit (ICU) to perform absolute and relative radiometric calibration as well spectral and geometric calibrations (Meini et al. 2016).

The ICU is composed of a high reflectance integrating sphere and a parabolic mirror that directs light source to illuminate the complete optical path. Lamps and LEDs are mounted inside the integrating sphere. The parabolic mirror is fixed to the inner part of the main cover. A dedicated solar port is used as entrance for sun radiation providing an additional calibration source.

The function of the ICU is to calibrate the instrument with both known spectral radiance and with filters characterized by spectral features to retain required data quality and to monitor the changes of the instrument on-orbit. The ICU input spectral radiance for calibration is generated by the sun (ICU_S Mode) or by internal sources (ICU_IS Mode), through a dedicated optical path.

In-flight calibration is performed by means of the ICU with a regular repetition rate, properly matched to the normal satellite maneuvering phases. The main functions of ICU_IS Mode include (1) instrument health check, (2) relative radiometric calibration, (3) spectral calibration, and (4) spatial calibration. The secondary function of ICU_IS Mode is for absolute radiometric calibration. The main function of ICU_S Mode is for absolute radiometric calibration. The secondary functions of ICU_S Mode include (1) relative radiometric calibration, (2) spectral stability check, and (3) spatial stability check.

Additional in-flight calibration reference sources are moon, flat-field and dedicated targets on the Earth. For these reference sources, specific spacecraft maneuvers are needed. These sources are observed through the instrument main port, giving a direct measurement of the main optical path transmittance and allowing simultaneously and independently a cross calibration of the ICU optical path transmittance.

9.2.7 EnMAP

The EnMAP hyperspectral instrument is equipped with an onboard calibration facility in order to achieve highly stable and precise data for scientific applications and achieve spectral calibration accuracy better than 0.5 nm in the VNIR and 1.0 nm in the SWIR and to determine the sensor linearity (Sang et al. 2008).

Figure 9.3 shows the onboard calibration units of the EnMAP instrument. The main sphere with white Spectralon (polytetrafluoroethylene) is installed with several light sources to provide various stable light illumination and the spectra for the different calibration tasks. Tungsten halogen lamps driven by precision current sources provide smooth broad spectrum radiation. White LEDs are added to enhance the short wavelength light output of the sphere, an approach that has been demonstrated to be viable for spectrometer calibration in space for relative radiometric calibration.

When performing in-flight spectral and relative radiometric calibration, the slit assembly (double entrance slit) of the spectrometers is homogeneously illuminated by the radiation light coming from the output port of the integrating sphere, as shown in Figure 9.3a. The light from the sphere is imaged onto the beam-splitter using a dedicated optical system with a mirror mounted to the shutter mechanism (see Figure 5.34 in Section 5.5.1).

For spectral calibration, a second smaller sphere with doped Spectralon couples light with a strongly structured spectral flux to the primary sphere. Many distinct and stable spectral features are created by combining light from tungsten filament lamps with rare earth oxide doped Spectralon as a diffuser material. All light sources are featured multiple times for redundancy and to monitor degeneration. For linearity measurements, constant irradiance, produced by focal plane LEDs, is recorded with different integration times (Kaufmann et al. 2016).

For absolute radiometric calibration the sun is used as a source of reference spectral irradiation and correct spectral composition. A sun-illuminated diffuser placed in front of the telescope reflects radiation entering the sun calibration port into the system aperture. Acting as a diffuse reflectance standard the full entrance pupil is simultaneously illuminated by the diffuser for all field points. Figure 9.3b illustrates the arrangement. The diffuser panel is mounted on a mechanism which allows it to be rotated in front of the telescope aperture for system calibration. Care is taken to keep residual specular reflections from the diffuser from entering the FOV of the telescope. The baffle design for the Sun calibration port is such that stray reflections on the panel are avoided. Spectralon has been selected as diffuser material since it has been demonstrated to be a stable

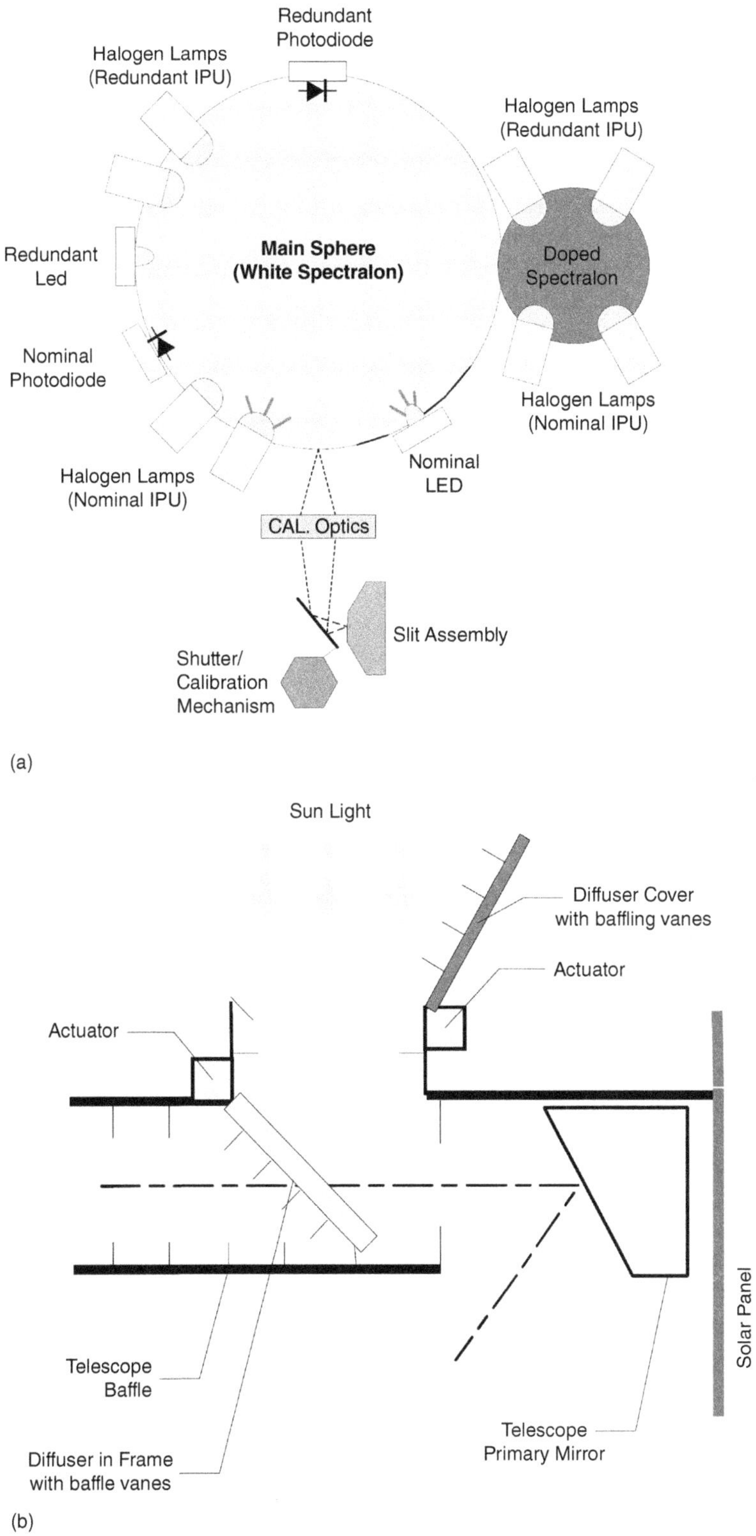

FIGURE 9.3 Onboard calibration units of the EnMAP hyperspectral instrument. (a) For absolute radiometric calibration, (b) for spectral and relative radiometric calibration. (Courtesy of DLR.)

standard of diffuse reflectance in a space environment. In the storage position of the mechanism the diffuser will be sheltered by a protection cover to prevent constant UV exposure causing premature degradation.

9.2.8 HISUI

The requirement for absolute radiometric accuracy is ±5% in both VNIR and SWIR regions, and ±2% between spectral bands. The requirement for spectral band position accuracy is 0.2 nm in VNIR region and 0.625 nm in SWIR region. To achieve these requirements HISUI hyperspectral imager is equipped with a partial-aperture onboard calibration unit, as shown in Figure 2.20. The calibration unit includes halogen lamps and multiple filters with known and stable spectral characteristics as well as a filter wheel. The filters are mounted on a filter wheel. Vicarious calibration at selected sites across the world, lunar calibration and cross calibration with other satellites will be used (Matsunaga et al. 2016).

To achieve this spectral calibration requirement, a trade-off study has been carried out and a calibration strategy has been selected (Tatsumi et al. 2010). The trade-off studies have been carried out on several candidates of wavelength standards for spectral calibration:

1. Solar Fraunhofer lines plus diffuser
2. Hg, Ar, or Xe discharge lamps that exhibit atomic emission lines
3. Light emitting diode (LED)/laser diode (LD) with single-mode spectrum (Aoyanagi et al. 2009)
4. Combination of spectral calibration filters and illumination lamp (Lockwood et al. 2006, Itten et al. 2008, Sang et al. 2008)

Fraunhofer structure in the solar spectrum has many strong features in the UV and visible wavelength range between 300 nm and 900 nm, but has much less features in SWIR region. The spectral bandwidth of Fraunhofer lines is relatively narrow, thus it is difficult to identify the absorption peak wavelengths using the broad sampling data.

For Hg, Ar, or Xe discharge lamps, there are many absorption spectral feature lines, but their emission level is low in wavelength between 2000 nm and 2500 nm. High voltage for discharge is required. Therefore, it is inconvenient for them to be used in space application.

LED/LD needs very precise temperature control in order to stabilize the wavelength of the emission as the wavelength standard. Emission level of the LED/LD is low in SWIR region.

The filter of special rare earth material defined by the National Institute of Standards and Technology's Standard Reference Material 2065 (NIST SRM 2065) exhibits sharp absorption peaks at the specific wavelengths. As a results of trade study, this filter is selected to determine the spectral stability and to trace spectral shifts of center-wavelength position of the HISUI spectral bands. The NIST SRM 2065 filter is certified for the locations of the 7 absorbance peaks in the spectral region from 0.970 μm to 1.946 μm and 13 additional absorbance peaks spanning the spectral region from 0.334 μm to 0.805 μm, which falls in the VNIR region. The transmittance function of wavelength of NIST SRM 2065 filter spectrum is shown in Figure 9.4.

However, as shown in Figure 9.4, the NIST SRM 2065 filter does not exhibit narrow band absorption peaks in the wavelengths beyond 2.000 μm. After having surveyed several transparent materials with absorption peaks around 2.400 μm, it was found that Mylar film exhibits absorption peaks at wavelength of over 2.000 μm (DuPont 2003). Figure 9.5 shows the transmittance spectrum of Mylar film. This film has the specific absorption band at 2.443 μm and two other weak absorption peaks at 1.6595 μm and 2.130 μm. There are no absorption peaks in the VNIR region.

As a result of the trade study, the combination of the NIST SRM 2065 filter and Mylar film is selected. A quartz tungsten-halogen lamp (QTH-lamp) is used as light source to illuminate them.

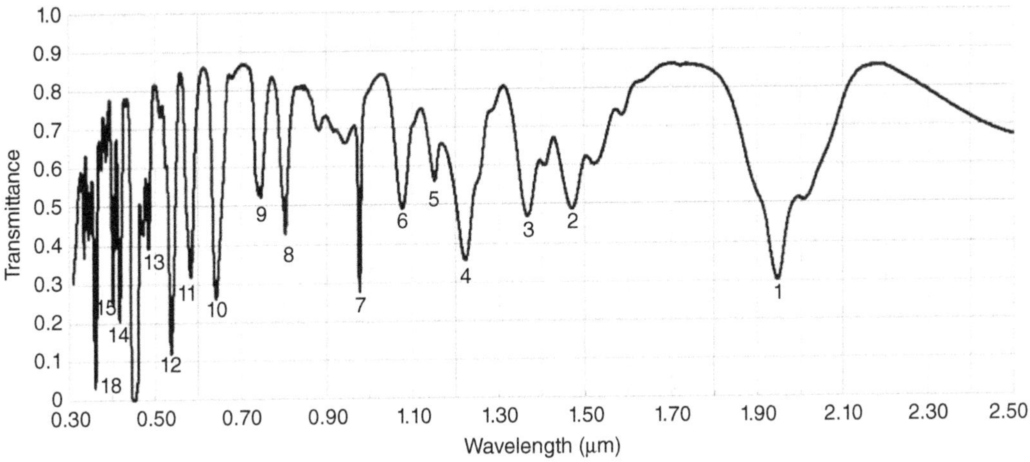

FIGURE 9.4 Transmittance as a function of wavelength of NIST SRM 2065 calibration filter.

Calibration optics of the HISUI hyperspectral instrument is composed of a small folding mirror, QTH-lamp and a filter wheel. The filter wheel is mounted with five filters, including the SRM 2065 filter with Mylar film and four band pass filters. The calibration light enters the sensor from the front of the TMA telescope, as shown in Figure 2.20. Calibration data will be periodically acquired in the night time with an interval of several tens of days.

9.3 ONBOARD RADIOMETRIC CALIBRATION TECHNIQUES

There are many approaches to onboard radiometric calibration. These approaches can be classified into three general categories: solar-based, man-made radiative source-based and vicarious calibrations.

9.3.1 Solar-Based Calibration

Sun can be used as a radiometric reference source. Diffusion can be on a panel or in an integration sphere. The advantages of using the sun is that its overall curve follows the baseline of the reflected signal of scenes. It does not consume power other than the power necessary to flip the injection optics. However, the sun's radiation is variable. In addition to the 11 years cycle and the annual

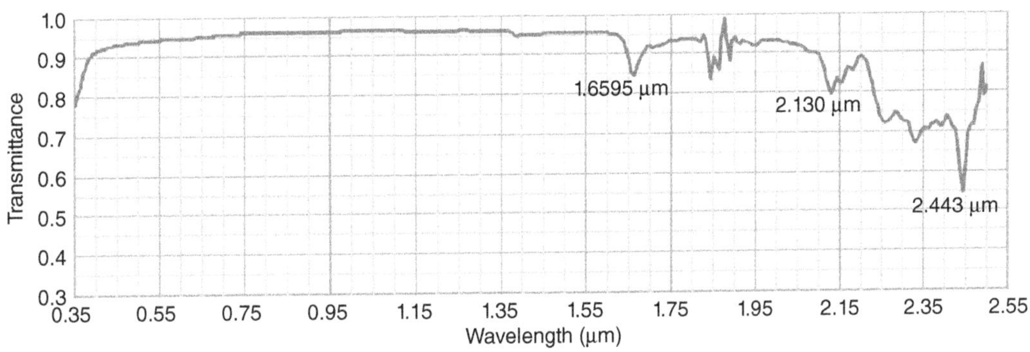

FIGURE 9.5 Transmittance curve of Mylar film.

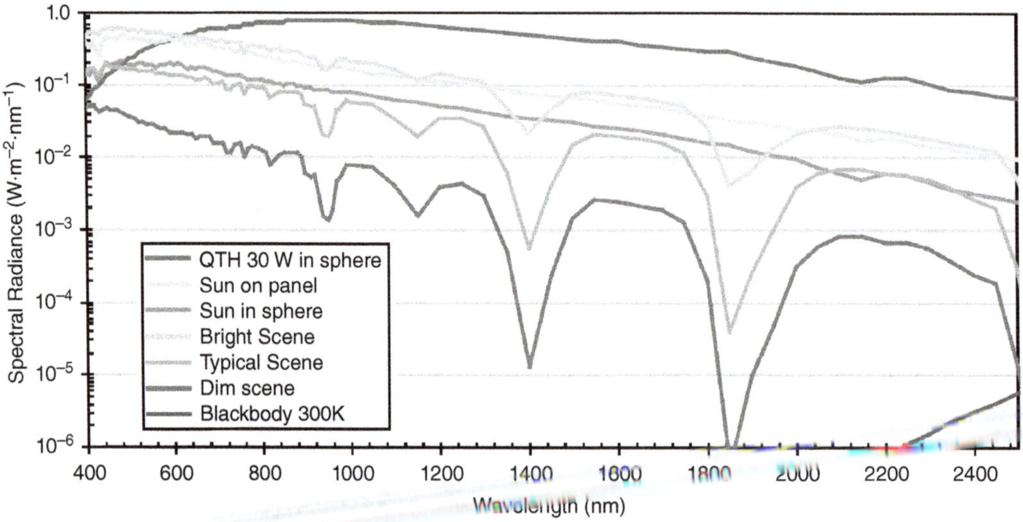

FIGURE 9.6 Comparison of radiance level of various reference sources and scenes.

cycle, there are random fluctuations at the surface of the sun that influence its radiant power. To be used as an absolute radiometric reference, data from external sources (e.g., solar observatories, some climatology satellites, etc.) are necessary.

The sun is also an important source of UV, which is known to have an impact on several optical materials. For instance, Spectralon, a common diffusing material is degraded by UV. To preserve the accuracy during the mission life it is necessary to monitor this degradation with a secondary reference (e.g., a lamp, radiometer, etc.).

Means of bringing the sun to the instrument must also be devised. This can be done by direct pointing of the instrument or of the platform, directly through a different optical port, or indirectly with a fiber optics. The ability to have a sun view (direct or indirect) must also be considered because of the inability to point the platform and potential obscuration by other equipment on the optical bench.

Figure 9.6 shows the spectral radiance of the sun diffused on a panel and diffused in an integration sphere in comparison with three scenes (typical, bright, and dim). The radiance of a diffused quartz tungsten halogen (QTH) lamp and of a blackbody at 300 K are also shown. From the figure, it can be seen that the diffused sunlight is less bright than the bright scenes. A diffused QTH lamp of 30 W has a higher radiance beyond 600 nm than the diffused sun.

9.3.1.1 Solar Diffuser

Solar diffusers (SDs) are a commonly used onboard calibration device. They are capable of providing a full system calibration including all the imaging optics while filling the full aperture and field-of-view of the sensor (Slater et al. 2001) if they are inserted in front of the frontmost optical component in the optical path of the instrument. Use of this technique also implies that the diffuser is illuminated by the same source as the Earth providing appropriate radiance levels for the visible through SWIR wavelengths.

Errors associated with using a SD for on-orbit calibration can be classified as the following three categories (Krause et al. 2002):

1. Errors that affect irradiance on the diffuser. These errors are either due to uncertainties in the knowledge of the:
   i. Exoatmospheric solar irradiance (ESI)
   ii. Geometry

2. Errors that affect the reflected radiance from the diffuser. These errors arise from:
   i. Uncertainty in the laboratory measurement of the diffuser
   ii. Assumption of spatial uniformity of diffuser
   iii. Degradations of the diffuser
   iv. Errors in the angular position of the diffuser
3. Errors due to stray light contributions

The uncertainty in total solar irradiance (TSI) of the sun has been determined to be ±0.15% as derived from a composite record compiled between 1976 and 1998 (Frölich and Lean 1998). This is negligible compared to the accuracy requirement of most sensors. However, the modeled solar spectral exoatmospheric irradiance has been determined to have an error of 1.0% in the VNIR (Krause et al. 2002). A comparable value for the SWIR spectral region is 3.0% (Thuillier et al. 2015). Implementations of solar monitoring, modeling, and reporting procedures are expected to further reduce the uncertainty in the modeled solar spectral exoatmospheric irradiance (Krause et al. 2002, Nieke et al. 2002). Perhaps of greatest importance is the selection of the solar exoatmospheric model that is applied to the data, an aspect of calibration that is often overlooked but is critical when comparisons of data acquired from different sensors is undertaken (Thome et al. 2001).

Geometry can affect the expected incident solar irradiance in a number of ways. The error in assuming the sun approximates a point source in the determination of the irradiance striking the panel is negligible. As well, accurate ephemeris tables allow the effect on the calculated irradiance of the Earth-Sun distance variations throughout Earth's orbit around the sun (~3.5% as described in Section 8.2.1) to be removed with negligible error. However, small errors in satellite pitch and yaw will add a noticeable cosine effect in the modeled irradiance striking the panel. A satellite roll error will not affect the sun-diffuser-sensor geometry, therefore not impacting the observed radiance assuming negligible scattered light effects (Krause et al. 2002). For radiation from a perfect point source striking a panel with a 45° angle of incidence (AOI), an error of 0.1° will result in a 0.17% error in the expected incident irradiance.

Determining the uncertainty in the expected reflectance of the SD has proven to be difficult as the quoted values vary wildly. For example, the Multi-Angle Imaging Spectroradiometer (MISR) is a scientific instrument on the Terra satellite launched by NASA on December 18, 1999. The MISR tests reported uncertainties in the determination of the bidirectional reflectance factor (BRF) of its Spectralon panel as 2.9% absolute and 1.4% relative uncertainties (Bruegge et al. 2001).

MODIS satellites use SD to perform on-orbit calibration of the reflective solar bands (RSBs). The SD panel was made of space-grade Spectralon with a near-Lambertian reflectance profile in the reflective solar spectral regions. Its BRF was characterized prelaunch by the instrument vendor using standard reference samples traceable to the NIST (National Institute of Standards and Technology) reflectance standards. MODIS SD BRF characterization was made at a number of wavelengths from 0.4 μm to 1.7 μm and different solar illuminating angles that consists of 3 elevation and 3 azimuth angles. The SD viewing angle was fixed to be the same as that used on-orbit. Due to measurement limitations, the SD BRF at 2.1 μm had to be derived from the total integrated scattering measurements and the BRF measurements at other wavelengths. A two-dimensional second-order polynomial function was used to fit the measured BRF values for each wavelength. The SD BRF for a given wavelength is determined by an interpolation of the polynomial values.

MODIS studies reported <1.0% absolute uncertainty in its Spectralon diffuser BRFs for all but the longest wavelength channel (2100 nm) (Pavlov et al. 1998, Walker et al. 1998). These are considerably larger than those reported by the MERIS team. They reported uncertainties in the absolute knowledge of their Spectralon diffuser panel BRDF to be 0.49%, 0.26% for the relative spatial uncertainty and 0.39% for the relative spectral values (Chommeloux et al. 1998).

The differences in the magnitudes of the uncertainties reported by the different mission teams appear to be the results of a number of factors. The MODIS and MISR calibrations were performed with reference to a NIST calibrated BRF sample. The sample had a reported uncertainty of 0.5% in its absolute BRF. Another 0.5% is added to the overall error due to the process of transferring that to the sample under test. The MERIS team used an alternative experimental method to determine the BRDF as opposed to BRF. As BRDF is a measure of the ratio of the incident irradiance to reflected radiance, a technique was used that did not require an absolute standard to be used, eliminating these two errors. As well, the effect of diffuser nonuniformities and spectral variations, included as part of the MODIS/MISR uncertainties, were reported separately by MERIS. MODIS reported the spatial nonuniformities as 0.7%, MISR as 1.0% whereas MERIS reported them as 0.5%. Spatial nonuniformities on the S13G/LO sample were reported at 0.3%. As for the spectral (wavelength) variations, MISR and MERIS were stated as 1%, MODIS was stated as 1.2% (Xiong et al. 2016).

The degradation of candidate diffuser materials in general, and Spectralon specifically, has been the subject of much investigation (Guzman et al. 1991, Bruegge et al. 1992, Hilsenrath 1994, Markham et al. 1997). Degradation of the reflectance of most materials, including Spectralon, is expected due to vacuum ultraviolet/ultraviolet (VUV/UV) radiation, atomic oxygen exposure and proton bombardment. UV luminescence of surface contaminants will also impact the observed reflectance. The degradation of a SD is wavelength dependent with the largest degradation occurring in the shortest wavelength region.

The inclusion of SDSM as part of the onboard calibration of the MODIS instrument has provided for the first time an actual measurement of the degradation of a Spectralon solar panel on an EO instrument. Monitoring results obtained after over 12 years of service of the Aqua satellite indicate a 20% degradation of the diffuser reflectance in the shortest wavelength region (412 nm) of the MODIS Aqua (Xiong et al. 2016). This degradation value improves to 8.0% at 531 nm, 4.0% at 645 nm and 1.4% at 858 nm.

The SD's measured degradation continues to decrease with increasing wavelength to the 936-nm band where no measurable degradation is observed. This is consistent with the ground-based measurements on Spectralon samples that showed little to no degradation beyond 900 nm. The negligible degradation observed in the SWIR wavelengths is significant in that an inter-band calibration technique may be capable of monitoring the degradation of the SD in the VNIR channels with respect to a stable diffuser BRDF in the SWIR channels. Of considerable interest is the stated uncertainty of 0.5% reported by the MODIS team in uncertainty of Spectralon degradation levels based upon their SDSM results (Guenther et al. 2002). This may be used as a best case for modeling accuracies of the panel degradation.

These results were based upon a panel deployment strategy that deployed the panel once a week for the first 2 years on-orbit followed by once every 2 weeks for the remainder of the reported 2.5 years. When not deployed for solar calibrations, the SD panel was stored away in a manner to reduce exposure to the degradation sources. Of considerable importance is the fact that the degradation is a well-defined function. This suggests that the temporal trend of the degradation for each spectral band can be predicted and applied to the calibration procedures.

MERIS and OLCI use erbium-doped Spectralon diffuser to perform onboard calibration. The BRDF of the radiometric diffuser had been extensively characterized prior to launch using a dedicated BDRF bench to an absolute accuracy of better than 1%. The impact of exposure to atomic oxygen, proton bombardment and UV/VUV radiation on the diffuser optical characteristics had been tested on ground. The degradation of the material was estimated by the evolution of hemi-reflectance data measured with a spectrophotometer equipped with an integrating sphere over the range 200–2500 nm. No evolution could be found after exposure to atomic oxygen and protons irradiation. However, decrease in the hemi-reflectance was observed on all samples exposed to UV/VUV radiation. This evolution occurs at short wavelengths below 550 nm. From 550 nm to 250 nm the hemi-reflectance reduces to 0.94 for an exposure of 20 h UV/VUV radiation, and from the same spectral range the hemi-reflectance reduces to 0.87 for an exposure of 100 h of UV/VUV radiation. The equivalent exposure duration over the 10 years of on-orbit MERIS lifetime is about 6.8 h. The small evolution of the diffuser hemi-reflectance, predicted to stay below 1.0% during MERIS mission life, has been confirmed by analysis of in-flight data. The remarkable stability of the diffuser has been obtained thanks to a specific post-production cleaning process applied to reduce the degradation (browning) of its scattering characteristics when exposed to space environment. The diffusers were thermally baked out in high vacuum (<10^{-6} mbar) during 68 h and at 90°C.

The diffusers mentioned above were all fabricated with space-grade Spectralon. Spectralon is one of three materials identified as best for use in on-orbit SDs (Krause et al. 2002). The other two are thermal control paints YB-71 and S13G/LO. YB-71 has been used for onboard calibrations on SeaWIFS and ETM+ (Markham et al. 1997). The reflectance of this material decreases quickly at wavelengths beyond 1000 nm (90–92% for wavelengths shorter than 900 nm, linearly decreasing to <65% by 2300 nm) (Markham and Barker 2003), especially when compared to that of space-grade Spectralon (>95% through 2500 nm). UV exposure testing shows slightly less degradation on Spectralon in VNIR than the similarly exposed YB-71 sample (Bruegge et al. 1992). If being considered, the shape of the BRDF of these paints should be compared with that of Spectralon. Since errors in the knowledge of the exact geometries of the panel to the incident solar irradiance as well as the sensor view angle are significant contributors to the total uncertainty of this method, it would be of considerable interest to perform sensitivity analysis on the BRDFs of the different materials for the proposed geometries.

S13G/LO is a silicon-based, white, diffuse reflecting, thermal control paint. Like Spectralon, the material exhibits near-Lambertian properties that have been modeled with an uncertainty of 2% in the VNIR and spatial nonuniformities stated as 0.3%. With uncertainties similar to that of Spectralon, the ease of use and small mass associated with the use of either paints make them an attractive alternative, particularly for a low-mass spaceborne instrument. However, the level of UV degradation that has been shown to occur below 700 nm was not well defined. The degradation of the BRDF will, unlike Spectralon, primarily occur during exposure to the sun early in the spacecraft lifetime. In addition, significant spectral features were noted in the SWIR region. These features appear to be enhanced by UV exposure.

Hyperion employed an onboard SD technique that used the S13G/LO diffuser panel, which was mounted on the backside of the telescope cover (Jarecke et al. 2001, 2002). The published papers on Hyperion note that the most uncertain part of the solar calibration is the correction for the spectral diffuse reflectance of the paint.

The sensitivity of the BRF to the uncertainty of the view angle is a function of the BRDF for a given illumination angle and the nominal view angle. The more Lambertian a surface is, the less sensitive it will be to an error in the reported view angle.

Errors due to stray light contributions can be significant and are the most difficult to analyze. They tend to be different for every sensor. Stray light in telescope optics must be taken into account and a first-order analysis of platform reflections and earthshine on the diffuser must be undertaken. Earthshine depends on time of day, date of year, location of sensor on orbit relative to Earth and Sun.

The accuracy of using a SD calibration method is extremely sensitive to the stray light. An analysis of the impact of stray light effects on the accuracy of the SD method indicates that errors involved in calibration with a SD system alone are in the order of 15% (Palmer 1991). This value was reiterated 10 years later by Slater et al. (2001). The analysis undertaken on the use of a S13G/LO diffuser suggests that if accurate analysis of stray light is performed and the diffuser door position is accurately known, absolute at-sensor radiance with 5% uncertainty can be achieved with 3% possible (Krause et al. 2002).

9.3.1.2 Solar Diffuser With Ratioing Radiometer

The SDSM equipped in MODIS satellites (Guenther et al. 1996), described in Section 9.2.3 for tracking the on-orbit degradation of the Spectralon, is an implementation of the technique of SD with ratioing radiometer, which was originally presented by Palmer and Slater (1991). At that time, they suggested that the inclusion of a ratioing radiometer with the SD can reduce the uncertainty of calibration from 15% to 2.1% with respect to the solar input. The prime manner in which the error is so drastically reduced is by removal of the effects of scattered light. Following work had suggested that estimations of the uncertainties are hard to make as they will be heavily dependent on the exact sensor configuration (Slater et al. 2001).

The ratioing radiometer technique used with the SD can be used in two operational modes (Slater et al. 2001). In the first mode, it is assumed that the stray light is negligible. In this mode it is possible to calculate the diffuser radiance directly from knowledge of the exoatmospheric solar irradiance, the BRDF of the diffuser and the angle of incidence. To use this method, it is essential that the reflectance of the panel remains unchanged between its prelaunch characterization and its initial calibration on-orbit. All subsequent calibrations are performed with respect to the initial on-orbit calibration (sun-to-diffuser ratio). It is also important that (1) stray light does not enter the field-of-view of the radiometer when it observed either the sun or the diffuser and (2) degradation of the diffuser is not accompanied by changes in the shape of the BRDF unless the angles of incidence of the radiometer and the sensor at the diffuser are equal.

In the second mode, the stray light is anticipated. The diffuser radiance can be calculated by referenced to a prelaunch calibration of the sun-to-diffuser ratio. This ratio may be determined by accurately measuring the solid angle of the radiometer's field-of-view and the diffuser reflectance and, if possible, conducting a solar radiation based calibration (SRBC). In this mode the assumptions noted as (1) and (2) in the no-stray-light mode are also of importance. However, the requirement that the diffuser does not change in reflectance from prelaunch calibration is not an issue.

In the case of MERIS and OLCI hyperspectral sensors, there are two SDs: diffuser-1 for calibration and diffuser-2 for checking. By monitoring the ratio of the instrument response to both diffusers, an estimate of the degradation of the frequently used diffuser-1 with respect to the reference diffuser-2 can be determined.

After almost 10 years in orbit, diffuser-1 of MERIS had been exposed to approximately 370 min of solar radiation, and showed a degradation of approximately 2% at 412 nm. According to the diffuser's qualification report (Delwart et al. 2007), the *browning* of the Spectralon diffusers follows a linear degradation trend for low to medium illumination durations. This was confirmed on-orbit, where the reduction in reflectivity is expressed per orbit.

As diffuser-2 had been exposed approximately 37 min, i.e., about 10 times less than diffuser-1, with the assumption that the degradation was due to exposure to solar illumination only. It was inferred that diffuser-2 had suffered a degradation of less than 0.2% after almost 10 years of the MERIS mission.

9.3.1.3 Integrating Sphere With Solar Illumination

An onboard integrating sphere illuminated by solar irradiance was proposed in the ARIES Phase A study report (Auspace Limited 1997) and introduced for consideration in the Canadian hyperspectral initiative (Duggan et al. 1998). In order to introduce the calibrated light into the optical chain of the hyperspectral instrument, the solar illuminated integrating sphere is located behind the primary mirror and project through a slit on the mirror. The light is then reflected back into the system by a collimating mirror placed on the back of the aperture cover while in the closed position. A second co-aligned monitoring integrating sphere is included for less frequent absolute calibration checks. This technique requires a complicated optical design with a reflective primary element. The technique was quoted to have a 4% error in absolute radiometric calibration.

9.3.2 Radiative Source-Based Calibration

Radiative sources for calibration can include QTH lamps, LEDs, laser diodes, and blackbodies.

9.3.2.1 Diffuser Panel With Lamps

The main issue relating to the use of an onboard calibration comprised of a lamp illuminating a diffusor panel is the stability of the lamp assuming that the panel is protected from the space environment and the associated degradations. Lamps are subject to significant degradation during launch and in the early mission lifetime (Slater et al. 2001). Effects due to gaseous convection in zero-g and vacuum conditions are among the problems encountered in the transfer of ground-based lamp calibrations to the space environment. Differences in the lamp color-temperature and the lack of Fraunhofer lines create additional problems in the use of lamp illuminated calibrations. When used, lamp-based onboard calibration systems are often configured to fill only part of aperture and are used for short-term (intra-orbit) relative calibrations when direct access to solar illumination is unavailable or awkward. Additional problems encountered with the use of lamps include providing stable power supplies, allowing for lamp stabilization, lamp aging, and difficulties in checking the stability of the lamp output in space. Their use as absolute calibration source has been avoided for these reasons.

Hyperion reported on the use of a number of low wattage QTH lamps to illuminate the SD in the closed position located on the back of the telescope cover. Prelaunch tests showed the lamps degrade from 1% to 6% over the initial 50 h of operation and 4–17% after 150 h. The operational lifetime of the lamps was 146 h. One of the initial 6 lamps selected for use on the flight unit failed for undetermined reasons during prelaunch thermal vacuum tests. Once on-orbit, the lamps exhibited an increase in brightness consistent with what was expected due to the loss in convective cooling between the filament and glass envelope experienced in the micro-g environment of space. This increase can be modeled by an increase in filament temperature of 100°C. An additional decrease of 3% with respect to the SD calibration had been observed in the first 140 days of on-orbit operation.

9.3.2.2 Integrating Sphere With Lamp Illumination

The SRCA of the MODIS satellites contains an integrating sphere with lamp illumination. The integrating sphere is illuminated by a number of low wattage incandescent lamps. However the SRCA only illuminates 25% of the entrance aperture making it of limited use for radiometric observations (Xiong et al. 2016). A radiometric calibrator of this type is subject to the same lamps' degradation problems as noted in Section 9.2.3 in the discussion on using lamps to illuminate panel diffusers.

The Compact Reconnaissance Imaging Spectrometer for Mars (CRISM) has an in-flight calibration system equipped with an integrating sphere illuminated by lamps. The integrating sphere provides a smooth, near-flat field of dispersed light as viewed through the VNIR and IR spectrometers. It is the primary in-flight reference for radiometric calibration. It samples all of the optics except the telescope, and is lined with sandblasted aluminum and has a slightly vignetted rectangular aperture. Illumination is provided by either of two small incandescent lamps, one controlled by the VNIR focal plane electronics that is located in the plane of the slit off to one side, and another one that is controlled by the IR focal plane electronics that is located orthogonal to the slit off to one side. The VNIR-controlled lamp is the primary bulb for in-flight calibration because its orientation generates less cross-slit brightness gradient, minimizing effects of small irreproducibility in the shutter's position viewing the sphere. To maintain lamp's brightness stability and minimize the lamp degradation during the mission life a closed-loop control is adopted. A silicon photodiode installed inside the sphere measures the brightness of the lamp and adjusts the current to match a brightness goal. There is one photodiode per lamp, controlled by the same focal plane electronics that control the lamp. A single closed-loop set point (~90% of maximum) is used in flight, which adjusts lamp's current up or down to reach a commandable brightness goal. Each lamp runs at ~2000 K at its set point. Convolved with the efficiency of the optics and detectors, signal in digital number/second as viewed at the detector peaks near 2200 nm. Under closed-loop control, either of the sphere lamp's brightness stabilizes in 3–4 sec after it is turned on. For the same sphere temperature, it repeatedly achieves the same current to within one part in a thousand. The brightness of the sphere exhibits slight and systematic variation dependent on its temperature (Murchie et al. 2007).

The EnMAP is also designed having an integrating sphere illuminated by lamps for onboard calibration, as shown in Figure 9.3. Homogeneous light from the output port of the sphere illuminates the spectrometer entrance slits and is imaged onto the field splitter unit using a dedicated optical system with a mirror mounted to the shutter mechanism. The calibration system features three positions for (1) Earth observation and absolute radiometric calibration (telescope input), (2) spectral calibration (sphere input), and (3) dark current measurement (shutter). The sphere (white Spectralon, i.e., polytetrafluoroethylene) is equipped with several light sources and monitoring photodiodes to provide various stable illumination and the spectra for different calibration tasks. Four Tungsten halogen lamps (including two redundant ones) are installed, which are driven by precision current sources and provide smooth broad spectrum radiation. Due to the spectral range of the lamps mismatch with the solar spectral, two LEDs (including a redundant one) are added to make up the short wavelength output of the sphere (Kaufmann et al. 2016).

9.3.2.3 Light Emitting Diodes and Laser Diodes

Light emitting diodes (LEDs) compared to QTH have the advantage of long life and high efficiency. They are also relatively mechanically robust. However, their tolerance to space environment, in particular radiation, need to be assessed. Laser diodes have similar structures a LED. The space qualification of LED should not be a problem because of their similarity to laser diodes like QTH, the output of LED, both its amplitude and its spectral shape, depends on the current that drives the LED.

It is possible to use an array of LED to directly illuminate the instrument without a diffuser; however, it is not sure that a good homogeneity can be achieved and maintained this way.

As most LEDs emit over a limited range of wavelength, several LEDs of various colors would be required to calibrate both the VNIR and SWIR spectrometers of a hyperspectral instrument, since even broadband LED, such as white LED, cannot cover the complete spectral range of a hyperspectral instrument. A tunable LED that covers a wide range of the visible spectrum has been devised by the NIST as a radiometric reference (Brown et al. 2002). This may have potential to be used as a reference for an onboard calibration. Simulations demonstrated that such a source would be able to approximate standard light-source distributions over the visible spectral range, from 380 nm to 780 nm, with deviations on the order of 2%. The tunable LED source can also simulate spectral distributions of special sources such as discharge lamps and display monitors.

9.3.2.4 Blackbody

Blackbody sources are commonly used in the infrared (especially in the mid-wave and long-wave infrared) for radiometric calibration. High-precision blackbodies are particularly suited for this task as they generate a simple spectral radiance profile close to the Planck function. Their radiance is primarily dependent on their temperature. A different temperature gives a different spectral radiance profile that does not overlap or cross the profiles obtained at other temperatures.

With a combination of geometry and coating, blackbodies having emissivity >0.9985 have been achieved. The temperature of the cavity can be monitored with precision platinum resistance thermometers (PRTs) with accuracy better than 0.03 K. As large opening can be achieved, a diffuser would not be necessary. With these devices, a very good radiometric accuracy can be achieved. Some blackbodies using these cavities have estimated radiometric accuracy better than 0.6% after 5 year in space. However, to obtain a significant signal, even in the SWIR, a very high temperature would be necessary. The required power would be prohibitive. Hence, their usage as bright radiometric reference is essentially confined to the thermal infrared.

9.3.3 Vicarious Calibration

9.3.3.1 Advances of Vicarious Calibration

An important method for in-flight calibration is often referred to as vicarious calibration (VC). Vicarious calibration refers post-launch calibration that makes use of well-characterized "invariant" natural targets of the Earth or lunar scenes. The invariant target approach enables the definition of a reference standard for the observed reflections.

Vicarious calibration techniques have been becoming popular on-orbit calibration procedures over the past decades. Onboard vicarious calibrations are devoted primarily to the monitoring of the radiometric responsivities or gain coefficients of the sensor spectral bands over time.

From a technique that allowed rough validation over two decades ago, it is now often considered as one of the primary methods of performing absolute calibration. This change in approach has come about for a variety of reasons including better understanding and reduced measurement uncertainties of the vicarious calibration targets, atmosphere characterization and modeling, as well as a better understanding of the difficulties in transferring and maintaining a calibration target in the harsh environment of space.

The POLDER sensors (both POLDER-1 and POLDER-2), SPOT 1-5 and AVHRR have no onboard calibration system at all and rely totally on vicarious calibration techniques. POLDER claimed an uncertainty of better than 3% in its radiometric calibration coefficients for the ocean color bands (443, 490, and 565 nm) based on a strategy of vicarious calibration techniques (Fougnie et al. 1999).

The HICO did not have an onboard calibration system and employed vicarious radiometric calibration (Franz et al. 2007). It acquired imagery from homogeneous sites (e.g., Railroad Valley Playa in Nevada) for vicarious calibration (Kappus et al. 2016). HICO relied on comparisons with the MODIS-Aqua and MODIS-Terra instruments, as reported by Gao et al. (2012), to determine corrections to the laboratory-determined radiometric calibration coefficients. In addition, on-orbit spectral calibrations were made by matching HICO-measured spectral data within and around the oxygen absorption band centered near 761 nm with theoretically calculated data (Gao et al. 2012).

Other sensors have grown to either partially or completely rely on vicarious calibration when their onboard calibration failed to provide the desired results. The onboard calibration of the Ocean Color Temperature Scanner (OCTS) aboard the ADEOS satellite was shown to have significant inaccuracies due to degradation of onboard lamps and inadequate results from the onboard SD. The MODIS onboard calibration plan foresaw that the accuracies achieved with the onboard calibration would degrade over time on-orbit, vicarious calibration is utilized to monitor the stability of the onboard calibration and data accuracy (Guenther et al. 1996, Xiong et al. 2002). The vicarious calibration techniques are used not just for confirming results obtained from onboard calibrations (Slater et al. 1996), but also for generation of updated calibration coefficients (Slater and Biggar 1996).

Many VCs and associated support campaigns have been carried out. This is significant because one of the downsides to undertaking VC-based ground measurements is the level of effort required to carry out the field campaign. Cooperation and coordination of these field campaigns will lead to improved understanding of the techniques and the processes affecting the vicarious calibration site.

In addition, cross comparisons between different sensors, although not actual calibrations, will provide independent verification of the sensor calibrations. The uncertainty of such cross comparisons is anticipated to be ±2% when they are conducted over spectrally flat targets and when there is close correspondence between spectral bands (Slater et al. 1996).

9.3.3.2 Ground-Based Vicarious Calibration

Ground-based VC can be used to calibrate spaceborne instruments that do not have an onboard calibration system, or that have a calibration system that failed or that is suspect. It characterizes the surface reflectance of relatively uniform areas within the GIFOV of the instrument with ground measurements and then to estimate the TOA radiance with an atmospheric model. The atmospheric correction is usually supported by a series of measurements to attempt to characterize the atmosphere and its constituents (water vapor and aerosol in particular) at the time of the satellite passage. Cross correlation between sensors is also used sometimes.

The efforts required to correctly characterize the ground site and the atmosphere are significant. It usually involves a team of scientists and several ground or airborne instruments. Since the site surface evolves in time and the atmosphere is ever changing, these efforts are to be redeployed every time a new vicarious calibration is necessary.

Because ground characterization is difficult and costly, the surface is subject to changes (e.g., erosion) and to seasonal variations, and atmospheric correction is notoriously uncertain, the accuracy achievable is likely to be relatively low. For broadband sensors such as Landsat/TM and NOAA/AVHRR, reasonable accuracy levels have been achieved. However, because of their higher spectral resolution, hyperspectral sensors are more sensitive to uncertainties on the aerosol content and to atmospheric correction errors. The achievable accuracy is likely to be less for spaceborne hyperspectral sensors than for broadband spaceborne sensors. This should only be considered in the case of a failure of the onboard calibration system or in the case that there is need to attempt to monitor the evolution of the calibration system.

Three methods have been developed for in-flight ground VC of multispectral imaging sensors in the solar reflective range (Slater et al. 1987, Slater et al. 1996, Dinguirard and Slater 1999) known as:

1. Reflectance based
2. Irradiance based (also known as improved reflectance based)
3. Radiance based

The reflectance-based method requires the accurate measurement of the spectral reflectance of the ground. The ground reflectance is converted to top-of-atmosphere (TOA) radiance using radiative transfer code (2-way) run with measurement data of spectral extinction depth and other meteorological parameters. Accuracies of 3.3% with respect to the solar-exoatmospheric irradiance were achieved in measurement techniques of the ground and atmosphere.

The irradiance-based method uses the same ground measurements as those acquired for the reflectance-based method and an additional measurement that allows the effect of diffuse irradiance to be removed. The additional measurement helps to reduce the uncertainties in the aerosol model used for scattering computations. The uncertainty in this technique is 3.5% with respect to the solar-exoatmospheric irradiance. Improvements in the panel BRF could lead to reduction of this value to 2.8%.

The radiance-based method measures the radiance of a ground target by flying on an aircraft over the target with a well calibrated, radiometrically stable, radiometer. By flying at a high altitude (i.e., >3 km) so that most of the atmospheric aerosols and water vapor are below the radiometer, the measured radiances are very close to the TOA radiances detected by the satellite sensor. This method eliminates the errors due to ground reflectance measurements and the radiative transfer code, although ground reflectance and meteorological measurements are still made in order to allow for radiative transfer corrections be applied to correct for absorption and scattering that will occur between the aircraft and the satellite. Estimates of the uncertainty of this technique are 2.8% with respect to the NIST standard, against which the airborne radiometer was calibrated with the largest contributor to the error being the calibration of the airborne radiometer. The error in correcting the radiance to TOA from the aircraft altitude is estimated at <0.1% (Dinguirard and Slater 1999), eliminating the need for acquisition at higher altitudes. Improvements in the transfer of the NIST standard to the radiometer and in the measurement error are expected to lead to the future reduction of this error to 1.8%. This method provides the most accurate results of the 3 techniques albeit at significantly greater costs than the other 2 techniques.

The uncertainties stated here, once combined with the unknowns in the solar-exoatmospheric irradiance or NIST standard, are consistent with the statement made by Fox (2001) that the estimates of attainable accuracies in VC derived radiances are 5–10%.

Undertaking VC field campaigns has in the past been labor intensive and far from systematic, standardized or permanent. Recent steps have been undertaken to make them more systematic and standardized using new techniques and approaches.

In order to address the above-mentioned issues, it has been proposed by an influential group of EO scientists (P. M. Teillet, K. J. Thome, N. Fox, and J. T. Morisette) that an automated web of in situ sensors referred to as Global Instrumented and Automated Network Test Sites or GIANTS, has been developed (Teillet et al. 2001b). Two systems have already been developed and deployed for the automated monitoring of atmospheric aerosol properties (Holben et al. 1998).

Ground temperature measurements are useful for tracking the TEB radiometric stability. MODIS uses the Dome Concordia, Antarctica (Dome C) site to track the stability and relative consistency of the 15 thermal emissive bands (Wenny and Xiong 2008). Ground temperature measurements of the site are made. Assuming the ground sensors are stable, any drift in temperature difference would indicate a drift in the MODIS calibration. It is seen that little to no drift is evident in either Terra or Aqua MODIS for the thermal emissive bands, which indicates good stability of the calibration. The relative bias between Terra and Aqua MODIS using the ground temperature as a comparison proxy shows that the calibration consistency of the two MODIS instruments is very stable, with no observable drift between the two sensors (Xiong et al. 2016).

9.3.3.3 Lunar Calibration

The moon is the only extraterrestrial object that is accessible to Earth-orbiting imaging systems beside the sun. The key advantage is its presumed high stability due to its age. It is a universal standard and easily available. Brightness of moon is within the dynamic range of most optical satellite systems (reflectance ~11%), and its spectral radiance is potentially knowable to a fraction of a percent (Kieffer et al. 1999). The radiometric stability of the moon is estimated as 10^{-9} per year (Kieffer and Widley 1996) and its irradiance levels, for selected lunar cycles, can be matched to within 0.1% (Godden and McKay 1997). Obviously these characteristics of the moon far surpass the stability of any onboard or terrestrial target and needs to be seriously considered as part of any calibration strategy.

The moon can be used for sensor calibration in one of two ways. It can be used to (Dinguirard and Slater 1999):

- Make direct measurement of absolute sensor radiometric calibration.
- Monitor the stability of an in-flight SD.

In using the moon to provide an absolute radiometric calibration source, the total integrated signal of the lunar face is compared to the expected lunar radiance. The model of the expected lunar radiance is extremely complex. The United States Geological Survey (USGS) has refined the model based on the extensive measurements acquired of the moon over a period of 7 years (Stone and Kieffer 2002, Stone et al. 2002). Acquisition of satellite data of the moon to perform such a calibration requires that the data be corrected to standard distances and for over-/under-sampling. Estimations of the uncertainty in the use of the lunar radiance model for absolute calibration are currently 2–3%.

The methodologies to use the moon to monitor sensor stability are considerably less complex. This procedure requires that the ratio of the total integrated signal of the lunar face to that of a solar panel measurement be monitored. As the sensor scans the moon in exactly the same manner as the SD, changes in that ratio can be attributed to changes in the reflectance of the SD. New reflectance values can then be applied based on the relative changes determined by the ratio. If the initial ratio is undertaken at a time when the absolute calibration of the sensor is known, then that baseline can be transferred to the time of any future lunar calibration. The group at the USGS has modeled the relative lunar irradiance. The relative irradiance, which is impacted by the phase and/or liberation angles of the moon, can be modeled to within 0.5–1% for lunar phase angles between 5° and 90° (Stone and Kieffer 2002).

Unlike performing an absolute lunar calibration, it is not required that the scan rate be exactly known; however, the scan rate of the two lunar measurements would ideally be identical. The sensor must scan the entire lunar disk. Extending the scan beyond the lunar disk will provide data to undertake a precise dark current correction. The signal of the entire disk of the moon, plus any scattered signal that falls outside the lunar disk, is integrated. It is this value that is to be used for comparison. In this way, the structure of the lunar surface does not impact the process and any change in sensor scattering that results in scattering outside of the lunar disk, although definitely a system degradation, will not, as it shouldn't, impact the radiometric calibration.

Lunar calibration are used in many spaceborne instruments, such as SeaWiFS, Hyperion, ALI, MTI, ACS, GMS-5, MODIS, GOES-x, SAGE-3, and VIIRS.

9.3.3.4 Stable Deserts

Stable deserts have been used for multi-temporal calibration of satellite sensors for some time. Areas as large as 100 km × 100 km have been identified with temporal instability of <1–2% over a year prior to application of atmospheric corrections (Cosnefroy et al. 1996). Changes in the observed signal levels of these sites can be used to perform a relative radiometric calibration of an instrument or the onboard calibration system with an uncertainty of <3% (Dinguirard and Slater 1999).

For example, Fourier Transform Hyperspectral Imager (FTHSI) onboard HJ-1A satellite (see Section 2.11 for details) utilized Dunhuang test site in Gobi Desert to validate the radiometer calibration. The Dunhuang test site is the first Chinese radiometric calibration site, which has been used by a number of vicarious calibration campaigns for different series of Chinese satellites. This test site is located at 94.4° east longitude, and 40.08° north latitude in the Gobi Desert. The entire field is a uniformly flat area, without vegetation coverage. The size of the uniform area is about 30 km × 40 km. The atmosphere is dry and clean, and the annual evaporation is great. The experimental results showed that using this test site the confidence intervals of calibration error for all FTHSI bands were between 2% and 12%, and the confidence intervals of 90 bands in 500–900 nm range were more precise, with corresponding confidence interval 2–6% (Gao et al. 2010).

9.3.3.5 Sun Glint

Sun glint is a phenomenon that occurs when sunlight reflects off the surface of the ocean at the same angle that a satellite sensor is viewing the surface. In the affected area of the image, smooth surface ocean becomes a silvery mirror, while rougher surface appears dark. This phenomenon is often undesired as it usually generates a signal that is beyond the operational dynamic range of the instrument.

The technique of using sun glint, applied to POLDER to perform inter-band calibration between the blue bands (<600 nm) and the red/NIR bands, requires that scenes are acquired over water that include a region of specular reflection. A ratio in two spectral bands of the difference in between the signal in the glitter area and the region outside the glitter is determined. The difference in the ratio is expected to be only dependent on the differential atmospheric absorption and scattering, which is easy to model. The view angle must be within a 3° cone of the specular angle and the surface wind speeds at the time of image acquisition need to be between 2 m/sec and 5 m/sec. This may be difficult to achieve with narrow swath width sensor. Accuracies of the inter-band calibration on POLDER with this technique were reported at 1–2% (Dinguirard and Slater 1999, Tobbé et al. 1999).

9.3.3.6 High Altitude Clouds

In the POLDER mission, clouds were used for validation of calibrations undertaken in other manners, including inter-band calibration of POLDER's 5 upper bands using sun glint techniques. The method makes use of the spectrally constant nature of the cloud reflected radiances. By using clouds at high altitudes (>10 km), the impact of aerosols and water vapor, which are concentrated at lower levels, are negligible. Corrections for Rayleigh scattering and ozone must still be performed. The technique is reported to have a 4% uncertainty for inter-calibration of spectral bands on POLDER

(Dinguirard and Slater 1999, Lafrance et al. 2002). As with the sun glint technique, the cloud images must be obtained at specific geometric conditions.

MODIS satellites use deep convective clouds (DCCs) to monitor sensor calibration stability. DCCs are invariant targets that can be easily found over the tropical region. The DCCs are considered to be the brightest targets among all terrestrial surfaces with nearly flat reflected solar spectra and minimal impact of water vapor and aerosols and, therefore, excellent invariant targets to track sensors radiometric stability (Doelling et al. 2004). A simple brightness temperature threshold is used to identify DCC pixels. The method is most successful when a large ensemble of DCC pixels is identified. A predetermined DCC BRDF model is often used to normalize measured radiances to that at nadir. In the case of MODIS, the mode of the monthly probability density function (PDF) is determined and used to track the nadir DCC reflectance over time. The results indicate that both Terra and Aqua MODIS instruments are stable to within 1% and the Aqua DCC trend appears to have a gradual decrease in recent years. The DCC results for band 2 (0.85 μm) are similar to that of band 1. For other unsaturated VIS bands 3 and 4, Terra and Aqua MODIS instruments are stable to within ±1%. In addition to onboard SD, moon, and DCC, other invariant desert sites are often used to monitor MODIS calibration stability of the reflective solar bands (Xiong et al. 2016).

9.3.3.7 Molecular Scattering Over Oceans

This technique has been used to provide an absolute calibration of the blue bands of multispectral imagers such as AVHRR, POLDER, SPOT, and VEGETATION. At these short wavelengths, the signal observed by a satellite over open ocean sites is mainly due to Rayleigh scattering. The TOA radiances over such surface targets can theoretically be calculated. Unwanted signal from aerosol scattering and surface conditions, such as foam, in water reflectance and ocean glint contribute to the image. In order to minimize their impact, images are selected of open oceans (clear water) with large viewing and solar illumination angles (increase atmospheric path), and westerly viewing angle to avoid sun glint. Non-Rayleigh scattering is determined from signal in a NIR band where molecular scattering is negligible. After removing the effect of foam, in-water reflectance and ocean glint through the use of models and meteorological data in the NIR band, the aerosol content is deduced and transferred to the blue bands using aerosol models. Use of the models in this manner means that the accuracy of the results is dependent upon the accuracy with which the spectral bands are known. This technique has been applied to sensors with the appropriate blue bands to within an estimated uncertainty of 2–3.5% (Dinguirard and Slater 1999). For SPOT 1-3 satellites, which did not have an appropriate blue band, a 5% uncertainty is estimated.

9.3.4 Offset Correction

9.3.4.1 Offset Signal of Detector Arrays

Almost all imaging detector arrays produce a dark offset signal, which is present at all times as an additive component to the radiance signal. If a detector array is linear, then the dark offset can simply be subtracted from the raw image frames (a combination of dark and radiance signal), after being determined as a result of dark calibration. The dark signal is almost always a function of temperature and also long-term effects such as radiation, residual radiance from bright objects, detector contamination, and other degradation effects. Periodic or semi-periodic dark signals may be present as well, such as electromagnetic interference (EMI) and random telegraph signals (RTS), respectively. Random fluctuations of the signal (in addition to the normal detector noise) may be present as well, such as single events upsets (SEU) due to high-energy particles.

Dark offset is usually a combination of thermally generated charge or dark current and electronic offset, the offset due to preamplifiers, etc. In a CCD array, the dark current is fairly easily separated from the electronic offset. This is done by setting the exposure time to zero (or nearly so), or extrapolating back to zero time from two or more frames taken with known exposure times.

Basically the electronic offset for a CCD array is common to all pixels read from the same output (and analog chain). However, due to charge transfer inefficiency effects, for example, the apparent electronic offset may not be common to all pixels. Also, even if the electronic offset is found to be common, it may be a function of temperature and also time in the long term, although not as variable as pixel dark currents. Note that changing exposure times result in changes to the CCD clock timing, which could have other implications, often in a complex fashion. It is best, if possible, to always use the same CCD (or other detector type) clock timing, other than for special diagnostic modes. An example of a diagnostic mode would involve readout of the whole CCD array, including the dark storage area, to determine if any such storage pixels are defective and causing problems with readout of image area pixels.

If frame-transfer smear subtraction is implemented, then it will be required to know what the dark charge component of the dark offset is. Effects of changing the exposure time, or other aspect of the CCD clock timing, will be required. The dark current may be low enough such that errors in the determination of electronic offset should be small, and hence errors in frame-transfer smear correction will be small, especially considering that the radiance signal will generally be much larger than the dark charge.

In SWIR detector arrays, which use photodiodes and per-pixel readout amplifiers and multiplexers, the dark current (or charge) is not easily separated from the electronic offset. Setting the exposure time to zero (or nearly so) may introduce nonlinearity that cannot be rationalized (sample-and-hold pedestal due to charge injection, for example). Testing of a candidate detector array will be required to confirm this. Also, each pixel will have a unique electronic offset, significantly affected by temperature and time.

Therefore, in order to keep the dark offset calibration process general, it is recommended that the dark current (or charge) and electronic offset be treated as a single item for the SWIR detector array, and characterized as such. This assumes a constant exposure time for scene acquisition. If different exposure times are required, it is recommended that dark calibration be performed separately for each exposure time.

Below are methods of determining and hence removing the dark offset signal (combined dark current and electronic offset).

9.3.4.2 Frame Method

Most commonly, the dark offset is determined by placing the detector array in a dark condition at a known and stable temperature. There are two basic methods of achieving a dark condition: (1) closing an aperture door or shutter or (2) pointing the hyperspectral imager at a known dark area on the Earth, such as a desert, ocean, or boreal forest on a moonless night. The second method is usually simple to achieve if the swath width is narrow and the dark frames can be acquired in a short period of time. A shutter can be closed at any time for any duration; however, a shutter is a mechanical component that can fail, resulting in mission termination. It is therefore desirable not to use a shutter.

Under a dark condition, several dark frames are collected and averaged to produce a reference dark frame that is relatively free of noise. Otherwise, when a single unaveraged dark frame is subtracted from raw scene data (containing dark plus radiance signal), the dark noise is increased by a factor of root-two (1.414). It must be ensured that no light is present, such as leakage of sunlight from the instrument housing.

Single events need to be detected and removed from the dark data. The simplest method of doing so is to examine each dark frame in comparison to the others or the averaged frame and reject it if there are any significant differences on a pixel-by-pixel basis. Given the high frame rate of a hyperspectral imager, the probability of a single event set of pixels is low enough such that the majority of the dark frames should be without any single events. This is especially true if dark calibration is performed outside the South Atlantic Anomaly (SAA) zone.

Dark calibration is normally performed at periodic intervals throughout the mission. A dark database is kept to monitor changes in dark offset and allow interpolation of dark reference data to be applied to scene data taken at any given time. In general, a detector array will not always be at the same temperature when the dark calibration is performed as when the scene acquisition occurs. Therefore, the temperature effect must be removed by correcting the dark database to correspond to the detector temperature at the time the scene was acquired. The temperature correction requires knowledge of dark offset versus temperature for each pixel. This must be determined as part of prelaunch calibration, where dark offset is determined at a variety of detector temperatures about the expected detector in-orbit operating temperature. The temperature dependence should be updated occasionally during the mission as the temperature dependence of the dark offset will inevitably change over time. An extreme example is the emergence of "hot" pixels. Although truly hot pixels will normally be classed an unusable, "warm" pixels will still be useable if temperature dependent effects are removed. Performing dark calibration at a variety of detector temperatures (at least two temperatures) may require that the temperature control set point be changed for the calibration.

9.3.4.3 Masked Pixel Method

With this method, a set of detector pixels is permanently masked (shielded from light) to achieve a dark condition usually near the edge of the detector array. For each frame (calibration and operational), dark offset data is available for the masked pixels. This allows dark offsets for the illuminated pixels to be better determined than without masked pixels, mostly due to temperature effects.

For example, in the OSIRIS instrument the SWIR detector array (InGaAs linear array with CTIA readout multiplexer) masked pixels were used. In the MERIS instrument, the CCD device is an area detector array with 780 (H) × 576 (V) pixels, but only 740 × 520 pixels are used as imaging area. The central 740 pixels of each line correspond to the cross-track line at a given wavelength. Each column of 520 detector elements corresponds to the spectral components of a pixel in the cross-track line at all wavelengths in the 390–1040 nm range. Five (5) masked columns are used on both sides of the imaging area to monitor the variation of the detector dark current and offset. In addition, fifteen (15) columns between the imaging area and the masked columns are implemented to optically isolate the masked pixels and protect them against charge contamination. Figure 9.7

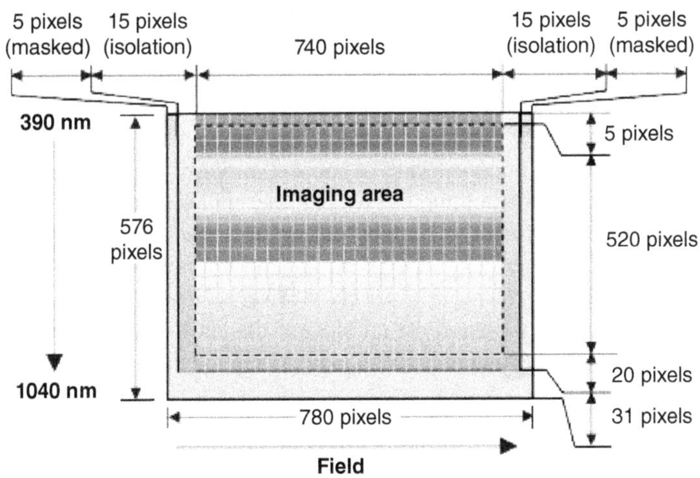

FIGURE 9.7 Masked pixel zones for monitoring variation of the detector dark current and offset in MERIS CCD detector array arrangement.

shows the masked pixels for monitoring variation of the detector dark current and offset in MERIS CCD detector array.

This masked pixel method proved essential in the removal of common but random dark offsets, on a frame-by-frame basis. It is believed that the dark offset variation, common to all pixels in the detector array, is caused by noise on the detector power or clock lines. Detector pixels can be quite sensitive to such noise in unexpected ways. The main drawback of the masked pixel method is rationalizing the measured dark offsets with the measured detector temperature and dark offsets of the unmasked pixels, both pre-flight and in-flight.

Although temperature sensors can be placed at the detector array to allow temperature effects on the dark offset to be corrected, the actual pixel temperatures may not be necessarily what the temperature sensor(s) report. This is especially true where large temperature gradients are present in the vicinity of the detector array, which is expected to be the case for SWIR detector arrays due to the low operating temperature. For example, the leads of the temperature sensor can have significant heat conduction, causing higher-than-actual temperatures to be reported. To minimize temperature measurement errors, the temperature sensor(s) and their leads should be mounted inside the detector package at the detector device. However, detector arrays will normally have temperature gradients over the pixels. Although the temperature of the edge pixels may only be known by measurement, it may be possible to determine the temperature of the center pixels. Unfortunately, the dark offset function of temperature is usually nonlinear, making determination of center pixel temperature and thus dark offset more difficult. As long as the thermal environment of the detector arrays and their surroundings does not change between prelaunch and post-launch environments, then the measured detector temperature versus dark offsets (of each pixel) should not change. This includes potential fracture or delamination of thermal interfaces due to launch vibration. Detector characterization and calibration with respect to dark offset (possibly other calibration) would likely have to be done in a thermal vacuum environment, as opposed to a more convenient purged, atmospheric pressure test setup.

Placement of the mask is important. For a frame transfer CCD array, the mask covering the storage area can simply be extended into the image area. A SWIR detector array requires a mask that is not normally present. Due to illumination angles at the detector array and the required spacing of the mask from the detector surface, the mask will need to cover (relative to the perpendicular) more than the desired number of fully dark pixels. The mask should also be as dark as possible to minimized stray light, and also should be placed under any cover glass of the detector array. The darkest edge of the detector array is the best place for a mask. For the VNIR, this would be the short wavelength (UV) end of the spectrum. For the SWIR, this would be the long wavelength end of the spectrum.

It is possible that the mask, which is usually a thin piece of metal, can cause a different radiation effect on the masked pixels than the unmasked pixels. Therefore, when any radiation testing is done, the test should be done with the mask in place on the test detector array.

The use of a mask helps, as a minimum, verify that thermal changes (or no changes) in particular have occurred during the mission, including initial post-launch check-out. It is quite common for dark offset correction procedures to change during the mission, not to mention calibration in general. It is, therefore, desired to have as much dark offset related information (e.g., temperature, masked pixel data) as possible throughout the mission, so that dark offset correction procedures can be improved and, if necessary, reapplied to previously processed data.

9.3.5 Summary of Radiometric Calibration Techniques

Table 9.1 summarizes radiometric calibration techniques discussed in this section. It compares the pros and cons of the techniques along with the uncertainties associated with the techniques. This includes an initial uncertainty in the application of the in-flight calibration method, and where applicable, the rate of degradation of the method itself.

TABLE 9.1
Comparison of Onboard Radiometric Calibration Techniques

Technique	Pros	Cons
Solar diffusor-Spectralon	• Provide good initial absolute calibration • Appear simple to implement • Well understood properties • Use the same light source as imagery	• Relatively heavy • Large panel needed to fill FOV of large aperture sensor • Subject to stray light errors • Degradation must be monitored
Solar diffusor-paint	• Lighter than Spectralon • Use the same light source as imagery	• Reflectance less than that of Spectralon • Reflectance reduced even more in SWIR • Spectral feature in SWIR • Higher degradation rate than Spectralon • Not as much flight experience as Spectralon • Subject to stray light errors
Redundant solar diffusor	• Good long-term monitoring of absolute • Use the same light source as imagery	• Dual panels, heavier • More complex
Solar diffusor with ratioing radiometer (absolute)	• Excellent long-term monitoring of absolute • Minimize stray light impact • Use the same light source as imagery	• Relative stability of moon equal or better with onboard complexity
Solar diffusor with ratioing radiometer (relative)	• Use the same light source as imagery	• Relative stability of moon equal or better with onboard complexity
Integrating sphere with solar illumination	• Power-free source	• Location of integrating sphere port to accommodate large FOV very constrained • Degradation of first strike wall quickly reduces integrating sphere efficiency • Can be very large
Diffusor panel with lamps	• Used for short-term (intra-orbit) relative calibrations when direct access to solar illumination is unavailable	• Use light source with different spectrum • Uncertain degradation during launch and transfer to orbit • Lamp reliability • Weak signal or high wattage lamps
Integrating sphere with lamp illumination	• Provide a smooth, near-flat field of dispersed light as viewed through the VNIR and IR • Photodiodes used to monitor and provide stable illumination	• Use light source with different spectrum • Location of integrating sphere port to accommodate large FOV very constrained • Uncertain degradation during launch and transfer to orbit • Lamp reliability
LED or diffused LED	• Absolute calibration • Long life time compared to QTH • Mechanically robust • Limited power tuning • Diffuser may be unnecessary	• Multiple LEDs required to cover the entire spectrum • Little space heritage • Spectral and radiometric stability dependent on stability of electrical current and temperature
Blackbody	• Absolute calibration • No diffuser required • Smooth radiance curve • High accuracy achievable • Tunable	• Require high power to reach a high temperature and a high enough signal • Long warming time

(*Continued*)

TABLE 9.1 *(Continued)*
Comparison of Onboard Radiometric Calibration Techniques

Technique	Pros	Cons
VC—ground based—reflectance	• High absolute accuracy • Accuracy should improve	• Ground deployment costs • Infrastructure costs • Ground instruments must be calibrated • Targets occasionally unavailable
VC—ground based—radiance	• Very high absolute accuracy • Accuracy should improve	• Ground deployment costs • Infrastructure costs • Cost of airborne radiometer calibration and data acquisition • Ground instruments must be calibrated • Targets occasionally unavailable
VC—ground based—irradiance	• High absolute accuracy • Accuracy should improve	• Ground deployment costs • Infrastructure costs • Ground instruments must be calibrated • Targets occasionally unavailable
VC—ground based—automated reflectance	• No ground deployment cost • Will provide superior temporal information • Normally available during each revisit period • No special operation requirement	• Cost of implementing infrastructure
VC—lunar calibration—absolute	• 1–2% uncertainty • Target available, predictable • Available 50% of the time	• Require lunar scan • Scan must be accurate/precise • May require participation of expert
VC—lunar—relative	• Uncomparable stability • < 1% uncertainty in model • Target availability, predictable • Available 50% of the time	• Require lunar scan • Scan must be accurate/precise • May require participation of expert
VC—stable deserts	• No ground deployment cost • Method has good relative accuracies even without atmospheric correction (1–2%/year)	Relative only
VC—sun glint	• No ground deployment cost • Good inter-band calibration (1–2%)	• Operator intensive • Must acquire meteorological data of remote area • Inter-band only • Require restrictive view/illumination geometry • Require specific meteorological (wind) conditions
VC—high altitude cloud	• No ground deployment cost	• Operator intensive • Must acquire meteorological data of remote area • Relatively large (4%) transfer error added to reference band uncertainty
VC—molecular scattering over ocean	• Provide accurate absolute calibration of blue bands • No ground deployment cost	• Only applicable to appropriate blue bands • Hard to find appropriate target images for narrow FOV sensors

9.4 ONBOARD SPECTRAL CALIBRATION TECHNIQUES

9.4.1 Onboard Spectral Calibration Strategies

By its nature, the high spectral resolution and large number of spectral bands inherent in hyperspectral instruments make the spectral calibration requirements significantly more stringent than those of their predecessors, multispectral sensors with relative broad bands and low number of bands. For those multispectral sensors that have the spectral bands defined by filters such as MODIS, MISR, ETM+, an SPOT, the calibration issue is simplified in yet other ways when compared to grating or prism-based instruments. Even for airborne hyperspectral sensors such as AVIRIS and CASI that can be calibrated on the ground or in the laboratory, the challenge of performing a complete spectral characterization of each of the individual spectral elements (pixels) remains a complex and time consuming process which continues to be updated and refined. Application of those techniques to the satellite hyperspectral sensors based on those uses on the ground, although feasible, often implies significant additional complexity and cost to the instrument design and operation. The development of new calibration concepts tested under laboratory conditions may result in techniques applicable to satellite instrumentation with significantly simplified design and operational complexity as well as reduced mass and volume from that of earlier techniques (etalon). These techniques still require further development in order to prepare them for spaceborne deployment. As an alternative, the development of spectral calibration procedures based on the acquisition and processing of scene-based imagery has made significant steps forward and appears to be capable of supplementing onboard hardware based techniques. Although some of the techniques can be enhanced or extended when ground truth data is available to supplement the satellite imagery, some of the techniques do not require such information.

The key specifications of a hyperspectral satellite that impact the spectral calibration are the wavelength range, spectral sampling interval, spectral resolution, and calibration accuracy (center wavelength and bandwidth). The SNR requirements potentially impact the ability to provide adequate spectral calibration. The radiometric calibration requirements impact on the radiometric accuracy. The relationship between the spectral and radiometric calibrations is described by Green (1998). It was suggested that knowledge of the center wavelength of each pixel as well as the knowledge of the FWHM of the pixels spectral response function (assumed to be Gaussian) must be known to within 1% of the FWHM itself. These conclusions were driven by the resulting radiometric errors that would occur adjacent to the solar and atmospheric spectral features when spectral misalignments occurred with respect to the atmospheric correction model applied to the data. In the regions between these solar and atmospheric features, the resulting radiometric error would be considerably smaller than that occurring in the vicinity of the feature. At sufficiently fine resolution, the top-of-atmosphere reflected radiance is highly structured. The smoothness of the "between feature" regions will depend on the resolution of the instrument.

It is undesirable to have spectral artefacts introduced to the resulting data cube during spectral calibration although the spectral calibration may be deemed adequate to allow the radiometric calibration to within the specified requirement. In other words, in a spectral region that is spectrally smooth, the system must insure that a spectral feature is not introduced despite the fact that all radiometric values are within the radiometric requirements. This ties once again the regions adjacent to the atmospheric and solar absorption features.

Both the nature and magnitude of cyclical and long-term drifts on both the sensor and calibration systems need to be considered. In terms of spectral calibration and its impact on the radiometric calibration, this situation is somewhat different in that long-term drift of the spectral parameters is not expected. Rather the deltas that occur in the spectral parameters between prelaunch spectral calibration and on-orbit commissioning as well as any cyclical (i.e., orbital effects) tend to be the focus of on-orbit spectral characterization efforts.

Mechanical changes can occur that impact the spectral system characteristics during launch or due to changes in the environment during transfer to orbit. Following that, it is possible the changes continue to occur due to thermal fluctuations that impact the opto-mechanical design and its alignment. Following that, any time and effort given to spectral on-orbit characterization tends to be to monitor or perhaps to refine the spectral calibration of the system. It is necessary to provide adequate monitoring and refinement of the system's spectral performance.

This section describes several onboard spectral calibration techniques employed by various satellite hyperspectral sensors. Where available, a review of the spectral performance of the system and the onboard spectral calibration capabilities is provided.

9.4.2 Filtered QTH Spectral Line Sources

Quartz tungsten halogen (QTH) lamps are popular VNIR light sources because of their smooth spectral curve and stable output. They do not have the sharp spectral peaks that arc lamps exhibit, and they emit little UV radiation.

QTH lamps with spectral filters had been proposed in the ARIES Phase A design. It was noted that the use of filtered QTH line sources require not only excessive use of power but also required significant warm-up periods, require considerable care, and are potentially subject to difficult to quantify degradations.

Hyperion used a number of low wattage QTH lamps to illuminate the SD in the closed position located on the back of the telescope cover. Prelaunch tests showed the lamps degrade from 1% to 6% over the initial 50 h of operation and 4–17% after 150 h. The operational lifetime of the lamps was 146 h.

As described in Section 9.2.8, HISUI hyperspectral instrument uses a QTH lamp as light source to illuminate the filters mounted on the filter wheel to calibrate the spectrum of the instrument.

9.4.3 Monochromator

An onboard monochromator approach has been used to calibrate spectral performance of hyperspectral instruments. MODIS satellites are the only examples of use this approach being seen so far. MODIS has a modified Czerny-Turner monochromator built into its SRCA, which performs on-orbit spectral and spatial characterization (Xiong et al. 2016). The monochromator contains a motor-driven grating/mirror assembly, which is used in conjunction with a filter wheel to diffract the light into one of the VIS, NIR, or SWIR bands.

The SRCA system has a lamp illuminated integrating sphere. The light coming from the integrating sphere is bounced off of a toroidal relay mirror passing through order filters mounted on the filter wheel and into the entrance slit of the monochromator, is then dispersed by a grating, focused through an exit slit and onto the scan mirror, the first element of the MODIS' main optical system. The SRCA system is self-calibrated by viewing a signal through a didymium glass filter located behind an auxiliary slit located adjacent to the exit slit. The centroid of each band can be determined to within 1 nm uncertainty with 0.5 nm precision.

The unique feature of SRCA is that it can generate calibration information at the same time that MODIS is collecting and recording spectral data, which allows long stretches of continuous data. This approach is very expensive and complexity being designed into the instrument's onboard calibration system. The probability that this method (or a similar one) be replicated on future missions is not considered likely (Teillet 1997).

9.4.4 Doped Spectralon Diffuser

A Spectralon diffuser doped with rare earth erbium oxide exhibits numerous spectral features in VNIR region, but few spectral features in SWIR region. It was deployed in the MERIS instrument onboard ENVISAT to perform spectral alignment of the sensor's spectral bands. The same doped Spectralon diffuser is also used in its successor Ocean and Land Color Imager (OLCI) onboard Sentinel-3A satellite.

Spectralons doped with two other rare earth holmium oxide and dysprosium oxide have been produced for use as wavelength calibration standards. The dysprosium oxide doped Spectralon standard has two more features in the SWIR region than the erbium oxide standard, but still does not have any distinguishable features beyond the 1700 nm while the VNIR features are not as clearly identifiable. Extension of the spectral registration of these features to the entire spectral range of hyperspectral sensor would require considerable faith in their interpolation/extrapolation.

The onboard spectral calibration of MERIS and OLCI makes use of an erbium doped Spectralon diffuser plate, which offers a number of spectral absorption features in their operational spectral range. The doped Spectralon panel is inserted in the instruments' field-of-view while illuminated by the sun during calibration procedures.

The absorption spectrum of the erbium doped Spectralon diffuser was measured with a Cary 5000 spectrophotometer in ultrahigh precision scan mode. Figure 9.8 shows the absorption spectrum of the erbium doped Spectralon together with the MERIS convolved elementary spectral pixels. The absorption peaks selected in the MERIS and OLCI spectral calibration are those centered at 408 nm (Peak 1) and 522 nm (Peak 3). Note that the two absorption peaks are both in the UV-VIS region.

In addition of the 2 absorption peaks of the erbium doped Spectralon diffuser, a number of other wavelength calibration bands have been used by taking advantage of the spectral programmability of the MERIS/OLCI instrument (Delwart et al. 2007). These calibration bands include:

1. Earth observation of the O_2-A absorption feature at 760 nm
2. Observation on the radiometric calibration diffuser of the sun Fraunhofer lines at 395, 486, 656, 854, and 866 nm

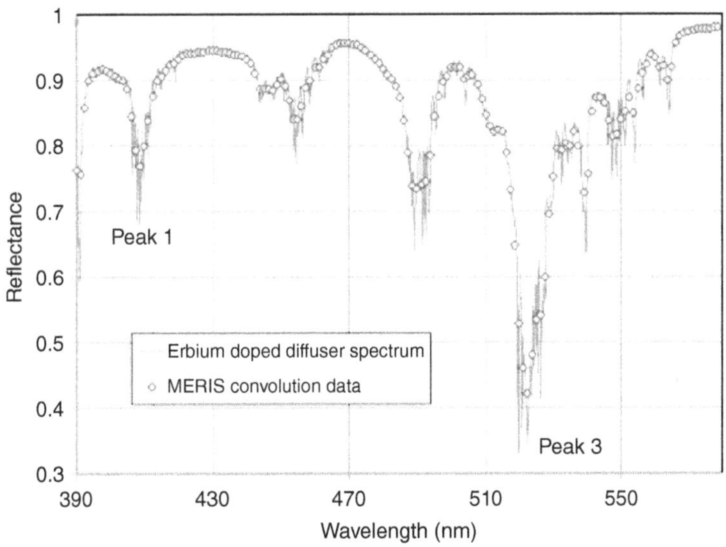

FIGURE 9.8 Erbium doped Spectralon spectrum and MERIS convolved elementary spectral pixels. (Courtesy of ESA.)

In total, the spectral calibration has been performed at 8 wavelengths covering most of the useful spectral range of the instrument. Calibrations are regularly performed at an average frequency of once every 3 months for the erbium/Peak 3 method and at lower frequency for other calibration spectral bands.

The in-flight spectral calibration measurements based on the erbium doped Spectralon diffuser were in consistent with the spectral variation observed across the FOV of the MERIS five cameras during the on-ground calibration campaign.

The results of the spectral calibration using O_2-A band (at 760 nm) and Fraunhofer lines (at 395, 486, 589, 656, 854, and 866 nm) are similar to those using the two erbium absorption peaks. This highlights the consistency of the measured in-FOV variation of the wavelengths. The O_2-A band and Fraunhofer lines perform even better in deriving absolute wavelength with an accuracy of about 0.2 nm compared to ±0.3 nm by the erbium measurements.

MERIS demonstrated an excellent spectral stability on-orbit as revealed by the regular in-flight spectral calibration campaigns performed. The spectral shift observed during the 10 years mission lifetime was small based on the three calibration methods described above: erbium doped Spectralon diffuser (Peak 1 and Peak 3), O_2-A absorption band and Fraunhofer lines. The spectral stability is within the measurement accuracy (0.1 nm). However, Cameras 2 and 4 had a slight modification of their spectral response over time (Bézy et al. 2016).

In EnMAP hyperspectral satellite, a second smaller integrating sphere with rare earth oxide doped Spectralon is equipped to perform spectral calibration. It couples light with a strongly structured spectral flux to the primary integrating sphere. Many distinct and stable spectral features are created by combining light from tungsten filament lamps with rare earth oxide doped Spectralon as a diffuser material (Kaufmann et al. 2016).

9.4.5 Atmospheric/Solar Lines

Techniques of using atmospheric and solar spectral features to perform spectral calibration or validation have been developed and reported. This is a kind of vicarious calibration techniques. Barry et al. (2002) developed a technique applied to Hyperion data sets. Gao et al. (2002) developed techniques that were applied to AVIRIS, PHILLS, and Hyperion data sets. Neville et al. (2003) developed a technique applied to AVIRIS, Short-wave infrared Full Spectrum Imager (SFSI) and Hyperion.

In the technique reported by Barry, frames of the reflected solar spectrum are acquired from the SD as the sun is rising through the Earth's limb. A 12-sec window existed for Hyperion to acquired such data as it was crossing the pole. The viewed spectrum consists of not only spectral features from the sun and the atmosphere, but also features contributed by the reflectance spectrum of the paint used on the SD as well. Nineteen (19) usable features were identified in the SWIR region. The only useful features in the VNIR region were the 762.5-nm atmospheric oxygen absorption line and the 520-nm solar Fraunhofer line. As with any other spectral calibration techniques, the observational data must contain clearly defined spectral features that can be identified and traced to a reference spectrum. For each spectral feature, a cubic spline was used to identify its exact location. The use of the spline is expected to have introduced a ±1.1 nm error. As this was performed for each spatial pixel in the field-of-view, a 2D surface of calculated wavelength shift is plotted for the SWIR detector array, with the x-axis representing the pixel location of the spectral feature based on original spectral calibration and the y-axis being the field-of-view location. A low order polynomial is then applied to reduce statistical noise in the data and produce a complete band to wavelength map of the focal plane. For the VNIR detector array, as there were only two points, a simple offset and tilt was determined for each spatial pixel.

Results of the analysis showed that the on-orbit measurements were within 5 nm, or half a band FWHM, of the prelaunch calibration in the SWIR. The VNIR results were slightly better. The accuracy of this technique is limited by the accuracy of the reference spectra and the error in the spline determination of the feature locations (±1.1 nm).

The techniques reported by Gao et al. (2002) and Neville et al. (2003) make use of superior spectral feature matching techniques to that described above. The data sets on which the analyses were applied in both cases are that of, unlike the Barry procedure, ground scenes.

Gao et al. (2002) suggested that for 10 nm spectral resolution (FWHM) instruments, 7 spectral features should be suitable for application of the described analysis (H_2O – 820 nm, 940 nm, 1140 nm, O_2 – 760 nm, CO_2 – 1580 nm, 2010 nm and 2060 nm). For instruments with spectral resolutions of 5 nm or better, the 430 nm Fraunhofer line becomes usable. For instruments with spectral resolutions of 2.5 nm, the additional Fraunhofer lines at 516, 656, 854, and 863 nm could be used. Use of any of these spectral features is, however, dependent on adequate signal-to-noise of the data sets. This includes adequate removal of pixel-to-pixel nonuniformities. Neville applied an iterative procedure to precondition the data to eliminate the impact of any scene variations on the measurement results. Gao et al. did not indicate whether such preconditioning of the data is performed in his technique. The results of applying both techniques, however, were found to be comparable.

Neville et al. (2003) inferred from analysis of the AVIRIS data set that the procedure is capable of making single pixel measurements down to ±0.05 nm. Standard errors (standard deviation of the mean) in the case of the AVIRIS results were estimated at ±0.002 nm. The technique can also be applied to determination of the bandwidths. However, the technique will function only if the band response function is known. It will not work if the band response function is to change in an unknown manner although this is not likely to be of concern. The AVIRIS bandwidth in the vicinity of the 942 nm water vapor band, for example, was found to increase 2.06% ± 0.05% by Neville et al.

As with the Barry technique, the results must be interpolated or extrapolated from the analyzed spectral features to each of the bands for each of the spatial pixels. The suitability of undertaking this process is evaluated by applying an atmospheric correction to the data set based upon the new sensor wavelength calibration. When this is done, the "spikes" disappear in the corrected scene spectra that occur when the atmospheric correction algorithm is spectrally misaligned with the data set. The only remaining spikes are found at the CO_2 lines and may be attributable in part to incorrect offset correction.

The results of the reported techniques were with respect to the spectral accuracies of the atmospheric models applied in the analysis, HITRAN 2000 by Gao et al. and MODTRAN 4.2 by Neville et al. It should also be noted that the analyses appear to indicate that the procedures are capable of providing similar results given different background scenes.

9.4.6 LED Spectral Line Sources

The use of LEDs as spectral line sources as part of an onboard calibration system has been proposed by Trottier (2002) and Moreau (2002). LEDs have higher efficiencies and longer lifetimes compared to QTH lamps and are not subject to degradation of their spectral output as is seen in the QTH lamp filters. LEDs whose wavelengths cover the entire spectral range of concern have been identified. However, use of a LED as a spectral calibration source requires a stable, high precision, low noise power supply source. LED as subject to spectral drift with temperature as well and must be carefully regulated in temperature and current.

LEDs may be viewed directly (although the uniformity of such an approach would be questionable) reflected from a diffuser panel as proposed by Trottier (2002) or as input to an integrating sphere as proposed by Moreau (2002). A tunable LED covering the 380–780 nm spectral range has been developed by NIST (Brown et al. 2002) as a radiometric reference. This is an integrating sphere with 40 LEDs, each driven by a selectable current, allowing the combination of spectral outputs to simulate various spectral sources. Such a technique, extended over a wider spectral range, may provide an interesting approach to spectral calibration.

9.4.7 Etalon

The use of tunable etalon filters for calibration of hyperspectral imagers has been proposed and tested using a CASI airborne hyperspectral instrument in laboratory conditions (Sinclair et al. 2002). The etalon provides a "comb" of spectral features at any one time when white light is passed through it. For the etalon tested, the comb contained 50 transmission features between 400 nm and 950 nm, each one with a spectral resolution (FWHM) of 1.2 nm or less. By adjusting the length of the etalon by tuning the piezoelectric transducer voltage applied to it, the position of the spectral features are stepped through the spectrum. The step size achieved in the tests was between 2.2 nm and 2.8 nm. The exact position of each of the spectral features on the sensor array is determined for each step position. The results are then used to determine the wavelength registration and bandwidth of each pixel.

When using a single frame of data per etalon step, the registration was found being good to within 4.7% of the bandwidth. The bandwidth was determined to within 5.5% of the required bandwidth. Given CASI's spectral bandwidth of 2–6 nm, this suggests accuracies of between 0.11 nm and 0.33 nm, which are approaching to the 0.1 nm accuracy suggested by Green (1998). These results improved to 3.3% for registration and 1.9% for bandwidth determination respectively if 6 frames of data were averaged. This suggests accuracies of between 0.04 nm and 0.11 nm, which reached the 0.1 nm accuracy suggested by Green (1998). The entire data acquisition was accomplished in 85 min. This acquisition time could have been reduced to ~50 min if only a single frame of data was acquired for analysis. This compares with days, or even weeks, that would be required to acquire a data set containing the equivalent information with a monochromator.

The miniature Fabry-Perot etalon used in these tests was only 1.5 cm in diameter and 2.5 cm long with a mass of under 20 g. A precision high stability variable voltage source is required to drive the Fabry-Perot etalon. A 250-W white light was used to illuminate the Fabry-Perot etalon that requires a collimation of the incident light of 2.9° or better. This could be replaced in an onboard calibration system by using the sun as a source. In such a case, care must be taken to avoid or properly deal with spectral features in the solar spectrum, i.e., the Fraunhofer lines. A small integrating sphere was also used in the laboratory setup in order to remove the effects of spatial nonuniformities in the Fabry-Perot etalon throughput. Two etalons would be required to cover the spectral range of the hyperspectral instrument being considered, one for the VNIR and one for the SWIR region. This technique appears very promising at that time; however, additional development work would be required to prepare the technique for use in space.

The use of etalons in space does in fact have a space heritage, albeit not for use in imaging instruments nor as part of a spectral calibration/validation system. Among the space instruments making use of etalons are the high-resolution Doppler imager (HRDI) and TIMED Doppler interferometer (TIDI). The HRDI was onboard the Upper Atmosphere Research Satellite (UARS) launched in 1992 and deactivated on December 15, 2005. The TIDI is a second generation of this instrument onboard the Thermosphere-Ionosphere-Mesosphere Energetics and Dynamics (TIMED) satellite launched on December 7, 2001 and still active at the time of the book being written (January 2020). These instruments observe the emission and absorption lines of molecular oxygen (and other atmospheric components) in small volumes (4 km in height by 50 km in width) above the limb of the Earth. From the Doppler shift of the lines, the horizontal winds can be determined, while the line shapes and strengths yield information about the temperature and atmospheric species makeup. The HRDI instrument was based upon a triple etalon Fabry-Perot interferometer. The TIDI designed is based on a fixed gap single etalon Fabry-Perot interferometer. In both cases it is the Fabry-Perot interferometer that acquires the actual scientific data. The HRDI has a spectral calibration requirement of a 0.1 nm. It was calibrated through the use of onboard low-pressure emission lamps.

TABLE 9.2
Pros and Cons of the Spectral Calibration Techniques

Technique	Pros	Cons
Filtered QTH spectral line sources	• Spectral location of the filters can be customized	• Uncertainty degradation during launch and transfer to orbit • Lamp reliability • Requires many lamps or filter wheel
Monochromator	• Complete spectral scan • High accuracy	• Large mass and volume • Slow • Low throughput
Doped Spectralon diffuser	• Relatively easy to use	• Limited spectral features (requires interpolation/extrapolation) • Solar panel subject to on-orbit degradation when using solar illustration
Atmospheric/solar lines—Hyperion technique	• Full field-of-view/aperture • Limited spectral features are located in the key positions	• Requires high precision, stable supply • Requires exposure of SD • Limited time to acquire solar light through Earth limb
Atmospheric/solar lines—Gao or Neville technique	• Full field-of-view/aperture • Limited spectral features are located in the key positions • Technique independent to any Earth target (applicable to any Earth scene)	• Accuracy tied to atmospheric model applied (may be a pro as well)
LED line sources	• Lightweight and low mass • Relatively low power consumption	• Requires stable, high precision, low noise power supply • LEDs may not be available at key spectral locations • Many LEDs required • Lack of space heritage
Etalon	• Complete spectral scan • Fast • Simple, lightweight, low mass	• Requires precision and stable power supply • Diffuser required to fill aperture • Lack of space heritage • Delicate alignment

9.4.8 Summary of Spectral Calibration Techniques

Rigorous spectral calibration of a hyperspectral instrument is a nontrivial task. Onboard calibration (recalibration) or validation of spectral registration and bandwidth on spaceborne instruments has been a particular challenge, especially as the spectral resolution is kept increasing and the atmospheric correction models demand better and better registration with the data set. Advanced developments lead to methods which appear to be able to provide the required accuracies.

Table 9.2 summarizes the pros and cons of the spectral calibration techniques described in this section.

9.5 CONCEPT DESIGN EXAMPLES OF ONBOARD CALIBRATION SYSTEM

This section presents three examples of concept design of onboard calibration of hyperspectral instruments.

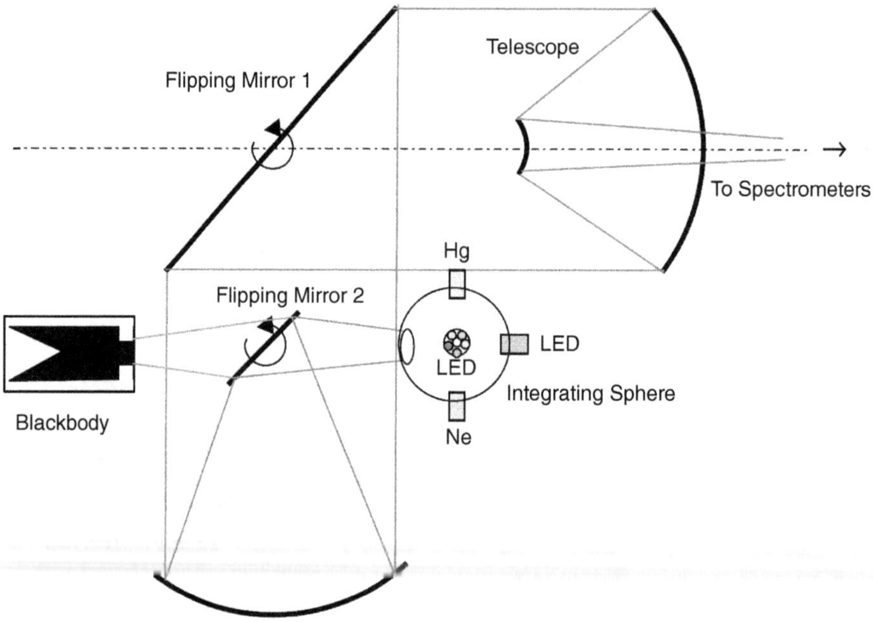

FIGURE 9.9 Concept design of onboard calibration using a integrating sphere and a blackbody placed in front of telescope.

9.5.1 Concept Design 1

This concept design places the calibration system in front of the telescope. The entire field-of-view of the hyperspectral instrument is calibrated. The flipping mirror 1 is used to switch between the observing scene and the calibration system, as shown in Figure 9.9.

This concept uses a 6-port integrating sphere and a blackbody as the reference sources. The passive cavity blackbody provides a dark reference because of its relative insensitivity to stray light. The integrating sphere provides bright reference sources.

One port of the sphere is used as output, two ports are used for spectral references with Hg and Ne lamps, three ports are used for the bright radiometric references. There are many options to fill the three ports of the integrating sphere. One option is to use various bundles of LED to cover the entire spectral range of the VNIR and SWIR spectrometers. LEDs are generally more reliable than QTH.

Alternatively, the three ports can be filled with a bundle of LEDs, a QTH lamp, and a spare QTH lamp. This option can be used in the situation that too many LEDs would be necessary to cover the entire spectral range or that QTH are judged being a better fitting than LEDs. The LED bundle would then serve as a source of blue-green light and as a monitor of the evolution of the QTH.

Switching between the integration sphere and the blackbody is done with flipping mirror 2. This flipping mirror causes an obscuration but since the telescope already has a 10% obscuration, if the obscuration in the calibration system matches the obscuration of the telescope, no energy will be further wasted.

9.5.2 Concept Design 2

Figure 9.10 shows a concept design of an onboard calibration system that uses an extended area LED array and blackbody. In this concept design, it is assumed that an extended area LED array incorporating LEDs of various wavelengths would be adequate to cover the required spectral range of the spectrometers. In this case no diffuser is necessary (a diffusing window can be placed directly on

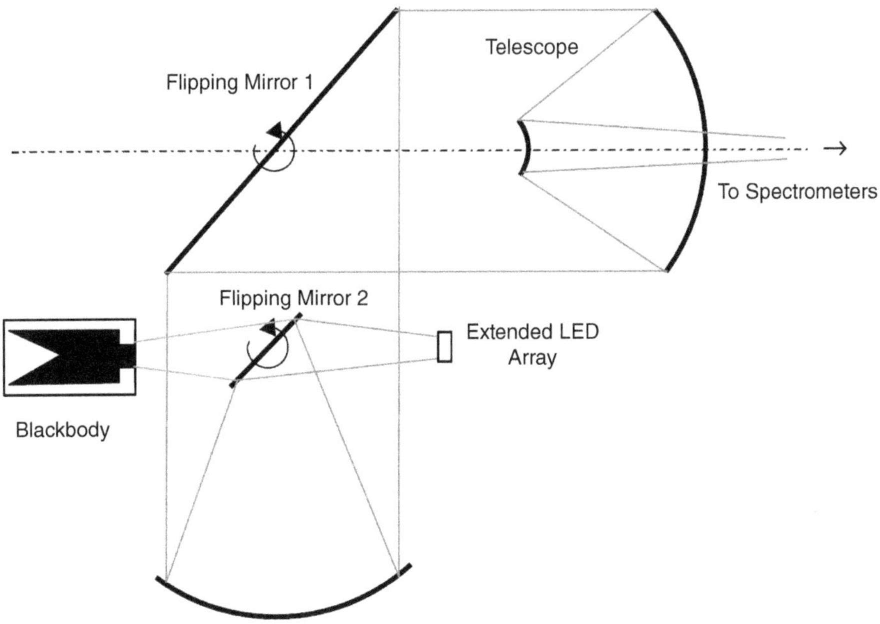

FIGURE 9.10 Concept design of onboard calibration using an area LED array and a blackbody placed in front of telescope.

the array). It is also assumed that the LED array can be used as spectral reference as well by turning on only one type of LED in the array at a time. A calibration system based on this concept can be very compact.

9.5.3 Concept Design 3

This concept design, as shown in Figure 9.11, is similar to the first except design but the feeding optics is different. In the first concept design, a large mirror is required to feed the telescope. If it

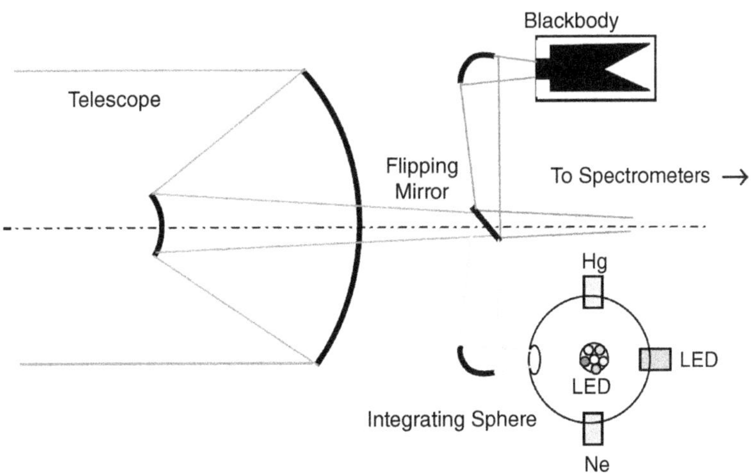

FIGURE 9.11 Concept design of onboard calibration using a integrating sphere and a blackbody placed in behind the telescope.

turned out that the associated mass is not available in the first concept design, this concept could be used instead.

In this concept design, calibration light beam is injected behind the telescope. Because the beam is much smaller at this location, smaller optics can be used to fill the field-of-view. A flipping mirror is used to switch the light of the observing scene to the spectrometers or to inject the radiance light from the bright or the dark reference sources sequentially. The drawback of this configuration is that the telescope is not part of the calibration process. Variation of transmittance of the surfaces of the telescope will not be captured by the radiometric calibration. For this reason, the first two concepts generate a more reliable calibration.

REFERENCES

Aoyanagi, Y., S. Satori, T. Totani, T. YaSunaka, A. Nakamura, and Y. Takeuchi. 2009. Development of space-borne small hyperspectral sensor HSC-III for micro satellite. *Proceedings of 7th IAA Symposium on Small Satellites for Earth Observation*. Berlin, Germany.

Auspace Limited. 1997. ARIES Phase A Design Report for ARIES Space Segment. *Auspace Limited Document* No. RPT-ARE-0001-AUS, March 3, 1007.

Barry, P. S., J. Shepanski, and C. Segal. 2002. Hyperion on-orbit validation of spectral calibration using atmospheric lines and an on board system. *Proceedings of SPIE* 4480:231–235.

Bézard, B., C. Tsang, R. W. Carlson, G. Piccioni, E. Marcq, and P. Drossart. 2009. Water vapor abundance near the surface of Venus from Venus express/VIRTIS observations. *Journal of Geophysical Research* 114:E00B39.

Bézy, J.-L., J.-P. M. Huot, S. M. Delwart, L. Bourg, R. Bessudo and Y. Delclaud. 2016. Medium resolution imaging spectrometer for ocean colour onboard ENVISAT. In *Optical Payloads for Space Missions*, ed. S.-E. Qian, 91–120. Sussex, UK: John Wiley & Sons.

Bovensmann, H., et al. 2002. SCIAMACHY in-flight instrument performance. *Proceedings of the Envisat Calibration Review*, ESA Special Publications SP-520, Nov. 2002, http://envisat.esa.int/calval/proceedings/sciamachy/scia_02.pdf.

Brown, S. W., C. Santana, and G. P. Eppeldauer. 2002. Development of a tunable LED-based colorimetric light source, *Journal of Research of the National Institute of Standards and Technology* 107:263–371.

Bruegge, C. J., N. Chrein, and D. Haner. 2001. A Spectralon BRF data base for MISR calibration applications. *Remote Sensing of Environment* 76:354–366.

Bruegge, C. J., A. E. Steigman, A. R. Rainen, and A. W. Springsteen. 1992. Use of Spectralon as a diffuse reflectance standard for in-flight calibration of earth-orbiting sensors. *Optical Engineering* 32(4):805–814.

CEOS. 1995. Leadership to ensure high-quality earth observation data: A strategic vision, committee on earth observation satellites, Working Group on Calibration and Validation.

Chommeloux, B, et al. 1998. Spectralon diffusers used as in-flight optical calibration hardware. *Proceeding of SPIE* 3427:382–393.

Cosnefroy, H., M. Leroy, and X. Briottet. 1996. Selection and characterization of Saharan and Arabian desert sites for the calibration of optical satellite sensors. *Remote Sensing of Environment* 58:101–114.

Delwart, S. and J.-P. Huot. 2002a. MERIS calibration review—smile and straylight corrections. *Proceedings of the ENVISAT Calibration Review Workshop Presentations*, Sept. 2002, ESTEC, Noordwijk, The Netherlands, http://envisat.esa.int/calval/workshop/meris/13_12_smile.pdf

Delwart, S. and J.-P. Huot. 2002b. MERIS calibration review—radiometric corrections. *Proceedings of the ENVISAT Calibration Review Workshop Presentations*, Sept. 2002, ESTEC, Noordwijk, The Netherlands, http://envisat.esa.int/calval/workshop/meris/08_radio_corr.pdf

Delwart, S., J.-P. Huot, L. Bourg, and C. Brockmann. 2003. Calibration and early results of MERIS on ENVISAT, *Proceedings of SPIE* 4881:337–344.

Delwart, S., R. Preuker, L. Bourg, R. Santer, D. Ramon, and J. Fisher. 2007. MERIS in-flight spectral calibration. *International Journal of Remote Sensing* 28(3):479–496.

Dinguirard, M. and P. N. Slater. 1999. Calibration of space-multispectral imaging sensors: A review. *Remote Sensing of Environment* 68(3):194–205.

Doelling, D. R., L. Nguyen, and P. Minnis. 2004. On the use of deep convective clouds to calibrate the AVHRR data. *Proceedings of SPIE* 5542:281–289.

Drossart, P., et al. 2004. VIRTIS Imaging Spectrometer for the ESA/Venus Express Mission. *Proceedings of SPIE* 5543: 175–185.

Dubuisson, P., R. Borde, D. Dessailly, and R. Santer. 2003. In-flight spectral calibration of the oxygen A-band channel of MERIS. *International Journal of Remote Sensing* 24(5):1177–1182.

Duggan, P., J. Harron, C. Hersom, and R. J. Soffer. 1998. HSI on-board calibration system design study. *Instrument Services Laboratory, CRESTech*, Document No. ISL-R-9809021, Sept. 15, 1998.

DuPont. 2003. Teijin Product Information, Mylar film Optical properties, 06/2003.

Fougnie, B., P.-V. Deschamps, and R. Frouin. 1999. Vicarious calibration of the POLDER ocean color spectral bands using in situ measurements. *IEEE Transactions on Geoscience and Remote Sensing* 37(3):1567–1574.

Fox, N. 2001. Traceability to SI for EO measurement. *CEOS CAL/VAL Newsletter*, Issue 9, pp. 1–9.

Franz, B. A., S. W. Bailey, P. J. Werdell, and C. R. McClain. 2007. Sensor-independent approach to the vicarious calibration of satellite ocean color radiometry. *Applied Optics* 46(22):5068–5082.

Frerick, J., B. Cozzoni, S. Slijkhuis, R. Debeek, and J. Skupin. 2002. SCIAMACHY in-flight calibration. *Proceedings of the ENVISAT Calibration Review*, ESA Special Publications SP-520, Nov. 2002, http://envisat.esa.int/calval/proceedings/sciamachy/scia_03.pdf

Frölich, C, and J. Lean. 1998. The Sun's Total Solar Irradiance: Cycles, Trends and Related Climate Change Uncertainties Since 1976. *Geophysical Research Letters*, 25(23): 4377–4380.

Gao, H. L., et al. 2010. HJ-1A HSI on-orbit radiometric calibration and validation research. *Science China Technological Sciences* 53:3119–3128.

Gao, B.-C., R. R. Li, R. L. Lucke, C. O. Davis, R. M. Bevilacqua, D. R. Korwan, M. J. Montes, J. H. Bowles, and M. R. Corson. 2012. Vicarious calibrations of HICO data acquired from the international space station. *Applied Optics* 51(14):2559–2567.

Gao, B.-C., M. J. Montes, and C. O. Davis. 2002. A curve-fitting technique to improve wavelength calibrations of imaging spectrometer data. *Proceedings of the 11th JPL Airborne Earth Science Workshop*, JPL Publication No. 03–4:99–105.

Green, R. 1998. Spectral calibration requirement for earth-looking imaging spectrometers in the solar-reflected spectrum. *Applied Optics* 37(4):683–690.

Green, R. O. and B. Pavri. 2002. Hyperion on-orbit calibration results from an AVIRIS underflight in Argentina. *Proceedings 11th Annual Airborne Earth Science Workshop, Jet Propulsion Laboratory, Pasadena, California.*

Green, R. O., B. Pavri, and T. Chrien. 2002. On-orbit radiometric and spectral calibration characteristics of EO-1 Hyperion derived with an underflight of AVIRIS and in situ measurements at Salar de Arizaro, Argentina. *IEEE Transactions on Geoscience and Remote Sensing* 41(6):1194–1203.

Godden, G. D. and C. A. McKay. 1997. A strategy for observing the moon to achieve precise radiometric stability monitoring. *Canadian Journal of Remote Sensing* 23(4):309–317.

Guenther, B., W. Barnes, E. Knight, J. Barker, J. Harnden, R. Weber, M. Roberto, G. Godden, H. Montgomery, and P. Abel. 1996. MODIS calibration: A brief review of the strategy for the at-launch calibration approach. *Journal of Atmospheric and Oceanic Technologies* 13:274–285.

Guenther, B., V. V. Salomonson, W. L. Barnes, and J. B. Young. 2002. On-orbit performance of the earth observing system moderate resolution imaging spectroradiometer; first year of data. *Remote Sensing of Environment* 83(1):16–30.

Guzman, C. T., et al. 1991. Requirements of a solar diffuser and measurements of some candidate materials. *Proceeding of SPIE* 1493:120–131.

Hilsenrath, E. 1994. Effect of space shuttle flight on the reflectance characteristics of diffusers in the near-infrared, visible and ultraviolet regions. *Optical Engineering* 33(11):3675–3682.

Holben, B. N., et al. 1998. AERONET—A federated instrument network and data archive for aerosol characterization. *Remote Sensing of Environment* 66(1):1–16.

Itten, K., et al. 2008. APEX—The hyperspectral ESA airborne prism experiment. *Sensors* 8(10):6235–6259.

Jarecke, P., K. Yokoyama, and P. S. Barry. 2001. On-orbit radiometric calibration the Hyperion instrument. *Proceeding of IEEE IGARSS* 2001:2825–2827.

Jarecke, P., K. Yokoyama, and P. S. Barry. 2002. On-orbit solar radiometric calibration of the Hyperion instrument. *Proceeding of SPIE* 4480:225–230.

Kappus, M. E. et al. 2016. Hyperspectral imager for the coastal ocean on the international space station. In *Optical Payloads for Space Missions*, ed. S.-E. Qian, 29–49. West Sussex UK: John Wiley & Sons.

Kaufmann, H. et al. 2016. Environmental mapping and analysis program—A German hyperspectral Mission. In *Optical Payloads for Space Missions*, ed. S.-E. Qian, 161–181. West Sussex UK: John Wiley & Sons.

Kieffer, H. H., J. M. Anderson, and K. J. Becker. 1999. Radiometric calibration of spacecraft using small lunar images. *Proceeding of SPIE* 3870:193–205.

Kieffer, H. H. and R. L. Widley, 1996. Establishing the moon as a spectral radiance standard. *Journal of Atmospheric and Oceaninc Technologies* 13:360–375.

Krause, K., S. F. Biggar, K. J. Thome, J. Eagan, and D. Kenyan. 2002. On-orbit radiometric calibration using a solar diffuser. *Proceedings of SPIE* 4483:135–145.

Kurucz, R. L., I. Furenlid, J. Brault, and L. Testerman. 1984. Solar flux atlas from 296 nm to 1300 nm. *National Solar Observatory Atlas* No. 1:1–239.

Lafrance, B., et al. 2002. Interband calibration over clouds for POLDER space sensor. *IEEE Transactions on Geoscience and Remote Sensing* 40(1):131–142.

Lockwood, R., et al. 2006. Advanced responsive tactically-effective military imaging spectrometer (ARTEMIS) design. *Proceedings of IEEE IGARSS 2006* 1628–1630.

Markham, B. L. and J. L. Barker. 2003. On-orbit performance of the Landsat-7 ETM+ radiometric calibration. *International Journal of Remote Sensing* 24(2):265–285.

Markham, B. L., W. C. Boncyk, D. L. Helder, and J. L. Barker. 1997. Landsat-7 enhanced thematic mapper plus radiometric calibration. *Canadian Journal of Remote Sensing* 23(4):318–332.

MatSunaga, T. et al. 2016. Hyperspectral imager suite (HISUI): Japanese spaceborne hyperspectral imager for resource and environmental mapping. In *Optical Payloads for Space Missions*, ed. S.-E. Qian, 215–222. West Sussex UK: John Wiley & Sons.

Meini, M., A. Bini, L. Giunti, E. Fossati, and R. Formaro. 2016. Hyperspectral payload for Italian PRISMA programme. In *Optical Payloads for Space Missions*, ed. S.-E. Qian, 183–213. West Sussex UK: John Wiley & Sons.

Melchiorri, R., G. Piccioni, and A. Mazzoni. 2003. VIRTIS-M flight lamps. *Review of Scientific Instruments* 74(8):3796–3801.

Moinelo, A. C., G. Piccioni, E. Ammannito, G. Filacchione, and P. Drossart. 2010. Calibration of hyperspectral imaging data: VIRTIS-M onboard Venus express. *IEEE Transactions on Geoscience and Remote Sensing* 48(11):3941–3950.

Montgomery, H., N. Che, and K. Parker. 2000. The algorithm for MODIS wavelength on-orbit calibration using the SRCA. *IEEE Transactions on Geoscience and Remote Sensing* 38(2):877–884.

Moreau, L. 2002. Preliminary onboard radiometric calibration concept technical memorandum. *ABB Bomem Inc. Technical Memorandum* No. BOM-THM-003, Issue 1, Rev. B, Feb. 11, 2002.

Murchie, S., et al. 2007. Compact reconnaissance imaging spectrometer for mars (CRISM) on mars reconnaissance orbiter (MRO). *Journal of Geophysical Research* 112(E05S03):1–57.

Neville, R. A., L. Sun, and K. Staenz. 2003. Detection of spectral line curvature in imaging spectrometer data. *Proceedings of SPIE* 5093:144–154.

Nieke, J., et al. 2002. Global imager's on-board calibration (VNIR-SWIR). *Proceeding of SPIE* 4483:231–241.

Palmer, J. M. 1991. Stray light effects on calibrations using a solar diffuser. *Proceedings of SPIE* 1991:143–154.

Palmer, J. M. and P. N. Slater. 1991. A ratioing radiometer for use with a solar diffuser. *Proceeding of SPIE* 1493:106–117.

Pavlov, M. M., M. L. Byers, and J. A. Walker 1998. Bidirectional reflectance factor (BRF) characterization on the MODIS flight solar diffuser. *Proceedings of SPIE* 3439:257–268.

Pearlman, J. S., P. S. Barry, C. C. Segal, J. Shapanski, D. Beiso, and S. L. Carman. 2003. Hyperion, a space-based imaging spectrometer. *IEEE Transactions on Geoscience and Remote Sensing* 41(6):1160–1173.

Piccioni, G., et al. 2007. VIRTIS: The visible and infrared thermal imaging spectrometer. *ESA-SP-1295, ESA Publications Division*, Vol. SP-1295, Noordwijk, The Netherlands.

Ramon, D., R. Santer, and P. Dubuisson. 2003. MERIS in-flight spectral calibration in O_2 absorption using surface pressure retrieval. *Proceedings of SPIE* 4891:505–514.

Sang, B., et al. 2008. The EnMAP hyperspectral imaging spectrometer: Instrument concept, calibration and technologies. *Proceedings of SPIE* 7086.

Sinclair, P., R. Berman, C. Hersom, and A. Hollinger. 2002. Spectral calibration of hyperspectral imagers using a white etalon. *Proceedings of 12th CASI Conference on Astronautics*, Ottawa, Canada, Canadian Aeronautics and Space Institute, November 12-14, 2002.

Skupin, J., et al. 2002. SCIAMACHY spectral performance. *Proceedings of the ENVISAT Calibration Review Workshop*, Sept. 2002, ESTEC, Noordwijk, The Netherlands.

Slater, P. N. and S. F. Biggar. 1996. Suggestions for radiometric calibration coefficient generation. *Journal of Atmospheric and Oceaninc Technologies* 13:376–382.

Slater, P. N., S. F. Biggar, J. M. Palmer, and K. J. Thome. 2001. Unified approach to absolute radiometric calibration in the solar-reflective range. *Remote Sensing of Environment* 77:293–303.

Slater, P. N., S. F. Biggar, K. J. Thome, D. I. Gellman, and P. R. Spyak. 1996. Vicarious radiometric calibrations of EOS sensors. *Journal of Atmospheric and Oceaninc Technologies* 13:349–359.

Slater, P. N., et al. 1987. Reflectance- and radiance-based methods for the in-flight absolute calibration of multi-spectral sensors. *Remote Sensing of Environment* 22:11–37.

Stone, T. C. and H. H. Kieffer. 2002. Absolute irradiance of the moon for on-orbit calibration. *Proceeding of SPIE* 4814:211–221.

Stone, T. C., H. H. Kieffer, and J. M. Anderson. 2002. Status of use of lunar irradiance for on-orbit calibration. *Proceeding of SPIE* 4483:165–175.

Tatsumi, K. et al. 2010. Onboard spectral calibration for the Japanese hyperspectral sensor. *Proceedings of SPIE* 7826.

Teillet, P. M. 1997. A status overview of earth observation calibration/validation for terrestrial application. *Canadian Journal of Remote Sensing* 23(4):291–298.

Teillet, P. M., K. J. Thome, N. Fox, and J. Morisette. 2001b. Earth observation sensor calibration using a global instrumented and automated network of test sites (GIANTS), *Proceeding of SPIE* 4540:246–254.

Teillet, P. M., et al. 2001. A generalized approach to vicarious calibration of multiple earth observation sensors using hyperspectral data. *Remote Sensing of Environment* 77:304–327.

Thome, K. J., S. F. Biggar, and P. N. Slater. 2001. Effects of assumed solar spectral irradiance on inter-comparisons of earth-observing sensors. *Proceedings of SPIE* 4550:260–269.

Thuillier, G., M. Hersé, D. Labs, T. Foujols, W. Peetermans, D. Gillotay, P. C. Simon, and H. Mandel. 2003. The solar spectral irradiance from 200 to 2400 nm as measured by the SOLSPEC spectrometer from the ATLAS and EURECA missions. *Solar Physics* 214(1):1–22.

Thuillier, G., M. Herse, P. C. Simon, D. Labs, H. Mandel, D. Gillotay, and T. Foujols. 1998. The visible solar spectral irradiance from 350 to 850 nm as measured by the SOLSPEC spectrometer during the ATLAS-1 mission. *Solar Physics* 177:41–61.

Thuillier, G., et al. 2015. The infrared solar spectrum measured by the SOLSPEC spectrometer onboard the international space station. *Solar Physics, Springer Verlag* 290(6):1581–1600.

Tobbé, B., et al. 1999. In-flight calibration of the polder polarized channels using the sun's glitter. *IEEE Transactions on Geoscience and Remote Sensing* 37(1):513–525.

Trottier, Y. 2002. On-board calibration system. *Spectral Applied Research Technical Memorandum No: SE-142431 Rev. 04*, Feb. 05, 2002.

Walker, J. A., M. M. Pavlov, and M. L. Byers. 1998. Characterization of the solar diffuser as the primary standard for onboard calibration of the moderate resolution imaging spectrometer. *Proceedings of SPIE* 3498:496–508.

Wehrli, C. 1986. World radiation center Davos-Dorf, Switzerland (WCRP). *Publication Series No. 7, WMO ITD* 149:119–126.

Wenny, B. N. and X. Xiong. 2008. Using a cold earth surface target to characterize long-term stability of the MODIS thermal emissive bands. *IEEE Geoscience and Remote Sensing Letters* 5(2):162–165.

Xiong, X., et al. 2002. Trending results of MODIS optics on-orbit degradation. *Proceeding of SPIE* 4814:337–346.

Xiong, X.-X., et al. 2016. Moderate resolution imaging spectroradiometer on terra and aqua missions. In *Optical Payloads for Space Missions*, ed. S.-E. Qian, 53–90. West Sussex UK: John Wiley & Sons.

10 Instrument Control and Onboard Data Handling

10.1 FUNCTIONS OF INSTRUMENT CONTROL AND ONBOARD DATA HANDLING SUBSYSTEM

A typical spaceborne hyperspectral imaging system consists of fore-optics (telescope, slit assembly, and possibly depolarizer, etc.), visible and near-infrared (VNIR) and shortwave infrared (SWIR) spectrometers and possibly a panchromatic (PAN) channel, focal plane arrays and their proximity electronics (PEs), an onboard calibration unit, a service module, temperature sensors, heaters, a processing formatting and control unit (PFCU), and a solid-state mass memory unit (MMU). Figure 10.1 shows these functional blocks, the data flow, and the connection to the instrument control and onboard data handling (ICODH) subsystem.

The main functions of the ICODH subsystem include the following:

- Controlling the Optical Head that contains SWIR, VNIR, and PAN spectrometers via the internal links.
- Receiving, formatting, and packetizing data from the SWIR, VNIR, and PAN spectrometers via the internal links.
- Interfacing the PFCU for sending science data packets.
- Controlling the cover and calibration unit, spectrometer shuttle, and heaters based on the signal received from the temperature sensors, through the Service Module via the internal links.
- Interfacing as a Remote Terminal of the spacecraft's Onboard Data Handling (OBDH) subsystem via the Spacecraft Bus (e.g., MIL STD 1553B) for telecommand/telemetry (TC/TM).
- Sending discrete analogue telemetry for monitoring temperatures and voltages in the units, and bi-level telemetry for monitoring the unit on/off status.
- Interfacing the spacecraft's Power Control and Distribution Unit (PCDU) to receive the supply voltage.
- Controlling the power supply distribution to the SWIR, VNIR, and PAN spectrometers.
- Performing the system built-in test.

Because of the central control function of the PFCU, a very high reliability is required for the operation. Therefore, redundant boards are often used, which are in cold standby redundancy for low-power operation. The switch over to the redundant board will be performed by the spacecraft.

The PFCU and MMU are normally physically separated from Optical Head and from Proximity Electronics. They can be mounted inside the spacecraft platform and fixed on the top-plate interface between the spacecraft and the Optical Head.

10.2 CONFIGURATION OF INSTRUMENT CONTROL AND ONBOARD DATA HANDLING SUBSYSTEM

The ICODH subsystem incorporates dual redundant PFCU and a solid-state MMU. Figure 10.2 depicts a typical configuration of such a subsystem.

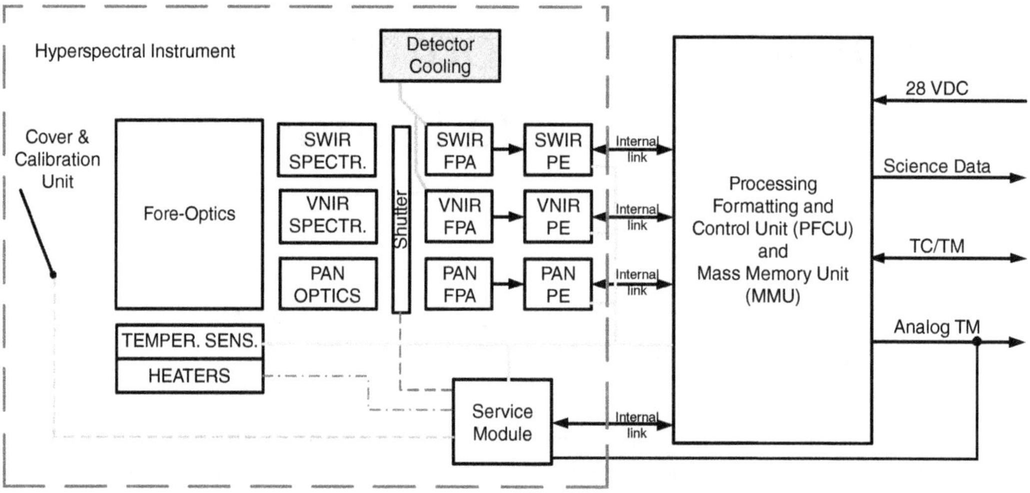

FIGURE 10.1 A functional block diagram of a typical hyperspectral imager and its instrument control and onboard data handling subsystem.

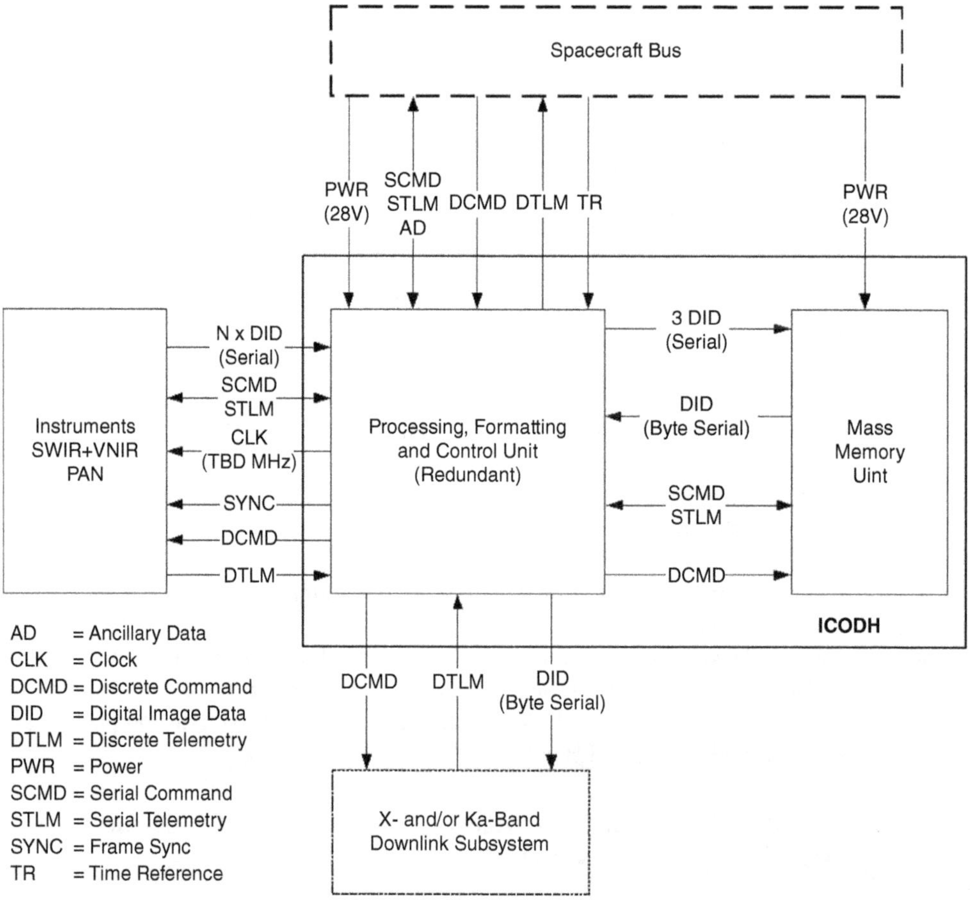

FIGURE 10.2 Configuration of the instrument control and onboard data handling subsystem.

The PFCU is the control center of a hyperspectral imaging system. It controls the imaging data acquisition of the spectrometers by sending commands and monitors the status of the instrument via the bidirectional Serial Command (SCMD) and Serial Telemetry (STLM) links. It provides the main clock (CLK) signal of the system to the detector arrays and a synchronous (SYNC) signal to synchronize the frames of the imaging data of the spectrometers. It issues Discrete Commands (DCMD) to control the actions of the instrument and receives signal from the instrument via Discrete Telemetry (DTLM) to report the status of the instrument. It receives acquired hyperspectral imaging data and PAN image data by the spectrometers via the multiple Digital Image Data (DID) links.

The PFCU interfaces with the spacecraft via the Spacecraft Bus to implement all the commands uploaded from the ground control center. It receives commands, telemetries, and ancillary data from the spacecraft via the bidirectional SCMD, STLM, and Ancillary Data (AD) links, and sends back housekeeping and status information of the instrument to the spacecraft. It receives discrete commands from the spacecraft via the DCMD link and passes them to the instrument for actions. It sends back the status information of the instrument to the spacecraft via DTLM. It receives time reference from the spacecraft via the Time Reference (TR) link and receives power supply of 28 V via the Power (PWR) link to supply power to the PFCU.

The PFCU sends the formatted science data acquired by the spectrometers to the MMU via the multiple DID serial links for storage. It receives stored science data via the fast byte serial link and encrypts the data before sending to the X- and/or Ka-band downlink subsystem for transmission to ground. It communicates with the MMU via the bidirectional SCMD and STML links. It sends discrete commands to the MMU via the DCMD link. The MMU receives power supply of 28 V via the PWR link to supply power to the MMU.

The PFCU controls the data downlink subsystem by first commanding the MMU sending stored data to it and then encrypting the data and sending them via the fast byte serial links to the X- and/or Ka-band downlink subsystem. It sends discrete commands to the downlink subsystem via the DCMD link and receives discrete telemetry from the downlink subsystem via the DTLM link.

10.3 PROCESSING, FORMATTING, AND CONTROL UNIT

10.3.1 Architecture of the PFCU

Figure 10.3 shows the architecture of the PFCU. The PFCU provides control and monitoring of the hyperspectral payload, acting as an agent to the satellite bus, controlling the MMU, command, and telemetry subsystem. It generally performs the following five tasks:

First, it receives the digital image data generated by the VNIR, SWIR, and PAN spectrometers. The data from each detector array is buffered and preprocessed to convert the instrument detector pixel ordering on the interface to the ordering required for subsequent ground processing or onboard compression. The selection of the SWIR and VNIR spectral bands allows a subset of the full spectrum to be stored. As the SWIR spectrometer generates more spectral bands than the VNIR spectrometer, its data rate and volume are higher than those of the VNIR spectrometer. In some hyperspectral imager system designs, the SWIR spectral bands are split into two streams (e.g., SWIR1 and SWIR2).

Second, if onboard data compression is enabled, the data are compressed in either lossless or near-lossless before formatting. PFCU formats and packetizes the data as per the Consultative Committee for Space Data Systems (CCSDS) standards and inserts the ancillary data required for ground processing.

The third function is the generation of a reference frequency to be used by the hyperspectral imager for all internal timings, together with synchronization pulses to synchronize the image frame timing across the VNIR, SWIR, and PAN channels. This function also generates timing and clocks for use within the PFCU, so as to maintain a synchronous design.

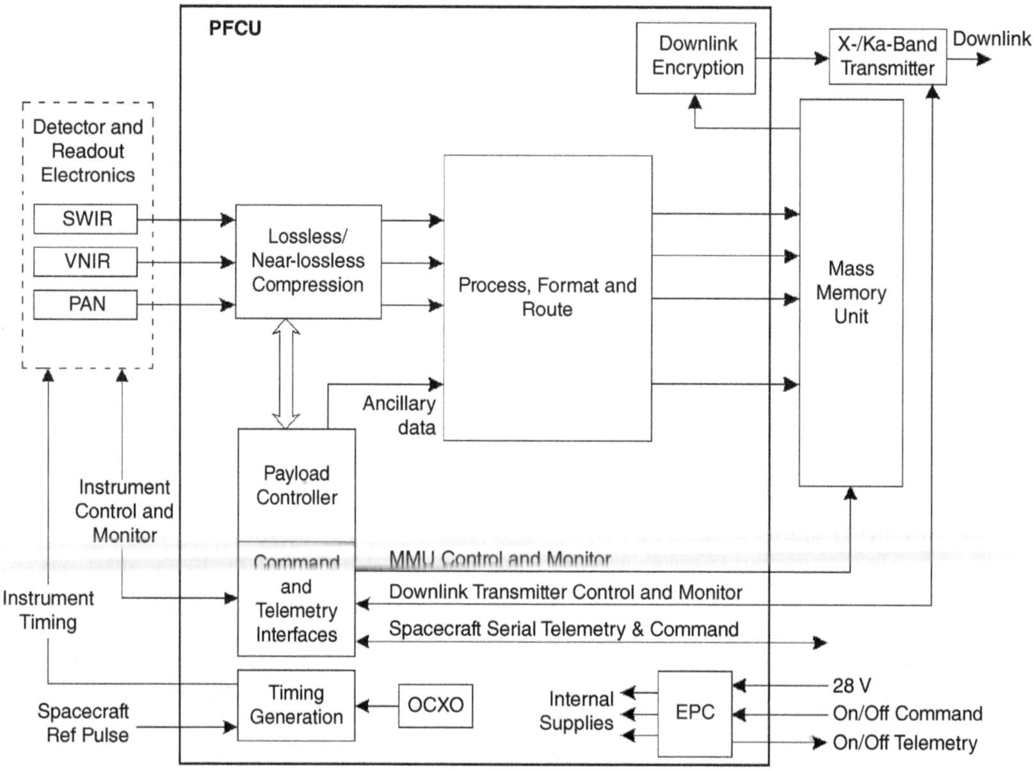

FIGURE 10.3 Architecture of Processing, Formatting and Control Unit. (EPC: electronic power conditioner, OCXO: ovenized crystal oscillator).

The fourth task is the replay of stored data for transmission to ground via an X- and/or Ka-band downlink subsystem when the spacecraft passes over a ground station. The PFCU conducts error detection and correction coding (channel coding) using CCSDS-recommended channel coding codes (Qian 2013b) if this is not already provided by the X- and/or Ka-band downlink transmitter.

The final primary task is the control and monitoring of the payload, and the execution of pre-stored imaging sequences synchronized to global positioning system (GPS) time (via Spacecraft Time). This ensures the appropriate sequence of events to image the required swath on the surface of the Earth and to subsequently transmit the data to a ground station.

10.3.2 Data Acquisition

10.3.2.1 Instrument Data Acquisition

There are three instrument channels if a PAN imager is also included. Each instrument channel has an independent acquisition chain. This gives a total of 3 instrument chains in a typical hyperspectral payload (SWIR, VNIR, and PAN). All of the instrument channel chains perform generically the same functions:

1. Data Reception from the instrument.
2. Data re-synchronization to the local (PFCU) processing frequency.

3. Data merging and re-ordering, if applicable, to convert from the instrument interface read-out order of Band Interleave by Pixel (BIP) or Band Interleave by Line (BIL) to Band Sequential (BSQ) for VNIR and SWIR.
4. Band Selection (removal of unwanted bands), if applicable, as selected by ground command.
5. Packet Generation (CCSDS function). This formats the instrument channel data into packets according to CCSDS-recommended standard. Provisionally, each image frame is mapped into a single packet or group of packets.
6. Segmentation Function (CCSDS function). This function segments the packets on each data channel into CCSDS Transfer Frames using the Advanced Orbiting System (AOS) format. Each channel will be carried on a separate CCSDS Virtual Channel.

10.3.2.2 Ancillary Data

Ancillary data is generated by the Payload Controller at the same time as the detector frame data are acquired and sent to MMU for recording—this comprises selected spacecraft/payload status and operational telemetry during the image acquisition period as required for interpretation of the detector frame data on the ground. The two sources of data are (1) payload ancillary data gathered via the Instrument Control and Monitor through the Command and Telemetry Interfaces to the Payload Controller, and (2) other payload elements and Spacecraft Bus ancillary data supplied over the serial telemetry and command.

The housekeeping data packetization is handled in the Payload Controller software leaving only the segmentation function to be done in hardware.

10.3.2.3 Virtual Channel Multiplexing

To optimize the number of MMU inputs and separate the storage files needed, some inputs will need to be combined after segmentation. This is handled by the multiplexing function that multiplexes selected CCSDS virtual channels (transfer frames) onto a single MMU input data stream. A possible candidate grouping for such merging is the PAN channel and the ancillary data channel.

10.3.3 Onboard Data Compression

Due to the extremely high data rate and large data volume generated by hyperspectral sensors, onboard data compression is sometimes adopted to reduce the data volume. Among the twenty-five (25) spaceborne hyperspectral sensors described in Chapter 2, at least five of them have been and will be deployed with an onboard data compression unit. These sensors include three earlier space exploration hyperspectral sensors: Visible and Infrared Thermal Imaging Spectrometer (VIRTIS; Coradini et al. 2007), Compact Reconnaissance Imaging Spectrometer for Mars (CRISM; Murchie et al. 2007, 2009), and Moon Mineralogy Mapper (M3; Green et al. 2011), one recently (December 2019) launched Hyperspectral Imager Suite (HISUI) aboard the International Space Station (Matsunaga et al. 2016) and the one to be launched German Hyperspectral Imager for Environmental Mapping and Analysis (EnMAP) (Kaufmann et al. 2016).

10.3.3.1 VIRTIS Lossless and Lossy Data Compression

In the VIRTIS sensor, either lossless or lossy data compression were used (Coradini et al. 2007). The lossless compression algorithm was derived from that developed for the VIMS/Cassini (Baines et al. 2000) and OMEGA/Mars Express (Bibring et al. 2004) imaging spectrometers. The lossless compression algorithm performed differential pulse code modulation (DPCM) as a preprocessing to take the advantage of spatial and spectral correlations of the spectral imaging data, then a Rice coding was applied on the residuals. The compression was mainly used for validation purposes early in the mission. Typical compression ratios of 2–4 were achieved depending on the entropy of the data to be compressed. The lossy compression algorithm was based on wavelet transforms. The following compression modes with related compression ratios were used:

- A DPCM was performed followed by Rice encoding of the residuals. The lossless compression ratio is low (about 1.5:1) for very noisy VIRTIS-M IR data and high (about 14:1) for synthetic IR ramp data. The nominal average compression ratio is about 2:1–4:1.
- Wavelet F1 was low lossy compression. The compression ratio was 8:1 (16-bit to 2-bit).
- Wavelet F2 was medium lossy compression. The compression ratio was 10.67:1 (16-bit to 1.5-bit).
- Wavelet F3 was high lossy compression. The compression ratio was 16:1 (16-bit to 1-bit).

10.3.3.2 CRISM Data Editing and Lossless Compression

In the CRISM sensor, a so-called data *editing* compression is applied to the raw data readout from the focal plane (Murchie et al. 2007). All CRISM raw data are read off the detector array in a 14-bit format and are compressed in real-time in hardware. Compression options, in succession, are as follows:

1. Raw data are read off the detector arrays as 14-bit values, subtracted offset on a line-by-line basis. The offsets are set by uploading a data structure to the Data Processing Unit (DPU).
2. The difference is scaled by a wavelength-dependent gain to yield a 12-bit format, on a line-by-line basis. The gains are set by uploading a data structure to the DPU.
3. One of the four commandable wavelength filters is then used to select spectral bands (rows) of the frame of 12-bit data and save only the selected spectral bands. The total number of spectra (rows) with useful signal is 544 (107 in the VNIR region and 437 in the IR region).
4. The selected spectral bands are binned spatially with one of the four selected options 1×1, 2×2, 5×5, or 10×10 in the spatial direction. No pixel is binned in the spectral direction.
5. Optionally, conversion from 12-bits to 8-bit using one of eight look-up tables specified on a line-by-line basis: the line-by-line choices are set by uploading a data structure to the DPU. The look-up tables (LUTs) are stored in nonvolatile memory and are not uploadable.
6. Optionally, DPCM + lossless Fast compression.

10.3.3.3 M3 Lossless Data Compression

In the M3 sensor, a complete electronic and computer signal chain was designed to operate the detector array and to digitize, compress, format, and store the measurements prior to downlink. There was an important mission constraint on the spacecraft downlink capacity, that is, the spacecraft downlink rate was 8.3 Mbit/sec to be shared among several high data rate instruments.

To meet the lunar science mapping objectives and to accommodate the constraint of the data downlink rate from the spacecraft to the Earth receiving station, two M3 instrument measurement modes were designed: (1) full spectral and spatial resolution measurement, and (2) reduced spatial and spectral sampling.

The full resolution measurement mode was designated "Target Mode." In this mode, data are collected at full resolution with 640 cross-track detector elements (600 are lunar surface imaging) and 260 spectral detector elements for every sample in the cross-track line with a nominal spatial sampling of 70 m. With a spacecraft velocity of 1544 m/sec and 12-bit sampling, this corresponds to a data rate of 44 Mbit/sec. During the nominal 120-min orbit with half illumination the potential per orbit data rate for M3 is 158 Gbit/orbit.

The reduced spatial and spectral sampling mode was designated "Global Mode" to collect full lunar coverage at reduced spatial and spectral sampling to allow efficient global reconnaissance with M3. With this mode, M3 spatially averages 2 × 2 spatial elements and designated portions of the spectral range by a factor of 2 and 4. The ranges of spectral averaging are spectral detector elements 1–32 (430–720 nm) by 4, 33–116 (730–540 nm) by 2, and 117–260 (1550–3000 nm) by 4. This resulted in a Global Mode data set with 86 spectral bands. The effective data rate reduction (compression) of Global Mode with respect to Target Mode is 12:1.

To further reduce the data rate of M3 data, a lossless compression algorithm was added to the electronics design for both Target Mode and Global Mode data. An average compression ratio of 1.5:1 was achieved.

With collection of data over half an orbit, the baseline data rate with Global Mode, and lossless compression, the data volume is 8.8 Gbit/orbit that is well within the 16-Gbit capacity of the M3 MMU. With these measurement modes, a mission mapping strategy was developed to measure the full surface of the moon in the first 6 months with Global Mode and acquire as much as 25% of the moon surface in the final 18 months of the nominal 2-year mission (Boardman et al. 2011).

10.3.3.4 HISUI Onboard Lossless Data Compression

HISUI consists of a hyperspectral imager and a multispectral imager and was delivered to the International Space Station on December 6, 2019. Onboard lossless data compression is used to reduce the data volume approximately to 70% (corresponding to a compression ratio 1.42:1) of the original data (Matsunaga et al. 2016).

The lossless compression algorithm is an updated version of the Fast, Efficient, Lossless Image Compression System (FELICS) (Howard and Vitter 1994), which is a hierarchical predictive coding method with resolution scaling. In order to improve the performance of image decorrelation and entropy coding of the FELICS, a two-dimensional (2D) interpolation prediction is introduced to decorrelate image data, and an adaptive Golomb-Rice encoder is used to replace the entropy encoder. The compression speed is almost 30 times faster than JPEG-LS or lossless JPEG2000, whereas the compression ratio is only 1% or 5% less than that obtained using JPEG-LS or lossless JPEG2000 (Hihara et al. 2015).

Thanks to the simplicity of the compression algorithm that very little hardware resource is required to implement the compression. The compression algorithm and the onboard calibration algorithm is implemented by a Field Programmable Gate Array (FPGA) chipset rather than an Application-Specific Integration Circuit (ASIC) circuitry. As a consequence, the circuitry for onboard compression, radiance calibration, and smile characteristics is embedded into the front-end of the electronics unit.

10.3.3.5 EnMAP Onboard Data Compression

The Environmental Mapping and Analysis Program (EnMAP) is a German hyperspectral satellite mission scheduled to be launched in 2020. EnMAP plans to use lossless data compression based on the CCSDS Standard 122—Image Data Compression (Qian 2013b). The onboard lossless compression will be performed individually for each spectral band offline during data take intervals. This wavelet transform based compression algorithm achieves an average compression ratio of 1.6 with a dedicated FPGA chipset. The MMU output is routed through a CCSDS coding unit directly to the 320-Mbit/sec X-band downlink system (Kaufmann et al. 2016).

10.3.3.6 Vector Quantization Based Near-Lossless Data Compression

Vector quantization (VQ) is an efficient coding technique for data compression of hyperspectral imagery because of its simplicity and its vector nature for preservation of the spectral signatures associated with individual ground samples in the scene. The author of this book has developed a series of VQ-based near-lossless data compression techniques for hyperspectral imagery (Qian 2010). Many innovative techniques for on-ground and onboard uses have been developed and patented

(Qian et al. 1996, 1997, 1998, 2000, Manak et al. 1998, Qian 2004, 2006, Qian and Hollinger 2004a, 2004b, 2004c, 2009). Reasonably high near-lossless compression ratios (>10:1) have been achieved for hyperspectral data in remote sensing applications (Qian et al. 2006).

A VQ compression procedure includes two main steps: codebook training (also referred to as codebook generation) and encoding (i.e., codevector matching). In the training step, similar vectors in a training sequence are grouped into clusters and each cluster is expressed by its centroid called the codevector. In the encoding step, each input vector is then compressed by replacing it with the best-matched codevector referenced by a single index of the matched codevector. The index (or address) of the matched codevector in the codebook is then transmitted to the decoder over a channel, and is used by the decoder to retrieve the same codevector from an identical codebook. This is the reconstructed reproduction of the corresponding input vector. Compression is thus obtained by transmitting the index of the codevector rather the input vector itself.

In VQ compression of hyperspectral data, the full spectral profile of each ground sample within a scene is defined as a vector, which is often referred to as a spectral vector. Since the number of unique targets in the scene of a datacube is limited, the number of trained spectral vectors (i.e., codevectors) can be much smaller than the total number of spectral vectors of the datacube. Thus, all the spectral vectors of the datacube can be expressed using a codebook with comparatively few codevectors and good reconstruction fidelity can be achieved.

VQ compression techniques make good use of the high correlation between spectral bands and in the spatial domain of a hyperspectral datacube, and thus achieve a high compression ratio. However, a big challenge to VQ compression techniques for hyperspectral imagery in terms of operational use is that it requires large computational resources, particularly in the codebook generation step. Since the size of hyperspectral datacubes can be hundreds of times larger than those of grey-level or multispectral images, the processing time required to train a codebook or to encode a datacube using the codebook could also be tens to hundreds of times larger.

To bring the traditional VQ-based compression techniques to practical usage for hyperspectral data, Qian et al. (1996) began introducing innovative solutions to significantly reduce the computational complexity over two decades ago. They proposed an efficient method of representing spectral vectors of hyperspectral datacubes, referred to as Spectral Feature Based Binary Code (SFBBC), which converts a spectral vector of $16N_b$ bits into a SFBBC code of $3N_b$-2 bits (N_b is the number of spectral bands, e.g., $N_b = 224$ in the case of AVIRIS). More importantly, with the SFBBC code, the Hamming distance, rather than the Euclidean distance, can be used in the codebook training and codevector matching process of the VQ compression. The Hamming distance is much faster than the Euclidean distance, as it is a simple sum of logical bit-wise exclusive-or operation. By using the SFBBC code the conventional VQ compression technique can be speeded up by a factor of 30 to 40 at a fidelity loss of signal-to-noise ratio (SNR) <1.5 dB (Qian et al. 1996).

Qian et al. (1998) further speeded up the VQ compression process by using remote sensing knowledge contained in the hyperspectral datacube to be compressed. A spectral index, such as the Normalized Difference Vegetation Index (NDVI), was introduced to benefit the compression algorithm. A novel VQ compression technique referred to as the spectral index based Multiple Sub-Codebook Algorithm (MSCA) was developed (Qian et al. 2000). The MSCA can accelerate both the codebook generation time (CGT) and the coding time (CT) by a factor of around $n = 8$ or 16 (n is the number of regions being segmented on the index-map), while the reconstruction fidelity is almost the same as that of the conventional VQ compression technique.

Three compression systems for hyperspectral imagery were created and tested by combining the SFBBC, MSCA, and spectral vector sub-sampling, the overall processing speed of the conventional VQ compression could be improved by a factor of around 1000 at an average SNR penalty of <1.0 dB (Qian et al. 2000).

Qian and Hollinger (2004a, 2004b) invented two fast near-lossless VQ compression techniques: Successive-Approximation Multistage Vector Quantization (SAMVQ) (Qian and Hollinger 2004a) and Hierarchical Self-Organizing Cluster Vector Quantization (HSOCVQ) (Qian and Hollinger 2004b).

Both of them can restrict the compression error introduced in the compression process to the level lower than the intrinsic noise of the original data. The SAMVQ compresses hyperspectral data in multiple stages manner using extremely small stage-codebook to successively approach a given threshold, while the HSOCVQ works by adaptively clustering vectors and splitting the clusters hierarchically so that each vector is encoded with an error less than a given threshold. Near-lossless compression is achieved when the threshold is set to smaller than the floor of the original data's intrinsic noise.

The near-lossless properties of the SAMVQ and HSOCVQ were examined by comparing the compression errors with the intrinsic noise of the original data (Qian et al. 2006). The experimental results show that compression errors are smaller than the intrinsic noise. A multidisciplinary user acceptability study was carried out, which involved eleven user groups covering a wide range of application areas and a variety of hyperspectral sensors, to assess the impact of VQ-based compression techniques on Earth-observation applications. The users examined the compressed data qualitatively and quantitatively using well-understood data and predefined evaluation criteria and ranked the compressed data according to the impact on the applications using a double-blind test approach. Most users accepted the compressed data at a compression ratio between 10:1 and 50:1 (Qian et al. 2005).

Two versions of hardware compressor prototypes that implement the SAMVQ and HSOCVQ techniques for onboard usage were built. The first was targeted for real-time use, while the second was for non-real-time use. Figure 10.4 shows the VQ near-lossless data compressor board for

FIGURE 10.4 A prototype VQ compressor board (on the board there are 4 compression engines and a Network Switch, each of which uses a FPGA chip).

real-time usage (Qian and Hollinger 2007). The prototype board is composed of multiple standalone compression engines (CEs), each with the ability to compress a subset of spectral vectors in parallel. These are autonomous devices, and once programmed they perform compression in a continuous mode, subset by subset. A CE is composed of a FPGA chip. The prototype board also has a Network Switch, a Fast Memory, and a PCI bus interface. The Network Switch which is also composed of a FPGA chip is used to serve the data flow transfer in and out of each of the CEs serving up to 8 CEs in parallel using a high-speed serial link.

The design of the hardware compressor is capable of accepting varying datacube sizes, numbers of spectral bands, and codebook sizes. The system is truly scalable, as any number of CE components can be used according to the mission requirements. The prototype compressor has been benchmarked with the initial target for a 250 Hz frame rate with 1024 cross-track pixels, 200 bands, and 12-bit data resolution. This corresponds to a data rate of 614 Mbps. The experimental results showed that the compression ratio and fidelity achieved by the prototype compressor were similar to those obtained by software simulation.

10.3.3.7 CCSDS Recommended Data Compression Standards

Three data compression standards designed for data acquired by satellites or spacecraft have been developed by the Data Compression Working Group (DCWG) within the CCSDS. (The author of this book was a member of the working group and participated in the development of the standards.)

Although JPEG and JPEG2000 exist as the international standards for image compression, they were not designed for compression of satellite data and images. The compression algorithms recommended by the CCSDS are intended to be suitable for use aboard spacecraft. In particular, the complexity of the algorithms is designed to be sufficiently low to make high-speed hardware implementation feasible. The algorithms permit memory-efficient implementation that does not require large intermediate frames for buffering. In addition, the CCSDS standards are intended to limit the effects of data loss that can occur on the communications downlinking channel. Consequently, the compression algorithms are appropriate for frame-based image formats (two dimensions of data acquired simultaneously) produced, for example, by charge-coupled device (CCD) arrays (called image frames) as well as strip-based input formats (i.e., images acquired one line at a time).

The three CCSDS data compression standards are the following:

1. Lossless Data Compression: CCSDS 121.0-B-2 (ISO-15887:2013) for lossless compression of 1D science data (CCSDS 2012, ISO 2013a).
2. Image Data Compression: CCSDS 122.0-B-2 (ISO-26868:2009) for lossy-to-lossless image compression (CCSDS 2017, ISO 2009).
3. Low-Complexity Lossless and Near-Lossless Multispectral and Hyperspectral Image Compression: CCSDS 123.0-B-2 (ISO-18381:2013) for lossless and near-lossless multispectral/hyperspectral image compression (CCSDS 2019, ISO 2013b).

The CCSDS standard 121.0-B was design for lossless data compression by exploiting redundancy in one dimension of science data. It consists of a preprocessor and the adaptive entropy encoder with lower computational complexity. This standard has been used in many space missions since its first issue in May 1997 and the second issue released in May 2012.

The CCSDS standard 122.0-B is a wavelet transform-based image compressor with lower implementation complexity. It consists of a discrete wavelet transform (DWT) module that performs the DWT decomposition of 2D image data, and a bit-plane encoder (BPE) that encodes the transformed data. The standard supports two choices of DWT, an integer and a floating-point

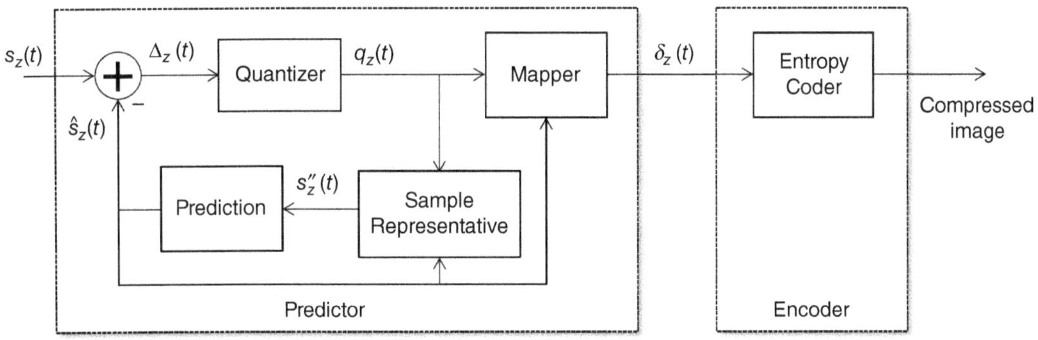

FIGURE 10.5 A black diagram of CCSDS 123.0-B-2 3D data compression standard. (Source of CCSDS.)

DWT, and thus is capable of providing both lossy and lossless compression. This standard has been used in many space missions since its first issue in November 2005 and the second issue released in September 2017.

The CCSDS standard 123.0-B defines a data compressor for multispectral and hyperspectral imagers and sounders that produce 3D datacubes. Issue 1 was released in May 2012 for lossless data compression of the 3D datacubes (CCSDS 2012b), which has been superseded. Issue 2 was released in February 2019 that extends the lossless multispectral and hyperspectral image compression standard to provide an effective method of performing either lossless or near-lossless compression of 3D image data (CCSDS 2019). The current issue of CCSDS 123.0-B compressor consists of a predictor followed by an encoder, as shown in Figure 10.5. The predictor uses an adaptive linear prediction method to predict the value of each image sample based on the values of nearby samples in a small 3D neighborhood. Prediction is performed sequentially in a single pass.

The major revision in the present issue of the CCSDS 123.0-B from the original issue (CCSDS 2012b) is that each prediction residual $\Delta_z(t)$, i.e., the difference between the predicted value and actual sample value, is quantized using a uniform quantizer. The quantizer step size can be controlled via an absolute error limit and/or a relative error limit, so that samples can be reconstructed with a user-specified error bound and/or samples predicted to have smaller magnitude can be reconstructed with lower error. Lossless compression of a band image is obtained simply by setting the absolute error limit to zero. The quantized prediction residual $q_z(t)$ is mapped to an unsigned integer mapped quantizer index $\delta_z(t)$, similar to the calculation of mapped prediction residuals in the original issue.

The first issue of CCSDS 123.0-B-1 standard for lossless data compression has been implemented by several user groups using FPGA chipsets (Keymeulen et al. 2014, Santos et al. 2016, Theodorou et al. 2016, Bascones et al. 2018, Fjeldtvedt et al. 2018). Keymeulen et al. (2014) reported a compression processing throughput of 520 Mbit/sec for compressing AVIRIS datacubes using Xilinx Virtex-5 FPGA with the parameter $P = 3$ (the number of previous bands used in prediction) and data input ordering BIP (band interleave by pixel). Theodorou et al. (2016) reported a compression processing throughput of 1790 Mbit/sec using Xilinx Virtex-5 FPGA with the parameter $P = 3$ and data input ordering BIP. Santos et al. (2016) reported a compression processing throughput of 2240 Mbit/sec of the SHyLoC implementation using Xilinx Virtex-5 FPGA with the parameter $P = 15$ for any data input ordering (BIP, BSQ, and BIL). Bascones et al. (2018) reported a compression processing throughput of 760 Mbit/sec using Xilinx Virtex-7 FPGA with the parameter $P = 15$ and data input ordering BIP. Fjeldtvedt et al. (2018) reported a compression processing throughput of 2350 Mbit/sec using Xilinx Zynq-7020 FPGA with the parameter $P = 15$ and data input ordering BIP. None of these implementations has been deployed onboard a spacecraft.

Since the original issue of CCSDS 123.0-B-1 standard has been superseded upon the release of the second issue of the standard in February 2019, it is expected that reports on new hardware implementation of the second issue of CCSDS 123.0-B standard will come up.

10.3.4 Transmission Chain

The transmission chain implements the downlink encryption function and consists of the following elements:

A data read back link from the MMU (for each of the redundant PFCUs).

- A data Downlink Encryption block using a selected method of the Advanced Encryption Standard algorithm.
- Forward Error Correction (FEC) encoder (if not provided by the downlink transmitters). Assuming an 8-PSK modulation, this will comprise an RS-block coder concatenated with a rate 5/6 Trellis (convolutional) code.
- An interface to the X- or Ka-band transmitter.

The downlink FEC encoding is provided by the transmitter. The transfer rates for the MMU read back channel is precisely controlled by the PFCU which generates the transfer clock for the operation.

10.3.5 Control and Monitoring

Since the Mission Control Center (MCC) is generally out of contact with the spacecraft, both the ICODH subsystem and the Spacecraft Bus must be pre-commanded. The use of time-tagged commands scripts is adopted to control the imaging sessions and the data transmission to ground stations.

The exact mix of command and telemetry types will depend on the requirements of the different instruments. There are different types of command and telemetry, such as standard pulsed digital, digital level, analogue, serial command, and serial telemetry, the latter including the ESA style (OBDH), MIL STD 1553B, and CAN-Bus.

10.3.6 Timing Generation

Timing is based on a stabilized reference clock Local Oscillator (e.g., 5 MHz) plus associated phase-locked loop frequency synthesizer and synchronous divide chains to generate both internal timing required by the ICODH subsystem and timing required by the different instruments.

Instrument timing generation function will, as a minimum, support the Start Pulse generation to the different instruments to ensure that they are synchronized. Provisionally, the SWIR, VNIR, and PAN are synchronized as a group. Additional timing may be required by individual instruments. It is expected that this instrument timing will need to be adjustable from the ground to compensate for on-orbit changes.

The function will also support a precise satellite time which is initialized to match the time in the Spacecraft Bus by command and maintained with the support of the Spacecraft Reference Pulse of 1 Pulse Per Second (PPS). This is used both for control and monitor timing, and for tagging instrument data packets with the time data when they were generated.

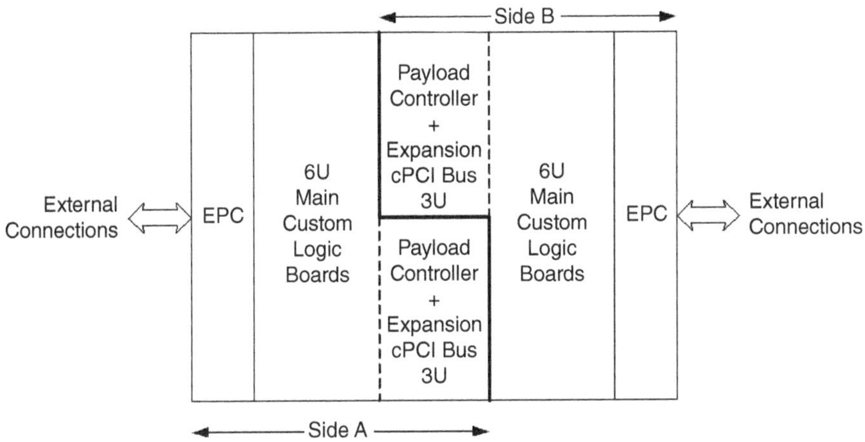

FIGURE 10.6 An example of the PFCU packaging layout.

10.3.7 Redundancy and Packaging

The PFCU comprises two standby redundant sides contained in a common housing. The two sides are electrically isolated from each other.

Figure 10.6 shows an example of the packaging layout of the PFCU. For each side, the processor and associated memory and I/O expansion boards use a standard board format with a compact Peripheral Connect Interface (cPCI) backplane and a half (e.g., 3U) card width. The main custom logic boards use a full card width (e.g., 6U) with custom backplane that contains a partial extension of the cPCI bus for interfacing purposes. This 6U format allows efficient packaging of the large devices and extensive memory required by this part of the unit. The electronic power conditioner (EPC) module is also configured with a 6U width.

The result is an L-shaped profile, with the two sides combined into an overall rectangular 6U chassis by rotating one side by 180° and interlocking it with the other side.

10.4 MASS MEMORY UNIT

The formatted data are stored in the MMU. These data include uncompressed raw (or compressed if compression mode is selected) image data from VINIR and SWIR detectors, PAN data, ancillary data, and the telemetry data from the spacecraft that are to be transmitted to ground. The formatter data bus is 16-bit wide. The data can be either stored as 16/32-bit words or larger words (128-bit wide) using an external multiplexer at the input and a de-multiplexer at the output. The maximum volume of data to be stored is defined by each mission.

The MMU is required to provide the following:

- At least a 4 Tb storage capacity (mission depended).
- 4 input ports at a rate of, for example, 350 Mbit/sec for SWIR1, SWIR2, VNIR, Pan + Ancillary data. This assumes that the SWIR is split into two inputs to even up data rates and that lower rate channels (e.g., PAN and Ancillary) can be combined.
- 1 output port at a peak rate of, for example, 1000 Mbit/sec required on the downlink output.
- Read-while-write to support transmission of previously acquired images while acquiring a new set of images. The total throughput required (write plus read) is up to 2400 Mbit/sec, assuming that the worst case (compression + acquire + transmit) is required.

10.5 PROXIMITY ELECTRONICS AND SERVICE MODULE ELECTRONICS

10.5.1 FPAs Proximity Electronics

An example of architecture for the VNIR and SWIR FPA Proximity Electronics is depicted in Figure 10.7. The function of the blocks in the figure is to provide all the needed control signals and biases for the detector arrays, to process the relevant output electrical signals (pixels) and to convert them to digital numbers. The interface between the spectrometers and the PFCU is based on dedicated serial lines using a custom protocol.

In a hyperspectral payload, typically a CCD array or a silicon complementary metal-oxide-semiconductor (CMOS) array is used for a VNIR spectrometer and a mercury-cadmium-telluride (HgCdTe) detector array is used for a SWIR spectrometer. Both the detector arrays are integrated on the same read-out integrate circuit (ROIC). This approach allows having a common modular design for both proximity electronics. The proximity electronics normally is placed as near as possible to FPAs, still maintaining a good thermal insulation with respect to the Optics Head to limit the heat flux toward the cold part of the instrument.

The operating concept of the VNIR and SWIR proximity electronics shown in the figure is based on the assumption that everything is under control from the PFCU. This approach assumes that up to 8 A/D channels are multiplexed across a single one. The specific arrangement shown the uses of separate interfaces for the prime and redundant PFCU. The proximity electronics units receive the

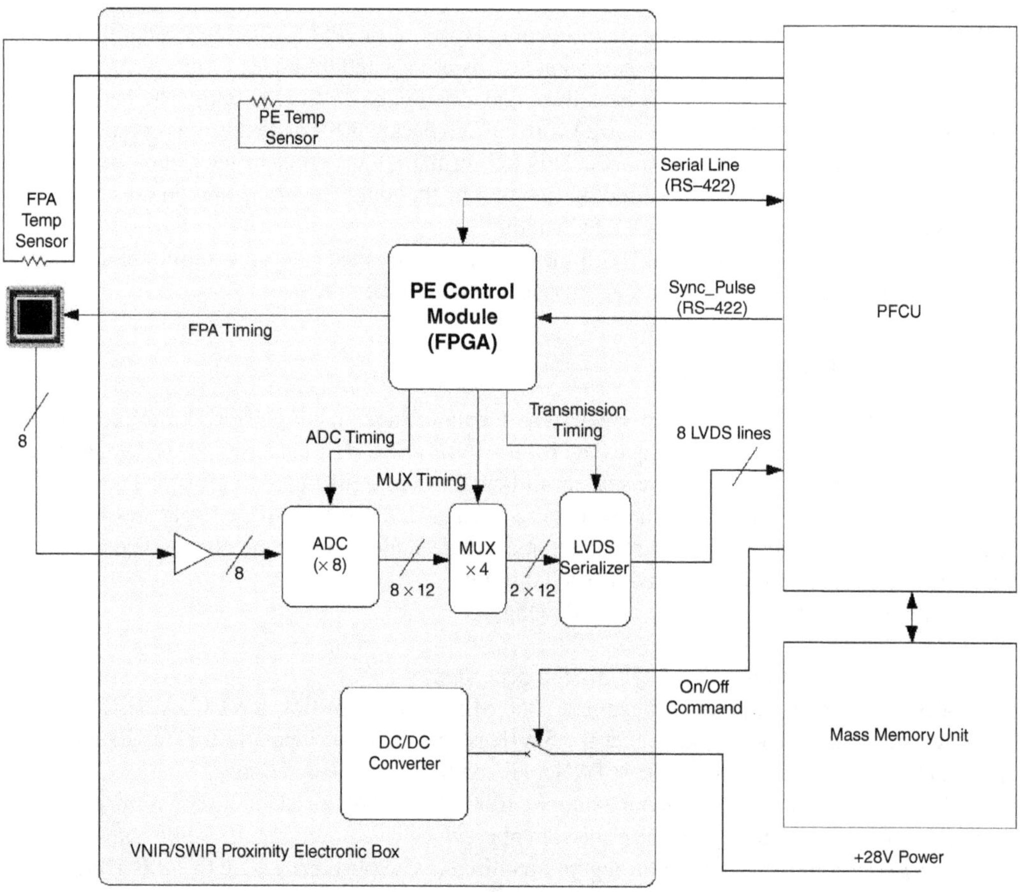

FIGURE 10.7 A functional block diagram of VNIR/SWIR proximity electronics.

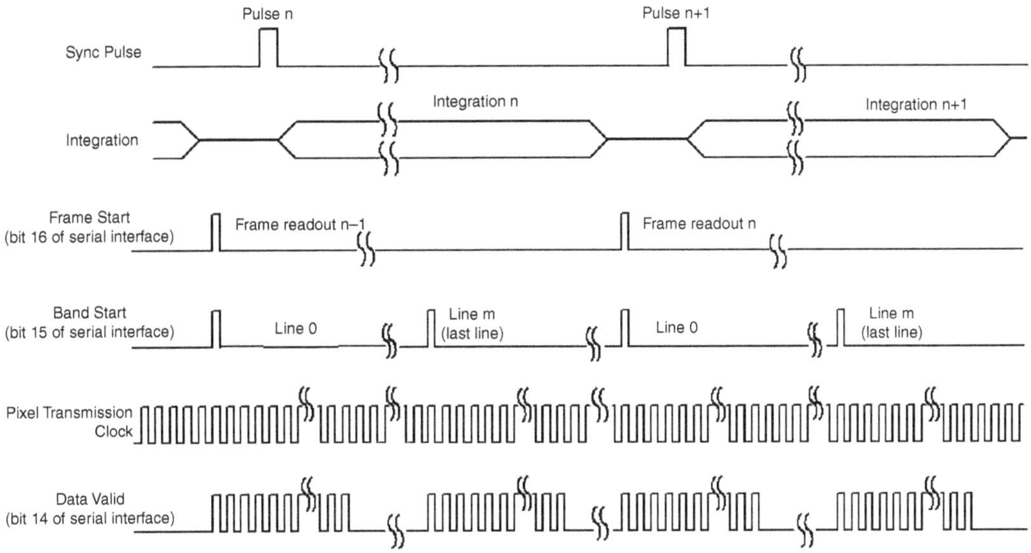

FIGURE 10.8 VNIR/SWIR proximity electronics serial link timing.

synchronisms (digital pulse) for image integration start, and provides the PFCU with the digitized pixels data. The frame synchronization and transmission concept is depicted by the serial link timing as shown in Figure 10.8.

10.5.2 Service Module Electronics

Figure 10.9 shows an example of architecture for the service module electronics. It is controlled by the PFCU via a serial link, such as RS-422. Commanded by the controller, it collects temperature from the various temperature sensors for monitoring the temperature-sensitive components. It provides power to drive the shutter actuator for imaging acquisition or in-flight calibration. The service module electronics provides power to heaters to keep the temperature for the VNIR focal plane array, SWIR focal plane array, and other components. Some hyperspectral payloads may equip with decontamination heaters. These heaters could also be used for annealing. The service module electronics provides power to the annealing/decontamination heaters to allow the decontamination/annealing of the detector arrays.

The service module electronics is in charge of the calibration unit management under the commands of the PFCU. In particular, if the calibration unit is based on a diffuser mounted on the internal surface of the cover that is used with some different inclination angles with respect to the sun in order to provide different levels of input radiance. A second diffuser panel can then be mounted on this diffuser in order to protect it when it is not in use.

10.6 ELECTRICAL INTERFACES

10.6.1 Cross-Strapping

Figure 10.10 shows a simplified diagram of the interfaces within the ICODH and between the ICODH and external entities. In this representation, the existence of one or more interfaces between entities is shown as a single line—the arrows on the line indicate the composite direction of all interfaces (a bidirectional line implies some mix of input, output, and bidirectional interfaces). Following subsections identify the types and count for each interface.

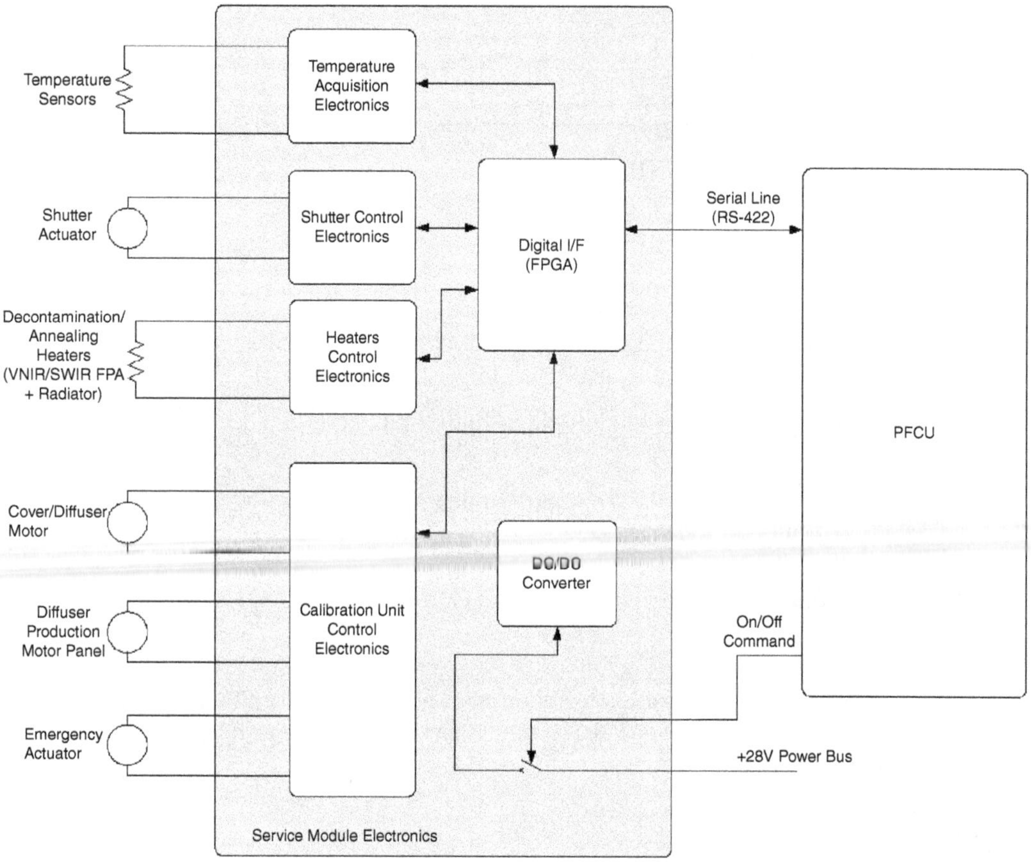

FIGURE 10.9 A service module electronics functional block diagram.

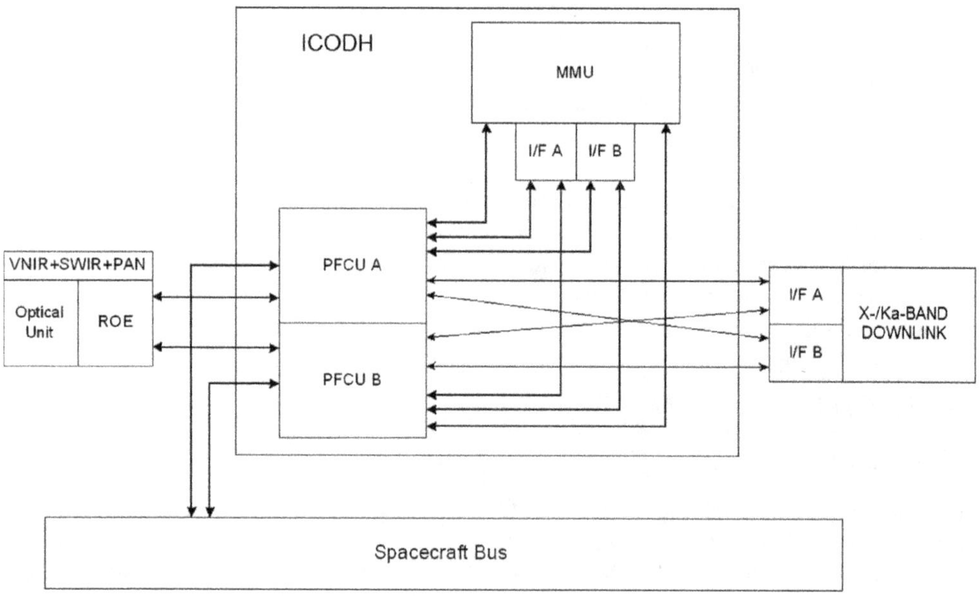

FIGURE 10.10 ICODH interface cross-strapping.

TABLE 10.1
An Example of Image Data Interfaces per Spectrometer

Instrument	Read-Out Channels	Total Data Rate Mbit/sec	Serial Interfaces
VNIR	8	350	2
SWIR	8	700	2
PAN	2	100	1

The diagram shows outline details of the proposed redundancy, using the nomenclature A and B to denote standby redundant functions.

10.6.2 Instrument Interfaces

The image data interfaces are in serial and are based on a matched pair of devices. Each serial interface comprises 4 low-voltage differential signaling (LVDS) (3 data + clock) sent over shielded twisted pairs and can support data rates up to 200 Mbyte/sec. In practice, other limitations in the electronics will limit the data rate further. Table 10.1 shows an example of the number of serial interfaces per spectrometer.

Serial command and telemetry interfaces can be used to control and monitor the spectrometers with 4 independent interfaces:

- SWIR
- VNIR
- PAN
- Hyperspectral Service Module

The serial interface comprises 5 signals as follows. The first 4 signals are from ICODH to the spectrometers, whereas the last signal is from the spectrometers to the ICODH.

1. Shift Clock: This is a burst-mode clock used to shift/sample the serial data. When active, it has a 50% duty cycle.
2. Gate: This indicates when data are presented to the spectrometers or should be provided by the spectrometers.
3. Direction: This signal indicates whether data are being sent to the spectrometers or are being sent to the ICODH.
4. Control Data: This is serial data to the spectrometers.
5. Monitor Data: This is serial data from the spectrometers to the ICODH.

Control and monitor signal data transfers can use a common timing approach, as shown in Figure 10.11. Each signal data transfer comprises two phases. In the first phase, control data are sent from the ICODH to the spectrometers, while in the second phase, the spectrometers send reply data to the ICODH The reply data is selected by the content of the control data.

The gap between the control phase gate and the reply phase gate is set by the time needed by the spectrometers to get the reply. Provisionally, this is assumed to be 4 clock periods.

The interface signals can be RS 422 operating in balanced differential mode over a shielded twisted pair of nominal impedance $Z_0 = 120\ \Omega$. Reverse termination is used (i.e., series resistors of $Z_0/2$ in each output of the transmitter).

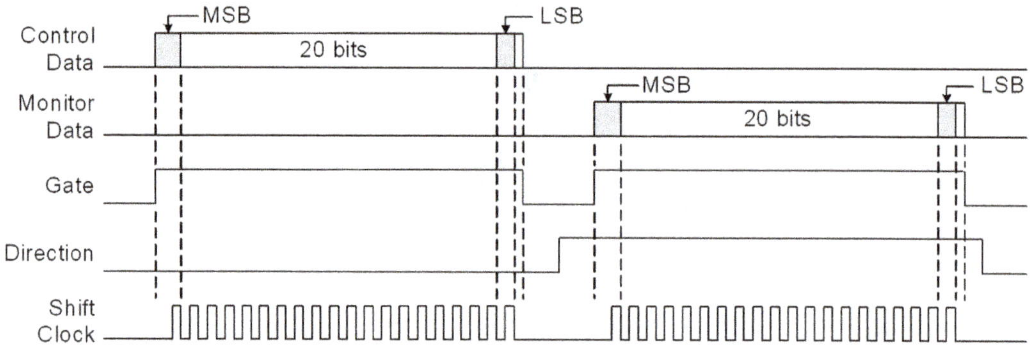

FIGURE 10.11 Serial command and telemetry interface timing.

The shift clock frequency will be a convenient division of the PFCU reference clock. This latter is nominally 5 MHz, suggesting shift clocks of either 1.25 MHz (1/4) or 625 kHz (1/8).

The message burst is a 20-bit (in this example) value divided as follows:

- 4-bit Function Code,
- 16-bit data.

The function code meaning depends on the spectrometers' interface, with the same code value being re-used for different applications on each interface. There is no a priori partitioning of codes between Command and Monitor operations (a Command operation in one spectrometer could be a Monitor operation in another).

For Command operations, the control message function code identifies the command type, while the data gives a command-specific parameter value. The reply in this case is the command confirmation and should be the value actually transferred to the controlling register (not just a re-shift of the input serial value). The reply function code must match that in the control message.

For Monitor operations, the control message function code identifies the parameter to be monitored and the data has no significance. The reply message gives the parameter value. Again, the function code must match that in the control message.

The PFCU provides the spectrometers with a reference clock frequency derived from a stable local oscillator. In addition, it will provide a Frame Sync pulse to each spectrometer channel.

10.6.3 MMU Interfaces

The input data interface to the MMU depends on the chosen supplier of the MMU. It is expected to support at least 400 Mbit/sec per interface with 4 active interfaces. There is 1 output data interface with byte-wide serial interface at the information rate of 1000 Mbit/sec.

A serial control and monitor link is used to provide multi-bit operations of the MMU, such as file management control and monitoring.

Other control and monitoring is done by discrete commands and telemetry similar to those for the spectrometers.

10.6.4 X-/Ka-Band Downlink Interfaces

Data transfer to the X-/Ka-band downlink can be done over a byte-wide serial interface at the information rate of 1000 Mbit/sec. It is assumed that downlink FEC coding is done in the transmitter.

The X-/Ka-band control and monitoring is done by discrete commands and telemetry similar to those for the spectrometers. In addition, a special high-power pulse command can be used to control redundancy switching in the radio frequency feed to the antenna.

10.6.5 Spacecraft Bus Interfaces

The spacecraft bus electrical interfaces are defined by the bus and are nominally as follows:

- Power is at 28 V DC nominal with separate power feeds to all payload units.
- Pulse commands are used to switch the PFCU on and off, this in turn controls the on/off switching of the other payload units.
- A MIL-STD 1553B serial interface is used to send multi-bits commands to the PFCU that control the payload operation. This same interface is used to transfer bus ancillary data to the PFCU.
- A 1-PPS signal provides a GPS reference pulse used to reference all onboard time to Coordinated Universal Time (UTC). This referencing is implemented in the PFCU.

REFERENCES

Baines, K. H., et al. 2000. Detection of sub-micron radiation from the surface of venus by Cassini/VIMS. *Icarus* 148:307–311.

Bascones, D., C. Gonzalez, and D. Mozos. 2018. FPGA implementation of the CCSDS 1.2.3 standard for real-time hyperspectral lossless compression. *IEEE Journal of Selected Topics in Applied Earth Observations and Remote Sensing* 11(4):1158–1165.

Bibring, J.-P., et al. 2004. Perennial water ice identified around the mars south pole. *Nature* 428: 627–630.

Boardman, J. W., C. M. Pieters, R. O. Green, S. R. Lundeen, P. Varanasi, J. Nettles, N. Petro, P. Isaacson, S. Besse, and L. A. Taylor. 2011. Measuring moonlight: An overview of the spatial properties, lunar coverage, selenolocation, and related Level 1B products of the Moon Mineralogy Mapper. *Journal of Geophysical Research* 116, E00G14:1–15.

CCSDS. 2012a. Lossless data compression. Issue 2.*Recommendation for Space Data System Standards (Blue Book)*, 121.0-B-2. Washington, D.C.: CCSDS, May 2012 (available at https://public.ccsds.org/Pubs/121x0b2ec1.pdf, accessed on May 23, 2019).

CCSDS. 2012b. Lossless multispectral & hyperspectral image compression. Issue 1. *Recommendation for Space Data System Standards (Blue Book)*, 123.0-B-1. Washington, D.C.: CCSDS, May 2012.

CCSDS. 2017. Image data compression. Issue 2. *Recommendation for Space Data System Standards (Blue Book)*, 122.0-B-2. Washington, D.C.: CCSDS, November 2017 (available at https://public.ccsds.org/Pubs/122x0b2.pdf, accessed on May 23, 2019).

CCSDS. 2019. Low-complexity lossless and near lossless multispectral and hyperspectral image compression. Issue 2. *Recommendation for Space Data System Standards (Blue Book)*, 123.0-B-2. Washington, D.C.: CCSDS, February 2019. (available at https://public.ccsds.org/Pubs/123x0b2.pdf, accessed on May 23, 2019).

Coradini, A., et al. 2007. VIRTIS: An imaging spectrometer for the Rosetta Mission. *Space Science Reviews* 128(1–4):529–559.

Fjeldtvedt, J., M. Orlandic, and T. A. Johansen. 2018. An efficient real-time FPGA implementation of the CCSDS-123 compression standard for hyperspectral images. *IEEE Journal of Selected Topics in Applied Earth Observations and Remote Sensing* 11(10):3841–3852.

Green, R. O., et al. 2011. The Moon Mineralogy Mapper (M3) imaging spectrometer for lunar science: Instrument description, calibration, on-orbit measurements, science data calibration and on-orbit validation. *Journal of Geophysical Research* 116, E00G19:1–31.

Hihara, H., et al. 2015. Onboard image processing system for hyperspectral sensor. *Sensors* 15: 24926–24944.

Howard, P. G. and J. S. Vitter. 1994. Fast progressive lossless image compression. *Proceedings of SPIE* 2186:1–12.

ISO. 2009. Space data and information transfer systems—Lossless Image Compression, *ISO*-26868:2009 (available at http://www.iso.org/iso/iso_catalogue/catalogue_tc/catalogue_detail.htm?csnumber=43849, accessed on May 23, 2019)

ISO. 2013a. Space data and information transfer systems—Lossless data compression, *ISO-15887*:2013 (available at https://www.iso.org/standard/62314.html, accessed on May 23, 2019).

ISO. 2013b. Space data and information transfer systems—Lossless multispectral and hyperspectral image compression, *ISO-18381:*2013 (corresponding to CCSDS 123.0-B-1) (available at http://www.iso.org/iso/home/store/catalogue_tc/catalogue_detail.htm?csnumber=62319, accessed on May 23, 2019)

Kaufmann, H., et al. 2016. Environmental mapping and analysis program – A German hyperspectral mission. In *Optical Payloads for Space Missions*, ed. S.-E. Qian, 161–181. West Sussex, UK: John Wiley & Sons.

Keymeulen, D., N. Aranki, A. Bakhshi, H. Luong, C. Sarture, and D. Dolman. 2014. Airborne demonstration of FPGA implementation of fast lossless hyperspectral data compression system. *Proceedings of NASA/ESA Conference on Adaptive Hardware and Systems*, pp. 278–284.

Manak, D., S.-E. Qian, A. Hollinger, and D. Williams. 1998. Efficient hyperspectral data compression using vector quantization and scene segmentation. *Canadian Journal of Remote Sensing* 24:133–143.

Matsunaga, T., et al. 2016. Hyperspectral Imager Suite (HISUI): Japanese spaceborne hyperspectral imager for resource and environmental mapping. In *Optical Payloads for Space Missions*, ed. S.-E. Qian, 215–222. West Sussex, UK: John Wiley & Sons.

Murchie, S., et al. 2007. Compact reconnaissance imaging spectrometer for mars (CRISM) on mars reconnaissance Orbiter (MRO). *Journal of Geophysical Research* 112, E05S03:1–57.

Murchie, S., et al. 2009. Compact reconnaissance imaging spectrometer for mars investigation and data set from the mars reconnaissance Orbiter's primary science phase. *Journal of Geophysical Research* 114, E00D07:1–15.

Qian, S.-E. 2004. Hyperspectral data compression using a fast vector quantization algorithm. *IEEE Transactions on Geoscience and Remote Sensing* 42(8):1791–1798.

Qian, S.-E. 2006. Fast vector quantization algorithms based on nearest partition set search. *IEEE Transactions on Image Processing* 15(8):2422–2430.

Qian, S.-E. 2010. A decadal R&D of near lossless data compression on-board satellites at Canadian Space Agency. *Journal of Applied Remote Sensing* 4, 041797:1–22.

Qian, S.-E. 2013a. *Optical Satellite Signal Processing and Enhancement.* Bellingham: SPIE Press.

Qian, S.-E. 2013b. *Optical Satellite Data Compression and Implementation.* Bellingham: SPIE Press.

Qian, S.-E., M. Bergeron, I. Cunningham, L. Gagnon, and A. Hollinger. 2006. Near lossless data compression on-board a hyperspectral satellite. *IEEE Transactions on Aerospace and Electronic Systems* 42(3):851–866.

Qian, S.-E., and A. Hollinger. 2004a. System and method for encoding/decoding multi-dimensional data using successive approximation multi-stage vector quantization (SAMVQ). *U. S. Patent* No. 6,701,021 B1, issued on March 2, 2004.

Qian, S.-E., and A. Hollinger, 2004b. System and method for encoding multi-dimensional data using hierarchical self-organizing cluster vector quantization (HSOCVQ). *U. S. Patent* No. 6,724,940 B1, issued on April 20, 2004.

Qian, S.-E., and A. Hollinger. 2004c. Method and system for compressing a continuous data flow in real-time using recursive hierarchical self-organizing cluster vector quantization (HSOCVQ). *U.S. Patent* No. 6,798,360 B1 issued on September 28, 2004.

Qian, S-E., and A. Hollinger. 2007. Current status of satellite data compression at Canadian Space Agency (invited paper). *Proceedings of SPIE* 6683(04):01–12.

Qian, S.-E., and A. Hollinger. 2009. Method and system for compressing a continuous data flow in real-time using cluster successive approximation multi-stage vector quantization (SAMVQ). *U.S. Patent* No. 7,551,785 B2, issued on June 23, 2009.

Qian, S.-E., A. Hollinger, M. Bergeron, I. Cunningham, C. Nadeau, G. Jolly, and H. Zwick. 2005. A Multi-disciplinary user acceptability study of hyperspectral data compressed using onboard near lossless vector quantization algorithm. *International Journal of Remote Sensing* 26(10):2163–2195.

Qian, S.-E., A. Hollinger, D. Williams, and D. Manak. 1996. Fast 3D data compression of hyperspectral imagery using vector quantization with spectral-feature-based binary coding. *Optical Engineering* 35:3242–3249.

Qian, S.-E., A. Hollinger, D. Williams, and D. Manak, 1997. 3D data compression system based on vector quantization for reducing the data rate of hyperspectral imagery. In *Applications of Photonic Technology II*, ed. G. Lampropoulos, 641–654. New York: Plenum Press.

Qian, S.-E., A. Hollinger, D. Williams, and D. Manak. 1998. 3D data compression of hyperspectral imagery using vector quantization with NDVI-based multiple codebooks. *1998 IEEE Geoscience and Remote Sensing. Symposium Proceedings* 3:2680–2684.

Qian, S.-E., A. Hollinger, D. Williams, and D. Manak. 2000. Vector quantization using spectral index based multiple sub-codebooks for hyperspectral data compression. *IEEE Transactions on Geoscience and Remote Sensing* 38(3):1183–1190.

Santos, L., L. Berrojo, J. Moreno, J. F. Lopez, and R. Sarmiento. 2016. Multispectral and hyperspectral lossless compressor for space applications (HyLoC): A low-complexity FPGA implementation of the CCSDS 123 standard. *IEEE Journal of Selected. Topics in Applied Earth Observations and Remote Sensing* 9(2):757–770.

Theodorou, G., N. Kranitis, A. Tsigkanos, and A. Paschalis. 2016. High performance CCSDS 123.0-B-1 multispectral & hyperspectral image compression implementation on a space-grade SRAM FPGA. *Proceedings of 5th International Workshop On-Board Payload Data Compression*, Frascati, Italy, pp. 28–29.

11 Ground Segment

11.1 OVERVIEW OF THE GROUND SEGMENT

After building and launching a hyperspectral satellite into orbit, the success of the hyperspectral mission relies on the ability of the ground segment personnel to operate the satellite from the ground and ensure that the data gathered are of good quality and are made readily available to users. To achieve this goal, a ground flight control center with an antenna to send commands to the satellite is needed, together with at least one ground receiving station to receive the acquired data downlinked from the satellite as it passes overhead. Facilities for data processing, storage, and distribution are also needed.

The ground segment(s) of a hyperspectral mission consists of all the ground-based elements of the satellite system used by operators and support personnel, as opposed to the space segment (Ley et al. 2009, Elbert 2014), as shown in Figure 11.1. The ground segment enables control of the hyperspectral satellite, and distribution of payload data and telemetry among interested parties on the ground. The primary elements of a typical ground segment include:

- Ground stations, which provide radio interfaces with the satellite
- Mission control center(s), from which the satellite is controlled
- Remote terminals, which are used by support personnel
- Ground networks, which connect the ground elements to one another

These elements may be located together or separated geographically, and they may be operated by different parties. Some elements may support multiple satellites simultaneously.

11.1.1 Ground Stations

Ground stations provide radio interfaces between the space segment and the ground segment for telemetry, tracking, and command (TT&C), as well as payload data transmission, and reception. There are typically backup ground stations from which radio contact can be maintained if the primary ground station is unable to operate due to any reason, such as a natural disaster. Such contingencies are considered in a continuity of operations plan.

Signals of the control command to be uplinked to the hyperspectral satellite are first extracted from ground network packets, encoded to baseband, and modulated, typically onto an intermediate frequency (IF) carrier, before being up-converted to the assigned radio frequency (RF) band. The RF signal is then amplified to high power and carried via waveguide to a parabolic dish antenna for transmission.

Received signals of the acquired raw data by the hyperspectral payload downlinked via a X- or Ka-band channel are passed through a low-noise amplifier (often located in the antenna hub to minimize the distance the signals travel) before being down-converted to IF. These two functions may be combined in a low-noise block downconverter. The IF signal is then demodulated and the data stream extracted via bit and frame synchronization and decoding. Data errors, such as those caused by signal event upsets (SEUs), are identified and corrected where possible. The decoded data stream is then packetized or saved to files for transmission on the ground network.

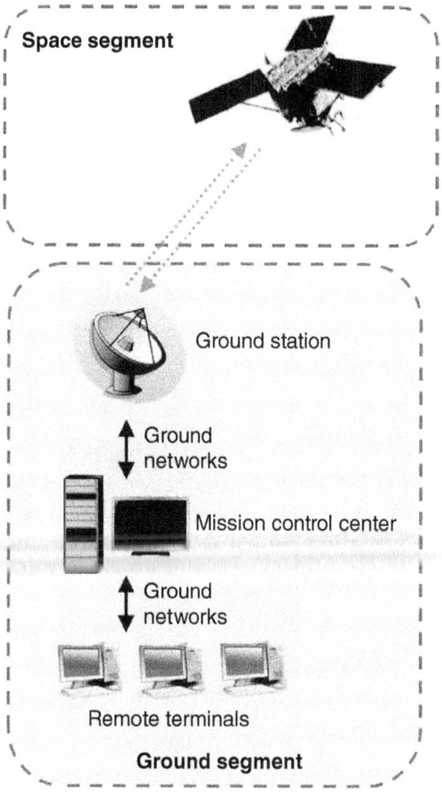

FIGURE 11.1 Elements of the ground segment and its networks.

Ground stations may temporarily store received telemetry for later playback to control centers, often when ground network bandwidth is not sufficient to allow real-time transmission of all received telemetry.

11.1.2 Mission Control Centers

Mission control centers process, analyze, and distribute satellite telemetry. They issue commands, data uploads, and software updates to the satellite. The mission control centers may also be responsible for configuration management and data archival. As with ground stations, there are typically backup control facilities available to support continuity of operations.

Mission control centers use telemetry to determine the status of the satellite and its payload. Housekeeping, diagnostic, science, and other types of telemetry may be carried on separate virtual channels. Flight control software performs the initial processing of the received telemetry.

Commands sent to the satellite are formatted according to the spacecraft database, and are validated against the database before being transmitted via a ground station. Commands may be issued manually in real time, or they may be part of automated or semiautomated procedures. Typically, commands successfully received by the satellite are acknowledged in telemetry, and a command counter is maintained in the satellite and on the ground to ensure synchronization. In certain cases, closed-loop control may be performed. Commanded activities may pertain directly to mission objectives, or they may be part of housekeeping. Commands (and telemetry) are often encrypted to prevent unauthorized access to the satellite or its data.

Mission control centers handle analytical tasks based on non-real-time data processing subsystems, such as:

- Mission planning and scheduling
- Orbit determination and maneuver planning
- Onboard memory management
- Short-term and long-term trend analyses
- Conjunction assessment and collision avoidance planning

11.1.3 Ground Networks

Ground networks handle data transfer between different elements of the ground segment. These networks often combine local area network (LAN) and wide area network (WAN) elements, for which different parties may be responsible. LAN interconnects computers using Ethernet. WAN is a telecommunications network that extends over a large geographical distance for computer networking. The design of ground networks is driven by requirements on reliability, bandwidth, and security.

Reliability is a particularly important consideration for critical systems, with uptime and mean time to recovery being of paramount concern. As with other aspects of the spacecraft system, redundancy of network components is the primary means of achieving the required system reliability.

Security considerations are vital to protect space resources and sensitive data. WAN links often incorporate encryption protocols and firewalls to provide information and network security. Antivirus software and intrusion detection systems provide additional security at network endpoints.

11.1.4 Remote Terminals

Remote terminals are interfaces on ground networks, separate from the mission control centers. They can be accessed by payload controllers, telemetry analysts, instrument and science teams, and support personnel, such as system administrators and software development teams. They may be receive-only, or they may transmit data via the ground network.

11.1.5 Hyperspectral Data Product Level

Table 11.1 lists the data product levels of hyperspectral imagery data. The ground segment normally provides product generation capabilities for Level 1A and 1B, as shown in the table. Further processing is necessary to produce higher-level information products that can be easily and immediately used by end users without further manipulation. This chapter assumes that this additional processing will be performed outside the ground segment.

The Level 1A Products are used to support the generation of calibration coefficients. The Level 1B Products are at-sensor radiance and are distributed to customers (Qian 2013).

Higher-level data products are considered customer specific processing because they are application or customer specific. It is difficult to satisfy different customer requirements with generic higher-level processing. For example, some customers would prefer to do their own atmospheric correction, which is required to generate Level 2 Products, since certain key input parameters (such as the aerosol model and visibility) are difficult to determine from the downlink data.

It is assumed that some customers will actually be value-added resellers (VARs) for different application market segments. In this case, these VARs would do the necessary processing to covert the Level 1B Products produced by the ground segment to higher-level products that can be directly used by end users.

TABLE 11.1
Hyperspectral Data Product Levels

Product Level	Description	Note
Level 0	Reconstructed unprocessed instrument data at full resolution, all communication artifacts, e.g., synchronization frames, communication headers, duplicate data are removed.	Received and extracted in the ground segment.
Level 1A	Reconstructed unprocessed instrument data at full resolution, time-referenced and annotated with ancillary information, including radiometric, spectral, and geometric calibration coefficients and geo-referencing parameters, e.g., platform ephemeris, computed and appended but not applied to the Level 0 data.	Used for generating calibration coefficients in the ground segment. Not deliverable to customer.
Level 1B	Reconstructed instrument data at full resolution, time referenced and annotated with ancillary information, converted into physical values, i.e., at-sensor radiance units.	Deliverable to customers.
Level 2	Derived physical variables, i.e., reflectance, at the same resolution and location as the Level 1 source data.	
Level 3	Level-3 Products consist of the Level-2 Products mapped to a uniform space/time grid.	
Level 4	Level-4 Products consist of derived data that could not be measured directly by the sensor. Level-4 Products are typically calculated through an extended analysis of the lower-level data, such as a modelling process.	

11.2 GROUND SEGMENT CONTEXT

Figure 11.2 shows an example of context of the ground segment of a hyperspectral mission. The ground segment interfaces with the space segment, calibration support data sources, and customers.

11.2.1 Interface to Space Segment

There are four links between the ground segment and the space segment. Each of them is described below.

11.2.1.1 Spacecraft Commands

The purpose of the Spacecraft Commands flow is for the ground segment to provide command and control signals to the hyperspectral satellite.

The required command and control functions include but are not limited to the following:

- Hyperspectral sensor data acquisition commands
- Hyperspectral sensor data compression commands
- Hyperspectral sensor housekeeping commands
- Panchromatic sensor data acquisition commands (if a panchromatic sensor is deployed)
- Panchromatic sensor housekeeping commands (if a panchromatic sensor is deployed)
- Altitude sensor data acquisition commands
- Altitude sensor housekeeping commands
- Onboard calibration parameter update commands
- Onboard memory management commands
- Downlink scheduling commands
- Ground motion compensation pointing mirror commands (if ground motion compensation strategy is adopted for increasing signal-to-noise ratio)

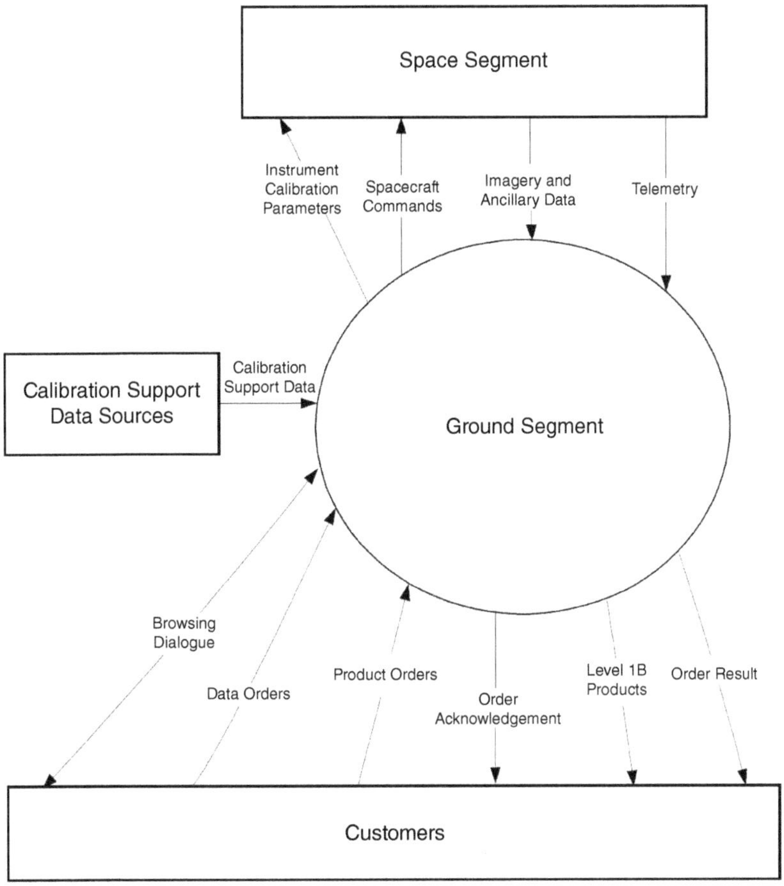

FIGURE 11.2 A ground segment context diagram.

- Spacecraft attitude control commands
- Orbit maintenance commands
- Pointing mirror alignment commands

Further, the pointing mirror alignment commands are replaced by the following:

- Gimbal control commands

11.2.1.2 Imagery and Ancillary Data

The purpose of the Imagery and Ancillary Data flow is for the hyperspectral satellite to provide imagery and ancillary data to the ground segment.

The imagery and ancillary data transmitted from the hyperspectral satellite have been formatted and packetized according to the Consultative Committee for Space Data Systems (CCSDS) standards, and consist of the following data:

- Hyperspectral imagery data, which may be either raw digital numbers (DNs) of the acquired radiance data, or compressed data.
- Ancillary data, which may include information such as satellite attitude and ephemeris, payload temperature, and solar or lunar calibration imagery.

11.2.1.3 Instrument Calibration Parameters

The purpose of the Instrument Calibration Parameters flow is for the ground segment to uplink the required radiometric calibration parameters for onboard radiometric correction to convert the acquired raw DNs to at-sensor radiance data. Onboard radiometric correction may not be required by most of hyperspectral missions. This function is described in this chapter for providing an option in case a hyperspectral mission requires this function.

The instrument calibration parameters normally specify the following items for each focal plane array in the hyperspectral imager:

- Radiometric offset coefficient
- Radiometric gain coefficient
- Higher-order radiometric coefficients, if deemed necessary (optional)

11.2.1.4 Telemetry

The purpose of the Telemetry data flow is for the hyperspectral satellite to provide payload health and status to the ground segment.

The telemetry data normally consist of data such as:

- Voltages
- Temperatures
- Operation statistics and modes
- Power
- Attitude
- Ephemeris

11.2.2 Interface to Calibration Support Data Sources

The purpose of the Calibration Support Data flow is for Calibration Support Data Sources to provide the ground segment with the support information necessary to perform the following calibrations for the hyperspectral sensor:

- Geometric calibration
- Vicarious radiometric calibration

For the geometric calibration, the calibration support data consist of ground control points (GCP) and a digital elevation model (DEM) over the region being imaged.

For the vicarious radiometric calibration, the calibration support data need to be collected over the region being imaged, and depend on the vicarious calibration method used:

- Irradiance-based calibration: ground reflectance, optical depth, and global-to-diffuse solar irradiance ratio measurements.
- Radiance-based calibration: airborne hyperspectral sensor measurements and ground reflectance and optical depth measurements.
- Reflectance-based calibration: ground reflectance and optical depth measurements.

11.2.3 Interface to Customers

11.2.3.1 Browsing Dialogue

A customer may search the Catalogue in the ground segment to review which data sets are currently held within the Archive in the ground segment. The Browsing Dialogue is comprises a set of queries and responses.

The browsing dialogue query normally contains search criteria such as:

- Geographic coverage
- Acquisition time (or range)
- Spectral range (or bands)
- Cloud cover percentage

The browsing dialogue response normally contains the following for each hyperspectral image (i.e., datacube) that matched the search criteria:

- Identifier of the image within archive
- Image size (in megabytes)
- Spectral bands
- Geographic coverage
- Acquisition time
- Cloud cover percentage
- Scaling factor to standard radiance units (if onboard radiometric conversion is performed)
- Compression ratio and/or quality factor (if onboard compression is performed)

The submission of a browsing dialogue query is initiated by a customer over the internet using a standard web browser. A browsing dialogue response is returned to the customer in response to the browsing dialogue query over the same web interface.

11.2.3.2 Data Orders

The purpose of the Data Order flow is to provide the ground segment with a description of the new data requested by a customer.

The data order normally provides the following information:

- Area of interest
- Desired date range
- Spectral regions, such as visible and near-infrared (VNIR), shortwave infrared (SWIR), or panchromatic (PAN)
- Customer information, such as name, organization, and address

The submission of a data order is initiated by a customer over the Internet using a standard web browser. A single data order will, if successful, result in the acquisition of a single scene.

11.2.3.3 Product Orders

The purpose of the Product Orders flow is to provide the ground segment with descriptions of the products desired by a customer.

The product orders normally specify the following:

- Identifier of an image within archive
- Product Level (1B – radiance)
- Customer information, such as name, organization, and address

The submission of a product order is initiated by a customer over the Internet using a standard web browser. A single product order will result, if successful, in the delivery of a single product.

11.2.3.4 Order Acknowledgment

The purpose of the Order Acknowledgement data flow is to acknowledge the receipt of data orders or product orders, and confirm their feasibility.

The order acknowledgement specifies:

- All information contained within the data order or product order.
- Indication of feasibility: positive or negative (only applicable to data orders).
- If an order is infeasible, a reason will be provided.

A single order acknowledgement will be sent for each data order or product order received by the ground segment.

11.2.3.5 Level 1B Products

The purpose of the Level 1B Product flow is to deliver the Level 1B data products to a customer who ordered them.

The Level 1B products are:

- The products saved in an application-friendly format, such as the Tagged Image File Format (TIFF) header and imagery within the same file or the Environment for Visualizing Images (ENVI) format (binary imagery file with separate ASCII header file).
- Absolutely radiometrically corrected.
- In radiance unit.
- Geo-referenced (geometric model supplied, but the imagery not resampled).

The Level 1B products only include hyperspectral imagery, which contains all the VNIR and SWIR bands. Products may include any combination of the VNIR, SWIR, and PAN, if the PAN instrument is deployed. The simultaneously collected higher spatial resolution PAN imager is treated as an additional band, processed in the same way as the hyperspectral imagery.

A Level 1B product is sent to the customer as soon as the processing has been completed to fulfill a Product Order.

11.2.3.6 Order Results

The purpose of the Order Result flow is to notify the customer that his/her data order or product order has concluded. This could mean that the order was successfully fulfilled, the order was rejected as being infeasible, or the order failed.

The order results contain:

- All the information contained within the data order or product order
- A final order status:
  a. Fulfilled successfully:
     1. For a data order, imagery with the requested characteristics was acquired and exists within the ground segment archive;
     2. For a product order, the product was created and is delivered online or on a storage media being shipped to the customer.
  b. Infeasible: possible reasons include (only applicable to data orders):
     1. No satellite passes over the area of interest within the date range
     2. Acquisition plan full within the date range over the area of interest
     3. The customer may try to readjust the order parameters so that the order can be successfully fulfilled.

   c. Failed: possible reasons include:
      1. No cloud-free images acquired within the date range
- If the final order status was successfully fulfilled, then the following additional information is provided:
   a. Data order: date acquired and identifier of the data set within the ground segment archive.
   b. Product order: product shipping date.

11.3 GROUND SEGMENT FUNCTIONAL DECOMPOSITION

Figure 11.3 shows the functional decomposition of an example ground segment of a hyperspectral satellite. This section provides a description of each of the functions and an overview of the interfaces that relate to these functions. Readers are referred to Section 11.4 for more detailed descriptions of the interfaces between the functions.

11.3.1 Order Management

The Order Management function is responsible for taking orders from Customers and managing their state as the remainder of the system, and performs the functions required to fulfill the order. The Order Management function provides the following functionality:

1. Maintains a user order database that contains a record of currently active user orders.
2. Manages the implementation of user orders and allows these orders to be monitored and tracked internally.
3. Accepts Data Orders and, in doing so, triggers the following activities:
   3.1 Creates a new user order in the user order database.
   3.2 Formulates a Feasibility Check, which is then forwarded to the Acquisition Planning function.
   3.3 Receives a Feasibility Result from the Acquisition Planning function and forwards this result to the Customer via an Order Acknowledgement message.
   3.4 For infeasible orders, terminates the user order and sends an Order Result message to the Customer.
   3.5 For feasible orders, submits a Collection Order to the Acquisition Planning function.
   3.6 Receives a Collection Order Result from the Acquisition Planning and forwards this result to the Customer via an Order Result message.
4. Accepts Product Orders and, in doing so, triggers the following activities:
   4.1 Creates a new user order in the user order database.
   4.2 Generates an Order Acknowledgement, which confirms reception of the Product Order.
   4.3 Submits a Product Order to the Product Processing function.
   4.4 Receives a Product Order Result from the Product Processing function and forwards this result to the Customer via an Order Result message.

11.3.2 Acquisition Planning

The Acquisition Planning function maintains a continuously feasible plan of acquisition-related activities and provides the major functions that depend directly on this plan.

The primary roles of the Acquisition Planning function are to schedule acquisition of imagery, downlink, and slew activities, representing the "always feasible" master plan for accepted acquisition-related activities.

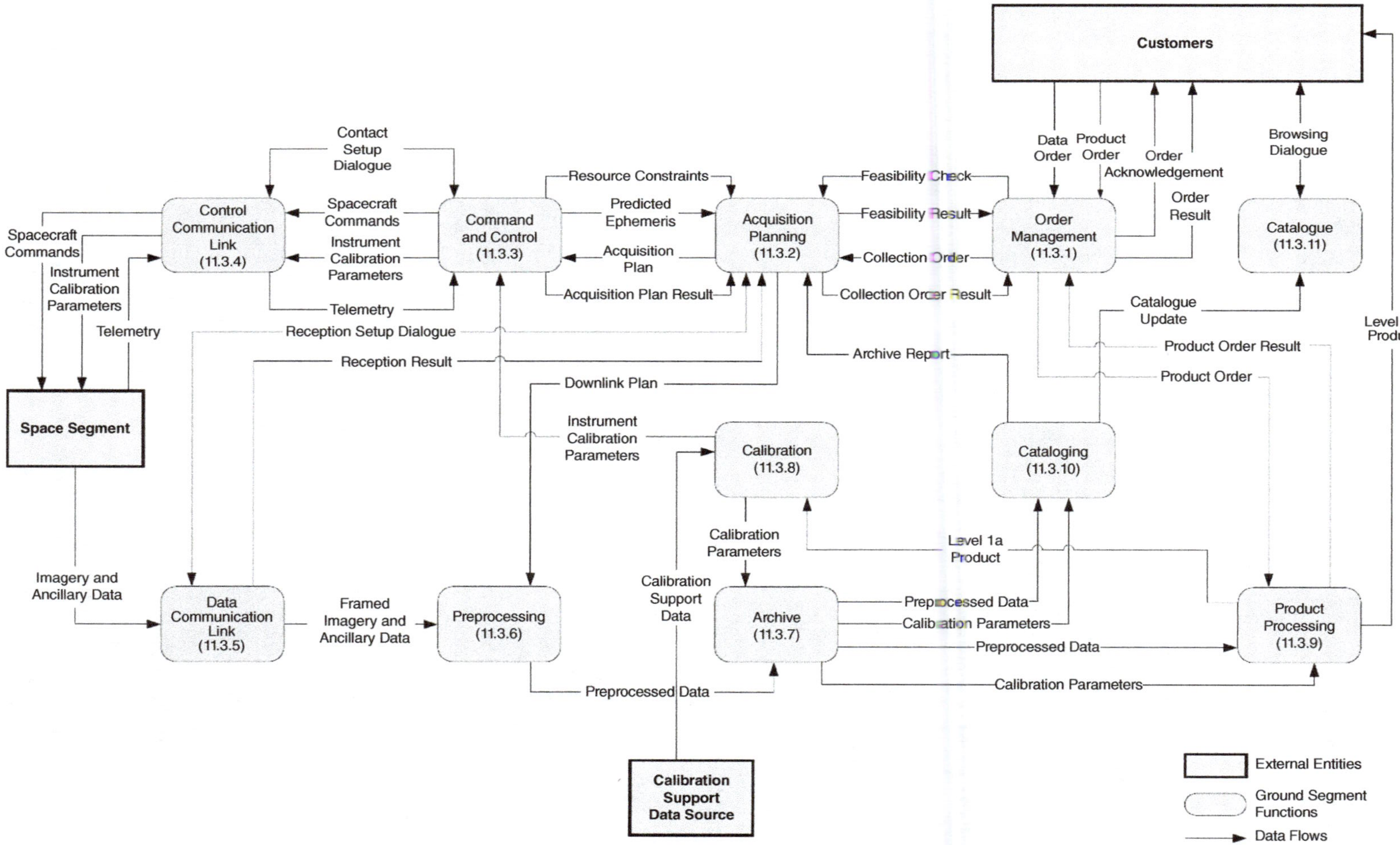

FIGURE 11.3 A block diagram of ground segment functional decomposition. (The numbers of the subsections that describe the functions are provided in the parentheses in the round boxes)

The major functions performed by the Acquisition Planning function are as follows:

1. Provides a feasibility checking service to support the Order Management function.
2. Performs constraints checking to determine the feasibility of a request.
3. Schedules the acquisitions of approved orders that have been received via Collection Order.
4. Performs rescheduling that is required due to resource availability changes (e.g., satellite availability or reception facility availability).
5. Periodically sends Acquisition Plans to the Command and Control function.
6. Establishes a Reception Setup Dialogue with the Data Communication Link in order to coordinate downlink activities with imagery acquisition operations.
7. Tracks the status of acquisition activities and reports status changes to the Order Management function via Collection Order Result messages.

11.3.3 Command and Control

The Command and Control function provides all the functions required to schedule, coordinate, control, and monitor the satellite and hyperspectral payload during the prelaunch, commissioning, and routine operation phases of the hyperspectral mission.

The Command and Control function receives Acquisition Plans from the Acquisition Planning function and Instrument Calibration Parameters from the Calibration function. Based on this information, the Command and Control function plans and executes daily satellite control operations, and monitors the state of the satellite. In support of daily operations, the Command and Control function:

1. Maintains a high-fidelity model of the hyperspectral payload and the satellite to ensure that all mission constraints are not exceeded.
2. Initiates a Contact Setup Dialogue with the Control Communication Link to establish communication with the satellite and the payload.
3. Generates Spacecraft Commands for uplink and forwards it to the Control Communication Link.
4. Receives and forwards Instrument Calibration Parameters to the Control Communication Link.
5. Receives, archives, and monitors Telemetry and tracking information from the Control Communication Link.

In the event that acquisition plans cannot be scheduled by the Command and Control function or executed by the spacecraft, an Acquisition Plan Result is sent to the Acquisition Planning function with a notification of the scheduling problem.

The Command and Control function also provides the following services to the Acquisition Planning function:

6. Delivery of information regarding the known ground segment and space segment Resource Constraints.
7. Delivery of Predicted Ephemeris derived from downlinked telemetry.

11.3.4 Control Communication Link

The Control Communication Link relays spacecraft commands generated by the Command and Control function to the hyperspectral satellite Space Segment. The link also relays telemetry data from the Space Segment to the ground. In support of these operations, the Control Communication Link provides the following functionality:

1. Exchanges Contact Setup Dialogue messages with the Command and Control function.
2. Sets up the associated control software/hardware for the acquisition, based on the Contact Setup Dialogue.

3. Relays Spacecraft Commands from the Command and Control function to the Space Segment.
4. Relays Instrument Calibration Parameters from the Command and Control function to the Space Segment.
5. Relays Telemetry data from the Space Segment to the Command and Control function.

11.3.5 Data Communication Link

The Data Communication Link relays acquired imagery data and ancillary data from the Space Segment to the ground. In support of these operations, the Data Communication Link provides the following functionality:

1. Exchanges Reception Setup Dialogue messages with the Acquisition Planning function.
2. Sets up the associated reception software/hardware for the acquisition, based on the Reception Setup Dialogue.
3. Receives downlinked Imagery and Ancillary Data and records them in a framed format online or on portable media.
4. Sends the Framed Imagery and Ancillary Data to the Preprocessing function.
5. Creates and sends Reception Results to the Acquisition Planning function, indicating the details and status of the acquisition.

11.3.6 Preprocessing

The Preprocessing function is responsible for any initial processing that needs to be performed prior to archiving the data. This Preprocessing function could be to:

1. Receive Downlink Plan from the Acquisition Planning function.
2. Segment the Imagery Data into images based on the Downlink Plan.
3. Segment the Ancillary Data according to the Downlink Plan.
4. Extract Attitude and Ephemeris Data.
5. Process Attitude and Ephemeris Data over a sufficiently long time interval to create Preprocessed Attitude and Ephemeris of sufficient quality for product generation.
6. Convert the Imagery, Ancillary, Ephemeris, and Attitude Data into a format suitable for archiving.
7. Decompress imagery data if the imagery data was compressed onboard.
8. Send the Preprocessed Data (which includes Imagery, Ancillary, Attitude, and Ephemeris Data) to the Archive.

Depending on the eventual design and image data structure and format, the Preprocessing function may not be required.

11.3.7 Archive

The Archive function is responsible for maintaining access to acquired data for the life of the mission. The Archive function provides the following functionality:

1. Provides interfaces allowing customers to submit and retrieve data to/from the archive; in particular:
    1.1 Receives and archives Preprocessed Data (which includes Imagery, Ancillary, Attitude, and Ephemeris Data) from the Preprocessing function.
    1.2 Receives and archives Calibration Parameters from the Calibration function.

2. Sends Preprocessed Data (which includes Imagery, Ancillary, Attitude, and Ephemeris Data) and Calibration Parameters to the Cataloging and Product Processing functions.
3. The recently submitted Preprocessed Data will be stored on a disk initially and, as future data arrives, will be transferred to removable media for long-term storage.
4. Manages storage of all archive data submitted using a combination of online and offline (portable media) storage.
5. Provides an interface to allow monitoring of archive health indicators and activity and to provide tools for examining the contents of the various data collections within the Archive. Archived data sets can be queried based on:
   5.1 Archive identifier
   5.2 Date range
   5.3 Data type

11.3.8 Calibration

The Calibration function is responsible for generating the Calibration Parameters that are used by both the space segment and the ground segment. The Instrument Calibration Parameters are transferred to the Command and Control function for being uplinked to the Space Segment. Calibration Parameters are also transferred to the Archive where they are distributed to the Cataloging and Product Processing functions for use during the generation of catalogue updates and products, respectively.

The Calibration function:

1. Receives Level 1A Product from the Product Processing function.
2. Receives Calibration Support Data from the Calibration Support Data Sources.
3. Generates radiometric, spectral, and geometric Calibration Parameters.
4. Sends the generated Calibration Parameters to the Archive.
5. Sends selected Instrument Calibration Parameters to the Command and Control function.

The specific functions within the Calibration function are primarily dependent on the space segment and the required accuracy of the Products. These functions typically include the following:

6. Generation of radiometric Calibration Parameters, providing a mapper from DNs to at-sensor radiance units.
7. Generation of spectral Calibration Parameters, providing the wavelength center of each band.
8. Generation of geometric Calibration Parameters, providing a mapper from image coordinates (line, column) to ground coordinates (latitude, longitude).

11.3.9 Product Processing

The Product Processing function generates products (normally Level 1A and Level 1B) for transfer to customers, and for Calibration Parameter generation. The Product Processing function:

1. Receives Preprocessed Data (which includes Imagery, Ancillary, Attitude, and Ephemeris Data) from the Archive. The Preprocessed Data may be recently collected or have been in the Archive for some time.
2. Produces Level 1A Products and delivers them to the Calibration function.
3. Receives Calibration Parameters from the Archive to be used for producing Level 1B data.
4. Receives Product Order from the Order Management function.
5. Produces Level 1B Products and delivers them to the Customers via Order Management function.
6. Sends Product Order Results to the Order Management function.

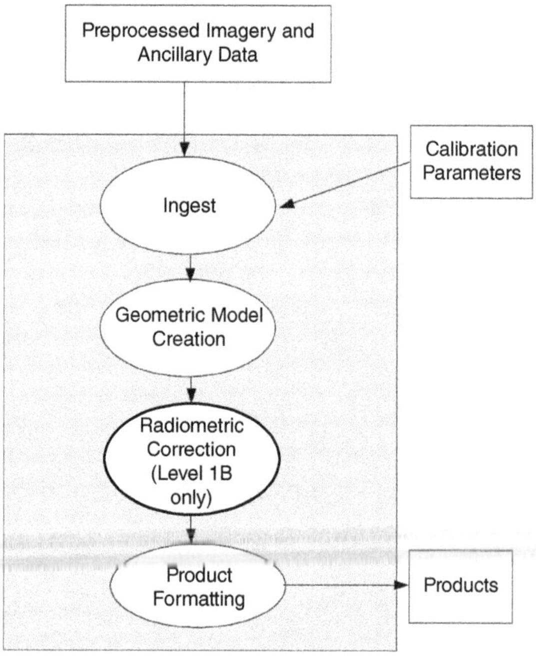

FIGURE 11.4 Processing steps of Product Generation in the Product Processing function.

The Product Processing function includes quality analysis tools for verifying that all products meet the specified product level specifications.

The generated product imagery is not geometrically or spectrally resampled, since this would affect the radiometric fidelity of the imagery; hence, geometric and spectral model information is included in the products.

Figure 11.4 shows the processing steps of product processing in the Product Processing function. There are four steps: ingest, geometric mode creation, radiometric correction, and product formatting.

In the ingest step, the Product Order specifies the Preprocessed Imagery Data in the Archive to be processed with a unique archive identifier. This imagery is retrieved from the Archive and the time interval for the imagery is read from the header. The corresponding Preprocessed Ancillary Data and Calibration Parameters appropriate for this time interval are then retrieved from the Archive. At this point, the Product Processing function has all the information it needs to create the Product.

The purpose of the geometric model is to describe the physical geometric conditions under which an imagery was acquired. This is captured within the geometric model by providing a complete bidirectional transformation between raw image coordinates (line, column) and ground coordinates (latitude, longitude). The accuracy of this transformation is the key to the geo-referencing procedure.

The image-to-ground transformation takes into account the geometry of the:

- Satellite, where attitude is constructed using the Preprocessed Attitude Data (such as star tracker and gyro data), and position is constructed using Preprocessed Ephemeris Data (such as GPS data).
- Payload, including the focal plane arrays.
- Earth's surface, including a reference datum and an elevation model.
- Earth's motion, including pole wander and rotation rate changes.
- Atmosphere, including the refraction of radiation at large view angles as it travels through the atmosphere.

Radiometric correction is the process by which pixel intensities in DNs are converted to at-sensor radiances. In this process, all known sensor effects, induced systematic effects, are removed from the imagery, such as:

- Detector responses and biases
- Optical throughput (due to lens and mirrors)

Radiometric correction is performed by applying correction factors (usually a gain and offset) to each pixel. These correction factors are included in the Calibration Parameters. Chapters 12 and 13 provide a detailed description of ground calibration and correction of hyperspectral imagery data from DNs to at-sensor radiances.

Radiometric correction will be performed on the ground for generating Level 1B products if onboard radiometric correction has not been applied.

Product formatting is the last step in the Product Generation function, and consists of the final formatting and assembly of the product components on the chosen data storage media:

- Imagery data
- Header (describes the imagery file)
- Metadata: all constituents other than the imagery and header, such as geometric and spectral model information.

11.3.10 Cataloging

The Cataloging function is responsible for populating the Catalogue by providing Catalogue Update entries. The Cataloging function provides the following functionality:

1. Receives Preprocessed Data (which includes Imagery, Ancillary, Attitude, and Ephemeris Data) and Calibration Parameters from the Archive.
2. Produces Catalogue Updates for each image collected. Each Catalogue Update includes:
   2.1 Browse image, which is a spatially and spectrally reduced resolution version of the full image, in a standard format.
   2.2 Metadata, which includes:
   - geometric coverage,
   - cloud cover assessment with a percentage and confidence,
   - acquisition date, and
   - available spectral bands.
3. Sends the Catalogue Updates to the Catalogue function.
4. Issues Archive Report to the Acquisition Planning function, containing:
   4.1 status of the acquisition,
   4.2 geometric coverage,
   4.3 cloud cover assessment with a percentage and confidence,
   4.4 acquisition date, and
   4.5 available spectral bands.

11.3.11 Catalogue

The purpose of the Catalogue function is to represent the contents of the Archive accurately and usefully. The Catalogue function provides the following functionality:

1. Maintains a repository of browse imagery and associated metadata that corresponds to the data stored in the Archive.
2. Receives Catalogue Update entities from the Cataloging function that consist of reduced resolution browse imagery and associated metadata.

3. Provides a browsing capability for Customers to search the contents of the Archive using the following information:
    - Geographic coverage
    - Acquisition date and time (or range)
    - Spectral bands (or range)
    - Cloud cover percentage (or range)
4. Provides the unique identifiers of the corresponding Archive data to the Customers.

11.4 OPERATION AND DATA FLOW BETWEEN FUNCTIONS

Section 11.3 has described the functions and provided an overview of the interfaces between these functions of an example ground segment of a hyperspectral satellite. This section provides a detailed description of these functions' operation and data flow between the functions. These functions, their relationship and data flow are shown in Figure 11.3.

11.4.1 Order Management to Acquisition Planning

11.4.1.1 Feasibility Check

The purpose of the Feasibility Check data flow is to determine whether a Data Order generated by a customer is consistent with the current or predicted spacecraft and ground segment constraints.

The Feasibility Check may specify the following items:

- Data Order Identifier (ID)
- Area of interest
- Desired date range
- Focal plane selection (i.e., VNIR and/or SWIR, and/or PAN)

11.4.1.2 Feasibility Result

The purpose of the Feasibility Result data flow is to notify the Order Management function of the results of the Feasibility Check.

The Feasibility Result may specify the following:

- Data Order ID
- Feasibility status, which can be any one of the following:
    a. The order is feasible.
    b. The order is not feasible due to conflict with existing planned operations during desired date range.
    c. The order is not feasible due to no passes over the area of interest during desired date range.
    d. The order is not feasible due to insufficient spacecraft resources (e.g., onboard memory storage constrain).

11.4.1.3 Collection Order

The purpose of the Collection Order data flow is to notify the Acquisition Planning function that a new acquisition must be scheduled that satisfies the imaging parameters contained in the corresponding Data Order.

The Collection Order may specify the following items:

- Collection Order ID
- Area of interest
- Desired date range
- Focal plane selection (i.e., VNIR and/or SWIR, and/or PAN)

It can be seen that the specified items of the Collection Order are identical to those of the Feasibility Check.

11.4.1.4 Collection Order Result

The purpose of the Collection Order Result data flow is to close the loop initiated by the Collection Order data flow. It occurs when imaging, archiving, and cataloging operations intended to satisfy the Collection Order have been completed.

The Collection Order Result may specify the following:

- Collection Order ID
- Collection Order status, which can be any one of the following:
    a. Success
    b. Failure due to spacecraft unavailability during the desired date range
    c. Failure due to no cloud-free acquisitions during the desired date range
    d. Failure due to acquisition errors during the desired date range
    e. Failure due to reception, archiving, or cataloging errors during the desired date range.

When the Collection Order status has a value of success, the following information is also included in the Collection Order Result:

- Archive ID

11.4.2 Acquisition Planning to Command and Control

11.4.2.1 Acquisition Plan

The purpose of the Acquisition Plan data flow is to periodically update the Command and Control function with a new set of imaging operation requirements for a single orbit that satisfy one or more Collection Orders.

The Acquisition Plan may specify the following:

- Orbit number
- Uplink station contact rise time
- Uplink station contact rise azimuth
- Uplink station contact set time
- Uplink station contact set azimuth
- Focal plane selection (VNIR and/or SWIR, and/or PAN)
- A list of events of acquisition of imagery, each of which consists of the following:
    a. Imagery start time
    b. Imagery stop time
    c. View angle
    d. Selection or deselection of image compression and onboard radiometric calibration
    e. Imagery compression ratio (for lossy compression if applicable)
- Downlink Station
- Downlink start time
- Downlink end time

11.4.2.2 Acquisition Plan Result

The purpose of the Acquisition Plan Result data flow is to provide final confirmation on the acceptability of an acquisition plan. It also serves as a mechanism for rejecting acquisition plans that are no longer valid given changes in resource constraint information.

The Acquisition Plan Result may specify the following:

- Orbit number
- Acquisition Plan status, which can be any one of the following:
  a. Accepted
  b. Rejected due to unavailability of uplink station
  c. Rejected due to unavailability of spacecraft

11.4.2.3 Resource Constraints

The purpose of the Resource Constraints data flow is to update the Acquisition Planning function with the most up-to-date information on the constraints that impact spacecraft operations. These constraints are used to assess collection order feasibility and formulate acquisition plans by the Acquisition Planning function.

The Resource Constraints may specify the following:

- Start time of the known unavailability of the space segment (due to high priority housekeeping tasks, space environment conditions, calibration events, etc.)
- End time of the known unavailability of the space segment
- Start time of the known unavailability of the Communication Link (due to conflicts with other missions, etc.)
- End time of the known unavailability of the Communication Link

11.4.2.4 Predicted Ephemeris

The purpose of the Predicted Ephemeris data flow is to update the Acquisition Planning function with revised predictions of the satellite orbit. This information is used by the Acquisition Planning function to check the feasibility of Collection Orders, plan imaging events, and schedule uplink and downlink contact times.

The Predicted Ephemeris may specify the following:

- A Two-Line Element (TLE) set.

11.4.3 Command and Control to the Control Communication Link

11.4.3.1 Contact Setup Dialogue

The purpose of the Contact Setup Dialogue data flow is to schedule uplink activities between the Command and Control function and the hyperspectral satellite Space Segment via the Control Communication Link.

The nature of the Contact Setup Dialogue interface varies depending on the type of the mission. The Command and Control function must interface with a set of TT&C facilities embedded within the Control Communication Link. These facilities could either be shared with one or more existing space missions or be dedicated for the sole use of a hyperspectral mission.

Direct uplink via an independent set of TT&C facilities serves as the primary means of communication with the Space Segment.

To orient the Control Communication Link TT&C antenna during uplink activities, the Command and Control function must provide the following information:

- Uplink station contact rise time
- Uplink station contact rise azimuth
- Uplink station contact set time
- Uplink station contact set azimuth

The Contact Setup Dialogue must also provide a handshaking protocol that allows the Command and Control function to establish a communication channel with the space segment (via the Control Communication Link) that will permit the exchange of uplinked spacecraft commands and downlinked spacecraft telemetry.

In the event that the TT&C facilities within the Control Communication Link are shared with other space missions, the Contact Setup Dialogue must also include a mechanism for requesting and scheduling time on the system. The nature of the scheduling mechanism will be driven primarily by the constraints imposed by the existing TT&C facilities.

11.4.3.2 Spacecraft Commands

The purpose of the Spacecraft Commands data flow is to upload commands generated by the Command and Control function to the Space Segment via the Control Communication Link.

Like the Contact Setup Dialogue, the nature of the Spacecraft Commands interface depends on the type of the mission. The interface is constrained by the requirements of the Control Communication Link TT&C facilities.

The Spacecraft Commands interface protocol requires spacecraft bus commands in addition to payload commands. The two primary sources of these additional commands are the required orbit maintenance and attitude control functions. The command and control functions required are listed in Section 11.2.1.1.

11.4.3.3 Instrument Calibration Parameters

The purpose of the Instrument Calibration Parameters data flow is to forward radiometric calibration parameters to the Control Communication Link for subsequent uplink to the Space Segment.

In some hyperspectral missions, onboard radiometric correction may be performed in Space Segment to generate at-sensor radiance data or to improve the efficiency of the onboard data compression prior to data downlink. To do so, instrument calibration parameters need to be uplinked to the space segment from the ground segment.

The Instrument Calibration Parameters may specify the following for each detector element in the hyperspectral sensor:

- Radiometric offset coefficient
- Radiometric gain coefficient
- Higher-order radiometric coefficients, if deemed necessary (optional)
- Radiometric correction coefficients for the panchromatic imager, if it is deployed

11.4.3.4 Telemetry

The purpose of the Telemetry data flow is to provide the Command and Control function with real-time information on the health of the Space Segment via the Control Communication Link. While the exact format of the telemetry may differ with different hyperspectral satellite missions, the general content is the same.

The telemetry information may include but is not limited to the following:

- Voltages
- Temperatures
- Sensor Status Indicators
- Power
- Attitude
- Ephemeris

11.4.4 Acquisition Planning to Data Communication Link

11.4.4.1 Reception Setup Dialogue

The purpose of the Reception Setup Dialogue data flow is to schedule downlink activities between the hyperspectral satellite Space Segment and the Data Communication Link.

The nature of the Reception Setup Dialogue interface varies depending on the type of the mission. The Acquisition Planning function must interface with a set of downlink facilities embedded within the Data Communication Link. These facilities could either be shared with one or more existing space missions or be dedicated for the sole use of a hyperspectral mission.

Direct downlink via an independent set of downlink facilities serves as the primary means of receiving data from the hyperspectral satellite Space Segment.

To orient the Data Communication Link antenna during downlink activities, the Acquisition Planning function must provide the following information:

- Downlink station contact rise time
- Downlink station contact rise azimuth
- Downlink station contact set time
- Downlink station contact set azimuth

In the event that the downlink facilities within the Data Communication Link are shared with other space missions, the Reception Setup Dialogue must also include a mechanism for requesting and scheduling time on the system. The nature of the scheduling mechanism will be driven primarily by the constraints imposed by the existing downlink facilities.

11.4.4.2 Reception Result

The purpose of the Reception Result data flow is to notify the Acquisition Planning function of the result of a scheduled reception activity.

The Reception Result may specify the following:

- Orbit number
- Reception status, which can be any one of the following:
  a. Success
  b. Failure due to missed satellite contact
  c. Failure due to data reception error
- Reception start time
- Reception end time

11.4.5 Acquisition Planning to Preprocessing

11.4.5.1 Downlink Plan

The purpose of the Downlink Plan data flow is to provide the Preprocessing function with the information necessary to split the downlinked data into separate images and extract the corresponding ancillary data.

The Downlink Plan may specify the following:

- Orbit number
- Focal plane selection (VNIR and/or SWIR, and/or PAN)
- For each image in the downlink, the following information is provided:
  a. Collection Order ID
  b. Image start time
  c. Image stop time

   d. Image start data index
   e. Image stop data index
- Ancillary start time
- Ancillary stop time
- Ancillary start data index
- Ancillary stop data index

11.4.6 Data Communication Link to Preprocessing

11.4.6.1 Framed Imagery and Ancillary Data

The purpose of the Framed Imagery and Ancillary Data flow is to provide imagery and ancillary data recorded by the Data Communication Link to the Preprocessing function prior to submission to the Archive.

The Framed Imagery and Ancillary Data may consist of the following:

- Hyperspectral imagery data
- Panchromatic imagery data, if applicable
- Ancillary data that consists of the following:
   a. Attitude data
   b. Ephemeris data
   c. Temperature data

11.4.7 Preprocessing to Archive

11.4.7.1 Preprocessed Data

The purpose of the data flow of Preprocessed Data from the Preprocessing function to the Archive is to submit imagery, attitude, and ephemeris data extracted from the downlinked data to Archive.

The Preprocessed Data may consist of each of the following three data types:

- Imagery
- Attitude time state
- Ephemeris time state

Imagery data may be accompanied by the following header information:

- Collection Order ID
- Image start time
- Image stop time
- Onboard data compression/radiometric calibration flags

The imagery data header may specify the following information:

- Focal plane (VNIR and/or SWIR, or PAN)
- Calibration flag (not applicable, radiometric, spectral, or geometric)
- Integration time

Both attitude and ephemeris data may be accompanied by the following header information:

- Time state start time
- Time state stop time

Note that the flow of Preprocessed Data also occurs between the Archive and the Cataloging functions, as well as the Archive and the Product Processing functions. In all the three cases, the content and format of the data flows are the same.

11.4.8 Archive to Cataloging

11.4.8.1 Preprocessed Data

The purpose of the data flow of Preprocessed Data from the Archive to the Cataloging function is to provide the data necessary for producing Catalogue updates (browse images and associated metadata).

The content of the flow of Preprocessed Data is described in detail in Section 11.4.7.1.

11.4.8.2 Calibration Parameters

The purpose of the Calibration Parameters data flow is to retrieve radiometric, geometric, and spectral calibration parameters from the Archive for use by the Cataloging function.

There are two types of radiometric calibration parameters: correction coefficients and a list (or a map) of dead detector pixels. The Calibration Parameters specify the following for each detector pixel in the hyperspectral sensor:

- Radiometric offset coefficient
- Radiometric gain coefficient
- Higher-order radiometric coefficients, if deemed necessary (optional)
- Radiometric correction coefficients for the panchromatic imager, if a panchromatic imager is deployed

The Calibration Parameters specify the following dead detector pixel information:

- Applicable date
- For each known dead detector pixel the following information is provided:
  a. Focal plane (either VNIR or SWIR)
  b. Band number
  c. Column number
- A dead detector pixel list for the panchromatic imager

Spectral calibration may be performed using a two-dimensional polynomial mapping from band number to the center wavelength of the detector arrays. A separate polynomial is assumed to be required for the VNIR and SWIR focal planes.

The Calibration Parameters specify the following spectral calibration parameters for each hyperspectral focal plane:

- Applicable date
- Focal plane (either VNIR or SWIR)
- Two-dimensional polynomial fit quality
- Polynomial coefficients

The Calibration Parameters specify the following geometric calibration parameters:

- Applicable date
- Attitude sensor to hyperspectral sensor alignment parameters

- Hyperspectral sensor pixel location parameters
- Attitude, ephemeris, and hyperspectral sensor timing parameters

The Calibration Parameters may specify the following additional geometric calibration parameters if a panchromatic imager is deployed:

- Applicable date
- Attitude sensor to panchromatic sensor alignment parameters
- Panchromatic sensor pixel location parameters
- Panchromatic sensor timing parameters

11.4.9 Cataloging to Acquisition Planning

11.4.9.1 Archive Report

The purpose of the Archive Report data flow is to close the loop initiated by an Acquisition Plan following the acquisition, reception, preprocessing, archiving, and cataloging of the required data.

The Archive Report specifies a list of image archive records, each of which consists of the following:

- Collection Order ID
- Archive status, which can be any one of the following:
  a. Success
  b. Failed due to excessive cloud cover
  c. Failed due to reception or archiving error

When the Archive status has a value of success, the following information will also be included with each image archive record:

- Archive ID

11.4.10 Cataloging to Catalogue

11.4.10.1 Catalogue Update

The purpose of the Catalogue Update data flow is to provide summary information to the Catalogue such that the customers can determine the availability of archived imagery. The Catalogue Update information may consist of two types of data:

- Reduced resolution browse imagery
- Associated image metadata

The metadata that accompanies a browse image may be composed of the following:

- Geographic corners
- Acquisition date
- View angle
- Cloud cover percentage

The Catalogue Update metadata may specify the following information:

- Focal plane (VNIR and/or SWIR, or PAN)
- Integration time

11.4.11 Archive to Product Processing

11.4.11.1 Preprocessed Data

The purpose of the data flow of Preprocessed Data from the Archive to the Product Processing function is to provide the data necessary to produce image products requested by the Order Management function.

The content of the flow of Preprocessed Data is described in detail in Section 11.4.7.1.

11.4.11.2 Calibration Parameters

The purpose of the Calibration Parameters data flow is to retrieve radiometric, spectral, and geometric calibration parameters from the Archive function for use by the Product Processing function.

This interface is identical to the Calibration Parameters interface described in Section 11.4.8.2.

11.4.12 Order Management to Product Processing

11.4.12.1 Product Order

The purpose of the Product Order data flow is to provide a mechanism for the Order Management function to request that an image product be generated and delivered using data already present in the Archive.

The Product Order may specify the following:

- Product Order ID
- Archive ID
- Product Level (1A or 1B)
- Customer information (name, organization, and address, etc.)

11.4.12.2 Product Order Result

The purpose of the Product Order Result data flow is to close the loop initiated by the Product Order data flow. A Product Order Result is generated after Product Processing activities corresponding to a given Product Order have been completed, and the product has been delivered/shipped to the customer.

The Product Order Result specifies the following:

- Product Order ID
- Product status, which can be one of the following status:
    a. Success
    b. Failure

When the product status has a value of success, the following information is included in the Product Order Result:

- Product shipping data

11.4.13 Product Processing to Calibration

11.4.13.1 Level 1A Product

The purpose of the Level 1A Product data flow is to provide the Calibration function with the required imagery and metadata for computing the satellite and ground segment calibration parameters.

A Level 1A Product consists of:

- Imagery:
    a. In an application-friendly format such as TIFF (header and imagery within the same file) or in the ENVI format (binary imagery file with separate ASCII header file)
    b. Raw in DNs

- Header (describes the imagery data file)
- Metadata: all constituents other than imagery and header, such as the geometric model, etc.

The format of the Level 1A Product delivered to the Calibration function is identical to that of the Level 1B Product that is delivered to the customers. The absence of radiometric correction is the only difference that distinguishes a Level 1A product from a Level 1B product.

11.4.14 Calibration to Command and Control

11.4.14.1 Instrument Calibration Parameters

The purpose of the Instrument Calibration Parameters data flow is to provide radiometric calibration parameters to the Command and Control function in preparation for uplink to the Space Segment. Radiometric correction may be performed onboard to generate at-sensor radiance or to improve the efficiency of the onboard data compression prior to data downlink.

This interface is identical to the Instrument Calibration Parameters interface described in Section 11.4.3.3.

11.4.15 Calibration to Archive

11.4.15.1 Calibration Parameters

The purpose of the Calibration Parameters data flow is to submit radiometric, spectral, and geometric calibration parameters calculated by the Calibration function to the Archive.

This interface is identical to the Calibration Parameters interface described in Section 11.4.8.2.

11.5 OPERATIONAL SCENARIOS

In general, there are three operational scenarios in the ground segment:

1. Ordering New Data
2. Ordering Level 1B Products
3. Conducting Instrument Calibration and Generating Calibration Coefficients

As described in Section 11.1.5, Level 0 data are unprocessed raw instrument data in DNs. Level 1A data are the same as Level 0 plus being time-referenced and annotated with ancillary information, including radiometric, spectral, and geometric calibration coefficients and geo-referencing parameters. Both Level 0 and Level 1A data are normally not deliverables to customers. A customer can only order Level 1B Products, which have been converted into at-sensor radiance units from DNs after removing all the instrument effects.

11.5.1 Ordering New Data

Ordering New Data will result in new data being acquired by the space segment and stored in the Archive. Once the new data is in the Archive, a customer can browse the Catalogue and order the Level 1B Products derived from it.

Figure 11.5 shows the trace diagram for the operational scenario of Ordering New Data. Each of the functions of the ground segment that participates in this scenario is shown as a box across the top of the figure. Time proceeds from top to bottom and each flow is shown as a directional arrow between the two functions. Flows are labelled with a letter (A, B, C, etc.) on the left edge of the diagram and the label is used in the description that follows.

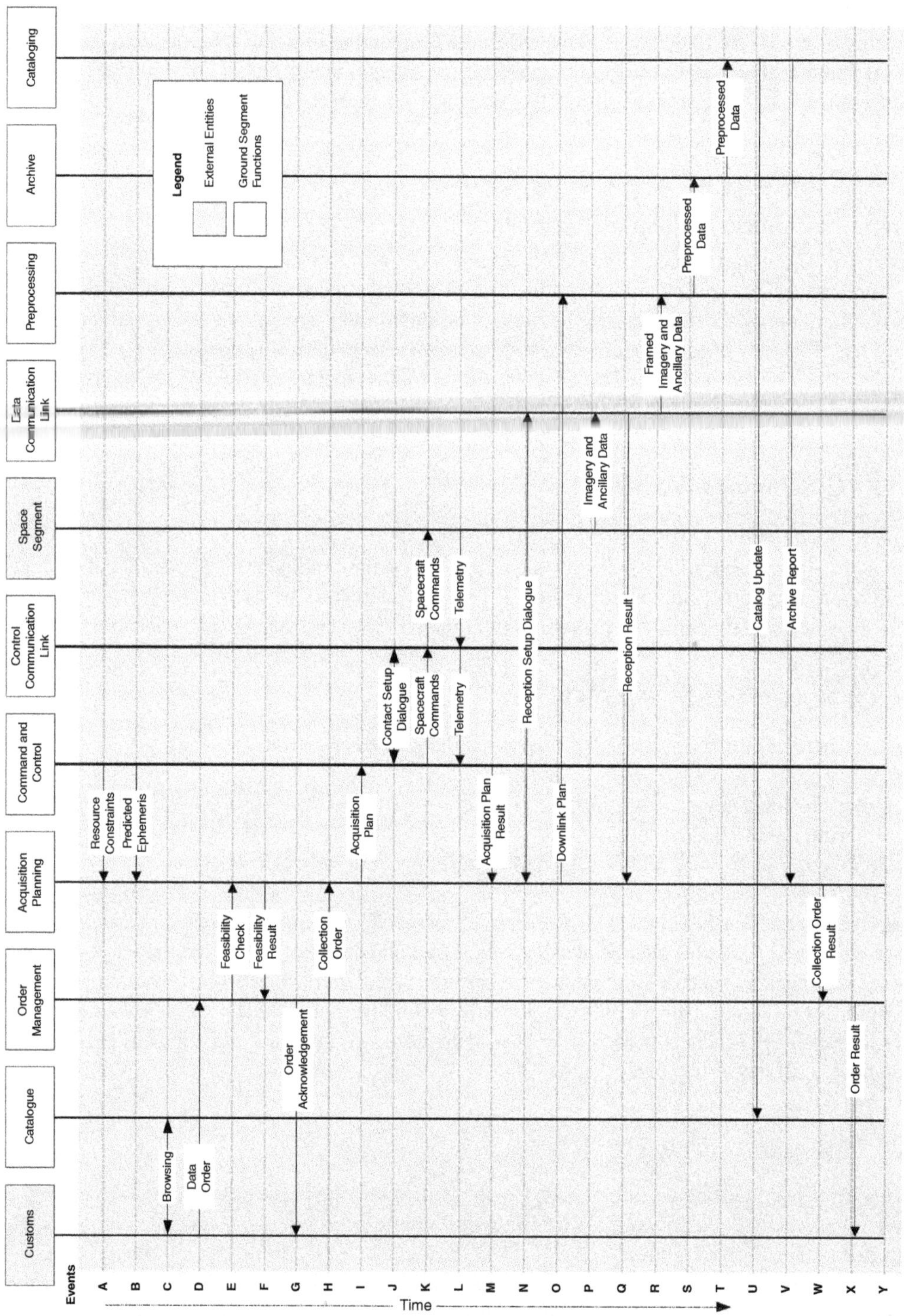

FIGURE 11.5 The trace diagram for new data ordering scenario.

11.5.1.1 Preliminary Events

On a periodic basis, the Command and Control function provides both Resource Constraints (A) and Predicted Ephemeris (B) to the Acquisition Planning function. Neither of these are synchronously linked in this scenario; however, they must have been provided prior to the start of the scenario.

The Resource Constraints provide updates on the availability of the space segment resources. For example, if an orbit maneuver needs to be performed, there will be a period of time following the maneuver when imaging will not be possible.

Propagating previously measured ephemeris into the future generates the Predicted Ephemeris. The Predicted Ephemeris is used by the Acquisition Planning function to build a geometric model for planning purposes.

11.5.1.2 Customer Interaction

This operational scenario begins with a customer initiating a Browsing Dialogue (C) with the Catalogue function for viewing existing data stored in the Archive that meet her/his needs. When the data is found, the customer then can submit a Product Order to the Order Management function to order Level 1B Data Product (see Section 11.5.2). When no data of interest is found, and if the custom decides to order new data, he/she submits a New Data Order (D) to the Order Management function.

The Order Management function then requests a Feasibility Check (E) for the order from the Acquisition Planning function. The Acquisition Planning function responds to the request with a Feasibility Result (F), indicating whether or not the order can be fulfilled successfully with some acceptable probability. The Order Management function then provides an Order Acknowledgement (G) to the customer to indicate successful receipt of the order and preliminary indication of its feasibility.

If the Feasibility Result is negative, the customer is sent a negative Order Result (not shown in Figure 11.5), and if positive, the order continues processing.

11.5.1.3 Planning, Command, and Control

Following a successful Feasibility Check, the Order Management function sends a Collection Order (H) corresponding to the New Data Order to the Acquisition Planning function. The Acquisition Planning function is responsible for determining a suitable opportunity to acquire new imagery data and inserting the order into the overall Acquisition Plan for a particular orbit. Once an Acquisition Plan (I) is complete (typically a fixed time prior to the uplink opportunity), the plan is transferred to the Command and Control function.

The Command and Control function is responsible for performing the high-fidelity constraint checks on the plan to ensure the health of the space segment. Between the Resource Constraints provided and the models in the Acquisition Planning function, an Acquisition Plan should have a very high probability of passing all constraint checks. Following the constraint checks, the Command and Control function converts the Acquisition Plan to a series of Spacecraft commands, inserting additional non-imaging commands as necessary and sends to Data Communication Link function to be uplinked to the space segment.

11.5.1.4 Space Segment Control

Some time prior to the uplink contact time, the Command and Control function initiates a Contact Setup Dialogue (J) that varies depending on the selected mission. The Contact Setup Dialogue involves programming the antenna system to track and communicate with the satellite.

Once the Control Communication Link function has established the link with the Space Segment, the Command and Control function begins sending Spacecraft Commands (K) to the Space Segment. In return, Telemetry (L) is provided indicating the state of the Space Segment. Using this mechanism, the imagery acquisition commands are sent to the Space Segment.

Following a successful transmission of the Spacecraft Commands to the Space Segment, an Acquisition Plan Result (M) is sent back to the Acquisition Planning function, indicating that the imagery data will be acquired.

11.5.1.5 Data Reception

Different hyperspectral missions may choose different transmission methods to downlink the imagery data from the payload to the Ground Segment. Prior to a downlink contact, the Acquisition Planning function initiates a Reception Setup Dialogue (N) with the Data Communication Link function. The reception antenna is programmed using the ephemeris and times in the Reception Setup Dialogue.

Immediately following the Reception Setup Dialogue, a Downlink Plan (O) is sent to the Preprocessing function to enable it to know what data to expect.

The compressed and/or raw Imagery and Ancillary Data (P) is then transmitted from the Space Segment to the Data Communication Link function at the appropriate time. The Data Communication Link function converts the data into a framed format and sends the Framed Imagery and Ancillary Data (R) to the Preprocessing function.

Following a successful reception, the Reception Result (Q) is sent back from the Data Communication Link function to the Acquisition Planning function.

11.5.1.6 Preprocessing, Archiving, and Cataloging

When the Framed Imagery and Ancillary Data (R) is transferred to the Preprocessing function, the necessary preprocessing is performed including image segmentation, and ephemeris and attitude extraction. This preprocessing is based on the Downlink Plan that was sent to the Preprocessing function earlier by the Acquisition Planning function.

Following preprocessing, the Preprocessed Data (S) is submitted to the Archive function where the data is tracked and a copy is made to a storage media (e.g., DVD).

Based on the newly acquired data submitted to the Archive, the Cataloging function retrieves the Preprocessed Data (T) from the Archive and creates a reduced-resolution imagery data along with the essential metadata. Together these data form the Catalogue Update (U) that is sent to the Catalogue as a representation of the contents of the Archive.

In addition, an Archive Report (V) is sent back to the Acquisition Planning function. This report contains information about the imagery such as the cloud cover, the actual geographic coverage and any data dropouts.

11.5.1.7 Completing the Order

Based on the Archive Report, the Acquisition Planning function determines if the customer's data collection needs were met. If the data collection was not successful (e.g., too much cloud coverage), the imagery acquisition may be replanned. If the data collection was successful, a Collection Order Result (W) is provided back to the Order Management function indicating the archive identifier of the new imagery.

The Order Management function then provides an Order Result (X) back to the customer which includes the archive identifier, thereby completing the scenario.

Note that the customer can then order the new imagery data as a Product by submitting a Product Order. The scenario associated with product ordering is described in the next section.

11.5.2 Ordering Level 1B Data Product

This scenario assumes that the data of interest to a customer has previously been acquired and stored in the Archive, as described in Section 11.5.1, Ordering New Data.

Figure 11.6 shows the trace diagram for Ordering Level 1B Data Product scenario. Same as in Figure 11.5, each of the functions of the Ground Segment that participates in this scenario is shown as a box in the top row. Time proceeds from top to bottom and each flow is shown as a directional arrow between the two functions. Flows are labelled with a letter on the left edge of the diagram.

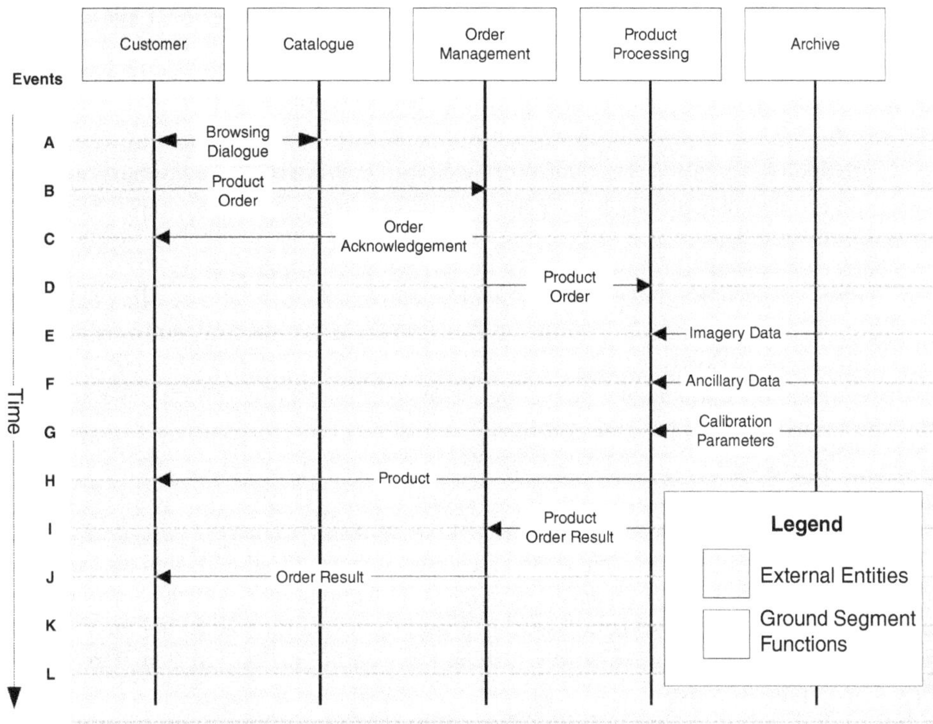

FIGURE 11.6 The trace diagram for Ordering Level 1B Data Product scenario.

11.5.2.1 Customer Interaction

This scenario begins with a customer initiating a Browsing Dialogue (A) with the Catalogue function for checking the existing imagery data that meets her/his needs. In this scenario, the customer finds such an imagery data set and submits a Product Order (B) to the Order Management function.

The Order Management function receives the Product Order and provides an Order Acknowledgement (C) to the customer to indicate successful receipt of the order.

11.5.2.2 Product Processing

The Order Management function sends a Product Order (D) corresponding to the received Product Order to the Product Processing function. The Product Processing function then retrieves the Imagery Data specified in the Product Order from Archive (E). The Product Processing function then retrieves the Ancillary Data and Calibration Parameters from the Archive that match the acquisition parameters of retrieved Imagery Data (F and G).

The Product Processing function then creates the Level 1B Product using the Imagery Data, Ancillary Data, and Calibration Parameters, and sends the Level 1B Product to Customer (H). Immediately following the product delivery, the Product Processing function sends a Product Order Result message (I) to the Order Management function to indicate the completion of the product processing.

11.5.2.3 Completing the Order

The Order Management function then provides an Order Result (J) back to the customer, which includes the shipping date and address for the product delivery, completing the scenario.

11.5.3 Onboard Instrument Calibration and Generating Calibration Coefficients

To conduct onboard instrument calibration, the same operational scenario as Ordering New Data described in Section 11.5.1 can be used with the following exceptions:

- A Calibration Specialist acts as a customer as in the Ordering New Data scenario and takes care of the onboard instrument calibration.
- The Calibration Specialist is responsible for determining when calibration events must be scheduled and executed. This decision is based on the following factors:
  a. The mission calibration plan, which contains a schedule about how often calibration events need to occur based on the predicted radiometric, spectral, and geometric stability of the payload.
  b. Historical calibration data, indicating the actual radiometric, spectral, and geometric performance of the payload over time.
- The calibration data need not be cataloged.

After the execution of the onboard instrument calibration using the above described operational scenario, the Calibration Specialist knows the identifier of the calibration data in the Archive. The Calibration Specialist can then initiate the generation of Calibration Coefficients from the newly collected calibration data. The scenario associated with this event is depicted in Figure 11.7, which shows a standard procedure by which the Calibration Coefficients are created from calibration imagery stored in the Archive.

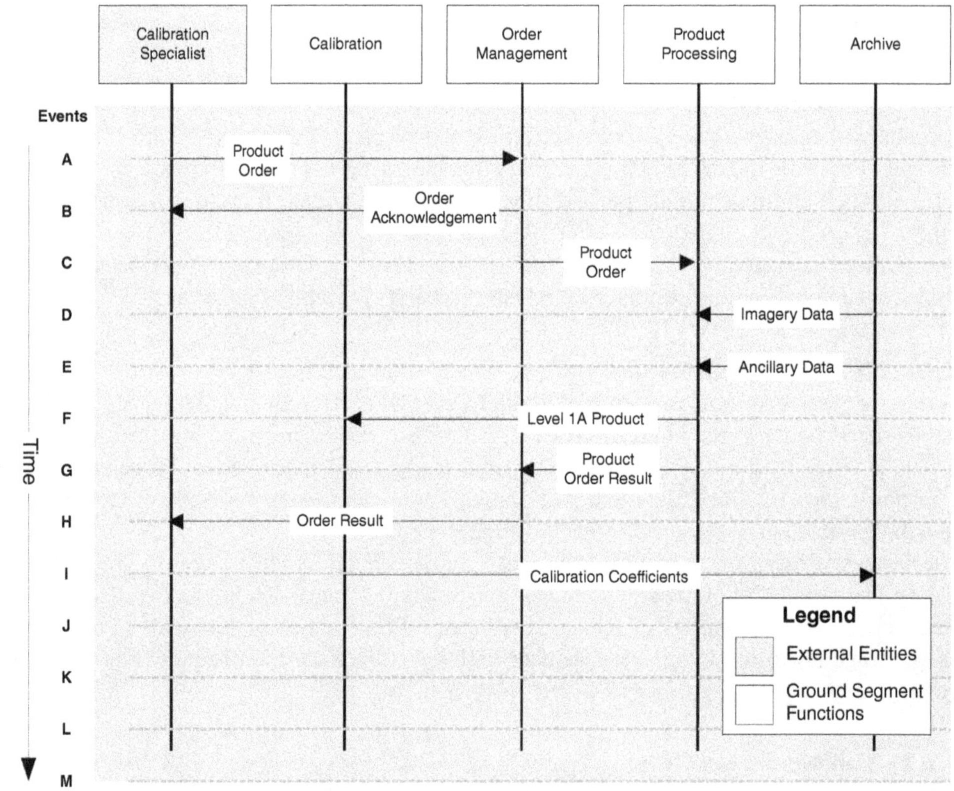

FIGURE 11.7 The trace diagram for the Calibration Coefficient Generation.

As shown in Figure 11.7, the Calibration Specialist submits a Product Order (A) to the Order Management function for the calibration data of interest. The Product Order function particularly indicates that the Level 1A Product needs be sent to the Calibration function.

The Order Management function receives the Product Order and provides an Order Acknowledgement (B) to the Calibration Specialist to indicate successful receipt of the order.

The Order Management function sends a Product Order (C) that corresponds to the Product Order to the Product Processing function. The Product Processing function then retrieves the Imagery Data specified in the Product Order from Archive (D). The Product Processing function then retrieves the Ancillary Data from Archive (E) that match the acquisition parameters of the retrieved Imagery Data.

The Product Processing function then creates the Level 1A Product using the Imagery Data and Ancillary Data and transfers the Level 1A Product to Calibration function (F). Immediately following the product delivery, the Product Processing function sends a Product Order Result message (G) to the Order Management function to indicate the completion of the product processing.

The Order Management function then provides an Order Result (H) back to the Calibration Specialist, which indicates that the Level 1A Product has been produced and transferred to the Calibration function.

The Calibration Specialist then coordinates the generation of the Calibration Coefficients using tools available within the Calibration function. The generated Calibration Coefficients are then sent to Archive (I) for later use in Level 1B Product processing.

REFERENCES

Elbert, B. 2014. *The Satellite Communication Ground Segment and Earth Station Handbook*, 2nd ed. Norwood, MA: Artech House.

Ley, W., K. Wittmann, and W. Hallmann (eds.). 2009. *Handbook of Space Technology*. West Sussex, UK: John Wiley & Sons.

Qian, S.-E. 2013. *Optical Satellite Signal Processing and Enhancement*. Bellingham, WA: SPIE Press.

12 On-Ground Calibration and Characterization

12.1 RATIONALE OF ON-GROUND CALIBRATION AND CHARACTERIZATION

The calibration and characterization of spaceborne hyperspectral sensors is a critical prerequisite for creating high-quality scientific data and for operational usages. The goal of calibration is to develop a quantitative understanding and characterization of a satellite system and its biases in time and space. Without accurate calibration, errors in the acquired data can spread through all subsequent image processing operations, including spectral, spatial and multi-temporal analyzes, thematic classifications, and the generation of vegetation indexes (Bannari et al. 1999). To get the maximum information from satellite data-derived products, sensors must constantly be calibrated, the data must be validated, and the stability and quality of data must be ensured. Calibration can be considered as a process that encompasses the entire system, from a satellite instrument to data products (Teillet et al. 1997).

The challenge is to ensure that the measurements and methods yield self-consistent and accurate geophysical parameters, even though the measurements are made with a variety of different satellite sensors under different observational conditions and the geophysical parameter retrieval methodologies vary. Without calibration we are unable to detect changes observed in satellite data to real environmental changes occurring at ground level and rule out the influence of other factors, such as instrument errors or influences of the atmosphere. This problem is exacerbated by the myriad of satellite sensors operated by multiple countries and organizations. Calibration allows the traceability of sensor data to the same physical standards. It is routinely required as sensors decay throughout their lifetime in space. Calibration is thus critical to the compilation of reliable long-term data sets for studying the effects of climate change and the fluxes of carbon to and from the oceans and land.

The satellite calibration/validation methods and instrumentation involved can be grouped into three categories (Dinguirard and Slater 1999):

1. On ground prior to launch
2. Onboard the spacecraft post-launch
3. Vicarious or indirect approaches using Earth scenes imaged in-flight.

On-ground prelaunch calibration methods encompass a vast array of careful instrument characterizations in the laboratory and occasionally outdoors, whereas onboard calibration and vicarious calibration are devoted primarily to the monitoring of the stability of spectrum, radiometric responsivities or gain coefficients of sensor over time (Guenther et al. 1996, Teillet 1997).

Calibration of spaceborne hyperspectral sensors is a nontrivial task. To increase the accuracy of the calibration, it is advisable to use independent methods. Different methods have been developed for relative and absolute calibration of optical sensors: calibration in laboratory before launch in a well-controlled environment; onboard calibration using lamps, a sphere, a solar diffusion panel, or a solar sensor; calibration through lunar observation; vicarious calibration using ground sites with simultaneous ground reference data; calibration using pseudo invariant features on the ground without independent reference data, inter-instrument, and inter-band calibration (Henry et al. 1993, Teillet et al. 1997).

Because of the inaccessibility of the satellite on orbit, the vicarious calibration method based on a ground site using simultaneous ground reference data is often considered an essential step to ensure the best accuracy versus investment compromise. This method has the advantage of reproducing the real conditions of image data acquisition. Its accuracy depends closely on the radiometric stability of the calibration site, site reflectance measurements, and knowledge of the atmospheric parameters at the time of image acquisition. In the best conditions of site and measurement stability, it ensures calibration accuracy in the range of ±2% to ±3% (Thome et al. 1997, Dinguirard and Slater 1999).

The subject of this chapter is the on-ground calibration and characterization. Characterization is the second subject addressed in this chapter. While the goal of calibration is to obtain instrument coefficients for converting raw sensor data into physical data, the goal of characterization is broader. The main objectives are to gather information on the system to be characterized, understand its behavior under various conditions, validate the quality of its data, and collect information that may later serve to help diagnose the instrument if some unexpected phenomena happen during its operation.

The subjects of the onboard calibration and vicarious calibration are described in Chapter 9. The subjects of conversion of acquired raw data to at-sensor radiance and data correction are described in Chapter 13.

12.2 CALIBRATION AND CHARACTERIZATION TYPES OF HYPERSPECTRAL SENSORS

The calibration is a fundamental step for the rigorous validation of products derived from satellite sensors. It is a process of quantitatively defining the sensor system's response to known controlled signal inputs, such as the standard calibration lamps, reference panels and ground reference scenes. The main aspects that need to be calibrated and characterized are the sensor system's response to electromagnetic radiation. Calibration and characterization of hyperspectral sensors includes following types:

1. Radiometric (sensitivity to input signals, offset, dark current, spectral cross-talk)
2. Spectral (spectral registration, spectral bandwidth)
3. Geometric
4. Spatial co-registration
5. Spatial resolution and modulation transfer function (MTF)
6. Linearity/nonlinearity
7. Uniformity (radiometric-spatial, radiometric-spectral)
8. Rectilinearity (spectral curvature, keystone)
9. Polarization sensitivity
10. Stray light (in-field, out-of-field)
11. Signal-to-noise ratio (SNR)

12.2.1 Radiometric

Ground radiometric calibration requires a bright light source which can uniformly fill the field-of-view (FOV) and aperture of the instrument. For example, two integrating spheres can be used, one with a barium sulfate coating or a Teflon coating for the visible and near-infrared (VNIR) wavelengths and another with a gold coating for the shortwave-infrared (SWIR) wavelengths. The integrating sphere has to be illuminated by sources appropriate for the spectral region of the instrument under test. The integrating spheres and their sources should be calibrated by source traceable to a national standard (e.g., US National Institute for Standards, United Kingdom's National Physics Laboratory, and the Canadian National Research Council).

Solar-based calibration from the ground was suggested by Slater et al. (2001) to use the same spectral distribution as on-orbit solar calibration. This would require a large heliostat, a parasol concurrent solar monitoring, and the use of atmospheric correction algorithms in calibration calculations. The diffuser would be used in the same manner as on-orbit. This approach has the same pros and cons as onboard solar-based calibration.

12.2.2 Spectral

Ground spectral registration can be calibrated using a series of atomic gas discharge lamps. Spectral bandwidth can be measured by a ground-based scanning etalon or a monochromator that could be stepped in fractional bandwidth increments. If the onboard calibration system includes a wavelength calibration component, such as a scanning etalon, that system would also have to be calibrated.

12.2.3 Geometric

Geometric or spatial characterization includes a broad range of tests that have for goal to characterize the relationship between the spatial distribution of radiation in the domain of the target and the spatial distribution of radiation in the focal plane domain of the instrument. Such activities include:

- Mapping of the distribution of intensity as function of position in the FOV
- Determination of the angular extend of the FOV of the entire instrument and of individual pixels
- Measurement of distortion and keystone
- Estimation of the spatial resolution and of the MTF
- Measurement of co-registration between different focal planes in the system

Some of these activities are described in more details in the next sub sections.

12.2.4 Spatial Co-Registration

Geometric registration within a single focal plane and co-registration between different focal planes can be characterized by scanning a highly collimated point source across the instrument FOV in fractional pixel increments. The FOV can be mapped at the same time using the same setup.

Co-registration of the VNIR and SWIR channels, and possibly panchromatic camera, should be adjusted and characterized at temperature extremes before and after vibration testing. The instrument design should incorporate reference surfaces, the instrument can be accurately positioned with respect to the test equipment.

12.2.5 Spatial Resolution and Modulation Transfer Function

The point spread function (PSF) and the modulation transfer function (MTF) are two ways of quantifying the spatial resolution or the blur function of the instrument. The PSF is the Fourier transform of the MTF. The optics and the detector array are internal elements that contribute to the total blur. External contribution to the total blur may include motion between the target and the instrument and atmospheric diffusion.

The spatial resolution can be estimated by observing specialized standard targets designed for that purpose. Another way is to image a knife edge (sharp contrasting target) and derive the PSF and the MTF from that measurement.

The PSF and MTF should be measured along two orthogonal directions because they can be different in the along-track and in the cross-track directions. Instrument motion, during operation, will add more blur along the direction of motion but that component can usually be added analytically to the budget, at least for the known parts of the motion vector.

12.2.6 Linearity/Nonlinearity

Linearity can be calibrated by varying the integration time of the instrument, although this relies on the stability of the instrument's clock and assumes an independence to photon rate. This photon rate assumption is false for photomultiplier tubes, but is believed to be good for bulk detectors. Measuring linearity by varying integration time does not require a target; closing the instrument shutter and integrating dark current only may even yield more accurate results by removing source stability issues, assuming the dark current is sufficiently stable.

In MEdium Resolution Imaging Spectrometer (MERIS), for example, radiometric response nonlinearity was characterized extensively at a number of stages in the instruments development phases. The following sources of nonlinear response in the instrument were identified:

1. The charge-coupled device (CCD) three phases clock voltage in dither clocking mode
2. The CCD transimpedance amplifier
3. The detection unit (DU) output amplifier
4. The video electronic unit (VEU) amplifier
5. The VEU analogue to digital converter (ADC)

The experimental results showed that the major contribution originated from the detectors, in particular the CCD's output transimpedance amplifier. At CCD detector level, the characterization consisted of measuring the detectors output in mV against a varying input irradiance obtained by moving the detector away from a stable and carefully monitored illumination source, and relying on the well-known quadratic law behavior of light propagation from a point source. The integral nonlinearity of the ADC was measured at VEU level, and showed to be negligible for all twelve gains settings.

In order to combine the unit level characterization to an instrument level nonlinearity function, verifications were performed between unit pairs. The results showed that these units did not introduce any noticeable nonlinearity contributions.

The MERIS had a nonlinearity requirement of <1%. This means that the main focus of the characterization activity should be on low light levels, a domain important for ocean color, as the majority of ocean surfaces have a relatively low light signal levels in the NIR spectral region outside of sun glint areas. The nonlinearity of the MERIS instrument showed to be slightly above the 1% requirement for the low light levels. An intensive characterization activity was performed both at unit level and at the different integration stages to understand the sources of this nonlinearity. A characterization data set of sufficient quality was delivered to correct the residual nonlinear response at instrument level to a relative radiometric accuracy of better than 0.2 % at typical oceanic scene levels.

12.2.7 Uniformity

Uniformity can be characterized to some extent by radiometric, spectral, linearity, and geometric calibrations. Spatial uniformity is the variation in radiometric response as a function of cross-track spatial position. Spectral uniformity is the variation in radiometric response as a function of wavelength.

The most accurate way to calibrate spatial uniformity of a hyperspectral instrument on ground is to scan a long uniform diffuse source across the instrument (Harron et al. 1992). The redundant data allows iterative calculation of both the source and camera uniformities, and an accuracy of 0.2% is possible. Alternatively, the geometric registration data can be reused, or the instrument could view a fixed highly uniform source.

The spectral uniformity can be measured by directly viewing a cavity blackbody with well-characterized emissivity and temperature assuming a sufficiently high flux can be achieved in the spectral range of the instrument. The blackbody does not need to fill the instrument aperture, since absolute radiometry is not needed, but stray light from the blackbody aperture must be mitigated. No reflective surface should be used between the blackbody and the instrument to avoid modifying the spectral distribution.

Since the greatest contributor to nonuniformity is likely to be the photo response nonuniformity (sometimes called fixed pattern noise even if it is not random) of the focal plane array, it may be advantageous to characterize this without the slit and dispersing element. This could be done by removing these components, or by performing this test partway through integration, or by designing a calibration port into the instrument for this purpose.

Radiometric calibration will naturally remove the nonuniformity but characterizing it for diagnostic purpose is useful.

12.2.8 Rectilinearity

Deviation in spectral rectilinearity is called spectral curvature (or smile), and deviation in spatial rectilinearity is called keystone. Spectral curvature is the variation of spectral registration and bandwidth as a function of cross-track spatial position. This can be measured using an etalon calibration system or a spectrally tunable laser. Keystone is the variation of cross-track spatial registration and MTF as a function of wavelength. Keystone refers to deviations of any shape, not just the common trapezoidal keystone shape. It can be measured by scanning a checkered black-and-white pattern with the instrument.

Monochromators have been used to measure spectral curvature, but this technique is slow and its accuracy is limited by the slit width, repeatability, and angular nonuniformity of the monochromatic. Keystone can be measured by reusing the geometric calibration data.

Neville et al. (2003, 2008) developed a technique to detect and correct smile using atmospheric absorption feature matching. The same authors also developed a technique to detect and correct keystone using inter-band correlation of spatial features (Neville et al. 2004).

12.2.9 Polarization Sensitivity

Most off-axis or asymmetrical optical systems tend to polarize light, and this polarization will change as the optical surfaces degrade. Polarization sensitivity can be calibrated with polarization filters, or with polarized laser standards. Glan-Thompson polarizers may be used for this purpose. They can yield extinction ratios greater than 100,000:1 over a wide range of angles and for a broad spectral range. By placing the polarizer at different angles in front of an unpolarized source, the polarization axes and sensitivity of the instrument can be mapped. The polarization sensitivity is usually wavelength-dependent so the test must be performed with polarizers that are efficient over the entire spectral range or with monochromatic sources and appropriate polarizers for the source being used at a given time.

MERIS was a spaceborne hyperspectral sensor for ocean color applications. It had a very stringent requirement for polarization sensitivity given the fact of the significant impact of the polarization sensitivity on the radiometric measurement due to extremely low albedo of the ocean scene. In the MERIS design, a depolarizer (also referred to as polarization scrambler) was deployed and inserted in front of the telescope just behind the UV filter in the optical path of each camera, to minimize the polarization sensitivity.

Polarization measurements are difficult to make, in particular when quantifying a depolarization efficiency at instrument level. In the case of MERIS, the measurement approach chosen by

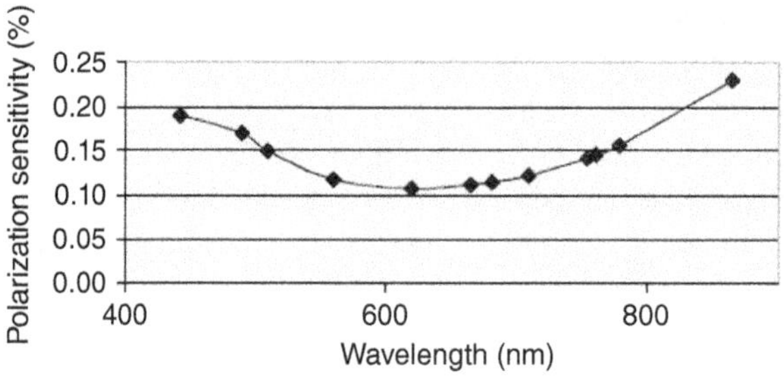

FIGURE 12.1 MERIS polarization sensitivity.

MERIS team was to verify it at component level. The compliance of the scrambling window spectral behavior was tested to validate the design when the system was illuminated with monochromatic polarized light at particular orientations. The advantage of this test is that the expected signal modulation is high enough to be detected, while direct measurement of the depolarization is buried into background noise and parasitic effects such as stray light.

The MERIS polarization sensitivity was relatively low (<3.0%) without applying the depolarizer. The polarization sensitivity was reduced to better than 0.25% by the use of a full pupil depolarizing scrambling window, as shown in Figure 12.1. This key instrument design allowed the users to make the assumption that the MERIS instrument is insensitive to polarization states, which simplifies and makes the retrieval of ocean color products much more robust.

It is worth to note, however, that the introduction of a depolarizer in the optical chain will introduce a slight degradation of the instrument's PSF, and introduce ghosts images that need to be carefully baffled.

12.2.10 Stray Light

Stray light can come from objects inside or outside the FOV. Stray light is primarily a function of coarse instrument geometry which is unlikely to drift over time, but degradation of baffle coatings and knife edges can have an impact.

One way to characterize stray light is to scan a collimated source at the instrument from many angles at many locations on the instrument, and measuring the variations in instrument response. In-field stray light requires a narrow source like a laser to look for fine ghosting effects. Several lasers need to be used to cover the full spectral range of the instrument. Out-of-field stray light does not need a narrow source, so time can be saved by using a wide source like a spotlight.

These tests are best done outside the thermal vacuum chamber to ensure that all instrument surfaces are included in the test. Temperature and vacuum is not expected to have a significant impact on stray light.

If a significant amount of stray light is found, this is usually addressed by redesign and rework, not calibration. Stray light being scene dependent, it cannot be calibrated without knowing the observed scene.

For example, the MERIS stray light contribution had been analyzed in detail using the nonsequential raytracing approach. The first analysis showed that great care had to be taken with regards to baffling, light trapping, the slit, the choice of the absorbing black paint, and the design of the edges of baffles and aperture stop—not only in the spectrometer where most of the stray light occurs, but also through both optical paths—observation and calibration.

The original overall stray light level was slight >2%, but this was reduced to less than 1% by proper design improvements, mainly in the spectrometer, trapping all diffraction orders except order −1, and by proper design and positioning of baffles and aperture stops.

The contribution of MERIS stray light to the calibration signal, relative to the light field at the instrument entrance, is between 0.5% and 2% depending on wavelength and location in FOV. The final impact on the instrument's radiometric gains, however, is <0.3% when stray light is included in an iterative calibration processing.

12.2.11 Signal-to-Noise Ratio

Although signal-to-noise ratio (SNR) cannot be calibrated, it is important to characterize it to validate the instrument performance and to refine the error budgets. Noise can be measured by accumulating many frames from a static scene for a long period of time and calculating the standard deviation in readings. Dividing the average of the measurements by the noise yields the SNR. The scene brightness should be representative of operations and comparable with the instrument design specifications. It must also be stable during the measurement period.

Signal-to-noise ratio is one of the many measures of radiometric sensitivity. The radiometric sensitivity can also be measured in absolute value such as in Noise Equivalent Spectral Radiance (NESR) by calibrating the measured noise into the desired units.

The test should be repeated for different levels of illumination to estimate the dependency of the noise with the source radiant power and determine for which levels of illumination the scene shot noise starts to dominate the total noise budget. The dark noise can be estimated by repeating the test with a source of negligible illumination: a shutter for short-wavelengths or a cold target for thermal infrared wavelengths.

12.3 ANALYSIS OF SPECTRAL AND SPATIAL ERROR SOURCES

12.3.1 Background of Spectral and Spatial Error Sources

It is difficult to separate the various causal agents of spectral and spatial errors in an imaging spectrometer into neat and direct factors. Mechanical and environmental factors combine to impact the spectral and the spatial performance. This is especially true for hyperspectral sensors operating in pushbroom mode since, as Mouroulis and Green (2018) pointed out, a pushbroom hyperspectral sensor is actually an integration of as many individual spectrometers as there are cross-track pixels in a swath (e.g., 1000 typically) into a single instrument. Optical components, especially those with direct views of space, have to work under harsh space conditions. Thermal cycling, irradiation by energetic particles and exposure to UV radiation, contamination, atomic oxygen, and micrometeorite impacts degrade the instrument performance over time.

Many reported works have stressed the importance of uniformity of spatial and spectral response in an imaging spectrometer if accurate information is to be obtained from the acquired data. As Green (1998) stated, the spectral response function of each pixel in the spectrometer must be known to a high accuracy. However, it is difficult to expect that a hyperspectral spectrometer exhibits the same response along the spatial and spectral directions and that it will be immune from degrading effects over time. But, it is realistic to design and build an instrument in which the variation of the response function is minimized and its behavior over time can be calibrated, with enough accuracy to satisfy the requirements.

Adopting the description by Mouroulis et al. (1998, 1999, 2000), a perfect imaging spectrometer would have a spectrum in which the point spread responses across the spatial and spectral domain line up. Although the point spread responses' widths may vary along the spectral direction, the responses are uniform along the spatial direction. Of course, in reality, this is not the case. In a real imaging spectrometer, the key geometric artefacts critical to performance and calibration are spectral curvature (smile) and spatial distortion (keystone).

Smile, defined as the spatial variation of the peak location of instrument point responses along the slit image, results in curved monochromatic images of the slit. Coupled with smile is the variation of the width of the monochromatic point response across the slit. According to Mouroulis (1999), for accurate calibration of hyperspectral imagers, the location of the peak of the response function, its half width and shape should be known to within a small fraction (typically 0.1) of a pixel along the length of the slit, as a function of wavelength. Smile detection and correction is described in Section 13.2.

Keystone is the variation of location of the instrument's PSF across the spectral dispersion or spectral direction. In other words, in an ideal instrument, the spectrum at one image point would be parallel to that at any other image point. Keystone is an estimation of this lack of parallelism. It is the variation of slit magnification along the spectral direction and controls the amount of mixing of spectral information and spectral contamination between pixels across the spectral range of the instrument (Mouroulis et al. 2000). Its impact on the performance of the instrument depends on the target sensed. Keystone detection and correction is described in Section 13.3.

Furthermore, one can easily understand that nonuniformity in an instrument point responses along the dispersion direction can cause errors similar to those caused by smile and keystone. It is not the size of the point spread that is critical but its shape and its spatial and spectral variation across the image plane. Mouroulis et al. (1998) have quoted values for the allowed errors at the level of a small fraction of a pixel.

As reported by Pitts et al. (2003) in the review of the performance of the airborne A-band spectrometer, shifts in the spectral registration were induced by thermal and mechanical instabilities of the instrument. It was suggested that the thermal design must consider the thermal behavior of the thin film coatings and surface optical treatments. In addition, good contamination control practices such as material selection and thermal vacuum bake-out should be employed in order to minimize self-contamination. At a minimum, materials chosen should satisfy the NASA's out-gassing requirements of a total mass loss (TML) of 1.0% or less and a collected volatile condensable materials (CVCM) of 0.10% or less. In some cases, when the material will be in the line of slight of a sensitive optical surface, more stringent criteria should be adopted. Sensitive elements should be adequately shielded from irradiation by energetic particles and UV radiation and due consideration given to the potential from damage to the optical system from micrometeorites.

12.3.2 Thermal Drifts

It is important that a spaceborne hyperspectral instrument be designed as athermal as possible and, as commonly done, the thermal behavior of the instrument be taken into account during the detailed design phase of the mission development. Distortions as a result of thermal variations could introduce tilts and decentration as well as defocus that would impact the smile, keystone, and variations in point spread response across the image plane. For example, tilts in the various mirrors introduced by uneven response to temperature of mounting fixtures or of the mirror substrate itself could conceivably shift the spectral dispersion nonlinearly across the image plane, depending on which optical element is impacted. Decentration and thermal-induced defocus could result in a loss of light due to vignetting. According to Yoder (2008), thermal shocks can significantly impact the performance of an instrument. Even slow changes in temperatures are important. If a mirror substrate exhibits inhomogeneous thermal expansion, the surface figure of the mirror can distort. Glass ceramics and metal materials behave different at extreme temperatures and, since glass and glass ceramics materials are generally used for high-performance systems, it is an issue that has to be looked into (Mouroulis et al. 1998). The addition of material such as a thin optical reflectance coating can produce a preload, either tensile or compressive, onto an optic. Optical coatings are usually not a problem for optics with diameter to center thickness ratio (which is referred to as aspect ratio) of less than ten (10). But, for mirrors with larger aspect ratios, the coatings can produce significant force to cause bending even at ambient temperatures. Even cooled, low aspect ratio optics with coatings of electrolysis nickel will tend to bend.

All the above phenomena are potential problems for any sort of optical instruments. Hyperspectral sensors based on dispersive spectrometers are even more at risk because their spectral registration is highly sensitive to the alignment of the optical elements of the spectrometer. It may be necessary to control their temperature to avoid spectral errors. Spectral calibration should be repeated frequently and ideally for potentially different instrument temperatures (e.g., at different positions on the orbit, different seasons, etc.).

12.3.2.1 Mirror Substrates and Prism

Traditionally, reflective space systems employ metal or/and ceramic glass material for the mirror substrates. It all depends on the thermal and optical performance requirements. A commonly used material for mirrors is aluminum. However, bimetallic effects do limit its usefulness in high performance systems. The Hyperion hyperspectral imager used all aluminum substrates, including the grating. The housing, also aluminum, was maintained at 20°C ±2°C to minimize temperature gradients (Folkman et al. 2001). Using the same material for the mirror substrates and the housing allows the instrument to expand and contract thermally with minimal variation in optical performance.

Another commonly used material is the glass ceramic Zerodur (Schott 2019) because of its extremely low coefficient of thermal expansion (CTE), good long-term stability and excellent surface quality. Zerodur was used for the parabolic mirrors on the ODIN Osiris UV/Visible Spectrograph (Warsaw et al. 1996). Other commonly used glass material is fused silica that has a higher CTE than Zerodur. Its high resistance to ionizing radiation makes it an attractive substrate to use when added shielding for refractive elements is required. Both Zerodur and fused silica can be easily coated, but there is a limit to the degree of weight reduction while maintaining the surface figure (Mouroulis et al. 1998). There is also the issue of the glass-to-metal interfaces, a limitation for wide temperature range operation. An increasing popular material for space is silicon carbide (SiC) because of its attractive thermal properties. This material has a high specific stiffness and thermal stability with excellent light weighting capabilities. Some forms of it can be polished to a high quality mirror finish.

Thermal effects are not that straightforward to model using an optical analysis program such as Zemax EE, which contains the tools required for sequential optical system design of imaging systems, as well as capabilities, such as sequential surfaces, polarization and nonsequential ray-tracing, and physical optics. The best that can be done is to investigate steady-state instrument performance, the first-order effects based on some general assumptions. A more detail analysis would have to wait for the detailed design phase of the instrument when the impact of the mounts and housing can be better assessed. At the early stage of the instrument development, the analysis can only be done based on some general assumptions on to the thickness of the mirrors, their configuration and mounting approach. Any error estimates can only be viewed as a rough order of magnitude since different assumptions impact the results.

To investigate the thermal behavior of the VNIR and SWIR spectrometers of a hyperspectral instrument, a modeling was carried out in the Hyperspectral Environment and Resource Observer (HERO) mission analysis study (Bergeron et al. 2008). In the modeling it was first assumed that all the mirrors and the gratings (or prisms) would be made of the same material and that, as a rough estimate, the mirrors would have flat back mounting surfaces and have aspect ratio of 10 (Yoder 2008). It was further assumed that the mirror substrates would be either all aluminum (CTE of 23.6×10^{-6}/°C) or fused silica (CTE of 0.58×10^{-6}/°C) as per Appendix B of Ref. (Yoder 2008). These two substrates were selected because their CTEs are about 10 times that of SiC (CTE of 2.4×10^{-6}/°C) and Zerodur (CTE of 0.05×10^{-6}/°C) and scaling could be applied.

The radii of curvatures of the mirrors were allowed to change as a function of temperature as well as the relative position between elements and the lines-per-millimeter dispersion of the gratings. Uneven dimensional changes of the mirrors were not modeled. The shifts in positions of the mirrors were hand calculated. The shifts in positions of the mirrors were calculated. Simple on-axis contraction and expansion of the mirror surfaces were estimated. The model also included the small index of refraction variation in temperature (Schott 2004) of the radiation-hardened prism in the VNIR region and the change in the shape and size of this prism. No tilt or decentrations were modeled. It was assumed that the mirrors would be mounted symmetrical and that the mounts would response to the temperature changes in a uniform manner. Shifts in the centroids of spots in the image plane with respect to the assumed nominal temperature of 20°C were calculated.

As expected, the materials with the higher CTEs resulted in greater centroid shifts as well as spot blurs. But, no matter what material was used, there was a consistent shift of complete spectrum along the spectral dispersion direction. For example, with aluminum mirrors and grating, the spectrum shift was found to be of the order of 0.58–0.65 pixel for the SWIR bands and 0.47–0.52 pixel for the VNIR bands over 10°C temperature change. Coupled with the shift along the spectral dispersion direction, there was also a shift along the spatial direction with obvious ramifications to smile and keystone. The amount of shifting depended on where in the slit the rays originated from. For example, for the low SWIR bands, spots at the edge of the slit shifted more (about a 0.25 pixel) whereas at the center, the shift was extremely small.

To check the sensitivity of the model, the calculations were repeated under two conditions. First, the mirrors and grating relative positions remained fixed while the radii of curvatures of the various surfaces were allowed to respond thermally. Second, the relative positions of the mirrors and gratings were allowed to change as a function of temperature while the radii of curvatures of the various mirrors and gratings remained fixed. In both cases, the prism in the VNIR channel was allowed to respond thermally. As expected, it was found that thermal-induced changes in radii of curvature had greater impact than relative shifts in element positions. Both effects are dependent on the model adopted and could be reduced if a different, for example, mirror backing shape with a different mirror aspect ratio had been assumed.

The modeling was repeated with fused silica as the mirror and grating substrates. The shifts in the spot centroids were much smaller (in the order of 0.11–0.15 pixel), even over the larger temperature range. The spatial variation of the centroid shift was still evident, although much less.

To further investigate the behavior of the hyperspectral instrument, each optical element was one-by-one tilted and decentered and the shift in the spectrum was calculated. In theory, the mirror mounts should be symmetrical. There would not be any tilt or decenter introduced into the system as a result of thermal variations. However, material ageing and other unforeseen factors could conceivably introduce small tilts and decenters into the system. The starting point was high precision tilts of 1-arc seconds and decenters of 1 μm (Bass et al. 1995). The tilts and decenters were each scaled up or down to yield approximately the same shift in centered values for each optical element. As expected, the most sensitive element to tilt and decentered was the primary mirror. A tilt of 2 arc seconds about either the *X*- or *Y*-axis or a decentration of 2 μm resulted in either a dispersion or spatial shift of 0.2 pixel. The other elements had various levels of sensitivities, with the least sensitive (tilts of 3 arc seconds to 20 arc seconds or decenters of 3–200 μm) for a 0.2 pixel. It is recommended that when the instrument is in the detailed design phase, considerable time be spent in material selection and mount design so that the instrument will be as dimensionally stable as dictated by the performance requirements.

12.3.2.2 Thin Film Coating Instabilities

Temperature changes can theoretically cause thermal contraction and expansion in thin film and dielectric coatings. When the temperature changes are extreme, difference between the coating material coefficients and the substrate can have disastrous consequences as reported by Heath et al. (1998). Temperature effects can also result in dispersion in the materials. Too large of a change in the refractive index can cause the optical performance of the coating to be greatly changed and performance impacted. In antireflective (AR) coatings, the swelling and/or contraction of the layers could change the transmission properties and modify the wavelength *ringing* common in complex coatings. Even though terrestrial testing cannot fully duplicate the space environment, requiring that all the coatings pass a comprehensive thin film environmental standard such as MIL standard goes a long way to verify the stability of any coatings.

Space optics should use coatings with sufficient interface adhesion to withstand thermal expansion and with low susceptibility to thermally induced dispersion. In addition, any coatings must be able to withstand the terrestrial environment that could involve lengthy exposure to high water vapor levels during spacecraft integration. Fortunately, by careful consideration of the actual properties of the thin films and the substrate (taking into account the deposition method used), coatings can be designed to withstand the thermal environment and be environmental stable. For example, filters fabricated with refractory metal oxides using ion assist deposition (IAD) were found to be extremely stable over extreme temperatures. Plasma-assist AR coatings were measured to have less than a 0.05% change in the spectral performance from room temperature to 80°C. If color glass filters are used for order sorters, it should be noted that steep long pass filters, in addition to a reduction in transmission, exhibit a linear shift of the edge wavelength of about 0.020–0.38 nm per °C toward the longer wavelengths, depending on the filter. The longer the edge wavelength, the larger is the temperature coefficient (Schott 2004).

Spaceborne hyperspectral instruments usually require protective and enhanced mirror coatings, AR coatings for the VNIR prism, order sorters with AR coatings and perhaps a dichroic mirror to separate the VNIR and SWIR channels (see Section 5.5.2). It is expected that any thermal drifts in these coatings will not contribute significantly to spectral and spatial registration errors, provided that thermal issues are taken into account in the detailed design phase of the mission.

12.3.2.3 Gratings

In replicated gratings, thermal cycling should have minimal impact. Although the CTE of the resin and the reflective coating can be higher than the grating substrate, the coatings are thin enough that they are fixed rigidly to the substrate. There have to be thin and of relatively low expansion to not distort during the curing process (Denton and Gee 1999). There is no evidence of deterioration in standard replicated gratings with age or when exposed to thermal variations from −196°C to 50°C. Those gratings that have to withstand higher temperatures can be made with a special epoxy such that the grating grooves do not distort. Well-protected and temperature stabilized components should survive the low Earth orbit (LEO) space environment with a minimum of changes.

A single-blaze, concave, electron-beam lithography grating has been used in the Hyperion onboard NASA's New Millennium Earth Orbiting-1 (EO-1) Mission. That grating was made with a spun-coat of polymethyl methacrylate (PMMA) with an aluminum layer (30–50 μm) evaporated on top. The grating substrate is aluminum (Folkman et al. 2001). The grating passed environmental testing including thermal cycling (−50°C to +50°C), vibration, out-gassing, and tape-test adhesion. The test results reported by Mouroulis et al. (1998b) show higher efficiencies and lower scatter as compared to standard ruled or holographic gratings. No information on how the gratings performed as a function of temperature is found.

An alternative technique to fabricate blazed gratings on curved surfaces with substantial sag using X-ray lithography is reported by Mouroulis et al. (2003). The test gratings were found to have excellent efficiencies with no inherent ghosts and smooth grooves, important for hyperspectral imagers. Imperfections in the diffraction grating can cause artefacts such as side-lobes in the line-spread function of a spectrometer (Hearn 1998). There is no information on how the gratings would withstand the space environment.

12.3.3 Radiation

Optics is known to degrade exponentially when exposed to a space environment. This is because optical materials when exposed to radiation experience transmission loss, compaction, surface degradation, and index of refraction perturbations. These effects are cumulative with time and are strongly dependent on energy and type of radiation (particles, photons, ions), dose rate, temperature, and glass chemical makeup. Furthermore, the radiation penetration range depends not only on the energy and type of particle, but also the geometry of the system.

There are three basic damage mechanisms in optical materials: ionization, knock-on quasi-elastic scattering, and nuclear reactions. Of these three, ionization is the dominant mechanism for space optics. Since the defects and disordered micro-regions created in glass via irradiation are metastable, the net damage is the result of a balance between defect creation and defect annealing. Hence, in space optics, one usually talks about long-term effects. It normally takes at least a few months for optics to accumulate enough of a radiation dose for the degradation to be noticeable. A secondary effect of charged particle irradiation is the electrical discharge of buildup charges within the surface layers of highly irradiated glass materials resulting in fissures called Lichtenberg patterns (Boies et al. 1994). Electrically grounded conductive coatings are commonly applied to minimize this effect.

12.3.3.1 Mirror Substrates

Metal mirrors undergo a reduction in reflectivity as a result of the action of low-energy protons. Surface bubbles caused by these low-energy protons will increase the bidirection reflectance distribution function (BRDF) of the mirror (Silverman 1995) through enhanced surface roughness. Again, as for the transparent materials, shielding will mitigate the effects. Fused silica and Zerodur, both materials commonly used for mirror substrates, are known to undergo compaction under radiation. However, the doses involved are higher than one would expect in a LEO orbit (Doyle et al. 1996). Tests on protected-silver Zerodur mirror showed that there was a densification of a thin layer within a few millimeters of the mirror surface with a square root of the dose dependence. The changes in the radius of curvature of the test mirrors were found to follow a linear relationship with accumulated dose rate. A reduction in radius of curvature of 1.53 μm per year in a geostationary Earth orbit (GEO) was estimated. However, the tests were done for high energetic electron flux, not that critical for LEO instruments. It was reported that Zerodur does not under any dimensional changes due to radiation in LEO (Doyle et al. 1996).

12.3.3.2 Prism Glass

Some hyperspectral instruments use prisms as spectral dispersive elements, such as PRISMA and EnMAP (see Sections 2.22 and 2.24). Schott Glass (Schott 2004) recommends that radiation hardened glass be used for prisms if the total anticipated dose exceeds 1 kRad. Cerium has been added to the glass melt to suppress radiation darkening in the visible and to provide some degree of resistance to electron, proton, and gamma radiation. Unfortunately this benefit comes at a cost. The cerium-doped glass exhibits significant lower transmissions in the short visible wavelengths (from 400 nm to 460 nm) than its non-radiation hardened equivalent. Over 10 mm thick, the transmission, at a wavelength of 400 nm, is only 27%. The glass is still not immune from radiation-induce transmission losses, under intense irradiation. However, according to Fruit et al. (1999), tests have shown that cerium-doped glass displays no significant changes in transmissions up to radiation dose 400 kRad. It was estimated that an instrument in a LEO of <1000 km altitude can expect to see, internal to the instrument, 1–5 kRad of mostly proton radiation. Hence, it is safe to assume, therefore, that a prism with material LaK9G15 will not undergo radiation-induced transmission losses. The index of refraction of LaK9G15 is, nevertheless, more strongly affected by irradiation than its non-radiation hardened version. Its index of refraction dose coefficient has been estimated to be 3.25×10^{-11} per rad. Even at that, the dose coefficient is still small and, at the radiation levels expect inside the instrument, radiation-induced index of refraction changes will not impact the instrument performance.

12.3.3.3 Gratings

Cotel et al. (2014) reported a wide range of diffraction gratings used in many space missions since 1968, including NASA's Lyman-Far Ultraviolet Spectroscopic Explorer (FUSE) and Hubble Space Telescope. These diffraction gratings include following three types:

1. Replica ruled 1200 groove/mm grating blazed at 250 nm, in dimensions 40 mm × 40 mm, aluminum coated
2. Master holographic 3600 groove/mm grating optimized for 50–150 nm, in dimensions 40 mm × 40 mm, platinum coated
3. Master holographic ion-etched 1200 groove/mm grating blazed at 250 nm, in dimensions 40 mm × 40 mm, aluminum coated

These gratings were undertaken the space qualification tests in the Long Duration Exposure Facility (LDEF) mission, which was deployed on orbit in 1984. The testing gratings were exposed in space vacuum for a long duration (69 months, about 34,000 orbits) with thermal cycle from –30°C to +70°C and in space environment with cosmic dust and sun irradiation during 10 months. The optical performances of these grating types were tested in space environment. Radiation impact on grating performance was not reported.

As part of the French Cooperative Passive Payload (FRECOPA) on LDEF, replicated holographic and ion-etched holographic gratings were tested (Bonnemason 1991). Two sets of gratings were directly exposed to space for 10 months. One set was shielded from direct solar radiation. The gratings were mounted on the trailing edge of the LDEF with a cumulative solar flux of 1600 h. Loss of efficiency was found to be wavelength dependent with the greatest loss occurring in the ultraviolet. The ion-etched, unshielded holographic grating experienced the greatest losses. The testing results also showed that there was no glass blank distortion or groove line deformation that could potentially impact spectral and spatial registration. The main problem was an increase in scatter. Whether the grating was shielded from direct solar radiation or not, the results were the same. There was a dramatic increase in scatter pre- and post-exposure to the space environment. No explanation was given for the increase although some of the scatter could have been introduced during retrieval and testing. In any case, increased scatter will impact the shape of the spectral response of the instrument. If the scatter sites were localized on the grating, it would further complicate the calibration.

12.3.3.4 Color Glass Filters and Thin Film Coatings

It is well-known that color glass filters are sensitive to irradiation. Tests carried out by Henson and Torrington (2001) have shown that these filters, commonly used as order sorters in imaging spectrometers, exhibit shifts in their cut-on and cut-off wavelengths and reductions in their transmissions when exposed to gamma radiation. Although the effect is mostly a radiometric factor, uneven reduction in transmission across the order sorter filters will impact the shape and size of the PSF. Under a 30-kRad-absorbed dose for the filters about 3 mm thick, the tested filters experienced a loss in transmission ranging from 2% to 9%. Under an irradiation of 1 MRad, a maximum 1 nm shift in cut-on wavelength exhibited. The transmission reduction was more pronounced near the filter cut-on wavelength than at higher wavelengths with the effect strongly dependent on the filter composition, and the dose rate. Guidelines for using color glass filters in harsh environments are provided by Appourchaux et al. (1994). Besides the obvious shielding requirement, the thickness of the glass should be minimized, since, at least as far as gamma radiation is concerned, transmittance loss is more pronounced the thicker the glass is. And the more tinted the color, the more resistant to irradiation. Adopting these guidelines should minimize any shift in the cut-on and cut-off wavelengths of the filters.

The principal mechanism of radiation damage in dielectric coatings is ionization and atomic lattice displacement. The charge particles interact with the electrons in the dielectric resulting in browning or other discoloration. However, there are a number of thin-film materials that are chemically stable and robust enough to withstand the rigors of spacecraft integration and the space environment. However, if an unsuitable soft material has to be used, it can be embedded within the layers of the coating for protection. Examples are the hybrid silver coatings discussed in the literature by Thomas and Wolfe (2000) and Stevenson and Sadkhin (2000).

Damage of thin film coatings directly caused by radiation is not expected to be a problem. However, it is recommended that only hard, dense coatings be used and that sensitive elements be shielded. As pointed out by Boies et al. (1994), mirror coatings do not provide any protection, at least from electron radiation, to the underlying substrate.

12.3.4 Contamination and Atomic Oxygen

The mirrors of a hyperspectral instrument, especially the primary mirror, could be seriously impacted by contamination. This contamination will not only increase scatter and reduce transmittance, but also could result in a spectral shift. Depending on the type of contamination and whether it is fixed or not on the mirror by UV radiation, it could result in an added *thin coating* being deposited on the mirrors. Selective absorption of one wavelength over another could occur with obvious ramifications on the uniformity of instrument PSF.

The complex nature of the space environment makes it difficult to predict the degree of performance degradation that one can expect for any optical instrument in space. But, contamination could very well be a significant factor in the degradation of performance of LEO platform optical instruments.

There are two main types of contamination of concern to space optics: particulate and molecular. Particulate matter generated from on-orbit erosion of materials and by the natural environment can physically obscure the FOV and block light, interfere with moving parts and abrade and scratch surfaces. Under the combined action of atomic oxygen (AO), UV radiation and thermal cycling, materials may lost mechanical integrity and be disintegrated becoming sources of particulate contamination for near-by surfaces.

It is difficult to predict the amount and type of molecular contamination that an instrument will be exposed to. Normally, one would expect that molecular contamination deposited during thermal cycling onto a surface would boil off within a few days. But the combined action of UV radiation and AO can fix the contamination layer. Factors such as the temperature and space history of the contamination source, the constituents of the contamination, the texture of the "fixing" surface, geometry factors plus the synergetic action of the space environment all interplay in the process. For example, whether the contamination is a silicone or hydrocarbon makes a big difference. Moreover, deposition of the contamination on any surface will not be uniform. It starts as contamination islands which gradually grow in size and, depending of the surface texture, grow into a thin film. As time goes on, distinct layers are deposited.

The performance of optics in an atomic oxygen environment is mixed. Most glasses resist atomic oxygen erosion well but their resistance can be highly variable. The Evaluation of Oxygen Interaction with Materials (EOIM-III) experiment flown on Space Shuttle STS-46 found that atomic oxygen greatly roughens metal mirrors (Stuckey et al. 1993). In the Optical Properties Monitor (OPM) experiment flown on the Russian Mir Space Station (Wilkes and Zweiner 1999), scattering of sample optical mirrors were found to gradually increased in both the visible and infrared over the lifetime of the mission. The cause of the scattering was found to be the result of contaminants as well changes in the coating structure. In the case of the MgF_2 and aluminium mirror the increase in scattering was mostly caused by contamination but also by the partial recrystallization of the MgF_2 thin film.

A postflight analysis of one of the samples found a layer of contaminate fixed onto the mirror causing not only a wavelength-dependent loss in mirror-reflectivity but also increase scatter.

12.3.5 Micrometeorite Damage

Dielectric coatings on optical components are often brittle. The coatings can suffer conchoidal cratering, star cracking, and interlayer delamination. The impact on performance depends on the size of the cracks or pits. When the pit or crack is much smaller than the wavelength of interest, there is little change in the coating performance. When the crack or pit size is of the order of the wavelength of interest, coating performance degrades.

Results from Long Duration Exposure Facility (LDEF) mission (Silverman 1995) and the Mir Environmental Effects Payload (MEEP) showed mostly localized damage in the optical substrates tested. The major effect was an increase in optical scatter with only a small change in the overall transmission or reflectivity. The amount of scatter increased nonlinearly with the size of the impact craters and with the amount of area affected. The larger the optic, the more likely a large impact will be and the greater the increase in the scatter. Obviously, the front most optical component will be the optical element most in danger.

The degree of scatter that a surface will experience is directly related to the mission orbit, orbit inclination plus direction that the surface is facing. Obviously, surfaces that point in the direction of the satellite's motion (i.e., ram direction) would experience more impacts than those facing the Earth. For LDEF (Mouroulis et al. 1998b), model predications estimated a BRDF scatter value for a surface facing in the ram direction to be about 7.2×10^{-2} per steradian for a wavelength of 5 μm attributed to micrometeorite and space-debris damage. Since BRDF scales as the inverse of wavelength squared (Silverman 1995), the impact in the VNIR region would be more pronounced than in the SWIR region. For an Earth-facing surface, the estimation was less than 4.0×10^{-3} per steradian. For those instruments whose view angle is off from the nadir, designing a shroud or baffle to exclude direct impact on the optics by micrometeoroids and debris can greatly reduce the amount of scatter. However, contamination from indirect impacts can still degrade performance via scattering from secondary eject. It is predicted, however, to be much less important than that due to direct impacts.

It is generally recommended to not use materials with low melt temperatures since these materials tend to exhibit more blistering and bubbles in space. With coated optics, the underlying substrate could be exposed to atomic oxygen and UV radiation and thus result in additional degradation. As noted by Dooling and Finckenor (1999), one of the more important lessons learned from LDEF was that small defects on a surface of a material or coating could precipitate even more damage below the surface. Atomic oxygen entering via a small defect may ricochet and carve out a larger cavity of damage behind the defect. It is consequently not recommended to use a material known to be reactive to atomic oxygen in an exposed area prone to micrometeorite or debris impacts.

12.4 SPECTRAL CALIBRATION REQUIREMENT ANALYSIS

12.4.1 Spectral Linearity

To analyze the requirement of spectral linearity, a strawman hyperspectral instrument is used as an example. This hyperspectral instrument consists of three spectrometers:

- A prism VNIR spectrometer covering the spectral range from 400 nm to 800 nm
- A grating SWIR1 spectrometer covering the spectral range from 750 nm to 1400 nm
- A grating SWIR2 spectrometer covering the spectral range from 1350 nm to 2500 nm

The spectral sampling interval is 5 nm per band (detector pixel). The pitch size of the detector array is 30 μm × 30 μm.

The designs of these three spectrometers were analyzed for dispersion linearity at the detector level. This is important for determining how spectral calibration must be performed. A highly nonlinear dispersion of a spectrometer (such as due to use of a prism) would require spectral calibration

TABLE 12.1
Spectrometer Dispersion Linearity (at Temperature of 20°C, Nominal Dispersion 5 nm/pixel, 30 μm Detector Pitch Size)

Spectrometer	Wavelength (nm)	Spot Position (pixel)	Deviation (nm)
VNIR	400	0.00	0.00
	600	39.91	0.52
	800	80.03	0.00
SWIR1	750	0.00	0.00
	1075	64.97	–0.01
	1400	129.93	0.00
SWIR2	1350	0.00	0.00
	1075	114.93	–0.01
	2500	229.86	0.00

at several wavelengths. Higher order polynomials are normally used to fit the discrete calibration points, resulting in an equation that gives wavelength as a function of pixel number. If spectral smile is present, then the polynomial coefficients will be slightly different for each spectral band.

Table 12.1 lists the spot positions as determined by ZEMAX-EE Optical Design Program for the two ends and the center wavelengths of the three spectrometers, i.e., three wavelengths per spectrometer. ZEMAX-EE determines the wavelength positions of the spots by computing the centroid location of the spots. The position is then converted to units of detector pixel numbers in the detector array, as listed in the table. A simple linear fit is made for each spectrometer, where the line ends are the spot positions (vs. pixel number). The deviation column gives the departure of the middle point from the straight line, in wavelength nm units. As expected, the two SWIR spectrometers are very linear. However, the VNIR spectrometer that uses prism as the dispersive element causes a small but significant spectral nonlinearity at the detector, 0.52 nm at the center band. It is assumed that a spectral calibration error ≤1% of a detector pixel is required, or 0.05 nm in the case of a design with a spectral dispersion of 5 nm/pixel as analyzed here.

In order to accurately calibrate the spectral linearity of the spectrometers a least square linear regression with a polynomial is used. It was analyzed that both a first order polynomial fitting and a second order polynomial fitting were insufficient. The order of a polynomial required for adequate spectral calibration of the VNIR spectrometer is a third order and is determined by ZEMAX-EE:

$$y = a_0 + a_1 w + a_2 w^2 + a_3 w^3 \tag{12.1}$$

$$y = -3.5234 + 5.6473w + 0.5219w^2 - 02434w^3 \tag{12.2}$$

Spot positions for nine VNIR wavelengths were determined by ZEMAX-EE, as summarized in Table 12.2. The third-order polynomial least square fitting to the spot positions gives a spectral calibration equation maximum error is 0.00499, which is less than the desired 0.05 nm, as shown in Table 12.2. Therefore, in theory, a four-point spectral calibration would be sufficient to perform the spectral calibration of the VNIR spectrometer. These points would be approximately equally spaced in wavelength.

12.4.2 Temperature Sensitivity

Temperature sensitivity of the three spectrometers was also analyzed. As a worst but realistic case, all reflective optical elements of the spectrometers were assumed to be made of aluminum. Also

TABLE 12.2
VNIR Spectrometer Curve-Fitting to Dispersion

Wavelength (nm)	Centroid Y Position (mm)	Y from Fitting (mm)	Fitting Error (nm)	Absolute Fitting Error (nm)
400	−1.1965	−1.1966	−0.00241	0.00241
450	−0.8988	−0.8986	0.00499	0.00499
500	−0.5998	−0.5997	0.00167	0.00167
550	−0.2999	−0.3000	−0.00214	0.00214
600	0.0004	0.0003	−0.00301	0.00301
650	0.3010	0.3010	−0.00101	0.00101
700	0.6019	0.6020	0.00190	0.00190
750	0.9029	0.9030	0.00260	0.00260
800	1.2039	1.2038	−0.00268	0.00268
Maximum				**0.00499**

assumed that they are mounted in an aluminum housing. A typical value of one-tenth the mirror diameter was assumed as the thicknesses of the mirrors since no mirror thicknesses were specified for the strawman spectrometers. The mirror dimensions change with the change of the instrument temperature. Of particular importance is the change of their curvature and the focal length. Of lesser importance is the change of the distance between the optical elements with the change of aluminum mounting temperature.

Table 12.3 shows how the wavelength registration shifts with 1°C temperature change at various wavelengths. Assuming that the CTE of aluminum is 23.6×10^{-6} and the mirror aspect ratio of 0.1. For a 1°C temperature change, the registration shifts by about 0.5 pixel (or between 2.36 nm and 2.92 nm) for the three spectrometers. However, the end-to-end (total) dispersion changes very little (<0.016 nm). This implies that only a one-point spectral calibration for each spectrometer would have to be performed as a function of temperature. As the worst case, the temperature of the SWIR spectrometers could not vary >3°C in order to keep the total (end-to-end) dispersion change <0.05 nm. This temperature range could be increased if materials with lower CTE than that of aluminum were used.

TABLE 12.3
Spectrometers' Wavelength Registration Shifts after 1°C Temperature Change (at Temperature of 20°C)

Spectrometer	Wavelength (nm)	Shift wrt 21°C (nm)	Change in Total Dispersion (nm)
VNIR	400	2.36	0.001
	600	2.37	
	800	2.36	
SWIR1	750	2.89	0.006
	1075	2.89	
	1400	2.90	
SWIR2	1350	2.91	0.016
	1075	2.92	
	2500	2.92	

12.4.3 Simulation of Spectral Calibration

Simulation of spectral calibration was performed to determine the effects of detector noise and spectral source characteristics on the calibration accuracy. In short, radiometric noise ultimately translates into spectral calibration noise. The major source of radiometric noise is radiance and dark signal shot noise, plus detector readout noise.

Of most interest here is the use of atmospheric and solar absorption lines. Solar Fraunhofer lines are very narrow. They are the only absorption lines that can be resolved with the spectrometer in the 5 nm sampling class. There are several atmospheric absorption lines that are very prominent and have larger spectral width than solar lines. Simulation programs such as MODTRAN are typically used to generate expected spectra for calibration activities. Curve fits of measured spectra to expected spectra are done to determine spectral correction factors for the spectrometer. The expected spectra are generated by convolving (filtering) the MODTRAN data with the known spectrometer spectral PSF, and then pixelating the filtered signal, to produce a simulated detector output. Note that the PSF will normally vary with wavelength and spatial location at the detector array, although the variation will be minimal in a good spectrometer design. It is assumed that the measured data from the detector array is corrected for dark and responsivity variation (both absolute and relative).

To perform the spectral curve fitting of the measured curve with the expected curve, the following algorithm is used. It is based on the method proposed by Gao et al. (2002).

Let L_M be the measured (nonuniformity corrected) spectral radiance from the detector, including noise sources, and let L_E be the expected spectral radiance, often calculated using MODTRAN.

L_M and L_E are discrete functions of i and j, where i is the detector pixel number along the spectral axis, and j is the detector pixel number along the spatial axis (cross-track direction).

L_E is computed from a spectrum having much finer resolution, such as that produced by MODTRAN, by averaging the finer resolution spectra into more coarsely spaced pixels. Therefore the wavelength step size in the fine resolution spectra must have an integer number of steps per pixel. However, MODTRAN produces discrete data in steps of wave number, and therefore the wavelength step size is nonlinear. The raw MODTRAN data is resampled (and over-sampled) using linear interpolation, such that the resampled spectrum has equal wavelength steps and an integer number of steps per pixel. For the analysis here, the fine wavelength step size used to resample the raw MODTRAN data is 0.01 nm, with a spectral dispersion of 5 nm/pixel (30 μm spaced pixels, in distance). Therefore, there are 500 samples per pixel in the resampled fine spectrum. These samples can also be called "sub-pixels," which are associated with a sub-index k.

Let λ_1 be the starting wavelength of the spectral radiance curve matching and λ_2 be the ending wavelength of the spectral radiance curve matching, such that λ_1 and λ_2 define the lower and upper wavelength limits about the spectral region of interest, and correspond to pixel numbers i_1 and i_2, respectively, for a given j.

An error function is defined as:

$$E(i,k) = L_M(i) / L_E(i+k) - 1 \tag{12.3}$$

where k is the sub-pixel shift factor as described above.

For the analysis here, $k = 1$ corresponds to a wavelength increment of 0.01 nm, and $i = 1$ corresponds to a wavelength increment of 5 nm.

The standard deviation is calculated as $\langle E(k) \rangle$ overall i between i_1 and i_2, for a given spectral shift value of k. k is varied to search for a local minimum value, $\min(\langle E(k) \rangle)$, a point at which the spectral curves are said to be matched.

For initial simulation, two Gaussian line shapes (spectral PSF) with a spectrum of 5 nm FWHM are spectrally matched. Detector noise is added to the unshifted measured curve. The wavelength

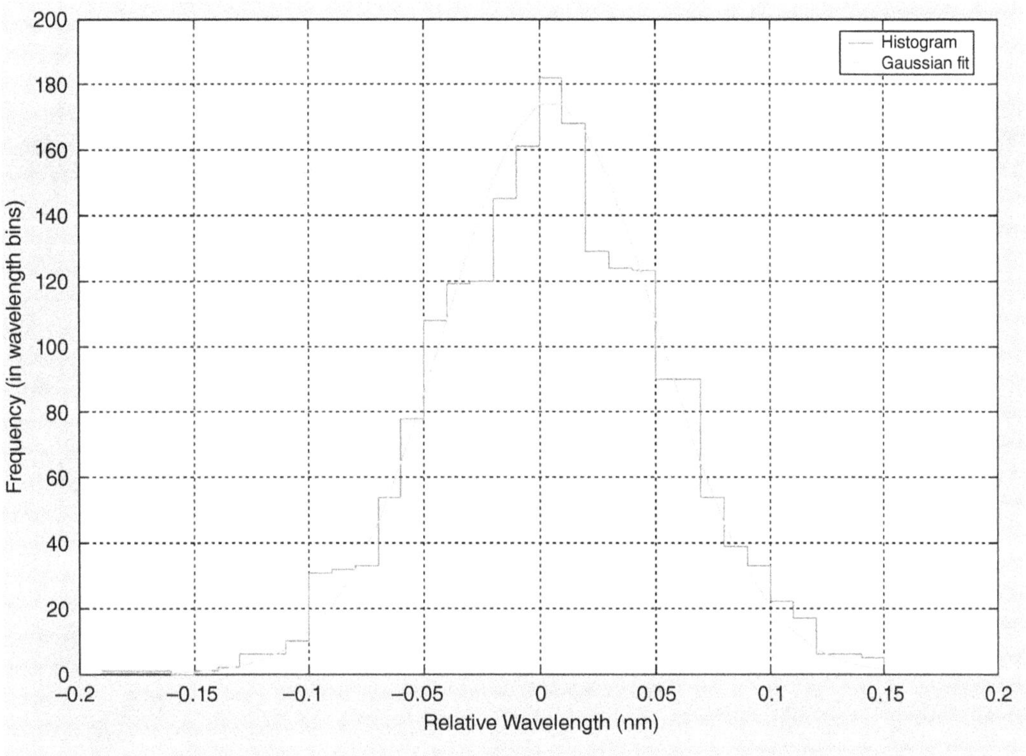

FIGURE 12.2 Histogram of simulated spectral calibration noise.

error is the deviation from zero wavelength shift of the matching error minimum. Running the simulation 2000 times generated the set of wavelength errors. The wavelength error statistics are shown in Figure 12.2 in the form of a histogram. A Gaussian (normal) curve is also shown on the same graph for comparison. A root mean square (RMS) wavelength error, $\left\langle \min\left(\langle E(k)\rangle\right)\right\rangle$ can also be calculated, as well as the mean error, $\min\left(\langle E(k)\rangle\right)$.

The RMS wavelength error is a function of many variables. Figure 12.3 shows how the RMS wavelength error varies with spectral bandwidth at FWHM (a Gaussian spectral PSF). Random variations in the curve is due to the finite number of samples of $\min\left(\langle E(k)\rangle\right)$ in calculating the RMS error, $\left\langle \min\left(\langle E(k)\rangle\right)\right\rangle$, for each value of spectral bandwidth. The plot is on a log-log scale to best show the wide range of x and y coordinates. The most important feature of this figure shown is the significant increase of RMS wavelength error as the spectral bandwidth at FWHM decreases below 1 nm. This is because all the spot energy is contained within a single pixel (5 nm wide). There is a very broad minimum error region (between 10^0 nm and 10^1 nm) shown in the graph, and therefore spectral bandwidth at FWHM is not critical, as long as it is greater than the critical minimum value of 1 nm. The *jog* RMS fit error in the graph near 0.4 nm is likely due to statistical variation in the simulation. With real absorption lines, other factors such as symmetry, and how quickly the signal changes with wavelength may be important.

Figure 12.4 shows how wavelength RMS error varies with SNR of the measured signal (i.e., detector output). Two curves are plotted, each with a different spectral bandwidth of 5 nm and 0.35 nm at FWHM. As expected, the wavelength RMS error decreases with increasing SNR. However, the curve with the small spectral bandwidth (0.35 nm) shows a dramatic increase in wavelength RMS error when SNR is lower than about 140. Also note that for the desired

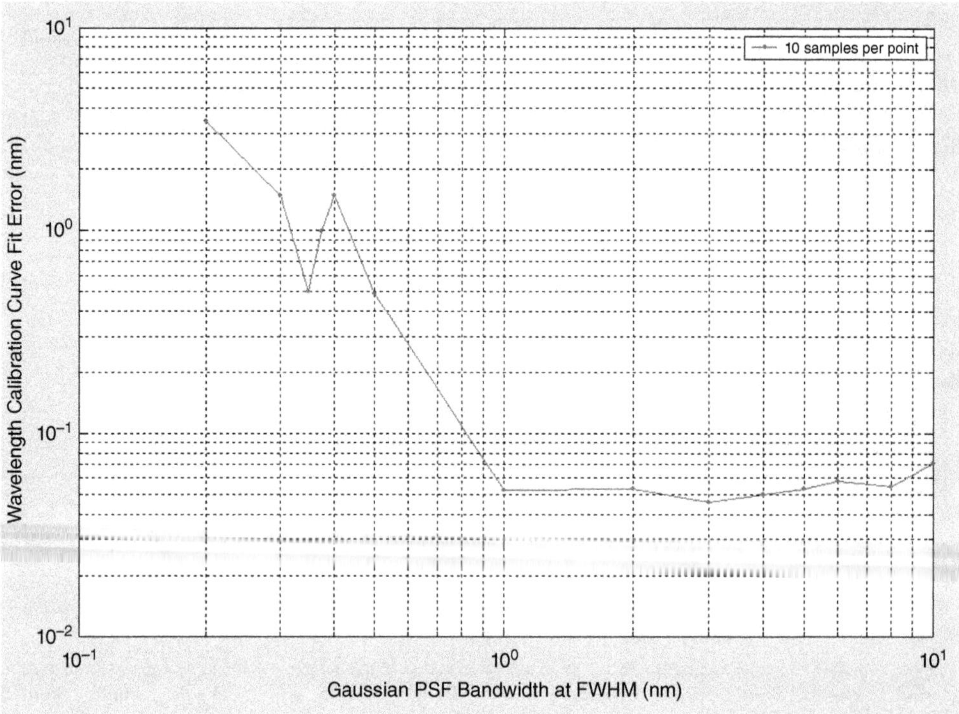

FIGURE 12.3 Wavelength calibration error versus spectral bandwidth at FWHM (5 nm/pixel spectral dispersion).

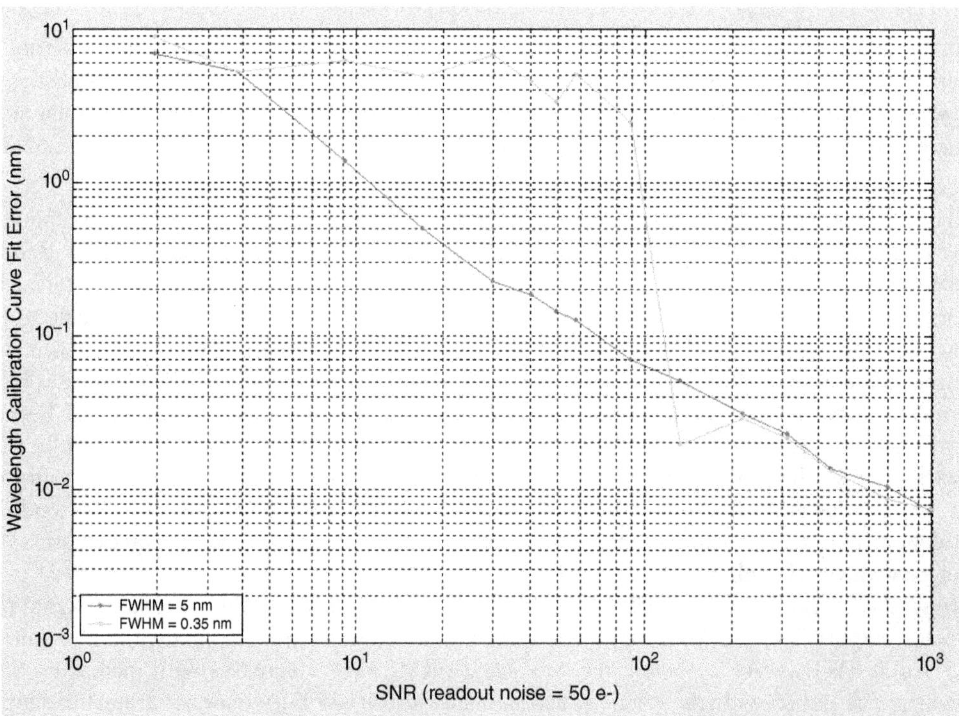

FIGURE 12.4 Wavelength calibration error versus SNR with two spectral bandwidths (5 nm and 0.35 nm).

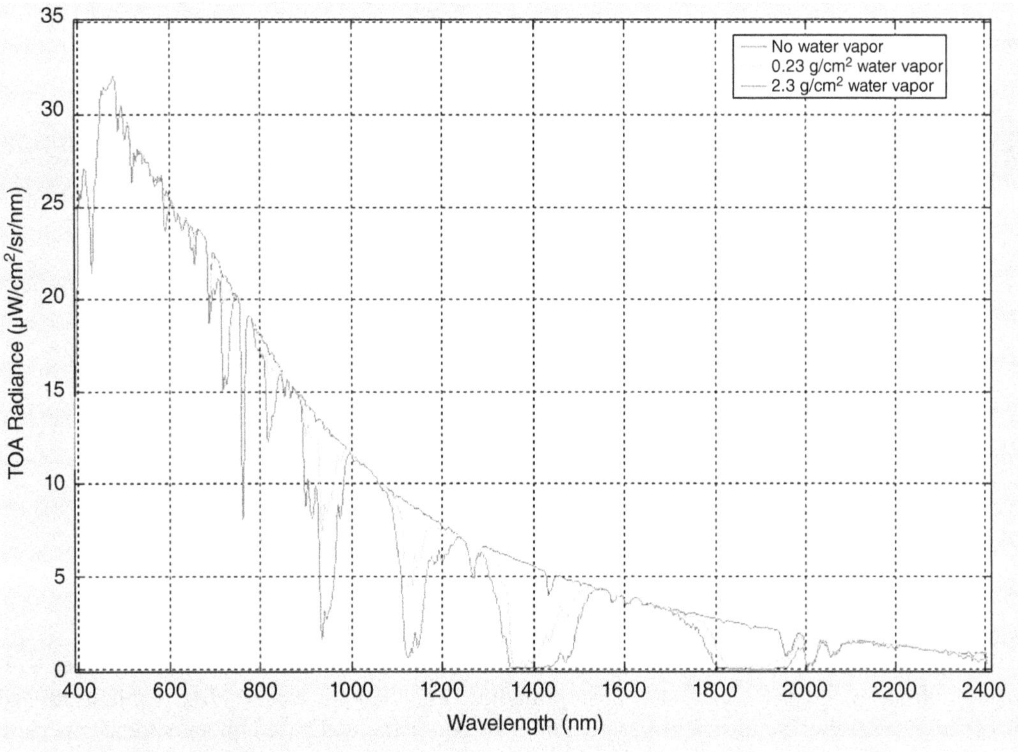

FIGURE 12.5 Overall top of atmosphere spectra with different level of water vapor generated by MODTRAN.

maximum error of 0.05 nm RMS, and SNR of 140 is required, which means a signal at least of 19,600 electrons is needed, not including dark or readout noise.

12.4.4 An Example of Scene-Based Spectral Calibration

In a scene-based spectral calibration, spectral features always occur in each scene. Potential spectral lines suitable for spectral calibration are examined. Figure 12.5 shows the overall top-of-atmosphere (TOA) solar reflective spectrum with different level of water vapor, as generated by MODTRAN algorithm. The solar radiation passes down through the atmosphere, reflects off the ground, and then passes backup through the atmosphere to a satellite instrument. The atmosphere absorbs certain wavelengths, as well as scattering light. The TOA spectral radiance as seen by a hyperspectral imager contains solar Fraunhofer absorption lines, atmospheric absorption lines, and spectral features of the scene on ground, the latter being of ultimate interest.

The presence of water vapor, carbon dioxide, and oxygen are the major causes of atmospheric absorption. Of these, water vapor has the greatest variability. Table 12.4 gives the parameters for MODTRAN algorithm used to generate the spectral curves shown in Figure 12.5. Therefore, Figure 12.5 shows the TOA spectra with and without water vapor. These spectra are convolved with a spectrometer Gaussian PSF of 5 nm bandwidth at FWHM. Such a bandwidth filters out most of the solar Fraunhofer lines, leaving only the most prominent lines providing enough absorption depth to be useful for spectral calibration, given the noise and other instabilities present. In addition, candidate atmospheric absorption lines should be free of water vapor effects. Otherwise good knowledge of water vapor is required when performing spectral calibration, so that accurate comparison spectra can be generated.

TABLE 12.4
MODTRAN Parameters Used in the Simulation

Satellite altitude	705 km
Ground elevation	130 m
Ground reflectance	60.0% (spectrally flat)
Aerosol model	Desert
Solar zenith angle	34.65°
Solar azimuth angle	282.65°
Platform heading	193.60°
Sensor tilt	0°
Pixel in swath	1000
EO source	IQSAL (5S)
Column 1	Wavelength in nm
Column 2, 3, 4	Radiance in μW/cm²/sr/nm
Column 2	Water vapor: 0 g/cm²
Column 3	Water vapor: 0.23 g/cm²
Column 4	Water vapor: 2.3 g/cm²

Figures 12.6 through 12.9 show candidate spectral absorption lines in different wavelength regions. Table 12.5 lists the candidate absorption lines and their wavelength, line depth, estimated SNRs and spectral calibration RMS error. The SNR was estimated based on an entrance aperture of 962 cm², 30 m ground sampling distance, and 5 nm/pixel spectral sampling at the corresponding wavelengths, but with the water vapor absorption lines ignored, as per Figure 12.5 (the case of no water vapor).

These are candidate absorption lines are selected because they have sufficient absorption depth, when filtered by a spectral of 5 nm FWHM, to allow accurate spectral calibration. In addition, these absorption lines are not too much affected by water vapor (H_2O), although simulation is required to determine how much knowledge of water vapor is required to allow accurate spectral calibration. There are many other absorption lines, but they are either too weak or too affected by water vapor to be useful.

For each absorption line given in Table 12.5, spectral calibration error was determined from the noise associated with the SNR, using a MATLAB® simulation program. As in simulations described previously, a simple inverted Gaussian spectral line model was used, with a FWHM bandwidth of

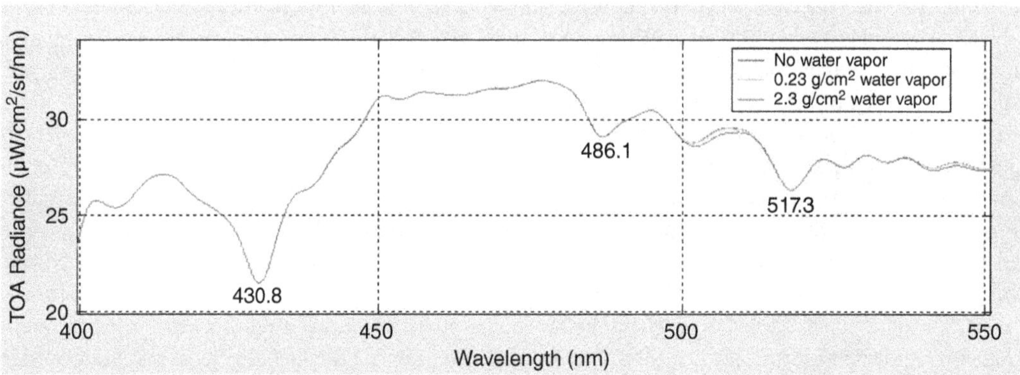

FIGURE 12.6 Spectral details of TOA solar Fraunhofer lines.

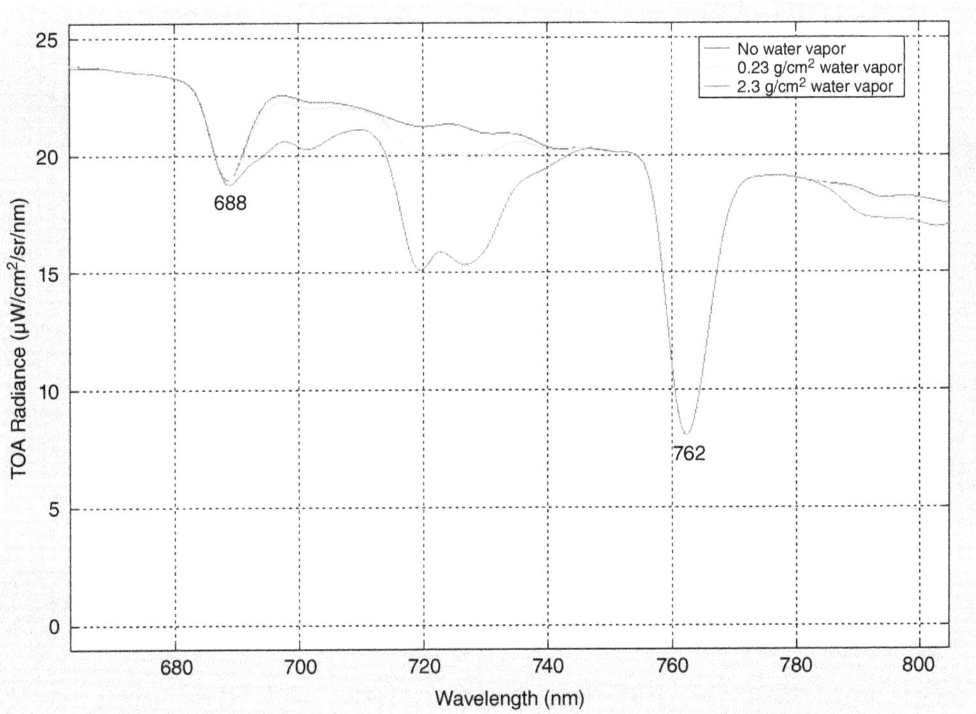

FIGURE 12.7 Spectral details of TOA atmosphere oxygen lines in VNIR region.

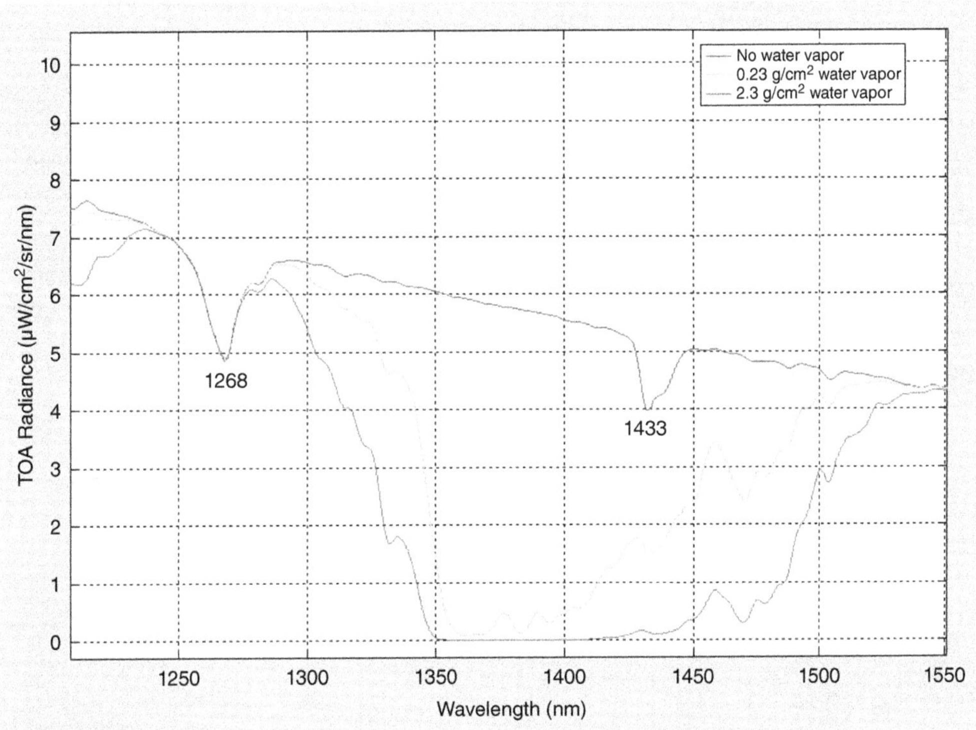

FIGURE 12.8 Spectral details of TOA atmosphere oxygen lines in SWIR region.

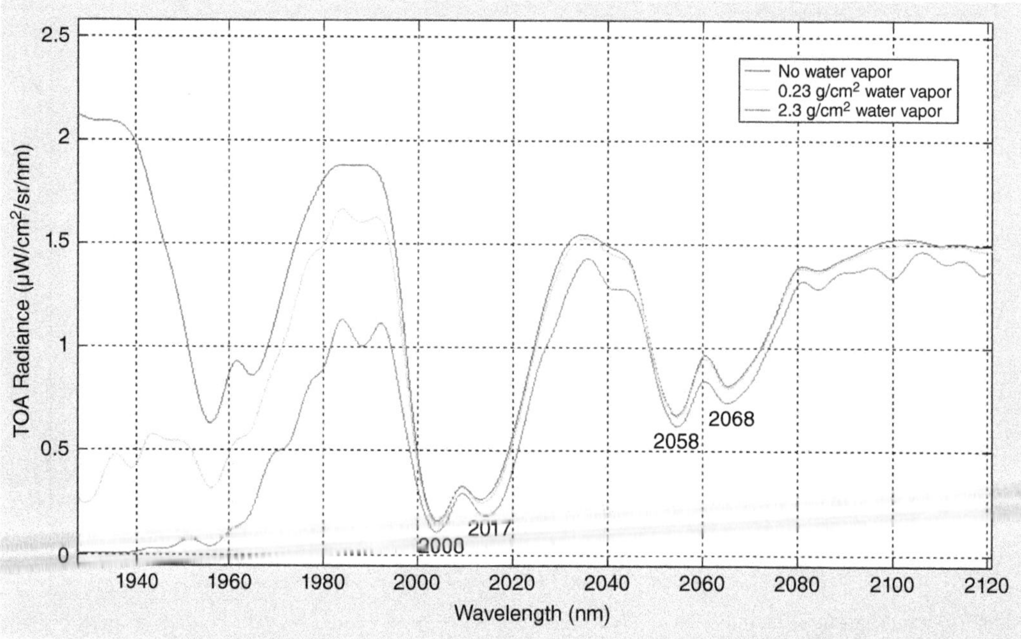

FIGURE 12.9 Spectral details of TOA atmosphere carbon dioxide lines in SWIR region.

5 nm and height equal to the line depth given in the table. For this table, the simulation of spectral curve matching was run 25 times for each spectral line, to give a reasonable estimate of the RMS error (standard deviation). As can be seen from the table, all errors due to noise are better than the desired accuracy of 0.05 nm, with a fair margin.

Figure 12.10 shows the effect on radiometric accuracy if there is uncalibrated or unknown spectral misregistration by 0.05 nm, and the desired accuracy limit. The error (absolute value) can reach about 2.8%. The error increases with smaller PSF since the rate of change of spectral radiance with wavelength is greater (line edges). Larger water vapor content will also cause higher radiometric errors, also due to higher rates of change in spectral radiance with wavelength.

TABLE 12.5
Candidate Absorption Lines of Atmosphere and Sun

Type	Element	Wavelength (nm)	Line Depth (%)	SNR	Spectral Calibration Error (nm RMS)	Comment
Solar	Fe, Ca	430.8	21	386	0.009	Strongest solar
Solar	H_2	486.1	4.6	460	0.005	
Solar	Mg	517.3	5.8	470	0.005	
Atmosphere	O_2 B	688	16	460	0.007	Affected by H_2O
Atmosphere	O_2 A	762	58	400	0.018	Very prominent
Atmosphere	O_2	1268	21	360	0.010	Somewhat affected by H_2O
Atmosphere	CO_2	1433	21	350	0.010	Very much affected by H_2O
Atmosphere	CO_2	2008, 2017	89	180	0.015	Somewhat affected by H_2O
Atmosphere	CO_2	2058, 2068	52	250	0.020	Affected by H_2O

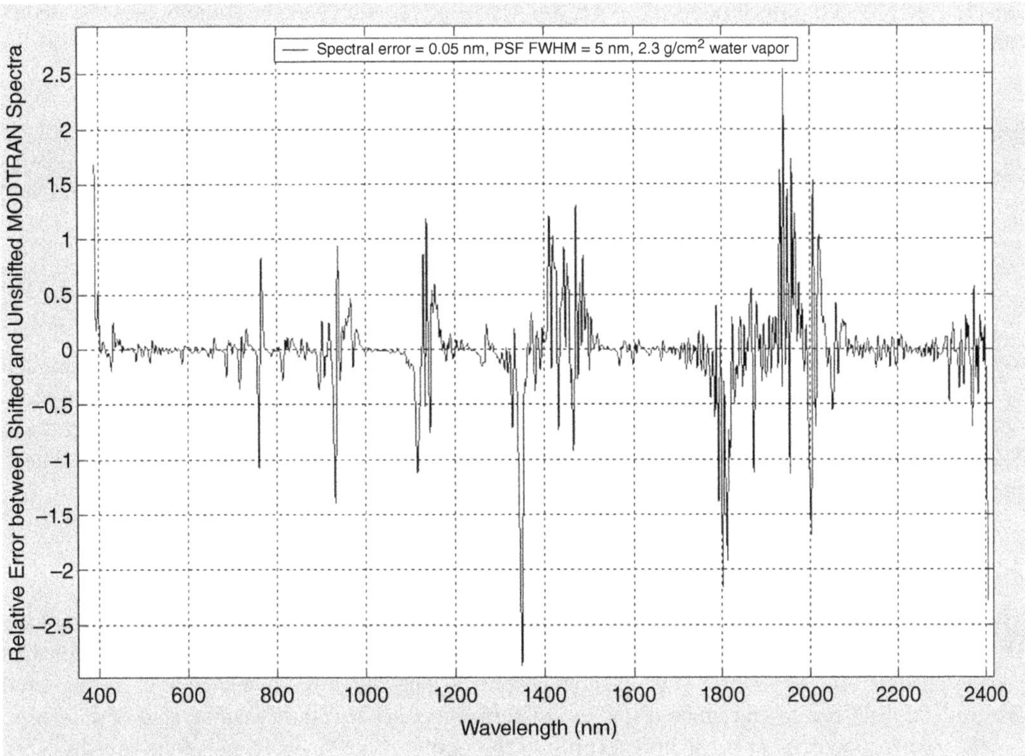

FIGURE 12.10 Radiometric error caused by spectral registration error.

Given the optical design analysis, it appears that spectral linearity is not affected by temperature; therefore, using only two absorption lines per spectrometer is likely sufficient to determine in-orbit spectral registration and dispersion. As can be seen from Table 12.5 and TOA spectra, the SWIR2 spectrometer may not have available absorption line near the short wavelength end (1350 nm) of its coverage. An absorption line near 1433 nm is a possibility, but it is very much affected by water vapor (see Figure 12.8). This absorption line is within a broad water vapor band range and is greatly attenuated if water vapor is present.

It is possible that spectral dispersion of each spectrometer only needs checking after launch, and a suitable ground-based campaign would be sufficient for post-launch determination of spectral calibration. Using a low water vapor ground site would minimize errors due to water vapor. During on-orbit operations, only checking spectral registration may be required.

12.5 PRELAUNCH CALIBRATION

The prelaunch characterization and calibration of spaceborne hyperspectral instruments should be planned and carried out in conjunction with their design and development to meet the mission requirements. The primary prelaunch measurements necessary to characterize the sensor are:

1. The absolute radiometric calibration
2. The spectral response
   a. Registration
   b. Bandwidth
3. The equalization coefficients (pixel relative sensitivity)

Besides the model parameters noted above, additional prelaunch characterizations must be undertaken. These normally include measurements to control and characterize:

1. Stray light
   a. In-band response
   b. Out of band response
2. Polarization
3. Linearity
4. Modulation transfer function
5. Dark current

With careful characterization of these effects, it may be possible to use the collected information to enhance the correction algorithms resulting in improved radiometric quality of the final data products. In a review of the calibration of spaceborne multispectral imaging sensors by Dinguirard and Slater (1999), they state that absolute radiometric calibration in the solar reflective range can, at best, provide 3–5% (1σ) with the highest uncertainties found in the extreme spectral range.

12.5.1 Considerations With Respect to Ground-Based Absolute Calibration

The absolute calibration coefficients are normally obtained by referencing to light source that are routinely checked against sources calibrated by national standards laboratories. The most prevalent illumination sources used for this purpose are large integrating spheres. They have the advantage of providing a full aperture, full field, end-to-end calibration at several different points over the dynamic range of the sensor.

The primary sources of error in determining the output of an integrating sphere are the lamp sources used to illuminate the spheres and the reflective coating of the sphere interior. With all sources of error being taken into account, it was stated that it would be difficult, time consuming and costly to achieve an uncertainty in the sphere output of ±3% (Biggar et al. 1993). Addition problems can also be noted for sphere-based calibrations. The relative spectral output of the lamps differs from that of the illumination source of the Earth scenes and the sun. This becomes a significant problem at the blue end of the spectrum where the lamp output falls dramatically. As well, for sensors with narrow spectral bandwidth (<10 nm), the presence of Fraunhofer lines within a bandwidth can change the response of the sensor by several percent with respect to its integrating sphere source calibration (Flittner and Slater 1991).

An alternative technique has been developed that allows the calibration to be undertaken by using sunlight as the illumination source. This technique, referred to as solar radiation based calibration (SRBC) was described by Slater et al. (2001). By using sun to illuminate the solar diffuser panel to be used for on-orbit calibration activities, the sun is used to provide the link between the prelaunch SRBC, in-flight solar diffuser based calibration and vicarious calibration activities. Slater et al. (2001) state that the technique is capable of achieving an uncertainty to better than 3% (1σ) with respect to the exoatmospheric solar irradiance. Although the results can be tied accurately to national laboratory standards, this provides a single standard to which the calibration of all Earth observation instruments, space, air, and ground based, can be tied. This is an important consideration giving the variations as high as 3% in the blue and 8% in the SWIR (Walker et al. 1991).

Biggar et al. (1993) reported the use of an early version of this technique with the SeaWiFS sensor. The method was refined and carefully analyzed (with error analysis) by Slater et al. (2001).

12.5.2 An Example—MODIS Prelaunch Calibration

MODIS radiometric calibration accuracy requirements are specified at the typical radiance of each spectral band. For the 25 reflective solar bands (RSBs), the accuracy requirements are ±2% for the reflectance and ±5% for the radiance products. To achieve these requirements prelaunch and in-flight calibrations are critical.

One of the objectives of prelaunch calibration and characterization was to examine the onboard calibration functions and to transfer the calibration reference from ground to onboard. MODIS prelaunch calibration and characterization were performed at different levels, including component, subsystem, instrument, and spacecraft levels, and through different phases, including ambient, pre-thermal vacuum (TVAC), TVAC, and post-TVAC testing. Extensive measurements were made to calibrate and characterize the instrument's radiometric, spectral, and spatial performance (Xiong et al. 2016).

Key radiometric performance was evaluated based on the measurements made in the TVAC environment that simulated the predicated instrument on-orbit operation environmental conditions. Three instrument temperature plateaus (cold, nominal, and hot) were used during the TVAC calibration and characterization. The radiometric calibration and characterization were made for each spectral band, detector array, subsample (for bands 1–7) and mirror side using both the primary and redundant electronic configurations, including limited cross-strapping configurations among the instrument power supply, signal processing, and data formatter.

A 100-cm-diameter spectral integrating source (SIS-100) was used in the radiometric calibration of the reflective solar bands. The SIS-100 can be operated with different lamp combinations or configurations, providing a range of radiance levels to calibrate the instrument's radiometric response. The SIS-100 was operated in laboratory ambient conditions and was located outside the testing chamber during instrument's TVAC measurements.

Instrument performance parameters, such as detector response, dynamic range, nonlinearity, SNR, short-term stability, and temperature sensitivity were derived from these measurements. The same parameters were also derived for the 15 thermal emissive bands (TEB) using a large aperture blackbody as the calibration source. A space view source was used for the measurements of instrument background and offset signals. Both the blackbody calibration source and space view source were operated inside the TVAC testing chamber.

MODIS data are collected using a double-sided scan mirror over a wide range of angles of incidence (AOI). The earth view data sector AOI ranges from 10.5° to 65.5°. The space view, blackbody, spectroradiometric calibration assembly (SRCA), and solar diffuser are centered at AOIs of 11.2°, 26°, 38°, and 52°, respectively. Since calibration is only performed at a fixed AOI, it is extremely important to characterize the sensor's relative response versus scan-angle (RVS). MODIS RVS characterization was performed in the laboratory ambient environment. The calibration source was located at a fixed position with MODIS placed on a rotating table such that its scan mirror's rotating axis was co-aligned with the table's rotating axis. This setup enables the calibration source to be viewed by the sensor at different AOI.

The polarization sensitivity was characterized for the MODIS spectral bands in the VIS and NIR regions. The polarization source assembly included a spectral integrating source and a prism polarizer. At each selected AOI, the polarization source assembly was rotated between ±180°, providing a calibration source with different polarizations.

Extensive prelaunch measurements were made to characterize the instrument's spectral and spatial performance. The spectral characterization was made using a ground-based spectral measurement assembly (SpMA). The SpMA is a double-grating monochromator with an internal source assembly. The sources include lamps and a glow bar, capable of covering the wavelength range of MODIS spectral bands. The slit width of the SpMA was adjusted, depending on the spectral band, its spectral resolution, and signal levels. The SpMA source spectrum was monitored by reference detectors with known spectral response during spectral characterization. The relative spectral response, in-band and out-of-band, was characterized at instrument level for each detector array. Key parameters, such as center wavelength and bandwidth were derived from prelaunch spectral characterization.

The spatial characterization was performed using an integration and alignment collimator. The collimator was used together with a narrow slit for measuring the along-scan responses and a reticle with three inclined slits for measuring the along-track responses. Key spatial performance parameters, including line spread function (LSF), MTF, instantaneous field-of-view (IFOV), and band-to-band registration, were determined in both along-scan and along-track directions.

12.6 ABSOLUTE AND RELATIVE RADIOMETRIC CALIBRATION

Absolute radiometric calibration is performed by calculating the ratio of the digital numbers output from a hyperspectral sensor with the value of an accurately known uniform-radiance field at its entrance pupil. In solar reflective range uncertainties are typically 3–5% (1σ), being highest at the extremes of the wavelength range.

Relative radiometric calibration is determined by normalizing the outputs of the detector elements to an average output from all the detector elements in the band. The result of the normalization is that all the detector elements give the same output when the focal plane of the hyperspectral sensor is irradiated with a uniform-radiance field. Absolute calibration performed for every single pixel individually automatically includes relative calibration. However, relative calibration can sometimes be easier to perform than absolute calibration.

For relative calibration, the absolute value of the radiance field need not be known. Typically, the RMS variation in the adjusted normalized outputs of the detector elements is in a range of 0.1–0.5%, depending on the SNR of the digitized output signal from the sensor. Relative calibration can also describe the ratio of the average outputs from two or more different bands of a hyperspectral sensor; it then is often referred to as inter-band calibration. A change in this ratio is an indicative of a temporal change in response of one or more spectral bands. Finally, multi-temporal calibration, or stability, is determined from the ratio of the average outputs from all the bands, for the same stable scene for two different dates.

The applications of hyperspectral satellites concern with geometrical and temporal measurements and scene classification. In the case of geometrical measurements, for example, mapping, relative calibration corrects the image for striping and similar cosmetic defects. It is an individual detector correction for striping, for example, and a multiple pixel correction for less-sensitive detector elements near the edge of a detector array, or a falloff in the image-plane irradiance. Temporal measurements require knowledge of changes in the average values of the sensor's response in each spectral band. If this is not known, then changes in the sensor's response are likely to be incorrectly attributed to changes in the observed scene. The results of in-flight calibration can be used to monitor sensor temporal changes.

Usually, relative calibration is used for detection of changes of a hyperspectral sensor over a few-day period because of its high precision, and absolute calibration is used in combination with relative calibration, for longer periods because of its greater long-term temporal stability. Scene classification concerns the statistical analysis of the digital number output in a scene. Relative calibration to remove striping, etc. is highly important.

Some applications, such as agriculture and monitoring natural disasters, benefit from the increasing number of satellite sensors in operation. It is desirable for some applications to compare data from multiple sensors having different spatial resolutions and spectral bands. It is essential for some other applications to know how the responses of the sensors compare to each other (Rao and Chen 1995). Such comparisons can be made relatively, but are more dependable if referenced to an absolute scale. These applications also require a good knowledge of the scene physics, for example, spectral and directional reflectance and atmospheric effects, as well as the pixel-level response of the sensor, stray light, and MTF.

A model linking the digital number output to the desired physical quantity has to be set up in order to take into account the different instrument parameters that have to be regularly checked on-orbit. Atmospheric radiative transfer codes have to be modified, or more general codes have to be used to account for the effect of the atmosphere on the measurement: absorption, scattering, MTF, adjacency, directional effects, etc. This type of model allows remote sensing data users to know when changes in the observed scene are due to natural changes, such as water stress for vegetation, and not due to changes in the observation conditions, bidirectional reflectance distribution function (BRDF), pixel size, etc.

The usual approach to sensor calibration starts with the formulation of a sensor calibration model (Schott 1997, Schowengerdt 1997). The simplest form of this model is a linear law linking the digital number output X to the radiance L_λ at the entrance pupil of the instrument

$$X = AL_\lambda \tag{12.4}$$

where A is the absolute calibration coefficient to be determined. This is done in prelaunch by accurate measurements, and then monitored on-orbit by onboard calibration devices using secondary or tertiary standard light sources (lamps or the sun) and vicarious methods, using images of specific well-known ground targets or the moon. The above equation assumes that the offset is either negligible or has been removed. If that is not the case, then the offset term has to be added to the equation:

$$X = AL_\lambda - X_o \tag{12.5}$$

If nonlinearity is suspected in the detection chain, then higher-order terms can be added.

REFERENCES

Appourchaux, T., G. Gourmelon, and B. Johlander. 1994. Effect of gamma-ray irradiations on optical filter glass. *Optical Engineering* 33(5):1659–1668.

Bannari, A., P. M. Teillet, and E. G. Richardson. 1999. Nécessité de l'étalonnage radiometrique et standardization des images numériques de télédétection. *Canadian Journal of Remote Sensing* 25(1):45–59.

Bass, M., E. Stryland, D. Williams, and W. Wolfe. 1995. *Handbook of Optics: Fundamentals, Techniques, and Design*. vol. 1, 2nd ed. Chapter 36, McGraw-Hill Inc, New York.

Bergeron, M., A. Hollinger, K. Staenz, M. Maszkiewicz, R. A. Neville, S.-E. Qian, and D. G. Goodenough. 2008. Hyperspectral environment and resource observer (HERO) mission. *Canadian Journal Remote Sensing* 34(S1):1–11.

Biggar, S. F., P. N. Slater, K. J. Thome, A. W. Holmes, and R. A. Barnes. 1993. Preflight solar-based calibration of the SeaWIFS. *Proceedings of SPIE* 1993:233–242.

Boies, M. T., J. Kinnison, and J. A. Schwartz. 1994. The effect of electron radiation on glass used for space-based optical systems. *Proceedings of SPIE* 2287:104–113.

Bonnemason, F. 1991. Ruled and holographic diffraction gratings experiment. *Proceedings of LDEF-69 Months in Space: First Post-Retrieval Symposium* 1301–1314.

Cotel, A., A. Liard, F. Desserouer, and P. Pichon. 2014. Overview of diffraction gratings technologies for space-flight satellites and ground-based telescopes. *Proceedings of SPIE* 10563: 105630P-1–105630P-8.

Denton, F. and A. E. Gee. 1999. Replication of optical surfaces: Capabilities and cautions. *Proceedings of Precision Fabrication and Replication*, Chapel Hill, North Carolina, USA, American Society for Precision Engineering.

Dinguirard, M. and P. N. Slater. 1999. Calibration of space-multispectral imaging: A review. *Remote Sensing Environment* 68:194–205.

Dooling, D. and M. M. Finckenor. 1999. Material selection guidelines to limit atomic oxygen effects on spacecraft surfaces. *NASA/TP*-1999-*209260*, June 1999.

Doyle, D., J. Bourrieau, P. Bricard, B. Johlander, H. J. Juranek, A. Litzelmannn, W. Pannhorst, and W. Rits. 1996. The effect on electron irradiation on the radius of curvature of a Zerodur mirror. *Proc. of SPIE* 2665.

Flittner, D. E. and P. N. Slater. 1991. Stability of narrow-band filter radiometers in the solar-reflective-range. *Photogrammetric Engineering and Remote Sensing* 57(2):165–171.

Folkman, M., J. Pearlman, L. Liao, and P. Jarecke. 2001. EO-1 Hyperion hyperspectral imager design, development, characterization and calibration. *Proceedings of SPIE* 4151:40–51.

Fruit, M., A. Gusarov, D. Doyle, and G. Ulbrich. 1999. Radiation impact on spaceborne optics: The dose coefficients approach. *Proceedings of SPIE* 3872:60–71.

Gao, B.-C., M. J. Montes, C. O. Davis. 2002. A curve-fitting technique to improve wavelength calibrations of imaging spectrometer data. *Proceedings of the 11th JPL Airborne Earth Science Workshop, JPL Publication* No. 03-4, 99–105.

Green R. O. 1998. Spectral calibration requirement for earth-looking imaging spectrometers in the solar reflected spectrum. *Applied Optics* 37:683–691.

Guenther, B., W. Barnes, E. Knight, J. Barker, J. Harnden, R. Weber, M. Roberto, G. Godden, H. Montgomery, and P. Abel. 1996. MODIS calibration: A brief review of the strategy for the at-launch calibration approach. *Journal of Atmospheric and Oceanographic Technology* 13(2):274–285.

Harron, J. W., J. R. Freemantle, A. B. Hollinger, and J. R. Miller. 1992. Methodologies and errors in the calibration of a compact airborne spectrographic imager. *Canadian Journal of Remote Sensing* 18(14):243–249.

Hearn, D. R. 1998. Characterization of instrument spectral resolution by the spectral modulation transfer function. *Proceedings of SPIE* 3439:400–407.

Heath, D. F., E. Hilsenrath, and S. Janz. 1998. Characterization of a hardened ultrastable UV linear viable filter and recent results on the radiometric stability of narrow band interference filters subjected to temperature/humidity, thermal/vacuum and ionizing radiation environments. *Proceedings of SPIE* 3501:401–411.

Henry, P., M. Dinguirard, and M. Bodilis. 1993. SPOT calibration over desert areas. *Proceedings of SPIE* 1938:67–76.

Henson, T. D. and G. K. Torrington. 2001. Space radiation testing of radiation resistant glasses and crystals. *Proceedings of SPIE* 4452:54–65.

Mouroulis, P. 1999. Spectral and spatial uniformity in pushbroom imaging spectrometers. *Proceedings of SPIE* 3753:133–141.

Mouroulis, P. and R. O. Green. 2018. Review of high fidelity imaging spectrometer design for remote sensing. *Optical Engineering* 57(4):040901.

Mouroulis, P., R. O. Green, and T. G. Chrien. 2000. Design of pushbroom imaging spectrometers for optimum recovery of spectroscopic and spatial information. *Applied Optics* 39(13):2210–2220.

Mouroulis, P., F. T. Hartley, D. W. Wilson, V. E. White, A. Shori, S. Nguyen, M. Zhang, and M. Feldman. 2003. Blazed grating fabrication through gray-scale X-ray lithography. *Optics Express* 11(3):270–281.

Mouroulis, P. and M. McKerns. 2000. Pushbroom imaging spectrometer with high spectroscopic data fidelity: Experimental demonstration. *Optical Engineering* 39(3):808–816.

Mouroulis P., J. T. Shea, and D. T. Thomas. 2001. Design, tolerancing and alignment of pushbroom imaging spectrometers for high response uniformity. *Proceedings of SPIE* 4441:106–116.

Mouroulis, P., D. A. Thomas, T. G. Chrien, V. Duval, R. O. Green, J. J. Simmonds, and A. A. Vaughan. 1998. Trade studies in multi/hyperspectral imaging systems. *NASA Code Y Internal Report.*

Mouroulis, P., D. W. Wilson, P. D. Maker, and R. E. Muller. 1998b. Convex grating types for concentric imaging spectrometers. *Applied Optics* 37:7200–7208.

Neville, R. A., L. Sun, and K. Staenz. 2003. Detection of spectral line curvature in imaging spectrometer data, *Proceedings of SPIE* 5093:144–154.

Neville, R. A., L. Sun, and K. Staenz. 2004. Detection of keystone in image spectrometer data. *Proceedings of SPIE* 5425:208–217.

Neville, R. A., L. Sun, and K. Staenz. 2008. Spectral calibration of imaging spectrometers by atmospheric absorption feature matching. *Canadian Journal of Remote Sensing* 34(S1):S29–S42.

Pitts, M., C. Hostetler, L. Polle, C. Holden, and D. Rault. 2003. An airborne A-band spectrometer for remote sensing of aerosol and cloud optical properties. *Proceedings of SPIE* 4882:353–362.

Rao, C. R. N. and J. Chen. 1995. Inter-satellite calibration linkages for the visible and near-infrared channels of the advanced very high resolution radiometer on the NOAA-7, -9, and -11 spacecraft. *Internation Journal of Remote Sensing* 16:1931–1942.

Schott, J. R. 1997. *Remote Sensing—The Image Chain approach*, New York, NY: Oxford University Press.

Schott, Glass. 2004. SCHOTT Radiation Resistance Glasses Catalogue.

Schott, 2019. Optical glass. http://www.schott.com

Schowengerdt, R. A. 1997. *Remote Sensing Models and Methods for Image Processing.* San Diego, CA: Academic Press.

Silverman, E. M. 1995. Space environment effects on spacecraft LEO materials selection guide: Part 1 and part 2. *NASA/CR-19291.*

Slater, P. N., F. S. Biggar, J. M. Palmer, and K. J. Thorne. 2001. Unified approach to absolute radiometric calibration in the solar-reflective range. *Remote Sensing of Environment* 77:293–303.

Stevenson, I. C. and G. Sadkhin. 2000. X-1 silver: A high durability silver coating for use in harsh environments. *Proceedings of 43rd Annual Technical*, Society of Vacuum Coaters.

Stuckey, W. K., C. S. Hemminger, G. L. Steckel, M. M. Hills, and M. R. Hilton. 1993. Spacecraft materials studies on the aerospace corporation tray on EOIM-III. *Proceedings of Third Post-Retrieval Symposium Part* 3:917–930.

Teillet, P. M. 1997. A status overview of Earth observation calibration/validation for terrestrial applications. *Canadian Journal of Remote Sensing* 23(4):291–298.

Teillet, P. M., D. Horler, and N. T. O'Neill. 1997. Calibration, validation, and quality assurance in remote sensing: A new paradigm. *Canadian Journal of Remote Sensing* 23(4):401–414.

Thomas, N. and J. Wolfe. 2000. UV-shifted durable silver coating for astronomical mirrors. *Astronomical Telescopes & Instrumentation*, June 2000.

Thome, K. J., B. Markham, J. Barker, P. N. Slater, and S. F. Biggar. 1997. Radiometric calibration of Landsat. *Photogrammetric Engineering & Remote Sensing* 63(7):853–858.

Walker, J. H., R. D. Saunders, J. K. Jackson, K. D. Mielenz. 1991. Results of a CCPR intercomparison of spectral irradiance measurements by national laboratories. *Journal of Research of the National Institute of Standards & Technologies* 96(6):647–668.

Warsaw, G. D., D.-L. Desaulniers, and D. Regenstein. 1996. Optical design and performance of the ODIN UV/visible spectrograph and infrared imager instrument. *Proceedings of 10th Annual AIAA/USI Small Satellite Conference.* Utah.

Wilkes, D. R. and J. M. Zweiner. 1999. Optical properties monitor (OPM) in-situ experiment flown on the Mir station. *Proceedings of SPIE* 3784:72–83.

Xiong, X.-X., et al. 2016. Moderate resolution imaging spectroradiometer on terra and aqua missions. In *Optical Payloads for Space Missions*, ed. S.-E. Qian, 53–90. West Sussex UK: John Wiley & Sons.

Yoder, P. R. 2008. *Mounting Optics in Optical Instruments*, 2nd ed. Bellingham, WA: SPIE Press.

13 Radiometric Conversion and Data Correction

13.1 CONVERSION TO AT-SENSOR RADIANCE AND TOP-OF-ATMOSPHERE REFLECTANCE

13.1.1 Process of Conversion From Raw Data to Radiance and Reflectance

A hyperspectral sensor acquires the intensity of the electromagnetic radiation of each pixel within its field of view (FOV) and records them as digital numbers (DNs). These DNs are normally 2-byte (16-bit) unsigned integer data that include the effects of surface reflectance, solar irradiance, scattering, atmospheric gas absorptions, variation in illumination due to topography, and instrument response function and artifacts. These DNs need to be calibrated (or corrected) and converted into physically meaningful units, like radiance or reflectance, before they can be used to retrieve scientific information for application purposes.

Radiance is the amount of radiation coming from an observed pixel within the FOV of a hyperspectral sensor. It includes radiation reflected from the surface, from neighboring pixels, and from clouds. The amount of radiation also depends on the surface cover. Sometimes pixels father away to the subject pixels also contribute to the radiance of the subject pixels. For example, in mountainous terrain considering a pixel in a valley bottom, it gets some radiation from slope pixels. Calibration to radiance compensates for variations in detector sensitivity. To derive a radiance image from a raw DN image, the raw data must be converted to radiance data by applying a series of gain values and offsets pixel-by-pixel, and band-by-band. The resulting pixel values are typically in unit of $W/(m^2 \cdot sr \cdot \mu m)$. This process is called radiometric calibration or conversion. Radiance images include the radiation effects of the sun. In fact, a radiance spectrum of a pixel closely matches that of the solar irradiance curve, also called the solar spectrum. For hyperspectral data analysis, the effects of solar irradiance should be removed by converting the radiance data to reflectance, which is the subject of Chapter 14.

Reflectance is the proportion of the radiation reflected off a surface to the radiation striking it. In hyperspectral data analysis, materials are identified by their reflectance spectra. So converting the data to reflectance is the first step toward identifying materials from hyperspectral imagery. Figure 13.1 shows the process of data conversion from raw DNs to radiance data and, ultimately, to reflectance data.

13.1.2 Radiometric Response Modeling of a Hyperspectral Sensor

In radiometric calibration of a hyperspectral sensor, the first problem to be solved is the definition of a way to model the physical entities involved. Spaceborne hyperspectral sensors are instruments that measure the radiance of a scene due to the reflection and scattering of input solar irradiance from the ground.

The measurement equation of a detector pixel measuring radiance in digital units can be written in a simplified equation as follows:

$$DN_{i,j} = G \cdot a_{i,j} \cdot \Omega \cdot L_\lambda \cdot \Delta\lambda \cdot \eta \cdot t \cdot \tau \tag{13.1}$$

where $DN_{i,j}$ is the digital number output by detector pixel i in band j, G is the gain of the instrument plus digitization, $a_{i,j}$ is the area of detector pixel i in band j, Ω is the instrument acceptance solid

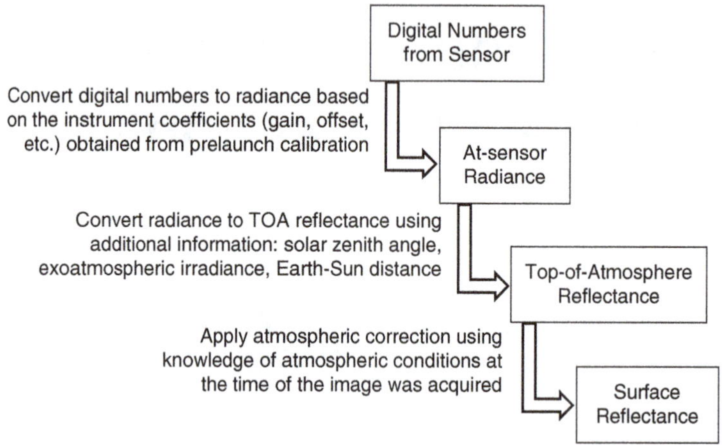

FIGURE 13.1 Process of data conversion from raw digital numbers to radiance data and reflectance data.

angle, L_λ is the spectral radiance at the instrument entrance aperture, $\Delta\lambda$ is the bandwidth, η is the detector quantum efficiency in electrons per incident photon, t is the integration time, and τ is the instrument optical transmission.

Instrument response nonlinearity, background, focal plane temperature effects, and response versus scan angle effects are not shown in Equation 13.1. These quantities are determined in pre-launch instrument characterization tests and are incorporated in instrument radiometric models and in the production of measured radiances. Equation 13.1 can be rewritten as:

$$DN_{i,j} = AL_\lambda \tag{13.2}$$

where

$$A = G \cdot a_{i,j} \cdot \Omega \cdot \Delta\lambda \cdot \eta \cdot t \cdot \tau \tag{13.3}$$

is the product of the gain and the instrument responsivity.

Because a hyperspectral sensor has a given spectral bandwidth b, one generally refers to the "effective" or "equivalent" radiance, which is the weighted average of $L(\lambda)$ across the given spectral bandwidth of the spectral sensitivity $s(\lambda)$ as follows:

$$L_\lambda = \frac{\int_{-\frac{b}{2}}^{\frac{b}{2}} L(\lambda)s(\lambda)d\lambda}{\int_{-\frac{b}{2}}^{\frac{b}{2}} s(\lambda)d\lambda} \; W/(\text{m}^2 \cdot \text{sr} \cdot \mu\text{m}) \tag{13.4}$$

If the radiance is independent of λ, i.e., $L(\lambda) = L_0$, then $L_\lambda = L_0$.

As this radiance is the solar flux reflected by the scene (ground + atmosphere), one can also express L_λ by:

$$L_\lambda = \frac{\rho_\lambda}{\pi} E_s \cdot cos\theta_s \cdot u(t), \tag{13.5}$$

where the Earth-atmosphere system is assumed being Lambertian, ρ_λ is the top-of-atmosphere equivalent reflectance of the scene, E_s is the equivalent exoatmospheric solar-spectral irradiance

$[W/(m^2 \cdot sr \cdot \mu m)]$, θ_s the solar zenith angle, and $u(t)$ is the term taking into account the temporal variation of the Earth-sun distance. With the normalization, E_s, and ρ_λ can be written as follows:

$$E_s = \frac{\int_{-\frac{b}{2}}^{\frac{b}{2}} E_s(\lambda)s(\lambda)d\lambda}{\int_{-\frac{b}{2}}^{\frac{b}{2}} s(\lambda)d\lambda} \tag{13.6}$$

$$\rho_\lambda = \frac{\int_{-\frac{b}{2}}^{\frac{b}{2}} \rho(\lambda)E_s(\lambda)s(\lambda)d\lambda}{\int_{-\frac{b}{2}}^{\frac{b}{2}} E_s(\lambda)s(\lambda)d\lambda} \tag{13.7}$$

Accordingly, the physical quantities related to the output signal are either radiance L_λ or TOA equivalent reflectance ρ_λ. Note that the applications that require to retrieve ρ_λ are more accurate (if sun-referenced), as the uncertainties in the knowledge of $E_s(\lambda)$ and $s(\lambda)$ are reduced by the definition of ρ_λ.

If the sensor is linear, the digital output X is proportional to the radiance ($X = AL_\lambda$), after dark-signal subtraction. However, in order to take into account a possible nonlinearity of the sensor at very low signal, a quadratic term can be adopted as follows:

$$X = AL_\lambda + BL_\lambda^2 \tag{13.8}$$

The prelaunch calibration methods provide the absolute calibration coefficients A and B if necessary.

For hyperspectral sensors using 2D detector arrays, such as a CCD or HgCdTe (mercury-cadmium-telluride) detector array, in order to take into account the sensitivity difference between detector elements, the instrument model has to express the behavior of each detector element. For Hyperion, for example, the model is as follows:

$$x_{j,k} = A_{j,k}L_{j,k} + C_{j,k} \tag{13.9}$$

where $x_{j,k}$ is the digital number output, $A_{j,k}$ is the gain and instrument responsivity, $L_{j,k}$ is radiance at detector element (j, k), and $C_{j,k}$ is the dark signal. The symbol k identifies the spectral band, while j is the pixel number in the cross-track line.

13.1.3 Conversion to At-Sensor Radiance

The procedure of converting raw data to at-sensor radiance is sensor dependent. This section provides examples of three spaceborne hyperspectral sensors, i.e., Medium Resolution Imaging Spectrometer (MERIS), Hyperion, and Environmental Mapping and Analysis Program (EnMAP).

13.1.3.1 MERIS Conversion to At-Sensor Radiance

For MERIS hyperspectral imager onboard ENVISAT, the conversion from DNs to at-sensor radiance was performed in the processing of the creation of the Level 1B data product. The processing includes:

1. reading the Level 0 product (raw DNs);
2. checking the data packets;

3. extracting measurement data from the packets;
4. correcting, calibrating, and geolocating the Earth imaging data into spectral radiance at-sensor values;
5. ingesting ancillary data; and
6. creating Level 1B products that include radiances, geolocation, and other annotations.

Before the conversion of raw data to radiance, the data packets in the data stream corresponding to the observation modes of MERIS are checked. Information in the packet header and data field header is used to detect anomalies in data stream, such as transmission error, format error, or sequence error.

Acquired DNs may be outside the range of the useful measurements $0 - L_{sat}$ caused by various phenomena. These DNs are invalid. The corresponding detector elements may be affected temporarily or permanently. When possible, these invalid pixels should be replaced with good estimates. These phenomena can be as follows:

1. Saturation due to the radiance level being above L_{sat} caused by, for example, sun glint, cloud, bright land, or snow/ice, which affects detector elements temporarily (typically several columns in several bands over several frames).
2. Recovery from saturation. After saturation, components of the acquisition chain need some time (a few columns) to recover; in the meantime the measurement is affected.
3. Blooming. Samples in bands and columns of the detector array close to a saturated one may be temporarily affected by photons or photoelectrons diffusing from the saturated pixel.
4. Glitches. High-intensity impacts (e.g., laser) generate isolated high-value samples.
5. Dead detector elements. Due to manufacturing defects or due to ageing in space, the response of some CCD elements to light may die, i.e., permanently deviate too much from the useful measurement range. Such dead detector elements need to be known.

Samples affected by saturation/recovery/blooming (1, 2, 3) are flagged. Glitches (4) are neither detected nor corrected due to unavailability of a simple model for detection. Samples corresponding to dead detector elements (5) are replaced with a cosmetically interpolated value after radiometric conversion.

The radiometric conversion is to estimate the spectral radiance using an inverse model of the MERIS acquisition and parameters stored in the characterization database and the MERIS samples DNs. The MERIS acquisition model is described as follows:

$$X_{b,k,m,f} = NonLin_{b,m}\left\{g\left(T_f^{VEU}\right)\left[A_{b,k,m}\left(L_{b,k,m,f} + G_{b,k,m}\left(L_{*,*,*,*}\right)\right) + S_{b.k,m,f}\right] + g_c\left(T_f^{CCD}\right)C_{b,k,m}^0\right\} + \varepsilon \tag{13.10}$$

where

- $X_{b,k,m,f}$ is the MERIS raw sample (not corrected onboard).
- *b, k, m,* and *f* are pixel number, spectral band number, hyperspectral camera number, and frame number, respectively. There were m = 5 cameras in MERIS, each covering 1/5 of the swath.
- $NonLin_{b,m}$ is a nonlinear function for pixel *b* of camera *m*.
- $g(T)$ and $g_c(T)$ are temperature-dependent gain terms (close to 1).
- T_f^{VEU} is the amplification unit temperature.
- T_f^{CCD} is the detector temperature.
- $A_{b,k,m}$ is the absolute radiometric gain for pixel *b* at band *k* of camera *m*.

- $L_{b,k,m,f}$ is the input radiance at the entrance pupil of a sample of the MERIS instrument.
- $S_{b,k,m,f}$ is the smear signal due to continuous sensing of light by MERIS.
- $G_{b,k,m}$ is a linear process representing the stray-light contribution to the signal. For a given sample, some stray light is expected from all the other simultaneous samples in the module, spread into the sample by specular (ghost image) or scattering processes, and from the samples in the previous and following frames.
- $C^0_{b,k,m}$ is the dark signal corrected onboard for temperature effects by the Offset Control Loop.
- ε is a random process representative of the instrument errors and parasitic processes not accounted for in the other terms of the model.

All terms without the subscript term f (frame) evolve in time due to ageing, but with a much slower rate that allows to represent them; for a given Level 1B product, they are fixed quantities retrieved from the databases.

For a given set of target, physical parameters, illumination, and observation angles, the radiance of a scene sensed by MERIS $L_{b,k,m,f}$ is proportional to the extraterrestrial sun spectral flux. Because there is no absolute spectral measurement of the sun irradiance simultaneous to MERIS acquisition, all results are produced with reference to the sun's spectral flux model that is included in the product header.

The term $A_{b,k,m}$ reflects all the amplification gains inside the instrument, which depend on:

- instrument programming (band settings, amplification programmable gains),
- components ageing,
- components temperature, and
- power supply voltage.

The terms, $g(T)$ and $g_c(T)$, are used to correct residual temperature for limitation or failure of the onboard temperature regulation. In normal operation, T depends on the time elapsed, because the sun zenith angle has decreased below a specific threshold (80°) and can be predicted.

The stray light term $G_{b,k,m}\left(L_{*,*,*,*}\right)$ in the MERIS acquisition model in Equation 13.10 may be strong enough to affect the least significant bits of the raw DNs. This may happen in particular when MERIS is observing a scene with some high-radiance areas (e.g., sun glint patch, partly cloudy, etc.). As the linear transform term, $G_{b,k,m}$ is assumed to be known well enough from the instrument characterization, it is possible to compute an estimate of the stray light and correct for it. Stray-light correction is handled separately from radiometric processing due to the specific nature of this step (deconvolution) and to the fact that it can be switched on and off.

13.1.3.2 Hyperion Levels 0 and 1 Processing to Generate At-Sensor Radiance

A Hyperion science data acquisition is referred to as a science Data Collection Event (DCE). A science DCE includes five data collections in following order and creates five corresponding files:

1. pre-image dark collect,
2. image collect,
3. post-image dark collect,
4. internal calibration lamp collect, and
5. post-lamp dark collect.

The time duration of both the pre- and post-image dark collects is 1 sec. The lamp-collect time duration is 3 sec and a typical image-collect time duration is 30 sec with the desired scene being contained in the centered 24 sec. The process of a DCE takes about 15 min with the longest time being the 10-min wait after the instrument is commanded from Idle mode to Standby mode to allow the analog signal processors (ASPs) of the visible and near-infrared (VNIR) and shortwave infrared (SWIR) spectrometers to achieve thermal stability before the pre-image dark collect. Another wait is 3 min after the internal calibration lamp is turned on to allow the lamp output to stabilize. The instrument is commanded into Imaging mode from Standby mode to acquire science data of the scene (a typical duration of 30 sec) and returns to Idle mode after the post-lamp dark collect. The remainder of the time, the instrument is in Idle mode. The instrument is in Standby mode only during DCEs (Barry, 2001).

The Hyperion science data and ancillary data obtained during a DCE are stored onboard in the Wideband Advanced Recorder Processor (WARP). The former include the five data files described above. While the latter contain information of the instrument telemetry obtained during the DCE. The science data and ancillary data are downlinked to one of several ground stations using an X-band downlink channel.

Level 0 processing turns the downlinked raw data into a set of Level 0 files with prescribed file names. Level 0 processing of EO-1 Hyperion science data retains the collected data in raw DNs and does not perform any artifact correction. A set of tasks performed on all collected scene data is as follows:

1. *Decode the data:* Perform Reed-Solomon decoding on downlinked science data. Extract the science data from the telemetry data and flag corrupted data.
2. *Separate the data into files:* Separate the data into files according to data-type boundaries, data stream boundaries, and DCE image number boundaries. Lunar calibration, ground image, solar calibrations are examples of data-type boundaries. Pre-image dark, image, post-image dark are examples of data stream boundaries.
3. *Perform checks to verify data integrity and instrument performance:* If a check fails, then the program terminates with a descriptive error message.
4. *Combine the VNIR and SWIR data sets:* Arrange VNIR-SWIR pixel order. Confirm that the VNIR and SWIR file pair up properly. Verify that the two filenames have the same year and day stamp, represent the same DCE image number, same event type, and same data type (i.e., pre-image dark calibration, image, post-image dark calibration, or lamp calibration). Concatenate the VNIR and SWIR science data together.
5. *Create HDF:* Send the science data into a formatted output file (HDF format) and append "HDF attributes" to the output file. These attributes list various properties about the data.

Hyperion Level-1 processing converts raw DNs to at-sensor radiance using the Level 1B code that contains the instrument responsivity function such as gains and offsets. This level processing produces the radiometric calibrated science data. Level-1 processing using the Level 1B code includes 10 steps as described below.

Step 0 is to identify and flag saturated pixels whose intensity is >4095 in the three data files: pre-image dark collect, image collect, and post-image collect (Hyperion used 12-bit digital quantization in the data processing chain; thus, the maximum value is $2^{12} - 1 = 4095$). A metadata file (MD15) is created that reports the location of saturated pixels before (.L0) and after co-registration (.L1B) (see Step 7).

Step 1 applies smear correction to SWIR data in the three data files (pre-image dark collect, image collect, and post-image collect). A metadata file (MD9) is created.

Step 2 removes echo from smear-removed data in the three data files. A metadata file (MD8) is then created.

Steps 3 and 4 are to subtract off the dark signal or offset ($O_{j,k}$). These two steps are required, because each acquired image of a scene includes not only the scene signal but also a signal caused by thermally generated electrons in the bulk material. To enable removal of this signal from the image, pre- and post-image dark frames are taken as part of each DCE. Each dark file is collected for 1 sec, corresponding to approximately 220 frames, because the frame rate of the instrument is 223.4 Hz. Level 1B processing codes use a dark value that is interpolated in time between the pre- and post-image dark, although the Level 1B code includes the full wait time between the pre-image dark collect and the start of the image collect, when performing dark interpolation calculations. Step 3 consists of (1) averaging frames of smear- and echo-corrected pre- and post-image dark files, (2) providing averaged pre-image (MD3A) and post-image (MD3B) dark files, (3) computing average value of corrected pre- and post-image dark files, and (4) generating the log file to report average values (MD5A for pre-image dark and MD5B for post-image dark). Step 4 is to subtract interpolated dark file values from echo- and smear-corrected image file values.

Step 5 is the conversion of raw DNs to at-sensor radiance. In this step, a calibration file (MD7) containing gains for each detector pixel $G_{j,k}$ measured in prelaunch characterization of the instrument is used. The dark-subtracted image is multiplied by the gain file to obtain radiometrically corrected data. Now the data are at-sensor radiance (Level 1). VNIR radiance data (bands 1 to 70) and SWIR radiance data (bands 71 to 242) are multiplied by 40 and 80, respectively, to retain precision of the maximum radiance within the range of a 16-bit signed integer.

Step 6 is to fix strips by repairing known bad pixels and output the Level-1A data file in signed integer format (with an extension name .L1A). The bad-pixel list was defined preflight. Pixels that are known to be dead are replaced with an average of neighboring FOV pixels or with 100% of the neighboring pixel if on the edge of the FOV prior to generating the Level-1 data file.

Step 7 is the spatial co-registration of VNIR and SWIR images. SWIR image data are shifted to align with the VNIR pixels. In the cross-track dimension (X) of the FOV, the SWIR pixels are shifted by −1. Within this context, FOV pixel #1 is removed and pixel #256 is padded with zeros. In the along-track dimension (Y), FOV pixels 129–256 are shifted by +1 pixel. For frame 1, FOV pixels 129–256 are padded with zeros and the last frame is removed. Spatial co-registration of the VNIR and SWIR data is the last step in Level-1B data processing. A metadata file is then generated to indicate that co-alignment has been performed on the Level-1 data product. The Level-1B saturated pixel report contains the location of both L0 and L1B saturated pixels (the co-alignment process changes the location of SWIR pixels).

Steps 8 and 9 are quality assessment (QA) and creation of Level 1 Digital Linear Tape (DLT). Step 10 is to ship data files to NASA GSFC for distribution to the science team.

Hyperion Level 1 data sets consist of radiometrically corrected images formatted as HDF files, and metadata in binary and ASCII formats. For data processed using the Level-1B code, the data type is a 16-bit signed integer with units of $W/(m^2 \cdot sr \cdot \mu m)$.

The Hyperion absolute radiometric accuracy is verified by cross-checking with radiance measurements from three different paths: solar calibration, internal lamp calibration source, and Lake Frome scene-based calibration effort.

For the solar radiometric calibration comparison, the measured radiance of the sun's irradiance was compared with solar irradiance models in the literature. The agreement was within 2% in the VNIR region and 5–8% in the SWIR region.

The in-flight calibration lamp was used as a ground to on-orbit transition. The lamp results indicated that the change in responsivity was <3% in the VNIR and 5–8% in the SWIR.

Scene-based calibration using Lake Frome in Australia was incorporated into the performance verification of the Hyperion imaging spectrometer instrument. The ground reflectance measurements and atmospheric correction leading to TOA radiances are consistent with the Hyperion ground and solar calibration at the 5% level in the 450–850 nm spectral range. The SWIR agreement is 10–15%.

13.1.3.3 EnMAP Levels 0 and 1 Processing to Generate At-Sensor Radiance

EnMAP Level 0 (LO) image products are raw data (DNs) and will be archived for long term, while Level 1B (L1B) image products are at-sensor radiance after systematical and radiometric correction. Level 1C (L1C) image products are at-sensor radiance after geometrical correction on top of L1B data. Level 2A (L2A) products are atmospherically corrected data and will be processed on demand (Palubinskas et al. 2017).

The Level-1B processor corrects a hyperspectral image cube for systematic effects of the hyperspectral sensor, for example, radiometric nonuniformities, and converts the raw DNs to physically meaningful at-sensor radiance values based on the calibration tables. The L1C processor creates ortho-images based on direct geo-referencing techniques implementing a line of sight model that uses onboard measurements for orbit and attitude determinations as well as the sensor-look direction vectors based on the valid calibration values. The L2A processor performs atmospheric correction and haze detection of the images by estimating the aerosol optical thickness and the columnar water vapor. Output products will be the surface reflectance and masks of land, water, haze, cloud, and snow.

The Level-0 processor collects and preprocesses the information from the different data streams to generate an internal product, which is not available for users. Level-0 processing steps include three tasks: transcription and screening, quicklook, and quality assessment. The details are described in the following sections.

1. Transcription and screening
   1.1 General
      1.1.1 Unpack source packets for each band file. This process removes packet header, interprets header information, and prepares files for further processing.
      1.1.2 Decompress unpacked band files of the spectral image. Decompression will be performed for each file separately.
      1.1.3 Check whether all status information, temperatures, currents, and voltages are available in the virtual channel (VC) and whether they are in the working range.
      1.1.4 Extract dark current (DC) measurements before and after scene data acquisition for each detector array and band file separately and reformat the data (e.g., to band sequential format). A separate product for the ingestion in the archive will be created.
   1.2 Scene Data
      1.2.1 Cut scene image into tiles after cutting off the DC measurements. Each image tile has the size of 1024×1024 pixels (approximately 30×30 km^2). Due to a time delay of about $\Delta t = 88$ ms between image data collection of the VNIR spectrometer and the SWIR spectrometer, the ground scene collected by the VNIR spectrometer is shifted by about 20 pixels (600 m on ground) in along-track direction.
      1.2.2 Generate quality masks for abnormal, cloud and haze pixels, and detector map and other derived metadata.
   1.3 Calibration Data
      1.3.1 Evaluate calibration measurements (e.g., full aperture sun diffuser, deep space, internal lamp, or LED measurements), and derive additional calibration information.
2. Quicklooks
   2.1 For each focal plane, two quicklooks will be generated. One quicklook in sensor geometry to be delivered together with L1B products and an another orthorectified quicklook to be delivered with L1C and L2A products.
   2.2 The quicklook files of the VNIR region will be transformed to 8-bit pixel values and radiometrically adjusted for optimal displaying with bands 550 nm as blue, 670 nm as green, and 850 nm as red.

2.3 The quicklook file of the SWIR region will be transformed to 8-bit pixel values and radiometrically adjusted for optimal displaying with bands 1050 nm as blue, 1650 nm as green, and 2200 nm as red.

3. Quality Assessment
   3.1 Generation of a mask for dead and abnormal pixels.
      3.1.1 A mask of the same size as the VNIR and SWIR detector arrays will be generated for marking dead and abnormal pixels in the data. A pixel will be marked as dead or abnormal if one of the following criteria is met:
      - no response (cold pixel),
      - saturated, even with low radiance input (hot pixel),
      - frozen output, and
      - flickering pixel.

      3.1.2 These criteria will be checked within the calibration process and the dead pixel mask may be changed, if necessary.

   3.2 Screening of raw data
   The screening results of the raw data will be written in the metadata file for quality assessments and further processing.

   3.3 Generation of other quality masks and parameters

Within a classification process, haze, cirrus, shadow, snow, land, defect masks, and cloud cover index will be created. Similar to the quicklook files, each mask will be generated in focal-plane geometry and orthorectified format.

The L1B processor converts the raw hyperspectral image data extracted by the Level 0 processor to at-sensor radiance values. The resulting Level-1B Data Product will be an at-sensor radiance datacube that has been corrected for systematic and radiometric errors and spectrally characterized. The L1B process includes the following steps:

1. Flag dead pixels, abnormal pixels, and saturated pixels using information from radiometric calibration tables. For the current saturated pixels, not only the current saturated pixels will be considered, but also pixels saturated in a previous frame(s) or neighboring pixels of the saturated pixel affected by blooming will be taken into account.
2. Correct nonlinearity response of the detector array in both spatial and spectral directions using the look-up tables (LUTs) for nonlinearity correction: $DN_{lin} = LUT(DN)$.
3. Subtract the offset for the SWIR spectrometer (the offset for the VNIR spectrometer is already internally corrected).
4. Subtract dark current from the DNs measured before and after each data take.
5. Correct response nonuniformity (RNU) of the detector array (flat fielding) using LUTs or polynomials: $DN_{cor} = RNU(DN_{lin} - DNDS - DN_{stray})$, where RNU is the response nonuniformity stored in the LUTs, DN_{lin} is the DN after nonlinearity response correction, and DN_{stray} is the DN after stray light correction.
6. Correct stray light by applying corresponding 2D matrices that contain weighting coefficients. The stray-light contribution to a particular pixel will be removed by subtracting the pixel's DN multiplied by the weighting coefficients from that pixel's DN.
7. Perform spectral referencing of each pixel to its spectral response function (SRF) using the center wavelength and FWHM bandwidth.
8. Perform radiometric referencing using an absolute calibration gain. Each pixel will be referenced to the corresponding radiometric coefficient from the Radiometric Coefficient Table.

The resulting Level-1B Data Product will be a hyperspectral cube at-sensor radiance after corrections for systematic errors and radiometric and spectral errors.

13.1.4 Conversion to Top-of-Atmosphere Reflectance

The at-sensor radiance is the variable measured by hyperspectral sensors. One can think of radiance as how much light the sensor "sees" from a target being observed. When observing on orbit through the atmosphere, some light scattered by the atmosphere is seen by the instrument and included in the observed radiance of the target. The atmosphere also absorbs light, which will decrease the observed radiance. Radiance most often has units of $W/(m^2 \cdot sr \cdot \mu m)$.

Radiance depends on the illumination (both its intensity and direction), the orientation, and position of the target and the path of the light through the atmosphere plus its conditions. The at-sensor radiance of a target material varies with the sensor, time, and scene. To reduce this variability of the at-sensor radiance, it can be achieved by converting the at-sensor spectral radiance to exoatmospheric top-of-atmosphere (TOA) reflectance, also known as in-band planetary albedo. TOA reflectance is a unitless term that provides the ratio of radiation reflected to the incident solar radiation on a given surface. TOA reflectance still contains effects of the atmosphere. Atmospheric correction is the process of removing the effects of the atmosphere to produce surface reflectance values. Surface reflectance can significantly improve the interpretability and usage of an image. Ideally the atmospheric correction process requires knowledge of the atmospheric conditions and aerosol properties at the time of image acquisition. This subject is described in detail in Chapter 14.

Comparing spectral images from different sensors, there are three advantages in using TOA reflectance instead of at-sensor spectral radiance. First, it removes the cosine effect of different solar zenith angles due to the time difference between data acquisitions. Second, TOA reflectance compensates for different values of the exoatmospheric solar irradiance arising from spectral band differences. Third, the TOA reflectance corrects for the variation in the Earth-sun distance between different data acquisition dates. These variations can be significant geographically and temporally. The TOA reflectance of the Earth is computed according to the following equation:

$$\rho_\lambda = \frac{\pi \cdot L_\lambda \cdot d^2}{E_s \cdot \cos\theta_s} \tag{13.11}$$

where

ρ_λ is the TOA reflectance,
π is the mathematical constant equal to ~3.14159,
L_λ is the spectral radiance at the sensor's aperture $[W/(m^2 \cdot sr \cdot \mu m)]$,
d is the Earth-sun distance in astronomical units,
E_s is the mean exoatmospheric solar irradiance $[W/(m^2 \cdot \mu m)]$, and
θ_s is the solar zenith angle in degrees.

Note that the cosine of the solar zenith angle is equal to the sine of the solar elevation angle. The solar elevation angle at the scene center is typically stored in the header file of Level 1 data products. The TOA reflectance calculation requires the Earth-sun distance d. Its numbers are tabulated in the Nautical Almanac Office. Table 6 in Chander et al. (2009) presents d in astronomical units throughout a year generated using the Jet Propulsion Laboratory Ephemeris 10 (DE405) data.

The mean exoatmospheric solar irradiances E_s using the Thuillier solar spectrum can be found in Tables 2–5 in Chander et al. (2009). The Committee on Earth Observation Satellites (CEOS) Working Group on Calibration and Validation (WGCV) recommends using this Thuillier spectrum for applications in optical-based Earth observation that use an exoatmospheric solar irradiance spectrum. It is believed to be the most accurate and an improvement over the other solar spectra (Thuillier et al. 2003).

13.2 SMILE DETECTION AND CORRECTION

13.2.1 Spectral Distortion—Smile

As described in Chapter 2, all dispersive element-based (grating or prism) spaceborne hyperspectral sensors use 2D-detector arrays and operate in pushbroom mode. In a whiskbroom hyperspectral sensor, all ground samples in a scene have their spectra recorded by a single linear detector array one after another, whereas a pushbroom hyperspectral sensor has effectively many (e.g., 1000 in PRISMA [PRecursore IperSpettrale della Missione Applicativa] and EnMAP) linear arrays (or spectrometers) to simultaneously record spectra of 1000 ground samples in a cross-track line. These 1000 linear arrays sense the ground samples independently; thus, such acquired hyperspectral imagery exhibits both spatial and spectral distortions, which need to be detected and corrected. The spatial distortion is often referred to as "keystone," which is the subject of Section 13.3. The spectral distortion is called spectral line curvature, sometimes also referred to as slit curvature, or as "smile/frown" in the remote sensing community. These terms generally refer to the misregistration of the detector elements (or pixels) across the FOV of the hyperspectral sensor in the spectral dimension.

Spectral misregistration in a pushbroom hyperspectral sensor leads to the geometric nonlinearity of the monochromatic image via a straight entrance slit when it appears in the focal plane of the spectrograph. In a more general sense, these spectral misregistrations can be caused by aberrations in the spectrograph collimator and imaging optics, or by a misalignment of the detector array in the instrument's focal plane, even though the latter may not produce a curved slit image. In the strict sense of the term spectral line curvature, these geometric distortions are caused by the dispersion element, prism, or grating. In the case of a grating spectrograph, the spectral line curvature derives from the grating equation and can be shown to vary as the product of the wavelength, the angular dispersion, and the square of the off-axis field angle (Schroeder 2000). The impacts of these misregistrations, regardless of the source, are similar and can be rectified by a common procedure. All such spectral misregistrations are referred to as spectral line curvature or smile.

In a pushbroom hyperspectral sensor that employs a 2D detector array, the spectrum is dispersed in one dimension, for example, along the columns, and the spatial field is oriented along the rows. Ideally, the output of this type of hyperspectral sensor consists of 2D (spectral–spatial) frames of data in which all the entries in a given row represent a single band-center wavelength and bandwidth of the ground samples in the cross-track line. However, because the detector elements are arranged on a rectilinear grid, the presence of spectral smile prevents the registration of the monochromatic image of the entrance slit onto the straight rows of detector elements.

Figure 13.2 shows an illustration of the smile distortion of a hyperspectral sensor. In this figure, an example of a hyperspectral sensor with a slit of 30-μm wide × 30-mm long is assumed. A 2D-detector array of the size of 1000 × 256 pixels with a pitch size of 30 μm × 30 μm is placed on the focal plane. The 256 pixels are for spectrum in the vertical axis of the detector array, whereas the 1000 pixels in the horizontal axis are for the FOV of the hyperspectral sensor covering 1000 ground samples in a cross-track line. In this configuration, the locations of the ground samples are denoted as Position [mm] on the slit in the horizontal axis of the figure. The vertical axis shows the values of the spectral distortion curves corresponding to different wavelengths of the hyperspectral sensor versus the position of the ground samples on the slit.

In the ideal case, all the smile distortion values should be zero (i.e., no spectral distortion) and show horizontal straight lines overlapped with the line 0. In the example, the smile distortion values vary at different spectral bands and show bend lines (smile curves). Some curves show large distortion at the two ends, while other curves show large distortion in the center. The maximum smile/frown distortion in this example is 7.7 μm, which represents a factor of 0.26 of the pitch size (30 μm) of the detector pixels. Detailed description of smile can be found in Qian (2013).

Spaceborne hyperspectral sensors all exhibit smile distortion. For example, Hyperion exhibited smile distortion up to 3.2 pixel (equivalent to 32 nm in terms of the 10 nm spectral

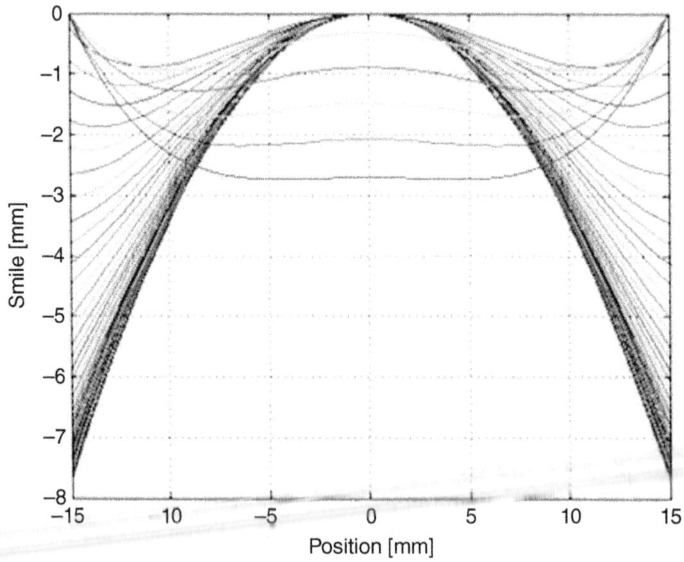

FIGURE 13.2 Illustration of spectral smile/frown distortion.

sampling interval) in the VNIR spectral region and up to 0.8 pixel (equivalent to 8.0 nm) in the SWIR region (Pearlman et al. 2003). The Hyperspectral Imager for the Coastal Ocean (HICO) had a 0.30-pixel smile (Kappus et al. 2016), while the Compact Reconnaissance Imaging Spectrometer for Mars (CRISM) exhibited a large smile distortion of 1.20 pixel (Murchie et al. 2007). The hyperspectral sensor of PRISMA was designed to maintain a smile distortion within 0.10 pixel (Meini et al. 2016), while the hyperspectral sensor of EnMAP is designed to maintain a smile distortion within 0.20 pixel (Kaufmann et al. 2016).

Accurate calibration and correction of smile is required before the acquired hyperspectral data sets are delivered to the user community for deriving application products and for quantitative analysis. The location of the peak of the Spectral Response Function (SRF), its shape, and the half-width must be within a small fraction of the nominal pixel bandwidth, typically less than a few percent. This requirement means that distortion of a pushbroom hyperspectral sensor along the spatial direction as well as the spatial variation of the optical Point Spread Function (PSF) must be kept to a minimum, typically a small fraction of a pixel before the calibration.

13.2.2 Smile Correction Using Atmospheric Absorption Feature Matching

The degree of spectral accuracy required generally depends on the spectral gradients exhibited by the measured spectra. Steeper gradients require higher spectral accuracy. The steepest gradients are most often caused by atmospheric absorption and in some cases by Fraunhofer lines in the exoatmospheric irradiance. When a radiance spectrum measured by a hyperspectral sensor is converted to surface reflectance by factoring out the irradiance spectrum and compensating for the atmosphere, spike artifacts appear in the neighborhoods of the sharp spectral features in the exoatmospheric irradiance and atmosphere absorption transmission spectra. These spike artifacts are caused by inaccuracies in the sensor's spectral specifications.

Neville et al. (2003, 2008) developed a method to measure the band centers and bandwidths of hyperspectral sensors, which uses only image data acquired by the sensor while in flight. The method uses spectral features present in the at-sensor radiance that are common to all pixels in the scene. These features are principally atmospheric absorption lines. This is done for each pixel in the FOV (i.e., cross-track line); thus, allowing the measurement of the sensor's smile distortion.

The band-center and bandwidth determinations are made by correlating the sensor-measured radiance with a modeled radiance.

The atmospheric radiative transfer code, MODTRAN 4.2, was used to perform the radiative transfer calculations. From the point of view of the characterization of the hyperspectral sensor to some specific level of accuracy, the radiative transfer reference data must be accurate to a commensurate level. From the pragmatic view of eliminating the spikes in the derived reflectance data, it is only necessary that the wavelength calibration of the measured image data matches the atmospheric reference data. In this method, the absorption features in the measured radiance are compared with those in a modeled at-sensor radiance calculated using MODTRAN 4.2.

The sensor-generated radiance is used to calculate the reflectance spectra in an iterative process in which both the calculated surface reflectance spectra and the sensor spectral parameters are sequentially improved until a stable result is achieved. As a first step, an atmospheric correction is performed to extract the surface reflectance, using as a starting point for estimating the sensor's nominal band-center wavelength and bandwidth values. Because these are in general incorrect, spikes occur in the neighborhoods of the atmospheric absorption features. These are suppressed by spectral smoothing over the affected regions to obtain approximate reflectance spectra for the scene. These locally smoothed reflectance spectra are then input to the radiance modeling process to achieve the modeled at-sensor radiance spectra.

To minimize the impact of approximations occurring in the surface reflectance, these spectra are filtered using a high-pass filter, thereby removing most of the influence of the spectral signature of the underlying scene material. These consist for the most part of features that vary slowly in comparison to the atmospheric features. A correlation procedure is then followed to obtain the first estimate for the correct band-center and bandwidth values. The procedure is repeated using these "corrected" values for band center and bandwidth to perform the atmospheric correction on the measured radiances, yielding an improved value for the derived surface reflectance. This process is repeated, updating the wavelength center values with each iteration, until a stable result is obtained or until a prescribed number of iterations has been performed. In practice, it is found that three to five iterations are sufficient.

The implementation of the method was carried out using a software tool referred to as the Imaging Spectrometer Data Analysis System (ISDAS) (Staenz et al. 1998). A LUT approach is adopted for the atmospheric correction (Staenz and Williams 1997). A "raw" LUT is created for the time, date, location, elevation, and appropriate atmospheric conditions using the MODTRAN 4.2 radiative transfer code. In the normal atmospheric correction procedure, a sensor-specific LUT is created by convolving the band response profiles with the raw LUT values. This convolved LUT is then sampled at the sensor's band wavelength values. In the band-center-bandwidth measurement procedure, many convolved LUTs are created. First, the convolution is repeated for a range of bandwidths selected by the operator. Each of these LUTs is then sampled for band-center wavelengths that are shifted by a range of operator-selected values.

The shift range is adjusted to encompass the anticipated smile amplitude, and the shift increment is adjusted to be commensurate with the pixel-to-pixel variation in wavelength shift. These are determined by iteration. The bandwidth increment and range values are chosen in a similar manner. If the pixel-to-pixel uniformity (responsivity) coefficients are not sufficiently accurate to eliminate striping in those bands involved in the feature analyses, errors in the wavelength estimates will occur. To cancel these effects and those caused by the discrete shift increment, a low-order polynomial is fitted to the final band-center and bandwidth values. Normally, a second-order polynomial is adequate and is in keeping with the theory for the grating spectrometer case, where a cosine dependence is predicted. This assumes that the angular range (at the spectrometer grating) is small.

The selection of absorption features for this procedure depends on the wavelength range, the spectral resolution, and the signal-to-noise ratio (SNR) of the particular sensor under study.

The deeper and better resolved the absorption feature that appears in the measured spectra, the lower the error estimates of the resulting wavelength measurements. To reduce the potential influence of the background-scene reflectance, features are chosen that are spectrally narrow or consist of a number of narrow subfeatures that are still resolvable by the instrument under study. In addition, both the measured spectra and the modeled spectra are filtered with a high-pass filter before the correlations are performed, again to reduce the influence of the scene reflectance. Ideally, the set of selected absorption features spans the spectral range of the instrument so that spectral correction measurements can be applied to the whole instrument range.

The comparison between the measured and modeled radiance spectra is performed separately for each selected absorption feature, thereby compiling for each cross-track pixel a set of band-center wavelength shifts and bandwidths, one value for each parameter for each feature in the absorption feature set. The wavelength and bandwidth estimates derived from each of these features and for each cross-track pixel are extended to all elements in the array by interpolation or extrapolation. For this purpose, the user can use either a low-order polynomial fit or a piecewise linear interpolation. Extrapolation to bands beyond the range spanned by these features can be problematic. This is most often accomplished using the band-center and bandwidth shifts measured for the nearest atmospheric feature.

Neville et al. (2008) measured the smile distortion of the Hyperion hyperspectral sensor using their proposed method on three data sets. These data sets were acquired at three scenes on different dates and locations: (1) Coleambally, Australia, January 12, 2002; (2) Ranger Mill, Australia, March 30, 2002; and (3) Saskatchewan, Canada, May 20, 2002. The Coleambally scene consists of a mix of densely vegetated fields and fields with bare soil, the Ranger Mill region is tropical savanna in which the vegetation is sparse compared to the vegetated fields at Coleambally, and the Saskatchewan site consists of fields of bare soil partially covered by crop residue. Measurements performed on all three data sets are shown in Figures 13.3 and 13.4.

Figure 13.3a shows the results for the oxygen feature at 761 nm, and Figure 13.3b, the results for the water vapor feature at 823 nm, both acquired by the VNIR spectrometer of the Hyperion instrument. In Figure 13.3a, both the measurements and the corresponding second-order polynomial curve fits are displayed, whereas for clarity Figure 13.3b shows only the fitted curves. Figure 13.3a and b also presents the prelaunch laboratory Hyperion calibration smile curves for the 763.14-nm and 824.16-nm bands, respectively.

Figure 13.4a presents the measurement results for the water vapor feature at 1136 nm overlaid by the corresponding second-order polynomial curve fits, while Figure 13.4b displays only the second-order polynomial fits for the carbon dioxide features at 1575–1607 nm. Included in these figures are the prelaunch laboratory Hyperion smile calibrations for the 1134.45-nm and 1578.37-nm bands in the SWIR region.

In total, 10 atmospheric absorption features, 3 in the VNIR region and 7 in the SWIR region, were extracted and used to establish a new wavelength calibration grid for correcting smile in both the VNIR and SWIR regions of the instrument.

The mean shift values for each of the four absorption features and for each of the three data sets were identified by calculating the means-over-pixels of the differences between the curve fits to the measured values and the laboratory calibration values. The maximum measured smile amplitude for the Hyperion data sets is 4.25 nm. Shapes extend to the asymmetry of the spectral curves, indicative of some type of opto-mechanical misalignment. The results contained in Figures 13.3 and 13.4 show significant differences between the measured and laboratory-calibrated values, both in the overall shifts and in the smile amplitudes. In addition, for the features in SWIR region, the laboratory-derived smile curves exhibit shapes that are at variance both with the theory and with the measured shapes. In all but the 823-nm feature, the measured shifts relative to the laboratory values are positive, i.e., the measured wavelength for a specific sensor band has shifted toward longer wavelengths.

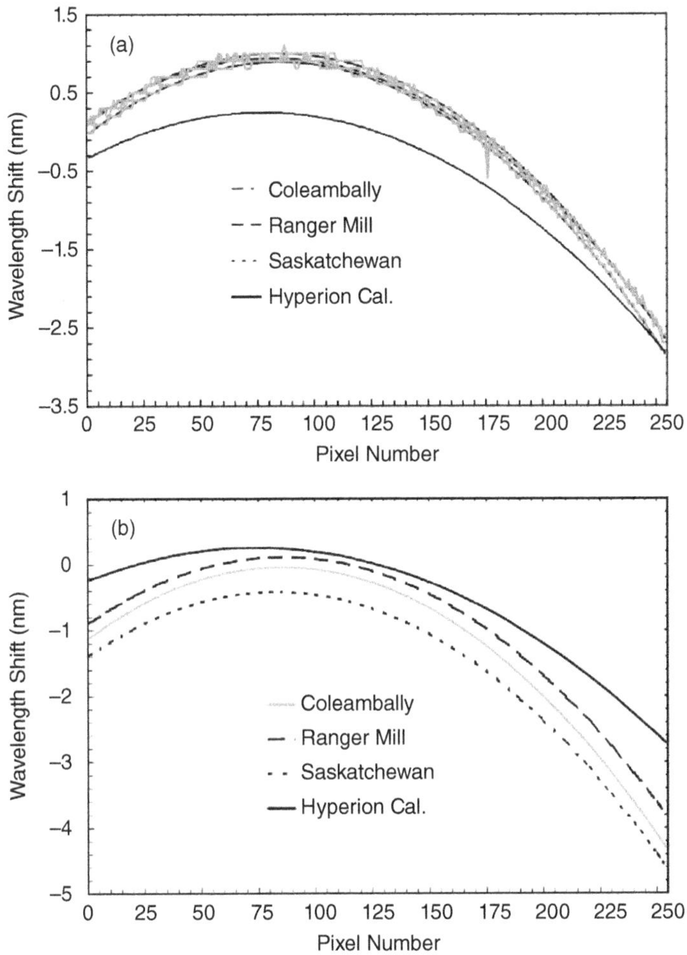

FIGURE 13.3 (a) Spectral smile measurements at the 761-nm oxygen absorption band, with their second-order polynomial fits, for three Hyperion data sets. The laboratory Hyperion smile calibration curve for the 763.14-nm band is included for comparison. (b) Polynomial fits to the 823-nm water vapor absorption band for three Hyperion data sets. The laboratory Hyperion smile calibration curve for the 824.16-nm band is included. (Reprinted from Neville et al. 2008; Copyright Government of Canada.)

13.3 KEYSTONE DETECTION AND CORRECTION

13.3.1 Spatial Distortion—Keystone

In addition to the spectral distortion described in the previous section, dispersive element-based hyperspectral sensors that use 2D-detector arrays also suffer from spatial distortion called keystone. This spatial distortion exhibits interband spatial misregistration along the spectral direction. Keystone is the variation of slit magnification with wavelength, but it is measured as an absolute length value or a fraction of the pixel pitch size of a detector array. The keystone distortions may be caused by geometric distortion, as can be seen in camera lenses, or by chromatic aberration, or a combination of both. In general terms, it refers to the cross-track spatial misregistration of the imaged pixels of the ground samples at the various spectral bands of the spectrograph.

Figure 13.5 depicts the keystone distortion of a hyperspectral sensor. As described in the previous section, a pushbroom imaging spectrometer employs a 2D-detector array. The spectrum is

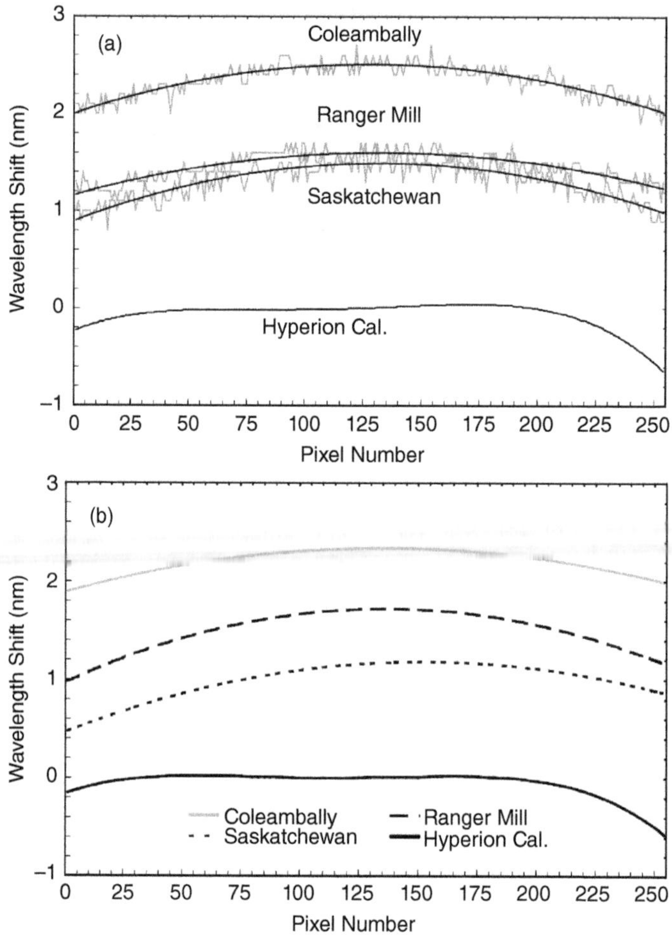

FIGURE 13.4 (a) Spectral smile measurements at the 1136-nm water vapor band, with their polynomial fits, for three Hyperion data sets. The laboratory calibration for the Hyperion 1134.45-nm band is included. (b) Polynomial fits to the measured wavelength shift values at the carbon dioxide bands at 1575–1607 nm. The laboratory calibration curve for the 1578.37-nm band is included for comparison. (Reprinted from Neville et al. 2008, Copyright Government of Canada.)

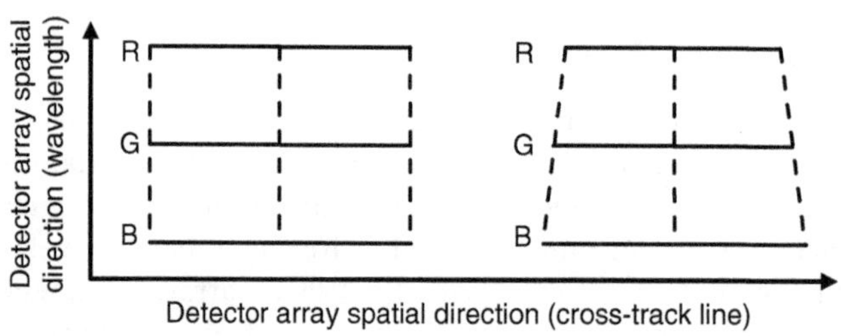

FIGURE 13.5 Schematic of spatial distortion (keystone) of a pushbroom imaging spectrometer. On the left, the lines for spectral bands blue (B), green (G), and red (R) are all straight, parallel, and aligned with the detector array. On the right, the lines for spectral bands G and R are shorter than the line for spectral band B because of keystone. The amount of keystone is measured by the difference in length between the lines B and R.

dispersed in one dimension, for example, in the vertical axis along the columns of the detector array, and the spatial FOV is oriented along the rows of the detector array in the horizontal axis. Ideally this type of imaging spectrometer generates 2D (spectral-spatial) frames of data in which all the entries in a given column record all spectral components (e.g., R, G, and B) of the same ground sample as shown on the left graphics. Due to geometric distortion, as can be seen in camera lenses, or chromatic aberration, or a combination of both, interband spatial misregistration occurs as shown on the right graphics. On the left graphics, the lines for spectral bands blue (B), green (G), and red (R) are all straight, parallel, and aligned with the detector array. On the right graphics, the lines for spectral bands G and R are shorter than the line for spectral band B, because of spatial misregistration. The amount of misregistration is measured by the difference in length between the lines B and R. The shape of this spatial distortion looks like a keystone and, therefore, it is often called keystone. It refers to the across-track spatial misregistration of the ground sample pixels in the various spectral bands of the detector array of the spectrograph.

Spaceborne hyperspectral sensors all exhibit keystone distortion. For example, the Hyperion VNIR spectrometer exhibited keystone, which caused spatial shifts ranging from –0.075 to 0.300 pixels, with the maximum keystone at the top left corner of the focal plane frame. The SWIR spectrometer showed smaller shifts than the VNIR spectrometer, ranging from –0.075 to 0.100 pixels. For the instrument as a whole, the spatial shift ranges from a minimum of 0.075 pixels to a maximum of 0.300 pixels, resulting in a total span of 0.375 pixels (Khurshid et al. 2006). HICO revealed 0.4-pixel keystone (Kappus et al. 2016), while CRISM exhibited a keystone distortion of 0.8 pixel (Murchie et al. 2007). The hyperspectral sensors of PRISMA and EnMAP are designed to maintain a keystone distortion within 0.1 pixel (Meini et al. 2016) and within 0.2 pixel (Kaufmann et al. 2016), respectively.

The simple description of keystone effect is that the spectral bands do not spatially align with the pixels properly on the ground and, therefore, do not measure the same real estate on the ground on a per pixel basis. The effect will basically diminish if the ground scene consists of large homogeneous targets. Keystone distortion matters for hyperspectral image applications, when the images are acquired by hyperspectral sensors with relatively large GSDs (i.e., the acquired spectra are from sub-pixels) and targets in the scene are different. Spatial distortion caused by keystone means that the fraction of mixed spectra from the sub-pixel object materials within a single ground sample varies with wavelength or spectral bands. This can make spectral unmixing ineffective. Keystone distortion will cause acquired adjacent pixels of different materials being mixed up varying with spectral bands. This will affect the recovered spectra of the pixels at the boundaries of the pixels of different materials. More specifically, if two neighboring ground samples contain different materials (i.e., different spectra), then the spectrum measured by the sensor will be a mixture of the spectra of the two different materials in which the fractional content of each material (spectrum) will vary with the band number. If the keystone is sufficiently large to span a whole pixel, then the measured spectrum of one ground sample will be the spectrum of its neighbor sample. This will make it extremely difficult, if not impossible, to identify material species in a spatially variable scene. The situation is complicated further by the fact that the degree of the misregistration caused by the keystone varies with cross-track pixel number. Spectral matching and spectral unmixing techniques will be unlikely capable of analyzing accurately data from such a sensor.

The keystone distortion presents a potentially serious problem to any hyperspectral image analysis and spectral interpretation. The spatial misregistration of the spectral bands of imaging spectrometer data will lead to confusion in the identification of the spectra acquired by the sensor. Indeed, because of the variation of the shifts as a function of cross-track pixel number, even the intra-scene comparison of the measured spectra will produce inconsistent results. The degree of the effect will depend both on the spatial variability of the scene as well as on the spectral variability of the scene constituents. While it is possible to extract examples from real data to make this point, it would be difficult to quantify the overall impact for a particular scene. In addition, the results so obtained would be both scene and sensor dependent. Keystone of a hyperspectral sensor needs to be corrected before the imagery data are delivered to users for applications.

13.3.2 Measuring Keystone Using Interband Correlation of Spectral Features

Neville et al. (2004) developed a method to measure the spatial misregistration of the spectral bands of a hyperspectral sensor based on the interband correlation of spatial features in a scene by using data acquired by the sensor of the scene in flight. This is done for each pixel in the FOV (i.e., cross-track line) and for all bands of each pixel, thus, allowing the measurement of the sensor's keystone and related interband spatial shifts. The determinations of interband spatial relationship are made by correlating these features as detected by the various bands.

A Sobel filter is used to detect the edges of distinct features in the cross-track direction for all spectral band images of a hyperspectral datacube. This is followed by a thresholding to extract the most prominent edges. The spatial locations of these edges are determined for each spectral band image relative to a reference band image by correlating the edge patterns in each band image relative to those in the reference band image. This is done by calculating the correlation coefficient over a user-defined sliding correlation window. Sub-pixel shifts are determined by interpolating the correlation values. For each pixel in the cross-track line, the shifts so determined are averaged together. These are assembled to give a 2D matrix of shifts, one for each cross-track pixel and one for each spectral band. In the final step, a low-order polynomial is fitted to the measured shift values, both in the cross-track dimension and the spectral dimension. Where the hyperspectral sensor is comprised multiple optical and/or detector modules to span an extended spectral range, the various modules are inter-registered using a similar technique. This method is implemented in the software tool ISDAS (Staenz et al. 1998).

Features appropriate for the edge detection generally must span a number of pixels in the along-track dimension of the scene, but need not be strictly linear. However, they must offer contrast that is well above the sensor noise level and the random scene modulations. The 3×3 Sobel kernel used to extract the edges effectively performs a weighted average in the along-track dimension, thereby reducing the influence of noise.

One of the main difficulties is that modulation of the scene radiance is wavelength dependent. Consequently, a feature that may offer high contrast in one part of the spectrum is unlikely to exhibit the same degree of contrast throughout the spectral range under consideration. Often the contrast is maintained for only a limited spectral range, disappearing and even reversing in other parts of the spectral range. To circumvent this problem, a number of reference bands are selected at intervals across the spectrum. All the spectral bands of a given sensor module are correlated with each of these intermediate reference bands. The corresponding intermediate shift curves, which result from the polynomial fits across the spectral bands, are individually adjusted to match a single primary reference band, yielding a shift matrix for each instrument module.

To determine the inter-module misregistration, a temporary keystone correction is performed on the edge-enhanced images, and the combined sets of features from all bands in the spectral range covered by each module are used in determining the shifts between the modules. The final keystone calibration matrix contains the shifts for each cross-track pixel for all the bands in the instrument. These shifts are given relative to a single spectral band.

The keystone distortion of the Hyperion hyperspectral sensor was measured by Neville et al. (2004) using their proposed method. Hyperion is a spaceborne pushbroom hyperspectral sensor consisting of two spectrometers, one for the VNIR region and another for the SWIR region. Together they span the spectral range of 427.55–2355.24 nm with 192 bands, not counting the duplicate overlap bands. The FOV in cross-track direction consists of 254 pixels (Pearlman et al. 2003).

Figure 13.6 shows the contour plot of the spatial shifts as measured relative to the reference band 120. The SWIR spectrometer, containing bands 49–192, exhibits very small shifts, ranging from −0.06 to 0.07 pixels. Their pattern indicates that they result primarily from a clockwise rotation. The VNIR spectrometer, containing bands 1–48, has spatial shifts ranging from −0.05 to 0.49 pixels. The pattern in the right half of the VNIR array is consistent with a clockwise rotation with shifts ranging from −0.05 pixels to 0.16 pixels, and a mean RMS keystone of 0.034 pixels. The pattern in

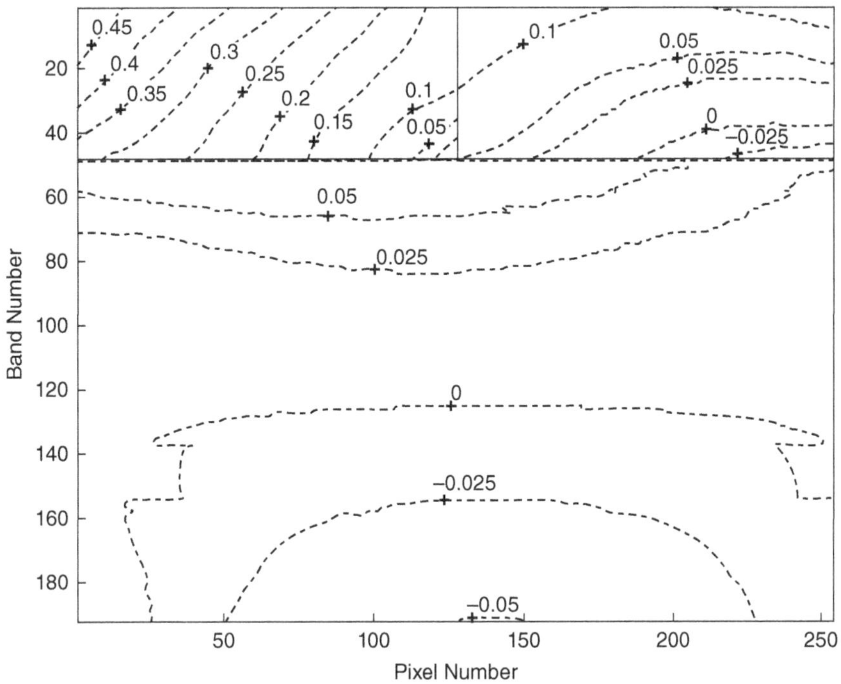

FIGURE 13.6 A contour plot of the spatial shifts exhibited by the spectral bands in the VNIR and SWIR spectrometers of Hyperion as measured relative to the reference band number 120.

the left half is consistent with a basic keystone distortion, all the shifts being positive, i.e., shifting to the right, with the pixel shift values increasing progressively as the pixel number decreases in the FOV. This distortion results in a compression of the cross-track scale for the left half of the VNIR spectrometer. For the left half of the VNIR array, the shifts range from 0 to 0.49 pixels and the mean RMS keystone value is 0.041 pixels. For the entire Hyperion instrument, the spatial shift ranges from a minimum of –0.06 pixels to a maximum of 0.49 pixels, giving a total span of 0.55 pixels and a mean RMS keystone measure of 0.072 pixels.

Upon the keystone of a imaging spectrometer is detected, the correction of keystone becomes a straightforward task. The keystone-induced spatial distortion of the acquired datacubes can be corrected by simply shifting back for each spectrum of the datacube at each spectral band an exact amount of distortion (shift) specified by the LUT or contour plot of the detected keystone of the instrument.

13.4 DE-STRIPING AND NOISE REDUCTION

The at-sensor radiance data obtained after the procedure of radiometric conversion may still contain instrument artifacts, such as stripes, random and system noise. Normally, these instrument artifacts are removed before the correction of keystone and smile. A software package developed at the Canada Centre for Remote Sensing (CCRS), referred to as ISDAS, contains a suite of preprocessing algorithms to remove or collect these instrument artifacts (Staenz et al. 1998, CCRS 2014).

13.4.1 De-striping

As depicted in Figure 1.2 in Chapter 1, a pushbroom hyperspectral sensor uses a 2D-detector array. One dimension is for the spatial domain corresponding to each of the ground samples (A, B, …, G) in the cross-track line, and another dimension is for the spectral domain to record the spectral components ($\lambda_1, \lambda_2, \ldots, \lambda_m$) of each ground sample. With the flight of the satellite in along-track direction, the next cross-track line on the ground is imaged, and so on. A fixed element (e.g., band m, column D) of the 2D detector array is always used to record a fixed-band (e.g., m) spectral component on a fixed column (e.g., D) ground sample in the cross-track line. This image acquisition configuration is prone to exhibiting along-track striping artifacts due to dead detector elements or non-uniform response of the detector elements. These defective columns must be corrected before performing further processing and quantitative analysis.

Sun et al. (2006) developed a technique that can automatically and accurately remove striping artifacts in hyperspectral imagery. The technique is called spectral moment matching (SpcMM). It uses spectral autocorrelation to estimate the expected mean and standard deviation of a sub-scene, which is composed of the image pixels acquired by an identical detector element (an along-track image column of a hyperspectral image cube). The rationale of the algorithm is the observation that there are usually highly correlated groups of bands in a hyperspectral image cube; the statistics of the sub-scenes measured by the corresponding detector elements in a set of highly correlated bands are usually very similar. The possibility of introducing undesired side effects into the de-striped images by the SpcMM is minimized due to the proper estimation of the expected mean and standard deviation for each along-track column.

The SpcMM algorithm consists of estimating the means and standard deviations of each along-track column in an image cube as follows:

1. Calculate the original means and standard deviations for each along-track column in the scene on a per-band basis.
2. Determine the specific number of spatial correlated bands for each along-track column.
3. Calculate the reference mean and standard deviation by averaging the original means and standard deviations of the corresponding along-track columns calculated in step 1 over the correlated bands.
4. Compute the gain and offset from the reference and original mean and standard deviation for de-striping an along-track column.
5. Adjust each pixel value in the image cube by applying the gain and offset of the column.

The commonly used de-striping techniques before the SpcMM are based on histogram matching and spatial moment matching. A histogram matching technique assumes that the sub-scenes measured by separate detector elements for the same band have approximately the same histograms. This technique removes the stripes by matching the sub-scene histograms of the same band to a reference histogram.

The spatial moment matching was originally developed for de-striping multispectral remote sensing images (Gadallah et al. 2000). It has been widely adopted for de-striping Hyperion imagery (Asner and Heidebrecht 2003, Datt et al. 2003, Goetz et al. 2003). The basis of the spatial moment matching is similar to that of histogram matching except that only the moment statistics, such as the means and standard deviations of the sub-scenes in a given band, are adjusted to match those of one or multiple references. Compared with the histogram matching technique, the spatial moment matching is less constraining, but matching means and standard deviations implicitly assumes that each detector element responds linearly to the level of incoming radiation.

The common idea behind the early de-striping techniques before the SpcMM is to estimate the expected statistics either by spatially smoothing the measured statistics of the neighboring sub-scenes in the same band or by simply employing the measured statistics of a reference. These techniques for estimating expected statistics for each sub-scene are based on spatial autocorrelation between the spatially adjacent sub-scenes. Accordingly, these early moment matching techniques were referred to as spatial moment matching (SpaMM) by Sun et al. (2006).

A low spatial autocorrelation between adjacent sub-scenes does not guarantee that SpaMM detects and removes real stripes. This is especially true for a hyperspectral image cube having limited along-track columns due to the constraint of flight-line length (e.g., several thousand pixels), which are insufficient to assure that all the along-track image columns within one band are statistically the same. As a result, the sub-scene to sub-scene statistical variations in the means and standard deviations introduced by the natural variability of the scene can mask the statistical variation caused by striping artifacts. In these cases, traditional histogram matching and global spatial moment matching cannot perform effectively. Even for local moment matching, any kind of spatial smoothing over the measured statistics of adjacent image columns might give an erroneous estimate of the expected statistics, for example, leading to a misclassification of a calibrated detector as a stripe detector, thereby introducing new stripes where none existed in the original image.

Both the SpaMM and SpcMM de-striping methods have been incorporated into the software tool ISDAS. When local SpaMM is used, the spectral bands in the VNIR and SWIR regions are usually de-striped separately with different smoothing window sizes (Datt et al. 2003). The smoothing window sizes were set to 5 and 15 pixels for the VNIR and SWIR regions of Hyperion. On the other hand, SpcMM automatically de-stripes an entire image cube without the manual selection of defective bands or cross-track spatial regions, or the individual selection of band-specific window sizes for spatial smoothing.

Sun et al. (2006) have tested two Hyperion data sets to assess their proposed de-striping SpcMM technique and compare it with the SpaMM procedure. The first data set was acquired on July 8, 2003, over Seal Harbour of an abandoned mine site in Nova Scotia, Canada. The second data set was acquired on January 12, 2002, over Coleambally, a flat agricultural site in the southeast of Australia. The raw Seal Harbour image cube originally has 3350 lines in along-track direction. Only a sub-cube cropped from lines 1750–2100 was de-striped using the SpcMM and local SpaMM procedures to provide a comparison of de-striped images. The Seal Harbour sub-cube was used to focus the assessment on an ocean area (Figure 13.7a), which is supposed to be homogeneous and useful to show the differences between the de-striped images and the side effects introduced by the local SpaMM procedure.

Figure 13.7 shows the images of band 57 (925 nm) over Seal Harbour before and after applying the SpcMM and local SpaMM de-striping. Figure 13.7b indicates that SpcMM performed very well in removing the block stripes without disturbing the rest of the image where no apparent

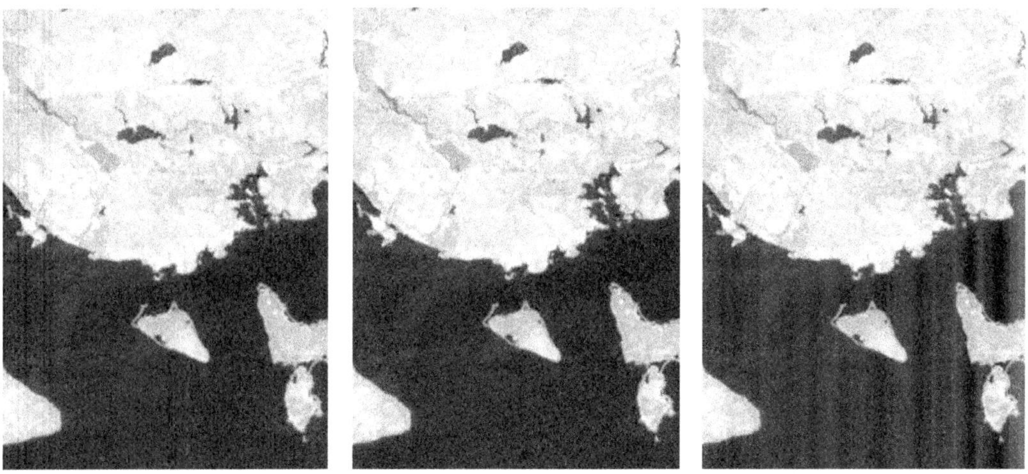

FIGURE 13.7 (a) Cropped original radiance image of band 57 (925 nm) acquired by Hyperion over Seal Harbor, Nova Scotia, Canada; (b) de-striped image using SpcMM; and (c) de-striped image using the local SpaMM procedure with a window size of 15 pixels for spatial smoothing. Note the stripes along the left side of the original image in (a) and the obvious wide stripes introduced by local SpaMM in (c). (Courtesy of Canada Center for Remote Sensing.)

stripe occurs, whereas the image depicted in Figure 13.7c reveals that the local SpaMM de-striping reduced the block stripes, but not to the degree achieved by the SpcMM procedure. In addition, it introduced some mid-frequency stripes in the right image area.

The whole Coleambally datacube consisting of 702 lines was de-striped with both SpcMM and local SpaMM procedures and utilized to show the impact of the de-striping results on the retrieved reflectance spectra. The Coleambally image cube shows a large number of nearly vertical edges between different terrain targets. This caused the frequencies of statistical variations between some adjacent along-track columns to be similar to that caused by the block striping artifacts in Hyperion SWIR bands. It is impossible for the local SpaMM procedure to remove the stripes without washing out these high-frequency variations. Detailed results can be found in Sun et al. (2006).

13.4.2 Random Noise Reduction

Systematic and random noise in acquired DNs of a hyperspectral sensor can be introduced by various components of the sensor system. It is an important step to reduce the noise in the acquired DNs. This will improve the data quality for the remote sensing applications. Because the noise occurs in various forms, it is difficult to be modeled.

The author of this book, in collaborating with his colleagues, has opened up a new research area since 2000 to reduce the noise of hyperspectral data using post-signal processing approaches for the purpose of increasing the SNR of the acquired satellite data. Wavelet shrinkage is introduced to separate noise from the signal in spectral derivative-domain with a hybrid spatial-spectral approach (Othman and Qian 2006). Some innovative techniques have been granted for patents in Europe, United States, and Canada (Qian and Othman 2011, 2013, 2014). These techniques can remove noise in hyperspectral data and increase SNR of the data by a factor of 2 without degrading the signal. Qian and his team further developed a series of de-noising techniques for hyperspectral image data by combining wavelet de-noising with Principal Component Analysis (PCA) and with Block-Matching Filtering (BMF) and have achieved SNR improvements by a factor of up to 5.6 (Qian et al. 2006, 2011, Chen and Qian 2009, 2011, Othman and Qian 2008, Qian and Lévesque 2009, Chen et al. 2011, 2014, Qian 2011, Luo et al. 2016). Most of these novel hyperspectral data de-noising techniques have been included in the book "Optical Satellite Signal Processing and Enhancement" written by Qian (2013). Readers who are interested in exploring more on this subject are recommended to read that book.

This section describes only the noise reduction offered by the ISDAS software tool. It is a scene-based noise reduction tool, referred to as "average-smooth," for reducing random noise of an image cube. The basic idea implemented in the tool is to replace each pixel spectrum with an averaged spectrum over a number of spectra that are most similar to the spectrum being processed (targeted spectrum). The most similar spectra of a targeted spectrum are determined based on a spectral screening technique, which can efficiently group pixel spectra according to their similarity (CCRS 2014).

Let $V_{M \times N \times L}$ represent a given radiance cube, then it can be mathematically described as follows:

$$V_{M\times N\times L} = \left\{ v_{i,j,k} : 1 \le i \le M, 1 \le j \le N, 1 \le k \le L \right\} \tag{13.12}$$

where M is the total number of cross-track lines (in along-track direction), N is the number of pixels in a cross-track line, L is the total number of spectral bands, and $v_{i,j,k}$ represents the radiance of a pixel in line i of column j for a spectral band k. The approach uses the gain coefficient frame ($G_{N \times L}$) for converting DNs to radiances provided as metadata by the data provider to generate a noise model on a per pixel basis. $G_{N \times L}$ is given as follows:

$$G_{N\times L} = \left\{ g_{j,k} : 1 \le j \le N, 1 \le k \le L \right\} \tag{13.13}$$

where $g_{j,k}$ is the gain coefficient for a pixel in column j of band k. The formula to calculate the noise $n_{i,j,k}$ for a single pixel of a given at-sensor radiance cube is given as follows:

$$n_{i,j,k} = \sqrt{\left(N_{f,g_{j,k}}\right)^2 + \frac{v_{i,j,k}g_{j,k}}{C_c}} \tag{13.14}$$

where N_f is the noise floor, and C_c is the charge conversion factor. Subsequently, the noise cube $N_{M \times N \times L}$ is generated and can be written as follows:

$$N_{M \times N \times L} = \left\{n_{i,j,k} : 1 \le i \le M, 1 \le j \le N, 1 \le k \le L\right\} \tag{13.15}$$

From $N_{M \times N \times L}$, a user-defined number of similar spectra are selected for each pixel by moving a window (or kernel) in the along-track direction with the assumption that all the pixels within the window are acquired under similar atmospheric conditions and viewing angles. For example, $S_1 = \{v_{1,k}: 1 \le k \le L\}$ and $S_2 = \{v_{2,k}: 1 \le k \le L\}$ are the two spectra, and $N_1 = \{n_{1,k}: 1 \le k \le L\}$ and $N_2 = \{n_{2,k}: 1 \le k \le L\}$ are their corresponding noise boundaries. S_1 and S_2 are similar if $|v_{1,k} - v_{2,k}| \le \min(n_{1,k}, n_{2,k})$, and S_1 and S_2 are similar in all spectral bands.

A user-defined number of similar spectra ($H = \{S_1, S_2, \ldots, S_H\}$) are selected from the moving window. The average spectrum for these H similar spectra $\bar{S} = \left\{\bar{v}_k : 1 \le k \le L\right\}$ is then calculated using the following formula:

$$\bar{v}_k = \frac{1}{H}\sum_{i=1}^{H} v_{i,k} \tag{13.16}$$

where $\bar{v}_k$ is the average spectrum for band k.

Finally, the original spectrum is replaced by the previously calculated average spectrum. The success of the noise reduction can be verified by randomly picking single-pixel spectra of different target types and comparing them before and after noise reduction (see an example shown in Figure 13.10). It should be noted that this procedure might affect the fraction mixture of a pixel in the scene due to the accuracy of the noise model. The effect this method might have on the information products needs to be investigated.

13.5 A CASE STUDY OF DATA CORRECTION AND ITS EFFECTIVENESS

Qian et al. (2006) reported a case study on how data correction (keystone and smile corrections) and random noise reduction of hyperspectral data can help improve the performance of onboard-data compression.

A hyperspectral satellite generates a huge data volume onboard, which can exhaust the onboard storage capacity and exceed the downlink capacity. To deal with this challenge, Qian and Hollinger (2004a, 2004b) developed two near-lossless data compression techniques for hyperspectral image data: Successive Approximation Multi-stage Vector Quantization (SAMVQ) and Hierarchical Self-Organizing Cluster Vector Quantization (HSOCVQ). These compression techniques well preserve spectral signature information for remote sensing applications (Hu et al. 2004, Qian et al. 2005). This is because the error introduced in the compression is less than the intrinsic noise of the original data caused by the instrument noise and other error sources such as calibration and atmospheric correction (Qian et al. 2005).

In the study, the effectiveness of random noise reduction of hyperspectral data using the average-smooth method described in Section 13.4.2 was evaluated using SAMVQ and HSOCVQ data compression as an example. This study was to examine whether the onboard data compression should be applied on radiance data right after radiometric conversion or on its noise-removed version.

This could help to decide whether or not the removal of random noise should be applied onboard before compression. The study evaluated the impact of removing random noise on data compression using three statistical based measures and a remote sensing application. The RMSE, SNR, and percentage error (%E) were used as the statistical based measures. A target detection of hyperspectral data was used as an example of the remote sensing application.

13.5.1 Test Data Set

The hyperspectral datacube tested was acquired using the SFSI-II (Neville and Powell 1992, Neville et al. 1996, Borstad et al. 2000) on June 7, 2002, flown at an altitude of 2900 m with a ground sample distance of 3.5 m × 3.5 m and 240 spectral bands between 1200 nm and 2450 nm with a spectral sampling interval of 5 nm. Man-made targets of five different materials: awning, polythene, white tarp, white cotton, and vinyl mat, were deployed in the scene. Figure 13.8 shows the man-made targets presented in the scene of the test datacube observed by the IKONOS panchromatic imager with a 1-m resolution. Seven pieces of awning with varying sizes ranging from 12 m × 12 m to 0.2 m × 0.2 m, four pieces of polythene, four pieces of white tarp, and four pieces of white cotton with varying sizes ranging from 6 m to 0.5 m were deployed. It is understandable that the IKONOS panchromatic imager can only see the targets of size down to 1 m × 1 m (with some blur or barely seeing it) due to its spatial resolution of 1 m. In addition, a 3 m × 3 m piece of white tarp was placed on a large vinyl turf mat (11 m × 14 m). The raw datacube is in 12-bit DN and contains 140 along-track lines. The size of the datacube is 140 lines × 496 pixels per line × 240 bands.

13.5.2 Data Processing Procedure

Figure 13.9 shows the block diagram of the data processing procedure for target detection of the hyperspectral test imagery using the spectral unmixing approach. There are three paths of processing procedure in the figure. In the upper path, which does not contain the onboard data compression,

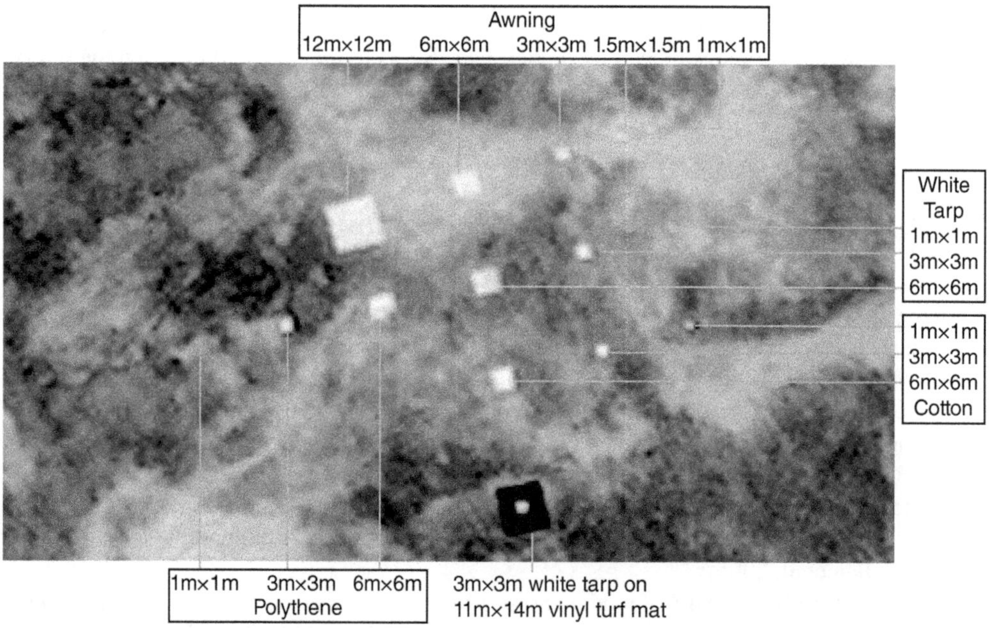

FIGURE 13.8 Man-made targets deployed on ground of the test SFSI-II datacube viewed in a 1-m resolution IKONOS panchromatic scene.

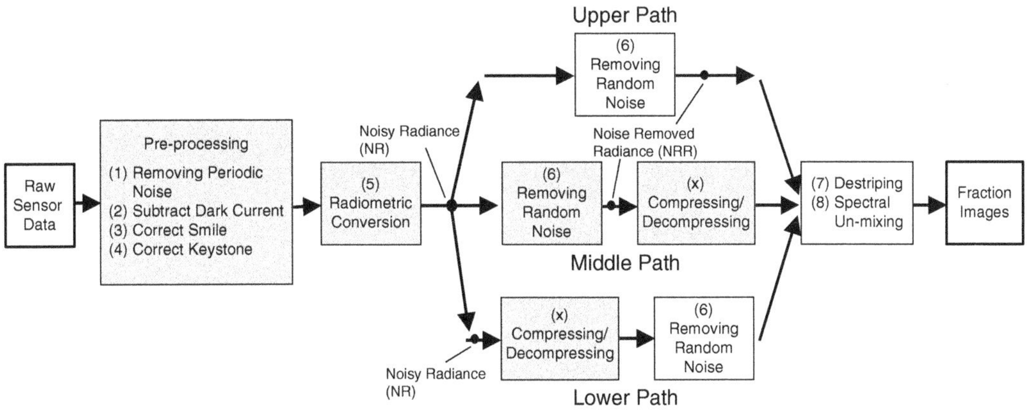

FIGURE 13.9 Block diagram of the data processing procedures.

there are eight (8) processing steps for generating the fraction images corresponding to the end-members from the acquired raw SFSI-II hyperspectral data. They are as follows:

1. *Removing periodic noise of raw data.* Periodic noise varies from aircraft to aircraft. It is probably caused by electrical emission and electromagnetic interference of an aircraft. In order to remove the periodic noise, the data are transformed into Fourier frequencies and a notch filter is applied to remove the peaks that exceed 20 DN values.
2. *Subtract dark current.* The average of the dark current recorded at the beginning and at the end of each flight line is subtracted from each pixel.
3. *Correct smile.* Correction of the spectral shift due to the slit curvature (Neville et al. 2003) using the smile detection/correction algorithm in the ISDAS software tool.
4. *Correct keystone.* Correction of the geometric distortion caused by misalignment of the lens in the focal plane (Neville et al. 2004) using the keystone correction algorithm in the ISDAS software tool.
5. *Radiometric conversion.* A scene-based calibration is performed using calibration coefficients derived from a previous SFSI-II survey to convert the raw DN into radiance. The radiance data are stored in 16-bit.
6. *Removing random noise of radiance data.* Random noise of the radiance data is removed by applying the average-smooth method described in Section 13.4.2. This is done by averaging together spectra from a number of different pixels that are equivalent in light of the known noise characteristics of the sensor. This noise removal processing can significantly improve the data quality for remote sensing applications. Figure 13.10 shows the typical radiance spectra of the test SFSI-II datacube over the site before and after the noise removal. For example, the H_2O band at 1470 nm and its water vapor value cannot be properly estimated before the random noise is removed. They can be well estimated in the noise-removed radiance (NRR) spectra.
7. *De-striping.* Stripes in band images of a radiance datacube are caused by the response difference between the adjacent detector elements. The intensity of the "bad" stripes is readjusted by applying a gain to them in order to correct them using procedure described in Section 13.4.1.
8. *Spectral unmixing.* Nine end-members and their corresponding regions of interest (ROI) are selected manually from the original radiance datacube. They are end-members for the five materials of the man-made targets and four ground features (forest, gravel road, sand, and grass). Each ROI is then used to extract the corresponding end-member spectrum from

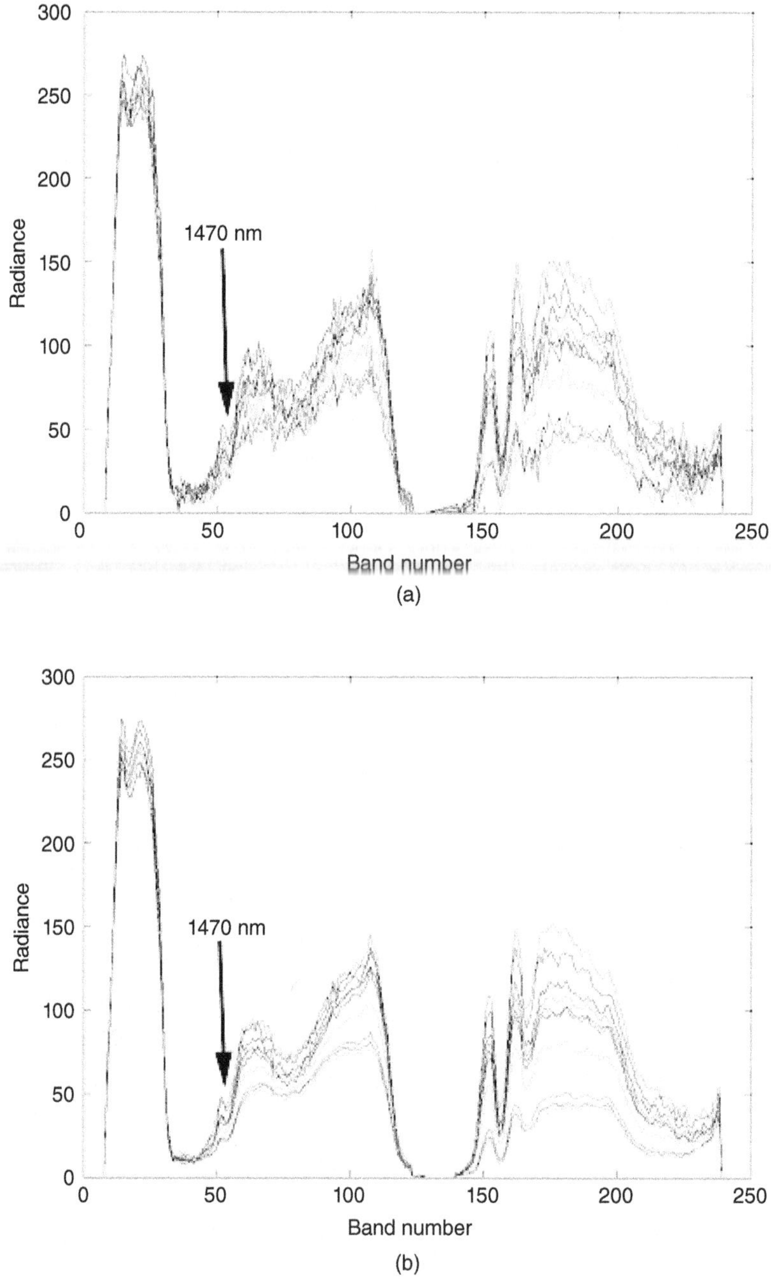

FIGURE 13.10 Selected spectra of the SFSI-II radiance over the target site (a) before and (b) after removal of random noise. Band numbers start at band 0 (1100 nm) and end at band 250 (2500 nm).

the radiance datacube to be evaluated using the constrained linear spectral unmixing algorithm in ISDAS. Fraction images corresponding to each end-member are produced from the radiance datacube and used for the target detection.

Processing steps 1–4 are referred to as preprocessing in this case study. The evaluation of the processing procedure of the middle and lower paths shown in Figure 13.9 is the subject of the case study. In the middle path processing procedure, the noisy radiance (NR) data undergo the removal of random noise before they are sent to compression. In this case, the input of the compression is the NRR data. In the lower path processing procedure, the NR data are sent to compression directly and undergone the removal of random noise after compression. The input of the compression is then the NR data.

The evaluation of the processing procedure of the upper and middle paths (i.e., without and with data compression on radiance data) has been carried out and reported in Qian et al. (2005).

13.5.3 Evaluation Results of Statistical Measures

The impact of removing random noise on data compression was first evaluated using the statistical based measures. Both the NRR and NR datacubes were compressed using SAMVQ at ratios 10:1 to 30:1 and HSOCVQ at ratios 10:1 to 20:1. The RMSE, SNR, and %E used are defined as follows:

$$\text{RMSE} = \sqrt{\left(\frac{1}{N_r \times N_c \times N_b}\right)\sum_{i=1}^{N_r}\sum_{j=1}^{N_c}\sum_{\lambda=0}^{N_b}\left[x_{i,j}(\lambda) - \hat{x}_{i,j}(\lambda)\right]^2} \tag{13.17}$$

$$\text{SNR} = 10\log_{10}\frac{(\text{Signal})^2}{\text{MSE}} \tag{13.18}$$

$$\%E = \left(\frac{1}{N_r \times N_c \times N_b}\right)\sum_{i=1}^{N_r}\sum_{j=1}^{N_c}\sum_{\lambda=0}^{N_b}\left|\frac{x_{i,j}(\lambda) - \hat{x}_{i,j}(\lambda)}{x_{i,j}(\lambda)}\right| \times 100\% \tag{13.19}$$

where $x_{i,j}(\lambda)$ and $\hat{x}_{i,j}(\lambda)$ are amplitudes of the radiance datacube and the reconstructed datacube, respectively, at spatial location (i, j) of band λ. N_r, N_c, and N_b are the total number of lines, the total number of pixels per line, and the total number of bands of the datacube, respectively.

Table 13.1 lists the RMSE, SNR, and %E of the reconstructed datacubes for both the NRR and NR datacubes. It can be seen that compression on the NRR datacube yields better reconstruction fidelity than that on the NR datacube for the same compression algorithm and same compression ratio. The compression on the NRR datacube attains a SNR gain (the column NRR–NR) between 7.4 dB and 9.7 dB and a percentage error reduction (the column NRR–NR) between 1.8% and 2.5%. This is due to the removal of random noise of the radiance data. Accordingly, the NRR datacube is more compressible than the NR datacube. The higher reconstruction fidelity of the compressed NRR datacubes is expected to produce more accurate fraction images for ground-target detection of hyperspectral data.

TABLE 13.1
Statistical Based Measures of Reconstructed Data With the Compression Applied on the Noise-Removed Radiance (NRR) Datacube and Noisy Radiance (NR) Datacube

Compression Algorithm and Ratio	RMSE			SNR (dB)			Percentage Error (%)		
	NRR	NR	NRR-NR	NRR	NR	NRR-NR	NRR	NR	NRR-NR
SAMVQ—10:1	8.93	21.92	−12.99	41.26	31.59	9.67	0.68	2.46	−1.78
SAMVQ—20:1	12.92	28.00	−15.08	38.05	29.46	9.67	0.95	2.96	−1.87
SAMVQ—30:1	15.22	33.35	−18.13	36.62	27.94	8.59	1.09	3.28	−2.19
HSOCVQ—10:1	17.43	32.63	−15.20	35.45	28.08	8.68	1.33	3.64	−2.31
HSOCVQ—20:1	20.85	36.05	−15.20	33.89	27.21	7.37	1.59	4.05	−2.46

13.5.4 Evaluation Results Using a Target Detection Application

Ground target detection of hyperspectral data was used as a remote sensing application to evaluate the impact of removing random noise of radiance data on data compression. The NRR and NR compressed datacubes using SAMVQ at ratios 10:1 to 30:1 and HSOCVQ at ratios 10:1 to 20:1 were decompressed to produce the reconstructed datacubes for evaluation. The reconstructed datacubes are of the same size and format as the original, but have undergone compression. A double-blind test approach was used to reduce self-deception and bias in the evaluation of the impact. The compressed datacubes were named using arbitrary numbers when they were sent back to the user. The user derived the fraction images for detecting the targets and evaluated the impact based on predefined criteria by comparing the products derived from the compressed datacubes with those derived from the original datacube. The user had no knowledge of the compression status of the datacubes evaluated. The following four criteria were predefined and used to assess the impact. One (1) point was scored if a criterion was met.

1. All man-made targets that are present in the fraction images derived from the original datacube are present in the fraction images derived from the compressed datacubes.
2. No targets other than the ones present in the fraction images derived from the original are present in the fraction images derived from the compressed datacubes.
3. The t-test of the distribution of the fraction images derived from the compressed datacubes is not significantly different from that of the fraction images derived from the original datacube at a significance level of 0.01.
4. The percentage standard error (SE%) of a target ROI is <5%. It is used to measure the relative average deviation of the fraction images derived from the compressed datacubes and is defined as follows:

$$SE\% = \frac{\frac{1}{n}\sqrt{\sum_{i=1}^{n}(f_i - \hat{f}_i)^2}}{\bar{f}} \times 100\% \tag{13.20}$$

where f_i is the fraction of an end-member for a pixel in an ROI derived from the original datacube, $\hat{f}_i$ is the fraction of the end-member for the same pixel derived from a compressed datacube, $\bar{f}$ is the mean fraction of the end-member for the ROI of the original datacube, and n is the size of the ROI.

The evaluation was performed on a ROI-by-ROI basis. The full score for a target ROI is 4. The full score for a compressed datacube is 20, as there are five targets. Table 13.2 lists the evaluation score per target, total score per datacube of the blind compressed datacubes with compression applied to the NRR and the NR datacubes.

When the compression was applied to the NRR datacube (i.e., the compression applied after removal of random noise), the compressed datacube SAMVQ 10:1 got a total score 18 of the full score of 20 and was ranked #1. For targets polythene, cotton, and vinyl mat, this datacube got a full score of 4. Both the compressed datacubes HSOCVQ 10:1 and 20:1 got a total score 15 of 20 and were ranked #2. Compressed datacubes SAMVQ 20:1 and 30:1 got a total score of 14 and 13, respectively. Accordingly, they were ranked #3 and #4. The user accepted all the five compressed NRR datacubes based on the predefined criteria.

When the compression was applied to the NR datacube (i.e., the compression applied before removal of random noise), both the compressed datacubes at compression ratio of 10:1 and 20:1 using SAMVQ got a total score of 5 of the full score of 20. The compressed datacube SAMVQ 30:1 got a total score of 3, while the compressed datacubes HSOCVQ 10:1 and 20:1 got a total score of 6 and 7, respectively.

TABLE 13.2
Application Evaluation Score per Target Material, Total Score per Compressed Datacube With Compression Applied to the Noise-Removed Radiance (NRR) Datacube and Applied to the Noisy Radiance (NR) Datacube (the Full Score per Target Material Is 4, the Full Score per Datacube Is 20)

Compression Algorithm and Ratio	Score per Target Material										Total Score per Datacube	
	Awning (ROI 1; Size = 71)		Polythene (ROI 2; Size = 29)		White Tarp (ROI 3; Size = 29)		Cotton (ROI 4; Size = 28)		Vinyl Mat (ROI 5; Size = 80)			
	NRR	NR	NRR	NR	NRR	NR	NRR	NR	NRR	NR	NRR	NR
SAMVQ—10:1	3	2	4	0	3	1	4	1	4	1	18	5
SAMVQ—20:1	3	2	3	0	2	1	3	1	3	1	14	5
SAMVQ—30:1	3	1	2	0	2	0	3	1	3	1	13	3
HSOCVQ—10:1	3	1	4	0	3	1	1	1	4	3	15	6
HSOCVQ—20:1	3	1	4	0	2	1	3	2	3	3	15	7

The total evaluation score of a compressed NR datacube is much smaller than that of a compressed NRR datacube for the same compression ratio and the same compression algorithm. These evaluation scores are consistent with the statistical measures of the compressed datacubes shown in Section 13.5.3. The evaluation results of this application show that removal of random noise of radiance data has significant impact on SAMVQ or HSOCVQ compression. Compression on the NRR radiance datacube by applying the removal of random noise to the radiance datacube before compression yields much higher evaluation scores and better user acceptability than with respect to the NR radiance datacube.

The noise removal algorithm was designed to remove random noise in original hyperspectral data. The SAMVQ and HSOCVQ algorithms are lossy data compression algorithms. They act like a low-pass filter, suppressing the high-frequency noise during the compression (Qian et al. 2004). The reason of the noise removal algorithm not working well on the compressed datacubes probably is that the random noise after compression using SAMVQ and HSOCVQ is different from that in the original (uncompressed) radiance data. The noise removal algorithm that was designed for removing the random noise in the original data cannot effectively handle the noise in the reconstructed data.

REFERENCES

Asner, G. P. and K. B. Heidebrecht, 2003. Imaging spectroscopy for desertification studies: Comparing AVIRIS and E0-1 Hyperion in argentina drylands. *IEEE Transactions on Geoscience and Remote Sensing* 41(6):1283–1296.

Barry, P. 2001. EO-1/Hyperion Science Data User's Guide, Level 1_B. TRW Space, Defense and Information Systems, Redondo Beach, CA.

Borstad, G. A., J. Lim, R. A. Neville, and P. Hauff. 2000. SFSI-2—A new source of airborne hyperspectral SWIR imagery. Proceedings of the 22nd Annual Canadian Remote Sensing Symposium, Victoria, British Columbia, 21–25 August 2000.

CCRS. 2014. Imaging spectrometer data analysis system (ISDAS) (available at ftp://ftp.ccrs.nrcan.gc.ca/ad/Lixin/ISDASv2/helps/IntroISDAS.htm, accessed on August 9, 2019).

Chander, G., B. L. Markham and D. L. Helder. 2009. Summary of current radiometric calibration coefficients for Landsat MSS, TM, ETM+, and EO-1 ALI sensors. *Remote Sensing of Environment* 113:893–903.

Chen, G., T. D. Bu, K. G. Quach, and S.-E. Qian. 2014. Denoising hyperspectral imagery using principal component analysis and block matching 4D filtering. *Canadian Journal of Remote Sensing* 40(1):60–66.

Chen, G. and S.-E. Qian. 2009. Denoising and dimensionality reduction of hyperspectral imagery using wavelet packets, neighbour shrinking and principal component analysis. *International Journal of Remote Sensing* 30(18):4889–4895.

Chen, G. and S.-E. Qian. 2011. Denoising of hyperspectral imagery using principal component analysis and wavelet shrinkage. *IEEE Transaction on Geoscience and Remote Sensing* 49(3):973–980.

Chen, G., S.-E. Qian, and S. Gleason. 2011. Noise reduction of hyperspectral imagery by combining PCA with block-matching 3D filtering. *Canadian Journal of Remote Sensing* 37(6):590–595.

Datt, B., T. R. McVicar, T. G. Van Niel, D. L. B. Jupp, and J. S. Pearlman. 2003. Preprocessing E0-1 Hyperion hyperspectral data to support the application of agricultural indices. *IEEE Transactions on Geoscience and Remote Sensing* 41(6):1246–1259.

Gadallah, F., F. Csillag, and E. Smith. 2000. Destriping multisensor imagery with moment matching. *International Journal of Remote Sensing* 21(12):2505–2511.

Goetz, A. E. H., B. C. Kindel, M. Ferri, and Q. Zheng. 2003. HATCH: Results from simulated radiances, AVIRIS and Hyperion. *IEEE Transactions on Geoscience and Remote Sensing* 41(6):1215–1222.

Hu, B., S.-E. Qian, D. Haboudane, J. R. Miller, A. B. Hollinger, and N. Tremblay. 2004. Retrieval of crop chlorophyll content and leaf area index from decompressed hyperspectral data: The effects of data compression. *Remote Sensing of Environment* 92:139–152.

Kappus, M. E. et al. 2016. Hyperspectral imager for the coastal ocean on the international space station. In *Optical Payloads for Space Missions*, ed. S.-E. Qian, 29–49. West Sussex UK: John Wiley & Sons.

Kaufmann, H. et al. 2016. Environmental mapping and analysis program – A German hyperspectral mission. In *Optical Payloads for Space Missions*, ed. S.-E. Qian, 161–181. West Sussex UK: John Wiley & Sons.

Khurshid, K. S., K. Staenz, L. Sun, R. Neville, H. P. White, A. Bannari, C. M. Champagne, and R. Hitchcock. 2006. Preprocessing of EO-1 Hyperion data. *Canadian Journal of Remote Sensing* 32(2):84–97.

Luo, G., G. Chen, L. Tian, K. Qin and S.-E. Qian. 2016. Minimum noise fraction versus principal component analysis as a preprocessing step for hyperspectral imagery denoising. *Canadian Journal of Remote Sensing* 42(2):106–116.

Meini, M., A. Bini, L. Giunti, E. Fossati, and R. Formaro. 2016. Hyperspectral payload for Italian PRISMA programme. In *Optical Payloads for Space Missions*, ed. S.-E. Qian, 183–213. West Sussex UK: John Wiley & Sons.

Murchie, S., et al. 2007. Compact reconnaissance imaging spectrometer for mars (crism) on mars reconnaissance orbiter (MRO). *Journal of Geophysics Research* 112(E05S03):1–57.

Neville, R. A., R. Marois, N. Rowlands, I. P. Powell. 1996. SFSI: The CCRS SWIR imaging spectrometer. *Proceedings of SPIE* 2819: 345–353.

Neville, R. A. and I. Powell. 1992. Design of SFSI: An imaging spectrometer in the SWIR. *Canadian Journal of Remote Sensing* 18(4):210–223.

Neville, R. A., L. Sun, and K. Staenz. 2003. Detection of spectral line curvature in imaging spectrometer data. *Proceedings of SPIE* 5093:144–154.

Neville, R. A., L. Sun, and K. Staenz. 2004. Detection of keystone in image spectrometer data. *Proceedings of SPIE* 5425:208–217.

Neville, R. A., L. Sun, and K. Staenz. 2008. Spectral calibration of imaging spectrometers by atmospheric absorption feature matching. *Canadian Journal of Remote Sensing* 34(S1):S29–S42.

Othman, H. and S.-E. Qian. 2006. Noise reduction of hyperspectral imagery using hybrid spatial-spectral derivative-domain wavelet shrinkage. *IEEE Transactions on Geoscience and Remote Sensing* 42(2):397–408.

Othman, H. and S.-E. Qian. 2008. Evaluation of wavelet deniosed hyperspectral data for remote sensing. *Canadian Journal of Remote Sensing* 34(S1):59–67.

Palubinskas, G., M. Bachmann, E. Carmona, B. Gerasch, H. Krawczyk, A. Makarau, M. Schneider, and P. Schwind. 2017. Image products from a new German hyperspectral mission EnMap. *Proceedings of IGFT* 2017—*Imaging & Geospatial Technology Forum* 2017, ASPRS Annual Conference.

Pearlman, J., P. Barry, C. Segal, J. Shepanski, D. Beiso, and S. Carman. 2003. Hyperion, a space-based imaging spectrometer. *IEEE Transactions on Geoscience and Remote Sensing* 41(6):1160–1173.

Qian, S.-E. 2011. Enhancing space-based signal-to-noise ratios without redesigning the satellite. *SPIE Newsroom.*

Qian, S.-E. 2013. *Optical Satellite Signal Processing and Enhancement.* Bellingham: SPIE Press.

Qian, S.-E., M. Bergeron and A. Hollinger. 2004. Evaluation of near lossless feature of two on-board data compression algorithms: SAMVQ and HSOCVQ. *WSEAS Transactions on Systems* 5(3):2153–2158.

Qian, S.-E. and A. B. Hollinger. 2004a. System and method for encoding/decoding multi-dimensional data using successive approximation multi-stage vector quantization. *U. S. Patent No. 6,701,021 B1*, issued on March 2, 2004.

Qian, S.-E. and A. B. Hollinger. 2004b. System and method for encoding multi-dimensional data using Hierarchical Self-Organizing Cluster Vector Quantization. *U.S. Patent No.6,724,940 B1*, issued on April 20, 2004.

Qian, S.-E., A. Hollinger, M. Bergeron, I. Cunningham, C. Nadeau, G. Jolly and H. Zwick. 2005. A multi-disciplinary user acceptability study of hyperspectral data compressed using onboard near lossless vector quantization algorithm. *International Journal of Remote Sensing* 26(10):2163–2195.

Qian, S.-E. and J. Lévesque. 2009. Target detection from noise-reduced hyperspectral imagery using spectral unmixing approach. *Journal of Optical Engineering* 48(2):1–11 (026401).

Qian, S.-E., J. Lévesque, and R. A. Neville. 2006. Effect of removing random noise of radiance data using smoothing on data compression onboard a hyperspectral satellite. *WSEAS Transactions on Systems* 1(5):219–224.

Qian, S.-E., J. Lévesque, and R. Rashidi Far. 2011. Assessment of noise reduction of hyperspectral imagery using a target detection application. *International Journal of Remote Sensing* 32(12):3267–3284.

Qian, S-E. and H. Othman. 2011. Method and System for Increasing Signal-to-Noise Ratio. *European Patent No. EP 1 982 305 B1*, granted on June 15, 2011.

Qian, S-E. and H. Othman. 2013. Method and System for Increasing Signal-to-Noise Ratio. *U.S. Patent No. 8,358,866 B2* granted on Jan. 22, 2013.

Qian, S-E. and H. Othman. 2014. Method and System for Increasing Signal-to-Noise Ratio. *Canadian Patent No. 2,640,683* granted on Oct. 21, 2014.

Qian, S.-E. and H. Othman. 2011. Method and system for increasing signal-to-noise ratio. European Patent No. EP 1 982 305 B1, granted on June 15, 2011.

Schroeder, D. J. 2000. *Astronomical optics*, 2nd ed. San Diego, CA: Academic Press.

Staenz, K., T. Szeredi, and J. Schwarz. 1998. ISDAS - A system for processing/analyzing hyperspectral data. *Canadian Journal of Remote Sensing* 24(2):99–113.

Staenz, K. and D. J. Williams. 1997. Retrieval of surface reflectance from hyperspectral data using a look-up table approach. *Canadian Journal of Remote Sensing* 23:354–368.

Sun, L., R. Neville, and K. Staenz. 2006. Automatic destriping of Hyperion imagery based on spectral moment matching. *Canadian Journal for Remote Sensing* 34(S1):S68–S81.

Thuillier, G., M. Herse, S. Labs, T. Foujols, W. Peetermans, D. Gillotay, P. C. Simon, and H. Mandel. 2003. The solar spectral irradiance from 200 to 2400 nm as measured by SOLSPEC spectrometer from the ATLAS 123 and EURECA missions. *Solar Physics* 214(1):1–22.

14 Atmospheric Correction

14.1 ATMOSPHERIC EFFECTS ON HYPERSPECTRAL DATA

A hyperspectral sensor measures the radiation of each pixel reflected from the surface within its field of view (FOV). Both the spatial and spectral information of a scene are available with hyperspectral data. There is growing interest in hyperspectral remote sensing for research and applications in a variety of fields, including geology, agriculture, forestry, coastal and inland water studies, environment hazards assessment, and urban studies. In hyperspectral data analysis, materials of the scene are identified by using their reflectance spectra. In order for hyperspectral imaging data to be used for quantitative remote sensing, such as land surfaces and ocean color, the measured radiation data must be converted to surface reflectance. Because the solar radiation on the sun-surface-sensor path in the 0.4–2.5 μm visible and near-infrared (VNIR) and shortwave infrared (SWIR) spectral regions is subjected to absorption and scattering by atmospheric gases and aerosols, the acquired hyperspectral imaging data are impacted by atmospheric effects. These effects include:

- absorption by atmospheric water vapor and gases (e.g., oxygen and ozone, etc.),
- atmospheric molecular scattering (Rayleigh effect), and
- aerosol absorption and scattering.

Accurate removal of these atmospheric effects is essential for hyperspectral applications. This process is referred to as atmospheric correction.

In addition, the influence of the solar zenith angle, sensor viewing geometry, and terrain have to be taken into account in the data processing to retrieve the surface reflectance.

Among the approximately 30 atmospheric gases, only 7 gases, namely oxygen (O_2), methane (CH_4), carbon dioxide (CO_2), ozone (O_3), nitrous oxide (N_2O), carbon monoxide (CO), and nitrogen dioxide (NO_2) produce observable absorption features in hyperspectral data. Gas absorption effects vary strongly with wavelength and gas properties, and can have a significant impact on the received solar flux.

Approximately half of the spectral region between 0.4 μm and 2.5 μm is affected by atmospheric water vapor absorption. The absorption effects from the other seven gases are generally located in much narrower wavelength intervals. Figure 14.1 shows examples of simulated atmospheric transmittances for atmospheric water vapor absorption and the gases.

Atmospheric water vapor absorption in the VNIR region is characterized by a number of bands of various strengths and spectral widths. Two very weak absorption bands are located at 0.60 μm and 0.66 μm. Slightly stronger and more significant bands are located at 0.73 μm, 0.82 μm, and 0.91 μm. At 0.94 μm and 1.14 μm, water vapor absorption is strong enough that measurements at the band centers and off-center locations can be used to derive the total column water vapor (CWV; Gao and Goetz 1990, Kaufman and Gao 1992). Water vapor absorption near 1.375 μm, 1.90 μm, and 2.5 μm is strong enough to make retrieval of the surface reflectance difficult or impossible. Therefore, these bands are used to obtain information about high-altitude moisture and cloud effects (Gao et al. 1998), because much of the signal at these wavelengths comes from the mid-to-upper troposphere. The large number of strongly absorbing water vapor bands makes the adjustment for water vapor a significant focus of atmospheric correction processes.

At 0.76 μm, a strong and narrow oxygen absorption line is present. Two weaker oxygen lines are located at 0.688 μm and 1.27 μm. Absorption peaks by methane (CH_4) are present from 2.2 μm to

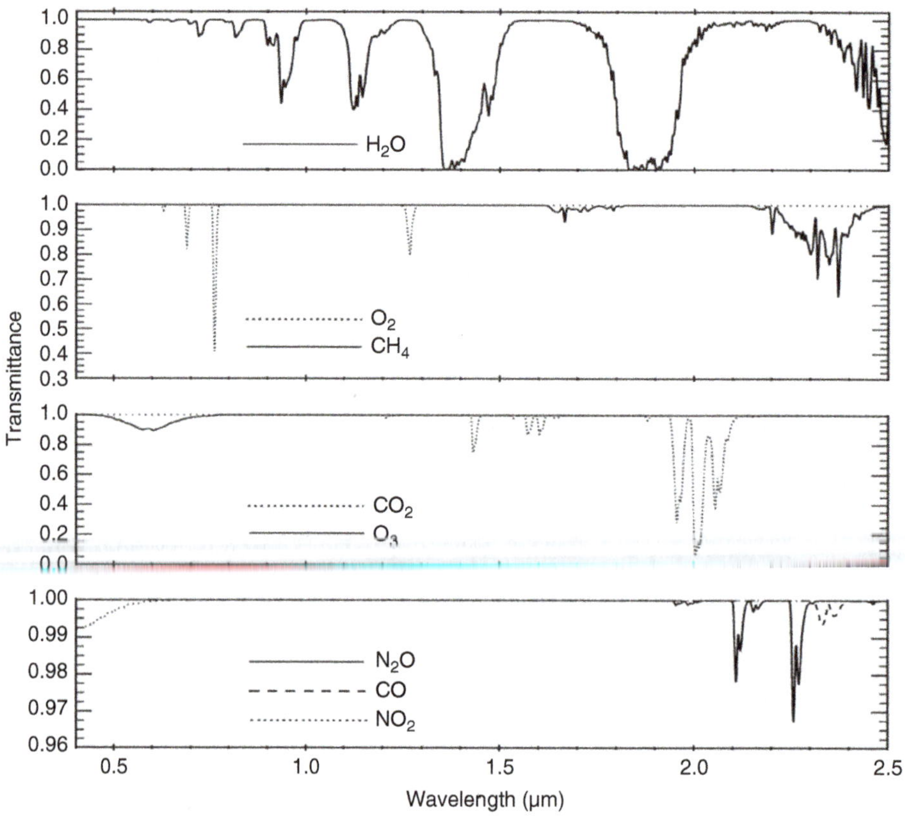

FIGURE 14.1 Transmittance spectra of atmospheric water vapor and sever gases. (Courtesy of Naval Research Laboratory.)

2.5 μm. CO_2 absorbs strongly from 1.9 μm to 2.1 μm. The effect of the CO_2 absorption is partially negated by a strong overlapping water vapor absorption band. CO_2 also exhibits a weak absorption line at 1.43 μm. Weak ozone absorption is present from 0.5 μm to 0.7 μm. Nitrous oxide (N_2O) presents two very week absorption peaks at 2.11 μm (about 2.2% deep) and 2.26 μm (3.3% deep).

While water vapor absorption clearly has the largest effect on the transmittance. Aerosol absorption, on the other hand, is relatively minor compared to molecular absorption, and is generally viewed as a smoothly varying continuous function. Typical absorption loss due to aerosols is limited to a few percent, with minimum absorption occurring for maritime aerosols and maximum absorption found in urban aerosols that tend to contain significant amounts of carbon (Griffin and Burke 2003).

Atmospheric gaseous molecules and aerosols scatter solar radiation. The short wavelength region between 0.4 μm and 0.7 μm is strongly affected by molecular scattering (Rayleigh scattering). Its effect decreases rapidly with increasing wavelength (λ^{-4}). Scattering from aerosols plays an important role of atmospheric effects and must be handled properly by atmospheric correction process if an accurate estimate of the surface reflection is to be obtained. Aerosol effects vary widely with atmospheric moisture, aerosol type, or makeup and optical depth or visibility. Under cloud-free conditions, which are usually desired for hyperspectral data collections, aerosols provide the bulk of the solar scattering. The aerosol scattering effect also decreases with increasing wavelength, but at a slower rate (typically λ^{-2} to λ^{-1}; Gao et al. 2009).

The application of atmospheric correction to retrieve surface reflectance can also lead to better characterization of the atmosphere. Secondary products of the physics-based atmospheric correction approaches can include a map of the integrated CWV and scene aerosol type and visibility. The spectral information itself can be used to provide quantitative information about atmospheric features such as clouds, dust, or smoke (Griffin et al. 2000).

14.2 STATISTICS-BASED ATMOSPHERIC CORRECTION APPROACHES

Various approaches exist for atmospheric correction of hyperspectral data. Most fall into two basic categories: statistical (or empirical) and physics based. The former use a priori knowledge of the reflectance characteristics of specific reference objects in a scene to develop statistical relationships between the at-sensor observations and the surface reflectance. Scene-based or empirical atmospheric correction methods are all statistics-based approaches. They were developed during the mid-1980s to remove atmospheric effects mostly from hyperspectral data acquired by airborne instruments for the derivation of relative surface reflectance spectra.

14.2.1 Empirical Line Method

The empirical line method (ELM) is a commonly used statistics-based atmospheric correction model (Roberts et al. 1985, Conel et al. 1987b). The ELM developed by Conel et al. (1987a) requires field measurements of reflectance spectra for at least one bright target and one dark target within the scene. The hyperspectral data of the scene are linearly regressed against the field-measured reflectance spectra to derive the gain and offset curves, which is akin to vicarious calibration. This process is equivalent to removing the solar radiance and the atmospheric path radiance from the measured signal. Gain and offset factors for each spectral band derived from this relationship are applied to all other pixels in the scene to remove the atmospheric component from the measurements for the derivation of surface reflectance for the entire scene.

This method produces spectra that are most comparable to the reflectance spectra measured in the field or in the laboratory (Aspinall et al. 2002). However, if changes occur in the atmospheric properties outside the area used for the empirical line approach, which is often the case, the reflectance spectra will contain atmospheric features. However, ELM does not properly account for topographic variations, which can result in elevation-dependent residual atmospheric absorptions.

14.2.2 Internal Average Relative Reflectance

Other scene-based statistical techniques are used to adjust or normalize the raw radiance data. These techniques tend to be more of a calibration tool than an atmospheric correction method. One such technique is the internal average relative (IAR) reflectance approach developed by Kruse (1988). The IAR approach calculates the average spectrum of an entire scene first. Then, spectrum of each pixel in the scene is divided by the average spectrum to estimate the relative reflectance spectrum for the pixel. This approach uses the properties of the data themselves to calculate an approximation of the reflectance. The resulting spectra represent reflectance relative to the average spectrum and resemble to laboratory spectra acquired of the same materials. However, the average spectrum of the scene itself may have spectral character related to absorptions of some particular materials. This can adversely affect the appearance of the IAR reflectance spectra and limit their usefulness.

14.2.3 Flat-Field Correction

The flat-field correction (FFC) approach developed by Roberts et al. (1986) is another scene-based statistical technique that adjusts or normalizes the raw radiance data. It assumes that there is an area in the scene that has spectrally neutral reflectance, i.e., the spectrum has little variation with wavelength. The mean spectrum of the flat field area is then used for the derivation of relative reflectance spectra of other pixels in the scene.

Both IAR and FFC approaches do not need any field measurements of reflectance spectra of surface targets. However, the derived relative reflectance spectra often contain absorption features that are not present in reflectance spectra of comparable materials measured in the field or in the laboratory (Clark and King 1987). The reason is that the mean spectrum of the entire scene or the flat field area often contains absorption effects of surface materials and is not 100% spectrally neutral. The use of such mean spectrum in the derivation of relative reflectance spectra of other pixels can introduce broad absorption bands in the resulting spectra.

14.2.4 Cloud Shadow Method

For atmospheric corrections over water surfaces, an empirical cloud shadow method was developed in the 1990s (Reinersman et al. 1998). This method calculates the differences between the total radiance values measured by a sensor over cloud shaded pixels and the neighboring pixels having similar optical properties. The differences are then used for the removal of the nearly identical atmospheric radiance contributions from the measured data. This method discounts the importance of adjacency effects.

Lee et al. (2005) improved the implementation of the cloud shadow method. In the method of Reinersman et al. (1998), the atmospheric effects, L_{sky}, is calculated from pairs of adjacent pixels that are under cloud shadow and under the sun with the atmospheric transmittance, t, and downwelling irradiance, E_d, calculated from radiance transfer with models for aerosols, which required the sensor to be well calibrated radiometrically and spectrally. In Lee et al. method, t and E_d are not explicitly derived. Instead, the product of t and E_d is estimated using the reflected radiance from the top of the cloud.

14.2.5 Dense Dark Vegetation Algorithm

Kaufman and Sendra (1988) proposed an algorithm for automatic atmospheric correction of satellite imagery. The algorithm is based solely on the satellite image being corrected and on climatology of the area. However, it requires that the image has a sufficient proportion of pixels with low reflectances, such as surfaces covered with dense dark vegetation. Once the presence of such pixels is established, the algorithm automatically chooses these pixels, derives the atmospheric optical thickness (a measure of the amount of haze), and conducts atmospheric correction of the image automatically. The algorithm is sensitive to the assumed reflectance of the dense dark vegetation. It is not very sensitive to the assumed aerosol characteristics, the accuracy of satellite calibration, or the knowledge of the exact fraction of the image covered by the dense dark vegetation.

14.3 RADIATIVE TRANSFER MODELING FOR PHYSICS-BASED ATMOSPHERIC CORRECTION

When true atmospheric correction is desired and direct information about a scene is unknown, statistics-based methods may not be applicable. For these cases, a more robust method is needed to retrieve the surface reflectance. Typically, physics-based radiative transfer models are chosen for this task. A theoretical modeling technique, which simulates explicitly the absorption and scattering effects of atmospheric gases and aerosols, is required in order to retrieve land surface reflectance spectra from hyperspectral data. The choice between empirical and radiative transfer based atmospheric correction is often a question of computation efficiency.

For physics-based models, the relationship between the surface reflectance, ρ, and the measured at-sensor radiance, L_t, needs to be set up. The radiances measured by a hyperspectral sensor on-orbit include surface-reflected solar radiances and atmospheric path radiances due to Rayleigh and aerosol scattering. These radiances are often converted to "apparent reflectances" (Tanre et al. 1986, Gao et al. 1993). The definition of apparent reflectance is given by:

$$\rho_t^*(\lambda,\theta,\phi,\theta_0,\phi_0)=\frac{\pi L_t(\lambda,\theta,\phi,\theta_0,\phi_0)}{\mu_0 F_0(\lambda)} \tag{14.1}$$

where L_t is the measured at-sensor radiance; λ, wavelength; θ, the sensor zenith angle; ϕ, the sensor azimuth angle; θ_0, is the solar zenith angle; ϕ_0, the solar azimuth angle; μ_0, the cosine of solar zenith angle; and F_0, the solar flux at the top of the atmosphere when the solar zenith angle is equal to zero.

Apparent reflectance is a measure of the reflectivity for the atmosphere and surface system. According to Tanre et al. (1986), when the surface is assumed to be Lambertian and the adjacency effect is neglected, $\rho_t^*(\lambda,\theta,\phi,\theta_0,\phi_0)$ can be expressed approximately as:

$$\rho_t^*(\lambda,\theta,\phi,\theta_0,\phi_0)=\left[\rho_{atm}^*(\lambda,\theta,\theta_0,\phi_0)+\frac{t_d(\lambda,\theta_0)t_u(\lambda,\theta)\rho(\lambda)}{1-s(\lambda)\rho(\lambda)}\right]T_g(\lambda,\theta,\theta_0), \tag{14.2}$$

where ρ_{atm}^* is the atmosphere path reflectance; t_d, the downward scattering transmittance; t_u, the upward scattering transmittance; ρ, the surface reflectance; s, the spherical albedo of the atmosphere; and T_g, the total gaseous transmittance in the sun-surface-sensor path.

The first term in the brackets, ρ_{atm}^*, is the contribution from atmospheric scattering to the measured apparent reflectance. The second term in the brackets, $t_d \cdot t_u \cdot \rho/(1 - s\rho)$, is the contribution from surface reflection to the measured apparent reflectance. The term T_g contains absorption bands of all atmospheric gases, as shown in Figure 14.1. Equation 14.2 treats the atmospheric scattering and gaseous absorption processes as two independent processes. The coupling effects between the atmospheric scattering and gaseous absorption are neglected. In the real atmosphere, the atmospheric scattering and gaseous absorption processes occur simultaneously. The coupling effects are small in regions where the atmospheric gaseous absorptions are weak and in regions where the scattering effects are small.

Solving Equation 14.2 for surface reflectance ρ and simplifying the notations for relevant quantities yields:

$$\rho=\frac{\frac{\rho_t^*}{T_g}-\rho_{atm}^*}{t_d t_u+s\left(\frac{\rho_t^*}{T_g}-\rho_{atm}^*\right)} \tag{14.3}$$

By simulating the atmospheric quantities T_g, ρ_{atm}^*, t_d, t_u, and s with radiative transfer models, the reflectance ρ (assumed for the horizontal Lambertian surface) can be retrieved from the measured radiance L_t using Equations 14.1 and 14.3. There are now limited case studies (e.g., Kaufman et al. 1997, Guanter et al. 2005) on the retrieval of aerosol models and optical depths from hyperspectral data for modeling aerosol absorption and scattering effects.

14.4 RADIATIVE TRANSFER MODELING BASED METHODS FOR LAND SCENES

14.4.1 Atmosphere Removal Algorithm (ATREM)

Atmosphere Removal (ATREM) proposed by Gao et al. (1993) is the first radiative transfer modeling based atmospheric correction algorithm for hyperspectral data. It simulates explicitly the absorption and scattering effects of atmospheric gases and aerosols and retrieves land surface reflectance

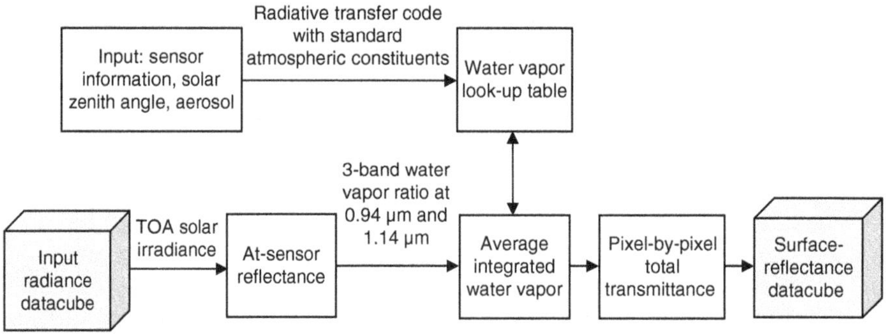

FIGURE 14.2 Schematic flow of the radiative transfer modeling based ATmospheric REMoval (ATREM) method.

spectra from hyperspectral data without the need for field measurements of reflectance spectra. The development of ATREM marked a major advance in atmospheric correction of hyperspectral data. The band model version of the ATREM source code (Version 3.1) was distributed to user community worldwide free of charge in the early 1990s. This made it possible for a large land remote sensing community to convert hyperspectral radiance data to surface reflectances for a variety of applications and hyperspectral research activities.

Figure 14.2 shows the schematic flow of the ATREM program. ATREM uses a narrow band spectral model to derive the gaseous transmittance and the Simulation of the Satellite Signal in the Solar Spectrum (5S) code (Tanre et al. 1986) to compute the scattering effect due to atmospheric molecules and aerosols by using a user-selected aerosol model. The individual gas concentrations are assumed to be uniform across the scene (except for water vapor) and are treated separately. In this manner, only a single scene transmittance spectrum is calculated for each uniform gas. ATREM provides a simple input parameter interface and relatively fast execution. Input parameters are limited to basic atmospheric parameters (aerosol visibility and model type, atmospheric model, and ozone concentration) and solar and viewing geometry.

Because water vapor is spatially variable, the integrated water vapor amount is derived on a pixel-by-pixel basis from the water vapor absorption bands 0.94 μm and 1.14 μm using a three-band ratio technique. The band ratio value is computed as the water vapor band radiance relative to the adjacent window radiance in the near-infrared (NIR) region. The methodology of retrieval of water vapor using a three-band ratio follows.

The water vapor bands are well characterized by the narrow band hyperspectral data. Typically, three to five bands are chosen to represent the strongest water vapor bands. Two more sets of bands located in the high-transmission regions, where absorption due to water vapor is small, are selected, one set on each side of the strongest water vapor absorption bands. A mean radiance is computed for each of the three band sets. The two window radiances are combined by using a spectrally weighted average, and a ratio of the water vapor band radiance to the overall window radiance is computed, producing a pseudo-transmittance value. This value is compared with a precomputed table of three-band ratios for the same band specifications over a range of CWV amounts.

Ratios are calculated for three water vapor bands: the default 0.94 μm and 1.14 μm bands and the weaker band 0.91 μm band. Each band has a distinct array of ratio values, with the 0.91-μm band having the smallest range of ratios for the given span of CWV values. The 1.14-μm band provides the greatest sensitivity to water vapor variations for amounts <3 cm. For CWV amounts greater than about 4 cm or 5 cm, the ratio values change very little. Thus, small variations in the ratio values can lead to large changes in the retrieved CWV. Significant errors are possible when dealing with very moist scenes.

Water vapor transmittance and CWV amounts are determined from the closest match to the measured three-band ratio values and are used to define the water vapor characteristics of the pixel. This procedure is repeated for every pixel in the image. The operation is fast, since it involves a simple ratio calculation and a look-up table (LUT) search for each pixel.

The transmission spectra of carbon dioxide (CO_2), ozone (O_3), nitrous oxide (N_2O), carbon monoxide (CO), methane (CH_4), and oxygen (O_2) in the 0.4–2.5 μm region are simulated based on the solar and observational geometry.

The measured radiances are divided by solar irradiances above the atmosphere to obtain the apparent reflectances. The surface reflectances are derived from the apparent reflectances using the simulated atmospheric gaseous transmittances and the simulated molecular and aerosol scattering data. The output of the ATREM algorithm includes a surface reflectance datacube and a water vapor image. The surface reflectance, ρ, defined in Equation 14.3 was referred to as the *scaled surface reflectance* by Gao et al. (1993) because of the assumption of horizontal Lambertian surface during the retrievals. The scaled surface reflectance is different from the *absolute* surface reflectance by a multiplicative factor, which is related to the slope and aspect of the surface and the bidirectional reflectance property of the surface.

Major upgrades to the ATREM code were made in the late 1990s and early 2000s. The band model was replaced with a line-by-line atmospheric transmittance model (Gao and Davis 1997) and the HITRAN (high-resolution transmission molecular absorption database) 2000 line database (Rothman et al. 2003, 2005). The 5S computer code was replaced with the newer 6S code (Vermote et al. 1997) for modeling atmospheric scattering effects. A module for modeling atmospheric NO_2 absorption effects in the 0.4–0.8 μm spectral region was added. The algorithm needs to have an input for column NO_2 amount (typically about 5×10^{15} molecules/cm^2).

14.4.2 Fast Line-of-Sight Atmospheric Analysis of Spectral Hypercubes (FLAASH)

FLAASH (Fast Line-of-sight Atmospheric Analysis of Spectral Hypercubes) is a MODTRAN-based atmospheric correction algorithm (Adler-Golden et al. 1998). It utilizes the full functionality of Moderate Resolution Transmittance 4 (MODTRAN4) radiative-transfer code (Berk et al. 1998), which includes a complete package of transmittance and scattering methods, including a full accounting for adjacency effects (the scattering from adjacent pixels into the current pixel in the FOV) associated with atmospheric scattering. Compared to previous versions of MODTRAN, the new correlated-*k* radiation transport algorithm in MODTRAN4 provides improved accuracy in treating molecular absorption over scattering paths. Figure 14.3 shows the schematic flow of the FLAASH algorithm.

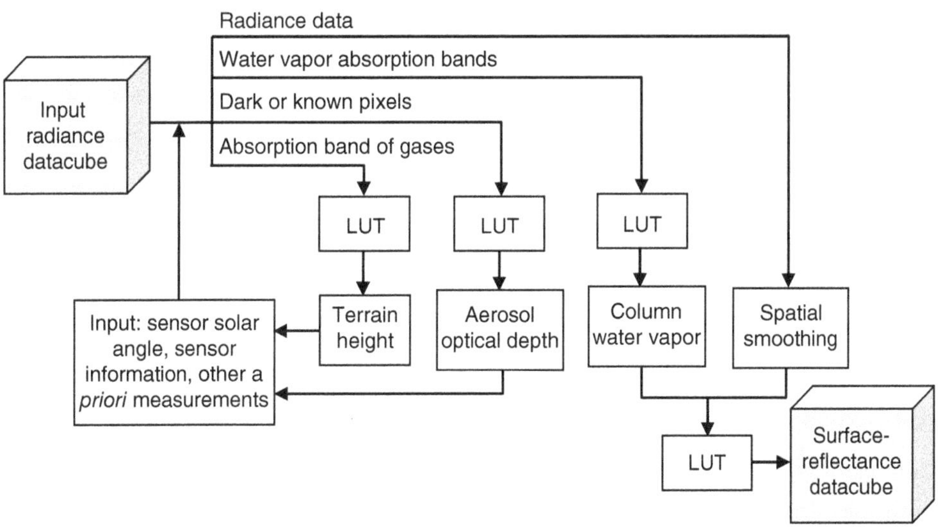

FIGURE 14.3 Schematic flow of the FLAASH atmospheric removal method.

To minimize the number of simulations (i.e., MODTRAN runs) required to generate the LUTs, a physics-based parameterization of the radiance-reflectance relationship is used. This relationship can vary across the scene due to variations in water vapor column density. Therefore, as in the ATREM the water vapor column is first determined for each pixel, the result is used as an input to the surface reflectance retrieval algorithm.

FLAASH also utilizes a three-band ratio technique but slightly different approach to estimate the water vapor transmittance of the intervening atmosphere. Its computations depend on separate within-band and window radiances and not the ratio of the two. This process requires a onetime calculation of a 2D LUT and an interpolation step for each pixel. The LUT is generated from MODTRAN calculations of the full gaseous absorption characteristics and aerosol/molecular scattering for each spectral band; therefore, taking the ratio of the two values is not required.

The radiance spectra are extracted from the hyperspectral radiance data and compared against the MODTRAN LUTs on a pixel-by-pixel basis to determine scaled surface reflectance. FLAASH offers the additional option of correcting for light scattered from adjacent pixels. Spatially averaged reflectance is used to account for the *adjacency effect*—radiance contributions that, because of atmospheric scattering, originate from parts of the surface not in the direct FOV of the sensor. FLAASH provides additional flexibility when compared to the other atmospheric correction programs in that it allows custom radiative transfer calculations for a wider range of conditions, including off-nadir viewing and all MODTRAN standard aerosol models.

FLAASH has multiple spectral resolution options to increase processing speed. An order of magnitude reduction in computation time can be achieved by using MODTRAN4's lower resolution (15 cm^{-1}) option. In addition to retrieving the surface reflectance and the CWV amount (for each pixel in a scene), it also offers the options of retrieving the aerosol optical depth (AOD; using a dark-pixel approach), the terrain elevation (using the 0.76 µm O_2 line), and of masking out clouds.

14.4.3 High-Accuracy Atmospheric Correction for Hyperspectral Data (HATCH)

The High-accuracy ATmospheric Correction for Hyperspectral data (HATCH) was developed by the same group as ATREM at University of Colorado in Boulder in view of the fact that the major techniques employed in the original ATREM were outdated as a result of new advancements in the area of atmospheric radiative transfer and by utilizing the latest advancements in the area of radiative transfer (Qu et al. 2003).

At the time when the original ATREM was developed, HITRAN 2000 was unavailable. HATCH utilizes HITRAN 2000, which provides more accurate molecular line parameters. In addition, ATREM employs the multiplication rule to handle transmittance in spectral regions where multiple gas absorptions are present. This results in less accurate transmittance calculation in these regions, such as in the spectral region at around 2.0 µm where both water vapor and CO_2 strongly absorb.

ATREM utilizes three-band ratio technique for water vapor amount retrieval. The prerequisite for this technique is that the surface reflectance spectrum must be linear in wavelength in the water vapor absorption region to be used. This introduces retrieval errors when the surface reflectance spectrum is not linear for wavelength, for instance, when the surface contains iron-rich soil and/or vegetation.

As a new atmospheric correction model, HATCH uses the new advancements in the area of radiative transfer. First, it utilizes HITRAN 2000 based correlated-*k* method, which allows accurate transmittance calculation in regions where multiple gases absorb strongly to accurately handle atmospheric transmittance.

Second, it utilizes multigrid discrete ordinate method, referred to as fast radiative transfer equation solver, to explicitly account for radiation in a nonhomogeneous scattering and absorbing atmosphere.

Third, a new technique called the *smoothness test* is introduced for retrieval of CWV amount, which eliminates the linearity assumption of the surface spectra in the conventional three-band ratio technique used by ATREM.

Finally, it utilizes an automatic spectral calibration of the sensors based on the acquired data and using the smoothness test technique.

The version developed for pushbroom hyperspectral sensors such as Hyperion is referred to as HATCH-2d. It takes into account the spectral calibration variation from one detector column to another of the 2D detector array. In performing atmospheric correction, HATCH-2d uses the corresponding individual spectral response function (center wavelength and full width half maximum of the spectral band) for each detector column instead of using one across the entire detector array. The automatic spectral calibration of the sensor is also performed individually for each detector column.

HATCH derives surface reflectance on a pixel-by-pixel basis, since the water vapor amount varies spatially.

14.4.4 Atmosphere Correction Now (ACORN)

Atmosphere Correction Now (ACORN) is a commercial atmospheric correction algorithm. It uses licensed MODTRAN4 technology (Berk et al. 2003) to produce surface reflectance without ground measurements. The software package provides an atmospheric correction of hyperspectral and multispectral data in the 0.4–2.5 μm spectral range (Kruse 2004).

ACORN uses LUTs calculated with the MODTRAN4 radiative transfer code to model atmospheric gas absorption as well as molecular and aerosol scattering effects, converting the calibrated sensor radiance measurements to apparent surface reflectance.

Water vapor is estimated from the data on a pixel-by-pixel basis using the water vapor absorption bands at 0.94 and/or 1.14 μm. A LUT for a range of water column vapor densities is generated using MODTRAN4 and then fitted in a least-squares sense against the hyperspectral data. A key feature of ACORN is full spectral fitting to solve for the overlap of absorptions between water vapor and liquid water in surface vegetation. Visibility is estimated from the hyperspectral data using nonlinear least-squares spectral fitting between the sensor radiance spectra and MODTRAN-modeled radiance with the AOD as the primary fitting parameter. The two-way transmitted radiance and atmospheric reflectance are calculated for each pixel using MODTRAN and the derived water vapor, pressure elevation, and AOD estimations. Apparent surface reflectance is derived from the total upwelling spectral radiance for a given atmosphere using a variant of the radiative transfer.

The main outputs of ACORN are a scaled surface reflectance datacube and a water vapor image. An image showing an estimate of leaf-water is also optionally produced. ACORN artifact suppression options include automated wavelength correction, removal of noisy channels, and *polishing* of residual errors.

14.4.5 5D LUT Approach in Imaging Spectrometer Data Analysis System (ISDAS)

Staenz and Williams (1997) developed a technique of retrieving surface reflectance from airborne and spaceborne hyperspectral data using a 5-dimension (5D) LUT procedure. This atmospheric correction technique has been integrated in the Imaging Spectrometer Data Analysis System (ISDAS) software tool suite (Staenz et al. 1998), which allows it to run with other tools that are relevant to surface reflectance retrieval, such as data preprocessing, post-processing, and visualization.

This technique can be applied in the VNIR and SWIR region up to 2.7 μm. The five dimensions of the LUT are as follows:

1. Pixel cross-track positions (defining the variations in viewing zenith and azimuth angles).
2. Terrain elevation.

3. Aerosol optical depth (AOD).
4. CWV content.
5. Spectral wavelength.

The LUTs are composed of simulated at-sensor radiance data generated by either the Modified Simulation of the Satellite Signal in the Solar Spectrum (M5S), LOWTRAN7, or MODTRAN2 or MODTRAN3 radiative transfer codes.

Surface reflectances are retrieved by first convolving the LUTs data along the wavelength dimension with the spectral response profiles of the sensor bands, then interpolating as necessary in the other four dimensions, finally, fitting the data to a parametric equation.

The technique also contains many additional features that include altitude-dependent layering of atmospheric aerosols and gases, scene-based estimation of atmospheric water vapor content, first-order removal of adjacency effects, and choice of different exoatmospheric solar irradiance functions. The LUT procedure can be applied to image cubes or individual spectra in forward (at-sensor radiance) and backward (surface reflectance) mode.

Performance tests showed good agreement between the LUT procedure and true radiative transfer code runs. Total relative errors of up to ±1.3% were achieved in case of MODTRAN2. These errors are a factor of about 0.4 lower than for M5S. Comparisons between reflectances retrieved from ground-based measurements and the LUT procedure indicate on average a relative agreement of better than 6% for the cases reported by Staenz and Williams (1997). These results were achieved using MODTRAN3. The LUT procedure was substantially faster than running a radiative transfer code on a pixel-by-pixel basis.

The 5D LUT Approach in ISDAS has been updated to use MODTRAN4 radiative transfer model. ISDAS atmospheric correction algorithm involves the following three steps:

1. Generate LUT.
2. Water vapor estimation.
3. Reflectance retrieval.

In step 1, an atmospheric LUT is generated by running MODTRAN 4.2 radiative transfer code for each combination among the first four dimensions of the LUT. ISDAS provides a dialog to help users to input the parameters required for creating a LUT.

In step 2, if a hyperspectral data set contains a number of spectral band images within the wavelength region of 850–1250 nm, then ISDAS will allow to estimate a 2D water vapor map based on a given at-sensor radiance data set and an atmospheric LUT. Water vapor content is estimated on pixel-by-pixel basis.

In step 3, after an atmospheric LUT and a water vapor map are generated in steps 1 and 2, the reflectance retrieval tool will derive an at-surface reflectance cube for the given at-sensor radiance cube.

Detailed description of the use of the ISDAS atmospheric correction software tool can be found in the reference (CCRS 2014).

14.4.6 Atmospheric and Topographic Correction (ATCOR)

A combined atmospheric and topographic correction (ATCOR) approach was proposed by Richter and Schlaepfer (2002). It uses a database of LUTs of the atmospheric correction functions, including path radiance, atmospheric transmittance, direct and diffuse solar flux, calculated with a radiative transfer code. In addition, the terrain shape obtained from a digital elevation model (DEM) is taken into account. A comparison of the surface reflectance derived from airborne imagery with the measured ground reflectance demonstrates that an accuracy of 1–3% can be achieved. Examples of processing of hyperspectral imagery in flat and rugged terrain are presented.

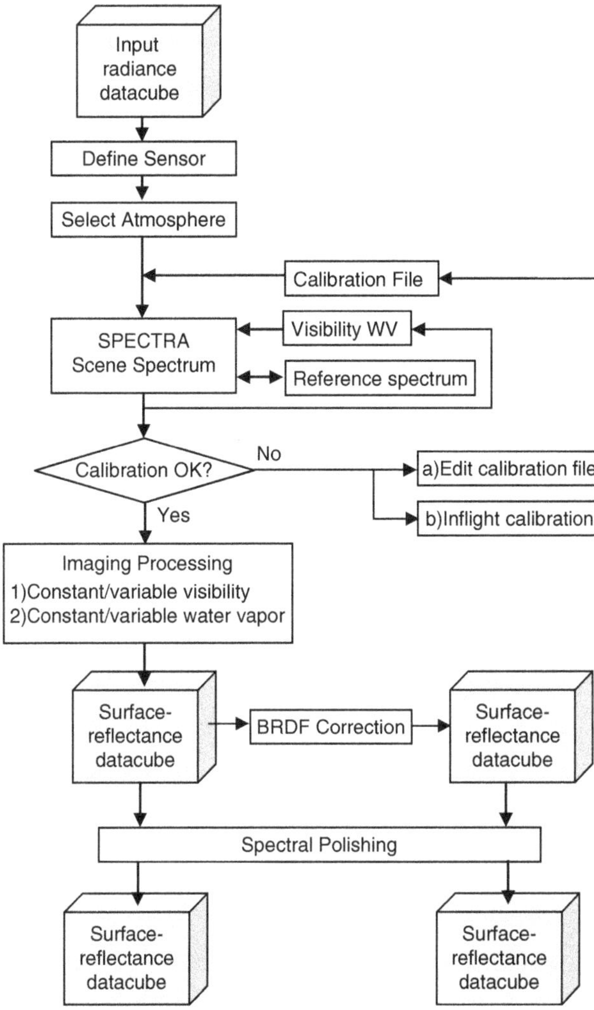

FIGURE 14.4 Typical workflow of the atmospheric/topographic correction process.

Figure 14.4 shows the typical workflow of the image-based atmospheric/topographic correction process, which is performed iteratively. The process requires the generation of the atmospheric LUTs pertaining to the image: flight and solar geometry, range of ground elevations, atmospheric parameters such as aerosol type and optical depth, water vapor, and ozone content, and the resampling with the spectral response function of the sensor channels.

First, a radiance image is input to the process with additional terrain information (DEM files). Then, the sensor has to be defined with the radiometric calibration file. After that a basic atmosphere/aerosol type combination needs to be selected (e.g., a summer atmosphere with a rural aerosol). The SPECTRA module checks the validity of the calibration file and estimates the visibility of the atmospheric water vapor column before processing the image cube.

Reflectance spectra of scene targets can be displayed as a function of visibility and water vapor and compared with field or library reference spectra: (a) if calibration problems exist in a few channels, a copy of the calibration file can be edited in these channels to match the reference spectrum; (b) if there are problems in many channels, the inflight radiometric calibration module should be used to generate a calibration file.

When the Calibration file is OK, the algorithm moves to the Image Processing step. Depending on the available sensor bands, there are two options to process the imagery: (1) constant or variable visibility, and (2) constant or variable atmospheric water vapor.

For large FOV sensors, an option is available to correct for cross-track illumination effects using bidirectional reflectance distribution function (BRDF). This is especially useful if the image recording took place in the solar principal plane.

In addition, a Spectral Polishing can be performed for the atmospherically corrected data and/or BRDF-corrected data. Finally, the algorithm also provides a classification function, which may be performed.

14.5 RADIATIVE TRANSFER MODELING BASED METHODS FOR WATER SCENES

14.5.1 Atmospheric Correction Over Optically Complex Water Scene

The atmospheric correction algorithms designed for remote sensing of land surfaces are not directly applicable for hyperspectral remote sensing of water color and ocean. Because inland water and ocean surfaces are much darker than land surfaces and the air-water interface is not Lambertian, very accurate modeling of atmospheric absorption and scattering effects and the specular water surface reflection effects is required in order to derive the so-called *water-leaving reflectance*, ρ_w, from hyperspectral imaging data. Water leaving reflectance is the reflectance resulted from scattering by water bodies, excluding contributions from specular reflection at the air-water interface.

The optical signal over water targets received by a hyperspectral sensor is generally a mixture of water-leaving radiance, atmosphere, and their interactions, the so-called top-of-atmosphere (TOA) radiance. Consequently, atmospheric correction is critically required to retrieve accurate water-leaving reflectance and alleviate atmospheric effects. Theoretically, the atmospheric signal in the visible wavelength is mainly made of Rayleigh scattering by gas molecules, Mie scattering by aerosols, and their interaction. The Rayleigh scattering can be exactly calculated from radiative transfer models (Gordon et al. 1988) and predicted a priori, while the aerosol scattering remains the largest uncertainty in the retrieval of water-leaving reflectance (Gordon and Wang 1994).

The signal measured by a satellite water color sensor can be decomposed into several terms (Gordon and Wang 1994, Gordon 1997):

$$\rho_t(\lambda) = \rho_r(\lambda) + \rho_a(\lambda) + \rho_{ra}(\lambda) + T(\lambda)\rho_g(\lambda) + t(\lambda)\rho_{wc}(\lambda) + t(\lambda)\rho_w(\lambda) \tag{14.4}$$

where $\rho_t(\lambda)$ represents the total reflectance measured by the sensor; $\rho_r(\lambda)$, $\rho_a(\lambda)$, and $\rho_{ra}(\lambda)$ are the contributions from atmospheric scattering due to air molecules (Rayleigh scattering), aerosols, and Rayleigh-aerosol interaction, respectively; $\rho_g(\lambda)$ is the reflectance related to the sun glitter and $\rho_{wc}(\lambda)$ related to the whitecaps; $T(\lambda)$ and $t(\lambda)$ are the direct and diffuse atmospheric transmittance from sea to the sensor, respectively. Ultimately, $\rho_w(\lambda)$ is the water-leaving reflectance.

Even under favorable observation conditions, $\rho_r(\lambda)$, $\rho_a(\lambda)$, and $\rho_{ra}(\lambda)$ together represent about 90% of $\rho_t(\lambda)$. This means that as much as 90% of the signal reaching the sensor originates from the atmosphere contributions and only 10% from the water in the visible spectrum (Huot et al. 2001). $\rho_g(\lambda)$ and $\rho_{wc}(\lambda)$ can be calculated for a given sun and sensor zenith angle and an empirical wind speed formulation (Frouin et al. 1996), respectively. During preprocessing, $\rho_t(\lambda)$ is corrected for gas absorption, Rayleigh scattering, white caps, and sun glint (Gordon, 1997). This term is referred as the Rayleigh corrected reflectance, $\rho_{rc}(\lambda)$:

$$\rho_{rc}(\lambda) = \rho_t(\lambda) - \rho_r(\lambda) - T(\lambda)\rho_g(\lambda) - t(\lambda)\rho_{wc}(\lambda) \tag{14.5}$$

For the sake of comparison of different atmospheric correction algorithms, the normalized water-leaving radiance, $L_{wn}(\lambda)$, which is related to $\rho_w(\lambda)$, is defined as follow:

$$L_{wn}(\lambda) = \frac{\rho_w(\lambda) F_0(\lambda) cos\theta_0}{\pi} \quad (14.6)$$

with F_0 being the extraterrestrial solar irradiance at a given wavelength, λ, and θ_0 is the sun zenith angle.

The traditional method for aerosol retrieval and atmospheric correction of remote sensing data over water surface is based on the assumption of zero water reflectance in the NIR region, which is also known as the "black-pixel assumption" method, in which the water-leaving reflectance is set to zero in the NIR wavelength (>700 nm) (Gordon and Wang 1994, Siegel et al. 2000). With this black-pixel assumption, the aerosol scattering at two NIR wavelengths can be derived and extrapolated to the visible wavelengths based on the property of aerosol wavelength dependence and then the atmospheric contribution can be subtracted from the TOA signal.

The NIR black-pixel assumption method has been used for atmospheric correction of ocean data acquired by the Coastal Zone Color Scanner (CZCS), Sea-Viewing Wide FOV Sensor (SeaWiFS), and Moderate Resolution Imaging Spectroradiometer (MODIS) (Gordon 1978, Gordon and Wang 1994, Wang 2007). It has been set to be the standard method of the National Aeronautics and Space Administration (NASA) ocean color products. However, the black-pixel assumption is only valid for clear water bodies (Morel and Prieur 1977), for example, open oceans (often referred to as case I water). For turbid waters (often referred to as case II water), such as coastal zones and inland lakes, this assumption is invalid due to the high reflection by suspended water particles in NIR region (Dekker et al. 1997, Lavender et al. 2005, Morel and Bélanger 2006).

Attempts to overcome the non-black-pixel fact at NIR bands by using SWIR bands of MODIS data (>1000 nm) have been reported (Gao et al. 2007, Wang 2007, Wang and Shi 2007, Wang et al. 2009, Werdell et al. 2010, Zhu et al. 2012). The studies showed that the method using SWIR bands is effective and flexible for turbid waters (Wang et al. 2009, Werdell et al. 2010, Zhu et al. 2012), suggesting that substituting SWIR bands for NIR bands in the traditional NIR black-pixel method has potential in the atmospheric correction processes of turbid case II waters.

Due to the extremely low albedo of water scenes, one issue that has recently come to light is that tropospheric NO_2 effect (Gao et al. 2009), given the fact that it can be as high as 10 times the global average in some urban areas. Typically, as shown in the graph on bottom in Figure 14.1, the atmospheric NO_2 transmittances at wavelength near 0.4 μm are 0.985–0.995 in the sun-surface-sensor path. When retrieving reflectance over bright land surface targets having reflectance ≥0.1 at wavelength near 0.4 μm and using radiative transfer modeling approaches, neglecting atmospheric NO_2 absorption effects does not introduce significant errors in the derived reflectance. However, when retrieving surface reflectance over darker water targets, such as inland waters and coastal waters, having reflectance ≈ 0.02–0.03 at wavelength near 0.4 μm using radiative transfer modeling approaches, the NO_2 absorption effects should be included in the models. Otherwise, errors on the order of 10–20% can be introduced in the retrieved surface reflectance at wavelength near 0.4 μm. One possible way to model the atmospheric NO_2 absorption effects is to use the NO_2 absorption cross-sections, such as those reported by Solomon et al. (1999), and the NO_2 column amount map derived from European satellite measurements, such as the measurements with GOME (Global Ozone Monitoring Experiment) type of spectrometers (Bracher et al. 2005).

14.5.2 Black-Pixel Assumption NIR Algorithm

The goal of the atmospheric correction is to derive the water-leaving reflectance, ρ_w, from the total radiance at the TOA after normalization of the illumination conditions and removal of the atmospheric effects. In Equation 14.4, the $\rho_g(\lambda)$ term is generally ignored because the sensors are equipped with a provision for tilting the scan plane away from the specular image of the sun. The $\rho_{wc}(\lambda)$ term is also not considered when the water surface is assumed to be calm for simplicity (i.e., low surface wind speeds). The $\rho_{ra}(\lambda)$ term can be ignored in the single-scattering case, as the amount of multiple scattering is small in such a case (Zhu et al. 2012). Accordingly, Equation 14.4 can be rewritten as follows by replacing the term $\rho_a(\lambda) + \rho_{ra}(\lambda)$ by a single-scattering term, $\rho_{as}(\lambda)$:

$$\rho_t(\lambda) = \rho_r(\lambda) + \rho_{as}(\lambda) + t(\lambda)\rho_w(\lambda) \tag{14.7}$$

To relate the derived water-leaving reflectance to the in situ measurement, all geometric influences on $\rho_w(\lambda)$ must be eliminated, which can be accomplished by normalizing the water-leaving reflectance, $[\rho_w(\lambda)]_N$, to a zenith view (Gordon and Clark 1981, Siegel et al. 2000) as follows:

$$\left[\rho_w(\lambda)\right]_N = \frac{\rho_w(\lambda)}{t_0(\lambda)} \tag{14.8}$$

where $t_0(\lambda)$ is the atmospheric diffuse transmittance in the solar direction. Therefore, if the values of $\rho_r(\lambda)$, $\rho_{as}(\lambda)$, and $t(\lambda)$ in Equation 14.7 can be calculated separately, $\rho_w(\lambda)$ can be derived.

Usually, the Rayleigh scattering reflectance $\rho_r(\lambda)$, can be exactly computed by solving the radiative transfer equations (Gordon et al. 1988). The diffuse transmittances of the atmospheric column $t(\lambda)$ can also be calculated approximately using the following equation (Gordon et al. 1983):

$$t(\lambda) = exp\left(\frac{-\left[\tau_{oz}(\lambda) + \frac{\tau_r(\lambda)}{2}\right]}{cos\theta}\right) \tag{14.9}$$

where $\tau_{oz}(\lambda)$ and $\tau_r(\lambda)$ are the ozone optical thickness and the Rayleigh optical thickness, respectively, and θ is the view zenith angle. Therefore, only the contribution from scattering by aerosol $\rho_{as}(\lambda)$ is unknown in Equation 14.7.

Because the aerosol model and concentration are highly variable in the spatial and temporal domains, $\rho_{as}(\lambda)$ cannot be predicted a priori like the Rayleigh scattering component, $\rho_r(\lambda)$. Estimation of the aerosol component $\rho_{as}(\lambda)$ is a key problem of atmospheric correction. According to the work by Gordon (1978) as well as by Gordon and Wang (1994), an atmospheric correction parameter, $\varepsilon(\lambda_i, \lambda_j)$, as the ratio of aerosol scattering reflectance at two wavelengths, λ_i and λ_j, can be defined as follows:

$$\varepsilon(\lambda_i, \lambda_j) = \frac{\lambda_{as}(\lambda_i)}{\lambda_{as}(\lambda_j)} = \frac{\omega_a(\lambda_i)\tau_a(\lambda_i)P_a(\theta, \theta_0, \lambda_i)}{\omega_a(\lambda_j)\tau_a(\lambda_j)P_a(\theta, \theta_0, \lambda_j)} \tag{14.10}$$

where $\tau_a(\lambda)$, $\omega_a(\lambda)$, and $P_a(\theta, \theta_0, \lambda)$ are the aerosol optical thickness, the aerosol single-scattering albedo, and the aerosol scattering phase function for a scattering angle, respectively; θ and θ_0 are the view zenith angle and the solar zenith angle, respectively. $P_a(\theta, \theta_0, \lambda)$ can be derived from local viewing geometric condition θ and θ_0. In Equation 14.10, the aerosol optical thickness, $\tau_a(\lambda)$, has a wavelength dependence that can be estimated as follows:

$$\tau_a(\lambda) = \sigma\lambda^{-\beta} \tag{14.11}$$

where β is the Ångström exponent and σ is the air turbidity index. Moreover, the aerosol scattering phase function $P_a(\theta, \theta_0, \lambda)$ is also dependent on the wavelength when aerosol particles are in uniform distribution, and the aerosol single-scattering albedo, $\omega_a(\lambda)$, can be considered as a constant for certain aerosol models. Accordingly, Equation 14.10 can be simplified as follows:

$$\varepsilon(\lambda_i,\lambda_j)=\frac{\rho_{as}(\lambda_i)}{\rho_{as}(\lambda_j)}=\left(\frac{\lambda_j}{\lambda_i}\right)^n \tag{14.12}$$

where n is a constant for a specific aerosol model.

As is well known, clear water reflectance is very low in the visible spectrum and becomes much lower as the wavelength increases. Accordingly, water-leaving reflectance of clear water targets has been observed to approach zero at the NIR wavelength (for turbid inland and coastal waters, the wavelength where the water-leaving reflectance approaches zero is longer, possibly in the SWIR range). Based on this fact, a reasonable assumption of zero water-leaving reflectance at NIR bands can be made. In that case, the aerosol scattering reflectance $\rho_{as}(\lambda)$ at two NIR wavelengths, NIR_1 and NIR_2, can be computed according to Equation 14.7 as follows:

$$\rho_{as}(\lambda)=\rho_t(\lambda)-\rho_r(\lambda)-\cancel{t(\lambda)\rho_w(\lambda)} \qquad \lambda=(NIR_1,NIR_2) \tag{14.13}$$

where $\rho_t(\lambda)$ is known, $\rho_r(\lambda)$ can be exactly calculated a priori, and $\rho_w(\lambda)$ is treated as zero at the NIR wavelength.

After the aerosol scattering reflectance, $\rho_{as}(\lambda)$, at two NIR wavelengths, NIR_1 and NIR_2 are calculated, the aerosol scattering reflectance at other required wavelengths can be derived by extrapolating back to the visible region as follows:

$$\rho_{as}(\lambda)=\left(\frac{NIR_2}{\lambda}\right)^n\times\rho_{as}(NIR_2) \tag{14.14}$$

where the value of n can be retrieved from Equation 14.15 as follows:

$$\varepsilon(NIR_1,NIR_2)=\frac{\rho_{as}(NIR_1)}{\rho_{as}(NIR_2)}=\left(\frac{NIR_2}{NIR_1}\right)^n \tag{14.15}$$

14.5.3 NIR Similarity Spectrum Algorithm

This algorithm includes two alternative assumptions: assumption on the water optical properties (i.e., spectrum similarity in NIR region), and assumption on spatial homogeneity of the atmosphere (Ruddick et al. 2000). The first assumption arises from the fact that the shape of water spectrum in NIR is largely determined by pure water absorption and hence invariant, while the magnitude of the spectrum is approximately proportional to the backscattering coefficient. Consequently, the ratio of any two NIR water-leaving reflectances, α, is constant. For MODIS-Aqua images, α is defined at two NIR spectral bands, as $\lambda = 748$ nm and 869 nm:

$$\alpha=\frac{\rho_{w(748)}/_{t(748)}}{\rho_{w(869)}/_{t(869)}}=1.945 \tag{14.16}$$

The second assumption is based on the fact that the atmosphere composition does not vary significantly in space and therefore the ratio of the aerosol reflectance, $\rho_a(\lambda)+\rho_{ra}(\lambda)$, referred as $\varepsilon(\lambda_1, \lambda_2)$, can be considered as spatially homogeneous over the area of interest. In clear waters, where $\rho_w(\lambda)$

in the NIR region of the spectrum is close to zero, $\rho_a(\lambda) + \rho_{ra}(\lambda)$ can be approximated by $\rho_{rc}(\lambda)$ in Equation 14.5. Accordingly, over clear water pixels ε(748,869) is calculated as follows:

$$\varepsilon(748,869) = \frac{\rho_{rc}(748)}{\rho_{rc}(869)} \tag{14.17}$$

Assuming spatial homogeneity of the atmosphere over the area of interest allows to estimate ε(748,869) over a clear water sub-scene and use it for the atmospheric correction of the entire image. A clear water sub-scene is defined as any pixel within the image for which $L_{rc}(\lambda)$ is nonnegative and below 0.015 mW/cm^2/μm/sr at 748 nm and 869 nm.

The NIR similarity spectrum algorithm is a three-step process. First, the atmospheric correction is performed with black-pixel assumption algorithm described in Section 14.5.2 to retrieve the scatterplot $\rho_{rc}(748)$ versus $\rho_{rc}(869)$ for the region of interest and subsequently to estimate ε(748,869) based on Equation 14.17 over the clear water sub-scene. Next, using Equations 14.16 and 14.17, $\rho_a(\lambda)$ is estimated at each wavelength by taking into account the nonzero water-leaving reflectance and allowing to determine the appropriate aerosol model for the region of interest. Finally, the black-pixel assumption algorithm is run again but forced with the previously selected aerosol model.

14.5.4 NIR-SWIR Algorithm With Turbid Water Index

Traditional algorithms for aerosol retrieval over water scenes are based on the assumption of zero reflectance in the NIR spectral region (wavelengths longer than 700 nm). However, it has been demonstrated that this approach leads to considerable errors in the presence of absorbing aerosols (Gordon, 1997, Bailey and Werdell 2006) or over case II turbid waters, where suspended sediments or high concentrations of phytoplankton and detrital particles may result in a non-negligible reflectance at NIR bands (Dekker et al. 1997, Lavender et al. 2005, Morel and Bélanger 2006).

Wang and Shi (2007) recommended using SWIR bands for atmospheric correction of water color hyperspectral data in areas of sufficient turbidity where the black-pixel assumption fails. They proposed using the NIR black-pixel assumption atmospheric correction algorithm unless a turbidity index, T_{ind}, exceeds a predefined threshold and, thereby, triggers the use of a SWIR-based atmospheric correction where two bands in SWIR region are used in lieu of two NIR bands (Wang 2007). Two bands at 1240 and 2130 (1640 nm band does not function properly) (Franz et al. 2006) in SWIR region of MODIS-Aqua are used. Shi and Wang (2009) concluded that these bands satisfy the black-pixel assumption in moderately (1240 nm) to extremely (2130 nm) turbid waters. This combined "NIR–SWIR" atmospheric correction approach has been evaluated extensively in several geographic locations, eg., the La Plata Estuary and the east coasts of the United States and China (Wang and Shi 2007, Shi and Wang 2009) and evaluated against an independent, globally distributed in situ data set (Wang et al. 2009).

The turbid water index, $T_{ind}(\lambda_i, \lambda_j)$, that is computed from the measured radiances at the NIR and SWIR bands is derived as:

$$T_{ind}(\lambda_i,\lambda_j) = \frac{\Delta\rho_{(RC)}(\lambda_i)}{\Delta\rho_{(RC)}(\lambda_j)} exp\left\{-\frac{\lambda_j - \lambda_i}{\lambda_k - \lambda_j} In\left(\frac{\Delta\rho_{(RC)}(\lambda_j)}{\Delta\rho_{(RC)}(\lambda_k)}\right)\right\} \tag{14.18}$$

where $\Delta\rho_{(RC)}(\lambda_l)$ is the Rayleigh-corrected (RC) TOA reflectance at a given wavelength, λ_l, i.e. (Shi and Wang 2007),

$$\Delta\rho_{(RC)}(\lambda_l) = \rho_t(\lambda_l) - \rho_r(\lambda_l) \tag{14.19}$$

and $\rho_t(\lambda_l)$ and $\rho_r(\lambda_l)$ are the measured TOA reflectance with gas absorptions corrected and the reflectance contributed from Rayleigh scattering, respectively.

Using the measured reflectances at the wavelengths, $\lambda_i = 748$ nm, $\lambda_j = 1240$ nm, and $\lambda_k = 2130$ nm, the turbid water index $T_{ind}(748,1240)$ from Equation 14.18 becomes:

$$T_{ind}(748,1240) = \frac{\Delta\rho_{(RC)}(748)}{\Delta\rho_{(RC)}(1240)} exp\left\{-\frac{492}{890} In\left(\frac{\Delta\rho_{(RC)}(1240)}{\Delta\rho_{(RC)}(2130)}\right)\right\}. \tag{14.20}$$

Detection of the turbid waters using the parameter $T_{ind}(748,1240)$ is based on the fact that for the productive inland and ocean waters, the water surface has a nonzero contribution at the NIR band 748 nm, while the black-water assumption is usually still true for the SWIR bands (Wang and Shi 2005). In Equation 14.20, the ratio of reflectances at 748 nm and 1240 nm provides the turbid water detection, while the exponential term normalizes for the aerosol component in the atmosphere. In fact, Equation 14.20 can also be approximated as (Shi and Wang 2007):

$$T_{ind}(748,1240) \approx 1 + \frac{t(748)\rho_w(748)}{\rho_A(748)}, \tag{14.21}$$

where $t(748)\rho_w(748)$ and $\rho_A(748)$ are the TOA water-leaving reflectance (Gordon and Wang 1994) and aerosol reflectance (including Rayleigh-aerosol interactions) at the wavelength 748 nm, respectively. Note that polarization effects are also included for both the Rayleigh $\rho_r(\lambda_l)$ and aerosol $\rho_A(\lambda_l)$ LUTs. Therefore, for the turbid waters, $T_{ind}(748,1240) > 1$ due to a nonzero ocean contribution at the band 748 nm, whereas for the open oceans, $T_{ind}(748,1240) \approx 1$ because of the black ocean at the NIR bands (i.e., $\rho_w(748) \approx 0$).

The turbid water index, $T_{ind}(748,1240)$, can be used to discriminate turbidity of inland waters and ocean where there are nonzero NIR reflectance. Thus, with the detection of the turbid waters, the SWIR band based atmospheric correction method should be used (Wang 2007). This combined NIR-SWIR method has shown significant improvements in MODIS ocean color products for cases with extremely turbid waters (Wang et al. 2007).

The turbid water index is quite effective and can be calculated prior to the atmospheric correction procedure. To use the $T_{ind}(748,1240)$ as a parameter for the algorithm to switch between the NIR and SWIR atmospheric correction methods, Shi and Wang (2007) found from various tests that a threshold of 1.3 is the most appropriate choice due to noise in $T_{ind}(748,1240)$ when derived from SWIR bands. A study for cases along the US east coast region has confirmed the threshold. Ocean color products chlorophyll *a* were derived from the MODIS water-leaving reflectances after atmospheric correction using the NIR black-pixel method and using SWIR band atmospheric correction method. The experimental results showed that, for cases with $1.1 < T_{ind}(748,1240) < 1.3$, mean chlorophyll *a* difference is usually within 5% between using the NIR and SWIR methods. The mean difference is >25% for cases of $T_{ind}(748,1240) \geq 1.3$. Due to considerably lower SNR values for the MODIS SWIR bands, it is generally better to use the NIR atmospheric correction algorithm for deriving ocean color products if it is applicable.

14.5.5 Self-Contained Atmospheric Parameters Estimation

As described in Section 14.5.4, the atmospheric correction for turbid case II water requires spectral bands in both NIR and SWIR region. It is known that MERIS is an imaging spectrometer working in VNIR region only as described in Section 2.7. In the atmospheric correction of MERIS data, to overcome the shortage of SWIR bands, an automatic atmospheric correction method for MERIS data over turbid waters, called Self-Contained Atmospheric Parameters Estimation for MERIS (SCAPE-M), was proposed by Guanter et al. (2010). This method does not use SWIR bands and does not require a priori assumptions on the water composition and its spectral response. It calculates reflectance of close-to-land water pixels through spatial extension of atmospheric parameters derived over neighboring land pixels. This method has been validated by comparing the derived aerosol optical thickness, surface reflectance, and water pigments to the in situ data acquired concurrently to the MERIS images with an excellent results (e.g., correlation coefficient, $R^2 = 0.74$ and root-mean-square error (RMSE) = 0.07 for reflectance, $R^2 = 0.75$ for product chlorophyll *a*, and $R^2 = 0.82$ for product phycocyanin).

SCAPE-M performs automatic cloud screening, aerosol and water vapor retrieval, and surface reflectance retrieval sequentially from MERIS Level-1B data over land surfaces and over inland and coastal water scenes. The atmospheric optical parameters required during the processing are calculated by interpolating from a LUT compiled with the MODTRAN4 atmospheric radiative transfer code (Berk et al. 2003). The Thuillier solar irradiance database (Thuillier et al. 2003) attached to MERIS images is used as a reference extraterrestrial solar constant. Elevation and topographic corrections are carried out by the use of a DEM projected onto each MERIS image to be processed.

Aerosol is retrieved over spatially integrated windows to improve the spatial coverage of the aerosol product. Aerosol optical thickness at 550 nm (AOT550) is retrieved at macro-pixels of 30 km × 30 km. For each macro-pixel, a maximum AOT550 threshold is estimated from dark pixels in the area, including inland water targets. Such a threshold is calculated as the maximum AOT that gives nonnegative reflectance at the dark pixels after atmospheric correction. The threshold value is refined by exploiting those pixels with sufficient green vegetation and bare soil pixels.

The TOA radiances at 5 land pixels, mixture of vegetation and bare soil, and with as much spectral contrast as possible are inverted assuming that surface reflectance can be provided by a linear combination of two end-members, pure vegetation, and bare soil spectra. End-member abundances and the AOT550 are retrieved concurrently. A default rural aerosol model is chosen. A user needs to set both the number of 5-pixel clusters and the number of vegetation end-members to be used in the inversion according to computation time. The final steps fill in the areas with unsuccessful AOT retrieval in the macro-pixel mosaic by interpolation, and smoothen the resulting mosaic-like AOT550 map using a cubic convolution.

AOT550s derived by SCAPE-M over 200 MERIS images are compared with AErosol RObotic NETwork (AERONET) data (Holben et al. 1998) from stations around the world (Guanter et al. 2007, 2008). A high correlation between MERIS- and AERONET-derived AOT was generally found. Correlation coefficient R^2 values were around 0.75, although some cases with low R^2 (0.337) or very high (0.925) were also reported. A systematic underestimation of MERIS-derived AOT500 by SCAPE-M with respect to AERONET measurements was above 0.4 in most of the sites. RMSEs were about 0.05 in most of the tested cases.

Surface reflectance is calculated by SCAPE-M from MERIS radiance data using the generated cloud mask, DEM, illumination and observation angles, terrain normal vector, AOT550, and CWV maps per pixel over land and water pixels. A Lambertian surface is assumed for land scenes in the modeling of radiative transfer effects. The same assumption has been applied to the processing of water scenes. The AOT550 and CWV maps derived from land surfaces are extended to inland water scenes by spatial inter- or extrapolation.

A smoothing of the mosaic of 30 km × 30 km macro-pixels is performed. A spatially continuous AOT map covering the imaged area is obtained after the smoothing process. Large water pixels noncompliant with the assumption of horizontal continuity of the atmospheric parameters are then screened out of the processing. In the default processing mode, inland water pixels are defined as those water pixels not further than 20 km from a land surface, or those inside closed water bodies with an area of <1600 km². These limits are empirically defined as the maximum distance within which the assumption of the continuity of AOT550 and CWV can normally be assumed.

The impact of adjacency effects over inland and coastal water pixels can be very strong in the NIR bands, for which water is very dark and land pixels normally present a high reflectance, and can be noticed in visible bands under certain conditions. SCAPE-M performs a simple adjacency correction on the final reflectance to reduce the impact. The correction includes modeling the at-sensor radiance as a linear combination of the photons coming directly from the target and those coming from areas adjacent to the target and scattered into the sensor direction. This is implemented by weighting the strength of the adjacency effect by the ratio of diffuse to direct ground-to-sensor transmittance. The adjacency-corrected surface reflectance, ρ, is then calculated as:

$$\rho = \rho_u + \frac{t_{dif}}{t_{dir}}(\rho_u - \bar{\rho}) \tag{14.22}$$

where ρ_u is the reflectance before the adjacency correction, $\bar{\rho}$ is the background reflectance calculated as the distance-weighted average reflectance of the neighboring pixels, and t_{dir}, t_{dif} are the atmospheric transmittance between the surface and the sensor for direct and diffuse radiation, respectively.

14.5.6 Modified NIR Black-Pixel Method

Due to the invalidity of the assumption of zero water-leaving reflectance in NIR region for turbid case II waters, the traditional NIR black-pixel method fails when applied to such cases. For water color hyperspectral sensors operating only in VNIR region, such as MERIS or ocean and land color imager (OLCI) onboard Sentinel-3A satellite, due to the shortage of SWIR bands, the algorithm using spectral bands in SWIR region cannot be extended to the data acquired by these sensors for case II waters.

On the other hand, water color hyperspectral sensors, such as MERIS and OLCI, provide two special NIR bands: an oxygen absorption band (O_2 A band at 761 nm) and a water vapor absorption band (vapor A band at 900 nm). These two bands are not within the atmospheric window and thus are often not used in terrestrial observations (Ding and Gordon 1995). Owing to oxygen or water vapor absorption, these two bands are of very low atmospheric transmittances $t(\lambda)$ (Dubuisson et al. 2009). A typical example provided by MODTRAN4 atmospheric radiative transfer code showed that $t(\lambda)$ is about 13% in both the 761 nm and 900 nm bands under standard middle-latitude summer atmosphere conditions (Berk et al. 2003). The water-leaving reflectance, ρ_w, is very small at these two bands (typically around 0.01–0.02), although it cannot be neglected over case II waters. The smaller values of both $t(\lambda)$ ($\lambda = 761, 900$) and water-leaving reflectance, ρ_w, at the two bands result in the product terms $t(761)\rho_w(761)$ and $t(900)\rho_w(900)$ in Equation 14.7 approaching zero, which allows the zero water-leaving reflectance assumption at these two NIR bands being reasonable, as in the traditional NIR black-pixel algorithm. Consequently, it is possible to follow the idea of the NIR black-pixel algorithm by using these two special bands. This extension of the NIR algorithm is referred to as the modified NIR black-pixel method (Zhu et al. 2012).

To evaluate this modified NIR black-pixel method, two issues need to be addressed. One is how to use the property of aerosol wavelength dependence to calculate the atmospheric correction parameter $\varepsilon(\lambda_i, \lambda_j)$, and the other is whether aerosol wavelength dependence is still valid when considering the aerosol multiple-scattering effect in the modified NIR black-pixel method.

Based on aerosol model simulation results, Gordon and Wang (1994) suggested that the variation of $\varepsilon(\lambda_i, \lambda_j)$ with wavelength over the entire range of 400–900 nm could be described better than in Equation 14.12 by using the following simple formula:

$$\varepsilon(\lambda_i, \lambda_j) = \frac{\rho_{as}(\lambda_i)}{\rho_{as}(\lambda_j)} = exp\left[c \times (\lambda_j - \lambda_i)\right] \tag{14.23}$$

where c is a constant like n in Equation 14.12, depending on the viewing geometric condition and aerosol models.

Equations 14.12 and 14.23 provide two means to interpolate the aerosol scattering reflectance from NIR to visible wavelength. To verify which equation is better, it was necessary to compare their behavior and robustness under different aerosol models. This process was carried out by a simulation experiment using the Second Simulation of the Satellite Signal in the Solar Spectrum (6S) software code (Vermote et al. 1997). Five aerosol models were examined, including the oceanic model, rural model, urban model, maritime model, and coastal model.

The simulation experiment quantitatively compared the performances of Equations 14.12 and 14.23 with the same geometry and illumination conditions. It is evident that the aerosol wavelength dependence still holds true and that both equations can fit the wavelength dependence of $\varepsilon(\lambda, 900)$ very well, with the correlation coefficient (R^2) larger than 0.90. Equation 14.23 presents a higher correlation coefficient than Equation 14.12 as a whole, with R^2 larger than 0.99 except in the oceanic aerosol model. These results suggest that Equation 14.23 is preferable to the aerosol single-scattering extrapolation. Similar results that support the conclusion were also observed when the geometry and illumination conditions were changed.

The multiple scattering within aerosol particles or between aerosols and gas molecules (i.e., the Rayleigh-aerosol interaction term $\rho_{ra}(\lambda)$ in Equation 14.4) may also affect the wavelength dependence of $\varepsilon(\lambda, 900)$, leading to the failure of Equations 14.12 and 14.23 when considering the multiple scattering conditions. Hence, it is necessary to verify the possibility of multiple scattering, even though its contribution is small in most cases. When multiple scattering is considered, the aerosol scattering term $\rho_{as}(\lambda)$ becomes $\rho_a(\lambda) + \rho_{ra}(\lambda)$ as follows:

$$\rho_A(\lambda) = \rho_a(\lambda) + \rho_{ra}(\lambda) \tag{14.24}$$

Like the single-scattering parameter $\varepsilon(\lambda_i, \lambda_j)$, the multiple-scattering parameter $\varepsilon_{(M)}(\lambda_i, \lambda_j)$ can also be defined as follows (Su 2000, Wang 2004):

$$\varepsilon_{(M)}(\lambda_i, \lambda_j) = \frac{\rho_A(\lambda_i)}{\rho_A(\lambda_j)} \tag{14.25}$$

Given the aerosol model and the aerosol optical thickness, the aerosol multiple-scattering reflectance, $\rho_A(\lambda)$, can be obtained by solving the atmospheric radiative transfer equations using 6S. In this case, $\varepsilon_{(M)}(\lambda,900)$ can be calculated for a certain geometry and illumination condition.

It was observed that $\varepsilon_{(M)}(\lambda,900)$ shows a strongly linear wavelength dependence within the spectral region of 400–900 nm for the rural model, urban model, maritime model, and coastal model, but not for the oceanic model. Such linear dependence is more significant for all aerosol models within the region of 560–900 nm.

The regression coefficients (R^2) between $\log_{10}[\varepsilon_{(M)}(\lambda, 900)]$ and $(900 - \lambda)$ for the rural, maritime, and coastal models are larger than 0.98, independent of the wavelength range. For the urban aerosol model, the relationship is more significant in the range of 560–900 nm than in the range of 400–900 nm. For the oceanic aerosol model, the relationship is still statistically significant, but R^2 is <0.8 and the slope is close to 0. These results suggest that the multiple-scattering parameter $\varepsilon_{(M)}(\lambda_i, \lambda_j)$ can also be expressed as a function of wavelength, especially within 560–900 nm. Therefore, it is reasonable to assume black pixels at the two special NIR bands and simply replace the aerosol single-scattering parameter with the multiple-scattering parameter if the multiple-scattering condition needs to be considered.

14.5.7 Direct Inversion Approach Using Neural Network

Two atmospheric correction methods for MERIS and MODIS data, respectively, using direct inversion approach by neural network (NN) were reported (Schroeder et al. 2007, Goyens et al. 2013).

These methods are based on an inverse modeling of extensive radiative transfer simulations within a coupled ocean-atmosphere system. The methods directly invert $\rho_t(\lambda)$ into the remote sensing reflectance, $R_{rs}(\lambda)$, defined as:

$$R_{rs}(\lambda) = \frac{\rho_w(\lambda)cos\theta_0}{\pi} \quad (14.26)$$

The direct inversion takes into account the spectral information in each spectral band simultaneously, not just in the spectral bands in NIR region like the methods described in Sections 14.5.2–14.5.6. Consequently, the NN method does not attempt to decouple oceanic and atmospheric signals to perform the atmospheric correction (see Equation 14.4). Another distinct difference is that the forward model parameterization of the NN includes only 8 aerosol models—compared to 80 models for the other algorithms. The radiative transfer simulations consist of reflectance data at both TOA and mean sea level. The radiative transfer simulations cover a large range of oceanic and atmospheric conditions for a variety of different sun and observing geometries. Specifically, the optically active water constituents cover the following concentration ranges: 0.05–50 mg/l for chlorophyll *a*, 0.05–50 g/l for suspended particulate matter (SPM), and 0.001–1 per meter absorption of color dissolved organic matter (CDOM) at 443 nm.

The simulated data were used to adapt the inverse model during a supervised learning procedure. The NN inverse model consists of a multilayer perceptron that learned the functional relationship between TOA and mean sea level reflectance. More details about the forward model parameterization can be found in (Schroeder et al. 2007).

When applied to MODIS data, the algorithm requires the TOA reflectance at all ocean color bands as input with further information about the sun and observing geometry as well as the surface pressure. In addition, MODIS Level-2 flags are used to mask pixels affected by land, high sun glint, and clouds. The associated NN outputs consist of mean sea level R_{rs} at eight MODIS bands up to 748 nm, including spectral aerosol optical thickness at four AERONET bands (440, 550, 670, and 870 nm). Additional NN flags are used for further quality control of the outputs that check for out-of-scope conditions compared to the simulated data ranges.

REFERENCES

Adler-Golden, S. M., A. Berk, L. S. Bernstein, et al. 1998. FLAASH, a MODTRAN4 atmospheric correction package for hyperspectral data retrievals and simulations. *Proceedings of the 7th Annual JPL Airborne Earth Science Workshop, JPL Publication* 97–21:9–14.

Aspinall, R. J., W. A. Marcus, and J. W. Boardman. 2002. Considerations in collecting, processing, and analyzing high spatial resolution hyperspectral data for environmental investigations. *Journal of Geographical Systems* 4:15–29.

Bailey, S. W. and P. J. Werdell. 2006. A multi-sensor approach for the on-orbit validation of ocean color satellite data products. *Remote Sensing of Environment* 102:12–23.

Berk, A., G. P. Anderson, P. K. Acharya, M. L. Hoke, J. H. Chetwynd, L. S. Bernstein, E. P. Shettle, M. W. Matthew, and S. M. Adler-Golden. 2003. MODTRAN4 version 3, revision 1 User's Manual. *Technical Report, Air Force Research Laboratory*, Hanscom Air Force Base, MA.

Berk, A., L. S. Bernstein, G. P. Anderson, P. K. Acharya, D. C. Robertson, J. H. Chetwynd, and S. M. Adler-Golden. 1998. MODTRAN cloud and multiple scattering upgrades with application to AVIRIS. *Remote Sensing Environment* 65(3):367–375.

Bracher, A., L. N. Lamsal, M. Weber, K. Bramstedt, M. Coldewey-Egbers, and J. P. Burrows. 2005. Global satellite validation of SCIAMACHY O3 columns with GOME WFDOAS. *Atmospheric Chemistry and Physics* 5:2357–2368.

CCRS. 2014. ISDAS atmospheric correction tool (available at ftp://ftp.ccrs.nrcan.gc.ca/ad/Lixin/ISDASv2/helps/ATM_correction.htm, accessed on December 6, 2019).

Clark, R. N. and T. V. V. King. 1987. Causes of spurious features in spectral reflectance data. *Proceedings of the 3rd Airborne Imaging Spectrometer Data Analysis Workshop, JPL Publication* 87–30:49–61.

Conel, J. E., R. O. Green, G. Vane, C. J. Bruegge, and R. E. Alley. 1987a. AIS-2 radiometry and a comparison of methods for the recovery of ground reflectance. *Proceedings of the 3rd Airborne Imaging Spectrometer Data Analysis Workshop JPL Publication* 87–30:18–47. Pasadena, CA: Jet Propulsion Laboratory.

Conel, J. E., R. O. Green, G. Vane, C. J. Bruegge, R. E. Alley, and B. J. Curtiss 1987b. Airborne imaging spectrometer-2: Radiometric spectral characteristics and comparison of ways to compensate for the atmosphere. *Proceedings of SPIE* 834:140–157.

Dekker, A. G., H. J. Hoogenboom, L. M. Goddijn, and T. J. Malthus. 1997. The relation between inherent optical properties and reflectance spectra in turbid inland waters. *Remote Sensing Reviews* 15:59–74.

Ding, K. and H. R. Gordon. 1995. Analysis of the influence of O2 A-band absorption on atmospheric correction of ocean-color imagery. *Applied Optics* 34:2068–2080.

Dubuisson, P., R. Frouin, D. Dessailly, et al. 2009, Estimating the altitude of aerosol plumes over the ocean from reflectance ratio measurements in the O2 A-band. *Remote Sensing of Environment* 113:1899–1911.

Franz, B. A., P. J. Werdell, G. Meister, E. J. Kwiatkowska, S. W. Bailey, Z. Ahmad, and C. R. McClain. 2006. MODIS land bands for ocean remote sensing applications. *Proc. of Ocean Optics XVIII*, Montreal, Canada.

Frouin, R., M. Schwindling, and P.-Y. Deschamps. 1996. Spectral reflectance of sea foam in the visible and near-infrared: In situ measurements and remote sensing implications. *Journal of Geophysical Research* 101:14361–14371.

Gao, B.-C., and C. O. Davis. 1997. Development of a line-by-line-based atmosphere removal algorithm for airborne and spaceborne imaging spectrometers. *Proceedings of SPIE* 3118:132–141.

Gao, B.-C. and A. F. H. Goetz. 1990. Column atmospheric water vapor and vegetation liquid water retrievals from airborne imaging spectrometer data. *Journal of Geophysics Research* 95(D4):3549–3564.

Gao, B.-C., K. B. Heidebrecht, and A. F. H. Goetz. 1993. Derivation of scaled surface reflectances from AVIRIS data. *Remote Sensing of Environment* 44:165–178.

Gao, B.-C., Y. J. Kaufman, W. Han, and W. J. Wiscombe. 1998. Removal of thin cirrus path radiances in the 0.4–1.0 μm spectral region using the 1.375-μm strong water vapor absorption channel. *Proceedings of the 7th Annual JPL Airborne Earth Science Workshop*, JPL Publication 97–21:121–130.

Gao, B.-C., M. J. Montes, C. O. Davis, A. F. H. Goetz. 2009. Atmospheric correction algorithms for hyperspectral remote sensing data of land and ocean. *Remote Sensing of Environment* 113:S17–S24.

Gao, B. C., M. J. Montes, R. R. Li, H. M. Dierssen, and C. O. Davis. 2007. An atmospheric correction algorithm for remote sensing of bright coastal waters using MODIS land and ocean channels in the solar spectral region. *IEEE Transactions on Geoscience and Remote Sensing* 45:1835–1843.

Gordon, H. R., 1978. Removal of atmospheric effects from satellite imagery of the oceans. *Applied Optics* 17:1631–1636.

Gordon, H. R. 1997. Atmospheric correction of ocean color imagery in the Earth observing system era. *Journal of Geophysical Research* 102:17081–17106.

Gordon, H. R., J. W. Brown, and R. H. Evans. 1988, Exact Rayleigh scattering calculations for use with the Nimbus-7 coastal zone color scanner. *Applied Optics* 27:862–871.

Gordon, H. R. and D. K. Clark, 1981. Clear water radiances for atmospheric correction of Coastal Zone Color Scanner imagery. *Applied Optics* 20:4175–4180.

Gordon, H. R., D. K. Clark, J. W. Brown, O. B. Brown, R. H. Evans, and W. W. Broenkow. 1983. Phytoplankton pigment concentrations in the middle atlantic bight: Comparison between ship determinations and coastal zone color scanner estimates. *Applied Optics* 22:20–36.

Gordon, H. R. and M. Wang. 1994. Retrieval of water leaving radiance and aerosol optical thickness over the oceans with SeaWiFS: A preliminary algorithm. *Applied Optics* 33:443–452.

Goyens, C., C. Jamet, and T. Schroeder. 2013. Evaluation of four atmospheric correction algorithms for MODIS-Aqua images over contrasted coastal waters. *Remote Sensing of Environment* 131:63–75.

Griffin, M. K. and H.-H. K. Burke. 2003. Compensation of hyperspectral data for atmospheric effects. *Lincoln Laboratory Journal* 14(1):29–54.

Griffin, M. K., S. M. Hsu, K. Burke, and J. W. Snow. 2000. Characterization and delineation of plumes, clouds and fires in hyperspectral images. *Proceedings of SPIE* 4049:274–283.

Guanter, L., L. Alonso, and J. Moreno. 2005. A method for the surface reflectance retrieval from PROBA/CHRIS data over land: Application to ESA SPARC campaigns. *IEEE Transactions on Geoscience and Remote Sensing* 43:2908–2917.

Guanter, L., L. Gómez-Chova, and J. Moreno. 2008. Coupled retrieval of aerosol optical thickness, columnar water vapor and surface reflectance maps from ENVISAT/MERIS data over land. *Remote Sensing of Environment* 112:2898–2913.

Guanter, L., M. C. González-Sampedro, and J. Moreno. 2007. A method for the atmospheric correction of ENVISAT/MERIS data over land targets. *International Journal of Remote Sensing* 28:709–728.

Guanter, L., A. Ruiz-Verdú, D. Odermatt, C. Giardino, S. Simis, V. Estellés, T. Heege, J. A. Domínguez-Gómez, and J. Moreno. 2010. Atmospheric correction of ENVISAT/MERIS data over inland waters: Validation for European lakes. *Remote Sensing of Environment* 114:467–480.

Holben, B. N., et al. 1998. AERONET-A federated instrument network and data archive for aerosol characterization. *Remote Sensing of Environment* 66:1–16.

Huot, J. P., H. Tait, M. Rast, S. Delwart, J. L. Bezy, and G. Levrini. 2001, The optical imaging instruments and their applications: AATSR and MERIS. *ESA Bulletin* 106:56–66.

Kaufman, Y. J. and B.-C. Gao. 1992. Remote sensing of water vapor in the near IR from EOS/MODIS. *IEEE Transaction on Geoscience and Remote Sensing* 30(5):871–884.

Kaufman, Y. J. and C. Sendra. 1988. Algorithm for atmospheric corrections to visible and near-IR satellite imagery. *International Journal of Remote Sensing* 9(8):1357–1381.

Kaufman, Y. J., A. Wald, L. A. Remer, B.-C. Gao, R. R. Li, and L. Flynn. 1997. The MODIS 2.1-μm channel - correlation with visible reflectance for use in remote sensing of aerosol. *IEEE Trans. on Geoscience and Remote Sensing* 35:1286–1298.

Kruse, F. A. 1988. Use of airborne imaging spectrometer data to map minerals associated with hydrothermally altered rocks in the northern grapevine mountains, Nevada and California. *Remote Sensing of Environment* 24:31–51.

Kruse, F. A. 2004. Comparison of ATREM, ACORN, and FLAASH atmospheric corrections using low-altitude AVIRIS data of Boulder, CO. *Summaries of 13th JPL Airborne Geoscience Workshop*, Jet Propulsion Lab, Pasadena, CA.

Lavender, S. J., M. H. Pinkerton, G. Moore, J. Aiken, and D. Blondeau-Patissier. 2005, Modification to the atmospheric correction of SeaWiFS ocean colour images over turbid waters. *Continental Shelf Research* 25:539–555.

Lee, Z. P., B. Casey, R. Parsons, W. Goode, A. Weidemann, and R. Arnone. 2005. Bathymetry of shallow coastal regions derived from space-borne hyperspectral sensor. Proceedings of Oceans 2005 MTS/IEEE Conference & Exhibition, 17–23 September 2005. Washington, DC.

Morel, A. and S. Belanger. 2006. Improved detection of turbid waters from ocean color sensors information. *Remote Sensing of Environment* 102:237–249.

Morel, A. and L. Prieur. 1977. Analysis of variations in ocean color. *Limnology and Oceanography* 22:709–722.

Qu, Z., B. C. Kindel, and A. F. H. Goetz. 2003. The high accuracy atmospheric correction for hyperspectral data (HATCH) model. *IEEE Transactions on Geoscience and Remote Sensing* 41:1223–1231.

Reinersman, P. N., K. L. Carder, and R. F. Chen. 1998. Satellite-sensor calibration verification with the cloud-shadow method. *Applied Optics* 37:5541–5549.

Richter, R. and D. Schlaepfer. 2002. Geo-atmospheric processing of airborne imaging spectrometry data, Part 2: Atmospheric/topographic correction. *International Journal of Remote Sensing* 23(13):2631–2649.

Roberts, A. D., Y. Yamaguchi, and R. P. Lyon. 1985. Calibration of airborne imaging spectrometer data to percent reflectance using field spectral measurements. Proceedings of the 19th International Symposium on Remote Sensing of Environment 2: 679–88. Ann Arbor, MI..

Roberts, D. A., Y. Yamaguchi, and R. Lyon. 1986. Comparison of various techniques for calibration of AIS data. *Proceedings of the 2nd Airborne Imaging Spectrometer Data Analysis Workshop JPL Publication* 86–35:21–30. Pasadena, CA: Jet Propulsion Lab.

Rothman, L. S., A. Barbe, D. C. Benner, et al. 2003. The HITRAN molecular spectroscopic database: Edition of 2000 including updates through 2001. *Journal of Quantitative Spectroscopy and Radiative Transfer* 82:5–44.

Rothman, L. S., D. Jacquemart, A. Barbe, et al. 2005. The HITRAN 2004 molecular spectroscopic database. *Journal of Quantitative Spectroscopy and Radiative Transfer* 96:139–204.

Ruddick, K., F. Ovidio, and M. Rijkeboer, 2000. Atmospheric correction of SeaWiFS imagery for turbid coastal and inland waters. *Applied Optics* 39:897–912.

Schroeder, T., I. Behnert, M. Schaale, J. Fischer, and R. Doerffer. 2007. Atmospheric correction algorithm for MERIS above case-2 waters. *International Journal of Remote Sensing* 28:1469–1486.

Shi, W. and M. Wang. 2007. Detection of turbid waters and absorbing aerosols for the MODIS ocean color data processing. *Remote Sensing Environment* 110:149–161.

Shi, W. and M. Wang. 2009. An assessment of the black pixel assumption for MODIS SWIR bands. *Remote Sensing of Environment* 113:1587–1597.

Siegel, D. A., M. Wang, S. Maritorena, and W. Robinson. 2000. Atmospheric correction of satellite ocean color imagery: The black pixel assumption. *Applied Optics* 39:3582–3591.

Solomon, S., R. W. Portmann, R. W. Sanders, J. S. Daniel, W. Madsen, B. Bartram, and E. G. Dutton. 1999. On the role of nitrogen dioxide in the absorption of solar radiation. *Journal of Geophysical Research* 104(D10):12047–12058.

Staenz, K., T. Szeredi, and J. Schwarz. 1998. ISDAS—A system for processing/analyzing hyperspectral data; technical note. *Canadian Journal of Remote Sensing* 24:99–113.

Staenz, K. and D. J. Williams 1997. Retrieval of surface reflectance from hyperspectral data using a look-up table approach. *Canadian Journal of Remote Sensing* 23(4):354–368.

Su, W. 2000. A new epsilon function for atmospheric correction algorithm. *Geophysical Research Letters* 27:3707–3710.

Tanre, D., C. Deroo, P. Duhaut, M. Herman, J. J. Morcrette, J. Perbos, and P. Y. Deschamps. 1986. Simulation of the Satellite Signal in the Solar Spectrum (5S). User's Guide. U.S. T. de Lille, 59655 Villeneuve d'ascq, France: Laboratoire d'Optique Atmospherique.

Thuillier, G., M. Herse, D. Labs, T. Foujols, M. Peetermans, D. Gillotay, P. C. Simon, and H. Mandel. 2003. The solar spectral irradiance from 200 to 2400 nm as measured by the SOLSPEC spectrometer from the Atlas and Eureca missions. *Solar Physics* 214:1–22.

Vermote, E., D. Tanre, J. L. Deuze, M. Herman, and J. J. Morcrette. 1997. Second simulation of the satellite signal in the solar spectrum (6S): An overview. *IEEE Transactions on Geoscience and Remote Sensing* 35(3):675–686.

Wang, M. 2004. Extrapolation of the aerosol reflectance from the near-infrared to the visible. The single-scattering epsilon vs. multiple scattering epsilon method. *International Journal of Remote Sensing* 25:3637–3650.

Wang, M. 2007. Remote sensing of the ocean contributions from ultraviolet to near-infrared using the short-wave infrared bands: Simulations. *Applied Optics* 46:1535–1547.

Wang, M. and W. Shi. 2005. Estimation of ocean contribution at the MODIS near-infrared wavelengths along the east coast of the U.S.: Two case studies. *Geophysics Research Letter* 32:L13606.

Wang, M. and W. Shi. 2007. The NIR–SWIR combined atmospheric correction approach for MODIS ocean color data processing. *Optics Express* 15(24):15722–15733.

Wang, M., S. Son and W. Shi. 2009. Evaluation of MODIS SWIR and NIR-SWIR atmospheric correction algorithms using SeaBASS data. *Remote Sensing of Environment* 113:635–644.

Wang, M., J. Tang, and W. Shi. 2007. MODIS-derived ocean color products along the China east coastal region. *Geophysical Research Letters* 34:L06611.

Werdell, P. J., B. A. Franz, and S. W. Bailey. 2010. Evaluation of shortwave infrared atmospheric correction for ocean color remote sensing of Chesapeake Bay. *Remote Sensing of Environment* 114:2238–2247.

Zhu, J, J. Chen, B. Matsushita, W. Yang, and T. Fukushima. 2012. Atmospheric correction of ENVISAT/MERIS data over case II waters: The use of black pixel assumption in oxygen and water vapour absorption bands. *International Journal of Remote Sensing* 33(12):3713–3732.

Index

A

B

C

D

G

I

N

P

Q

R

For Product Safety Concerns and Information please contact our EU
representative GPSR@taylorandfrancis.com
Taylor & Francis Verlag GmbH, Kaufingerstraße 24, 80331 München, Germany

www.ingramcontent.com/pod-product-compliance
Lightning Source LLC
Chambersburg PA
CBHW080735150926
51501CB00005B/169

* 9 7 8 0 3 6 7 2 1 7 9 0 7 *